Power System Analysis

SECOND EDITION

T. K. Nagsarkar

Formerly, Professor and Head
Department of Electrical Engineering
Punjab Engineering College
Chandigarh

M. S. Sukhija

Formerly, Founder Principal
Guru Nanak Dev Engineering College
Bidar, Karnataka

OXFORD

UNIVERSITY PRESS

OXFORD
UNIVERSITY PRESS

Oxford University Press is a department of the University of Oxford.
It furthers the University's objective of excellence in research, scholarship,
and education by publishing worldwide. Oxford is a registered trade mark of
Oxford University Press in the UK and in certain other countries.

Published in India by
Oxford University Press
Ground Floor, 2/11, Ansari Road, Daryaganj, New Delhi, 110 002, India

First Edition published in 2007
Second Edition published in 2014
Second impression 2019

ISBN-13: 978-0-19-809633-7
ISBN-10: 0-19-809633-X

Typeset in Times New Roman
by Anvi Composers, New Delhi 110063
Printed in India by Gopsons Papers Ltd, Noida 201301

To my parents
Late Ramakanta Nagsarkar
and
Late Shova Nagsarkar

T.K. Nagsarkar

Humbly dedicated to
the greatest Master for penning a
destiny of unmitigated joy

M.S. Sukhija

Preface to the Second Edition

Our title *Power System Analysis* was published in 2007 and since then it has been readily accepted as a standard textbook both at the graduate and postgraduate levels. This second edition is a gradational revision of the power systems analysis course. Without disturbing the simple and mellifluous flow of the language, the second edition aims to fulfil the following objectives:

- Reflect the overall changes in the energy sector scenario which have come about during the past five years or so
- Further fortify the understanding of principles of power system analysis
- Expose readers to current topics such as *voltage stability* which have acquired significance in the context of large integrated power systems operating close to their maximum capacity

New to the Second Edition

The purpose of revising the already accepted title *Power System Analysis* was to see it as an opportunity to bring about improvements. Therefore, the changes which have been incorporated in the second edition are largely based on the following:

- Suggestions from students and faculty members in various institutions
- Our continual references to look for 'chinks in the armour' in our work
- Periodic comments from the publisher's reviewers

Common Features

The most important change in the second edition is the inclusion of the following in each of the chapters:

- *Learning Outcomes* at the beginning of each chapter defines the results a reader will be able to achieve following a study of the chapter.
- End chapter *Summary*, along with *Significant Formulae*, provides a recapitulation of the chapter.
- Descriptive questions have been included in the chapter end *Exercises*.
- *Multiple Choice Objective Questions*, with one correct answer, have been added to enable a reader to quickly gauge his/her understanding and retentivity of the principles and laws of power system analysis.
- Existing terminology, principles, laws, symbology, etc., wherever required have been further clarified.

Extended Chapter Material

Chapter 1 (*Power Sector Outlook*) Data related to the energy sector has been updated along with a brief description of Vision 2020. Additionally, solar power and magneto-hydro-dynamic (MHD) power generation have been included.

Chapter 2 (*Basic Concepts*) Comparison of single- and three-phase power transmission, along with typical solved examples, have been included.

Chapter 3 (*Transmission Line Parameters*) Overview has been re-drafted to highlight the importance of transmission lines in power transmission and the electrical properties of commonly used conductor materials are shown in tabular form. A separate section is included to explain at length skin, proximity, and spirality effects which lead to non-uniform distribution of alternating current in a conductor.

Chapter 4 (*Transmission Line Model and Performance*) Overview has been re-written to bring out the effect of distributed line parameters. Significance of ultra-fast transients leading to overvoltages has also been stated. Significance of propagation constant and Ferranti effect have been explained at length along with the inclusion of several typical solved examples.

Chapter 6 (*Formulation of Network Matrices*) Overview has been expanded to emphasize the formulation of network matrices based on graph theory.

Chapter 7 (*Power Flow Studies*) A new section to explain the simulation of DC power flow solution has been added. Similarly, a detailed comparison of the various power flow solution methods has been provided in tabular form.

Chapter 10 (*Symmetrical Components and Unsymmetrical Fault Analyses*) A table summarizing the phase shift between primary and secondary voltages of three-phase transformers, according to different vector groups, has been added for easy simulation for fault analysis.

Chapter 12 (*Voltage Stability*) Keeping in mind the several collapses of major power networks worldwide due to voltage instability, this new chapter has been included. After explaining and defining voltage stability and voltage collapse, the former is classified as large disturbance and small disturbance voltage stabilities. Formulation of transient voltage stability problem, due to a small disturbance, has been discussed along with techniques for performing transient voltage stability studies. Appropriate typical solved examples explain the application of the techniques.

Chapter 13 (*Contingency Analysis Techniques*) A typical solved example has been included to further explain the method of performing contingency analysis.

Chapter 14 (*State Estimation Techniques*) Variations in Examples 14.1 and 14.2 have been introduced to further explain the method of weighted least squares and the application of line power flow estimator for computing the system variables respectively.

Chapter 15 (*An Introduction to HVDC Power Transmission*) A whole new section explaining the operation of HVDC converter, in detail, has been included. Additionally, typical solved examples have been added. These examples further clarify the principles of HVDC operation.

Academic Capacity Test (ACT) has been provided as an online resource for students. It consists of 100 multiple choice objective questions, randomly selected from all the chapters in the textbook. ACT has been designed to enable a reader to identify chapter-wise strengths and weaknesses.

Acknowledgements

Words may not be enough to express our sense of gratefulness to our innumerable readers who took time out to provide priceless inputs. In particular we would like to acknowledge our gratitude to our colleagues, reviewers, and the staff of OUP who provided invaluable constructive criticism. Finally, last but not the least, we are indebted to the countless faculty members and young readers for their eager acceptance of our work.

T.K. Nagsarkar
M.S. Sukhija

Preface to the First Edition

The enjoyable experience of writing our first book *Basic Electrical Engineering* and its ready acceptance by the engineering faculty has inspired us to author our second book *Power System Analysis*. Written with the objective of assisting students to acquire the concepts and tools of analyses, it will also equip them in finding solutions to power system engineering problems in practice. In view of this objective, the language and style of writing this book is completely student oriented.

The growth of modern-day integrated power systems has fortunately been accompanied by the development of fast interactive personal computers. The latter has necessitated the integration of personal computers into the curricula of power system engineering programmes and has enabled the teachers to augment the learning process through the simulation and designing of more practice-oriented problems and taking up more complex topics for analyses.

Matrix laboratory (MATLAB) is a very powerful matrix oriented software package for numerical computations. It is a handy tool for solving numerical problems requiring numerous matrix operations, and therefore, a boon for analysing power system problems. Keeping in mind the current developments and future requirements, the book has an intentional bias towards computer simulation of power systems and the application of MATLAB for analyses and solutions of complex problems. The text integrates, through examples, several executable MATLAB functions and scripts. This will help the readers to comprehend not only the translation of a simulation into an executable MATLAB solution but will also encourage them to modify and even develop their own programmes. The MATLAB commands and their effects have been explained wherever they have been used in the text. Each chapter includes several such examples and unsolved problems with answers.

About the Book

Power System Analysis aims at providing a comprehensive coverage of the curricula and will serve as a very useful textbook for electrical engineering students at the undergraduate level. The book provides a thorough understanding of the basic principles and techniques of power system analysis. Beginning with basic concepts, the book gives an exhaustive coverage of transmission line parameters, symmetrical and unsymmetrical fault analyses, power flow studies, power system control, and stability analysis. With the inclusion of some advanced topics such as state estimation, stability analysis, contingency analysis, and an introduction to HVDC and FACTS, it would also serve the requirements of teachers and students alike at the postgraduate level.

Content and Coverage

The book comprises 15 chapters and three appendices. Each chapter in this book commences with an overview, which briefly outlines the topics covered in the chapter, and ends with numerous unsolved problems which help the readers to assess their comprehension of the subject matter studied in the chapter.

Chapter 1 in addition to tracing the history of the growth of the power sector outlines its structure and its present state. Statistical data is included to provide a perspective of the Indian power sector and its future plans for meeting the load demand and making it more energy efficient. The concept of deregulation of the power industry is also covered.

Chapter 2 covers the representation of power system elements suitable for circuit analysis. A review of phasor notation, phase shift operator for three-phase systems, and the power in single-phase and three-phase circuits is presented. It describes per unit representation of power systems and its advantages in power system analysis.

Chapters 3 and 4 deal with the parameters of transmission lines and steady-state performance and analysis of transmission lines, respectively. Chapter 3 outlines the computation of the parameters of transmission lines. Chapter 4 covers their simulation as short, medium, and long lines. Power handling capability and reactive line compensation of lines are discussed. The phenomenon of travelling waves on transmission lines is also included in the chapter.

Chapter 5 covers the representation of synchronous machines, transformers, and loads in the steady state and transient analysis.

Chapter 6 introduces graph theory along with the commonly employed terminology in the formulation of network matrices. Since network matrices form the basis of power system analyses, the chapter comprehensively covers the formulation of bus admittance and bus impedance matrices of a power system network. The chapter also includes the formulation of nodal equations, both in the admittance and impedance frames of reference, and their solutions by direct and indirect methods. Sparsity techniques for storing non-zero elements, network reduction, and optimal numbering schemes are also covered in the chapter.

Chapter 7 on power flow studies of integrated power systems, under normal operating conditions, provides a detailed description of the formulation of power flow equations. Solutions of these power flow equations by the well-accepted Gauss, Gauss–Seidel, and Newton–Raphson methods have been presented in detail in this chapter. Fast decoupled method for solution of power flow problems suitable for online studies has also been presented.

Chapter 8 deals with the maintenance of active power balance and control of voltage magnitude and power frequency, within specified limits, when a system is operating in the steady state. Beginning with the basic control loop in a generator, the automatic voltage control (AVC) and load frequency control (LFC) loops are described and their steady-state and dynamic performances are outlined in detail. The LFC of a single control area is first discussed and then extended to a two-area control system. Tie-line bias control and its application to a two-area control system are also presented in detail.

Chapter 9 deals with the methodical computation of bus voltages and line currents under balanced three-phase fault conditions. Impedance matrix for three-phase faults and computations of three-phase fault currents are also included.

Chapter 10 covers the analysis of power systems under unsymmetrical fault conditions. Symmetrical components which transform unbalanced currents and voltages into sets of three balanced components are employed as a tool for transforming an unbalanced circuit into a balanced circuit. The latter transformation makes it feasible to analyse the faulted network on per phase basis. The application of symmetrical components to various types of unbalanced faults, along with computational algorithms, has been detailed. Series faults such as one-conductor or two-conductors open are discussed. Generalized formulation for unbalanced short circuit computation using bus impedance matrix is also given at the end.

Chapter 11 commences with the assumptions commonly made in stability studies, followed by the derivation of the swing equation based on an analogy with the laws of rotational mechanics. The chapter includes an analysis of the single machine transient stability problem based on the equal area criterion. The solution of swing equation by conventional step-by-step method as well as modified Euler method has been presented. An algorithm for studying the multi-machine transient stability of a power system is also included. Numerical solutions of the non-linear algebraic equations are explained. MATLAB functions and scripts have been included to demonstrate their utility in obtaining numerical solutions and plotting the swing equation. Linearization of the swing equation and its solution for performing steady-state stability analyses are explained in detail. Some methods for improving stability are discussed.

Chapter 12 covers the contingency analysis of a power system and it deals with the determination of line currents following a line outage or a switching operation. The concepts of compensating currents, distribution factors, and their computation by employing the Z-bus matrix are explained in detail. Contingency analysis of interconnected power systems by network equivalents is also presented.

Chapter 13 provides techniques to estimate the state of a power system using measured quantities such as P, Q, and line flows. Quantitative techniques to test the goodness of the state estimates from measurements and the elimination of bad data are also comprehensively described.

High voltage direct current (HVDC) systems operate in conjunction with ac systems to transmit bulk power in present day power systems.

Chapter 14 provides a historical perspective on the HVDC transmission, followed by a comparison of the ac and dc transmission systems. Typical configurations of HVDC converter stations and various types of dc links are also described.

Chapter 15 outlines the use of flexible ac transmission systems (FACTS) technology for enhancing transmission capability and improving grid reliability. The chapter highlights the restrictions in ac power flow in existing transmission systems due to parametric limitations and then explains how

FACTS technology can be employed to change the power flows by varying the parameters such as line reactance and voltage magnitudes. The various types of FACTS controllers employed for parametric variation have been described.

Three appendices are given at the end of the book. Appendix A defines the various types of matrices encountered in the modelling of complex engineering problems and outlines the matrix algebra involved in finding their solutions. Creation of matrices and the use of MATLAB commands have been demonstrated through examples. Numerous unsolved problems are provided at the end of the appendix. Appendix B comprises test data for power flow and the results for standard IEEE test systems. Solutions to end chapter exercises are provided in Appendix C.

We are confident that this book will enable students to achieve a better understanding of complex power system problems and help them solve these problems by utilizing the existing MATLAB functions and scripts. Students will become adept at developing their own functions and scripts independently, essentially after solving the problems at the end of each chapter. The faculty teaching the power systems engineering programme, on the other hand, will be able to enhance the teaching process by taking up modelling and analyses of problems which are encountered in the operation and control of power systems in practice.

We have developed this book over the past many years while teaching this subject to the students of electrical engineering at the postgraduate and undergraduate levels. The text has been written to match the evolving curriculum of the subject taught in universities in India and abroad. We acknowledge that the book is greatly influenced by earlier texts and literature written on diverse aspects of power systems by many outstanding professors and engineers. The details of the sources have been compiled in the bibliography at the end of the book. We wish to thank these individuals who have been instrumental in the development of this book. We are also grateful to our colleagues and friends who have provided valuable suggestions and criticism in formulating the contents of this book.

T.K. Nagsarkar
M.S. Sukhija

Brief Contents

Detailed Contents

Symbols and Acronyms

Δf frequency deviation

ΔI_{ij} current changes in the line connected between buses i-j.

ΔP, $\Delta \delta$, $\Delta |V|$ deviations from normal operating values of power, angle, and voltage magnitude, respectively

ΔP_{1-2} change in tie-line power

ΔP_G increase in the generated power

ΔP_i, ΔQ_i active and reactive power mismatch respectively at bus i

ΔP_L increase in the load demand

ΔP_{ref} change in reference power setting; deviation in speed changer or reference position

ΔP_T change in the turbine power

ΔP_{tie} tie-line power deviation

ΔP_V signal that regulates the control valves of steam or hydro turbine unit

ΔS_i complex power mismatch at bus i

Δt uniform step time interval

ΔV vector of the changes in bus voltages due to compensating currents

Δx elemental strip length

Δx_i^0 addition to be made to the initial estimate of the state variable x_i^0

Δy_A change in position of point A corresponds to a change in ΔP_{ref}

Δy_C change in governor output command ΔP_g

Δy_E a change in valve position command ΔP_V

$\Delta \delta_1$, $\Delta \delta_2$ small changes in voltage angles

$\Delta \delta_n$ change in δ during the nth interval

α electrical angular acceleration (rad/s^2); attenuation constant, reflection operator for waves approaching from the left; acceleration factor in power flow study; damping factor

α_{n-1} acceleration at time $t = (n-1)\Delta t$

α_0 temperature coefficient of the conductor resistance at 0°C.

α_m rotor angular acceleration (mech. rad/s^2)

α_R receiving end reflection coefficient

α_S sending end reflection coefficient

α' reflection operator for waves approaching from the right

β area frequency response characteristic (AFRC); phase constant; refraction operator for waves approaching from the left

β' refraction operators for waves approaching from the right

γ natural frequency of oscillations; propagation constant = $\alpha + j\beta$ = $\sqrt{yz}$

δ_m rotor angular position with respect to synchronously rotating reference (mech. rads)

δ_{n-1} angular position at time $t = (n-1)\Delta t$

δ_0 initial rotor angle position of the machine

δ_1, δ_2 — voltage angles of $|E_1|$ and $|E_2|$, respectively

δ_c — fault clearing angle

δ_{crit} — critical fault clearing angle

δ_i — voltage phase angle at bus i

δ_m — angle (mech. rads)

ε — permittivity constant (F/m); convergence tolerance

ε_0 — permittivity constant for free space $= 8.854 \times 10^{-12}$ F/m

ζ — vector of random errors or noise of measured physical quantities

ζ_i — error or noise of ith measured variable

$\overline{\zeta}$ — noise vector

η — efficiency

θ — angle in electrical radians measured from the rotor pole axis

θ_1, θ_2 — angles of phase displacement from the reference

θ_{im} — admittance angle of transfer admittance Y_{im}

θ_m — rotor angular position; angular displacement of rotor with respect to stationary reference axis (rad/s)

μ — mean; permeability constant of the medium

μ_0 — $4\pi \times 10^{-7}$

μ_r — relative permeability

3ϕ — three-phase

φ — phase angle between the voltage and current phasors

ϕ — total flux (Wb)

ϕ_{ar} — armature reaction flux per pole

ϕ_f — flux per pole; rotor field flux

ϕ_{fa} — flux linking the rotor field and the armature

ϕ_r — resultant air-gap flux

ρ — resistivity of conductor material (Ω m)

σ — standard deviation

σ_i^2 — variance

τ_d'' — short circuit sub-transient time constant

τ_d' — direct axis short circuit transient time constant

τ_{do}' — direct axis open circuit transient time constant

$\chi_{k\alpha}^2$ — chi-square distribution

Ψ — flux linkages (WbT)

Ψ_{12} — total flux between two points A_1 and A_2

Ψ_{int} — total internal flux linkages (WbT/m)

ω — angular speed (radians/sec)

ω_s — synchronous speed (elect. rad./s)

ω_{sm} — angular synchronous speed (mech. rad/s)

$|E_1|,$ — magnitudes of the end voltages of

$|E_2|$ — control areas 1 and 2, respectively

$|I|$ — effective or root mean square (rms) value of the sinusoidal current $\left(= I_m / \sqrt{2}\right)$

$|S_i|$ — absolute value of the ith measurement of real and reactive power flow

$|V_i|$ — magnitude of voltage at bus i

$|V_{R(FL)}|$ — receiving end voltage at full load

$|V|$ — effective or root mean square (rms) value of the sinusoidal; voltage magnitude $\left(= V_m / \sqrt{2}\right)$

$|V|_{\mathrm{ref}}$ — reference voltage

$|Y_{im}|$ — magnitude of transfer admittance between bus i and bus m

$|\hat{x}$ — indication that the equations are evaluated from the elements of state estimate vector $\hat{x}$

0^+ and 0^- — instant of time just after and just before the instant $t = 0$

A — cross-sectional area (m²)

A — element-bus incidence matrix or bus incidence matrix; symmetrical components transformation matrix

$\overline{A}$ — element-node incidence matrix

$a + jb$ — complex turns ratio of phase shifting transformer

a	phase shift operator; turns ratio	$D_s{}^b$	GMR of a bundled conductor
$ABCD$	constants of a two-port network	D_{sX}	self-GMD
AVC	automatic voltage control	$d\phi$	flux in the tubular element of thickness dx
B	magnetic flux density (Wb/m^2)		
B_1, B_2	tie-line bias parameters	e	emf induced in the stator coil
B_m	maximum value of flux density at the centre of the pole; magnetizing susceptance of a transformer	E	field of electric intensity (V/m); rms value of the induced phase emf
		E, E_{ar}, E_r	phasor voltages proportional to the flux phasors ϕ_f, ϕ_{ar}, and ϕ_r respectively
B_x	flux density at a distance x metres from the centre of the conductor		
		e, j	specified voltage and current vectors respectively
c_1	accuracy constant, in decimal, of the measurement (for real and reactive power flow measurements, typical values of c_1 are 0.01 or 0.02)	E, V	internal emf and terminal voltage respectively of generator
		$E[z]$	expected value of z
		e_1, e_2	emf induced in the primary and secondary windings respectively
c_2	constant to account for transducer, and analogue to digital converter accuracy (typical values of c_2 are 0.0025 or 0.005)	E_1, E_2	rms emf induced in the primary and secondary windings respectively
C_{ab}	capacitance between conductors a and b	e_i, f_i	real and imaginary parts of complex voltage of bus i
CCC	capacitor commuted converters	e_{km}	source voltage connected in series with the element
col	a row vector which provides the column number of Y_{bus}	B'_{im}, B''_{im}	elements of $[-B]$ matrix
$\cos\varphi$	power factor	elemz	a row vector which stores the primitive impedance of the lines
C_R	centre of the receiving end power circle	E_m	peak value of induced emf
C_S	centre of the sending end power circle	E_m	vector of measured voltage differences linked with each power flow measurement
d	distance between adjacent conductors of a bundled conductor		
D	uniform electric flux density (C/m^2)	E'	voltage behind transient reactance of the machine
D_1, D_2	distances of two points A_1 and A_2 respectively from the conductor surface	E''_G, E''_M	pre-fault internal voltages behind the sub-transient reactance of the machines
D_f	electric flux density on a cylindrical surface	f	frequency (Hz)
d/dt	operator p	$\hat{f}$	performance function
D_m	geometric mean distance (GMD) between conductor X and conductor Y	$\mathcal{F}$	magneto motive force (mmf) (AT)
		f_0	constant frequency; nominal frequency (50 Hz)
D_s	GMR of individual conductors	FACTS	flexible ac transmission systems

f_r	sub-synchronous frequency	I_{abc}	column vector of three-phase currents
FSD	full scale deflection of the instruments such as watt and var meters	$I_{abc,\,k\,(F)}$	column vector of three-phase post-fault currents at bus k
G	symmetric gain matrix	I_b	current flowing from bus p to bus q
G_C	equivalent conductance to represent core loss in a transformer	I_{base}	base current
G_{ii}, B_{ii}	self-conductance and self-susceptance of bus i	I_{bus}	vector matrix of injected bus current for the network
G_{im}, B_{im}	transfer conductance and transfer susceptance between bus i and bus m	I_C	compensating current vector
		I_{ch}	charging current
GMD	geometric mean distance	i_f	generator field current
GMR	geometric mean radius	IGCT	integrated gate-commutated thyristors
$G_{SG}(s)$	transfer function of speed governing mechanism	I_k	per unit current injections in bus k
$G_T(s)$	transfer function of a hydro-turbine	i_{km}	current through the element k-m
GTO	gate turn-off thyristors	I_{km}	short circuit current in the line flowing from bus k to bus m
H	coefficient matrix of dimension $m \times n$	IPFC	interline power flow controller
H	constant related to inertia (MJ/MVA or s); inertia constant (s); magnetic field intensity (AT/m)	I_{pu}	current in per unit
		I_q, I_d	quadrature axis and direct axis components of the armature current I_a
HVDC	high voltage direct current		
H_x	magnetic field intensity at a distance x metres from the centre of the conductor	$I_{S,\,bus\,(F)}$	post-fault symmetrical component bus current vector
I	identity matrix	I_S, I_R	currents at the sending end and receiving end respectively of a transmission line
I	current phasor		
$I(x)$	current flowing through the elemental length Δx	$I_{S,\,k(F)}$	column vector of symmetrical component post-fault currents at bus k
$I(x, s)$	Laplace transforms of the current at position x	IT	iteration count
$i(x, t)$	instantaneous current as a function of t and x	J_m	total moment of inertia of the rotor masses (kgm^2)
I_0	no-load current in transformer	J	Jacobian matrix of partial derivatives
I_{012}	column vector of symmetrical components of currents	J_1	sub-matrix of the partial derivative of P's with respect to relevant δ's
I_1, I_2	rms current in the primary and secondary windings respectively	J_2	sub-matrix of the partial derivative of P's with respect to relevant V's
I_a	current flowing from bus i to bus j	J_3	sub-matrix of the partial derivative of Q's with respect to relevant δ's
I_a	rms phase current		

J_4	sub-matrix of the partial derivative of Q's with respect to relevant V's	M	angular momentum of the rotor $= J_m \omega_m$; inertia constant
j_{km}	source current connected in parallel with the element k-m,	m	number of measurements
k	iteration count	MTO	MOS turn-off thyristors
K_A, T_A	transfer function gain and time constant respectively of the amplifier	MVA_{base}	base MVA
		$MVA_{base\text{-}1\phi}$	single-phase base MVA
K_E, T_E	gain constant and time constant respectively of the exciter	$MVA_{base\text{-}3\phi}$	three-phase base MVA
K_I	integral gain constant	m/n	redundancy factor
$K_{I, crit}$	critical gain setting	$(n-m)$	redundancy or degrees of freedom
K_{I1}, K_{I2}	integrator gains	n	number of state variables
K_{ij-k}	current-injection distribution factor	N	number of turns of a conductor
K_{SG}	static gain of speed governing mechanism	N_1, N_2	number of turns in the primary and secondary windings respectively
K_T	gain constant of the turbine	nbus	number of buses in the system
kV_{base}	base kV	N_S	speed in rpm
$kV_{base\text{-}LL}$	line-to-line base kV	P	number of poles of the alternator; average real power over one cycle in a single-phase circuit
$kV_{base\text{-}LN}$	line-to-neutral base kV		
K_w	winding factor		
K'_{rs-p},	modified generation shift	$p(t)$	instantaneous power in single-phase circuit
K'_{rs-q}	distribution factors	$p(z)$	Gaussian or normal probability density function
1LO	one-line conductor open		
2LG	double line-to-ground	$P, f, \delta,$	actual values of the power, frequency, angle, and voltage magnitude respectively
2LO	two-line conductors open	$|V|$	
L	inductance of a circuit (H)		
l	length (m); length of line; length of the conductor or the axial stator length	P_{1-2}	transfer of power over the tie-line connecting area 1 with area 2
L_{12}	inductance due to the external flux included between A_1 and A_2	$P_{3\phi}$	average three-phase power
		$P_{3\phi}$	total instantaneous three-phase power
L_{fa}	mutual inductance between the rotor field and stator armature		
LFC	load frequency control	P_a	accelerating power which accounts for any unbalance between P_m and P_e (W)
L_{ij-pq}	line-outage distribution factor		
L_{int}	inductance of conductor due to internal flux linkages	$P_{a, pu}$, $P_{i, pu}$, $P_{u, pu}$	per unit accelerating, mechanical, and electrical power, respectively
LL	line-to-line	P_d	coefficient of damping power
loc	a row vector which stores the start of each row of Y_{bus}	$P_{G(0)}$	constant generator power output
locs	gives the size of the loc vector	P_{Gi}, Q_{Gi}	generated active and reactive power respectively at bus i

P_i	mechanical power input from the prime mover minus mechanical losses (W)	R	speed regulation due to governor action; resistance
P_i, Q_i	injected active and reactive power respectively at bus i	$\boldsymbol{R}$	diagonal variance matrix
P_{i0}	initial power input to the machine	r_1, r_2	radii of conductors 1 and 2 respectively (m)
P_{im}, Q_{im}	active and reactive line power flows respectively measured at bus i	r_1, r_2	resistances of primary and secondary windings respectively
		R_a	armature resistance
P_{Li}, Q_{Li}	active and reactive power loads respectively at bus i	r_a, r_b	radii of conductors a and b respectively
P_{Gi}^{sp}, Q_{Gi}^{sp}	specified active and reactive power generation respectively at bus i	R_{dc}	direct current resistance
		R_e, X_e	resistance and inductance respectively of the field winding of an exciter
$P_{\angle i}^{sp}, Q_{\angle i}^{sp}$	specified active and reactive power load respectively at bus i	R_{eq}, X_{eq}	equivalent series resistance and leakage reactance respectively of the autotransformer
P_i^{sp}, Q_i^{sp}	specified injected active and reactive power respectively at bus i	R_f, L_f	resistance and self-inductance respectively of the field winding
P_{max}	maximum power transferred; peak value of the electrical power output of the generator	R_k	damper winding resistance
		R_{t2}	resistance of the conductor at $t_2°C$
P_{rated}	MW rating of the generator probability density function	r_1'	$= 0.7788r_1$, the geometric mean radius (GMR) of conductor 1
P_S	synchronizing coefficient ($= P_{max}\cos\delta_0$); sending end real power	S	apparent power in a single-phase circuit
P_u	electrical power output plus electrical losses (W)	$S_{3\phi}$	Apparent power or volt-ampere for a three-phase circuit
P_{u0}	initial power output of the machine	$\boldsymbol{S}_c$	vector of calculated power flows
P_{uA}	power output when the machine is under steady-state operation	S_{ci}	power flow computed from the estimated values of the state vector $\hat{\boldsymbol{x}}$
P_{uB}	power output when the machine is under fault cleared condition	S_{Gi}	complex power of generator at bus i
P_{uC}	power output of the machine during faulted condition	$\boldsymbol{S}_i$	complex power injected at bus i which is the difference between generator and load complex power at the bus
q	total charge (C)		
Q	reactive power in a single-phase circuit	SIL	surge impedance loading
$Q_{3\phi}$	average three-phase reactive power	S_{im}, S_{mi}	complex line powers measured at buses i and m respectively
		SLG	single line-to-ground
q_a, q_b	charges on the conductors a and b respectively	S_{Li}	complex power of load at bus i
r	mean radius at the air-gap	S_{li}	ith measurement of power line flow

S_m measurement vector of order m

SMES   superconducting magnetic energy storage

S_S sending end per phase complex power

SSG   static synchronous generator

SSSC   static synchronous series compensator

STATCOM
static synchronous compensator

SVC   static var compensator

T turns of the stator coil

t time(s)

T_a net accelerating torque (Nm)

TCDB   thyristor-controlled dynamic brake

TCPST   thyristor-controlled phase shifting transformer

TCR   thyristor-controlled reactor

t_{crit} critical fault clearing time

TCSC   thyristor-controlled series capacitor

TCSR   thyristor-controlled series reactor

T_e net electromagnetic torque that accounts for the total power output of the generator plus electrical losses (Nm)

T_m mechanical or shaft torque supplied by the prime mover minus retarding torque due to rotational losses (Nm)

T_{SG} time constant of speed governing mechanism

T_T time constant of the turbine

T_W time required in seconds for the water to flow in the penstock to reach the turbine

T'_{do} time constant of the open circuit direct axis rotor field winding

U unit matrix

UPFC   unified power flow controller

V velocity of propagation of travelling wave

$V_{S, k(F)}$ column vector of symmetrical component post-fault voltages at bus k

V voltage phasor

$v(t)$ time varying voltage quantity

$V(x)$ voltage at a point that is at a distance of x metres from the receiving end

$V(x, s)$ Laplace transforms of the voltage at position x

$v(x, t)$ instantaneous voltage as a function of t and x

v, i element voltage and current vectors respectively

V_{012} column vector of symmetrical components of voltages

V_{12} potential difference between two points X_1 and X_2

V_a, V_b, V_c three-phase voltage phasors

V_{a0}, V_{b0}, V_{c0} zero sequence components of three-phase voltages

V_{a1}, V_{b1}, V_{c1} positive sequence components of three-phase voltages

V_{a2}, V_{b2}, V_{c2} negative sequence components of three-phase voltages

V_{ab} voltage between conductors a and b

V_{abc} column vector of three-phase voltages

V_{abc}, I_{abc} phasor representation of three-phase balanced voltages and currents

$V_{abc, k(F)}$ column vector of three-phase post-fault voltages at bus k

V_B voltage of the infinite bus

v^b, i^b backward travelling waves moving in the direction of negative x

v_{ba} potential difference between two points a and b

V_{base} base voltage

V_{bus} voltage vector of bus voltages (bus voltages are specified with respect to a reference node, usually but not necessarily the ground)

$V_{bus, 0}$ n-dimensional vector constituting the pre-fault bus voltages

$V_{\text{bus}, F}$ n-dimensional vector of the post-fault bus voltages

V_c difference vector of calculated voltages

V_F pre-fault voltage at the point of fault

V_f Thevenin's equivalent voltage at the faulted point, which is the pre-fault voltage to neutral at the point of fault

v^f, i^f forward travelling waves moving in the direction of positive x

V_i^k complex voltage at kth iteration of bus i

V_k per unit voltage of bus k with respect to the reference

$V_{k, F}$ post-fault bus voltages of bus k

v_{km} voltage across element k-m

V_m difference vector of measurement voltages

V_m maximum value of voltage

$V_{pq\,012}$ column vector of symmetrical component voltage drops between p and q

V_{pq} column vector of voltage drops between p and q

V_{pu} impedance in per unit

V_{pu} voltage in per unit

$V_{R\,(\text{NL})}$ no load voltage at the receiving end

V_S slack bus voltage

$V_{Sk\,(0)}$ column vector of symmetrical component pre-fault voltages at bus k

$V_{S,\,\text{bus}\,(F)}$ post-fault symmetrical component bus voltage vector

$V_{S,\,\text{bus},\,0}$ pre-fault symmetrical component bus voltage vector

$V_{S,\,\text{bus}}$,   symmetrical component bus

$I_{S,\,\text{bus}}$ voltage and bus current vectors, respectively

V_S, V_R voltages at the sending end and receiving end respectively of a transmission line

VSC   voltage-sourced converter

V_t terminal voltage per phase

V_T Thevenin's voltage vector

v^t, i^t refracted travelling waves

V' column vector of new values of the bus voltages as a result of connecting the impedances between the two pairs of buses

V'_2, I'_2 secondary voltage and current respectively referred to the primary

W diagonal weighting matrix of order m

w_i weighting factor for ith measurement,

W_{KE} kinetic energy of the rotating mass

$W_{KE,\,0}$ stored area kinetic energy

x vector of state variables

$\widehat{x}$ state vector

X reactance

x_1, x_2 reactances of primary and secondary windings respectively

X_{12} tie-line reactance

X_{ar} fictitious reactance to represent the effect of armature reaction

X_{Cse} capacitive reactance in series with line

X_{Csh} shunt capacitive reactance

X_d reactance of the generator along the direct axis

X'_d direct axis transient reactance

X''_d direct axis sub-transient reactance

X_E total external reactance viewed from the generator terminals

X_{eq} equivalent reactance between the machine internal voltage and the infinite bus

X_L leakage reactance

X_{Lsh} shunt inductive reactance

X_q reactance in the quadrature axis

X_S synchronous reactance

Y_{bus} bus admittance matrix

y shunt admittance per unit length per phase to neutral $= G + j\omega C$

Y total shunt admittance per phase to neutral $= yl$

$Y_{bus, aug}$ augmented bus admittance matrix with ground as reference

$Y_{bus, mod}$ modified bus admittance matrix with with y_{Li} added to the diagonal terms of Y_{bus} matrix

$Y_{bus, red}$ reduced bus admittance matrix

$y_{E(0)}$ constant valve setting

Y_{ii} self-admittance of bus i

Y_{ij} transfer admittance between bus i and bus j

y_{km} self-admittance of the element between bus k and bus m

y_{Li} equivalent load admittance connected to ith bus and ground bus

Z loop impedance matrix

z vector of measurement variables

z, y primitive network impedance and admittance matrix respectively

Z complex impedance $(= R + jX)$

z series impedance per unit length per phase $= r + j\omega L$

Z total series impedance per phase $= zl$

$Z(s)$ transform series impedance

Z_1, Z_2, Z_0 Thevenin's equivalent impedance of the positive, negative, and zero sequence networks, respectively

Z_a, Z_b impedances of the connecting branches

Z_a, Z_b, Z_c series impedances in phases a, b, and c respectively

$Z_{ab}, Z_{ba},$ mutual impedances between
$Z_{bc}, Z_{cb},$ phases
Z_{ca}, Z_{ac}

Z_{base} base impedance

Z_{bus} bus impedance matrix for the network

Z_c characteristic impedance or surge impedance

Z_F fault shunt phase impedance matrix

$Zg(s)$ transform shunt impedance to ground

z_{km} impedance of line connecting buses k and m; self-impedance of the element

Z_L load impedance

Z_{pq012} symmetrical component transformed phase impedance matrix between p and q

Z_{ps} leakage impedance between primary and secondary; secondary is short-circuited and tertiary is kept on open circuit

Z_{pt} leakage impedance between primary and tertiary; tertiary is short-circuited and secondary is kept on open circuit

Z_S synchronous impedance

$Z_{S, (F)}$ fault shunt symmetrical component impedance matrix

$Z_{S, bus}$ symmetrical component bus impedance matrix

z_{true} vector of true values of the state variables

Z'_L the equivalent load impedance transformed to the primary side

Z'_{st} leakage impedance between secondary and tertiary; tertiary is short-circuited and primary is kept on open circuit

Z^A_{bus} bus impedance matrix of system A

Z^B_{bus} bus impedance matrix of system B

CHAPTER 1

Power Sector Outlook

Learning Outcomes

A focussed study of this chapter will enable the reader to:
- Learn the history and the growth of the power sector in India
- Obtain an overview of the existing situation of power generation, transmission, distribution, and consumption patterns
- Acquaint with Electricity Act 2003 and Electricity Act 2007 and their objectives
- Understand restructuring of the power sector and its importance in the Indian context
- Identify various systems and sub-systems of a power network, including categorization of various types of conventional (hydro, thermal, nuclear) and non-conventional (solar, wind, tidal, magnetohydrodynamic) primary sources of energy
- Know the significance of computers in online control for the efficient management of power systems and reliable supply of electrical energy

OVERVIEW

Electrical power is the most convenient form of energy since it is available to the consumer at the very instant it is switched on. The other benefits of electrical energy are the ease with which it can be generated in bulk and transmitted efficiently and economically over long distances.

1.1 HISTORY OF POWER SECTOR GROWTH

The first electric supply system was introduced by Thomas Edison in 1882 at the Pearl Street Station in New York, USA. Power was generated in a steam engine driven dc dynamo (generator), and dc power was distributed through underground cables for lighting purposes only. The scope of distribution was limited to short distances because of the low voltage of the distribution circuits.

In pre-independent India, the generation of electric power was mainly undertaken by the private sector and was limited to urban areas. The development of the power sector commenced with the commissioning of a 130-kW generator, in 1897, at Sidrapong in Darjeeling. In 1899, The Calcutta Electric Supply Company (CESC) established the first 1000-kW steam engine driven plant in Calcutta.

Post-Independence, the Government of India (GOI) took upon itself the task of developing the power sector in a rationalized manner so as to expand the electric supply industry for the benefit of the entire country. The early 1950s, therefore, saw the setting up of state electricity boards for the systematic growth of the power sector. Side by side a number of multi-purpose hydroelectric schemes were also commenced. In due course of time, work was started on building hydro, thermal, and nuclear power generating stations. In 1975, the GOI set up the National Hydroelectric Power Corporation (NHPC) and the National Thermal Power Corporation (NTPC) to signal their participation in the generation programmes and provide a stimulus to the growth of the power industry. Subsequently, Nuclear Power Corporation of India Limited (NPCIL) and Power Grid Corporation of India Limited (PGCIL) were established to provide an additional fillip to the power sector.

1.1.1 Installed Generation Capacity

Starting with an installed capacity of 1,713 MW at the end of 1950, installed capacity is being continually enhanced to meet the growing demand for power. As per the Central Electricity Authority (CEA) Monthly Power Sector Report (December 2012), the total installed generation capacity, constituting of hydro, thermal (including steam, gas, diesel), nuclear, and renewable energy sources (RES), as on 31 December 2012, stood at 210951.72 MW. The breakdown of installed capacity of different types of plants is given in Table 1.1. By the end of the 12 th Five Year Plan (2012–17), an additional total generation capacity of 79,790 MW (out of which hydro [9,204 MW], nuclear [2,800 MW], and thermal [67,786 MW]) is envisaged.

Table 1.1 Installed generation capacity (as on 31 December 2012)

Type of generation	Capacity in MW	% of installed capacity
Coal	120873.38	57.3
Thermal Gas	18903.05	9.0
Diesel	1199.75	0.6
Nuclear	4780.00	2.3
Hydro	39339.40	18.6
RES*	25856.14	12.3
Grand total	210951.72	100.0

* Renewable energy sources (RES) include small hydro projects, biomass gas, biomass power, urban and industrial waste power, and wind energy.

1.1.2 Gross Electricity Generation

The gross electricity generation at the national level, not considering the generation from the captive plants, grew from 5107 GWh in 1950 to 5,65,102 GWh during the second year of the tenth plan (fiscal year 2003–04). The electric energy generation target for the year 2011–12 was 8,55,000 GWh. Actual electric energy generation during the year was 8,76,400 GWh, and the growth in generation during 2011–12 was 8.05%. The details of generation and growth rates are given in Table 1.2.

Table 1.2 Annual electric energy generation targets and achievement

Category	Target 2011–12 (GWh)	Actual 2011–12* (GWh)	% of Target	Actual 2010-11 (GWh)	Growth (%)
Thermal	7,12,200	7,08,500	99.47	6,65,000	6.53
Nuclear	25,100	32,300	128.41	26,300	22.86
Hydro	1,12,100	1,30,400	116.40	1,14,300	14.15
Bhutan Import	5,600	5,300	94.60	5,600	−5.82
Total	8,55,000	8,76,500	102.51	8,11,200	8.05

* Generation excludes generation from plants up to 25 MW capacity.

The Central Electricity Authority Load Generation Balance Report 2012–13 predicted the anticipated power supply position in the country during the year 2012–13, taking into consideration the power availability from various stations in operation, fuel availability, and anticipated water availability at hydroelectric stations. A capacity addition of 17,956 MW during the year 2012–13 (comprising 15,154 MW of thermal, 802 MW of hydro, and 2,000 MW nuclear power stations) was envisaged. The gross energy generation in the country was assessed to be 9,30,000 GWh from the power plants in operation and those expected to be commissioned during the year.

1.1.3 Consumption of Electric Power

As is expected for a nation on the move, the consumption of electricity increased from year to year. The electricity consumption stood at 4,157 GWh by the end of 1950; it had increased to 3,22,459 GWh during the last year of the ninth five-year plan; and by the end of the second year (2003–04) of the tenth year plan, the electricity consumption registered was 3,60,937 GWh; an increase of 12% over the last two years. In 2009, the electricity consumption figure stood at 6,00,000 GWh which is expected to double by the next decade. The major consumers of electricity are: industrial sector (34.51%), domestic sector (24.86%), and agriculture sector (24.13%).

1.1.4 Rural Electrification

As a first step towards improving the quality of life in rural India, it was essential to undertake electrification of the villages. As per the report on the Status of Rural Electrification as on 31 March 2011, brought out by the Ministry of Power, Government of India, the number of villages electrified stood at 4,39,502, which represents a coverage of 74.02% of the villages, as against the 3061 villages electrified as on 31 March 1951. Based on the 1991 census data, 17 states/Union Territories have achieved 100% electrification of the villages. The details of rural electrification as on 31 March 2011 are presented in Table 1.3.

Table 1.3 Status of rural electrification as on 31 December 2011

Total number of villages	5,93,732
Villages electrified	4,39,502 (74.02%)
Villages to be electrified	1,54,320 (25.98%)
Total number of households	13,82,71,559
Electrified households	6,01,80,685 (43.5%)
Non-electrified households	7,80,90,874 (56.5 %)

1.1.5 Transmission and Distribution Lines

At the end of the second year of the tenth five-year plan, that is, 31 March 2004, the total length of the transmission and distribution lines stood at 63,44,858 circuit kilometres (ckm) as against 29,271 ckm on 31 March 1950. Table 1.4 provides the details of operating voltages and line lengths of transmission lines during the development of Indian transmission and distribution system through the five year plans.

Table 1.4 Transmission lines (all figures in ckm*) as on 31 December 2010

	6th plan	7th plan	8th plan	9th plan	10th plan	11th plan upto Dec 2010
± 500 kV HVDC						
Central	0	0	1,634	3,234	4,368	5,948
State	0	0	0	1,504	1,504	1,504
JV/Private	0	0	0	0	0	782
Total	0	0	1,634	4,738	5,872	8,234
765 kV						
Central	0	0	0	751	1,775	3,573
State	0	0	0	409	409	409
Total	0	0	0	1,160	2,184	3,982
400 kV						
Central	1,831	13,068	23,001	29,345	48,708	68,423
State	4,198	6,756	13,141	20,033	24,730	29,931
JV/Private					2,284	4,558
Total	6,029	19,824	36,142	49,378	75,722	1,02,912
220 kV						
Central	1,641	4,560	6,564	8,687	9,444	10,360
State	44,364	55,071	73,036	88,306	1,05,185	1,21,630
JV/Private						423
Total	46,005	59,631	79,600	96,993	1,14,629	1,32,413
Grand Total	52,034	79,455	1,17,376	1,52,269	1,98,457	2,47,541

*ckm is equal to 2 × route km.

The Government of India plans to quadruple the distribution network by adding 3.2 million ckm of distribution lines in the eleventh plan. Another 4.2 million ckm is planned to be added in the twelfth plan. Thus, by the end of the twelfth plan, the total distribution network in the country would have doubled, thereby greatly facilitating delivery of power to the expanding base of end-use customers. Table 1.5 indicates the details of future requirements of distribution network in the 11th and 12th plan periods as envisaged by the Working Group on Power for the 11th plan.

Table 1.5 System augmentation of distribution lines (all figures in ckm)

Particular	11th Plan	12th Plan
66 kV overhead	23,335	30,546
33 kV overhead	1,13,936	1,49,142
6.6/11/22 kV overhead	10,36,396	13,56,638
LT lines	20,80,106	27,22,857
Total	32,53,773	42,59,183

1.1.6 Per Capita Electricity Consumption

The per capita electricity consumption, which is an indicator of the development of a country, has been steadily increasing since 1950. For example, in 1950 this figure stood at 15.6 kWh, which increased to 559 kWh during the last year of the ninth five-year plan 2001–02. During the year 2003–04, the second year of the tenth plan, the per capita electricity consumption rose to 592 kWh. As per the highlights reported by the Central Electricity Authority, annual per capita consumption of electricity in the country during the years 2004–05 to 2010–11 is provided in Table 1.6.

Table 1.6 Annual per capita consumption of electricity

Year	Per capita consumption (kWh)
2004–05	612.5
2005–06	631.4
2006–07	671.9
2007–08	717.1
2008–09	733.5
2009–10	778.6
2010–11	818.8

1.2 VISION 2012 FOR THE POWER SECTOR

In order to develop the power sector and to alleviate the losses in the power sector, the GOI has prepared a scheme entitled 'Mission 2012: Power for All'. The all-inclusive blueprint lays down an integrated strategy for the development of the power sector. The objectives defined to achieve the vision 2012 are as follows:

- Sufficient power to achieve GDP growth rate of 8%
- Reliability of power
- Quality power

- Optimum power cost
- Commercial viability of power industry
- Power for all

Achieving the objective of electrifying all households by the target year requires an addition of a massive 1,00,000 MW generation capacity and amalgamation of the regional grids into a national grid, with the latter having an inter-regional transfer capacity of 30,000 MW. Therefore, in order to achieve the above objective, additional generating capacity should be created and the transmission and distributions networks should be enhanced.

Consequently, the programmes of GOI focus on the following:

- Access to electricity to be made available for all households in the next five years
- Availability of power on demand to be fully met by 2012
- Energy shortage and peaking shortage to be overcome by providing adequate spinning reserves
- Reliability and quality of power to be supplied in an efficient manner
- Electricity sector to achieve financial turnaround and commercial viability
- Consumers' interests to be accorded top priority

1.2.1 Strategies to Achieve Power for All

The following strategies to achieve 'Power for All' have been outlined for developing the power sector:

Power generation strategy will focus on an integrated approach including low cost generation, optimization of capacity utilization, controlling the input cost, optimization of fuel mix, technology upgradation, capacity addition through nuclear and non-conventional energy sources, high priority for development of hydro power, and a comprehensive project monitoring and control system.

Transmission strategy focuses on development of a National Grid including interstate connections, technology upgradation and optimization of transmission cost.

Distribution strategy is to concentrate on distribution reforms by focussing on system upgradation, loss reduction, theft control, consumer service orientation, quality power supply commercialization, decentralized distributed generation, and supply for rural areas.

Regulation strategy aims at protecting consumer interests and making the sector commercially viable.

Financial strategy aims at generating resources required for the growth of the power sector.

Conservation strategy is to optimize electricity utilization with focus on demand side management and load management and technology upgradation to provide energyefficient equipment/gadgets.

Communication strategy focuses on achieving political consensus, with media support, to enhance general public awareness.

1.2.2 Vision 2020 for the Power Sector

The Vision 2020 committee set up by the Planning Commission in June 2000 envisages efficient and environment-friendly energy resources which would become the growth engines to provide speedy and sustainable future economic development. In order to power the country's industries, transport vehicles, homes, and offices, the demand for power is estimated to grow by another 3.5 times or more in the next two decades. This, in turn, will require that the installed generation capacity be compulsorily tripled from 101,000 MW to 292,000 MW. Such an overall growth in power demand will need a matching supply of all forms of fuels leading to a doubling of the coal demand and tripling of the demand for both oil and gas. Such swelling demands for the nation's growing requirements of energy will further strain the social and physical environments, in addition to increasing vulnerability due to fluctuating international market prices.

In today's energy scenario, it would be prudent to take a focussed visionary approach to place greater dependence on renewable energy sources, which not only offer immense economic benefits but also offer social and environmental benefits. With India already being the fifth largest wind power generating country in the world, use of other renewable energy technologies such as solar power, solar thermal, biomass, and small hydro power is being explored.

1.3 POWER SECTOR REFORMS

Development of the power sector continues to be one of the greatest challenges in maintaining economic growth and further reducing poverty in India. About 45% of the households remain unconnected to the public power system, and those who are connected often receive infrequent and unreliable service.

The State Electricity Boards (SEBs) have been incurring losses and are unable to even make payments to the Central Power Sector Units (CPSUs) such as NTPC and PGCIL for the purchase of power. The accumulation of outstanding amounts to the CPSUs grew to over ₹ 40,000 crore, seriously hampering their capacity addition programme. The reform of the power sector is crucial, as financial losses amount to 1.5% of the GDP. To strengthen the financial health of the power sector, the GOI has taken up reforms for gradual elimination of losses. In India, the power sector reform process was initiated in 1991 and since then the Indian power sector has been witnessing major structural changes.

1.4 PERFORMANCE/POLICY INITIATIVES/DECISIONS

During the year 2003–04, the following decisions were taken for achieving the objectives of 'Power for All'.

- Electricity Act 2003 was enacted in July 2003.
- Accelerated electrification programme for 1,00,000 villages and one crore rural households launched. The scheme outlay of ₹ 6,000 crore comprised a grant component of ₹ 2,400 crore.
- 50,000 MW hydro initiative launched.
- Improvement in power supply position: Since the beginning of the ninth five-year plan (1996–97), the peak shortfall had reduced from 18% to about 11%. Supply shortfall had also reduced from 11.5% to 7.1%.
- Generation performance: During 2003–04, the generation, compared to the previous year, improved from 531 billion units (BU) to 558 BU (1 billion = 10^9). Overall plant load factor (PLF) of generating stations improved from 72.2% to 72.7% while in the central sector it improved from 77.1% to 78.7%.

Capacity addition target and achievement   Uninterrupted and reliable supply of electricity for 24 hours a day needs to become a reality for the whole country including rural areas. In order to fully meet the energy and peak demand by 2012, enough generating capacity has to be created with some spare generating capacity so that the system is also reliable. The sector is to be made financially healthy so that the state government finances are not burdened by the losses in this sector. The sector should be able to attract funds from the capital markets without government support. The consumer is paramount and he/she should be served well with good quality electricity at reasonable rates.

To ensure grid security, quality, and reliability of power supply, a reasonable spinning reserve at the national level has to be created in addition to enhancing the overall availability of installed capacity to 85%.

A capacity of about 1,00,000 MW is planned to be set up during the tenth and eleventh five-year plans, that is, between 2002 and 2012. This implies doubling the installed capacity which works out to adding about 430 MW every fortnight! Capacity addition plan for addition of 41,110 MW has been finalized for the tenth plan period. The central, state, and private sector's share of the capacity addition is expected to be 51%, 16%, and 25% respectively. About 7% is expected to come from renewable sources and 2% from the Tala project in Bhutan.

A hydro power initiative has also been launched by the Prime Minister in 2003, under which a capacity of 50,000 MW is to be added in the same period, that is, 2002 to 2012. Outlay for power sector for the tenth plan period has been enhanced to ₹ 1,43,399 crore, an increase of approximately 214% over the ninth plan outlay of ₹ 45,591 crore. Advance action plan has also been initiated to identify the capacity addition required in the eleventh plan. In the last two plan periods, barely half of the planned capacity addition was achieved. The optimistic expectations from the Investments in Power Projects (IPPs) have not been fulfilled and, in retrospect, it appears that the approach of inviting investments on the basis of government guarantees was perhaps not the best way.

1.5 ELECTRICITY ACT 2003

The Electricity Act 2003 envisages bringing in a market-oriented management in the power sector by introducing in it a spirit of competition which hitherto was non-existent. The major focus of the Act is to amend existing laws and enact new laws in the areas of generation, transmission, distribution, buying and selling, and utilization of power. The objectives of the Act are as follows:

- Generating a market-responsive competitive power industry
- Guaranteeing reliable and quality power supply to all areas
- Rationalizing tariff regime
- Bringing down the levels of cross-subsidization
- Safeguarding consumer interests

The salient provisions in the Act are as follows:

- No licences for setting up generating stations, except hydro stations, subject to their meeting specified technical standards
- No permission is necessary for establishing captive power plants

- 'Open access' permitted for transmission
- Multiple licences are permissible for transmission and distribution (T&D) in the same geographical area
- Establishing a spot market (stock market) for bulk power
- Function (generation, transmission, and distribution) based unbundling of state electricity boards
- Accountability through mandatory metering of electrical energy supplied to consumers

The Act, however, permits the state electricity boards to operate, for a limited period, with their integrated structure and allows them to select the order and segments of restructuring. The Electricity Bill 2003 passed by the Parliament in May 2003 is a unified central legislation and replaces the earlier three electricity Acts of 1910, 1948, and 1998 along with their amendments.

1.5.1 Electricity Act 2007

The Electricity (Amendment) Act was passed by the Parliament in June 2007. It would be out of the scope of this text to discuss the Electricity (Amendment) Act 2007 in its entirety. In Section 6 of the Electricity Act 2003, the following section shall be substituted:

"6. The concerned State Government and the Central Government shall jointly endeavour to provide access to electricity to all areas including villages and hamlets through rural electricity infrastructure and electrification of households."

Other important amendments to the Electricity Act 2003 are as follows:

- The term 'elimination' has been omitted in relation to cross-subsidies.
- Captive units will not require a licence to supply power to any user.
- Strict action against unauthorized usage of power.
- Power theft has been recognized as a criminal offence, punishable under Section 173 of the Code of Criminal Procedure, 1973.

1.6 RESTRUCTURING THE POWER SECTOR

It would be appropriate to develop the concepts associated with regulation and deregulation, in respect of the power sector in India, before describing the concepts of restructuring.

Regulation of the power sector implies that it must function within the laws and regulations which have been specified by the government.

Deregulation of the power sector implies that the government has specified rules and economic enticements for restructuring, controlling, and driving the electrical power sector.

Clearly, a regulated power sector means that it functions in a monopolistic and risk-free environment. On the other hand, in a deregulated scenario, the power sector operates in a competitive environment and is subject to market risks. Thus, regulation and deregulation symbolize opposite concepts without any one of them being absolutely black or white.

1.6.1 Features of a Regulated Power Sector

Monopolistic   There is only a single authority to generate, transmit, distribute, and sell electrical energy.

Responsibility to supply   The authority is obliged to supply energy to all areas irrespective of viability and profitability.

Government as a supervisor   The government acts as a regulator by legislating laws and rules within which the authority should operate and do business. This implies that

(a) the operating principle of the authority must be based on least-cost operation, that is, it must function such that it minimizes its overall revenue needs, and

(b) the government decides the rates to be charged by the authority.

The authority is expected to function within the government's specified regulatory guidelines and practices and it is assured a reasonable return on its investments.

1.6.2 Structure of a Regulated Power Sector

The electrical power sector in India, until recently, has been vertically integrated, with all the functions of power generation, transmission, and distribution being performed by a single entity, which complicated the separation of costs attributed to the three activities. Therefore, the electricity tariff rate charged to consumers is based on cumulative costs.

From the foregoing, the structure of a regulated authority may be conceptualized as one in which (i) information flow is present between the generating and transmission systems only, and (ii) the direction of flow of money is from bottom to top only, that is, from consumers to the authority.

1.6.3 Structure of a Deregulated Power Sector

Unlike the regulated power sector, a deregulated power sector is characterized by a competitive structure in which the various job functions, in a traditional set-up, are identified and segregated so that these job functions, whenever practical, can be thrown open to competition for improving efficiency and profitability. The procedure of restructuring is called unbundling.

Generally, the objective of the government in deregulating the power sector is to induce competition, by allowing several new players, in the production of electrical energy (generation) and retail marketing (distribution) of electricity, while maintaining a single transmission and distribution system in an area. Figure 1.1 presents a conceptual perspective of a deregulated power sector.

As can be observed from the figure, an unbundled power sector permits the entry of various players to undertake different tasks. In order to ensure smooth and uninterrupted functioning, a central operating authority, designated as Self-governing System Administrator (SSA) or Independent System Operator (ISO), is appointed for the entire system. The SSA is an independent authority. It does not possess its own generation facilities for business or indulge itself in market competition. The SSA ensures that there is a balance between generation and imports on one hand and consumption and exports on the other, at all times.

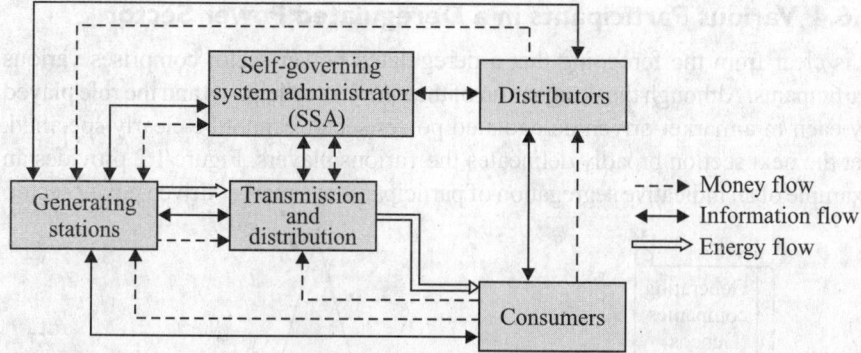

Fig. 1.1 Conceptualized perspective of a deregulated power sector

In the deregulated model, the flow of energy is from the generating stations to the consumers, via the transmission and distribution system as in the regulated power sector. In terms of the functioning of the deregulated power sector, one model is that various generating companies will sell and deliver power via the transmission and distribution (T&D) system to the distributors. The consumer transacts with the distributors. Another practice is that the consumer transacts directly with the generating company. The T&D system is operated by the SSA.

From the perspective of information flow, the consumer generally communicates with the distributor and the generating companies while T&D authorities and the distributors communicate with the SSA. In the model shown in Fig. 1.1, money flow is from the consumer to the distributor, generating companies, and T&D companies. Money is also exchanged between the SSA and the generating and T&D companies. The generating companies pay to the T&D authority for using their facility for supplying energy to the consumer. There is, however, no money flow between the distributor and the SSA.

Functionally, the consumer places a demand for energy with the distributor who in turn buys power from the generating company and transfers it to the customer via the regulated T&D system which is operated by the independent SSA. The SSA is accountable for maintaining a liaison with the various players and keeping track of the transactions being enacted.

Owing to the various players in the deregulated structure, the energy bill gets segregated into various amounts to be paid towards generation, transmission, and other costs. This is in contrast to the single energy bill in the regulated power sector. The different heads under which an energy bill gets segregated into, in the deregulated scenario, are as follows.

- Price of energy supplied.
- Price of energy delivered: This is analogous to the price of transportation of goods from one station to another.
- Price of quality of energy supplied: The quality of energy supplied is determined by the extent to which frequency is regulated and the voltage magnitude is controlled at the consumers' end. These services are individually priced and charged. The price for these services, however, may or may not be indicated in the energy bill separately.

1.6.4 Various Participants in a Deregulated Power Sector

It is clear from the foregoing that a deregulated power sector comprises various participants. Although the designation of the various participants and the role played by each in a market-driven deregulated power sector cannot be clearly specified, yet the next section broadly delineates the various players. Figure 1.2 provides an example of an indicative segregation of participants in a market-driven power sector.

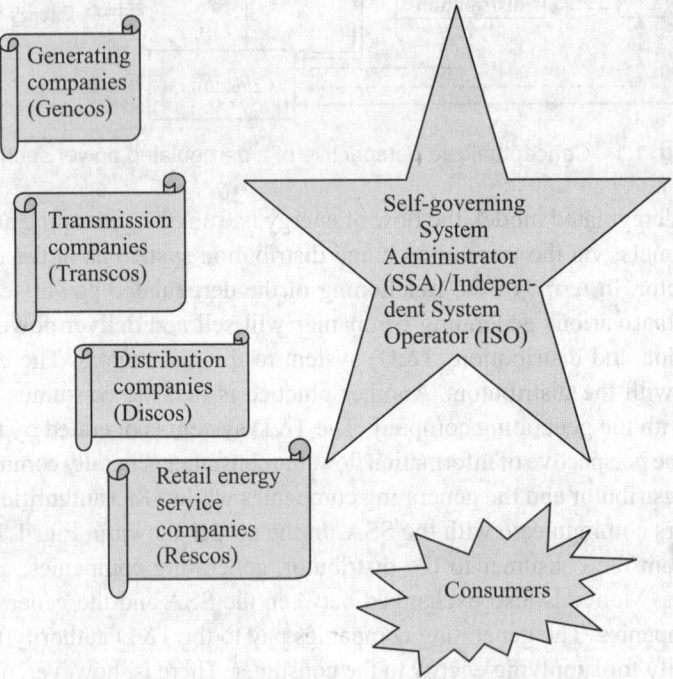

Fig. 1.2   Various components of a deregulated power sector

1.6.5 Role Description of Various Participants

Generating company (Genco)   Genco is a company which owns and operates generating stations to produce electrical power. The bulk energy produced by a Genco is sold at its site which is similar to a petroleum company selling bulk crude at its site.

Transmission company (Transco)   The role of a Transco is to transfer bulk power generated at the site of a Genco to where it is to be delivered. In respect of ownership, management, and maintenance of transmission lines, one of the following methodologies may be adopted in a deregulated power sector. Normally, Transco owns and maintains the transmission lines under monopoly contract. The operation of the transmission lines is undertaken by the SSA/ ISO. Since Transco is the sole franchisee of the transmission lines, it is paid for by the SSA/ISO for the use of the facility. Transco may also be assigned the administrative responsibility of carrying out the engineering functions to ensure that the transmission lines perform the task of transmitting power adequately.

Distribution company (Disco)   Disco performs the function of power delivery to individual consumers consisting of business houses, commercial organizations, and domestic customers. Disco is also an owner-operated organization under monopolistic contract. There are two models in which a Disco can function. In the first model, Disco owns and controls the distribution network and it earns its income by renting out the same or by billing for delivery of electric energy. In the second model, a Disco buys power in bulk either directly from a Genco or from the spot (stock) market and delivers it to the consumers.

Retail energy service company (Resco)   Rescos can be carved out of the several retail departments of the earlier vertically integrated utilities or these could be new entrants in the electrical power industry who believe that they are competent in the art of selling. The job of a Resco is to vend electrical power, that is, buy power from Gencos and sell it directly to the consumers.

Self-governing System Administrator (SSA)/Independent System Operator (ISO)   An SSA/ISO is an independent authority whose role is to ensure reliable and secure electrical power system. It does not involve itself in the electricity market trade. However, in order to ensure security and reliability of the electrical power supply, an SSA/ISO obtains various services such as reactive power and supply of emergency reserves from different players in the system. Normally, an SSA/ISO does not possess any generation capacity, apart from reserve capacity, in some cases.

Consumers   In the deregulated environment, consumers are categorized as a single entity and are perceived as the buyers. A consumer has the option to purchase electrical energy from a local Disco, or directly from a Genco, or from the spot (stock) market by bidding.

1.6.6 Mechanism of Competition

One of the basic objectives of deregulation is to promote competition within the electrical power industry. Figure 1.3 symbolically explains how the mechanism of competition is created in the electrical power industry.

In a deregulated electric power industry, competition is at the generation level and at the retail market level. Companies (Gencos) generate bulk power and put it on the market for sale at the wholesale level. Typically, customers who buy power in bulk from Gencos are large industrial consumers or other companies. For maximizing their profits, Gencos offer their power in the marketplace.

At the retail market level, distribution companies (Discos) buy power in bulk from Gencos and sell the same to individual customers, in small quantities, as per the requirement of the latter. Rescos and Discos compete for increasing their consumer base by offering competitive prices, good services, and other attractive service features.

A deregulated electrical power sector, therefore, is competitive at the wholesale and retail levels and in between is the monopolistic transmission and delivery system.

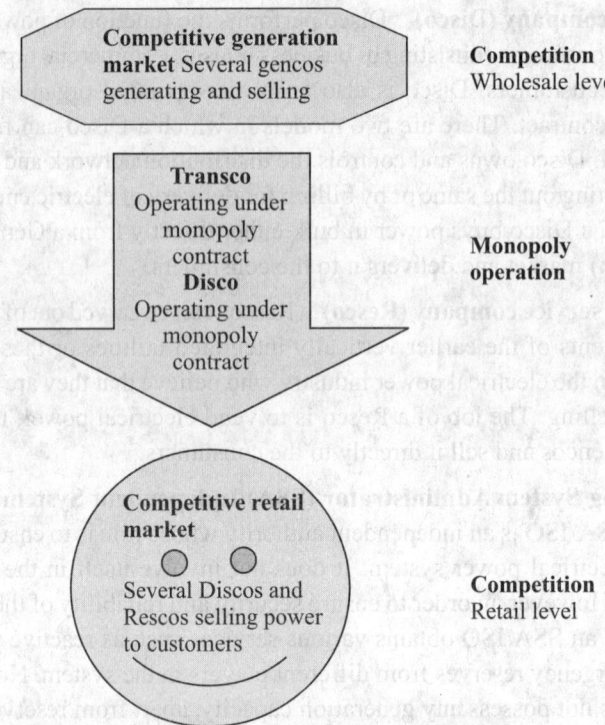

Fig. 1.3 Symbolic representation of the mechanism of competition in a deregulated environment

1.7 WHOLESALE POWER MARKET

It is logical to expect that there exists (i) a marketplace for the sellers and buyers to trade, and (ii) an adequate system for transportation and distribution. Similarly, in the context of a deregulated power industry, the following two additional systems are essential.

Power market A methodology for the bulk power producers (Gencos) to sell and Rescos and Discos to buy power.

System operation A viable transmission system, on real-time basis, which will transport power from the sellers' (Genco) site to the consumers' site.

The cornerstone of a deregulated power industry is a level playing field to all the entities. System operation is satisfactorily achieved by Transcos and Discos, which are regulated by an independent mechanism. The concept of a power market was both novel and alien to the power industry and required the introduction of new players.

1.7.1 Instruments of Sales Transactions

A market mechanism through which Gencos can sell their product (power) at competitive prices and transact business with buyers (Rescos and Discos) is an essential feature of a deregulated power industry. There are three fundamental models for transacting business and they are described as follows:

Poolco   It is the single government or semi-government buyer which also functions as the system operator. Its job is to buy power for all consumers. It operates by inviting bids from all Gencos and buying power, at the lowest quoted price, to meet the total demand.

Bilateral exchange   It functions on a multi-seller and multi-buyer approach. In this, bilateral agreements are reached confidentially, between a seller and a buyer, to exchange power at a mutually agreed price.

Power exchange (PX)   It is a trading exchange for electrical power, very similar to the monetary stock exchange, and is established by the government. In this mechanism, business is transacted through the PX. Similar to a monetary stock exchange, the PX constantly revises and declares the current price, called the market clearing price (MCP), at which transactions are done. A feature of the PX is that both the sellers and the buyers are really 'conversing' with the PX (marketplace) instead of individual sellers and buyers and thus are not aware of whom they are dealing with.

It may be mentioned that none of the three market instruments are mutually exclusive. It is not impossible to have several combinations of all three market devices to be operative at the same time. In practice, however, it is prudent for two of the three devices to be in operation concurrently.

1.8 RESPONSIBILITIES OF THE SSA/ISO

The SSA/ISO is the key to the successful operation of a deregulated power industry. It must function transparently to ensure a secure and reliable system, justifiable and impartial transmission tariffs, and provide other services. The SSA/ISO is expected to perform the following basic functions:

- Deliver power on request to the sellers and buyers, and to transmission services for the transportation of power.
- Determine and post prices for transmission usage, offer to reserve or sell track usage, invoice users and settle the same with the users, and promptly pass on revenues to owners of T&D system owners.
- Operate the system in a stable and economical mode.
- Assure and provide high quality of service.
- Provide ancillary services such as generation/load balance control area, respond quickly to correct generation/load imbalances by utilizing spinning reserve, maintain system voltages within specified limits by injecting or absorbing reactive power, and operate the system so as to minimize the transmission losses.
- The SSA/ISO should itself operate in a manner to reflect optimum economic efficiency and should deal fairly, equitably, and transparently with all the entities in the industry.

1.9 STATUS OF DEREGULATION OF THE POWER SECTOR IN INDIA

The physical infrastructure in the power sector in India has witnessed incredible growth since Independence. This expansion is attributed to

(a) government budgetary support,

(b) cross-subsidy,

(c) emphasis on utilizing indigenous resources for expansion, and

(d) centralized supply and grid expansion.

Until 1991, the power sector was vertically integrated with over 98% generation and over 95% distribution being undertaken by the state-or central government-owned utilities, like the electricity boards. Therefore, the power industry imbibed the inherent shortcomings such as operational inefficiencies, high T&D losses, and over-staffing of a vertically integrated system.

The government of Orissa was the first state to initiate the process of restructuring, in the mid-1990s, with the help of a loan from the World Bank (WB). The model of unbundling envisaged: (i) restructuring the state-owned electricity board (SEB) into three separate entities consisting of generation, transmission, and distribution, (ii) generation and distribution to be privatized, and (iii) setting up an independent regulatory authority to monitor and control these players.

Following in the footsteps of the Orissa state, various other states, such as Andhra Pradesh, Haryana, Rajasthan, and Uttar Pradesh, have obtained loans from international funding agencies, such as the WB and the Asian Development Bank (ADB), for unbundling their respective SEBs.

1.10 CONSTITUENTS OF A PRESENT-DAY POWER SYSTEM

An electrical power system is a complex network of several subsystems, which convert some form of latent energy into electrical energy and transform, transmit, and distribute it for consumption at the customers' terminals. Figure 1.4 is a single-line representation of a three-phase ac power system.

The various subsystems of an electrical power system may be classified as follows:

(a) Generation

(b) Transmission and distribution

(c) Loads

(d) Protection and control

It may be noted that transformers are used in all the subsystems. A transformer transfers power with very high efficiency from one voltage level to another voltage level. The insulation requirements limit the generated voltage to low values up to 30 kV. Thus step-up transformers are used for transmission of power at the sending end of the transmission lines. At the receiving end of the transmission lines, step-down transformers are used to reduce the voltage to suitable values for distribution to consumers of electric energy. Furthermore, depending on power handling capacity, two types of transformers, namely power transformers and distribution transformers, are shown in Fig. 1.4. Power transformers are usually rated from 250 kVA up to 1000 MVA, and distribution transformers are rated between 20 kVA and 250 kVA.

In Fig. 1.4, voltage levels at various subsystems are indicated. EHV designates extra-high voltage, usually above 220 kV and up to 800 kV. HV denotes high voltage, usually from above 66 kV to no more than 220 kV. MV means medium voltage, usually from above 1 kV but less than 66 kV. LV stands for low voltages, which are 1 kV or less.

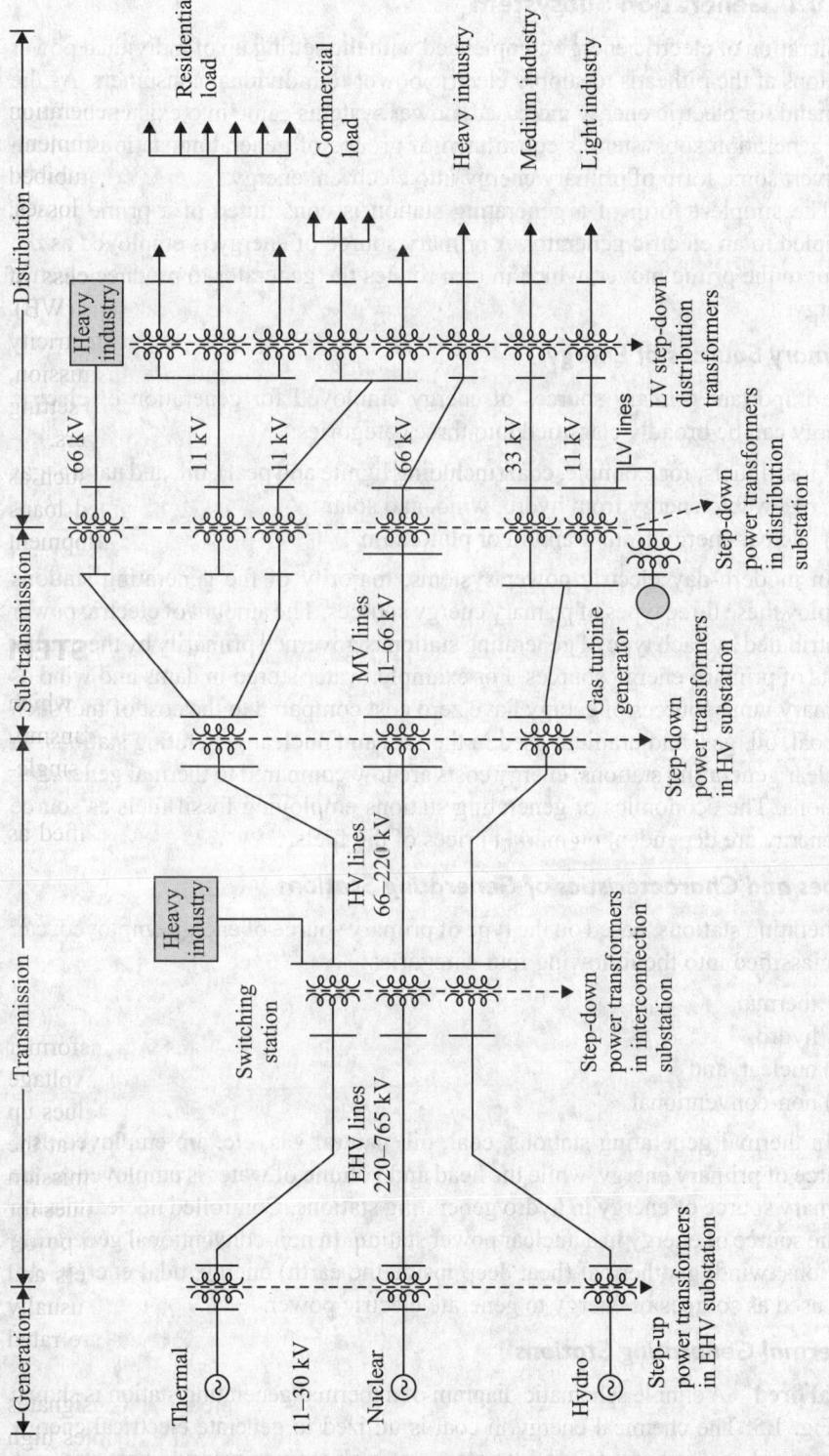

Fig. 1.4 Single line representation of an ac network

1.10.1 Generation Subsystem

Generation of electric energy commenced with the setting up of individual power stations at the pitheads to supply electric power to individual consumers. As the demand for electric energy increased, power systems came into existence. Thus, the generation subsystem is constituted of groups of generating stations, which convert some form of primary energy into electrical energy.

The simplest form of a generating station is constituted of a prime mover coupled to an electric generator. A primary source of energy is employed as the input to the prime mover, which in turn rotates the generator to produce electric energy.

Primary Sources of Energy

The important primary sources of energy employed for generation of electric energy can be broadly classified into three categories:

(i) fossil fuels, for example, coal (including lignite and peat), oil, and natural gas
(ii) renewable energy from hydro, wind, and solar
(iii) nuclear energy from uranium or plutonium

In modern-day electric power systems, majority of the generating stations employ these three types of primary energy sources. The amount of electric power contributed by each type of generating station is governed primarily by the market costs of primary energy sources. For example, water stored in dams and wind as primary input sources of energy have zero cost compared to the cost of fuel such as coal, oil, gas, and uranium used in thermal and nuclear generating stations. In nuclear generating stations, energy costs are low compared to thermal generating stations. The economics of generating stations employing fossil fuels as source of energy are dependent on market prices of the fuels.

Types and Characteristics of Generating Stations

Generating stations, based on the type of primary source of energy employed, can be classified into the following four categories:

(i) thermal,
(ii) hydro,
(iii) nuclear, and
(iv) non-conventional.

In thermal generating stations, coal, oil, natural gas, etc. are employed as a source of primary energy, while the head and volume of water is employed as the primary source of energy in hydro generating stations. Controlled nuclear fission is the source of energy in a nuclear power station. In non-conventional generating stations, wind, geothermal (heat deep inside the earth) energy, tidal energy, etc. are used as sources of energy to generate electric power.

Thermal Generating Stations

Coal fired A simple schematic diagram of a thermal generating station is shown in Fig. 1.5. The chemical energy in coal is utilized to generate electrical energy. Pulverized coal is burnt to produce steam, at high temperature and pressure, in a boiler. The steam so produced is passed through an axial flow steam turbine, where the internal heat energy of the steam is partially converted into mechanical energy.

The steam turbine, which is the prime mover, is coupled to an electric generator. Thus, mechanical energy produced by the rotating turbine is converted into electric energy.

The efficiency of the process of conversion of chemical energy into thermal energy and then into mechanical energy is poor. Due to heat losses in the combustion process, rejection of large quantity of heat in the condenser and rotational losses, the maximum efficiency of the conversion process is limited to about 40%. In order to increase the thermal efficiency of conversion of heat into mechanical energy, steam is generated at the highest possible temperature and pressure. To further increase the thermal efficiency, steam is reheated after it has been partially expanded by an external heater. This reheated steam is returned to the turbine where it is expanded in the final stages of bleeding.

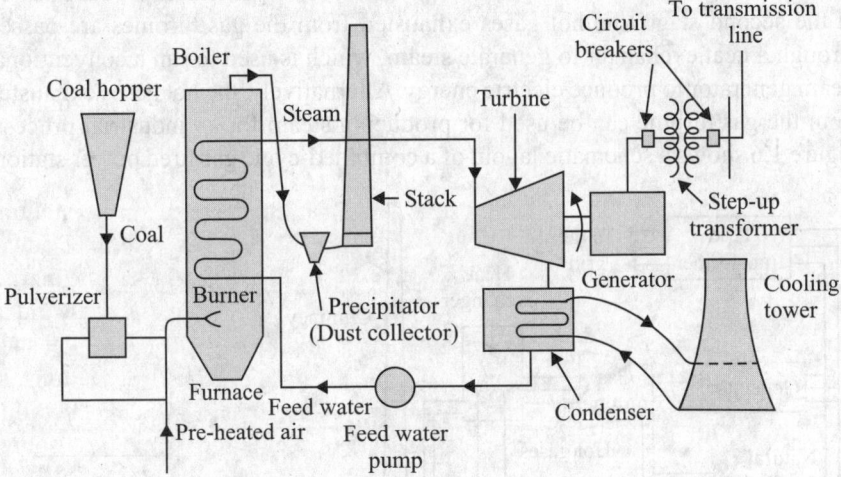

Fig. 1.5 Schematic diagram of a thermal generating station

Modern practice is to design and build generating units having large megawatt generating capacity, since their capital cost per kilowatt decreases as the megawatt capacity is increased. Increasing the unit capacity from 100 MW to 250 MW results in a saving of about 15% in the capital cost per kilowatt. It is also established that units of this magnitude result in fuel saving of the order of 8% per kilowatt-hour. Additionally the cost of installation per kilowatt is considerably lower for large units. Currently, the maximum capacity of turbo-generator sets being produced is nearly 1200 MW. In India, super thermal units of capacity 500 MW are being commissioned by BHEL.

Thermal generating stations also employ cogeneration in order to utilize the large amount of waste heat. In cogeneration, electricity and steam or hot water are simultaneously made available for industrial use or space heating. It is claimed that cogeneration results in an overall increase in efficiency of up to 65%. Cogeneration has been found to be particularly advantageous for industries such as paper, chemicals, textiles, fertilizers, food, and petroleum refining.

The waste gases produced by coal fired generating stations contain particles and gases such as oxides of sulphur and NO_x. These gases are released to the atmosphere resulting in pollution of air. Thermal pollution also results due to the large amount of heat released via the condenser to the cooling water.

Oil fired   In the oil fired steam station, oil is employed to produce steam, which is used to run the steam turbine. In the oil fired stations, the oil used is of two types: crude oil, which is the oil pumped from oil-wells, and residual oil, which is the oil left behind after the more valuable fractions have been extracted from the crude oil. Cost of transporting oil through pipelines is less than shipping coal by rail. However, residual oil fired stations have to be located close to the oil refinery because it is uneconomical to transport residual oil by pipelines because of its high viscosity.

Gas fired   The primary source of energy in such types of generating stations is natural gas. A gas turbine engine, which is similar to a turbo-prop engine used in an aircraft, is employed as a prime mover to run the generator. In order to achieve higher thermal efficiency, combined-cycle method is used to generate electricity. In the first stage, gas turbine engines coupled to electric generators produce power. In the second stage, the hot gases exhausted from the gas turbines are passed through a heat exchanger to generate steam, which is used to run a conventional steam generator to produce electric energy. Alternatively, the hot gases exhausted from the gas turbine can be used for producing steam for an industrial process. Figure 1.6 shows a schematic layout of a combined-cycle gas fired power station.

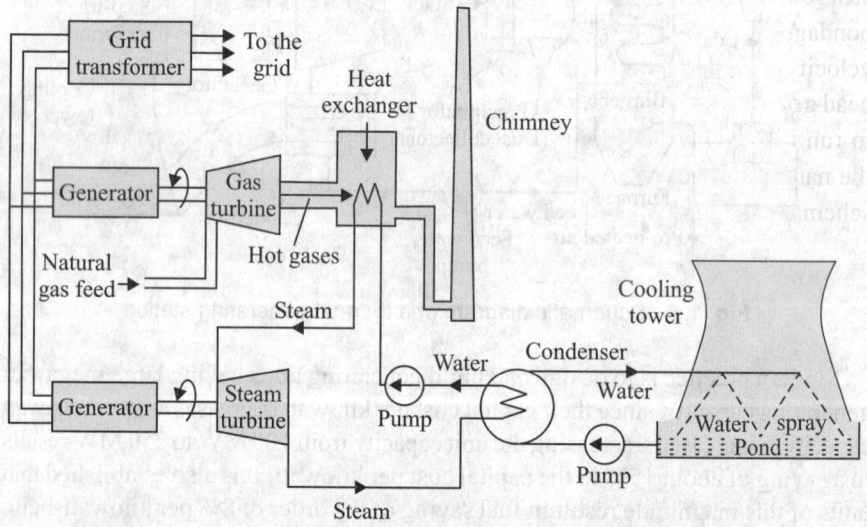

Fig. 1.6   Schematic layout of a combined-cycle gas fired power station

For the same amount of power generated, combined-cycle gas fired stations are more environment friendly. The flue gases emitted by these stations contain almost zero sulphur dioxide, 50% carbon dioxide, and 25% NO_x as compared to those produced in a coal fired power station. Compared with coal fired steam power stations, the installation cost of gas fired stations is lower and they can be quickly started. The operational cost of gas fired stations is high due to the high fuel cost when employed to supply power on their own. As such, they are used to supply peak load demand and for short periods.

The world over, gas turbines in conjunction with 100 MW generators are being used to generate electrical power. In India, a gas power station with an installed capacity of 180 MW (6 × 30 MW) is operational in Delhi.

Diesel oil fired　Diesel oil is used to run large internal combustion engines of the type employed in ships. The diesel oil fired stations exhibit characteristics similar to those of a gas fired station. However, the speed of a diesel oil fired station is considerably low and its fuel efficiency is higher than that of a steam power station. Since diesel is more expensive than oil, in an oil fired steam station, the use of diesel oil fired stations is limited to supplying stand-by power.

Hydro Generating Stations

In a hydroelectric generating station, the potential energy and quantum of water are utilized to generate electrical power. In other words, hydroelectric schemes function on flow of water and difference in level of water known as head. Due to the difference in head, considerable velocity is imparted to the water, which is used to drive a hydro turbine. This hydro turbine acts as a prime mover and is coupled to an electric generator to produce electrical energy.

Hydroelectric stations depend on the availability of a head of water. As such they are often sited in mountainous terrain and require long transmission lines to deliver power to the load centres. Hydroelectric schemes are classified on the basis of the head utilized to generate power: high head storage type, medium head pondage type, and run-of-river. In low head type of hydro generators, both the velocity of water and difference in levels are used to rotate the turbine. In high head generators, the difference in levels is used to impart high velocity to the water to run the turbine. As the name suggests, in the run-of-river hydro generators, the natural flow of river water is used to drive the turbines. Figure 1.7 shows a schematic diagram of the high head storage type hydroelectric scheme.

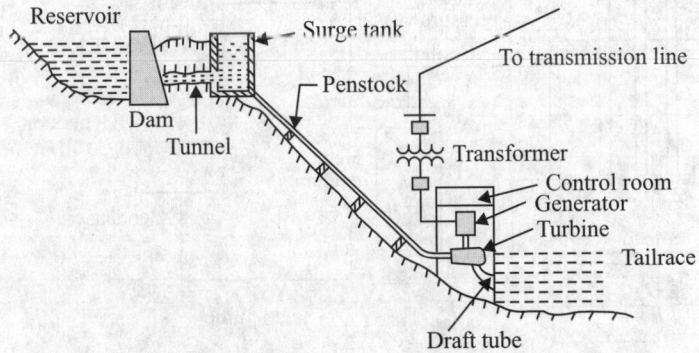

Fig. 1.7　Schematic diagram of a high head storage type of a hydroelectric scheme

The power P generated in a hydroelectric station is given as

$$P = 9.81 \rho Q h \eta \times 10^{-3} \text{ kW} \tag{1.1}$$

where Q is the discharge of water in m³/s through the turbine, ρ is the specific weight of water in 1000 kg/m³, h is the head of water in metres, and η is the generation efficiency.

The merits of a hydroelectric station are as follows:

- Minimal operational costs (since there is no fuel cost involved)
- No air pollution

- No waste products
- Minimal maintenance
- Quick start-up time (within five minutes)
- Long life (minimum fifty years)

The demerits of a hydroelectric station are as follows:

- High capital costs
- Long gestation period
- Ecological damage to the region

Nuclear Power Stations

The fuel in a nuclear power station is uranium. Of the two isotopes of uranium, uranium-235 and uranium-238, found in natural uranium, only uranium-235 is capable of undergoing fission. Fission in uranium-235 is brought about by bombarding it with neutrons. Due to the fission reaction, heat energy and neutrons are released. The released neutrons further react with fresh uranium-235 atoms to generate more heat and produce more neutrons. Thus the fission reaction is a chain reaction and is required to be conducted under controlled conditions in a nuclear reactor.

In a nuclear power station, the nuclear reactor constitutes the heart of the station and replaces the boiler in coal or oil fired stations. Figure 1.8 shows a schematic layout of a nuclear power station.

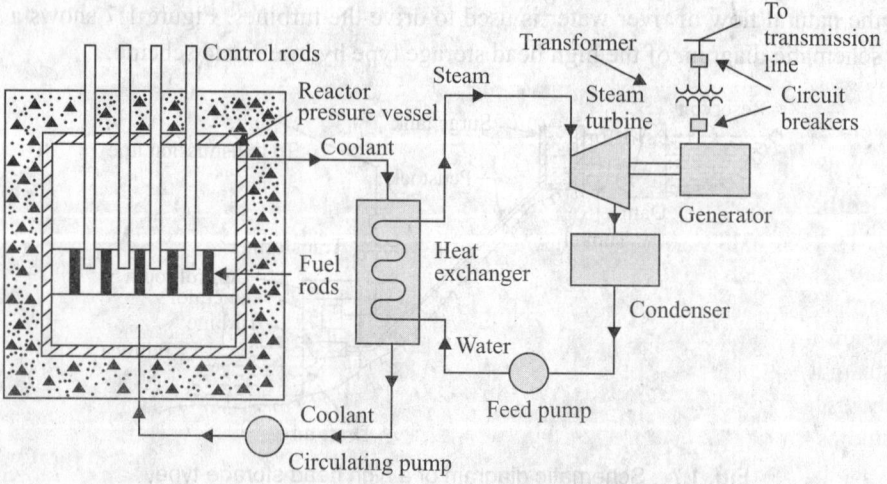

Fig. 1.8 Schematic layout of a nuclear power station

In the reactor pressure vessel, nuclear fuel rods are embedded in neutron speed reducing agents such as heavy water and graphite called moderators. These moderators reduce the speed of neutrons to a critical value. The nuclear reaction is controlled by inserting boron steel rods, which have the property to absorb neutrons. Thus, the rate of nuclear fission is controlled by controlling the neutron flux.

A primary coolant such as heavy water or carbon dioxide is used to transfer the heat generated due to the fission reaction to the heat exchanger. Steam is produced in the heat exchanger, which is used to run a conventional steam turbine.

The fuel requirements of a nuclear generating station are minimal compared to a coal fired generating station. In addition, the cost of transporting nuclear fuel is negligible. Another advantage of nuclear power stations is that they do not produce any air pollution. Nuclear stations, therefore, can be sited close to load centres. However, since radioactive fuel waste is produced in the nuclear reactor, safety considerations demand that nuclear stations be sited away from the populated areas. Nuclear stations require a high capital investment. The operational cost of such stations, however, is low.

Non-conventional/Alternative Generating Stations

Wind power stations   Wind as a source of energy has been used for centuries to grind grain and pump water. It is particularly attractive since it is non-polluting. However, it is unpredictable and unsteady. The expression for theoretical power generated, in watts, by wind of average velocity V metres per second is given by

$$P = 0.5\rho A V^3 \text{ W} \tag{1.2}$$

where ρ is the air density (1201 g/m^2 at normal temperature and pressure) and A is the swept area in square metres.

The success of wind power generating stations is governed by the initial capital cost, maintenance cost, useful life, and power output. Wind power generating stations are useful for meeting low power requirements in small isolated areas. In India, the gross potential of wind power has been assessed at approximately 45,000 MW, and the technical potential is estimated at 13,000 MW. Wind power stations have been set up in the states of Gujarat, Maharashtra, Orissa, Andhra Pradesh, and Tamil Nadu. The largest installation of wind turbines in the country so far is in the Muppandal–Perungudi area near Kanyakumari in Tamil Nadu with an aggregate installed capacity of about 500 MW. State-of-the-art technology is now available in India for manufacturing wind turbines of capacity up to 750 kW.

Geothermal power stations   Geothermal power generation involves conversion of the heat energy contained in hot rocks inside the core of the earth into electricity through steam. Water is used to absorb heat from the rock and transport it to the earth's surface, where it is converted to electric energy through conventional steam-turbine generator. Geothermal energy has been employed to generate steam in a limited way in Italy, New Zealand, Mexico, USA, Japan, etc. In India, the use of geothermal energy is still at the developmental stage with feasibility studies for a 1-MW station in Ladakh being undertaken. Though the efficiency of a geothermal station is less than that of a conventional fossil fuel plant, geothermal stations have become attractive due to their low capital cost and zero fuel cost. The total available geothermal power globally has been estimated at 2000 MW of which only about 500 MW has been tapped. In India, despite a number of hot springs, the availability of exploitable geothermal energy potential appears to be unattractive.

Tidal power stations   The gravitational effects of the sun and the moon and the centrifugal forces of the earth's rotation on its axis cause sea tides. In about 25 hours there are two high tides and two low tides. The minimum head required for generation is about 5 m. Tidal power stations use periods of high tides to fill reservoirs, through open sluice gates, behind embankments along the seashores.

During low-tide periods, when the tide is falling on the seaward side of the embankments, the sluice gates are closed and stored water is made to flow through turbines coupled to generators. This is known as *ebb generation*.

A tidal power station is usually sited at the mouth of an estuary or a bay. A barrage or an embankment is constructed at the site to store water. A two-way generation can be achieved in a tidal power station. As the tidal waves come in, water flows through the reversible turbines to generate power and fill the estuary/bay. As the tide falls, water flows out of the estuary/bay and the turbines. Since the turbines are reversible, power is again generated.

The disadvantage of tidal power stations is that, due to variation in high and low tide timings, they may be generating at peak demand on some days and idle for other days. Another disadvantage is the high cost of civil engineering works required.

With hundreds of kilometres of coastline, a vast potential source of tidal energy is available in India. It has been planned to set up a 600-MW tidal power station in India by constructing a dam at Kandala on the Gujarat coast. Other sites under exploration are at Bhavnagar, Navalakhi (Kutch), Diamond Harbour, and Ganga Sagar.

Solar power The earth receives radiation continually from the sun to the equivalent of 1.17×10^{17} W. This energy from the sun is utilized to generate electricity. A solar cell is a thin silicon wafer of thickness 0.25 mm and can have a round or square form. It has the property of converting light energy of the sun into current. Figure 1.9 shows the one-dimensional geometric view of a PN junction solar cell.

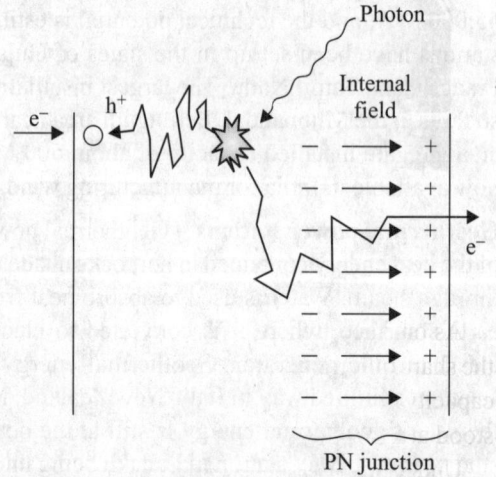

Light photons from the sun penetrate into the PN junction diode (cell) and impart enough energy to the valence electrons to make them jump into the conduction band. Due to the unaccounted number of photons penetrating the cell, an extremely large number of electrons enter the conduction

Fig. 1.9 Direct conversion of solar energy to electricity in a photovoltaic PN junction diode

band and are pushed out of the cell by the internal electric field which has already been produced during the manufacture of the PN junction diode. This flow of electrons leads to the flow of current. The process of direct conversion of solar light energy into electric current is called 'photovoltaic' (PV) effect.

The electrons will continue to flow out of the cell as long as light photons from the sun continue to penetrate the cell. As such, a cell never loses power, like a battery, since cells do not 'consume' electrons. Therefore, a cell may be viewed as a converter since it changes (sun) light energy into electric energy.

Figure 1.10 shows a view of a basic circular PV device. The metal contacts placed in front and at the back draw and deliver the electrons to the cell. In this manner, the same electrons continue to travel the same path and in the process deliver light energy to the load.

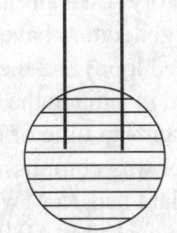

A typical silicon PV cell produces only 0.5 V DC. Therefore, a PV cell forms the basic device to form modules or panels to obtain higher voltages. **Fig. 1.10** Basic PV device Since a minimum of 12 V is required to charge a storage battery, a typical 65 × 140 cm panel will be made of 36 PV solar cells connected in series to produce 18 V. When loaded, the output voltage of the panel drops to 14 V which is the minimum voltage required to charge a storage battery. Thus, 36 solar cells panel has become the standard or basic module for the solar battery charger industry. Solar panels can be connected in series to obtain higher voltages of 24 V, 48 V, and more. Higher current capacity and therefore more power output can be obtained by connecting the basic modules of 36 cells each in parallel. Figure 1.11 provides a pictorial view of a basic module of 36 cells.

Fig. 1.11   Pictorial view of a basic solar PV module
(*Courtesy*: SunGift)

As on 31 March 2012, out of an installed capacity of about 200 GW in India, the share of renewable energy was 24,915 MW, which constitutes 12% of the total capacity. The share of solar energy in the installed renewable energy component stood at 905 MW (4%).

Under the National Action Plan on climate change, one of the eight missions that the GOI has set up in January 2010 is the Jawaharlal Nehru National Solar Mission (JNNSM). The aim of the Mission is to develop and promote the use of solar energy technology. The Mission aims to achieve, in three phases, a cumulative target of 20,000 MW and 2,000 MW, respectively, in grid and off-grid solar power generation by 2022.

MHD generation   The magnetohydrodynamic power generation technology (MHD) is the production of electrical power utilizing a high-temperature conducting fluid (plasma) moving through an intense magnetic field. The conversion process in MHD was initially described by Michael Faraday in 1893. However, the actual utilization of this concept remained unthinkable. The first known attempt to develop an MHD generator was made at Westinghouse Research

Laboratory (USA) in around 1936. Since the 1960s, many different types of MHD power generators have been classified according to the type of cycle (open loop or closed loop) and the type of fluid used. Open loop MHD generators were first realized in 1965 in the USA. It was a 32 MW facility, fuelled with alcohol, which had a start-up time of three minutes. In 1971, an MHD pilot plant using natural gas fuel was commissioned at the Institute of High Temperatures, USSR. This pilot plant had 75 MW of power (25 MW of MHD and 50 MW from steam). In 1984, a coal-fired MHD pilot plant was constructed in USA. Closed loop MHD generators are usually associated with nuclear reactors as heat source, where the working fluid can be a noble gas or liquid metal. In India, BHEL Tiruchirappalli started work in MHD technology in 1978 in close cooperation with BARC and High Temperature Institute, Moscow, which was a pioneer in large-scale MHD activities, and a 5 MW pilot plant was commissioned in Tiruchirappalli in 1985. Later in 1993 there was a proposal for installing a 200 MW retrofit in an existing thermal station, but it was later shelved.

Working principle   The MHD generator can be considered to be a fluid dynamo. This is similar to a mechanical dynamo in which the motion of a metal conductor through a magnetic field creates a current in the conductor, except that in the MHD generator the metal conductor is replaced by conducting gas plasma.

When a conductor moves through a magnetic field, it creates an electrical field perpendicular to the magnetic field and the direction of movement of the conductor. This is the principle, discovered by Michael Faraday, behind the conventional rotary electricity generator. Dutch physicist Hendrik Antoon Lorentz provided the mathematical theory to quantify its effects.

The flow (motion) of the conducting plasma through a magnetic field causes a voltage to be generated (and an associated current to flow) across the plasma, perpendicular to both the plasma flow and the magnetic field according to Fleming's Right Hand Rule. This is illustrated in Fig. 1.12.

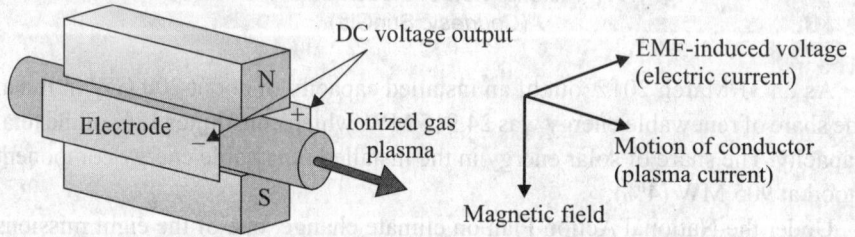

Fig. 1.12   Magnetohydrodynamic power generation
(*Courtesy:* www.electropaedia.com)

The MHD system   The MHD generator needs a high-temperature gas source, which could be the coolant from a nuclear reactor or more likely high-temperature combustion gases generated by burning fossil fuels, including coal, in a combustion chamber. Figure 1.13 shows the possible system components.

The expansion nozzle reduces the gas pressure and consequently increases the plasma speed through the generator duct to increase the power output. Unfortunately, at the same time, the pressure drop causes the plasma temperature to fall which also increases the plasma resistance.

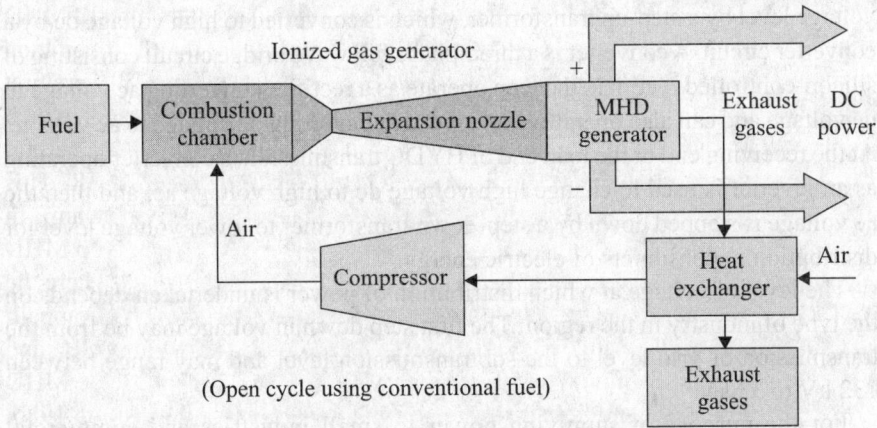

Ionized gas generator

(Open cycle using conventional fuel)

Fig. 1.13 Magnetohydrodynamic electricity generation
(*Courtesy*: www.electropedia.com)

The following are some of the merits of MHD power generation:

- Simple structure
- Works at high temperatures
- High Carnot-cycle efficiency
- Easy to realize combined cycle with other systems

The following are the disadvantages of MHD power generation:

- Simultaneous presence of high temperature and a highly corrosive and abrasive environment
- MHD channel operation under extreme conditions of electric and magnetic fields
- Expensive initial instalments

1.10.2 Transmission and Distribution Subsystem

The transformer and transmission line subsystems are designed to transmit bulk electric power for consumption at the load centres. In the generating stations, power is generated at voltage levels, which vary between 11 to 30 kV. The transformers at the generating station end step up the voltage to the level of transmission voltage suitable for transmission of bulk power. Since these transformers step up the voltage, they are also known as step-up transformers.

The power transmitted over a transmission line is proportional to the square of the transmission voltage. Therefore, ideally it is desirable to have the highest transmission voltages. As such, continuous efforts are undertaken to increase the transmission voltages. In the western countries, transmission of power is undertaken at transmission voltages of 765 kV. In India, the transmission voltage levels vary between 66 to 400 kV.

High voltage direct current (HVDC) transmission of bulk power over long distances is more economical than high voltage alternating current (HVAC) transmission when bulk power is to be transmitted over distances greater than 600 km. The dc voltages at which transmission takes place are 400 kV and above. At the generator end the ac voltage generated is stepped up to the transmission voltage level by a step-up transformer, which is converted to high voltage dc by a converter circuit. A converter is a three-phase full wave bridge circuit consisting of silicon-controlled rectifiers that can operate as a rectifier converting ac voltage to dc voltage and can also operate as an inverter converting dc voltage to ac voltage. At the receiving end or the load end of HVDC transmission, a converter operating as an inverter is used to change high voltage dc to high voltage ac, and then the ac voltage is stepped down by a step-down transformer to lower voltage level for distribution to consumers of electric energy.

The level of voltage at which distribution of power is undertaken depends on the type of industry in the region. The first step down in voltage may be from the transmission or grid level to the subtransmission level and may range between 132 kV to 33 kV.

For the purpose of supplying power to small industries and commercial and domestic consumers, the voltage is again stepped down at the distribution substation. The distribution of power is undertaken at two voltage levels; the primary or feeder voltage at 11 kV and the secondary or consumer voltage at 415 V for three-phase supply and 230 V for single-phase supply.

Subtransmission System

The portion of the transmission system that connects the high-voltage substations through step-down transformers to the distribution substations is called the subtransmission network. There is no clear demarcation between the transmission and subtransmission voltage levels. The voltage level of a subtransmission system ranges from 66 kV to 132 kV. Some heavy industrial consumers are connected to the subtransmission system.

A distribution subsystem constitutes the part of the electric power system between the step-down distribution substation and the consumers' service switches. A distributed system is designed to supply continuous and reliable power at the consumers' terminals at minimum cost. A typical distribution system is shown in Fig. 1.14.

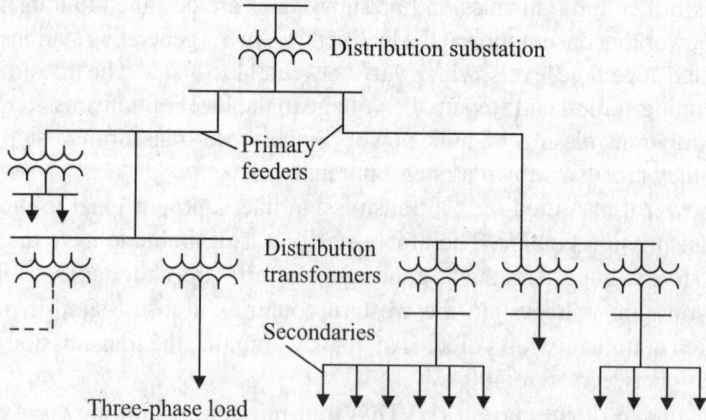

Fig. 1.14 Layout of a typical distribution system

At the distribution substation, the voltage level is brought down from 66 kV at the subtransmission level to 11 kV at the distribution level. Each distribution substation normally serves its own area, which is a subdivision of the area served by the distribution system. Distribution transformers are ordinarily connected to each primary feeder and its sub-feeders and laterals. Each transformer or banks of transformers serve to step down the voltage to utilize voltage of three-phase 415 V or single-phase 230 V and supply a consumer or a group of consumers over its secondary circuit. Each consumer is connected to the secondary circuit through service leads and a meter.

The subtransmission and distribution systems remained neglected for a long time and it was only in the mid 1990s when the development of their infrastructure was recognized as a core issue in the power sector. The reason for lack of initiative to update the subtransmission and distribution infrastructure was attributed to a generation-centric focus by both the central and state governments. The bias towards generation is obvious when it is observed that till 1993 the ratio of plan outlay for the development of the generation subsystem to the transmission and distribution subsystem was 3:1 as against the desired 1:1. However, during the period 1997–2002 (ninth plan period) this ratio improved to 1.3:1, which is mainly due to a reduction in investment by the government in the generation subsystem.

The shortcomings in the distribution infrastructure, on the other hand, have been identified as follows:

(a) Insufficient transformation capabilities
(b) High technical losses
(c) High non-technical losses, such as pilferage and commercial losses
(d) Inadequacy in addressing consumer concerns including poor service
(e) Absence of redundancies

Another very pressing issue related to the subsystem is that of unbearably high T&D losses. The collective T&D losses in the power industry in India increased from 17.5% to an astonishing 21.7% during the period 1970–85. On an all-India basis, the average T&D losses are estimated at 40%, which is very high when compared to the international average of approximately 6–7%. Table 1.7 provides the year-wise T&D losses in the power industry in India.

Table 1.7 Year-wise T&D losses in the Indian power industry

Year	T&D losses in % (All India)
2007–08	27.20
2008–09	25.47
2009–10	25.39
2010–11	23.97

Due to concerted efforts, the transmission and distribution (T&D) losses have come down but have stagnated at 23.97%.

The aggregate technical and commercial losses (AT&C) are of the order of 50% power generation. Billing for generated power is approximately 55% of the total power generated while the realization is only about 41%.

National Grid

Hitherto transmission networks were developed with a focus on self-sufficiency on regional basis. As such the period from the mid 1970s to the early 1990s saw the building up of strong state grids and the emergence of regional grids. Presently, the regional grid networks are adequately strong to meet the inter-state transmission requirements while the state grids can focus on meeting the intra-state needs of their respective states.

The spotlight is now on building a national grid for better utilization of hydro resources, saving the transportation cost of coal (since it is economical to transmit electrical energy), sharing of reserves, etc.

1.10.3 Load Subsystem

From the perspective of a power supplier, an item (component) consuming electrical energy is a load. Therefore, loads on a power system can be broadly categorized as follows: (a) industrial, (b) commercial, and (c) domestic.

Industrial loads, which are voltage and frequency dependent, are a combination of motor loads, lighting loads, etc. Induction motors comprise a high percentage of the industrial load and consume considerable reactive power. Both commercial and domestic loads are voltage dependent and are mainly constituted of lighting, heating, and cooling. A few terms related to load subsystems are described here.

Load curve of a utility is a plot of variation of composite load against time. If the variation of load is on a 24-hour basis it is called a *daily load curve*.

Peak or maximum demand is defined as the maximum load occurring in a 24-hour cycle.

Load factor (*LF*) is defined as the ratio of average load during a period to maximum load during the same period. Thus,

$$LF = \frac{\text{Average load in kW/MW}}{\text{Maximum load in kW/MW}} \text{ for a specified period} \tag{1.3}$$

If the specified period is a 24-hour cycle, the LF is called a daily LF. Multiplying Eq. (1.3) by 24 yields

$$\text{Daily LF} = \frac{\text{Average load in kW/MW} \times 24}{\text{Maximum load in kW/MW} \times 24}$$

$$= \frac{\text{Energy consumed in 24 h in kWh/MWh}}{\text{Maximum load} \times 24 \text{ in kWh/MWh}} \tag{1.4}$$

If the specified period is one year ($24 \times 365 = 8760$ h), the LF is called an annual LF. The annual LF is used to assess the performance of a generating station. Thus,

$$\text{Annual LF} = \frac{\text{Energy consumed in 8760 h in kWh}}{\text{Maximum load} \times 8760 \text{ in kWh/MWh}} \tag{1.5}$$

Higher the annual load factor, more economical is the plant operation. The desirable range of annual load factor of a system is between 55% to 70%.

Diversity factor is defined as the ratio of the sum of maximum demands of individual category of consumers, such as industrial, commercial, and domestic, to the maximum load on the system. Thus,

$$\text{Diversity factor} = \frac{\sum \text{Max. demand of individual category of consumers}}{\text{Max. demand on system}}$$

$$\tag{1.6}$$

It is a parameter which provides the diversification in load and is used to decide the installed capacity of a generating station. It is greater than unity, and therefore the installed capacity will be less than the sum of the maximum demands of individual category of consumers.

Utilization factor is defined as the ratio of maximum demand to installed capacity, that is,

$$\text{Utilization factor} = \frac{\text{Max. demand in MW}}{\text{Installed capacity in MW}} \tag{1.7}$$

Plant factor is defined as the ratio of annual energy generated to the possible annual energy that can be generated based on installed capacity. Thus,

$$\text{Plant factor} = \frac{\text{Annual energy generated in MWh}}{\text{Installed capacity in MW} \times 24 \times 365} \tag{1.8}$$

Example 1.1 Compute the (i) average load, (ii) maximum load, and (iii) daily average load factor, if the daily variation of load on a power company is as follows:

Interval number	Clock time in hours		Load in MW
1	00	06	4
2	06	09	8
3	09	11	12
4	11	14	18
5	14	18	15
6	18	20	12
7	20	22	8
8	22	24	4

Solution The MATLAB function lodata is used to draw the daily load curve. The input to the program consists of the following:

nintrvl   The number of intervals which is 8 in this case
load   It is a matrix whose order is equal to (number of intervals) × 3. The first two columns of the matrix represent the duration in clock hours and the third column represents the load in MW.

The output variables are energy, Pavg (average power), LF (load factor), and Pmax (maximum power demand).

```
% Program for plotting the load curve and computing avearge load and load factor
% Program developed by the authors
function [energy,Pavg,LF,Pmax]=lodata(nintrvl,load);
% Initialization
time = zeros(1,nintrvl);P=zeros(1,nintrvl);
Pmax = 0;X = 0;
hold on
% Plotting the load curve
for I =1:nintrvl;
 time(1,I) =load(I,2)-load(I,1);
 x=linspace(X,load(I,2),500);
 Y =load(I,3);
 plot(x,Y)
 X=X+time(1,I);
 if I < nintrvl;
 y=linspace(load(I,3),load(I+1,3),500);
 x=load(I,2);
```

```
plot(x,y)
% Computing the average load and load factor
energy=0;
for I =1:nintrvl;
energy=energy+(load(I,2)-load(I,1))*load(I,3);
if load(I,3) > Pmax;
    Pmax=load(I,3);
else
end
end
else
end
end
% Setting the axes and labelling the plot
axis([0 24 0 25]);
xlabel ('Time in hours');
ylabel ('Load in MW');
title ('DAILY LOAD CURVE');
hold off
% Output
energy
Pavg=energy/24
LF=Pavg/Pmax
Pmax
≫ nintrvl=8;load=[00 06 4;06 09 8;09 11 12;11 14 18;14 18 15;18 20 12;20 22
8;22 24 4];
% Input data
≫ [energy,Pavg,LF,Pmax]=lodata(nintrvl,load);        % Statement to call function
                                                      lodata for execution.
energy =
 234
Pavg =
 9.7500
LF =
 0.5417
Pmax =
 18
≫
```

The daily load curve is shown in Fig. 1.15.

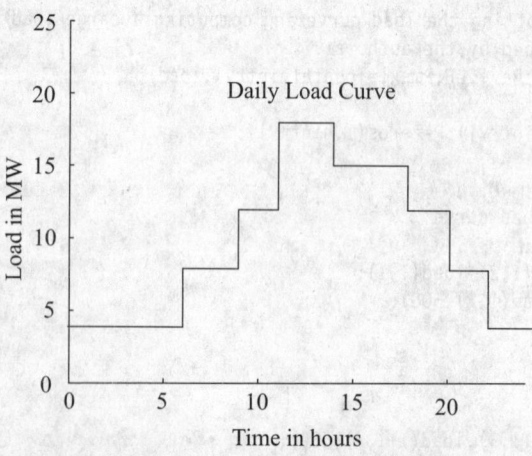

Fig. 1.15 Load curve of Example 1.1

1.10.4 Protection and Control Subsystem

Protection and control subsystem is constituted of relays, switchgear, and other control devices, which protect the various subsystems against faults and overloads, and ensure efficient, reliable, and economic operation of the electric power system.

1.11 ENERGY CONSERVATION

A unit of energy saved is a unit of energy generated at no extra cost. One of the objectives of the GOI under its Mission 2012: 'Power for All' is to evolve a conservation strategy to optimize the utilization of electrical energy. The focus will be on (i) demand management, (ii) load management, and (iii) technology upgradation to provide energy-efficient equipment and gadgets. It is possible to bring about an energy saving of the order of 20% in various sectors without sacrificing any of the end-use benefits of energy.

Keeping in mind the need and importance of energy conservation, the GOI has enacted the Energy Conservation Act (2001) under which a Bureau of Energy Efficiency has been established for the promotion of conservation and efficient use of energy.

1.12 COMPUTERS IN POWER SYSTEM ANALYSIS

Historically, digital computers were first employed for analysing power system problems, in a restricted manner, in the late 40s of the last century. With the advent of computers having capabilities to handle large volumes of data with adequately fast processing speeds, in the mid 1950s, their usage in analysing varied and more intricate problems related to larger and complex power system networks was a natural outcome.

The operation and control of present-day interconnected power networks, each constituting of substations, transmission lines, and transformers, has become so complex that from the perspective of economy and reliability of supply it is essential that these be monitored through a central point called an Energy Control Centre (ECC). An ECC is an online computer which undertakes signal processing based on remote data acquisition system and performs in both normal and emergency situations. The constituents of an ECC are as follows.

(a) An operator who acts as a human–machine interface
(b) A visual display unit (VDU) which enables the selection of presentation of the desired portion of the network, along with the data summaries and performance indices, through paging buttons
(c) Editing and special function keyboards to change operating conditions, system parameters, transformer taps, switch-in-out line capacitors, etc.
(d) Light pen cursor for operating circuit breakers, switches, etc. and for changing displays directly on the VDU

SUMMARY

- After Independence, the GOI, amongst other development plans, took upon itself to develop the power sector.

- Vision 2012 for the power sector, in addition to providing 'Power for all', also envisioned reliable and quality power at optimum cost along with the development of a competitive power industry in addition to providing sufficient power to achieve a GDP growth of 8%.
- Vision 2020 envisages efficient and environment-friendly energy resources which would become the growth engines to provide speedy and sustainable future economic development.
- In order to achieve the objectives of Vision 2012 for the power sector, Electricity Act 2003 was amended and enacted to bring about a market-oriented management, through restructuring and deregulation of the power sector, so as to introduce a spirit of competition. Electricity Act 2007 further amended Electricity Act 2003 to allow setting up of captive power units without obtaining licences and making power theft a criminal offence.
- A power sector is a complex network constituted of (a) generation, (b) transmission and distribution, (c) loads, and (d) protection and control subsystems.
- Primary sources of energy are (a) fossil fuels such as coal, oil, and natural gas, (b) renewable energy sources like hydro, wind, solar, and (c) nuclear energy.
- Based on the type of primary source employed, generation stations are categorized into (a) thermal, (b) hydro, (c) nuclear, and (d) non-conventional. In 2009–10, the total all-India generation was 771,173 GWh. Generation voltages range between 11 kV and 30 kV.
- DC transmission voltage is 500 kV while the AC voltages range between 66 and 400 kV. By 2017, it is planned to add transmission lines operating at 765 kV, along with HVDC Bipole lines. Distribution of power is undertaken at AC voltages up to 500 V.
- The load subsystem consists of (a) industrial, (b) commercial, and (c) domestic loads. The important terms used to define a load subsystem are: (a) load curve, (b) daily and annual load factor, (c) diversity factor, (d) utilization factor, and (e) plant factor.

EXERCISES

Review Questions

1.1 Trace the history of the growth of the power sector after Independence.

1.2 Write a short essay on 'Vision 2020' for the power sector.

1.3 Describe the models of regulation and deregulation associated with the power sector. Discuss the features of a regulated power sector and explain the structure of a regulated authority.

1.4 With the help of block diagrams, illustrate and explain the structure of a deregulated power sector.

1.5 Explain the mechanism of competition.

1.6 Write short notes on: (i) wholesale power market, (ii) instruments of sale transaction, and (iii) responsibilities of the SSA/ISO.

1.7 Draw a neatly labelled diagram of a power network and indicate the various subsystems along with their operation voltages.

1.8 Enumerate the various sources of energy and categorize generating stations based on primary source of energy employed.

1.9 Draw a block diagram of a thermal station and describe its main features. Explain cogeneration.

1.10 Write short notes on: (i) oil-fired, (ii) gas-fired, and (iii) diesel oil-fired generating stations.

1.11   Describe with the help of a diagram the salient features of a hydro generating station and itemize its merits and demerits.

1.12   Describe the working of a nuclear generating station.

1.13   List the various types of non-conventional generating stations and describe any two of them.

1.14   (a) Briefly describe the transmission and distribution subsystems.

   (b) Highlight the concerns of the T&D subsystem in the Indian power sector.

1.15   (a) Describe the components of the load subsystem.

   (b) Discuss the utility of load curve and define peak demand.

1.16   Define and write notes on: (i) load factor, (ii) diversity factor, (iii) utilization factor, and (iv) plant factor.

1.17   Write a note on the importance of computers in power system analysis.

Numerical Problems

1.1   A power utility with an installed capacity of 100 MW is supplying a composite load whose details are as follows:

Type of load	Average load in MW	Load factor
Industrial	48	0.8
Commercial	12.5	0.5
Domestic	24.5	0.7

   Calculate the diversity factor.

1.2   A generating station has a plant factor of 50% and a maximum demand of 450 MW. If the annual load factor is 60%, determine the additional load the station can supply. Assume that the generating station can be fully loaded.

1.3   The demand for power on a 100-MW generating station is as follows: 80 MW for 4 h, 50 MW for 8 h and 20 MW for 6 h. For the remaining part of the day it is switched off. Determine the annual load factor of the station. Assume that the station is under maintenance and repair for 45 days in a year.

1.4   The demand for energy on a utility is growing exponentially and can be expressed as $P = P_0 e^{at}$ where a and t are the growth rate and time respectively. Determine the growth rate if the energy consumption is expected to increase 1.5 times in 10 years.

1.5   Mathematically the demand on a utility is estimated to be $P = P_0 e^{a(t - t_0)}$, where P is the demand in the year t, P_0 is the demand in the base year t_0, and a is the per annum growth rate. If the maximum power demand in the base year was 250 GW, write a MATLAB function to plot the growth of demand for the next 15 years. What is the demand after 10 years?

1.6   The month-wise load on a generating station is as follows:

Month	Jan.	Feb.	Mar.	Apr.	May	Jun.	Jul.	Aug.	Sep.	Oct.	Nov.	Dec.
Load in MW	6	6	4	3	6	10	12	14	13	4	5	6

   Write a MATLAB function to plot the annual load curve for the generating station and determine the average load and load factor.

Multiple Choice Objective Questions

1.1   Who introduced the first electric supply system?
      (a) Edison   (b) Faraday
      (c) Tesla   (d) Marconi

1.2   Which of the following was set up for the development of nuclear power?
      (a) NHPC   (b) NTPC
      (c) NPCIL   (d) None of these

1.3   Which of the following is not an objective of 'Vision 2012' for the power sector?
      (a) Reliability of power
      (b) Quality power
      (c) Sufficient power to achieve 8% GDP growth rate
      (d) None of these

1.4   The process of reforming the power sector was started in
      (a) 1981   (b) 1991
      (c) 2001   (d) 2011

1.5   Electricity Act 2003 was enacted in the month of
      (a) April   (b) May
      (c) June   (d) July

1.6   Loss reduction and theft control was which part of the strategy to achieve 'Power for All'?
      (a) Power generation   (b) Transmission
      (c) Distribution   (d) Conservation

1.7   As per 'Electricity Act 2003' which of the following type of generating stations requires a licence?
      (a) Oil fired   (b) Hydro
      (c) Gas fired   (d) Solar

1.8   Which of the following industries employs cogeneration?
      (a) Fertilizer   (b) Petroleum refining
      (c) Paper   (d) All of these

1.9   Which of the following is not true of a hydroelectric station?
      (a) No ecological imbalance   (b) Long life
      (c) No air pollution   (d) None of these

1.10 Which of the following contains nearly zero sulphur dioxide?
      (a) Oil fired   (b) Coal fired
      (c) Gas fired   (d) Diesel fired

1.11 Which of the following is a reason to locate nuclear stations away from populated areas?
      (a) Primary coolant is heavy water.
      (b) High capital investment is required.
      (c) Radioactive waste is generated.
      (d) All of these.

1.12 The success of wind power stations is based on
      (a) initial capital cost   (b) power output
      (c) useful life   (d) all of these

1.13 Which of the following states has not set up wind power stations?
      (a) Orissa   (b) Punjab
      (c) Tamil Nadu   (d) Maharashtra

1.14 The voltage produced by a typical silicon PV solar cell is
(a) 0.25 V DC
(b) 0.25 V AC
(c) 0.5 V DC
(d) 0.5 V AC

1.15 Which of the following is not required for the flow of electrons in a PV cell?
(a) External field
(b) Internal field
(c) Penetration of a photon of light
(d) None of these

1.16 The minimum voltage required to charge a 12 V storage battery is
(a) 18 V
(b) 16 V
(c) 14 V
(d) 12 V

1.17 Which of the following will become a part of the transmission system by 2017?
(a) 765 kV
(b) HVDC Bipole
(c) 400 kV
(d) All of these

1.18 Currently the level of T&D losses stands at
(a) 17.5%
(b) 21.7%
(c) 25%
(d) 40%

1.19 A higher annual load factor indicates
(a) economical plant operation
(b) uneconomical plant operation
(c) no effect on economics of plant operation
(d) none of these

1.20 Which of the following is employed to determine the installed capacity of a generating station?
(a) Maximum demand
(b) Annual load factor
(c) Diversity factor
(d) Plant factor

Answers

1.1 (a)	**1.2** (c)	**1.3** (d)	**1.4** (b)	**1.5** (d)	**1.6** (c)
1.7 (b)	**1.8** (d)	**1.9** (a)	**1.10** (c)	**1.11** (c)	**1.12** (d)
1.13 (b)	**1.14** (c)	**1.15** (a)	**1.16** (c)	**1.17** (d)	**1.18** (c)
1.19 (a)	**1.20** (c)				

CHAPTER 2

Basic Concepts

Learning Outcomes

An intense study of this chapter will enable the reader to:
* Differentiate between single-phase and three-phase ac power systems
* Visualize mentally the physical structure of a power system and draw a schematic line diagram to properly locate generators, transformers, transmission lines, loads, and so on to show the entire interconnected network
* Appreciate the importance of phasor notation and apply it as a mathematical tool to analyse power system networks in the steady state
* Distinguish between complex, real, and reactive power and highlight their significance along with the need for power factor improvement
* Recognize the importance of phase shift operator and apply it for analysing three-phase networks
* Grasp the significance of per unit quantities and be able to compute the same and draw a per unit diagram of an interconnected power system

OVERVIEW

Modern electricity supply systems are invariably three-phase systems. The design of distribution systems is such that equal loading on all the three phases of the network is ensured by allotting, as far as possible, equal domestic loads to each phase of three-phase low-voltage distribution feeders; while industrial loads usually take three-phase supplies. Thus, normal operation is close to balanced three-phase working.

Power system engineers should be very conversant with steady-state circuits, particularly three-phase circuits. In the steady state, most power system voltages and currents are sinusoidal functions of time (at least approximately), and operate at the same frequency. In the steady-state operation, three-phase networks carry a balanced load and thus can be analysed by representing them on per-phase basis with one phase and the neutral.

2.1 REPRESENTATION OF POWER SYSTEMS

Since balanced three-phase networks are always solved as a single-phase or per-phase equivalent circuit comprising one of the three lines and a neutral return, a power system can be represented by one of the three phases and the neutral.

The components such as generators, transformers, and loads can be indicated by standard symbols, while a transmission line is represented by a single line between its two ends. The diagram is simplified further by omitting the neutral wire. Such a simplified diagram of an electric system is called a single-line or one-line diagram.

Detailed significant information about the system parameters is also indicated in the single-line diagram. This information includes ratings of synchronous generators, generation voltage, machine reactances, etc. It also includes information about the type of connection of three-phase transformers on the primary and secondary side, their ratings, and voltage ratios. The details of circuit breakers and their types such as oil, air-blast, and SF_6 breaker as well as the details of transmission lines such as sizes of conductors, configurations of the three phases above ground, and lengths can also be found from the diagram. The active and reactive power load demands at load buses and many other details are mentioned in the single-line diagram. The information to be provided on the single-line diagram depends on the type of analysis to be undertaken on the system. For example, to perform a load flow analysis, information regarding power generation, load demands, line, and transformer data is essential. However, to perform a transient stability study of the system, additional details about machine parameters, location of circuit breakers and relays, and the speed with which relays and circuit breakers operate to isolate the faulted part of the system are essential. Sometimes one-line diagrams include information on the current and potential transformers which are installed for connection to the relays and for metering purposes.

The formulation of power system problems for load flow, short-circuit, and transient stability analysis involves the formulation of suitable mathematical representations of all the components of the power system. For such computations, one-line diagram of the system is first drawn, then single-phase or per-phase equivalent circuits for the various electrical components (developed in later chapters of the book) are used to form the per-phase impedance diagram of the system. The per-phase impedance diagram is sometimes called the per-phase positive-sequence diagram, since it shows impedances to balanced currents in one phase of a symmetrical three-phase system.

For the sake of commonality and understanding of single-line diagrams of power systems, certain conventions for representing various components are used. These conventions are shown in Fig. 2.1. The standard symbol to designate a three-phase star with the neutral solidly grounded is shown in the figure. If a resistor or reactor is inserted between the neutral of the star point and ground, the appropriate symbol for resistance or inductance may be added to the standard symbol for the grounded star. Most transformer neutrals in transmission systems are solidly grounded. Generator neutrals are usually grounded through current limiting resistances or inductance coils during unbalanced short-circuits involving ground. It is essential to know the points at which a power system is earthed so as to be able to compute the magnitudes of currents in faults involving the ground. If a load flow study is to be made, the impedance diagram does not include the current limiting impedances connected between the neutrals of the generators and the ground. This is because under balanced condition of operation the neutrals of the generators are at the same potential as the neutral of the system. Since

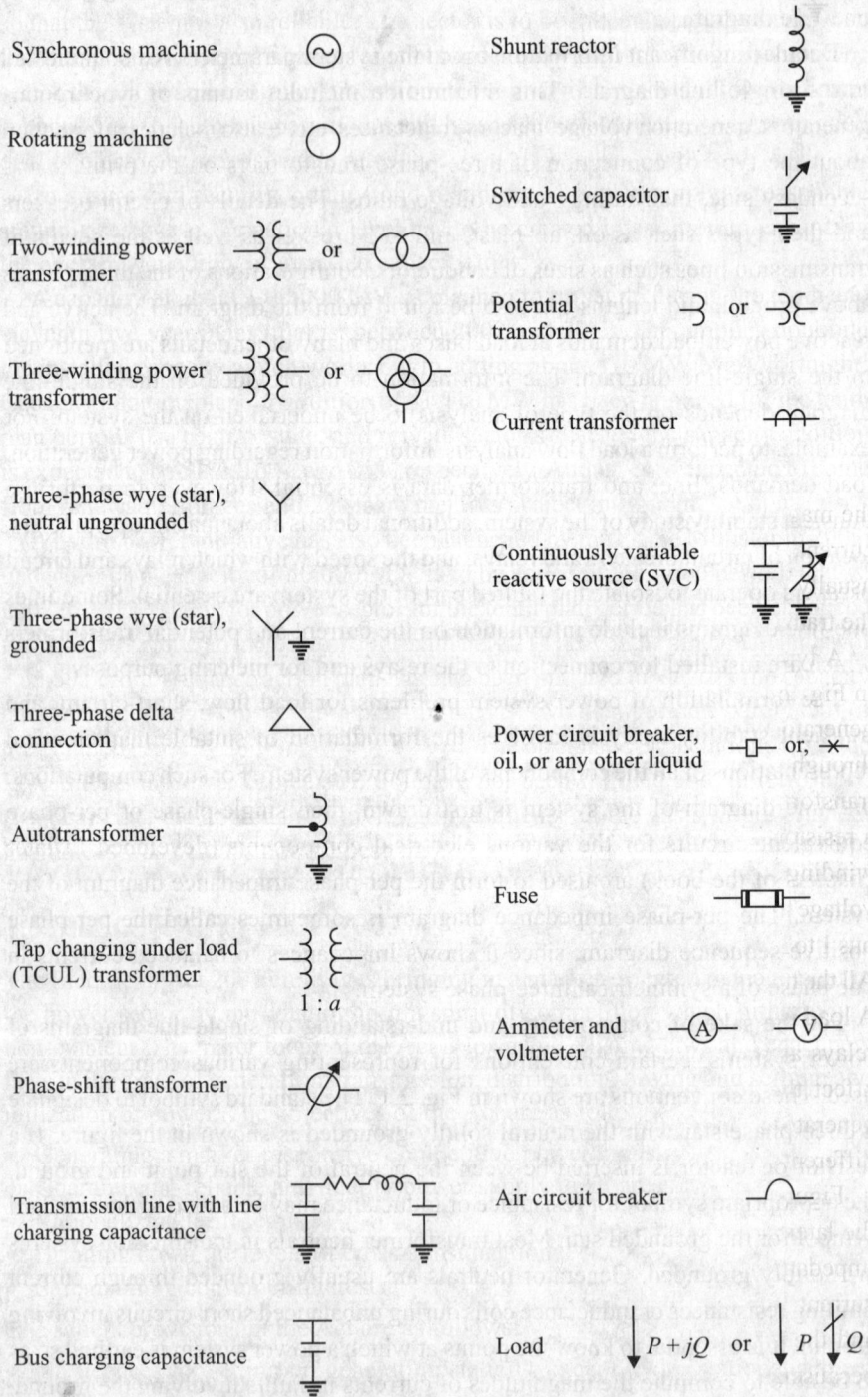

Fig. 2.1 Standardized symbols for components of power systems

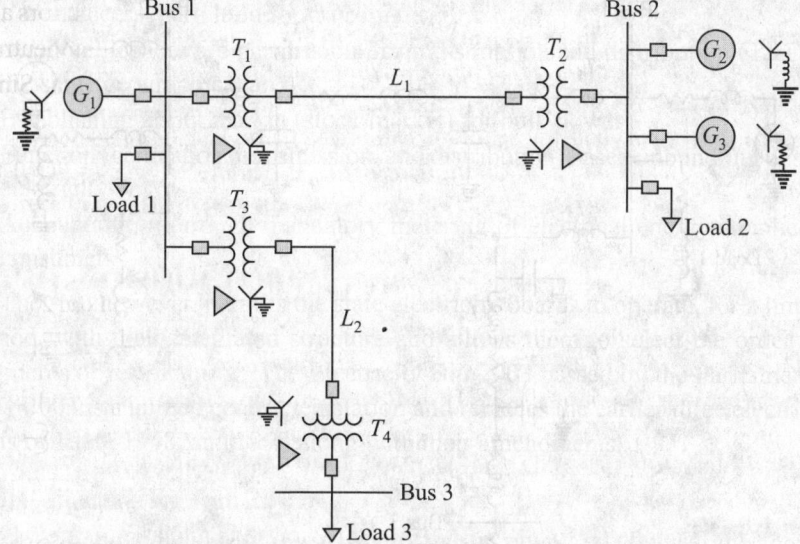

Fig. 2.2 Single-line diagram of a power system

the magnetising current in a transformer is negligible compared to its full load current, the shunt magnetising branch of the transformer equivalent circuit is usually omitted. In fault computations, resistances of various components and the transmission line charging capacitances are often neglected.

A typical single-line diagram of a simple electrical power system is shown in Fig. 2.2. The diagram given in Fig. 2.2 shows that the power system has two generator buses and one load bus. The generator connected to bus 1 is earthed through a resistance, and is connected to the transmission line L_1 through a step-up transformer T_1. Two generators, one grounded through a reactor and one through a resistor, are connected to bus 2 and through the step-up transformer T_2. Two-winding step-up transformers are connected to the generator buses to increase the voltage to the desired transmission voltage. Transmission line L_2 connects generator bus 1 to load bus 3 through a step-up transformer T_3. T_4 is a step-down transformer. All the transformers are delta-star connected with their star sides solidly grounded. A load is connected to each bus. The circuit breakers, switchgear, and protective relays are not shown in the diagram because emphasis is on the elements that affect the steady-state transmission of power. On the diagram ratings of the generators and transformers, information about loads and the impedances of the different components of the system is often indicated.

Figure 2.3 shows the combination of the equivalent circuits (developed in the later chapters) for the various components in Fig. 2.2 to form the per-phase impedance diagram of the system. The impedance diagram does not include the current limiting impedances connected between the neutrals of the generators and the ground as shown in the single-line diagram given in Fig. 2.2. This is because under balanced conditions, no current flows in the ground and the neutrals of the generators are at the same potential as that of the neutral of the system.

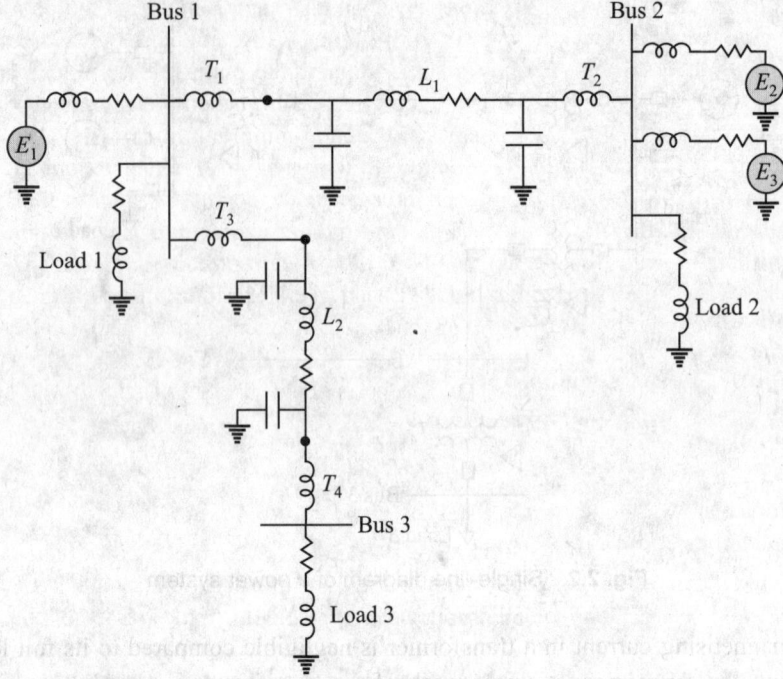

Fig. 2.3 Positive sequence impedance diagram of the system shown in Fig. 2.2

2.2 SINGLE-PHASE VERSUS THREE-PHASE SYSTEMS

As already mentioned, present-day electric power systems are three-phase systems. The advantages that a three-phase system offers in comparison to a single-phase system of equal rating are as follows:

(i) A single-phase power system delivers pulsating power because it drops to zero, three times in a cycle. On the other hand, in a three-phase system, power is also pulsating but the load receives constant power at any instant of time. Thus, induction motors show better operating characteristics.

(ii) The volume of conductor required in a three-phase system is less than that required in a single-phase, two-wire system of equivalent KVA rating. Consider a two-wire, single-phase ac generator system shown in Fig. 2.4. If the system supplies a load of P W at a power factor angle of φ and a maximum voltage of V_m volts, the current I_{1p} flowing through the wires is given by

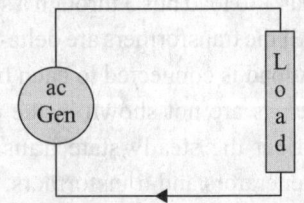

Fig. 2.4 Two-wire, single-phase ac generator system

$$I_{1p} = \frac{P}{\left(V_m/\sqrt{2}\right)\cos\varphi}\ \text{A} \tag{2.1}$$

If R_{1p} is the resistance of each conductor, it may be expressed as

$$R_{1p} = \rho\,\frac{l}{a_{1p}}\ \Omega \tag{2.2}$$

where ρ is resistivity in ohm-m, l is length of the conductor in metres, and a_{1p} is its area of cross section in m^2. Using Eqs (2.1) and (2.2) the total loss in the system P_{1p} is given by

$$P_{1p} = 2I_{1p}^2 R_{1p} = 2\left(\frac{P}{(V_m/\sqrt{2})\cos\varphi}\right)^2 \times \rho \frac{l}{a_{1p}} = \frac{4P^2}{V_m^2 \cos^2\varphi} \times \frac{\rho l}{a_{1p}} \text{ W} \quad (2.3)$$

From Eq. (2.3) area of cross section is written as

$$a_{1p} = \frac{4P^2}{V_m^2 \cos^2\varphi} \times \frac{\rho l}{P_{1p}} \text{ m}^2$$

Volume of conductor required is given by

$$V_{1c} = a_{1p} \times l = \frac{4P^2}{V_m^2 \cos^2\varphi} \times \frac{\rho l^2}{P_{1p}} \text{ m}^3 \quad (2.4)$$

Figure 2.5 shows a commonly used three-phase, three-wire ac system with the neutral earthed.

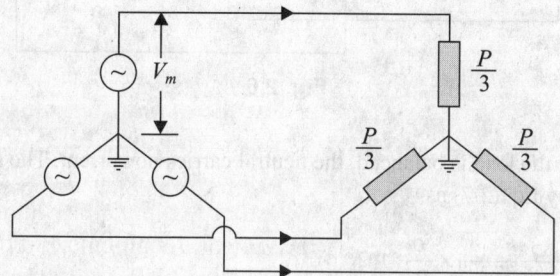

Fig. 2.5 Three-phase, three-wire ac system

In this case, assume that the load supplied is P W at a voltage of V volts/phase. The current is given by

$$I_{3p} = \frac{P}{3(V_m/\sqrt{2})\cos\varphi} = \frac{\sqrt{2}P}{3V_m\cos\varphi} \text{ A} \quad (2.5)$$

If R_{3p} is the resistance of each conductor, the total power loss can be expressed as

$$P_{3p} = 3I_{3p}^2 R_{3p} = 3\left(\frac{\sqrt{2}P}{3V_m\cos\varphi}\right)^2 \times \rho \frac{l}{a_{3p}} = \frac{2P^2}{3V_m^2\cos^2\varphi} \times \frac{\rho l}{a_{3p}} \text{ W} \quad (2.6)$$

The cross section of the conductor from Eq. (2.6) is given by

$$a_{3p} = \frac{2P^2}{3V_m^2\cos^2\varphi} \times \frac{\rho l}{P_{3p}} \text{ m}^2 \quad (2.7)$$

Hence, the volume of conductor required is expressed as

$$V_{3c} = 3 \times a_{3p} \times l = \frac{2P^2}{V_m^2\cos^2\varphi} \times \frac{\rho l^2}{P_{3p}} \text{ m}^3 \quad (2.8)$$

For the same efficiency, the losses in the two systems must be equal, that is, $P_{1p} = P_{3p}$. Therefore, dividing Eq. (2.8) by Eq. (2.4)

$$\frac{V_{3c}}{V_{1c}} = \frac{2}{4} = 0.5$$

From the foregoing discussion, it becomes clear that the volume of conductor required for a three-phase, three-wire ac system is 50 per cent of the conductor required for a single-phase, two-wire ac system.

Example 2.1 A three-phase, four-wire ac system supplies a balanced three-phase load of P W at a power factor of cos φ as shown in Fig. 2.6. If the maximum voltage between any conductor and neutral is V_m, compare the volume of conductor required with a single-phase, two-wire ac system.

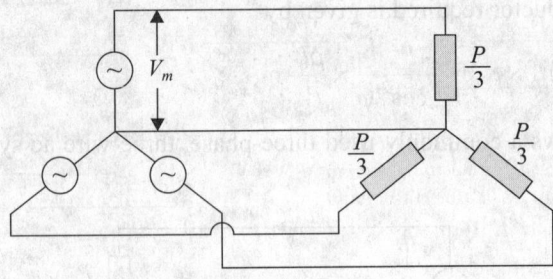

Fig. 2.6

Solution Since the load is balanced, the neutral carries no current. The area of the conductor is given by Eq. (2.7) as

$$a_{3p} = \frac{2P^2}{3V_m^2 \cos^2 \varphi} \times \frac{\rho l}{P_{3p}} \text{ m}^2$$

Usually, the area of the neutral conductor is half that of the line conductor. Thus the total volume of conductor required for a three-phase, four-wire system, V_{3c}, may be written as

V_{3c} = volume of conductor for three-phase conductors
     + volume of conductor for the neutral conductor

$$= 3a_{3p}l + 0.5a_{3p}l = 3.5 \times l \times \frac{2P^2}{3V_m^2 \cos^2 \varphi} \times \frac{\rho l}{P_{3p}} = \frac{7P^2}{3V_m^2 \cos^2 \varphi} \times \frac{\rho l^2}{P_{3p}} \text{ m}^3$$

$$(2.1.1)$$

Dividing Eq. (2.1.1) by (2.4) leads to

$$\frac{V_{3c}}{V_{1c}} = \frac{7}{3 \times 4} = 0.583$$

2.3 PHASOR NOTATION

In electrical power systems operating in the steady state, voltage and current variables approximate closely to sinusoidal quantities varying with time at the same frequency. Therefore, analyses of power systems in the steady state, employing phasor notation and impedances (or admittances), and complex power are of great significance to the power engineer.

A phasor is a complex number that contains the amplitude and phase angle information of a sinusoidal function. The concept of a phasor can be developed using Euler's identity that relates the exponential function to the trigonometric functions as follows:

$$e^{\pm j\theta} = \cos\theta \pm j\sin\theta \tag{2.9}$$

or $\quad \cos\theta = \text{Re}\{e^{j\theta}\}$ (2.10a)

and $\quad \sin\theta = \text{Im}\{e^{j\theta}\}$ (2.10b)

where Re and Im denote respectively the *real* and *imaginary* parts of the quantity in the parenthesis.

A sinusoidal voltage quantity $v(t)$, having an angular speed of ω rad/s, a maximum voltage of V_m, and a phase displacement of θ_1 can be conventionally written using Eq. (2.10a) as

$$v(t) = V_m\cos(\omega t + \theta_1) = \text{Re}\{V_m\, e^{j(\omega t + \theta_1)}\}$$

$$= \text{Re}\{V_m e^{j\theta_1} e^{j\omega t}\} \tag{2.11}$$

It may be observed that the factor $e^{j\omega t}$ in Eq. (2.11) contains the information about the angular frequency. Once the sinusoids of the same frequency are involved, this information is redundant. Then the representations of $V_m\cos(\omega t + \theta_1)$ in Eq. (2.11) are simplified, by dropping the term $e^{j\omega t}$, and may be expressed in polar form as

$$v(t) = V_m\cos(\omega t + \theta_1) = V_m e^{j\theta_1} \text{ or simply } V_m \angle\, \theta_1 \tag{2.11a}$$

The term $(V_m e^{j\theta_1})$ in Eq. (2.11a) is a complex number, which contains the amplitude and phase angle information of the sinusoidal function. This complex number, as per definition, is the conventional circuit theory phasor representation of the sinusoidal function. However, in power system engineering, it is convenient to use effective phasor representation because ac power can be directly equated to dc power. The effective voltage phasor V is defined as

$$V = \frac{V_m}{\sqrt{2}} e^{j\theta_1} = |V| e^{j\theta_1} \tag{2.12}$$

where $|V| = V_m/\sqrt{2}$ is called the effective or root-mean-square (rms) value of the sinusoidal voltage. Similarly, sinusoidal current $i(t)$ can be expressed as

$$i(t) = I_m\cos(\omega t + \theta_2) = I_m e^{j\theta_2} \text{ or } I_m \angle\, \theta_2 \tag{2.13}$$

and the rms value of the sinusoidal current I is defined as

$$I = \frac{I_m}{\sqrt{2}} e^{j\theta_2} = |I| e^{j\theta_2} \tag{2.14}$$

where $|I| = I_m/\sqrt{2}$ is the effective or rms value of the sinusoidal current and θ_2 is the phase displacement from the reference. The representation of rms voltage and rms current phasors in rectangular form is as follows:

$$V = |V|(\cos\theta_1 + j\sin\theta_1) \tag{2.15a}$$

$$I = |I|(\cos\theta_2 + j\sin\theta_2) \tag{2.15b}$$

2.4 COMPLEX POWER IN SINGLE-PHASE AC CIRCUITS

Although the fundamental theory of the transmission of energy describes the travel of energy in terms of the interaction of electric and magnetic fields, the power-system engineer is almost always more concerned with describing power, which is the rate of change of energy with respect to time, in terms of voltage and current. The power (in watts) being absorbed by a load at any instant is the product of the instantaneous voltage drop across the load (in volts) and the instantaneous current into the load (in amperes). Figure 2.7 shows a single-phase sinusoidal voltage supplying a load.

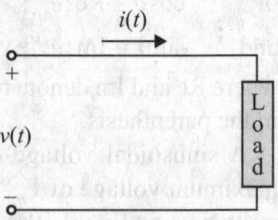

Fig. 2.7 A sinusoidal source supplying a load

If instantaneous voltage is represented by

$$v(t) = V_m\cos(\omega t + \theta_1)$$

and instantaneous current is given by

$$i(t) = I_m\cos(\omega t + \theta_2)$$

then using associated reference directions shown in the figure, instantaneous power $p(t)$ is given by

$$p(t) = v(t)\,i(t) = V_mI_m \cos(\omega t + \theta_1)\cos(\omega t + \theta_2)$$

$$= \frac{V_mI_m}{2}[\cos(\theta_1 - \theta_2) + \cos(2\omega t + \theta_1 + \theta_2)]$$

$$= \frac{V_mI_m}{2}[\cos(\theta_1 - \theta_2) + \cos(2(\omega t + \theta_2) + (\theta_1 - \theta_2))]$$

$$= \frac{V_mI_m}{2}[\cos(\theta_1 - \theta_2) + \cos 2(\omega t + \theta_2)\cos(\theta_1 - \theta_2)$$

$$- \sin 2(\omega t + \theta_2)\sin(\theta_1 - \theta_2)]$$

$$= \frac{V_mI_m}{2}\cos(\theta_1 - \theta_2)[1 + \cos 2(\omega t + \theta_2)]$$

$$- \frac{V_mI_m}{2}\sin 2(\omega t + \theta_2)\sin(\theta_1 - \theta_2)$$

$$= |V||I|\cos(\theta_1 - \theta_2)[1 + \cos 2(\omega t + \theta_2)]$$

$$- |V||I|\sin 2(\omega t + \theta_2)\sin(\theta_1 - \theta_2)$$

$$= |V||I|\cos\varphi[1 + \cos 2(\omega t + \theta_2)]$$

$$- |V||I|\sin 2(\omega t + \theta_2)\sin\varphi \tag{2.16}$$

where $\varphi = (\theta_1 - \theta_2)$ is the phase angle difference between the V and I phasors.

Figure 2.8 shows instantaneous variation of $v(t)$, $i(t)$, and $p(t)$ against time t. From Fig. 2.8 and Eq. (2.16), it is clear that $p(t)$ has two components, a constant component represented by $VI\cos\varphi$ and a time-varying component given by $VI\cos\varphi\cos(2\omega t + \theta_1 + \theta_2)$. The time-varying component has an angular speed equal to 2ω, which is twice that of the voltage or current wave.

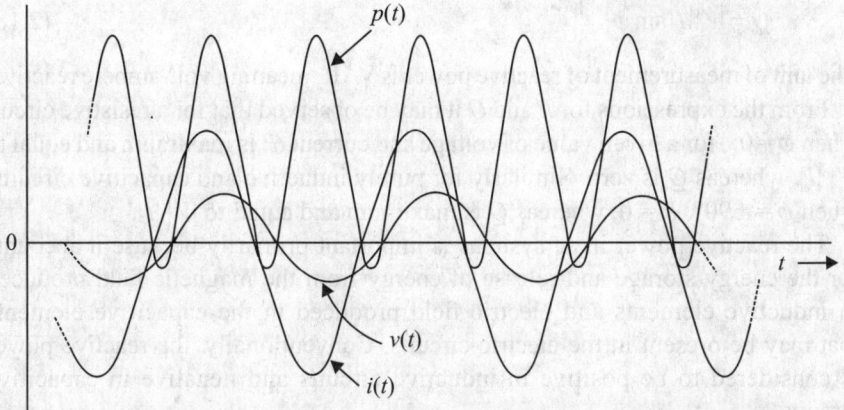

Fig. 2.8 Plot of $p(t)$ versus t

From Eq. (2.16), observe that for a fixed value of φ, the instantaneous power consists of two terms. The first term $p_1(t) = VI \cos\varphi \, [1 + \cos 2(\omega t + \theta_2)]$ has two components: a constant part and a time-varying part. The time-varying part has a frequency which is twice that of the voltage and current; and this component does not contribute towards the average value of power delivered to the one-port network as the average value of a cosine curve over one complete cycle is zero. Therefore, the first term $p_1(t)$ has an average value $p_{av} = |V||I| \cos\varphi$ with zero as the minimum value and $2|V||I| \cos\varphi$ as the maximum value. It varies about the average value sinusoidally at twice the supply frequency. This power is termed as instantaneous real power, or active power, or simply power. The second term of Eq. (2.16) $p_2(t) = -|V||I| \sin 2(\omega t + \theta_2) \sin \varphi$, has a peak value $q = |V||I| \sin\varphi$, and varies about its mean value of zero sinusoidally at twice the supply frequency. This power is referred to as instantaneous reactive power. This is a component of power that oscillates between the source and the energy storing elements of the circuit, that is, the reactive elements of the load.

If P is defined as average power over one cycle, that is, $T = 2\pi/\omega$, and power factor angle $\varphi = (\theta_1 - \theta_2)$, then the average real power P over one cycle, that is, $T = 2\pi/\omega$, is defined as the average value of instantaneous power $p_1(t)$ over one cycle and is obtained as

$$P = \frac{\omega}{2\pi} \int_0^{2\pi/\omega} p_1(t) dt = \frac{\omega}{2\pi} \int_0^{2\pi/\omega} |V||I| \cos\varphi \, [1 + \cos 2(\omega t + \theta_2)] \, dt$$

$$= |V||I| \cos\varphi \, \frac{\omega}{2\pi} \left(t - \frac{\sin 2(\omega t + \theta_2)}{2\omega} \right)_0^{2\pi/\omega}$$

or $\qquad P = |V||I| \cos\varphi$ $\hfill$ (2.17)

The unit of measurement of active power is watt. In Eq. (2.17), the cosine of the phase angle between the voltage and current phasors, $\cos\varphi$, is called the power factor.

The reactive power, also known as reactive volt-ampere, or quadrature power is defined as the average value of the instantaneous reactive power $p_2(t)$ and may be written as

$$Q = |V||I|\sin\varphi \qquad (2.18)$$

The unit of measurement of reactive power is VAR, meaning volt-ampere reactive.

From the expressions for P and Q it may be observed that for a resistive circuit when $\varphi = 0°$, for a given value of voltage and current, P is maximum and equal to $|V|\,|I|$, whereas Q is zero. Similarly for purely inductive and capacitive circuits, when $\varphi = \pm 90°$, $P = 0$, whereas Q is maximum and equal to $|V|\,|I|$.

The reactive power in ac systems is important primarily because it accounts for the energy storage and release of energy from the magnetic field produced in inductive elements and electric field produced in the capacitive elements that may be present in the electric circuits. Conventionally, the reactive power is considered to be positive in inductive circuits and negative in capacitive circuits.

The product of the voltage and current of an ac circuit is termed the *apparent power*. It is usually denoted by the symbol S. Thus,

$$S = |V||I| \text{ volt-ampere or VA} \qquad (2.19)$$

Equations (2.17) and (2.18) can now be written as

$$P = |V||I|\cos\varphi = S\cos\varphi \text{ W} \qquad (2.20)$$

$$Q = |V||I|\sin\varphi = S\sin\varphi \text{ VAR} \qquad (2.21)$$

Then, $\quad S = \sqrt{P^2 + Q^2} \text{ VA} \qquad (2.22)$

Equation (2.22) suggests that power can be interpreted by means of the phasor diagram shown in Fig. 2.9. The current component $I\cos\varphi$ in phase with the voltage V supplies active power, whereas the current component $I\sin\varphi$ in quadrature with the voltage V supplies reactive power. The component $I\cos\varphi$ is termed as active power component of the current, while the component $I\sin\varphi$ is termed as reactive power component of the current. The complex power S is defined as a complex number whose real part represents active power and the imaginary part represents reactive power.

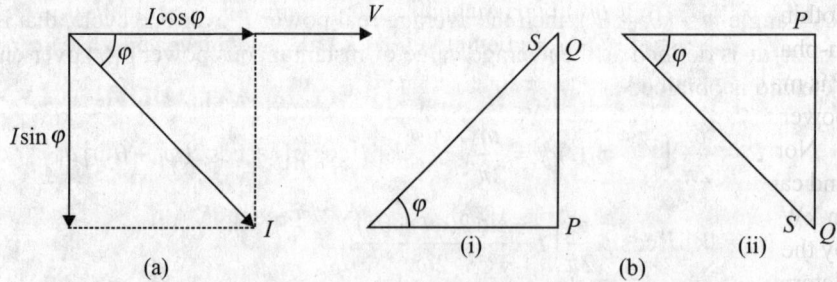

Fig. 2.9 (a) In-phase and quadrature components of current
(b) Complex power triangle (i) inductive load and (ii) capacitive load

The complex power phasor diagram is shown in Fig. 2.9(b). From the phasor diagram of Fig. 2.9(b), the complex power S can be written in the form

$$S = P \pm jQ \qquad (2.23)$$

where Q is positive for inductive load and negative for capacitive load. Alternatively, complex power in terms of voltage phasor V and current phasor I can be expressed as

$$S = VI^* \tag{2.24}$$

where I^* is the complex conjugate of I.

Let $|V| = Ve^{j0}$, $|I| = I\angle - \varphi = |I|e^{j(0 - \varphi)}$, I lags V by φ.

Then, $S = VI^* = |V|e^{j0}|I|e^{-j(0 - \varphi)} = |V||I|e^{j\varphi} = |V||I|(\cos\varphi + j\sin\varphi)$

or $S = P + jQ = |S|\angle\varphi = |S|(\cos\varphi + j\sin\varphi) \tag{2.25}$

Sign convention for power   If the current entering I at a node lags the applied node voltage V by an angle φ (between 0° and 90°), then $P = |V||I|\cos\varphi$ and $Q = |V||I|\sin\varphi$ entering the node are both taken to be positive, indicating that watts and VARs are being absorbed by the node. When I leads V by an angle φ, P is still positive and $Q = |V||I|\sin\varphi$ is treated as negative, indicating that negative VARs are being absorbed or positive VARs are being supplied by the circuit (capacitive) connected to the node.

Units of power   From the foregoing, it is observed that complex power S, apparent power $|S|$, real power P, and reactive power Q are products of voltage and current. The measurement unit of all power quantities is watt. However, to distinguish between the various power quantities and to avoid confusion, complex and apparent power S and $|S|$ respectively, have been assigned the units of volt-amperes (VA), real power P the units of watts (W), and reactive power Q the units of volt-amperes reactive (VAR). Since power systems handle large magnitudes of power quantities, practical units of mega (MVA, MW, MVAR) or kilo (kVA, kW, kVAR) have been adopted.

2.4.1 Need for Power Factor Improvement

In Fig. 2.9(a), current I is lagging voltage V by angle φ. This can be resolved into two components: in-phase component $|I|\cos\varphi$, in phase with V and the quadrature component $|I|\sin\varphi$, lagging V by 90°. The quadrature component contributes nothing to the active power, the entire active power is being contributed by the in-phase component. When $\cos\varphi = 1$, for given values of V and I, the active power P is maximum and is numerically equal to $|V||I|$ volt-amperes. The value of active power decreases with the decrease in power factor.

Normally ac apparatus, such as generators, transformers, transmission lines, and cables, is generally rated in kVA or MVA, rather than the active power output in kW or MW. The allowable output of the device is limited by heating caused by the losses in the device; the operating voltage and current of the device in turn determine these losses. Consequently, the capacity of the electrical equipment installed to supply a given load is essentially determined by the kVA or MVA of that load rather than by the load power in kW or MW. On the other hand, the boiler and turbine sizes and fuel requirements in a thermal generating station (turbine size and water requirements in a hydroelectric station) are determined essentially by the kW or MW power output of the generator and not by its kVA or MVA rating. Thus it is quite logical to find the electric utilities fix the electrical rate structures dependent mainly on active power and one of the two quantities either volt-amperes or reactive power. All three quantities are of economic importance.

Thus, from the point of view of the supply company, a low power factor is a very serious matter, and since the value of the power factor is decided by the nature of the load, that is, by the consumer devices, such as motors, the company encourages consumers to make their power factor as high as possible. The usual method is to frame the tariff on a kVA basis by imposing a high charge for the kVA and a low charge for the kW of energy consumed.

Most electric loads are reactive in nature and have lagging power factors less than unity. Particularly the industrial loads that are mostly induction motor drives have low power factor. The power factor may even be much less than 0.8 if induction motors are not fully loaded and in the presence of arc furnaces. Transmission lines, transformers, and generators of the electric power utility have to carry the lagging reactive power requirement of the load so that their full active power capability is not exploited. In addition, reactive current causes additional ohmic losses and large voltage drops. These factors create operational difficulties associated with the maintenance of the required voltage at the alternator terminals or at the end of a transmission line and cause financial loss to the utility. The utility, therefore, encourages its industrial consumers to improve the power factor of the load by installing static shunt capacitors that draw leading current to neutralize the lagging current drawn by low power factor of the load. If the industries continue to draw lagging power factor, then penalty is imposed through enhanced tariff based on the reactive component of the consumer's load.

Industrial consumers thus find it economical to improve the power factor of their individual motors and/or the total installation by installing static shunt capacitors. The limit to which the power factor must be improved is dictated by the balance of the yearly tariff saving against the yearly interest and depreciation cost of installing the capacitors.

To summarize, a low power factor necessitates a high capital cost for alternators, switchgear, transformers, and cables. However, the disadvantage of a low power factor does not end here since there are operational difficulties such as low voltage at the consumer end.

2.5 PHASE SHIFT OPERATOR a

Electrical power systems in operation, the world over, are three-phase systems. At the generating station, three-phase balanced ac voltages are generated. The generated voltage of each phase has the same amplitude but displaced from each other by 120°. The rms values of phase voltages may be represented by the phasors V_a, V_b, and V_c. In polar and rectangular form the three-phase rms voltages can be written in terms of rms phase voltage V as

$$V_a = |V|\angle 0° = |V|e^{j0} = |V|(1 + j0) \tag{2.26a}$$

$$V_b = |V|\angle{-120°} = |V|e^{-j120°} = |V|(-0.5 - j0.866) \tag{2.26b}$$

$$V_c = |V|\angle{-240°} = |V|e^{-j240°} = |V|(-0.5 + j0.866) \tag{2.26c}$$

where j is a positive phase operator, or projection on the imaginary axis.

In order to facilitate analyses of power systems, it is appropriate to take advantage of Euler's exponential and define a phase shift operator a, which has a unit magnitude and shifts a phasor through 120°, without effecting its magnitude, in the anti-clockwise direction. Thus,

$$a = e^{j(2\pi/3)} = 1\angle 120° = \left(\cos\frac{2\pi}{3} + j\sin\frac{2\pi}{3}\right)$$

or $\qquad a = (-0.5 + j0.866)$ $\hfill$ (2.27a)

Similarly,

$$a^2 = a \times a = e^{j(4\pi/3)} = 1\angle 240° = \left(\cos\frac{4\pi}{3} + j\sin\frac{4\pi}{3}\right)$$

$$= (-0.5 - j0.866) \hfill (2.27b)$$

$$a^3 = a \times a^2 = 1\angle 120° \times 1\angle 240° = 1\angle 0° = 1 + j0 \hfill (2.27c)$$

$$a^4 = a \times a^3 = a = 1\angle 120° = (-0.5 - j0.866)$$

The phase shift operator a *is* also known as the phasor a or simply a-operator. The concept of a-operator is similar to the complex number $j = e^{j(\pi/2)}$ which shifts a phasor through 90° in the anti-clockwise direction without effecting the magnitude of the phasor.

The phasor representation of three-phase balanced voltages mentioned in Eq. (2.26) may be written in a vector form as

$$V_{abc} = \begin{bmatrix} V_a \\ V_b \\ V_c \end{bmatrix} = \begin{bmatrix} 1 \\ a^2 \\ a \end{bmatrix} V_a \hfill (2.28)$$

Similarly balanced three-phase line currents may be represented by the phasors I_a, I_b, and I_c. In polar and rectangular form the three-phase rms currents can be written in terms of rms phase current I as

$$I_a = |I|\angle\varphi° = |I|e^{j\varphi} = |I|(\cos\varphi + j\sin\varphi) \hfill (2.29a)$$

$$I_b = |I|\angle(\varphi - 120°) = |I|e^{j\varphi - j120°} = |I|e^{j\varphi}e^{-j120°} = a^2 I_a \hfill (2.29b)$$

$$I_c = |I|\angle(\varphi - 240°) = |I|e^{j\varphi - j240°} = |I|e^{j\varphi}e^{-j240°} = a I_a \hfill (2.29c)$$

where φ is the phase angle of lag between the currents and the voltage reference. In terms of a vector, the rms phasor currents may be expressed as

$$I_{abc} = e^{j\varphi}\begin{bmatrix} I_a \\ I_b \\ I_c \end{bmatrix} = \begin{bmatrix} 1 \\ a^2 \\ a \end{bmatrix} |I_a|e^{j\varphi} \hfill (2.30)$$

Example 2.2 Calculate the following linear combinations of phasor a
(a) $1 + a$, (b) $a^2 - 1$, (c) $a + a^2$, and (d) $1 + a + a^2$.

Solution
(a) $1 + a = 1 + (-0.5 + j0.866) = (0.5 + j0.866) = 1\angle 60° = -a^2$
(b) $a^2 - 1 = (-0.5 - j0.866) - 1 = (-1.5 - j0.866) = \sqrt{3}\angle -150°$
(c) $a + a^2 = (-0.5 + j0.866) + (-0.5 - j0.866) = -1.0 + j0 = 1\angle 180°$
(d) $1 + a + a^2 = 1 + (-0.5 + j0.866) + (-0.5 - j0.866) = 0$

2.6 POWER IN BALANCED THREE-PHASE CIRCUITS

The power in a three-phase circuit is defined as the summation of the power delivered to all the three phases. In a balanced three-phase circuit, though the power in the individual phases varies sinusoidally at twice the supply frequency, it will be seen that the total of the instantaneous power in the three-phase circuit is a constant and independent of time. The concepts of instantaneous, active, reactive, apparent, and complex power and power factor as applied to three-phase circuits will be presented in this section.

2.6.1 Instantaneous Power in Three-phase Circuits

Power in a three-phase balanced network, having phase sequence *A-B-C*, can be obtained by taking the sum of the products of instantaneous voltages and currents. If p is the total instantaneous power, then

$$p_{3\phi} = v_A i_A + v_B i_B + v_C i_C \tag{2.31}$$

The instantaneous values of phase voltages v_A, v_B, v_C and phase currents i_A, i_B, i_C can be written as

$$v_A = \sqrt{2}\,|V_P|\sin \omega t$$

$$v_B = \sqrt{2}\,|V_P|\sin(\omega t - 120°)$$

$$v_C = \sqrt{2}\,|V_P|\sin(\omega t - 240°)$$

$$i_A = \sqrt{2}\,|I_P|\sin(\omega t - \varphi)$$

$$i_B = \sqrt{2}\,|I_P|\sin(\omega t - \varphi - 120°)$$

$$i_C = \sqrt{2}\,|I_P|\sin(\omega t - \varphi - 240°)$$

where V_P and I_P are the rms values of phase voltage and phase current, respectively, while φ is the phase angle between the phase voltage and corresponding phase current. The instantaneous total power in the three phases then becomes

$$p_{3\phi} = v_A i_A + v_B i_B + v_C i_C$$

$$= \sqrt{2}\,|V_P|\sin\omega t \; \sqrt{2}\,|I_P|\sin(\omega t - \varphi)$$

$$+ \sqrt{2}\,|V_P|\sin(\omega t - 120°)\; \sqrt{2}\,|I_P|\sin(\omega t - \varphi - 120)$$

$$+ \sqrt{2}\,|V_P|\sin(\omega t - 240°)\; \sqrt{2}\,|I_P|\sin(\omega t - \varphi - 240°) \tag{2.32}$$

On simplifying Eq. (2.32), the instantaneous power becomes

$$p_{3\phi} = 3|V_P||I_P|\cos\varphi \tag{2.33}$$

Hence, the average three-phase power is given by

$$P_{3\phi} = \frac{1}{T}\int_0^T p\,dt = 3|V_P||I_P|\cos\varphi \tag{2.34}$$

In a star-connected system, line voltage $|V_L| = \sqrt{3}\,|V_P|$ and $|I_L| = |I_P|$. Substituting for $|V_P|$ and $|I_P|$ in Eq. (2.33), the power in the star-connected network becomes

$$P_{3\phi} = \sqrt{3}\,|V_L||I_L|\cos\varphi$$

Similarly, for a delta-connected system, $V_L = |V_P|$ and $|I_L| = \sqrt{3}\,|I_P|$. Hence,

$$P_{3\phi} = \sqrt{3}\,|V_L||I_L|\cos\varphi$$

Therefore, it can be concluded that power in a balanced three-phase network, star or delta, is given by

$$P_{3\phi} = \sqrt{3}\,|V_L||I_L|\cos\varphi \qquad (2.35)$$

It may be noted that unlike single-phase systems, where the instantaneous power is pulsating, the total instantaneous power for a three-phase system is constant. Thus a three-phase supply system has a distinct advantage over a single-phase system, especially in the running of large industrial drives.

2.6.2 Reactive and Apparent Power

The reactive power in a three-phase system is equal to the sum of the reactive powers of the individual phases. For a balanced three-phase system, the reactive power Q is

$$Q_{3\phi} = 3|V_P||V_P|\sin\varphi \qquad (2.36a)$$

$$= \sqrt{3}\,|V_L||I_L|\sin\varphi \qquad (2.36b)$$

The apparent power or the volt-ampere for a three-phase circuit is the sum of the volt-amperes of the three individual phases. For a balanced three-phase system, the apparent power is given by

$$|S_{3\phi}| = 3|V_P||I_P| \qquad (2.37a)$$

$$= \sqrt{3}\,|V_L||I_L| \qquad (2.37b)$$

It may be observed that in a balanced three-phase system

$$P_{3\phi} = 3|V_P||I_P|\cos\varphi = S\cos\varphi$$

$$Q_{3\phi} = 3|V_P||I_P|\sin\varphi = S\sin\varphi$$

$$|S_{3\phi}| = \sqrt{P_{3\phi}^2 + Q_{3\phi}^2} \qquad (2.38)$$

2.6.3 Complex Power

For convenience in calculations, the concept of complex power is useful. It is a complex number the real part of which is the total active power and the imaginary part is the total reactive power in the circuit. Thus,

$$S_{3\phi} = P_{3\phi} + jQ_{3\phi}$$

$$= S_{3\phi}(\cos\varphi + j\sin\varphi) = S_{3\phi}\angle\varphi$$

Complex power in terms of V_P and I_P can be expressed as

$$S_{3\phi} = 3V_P I_P^* \qquad (2.39)$$

Let $\quad V_P = |V_p|e^{j0},\ I_P = |I_p|e^{j(0-\varphi)},\ I_P$ lags V_P by φ.

Then, for a balanced load,

$$S_{3\phi} = 3V_P I_P^* = 3|V_P|e^{j0}\,|I_P|e^{-j(0-\varphi)}$$

$$= 3|V_P||I_P|e^{j\varphi} = 3|V_P||I_P|(\cos\varphi + j\sin\varphi)$$

$$= P_{3\phi} + jQ_{3\phi} = |S_{3\phi}|\angle\varphi = |S_{3\phi}|(\cos\varphi + j\sin\varphi) \qquad (2.40)$$

2.7 PER UNIT QUANTITIES

Analyses of power systems employing actual values such as electrical units of voltampere (VA), voltage, current, and impedance do not adapt themselves easily to computations. However, these quantities are often expressed as per unit of a base or reference value specified for each. Electric power engineers often prefer to express impedances, currents, voltages, and power in per unit (pu) values rather than their actual units.

2.7.1 Per Unit Normalization

The numerical per unit value of any quantity is its ratio to the chosen base quantity of the same dimensions. Thus, per unit (pu) quantity is a normalized quantity with respect to a chosen base value. The base values for the quantities such as voltages, currents, impedances, and power are picked to compute per unit values of the quantities. The base variables are picked to satisfy the same kind of relationship as the actual variables. For example, corresponding to the equation between actual complex voltage V, complex current I, and complex impedance Z,

$$V = ZI \tag{2.41}$$

the equation for the base quantities, which are real numbers, is

$$V_{base} = Z_{base} I_{base} \tag{2.42}$$

Dividing Eq. (2.41) by Eq. (2.42) gives

$$\frac{V}{V_{base}} = \frac{Z}{Z_{base}} \frac{I}{I_{base}} \tag{2.43}$$

or $\qquad V_{pu} = Z_{pu} I_{pu} \tag{2.44}$

2.7.2 Power and Circuital Formulae in pu Values

Complex power is given by

$$S = P + jQ = VI^* \tag{2.45}$$

and the base power is given by

$$S_{base} = V_{base} I_{base} \tag{2.46}$$

Dividing Eq. (2.45) by Eq. (2.46) gives

$$S_{pu} = V_{pu} I_{pu}^* \tag{2.47}$$

It may be noted that Eqs (2.42) and (2.46) involve four base quantities V_{base}, I_{base}, Z_{base}, and S_{base}, and specifying any two base quantities determines the remaining two base quantities. For example, if S_{base} and V_{base} are chosen as base quantities, then

$$I_{base} = \frac{S_{base}}{V_{base}} \tag{2.48}$$

and $\qquad Z_{base} = \frac{V_{base}}{I_{base}} = \frac{V_{base}^2}{S_{base}} \tag{2.49}$

The per unit values of other variables can be written as follows:

$$S_{pu} = P_{pu} + jQ_{pu}$$

where

$$P_{pu} = \frac{P}{S_{base}} \text{ and } Q_{pu} = \frac{Q}{S_{base}} \tag{2.50}$$

Also, $Z_{pu} = R_{pu} + jX_{pu}$

where $R_{pu} = \frac{R}{Z_{base}} \text{ and } X_{pu} = \frac{X}{Z_{base}} \tag{2.51}$

Similarly,

$$Y_{pu} = G_{pu} - jB_{pu} = \frac{1}{Z_{pu}} \tag{2.52}$$

where

$$G_{pu} = \frac{G}{Y_{base}} \text{ and } B_{pu} = \frac{B}{Y_{base}} \tag{2.53}$$

2.7.3 Per Unit Representation of Power Systems

In an electrical network, the same base volt-ampere is used in all parts of the system, and one base voltage is selected arbitrarily at any point in the system. Base voltages at all other points must be related to the arbitrarily selected base voltage by the turn ratios of the connecting transformers.

For single-phase systems, or balanced three-phase systems (when solved as a single line with a neutral return), the base quantities in the impedance diagram are per phase volt-ampere in MVA and line-to-neutral voltage in kV. The following formulae relate the various quantities:

$$\text{Base current } I_{base} = \frac{MVA_{base}}{kV_{base}} kA \tag{2.54}$$

$$\text{Base impedance } Z_{base} = \frac{kV_{base}}{I_{base}} = \frac{kV_{base}}{\dfrac{MVA_{base}}{kV_{base}}} = \frac{(kV_{base})^2}{MVA_{base}} \Omega \tag{2.55}$$

$$\text{Base active power} = MVA_{base} \ MW \tag{2.56}$$

$$\text{Base reactive power} = MVA_{base} \ MVAR \tag{2.57}$$

If Z is the actual value of impedance (in ohm) of the element, then

$$Z \text{ in pu} = \frac{Z \text{ in ohms}}{Z_{base}} = \frac{Z \text{ in ohms}}{\dfrac{(kV_{base})^2}{MVA_{base}}}$$

or $$Z \text{ in pu} = [Z \text{ (in ohms)}] \times \frac{MVA_{base}}{(kV_{base})^2} \tag{2.58}$$

$$Z \text{ in per cent} = (Z \text{ in ohms}) \times \frac{MVA_{base}}{(kV_{base})^2} \times 100 \tag{2.59}$$

Similarly,

$$Y_{base} = \frac{MVA_{base}}{(kV_{base})^2} \text{ mho}$$

Admittance Y in pu = admittance Y (in mho) $\times \dfrac{(kV_{base})^2}{MVA_{base}}$ (2.60)

Other quantities can be similarly converted into pu quantities. Thus,

$$\text{Per unit current} = \frac{\text{Actual current in kA}}{I_{base}}$$ (2.61)

$$\text{Per unit voltage} = \frac{\text{Actual voltage in kV}}{kV_{base}}$$ (2.62)

In these equations, the subscript $_{base}$ denotes base quantities.

The equation derived for pu impedance [Eq. (2.55)] is correct only for a single-phase system. However, in three-phase systems, data is usually given as total three-phase MVA and line-to-line kV, and it is preferred to work with this data. Let this problem be investigated by rewriting Eqs (2.58) and (2.60) using the subscript LN for line-to-neutral and 1ϕ for single-phase systems, then

$$Z \text{ in pu} = (Z \text{ in ohms}) \times \frac{MVA_{base-1\phi}}{(kV_{base-LN})^2}$$ (2.63)

and

$$Y \text{ in pu} = (Y \text{ in mho}) \times \frac{(kV_{base-LN})^2}{MVA_{base-1\phi}}$$ (2.64)

Now, using subscript LL to indicate line-to-line and 3ϕ for three-phase, the following expressions may be written for a balanced system

$$kV_{base-LN} = \frac{kV_{base-LL}}{\sqrt{3}}$$ (2.65)

and

$$MVA_{base-1\phi} = \frac{MVA_{base-3\phi}}{3}$$ (2.66)

Substituting Eqs (2.65) and (2.66) in Eqs (2.63) and (2.64) yields

$$Z \text{ in pu} = (Z \text{ in ohms}) \times \frac{\dfrac{MVA_{base-3\phi}}{3}}{\left(\dfrac{kV_{base-LL}}{\sqrt{3}}\right)^2}$$

$$= (Z \text{ in ohms}) \times \frac{MVA_{base-3\phi}}{(kV_{base-LL})^2}$$ (2.67)

and $\quad Y \text{ in pu} = (Y \text{ in mho}) \times \dfrac{(kV_{base-LL})^2}{MVA_{base-3\phi}}$ (2.68)

Thus, it may be noted that except for the subscripts, Eqs (2.63) and (2.67) are identical. Subscripts have been used in these expressions to emphasize the distinction between working with per-phase and three-phase quantities. These equations can be used without the subscripts, but one must use

(a) line-to-neutral kV with MVA per phase and
(b) line-to-line kV with three-phase MVA.

Example 2.3 A three-phase transmission line transmits 50 MW at 0.8 power factor lagging at 132 kV. If the impedance of the transmission line is $(40 + j100)\Omega$, calculate the pu values of (i) complex power, real power, and reactive or VAR power, (ii) voltage, (iii) current, and (iv) impedance, resistance, and reactance. For the transmission line, assume $MVA_{base} = 100$ and $kV_{base} = 132$.

Solution
From Eq. (2.55)

$$Z_{base} = \frac{(kV_{base})^2}{MVA_{base}} = \frac{(132)^2}{100} = 174.24 \, \Omega$$

From Eq. (2.54)

$$I_{base} = \frac{100 \times 1000}{\sqrt{3} \times 132} = 437.40 \, kA$$

(i) Complex power $S = \frac{50}{0.8} = 62.5$ MVA

$$S_{pu} = \frac{62.5}{100} = 0.625; \quad P_{pu} = \frac{50}{100} = 0.5;$$

$$Q_{pu} = \frac{62.5 \times 0.6}{100} = 0.375$$

(ii) From Eq. (2.62),

$$kV_{pu} = \frac{\text{Actual voltage}}{kV_{base}} = \frac{132}{132} = 1.0$$

(iii) Actual current $= \frac{50 \times 1000}{\sqrt{3} \times 132 \times 0.8} = 273.37$ kA

From Eq. (2.61)

$$I_{pu} = \frac{\text{Actual current}}{I_{base}} = \frac{273.37}{437.40} = 0.625$$

(iv) Actual impedance $Z = \sqrt{(40)^2 + (100)^2} = 107.70 \, \Omega$

$$Z_{pu} = \frac{\text{Actual impedance}}{Z_{base}} = \frac{107.70}{174.24} = 0.618$$

$$R_{pu} = \frac{\text{Actual resistance}}{Z_{base}} = \frac{40}{174.24} = 0.23$$

$$X_{pu} = \frac{\text{Actual reactance}}{Z_{base}} = \frac{100}{174.24} = 0.574$$

2.7.4 Conversion of pu Quantities from One Base to Another Base

The per unit (pu) impedance of a generator or a transformer, as supplied by the manufacturer, is generally based on the rating of the generator or the transformer itself, while the selected base volt-ampere and base voltage for a power system are different from the ratings of the components. Therefore, it is imperative to change the pu values computed on rated base quantities to the common system base quantities.

The expressions given by Eq. (2.63) or Eq. (2.67) may be used for the determination of per unit impedance of a circuit element from its actual value in ohms. From this equation it may be noted that per unit impedance is directly proportional to (MVA_{base}) and inversely proportional to $(kV_{base})^2$.

If Z_{old} is the pu impedance on a given set of base quantities MVA_{old} and kV_{old}, and Z_{new} is the pu impedance on a new set of base quantities MVA_{new} and kV_{new}, then

$$Z_{old} = (Z \text{ in ohms}) \times \frac{MVA_{old}}{(kV_{old})^2} \qquad (2.69)$$

and

$$Z_{new} = (Z \text{ in ohms}) \times \frac{MVA_{new}}{(kV_{new})^2} \qquad (2.70)$$

From Eqs (2.69) and (2.70), the relationship between the old and new per unit values is

$$Z_{new} = (Z_{old}) \times \frac{(kV_{old})^2 \times (MVA_{new})}{(kV_{new})^2 \times (MVA_{old})} \qquad (2.71)$$

If the voltage bases are same, Eq. (2.71) reduces to

$$Z_{new} = (Z_{old}) \times \frac{(MVA_{new})}{(MVA_{old})} \qquad (2.72)$$

2.7.5 Advantages of pu Values

Following are the advantage of working with pu values to analyse power systems.

(a) Per unit values of equipment (generators, transformers, etc.) that have widely varying ratings normally fall within a narrow range while their actual ohmic values differ from equipment to equipment having different ratings. Therefore, pu values can also be selected, when ohmic values of impedances are not available, from tables which provide average values of various categories of equipment. Also, since the range of pu values of different categories of equipment is known, it is easier to detect errors in data or computation.

(b) Per unit values of quantities, when the base values are appropriately selected, are independent of the side of the transformer to which they are connected. On the other hand, ohmic values have to be referred to one side of the transformer.

(c) Per unit values are independent of the type of the power system, that is, whether the power system is single phase or three phase.

(d) Although the voltage bases on the two sides of a transformer in a three-phase circuit must have a definite relationship, the pu values of impedances are independent of the way in which transformers are connected.
(e) The equipment, supplied by manufacturers, provides the equipment parameters in pu on the name plate rating.
(f) Analyses of power systems are simplified considerably by using pu values.

2.7.6 Steps to Compute pu Values

Analyses of power system problems are greatly simplified by using single-line impedance diagrams in which system parameters are expressed in per unit. The procedure is summarized in the following steps.

Step 1 Select a common volt-ampere base for the entire power system and a voltage base for one part of the system. Usually, a voltage base is selected for a transmission line in the power system.

Step 2 Compute voltage bases for all parts of the power system by correlating the transformation ratios of the transformer banks.

Step 3 Convert pu values (which are provided on the name plate rating of the equipment) to the common system volt-ampere base and the applicable voltage base. In case the parameters are provided in actual ohmic values, compute base impedance for the part of the power system in which the equipment is connected and calculate the pu values.

Step 4 Draw a single-line diagram of the power system indicating values of all parameters in pu. Proceed to analyse the power system.

Step 5 Convert to actual values where required.

Example 2.4 Show that the per unit impedance of a transformer is independent of its primary or secondary side.

Solution Assume that the volt-ampere rating of the transformer is kVA, its primary and secondary voltages are kV_P and kV_S, and its corresponding primary and secondary winding impedances are Z_P and Z_S ohms, respectively. Next, suppose the transformer rating is selected as the base power, that is, $(kVA)_b$.

Then, the primary side base current $I_{bP} = \dfrac{kVA}{kV_P}$ and the base impedance is given by

$$Z_{bP} = \frac{1000 \times kV_P}{I_{bP}} = \frac{1000(kV_P)^2}{kVA}\,\Omega$$

Hence, the pu value of the primary side impedance is written as

$$Z_{(pu)P} = \frac{Z_P}{Z_{bP}} = \frac{Z_P \times kVA}{1000(kV_P)^2} \tag{2.4.1}$$

By following a similar procedure, the pu value of the secondary can be derived as shown below.

Primary side base current $I_{bS} = \dfrac{kVA}{kV_S}$ and base impedance is $Z_{bS} = \dfrac{1000(kV_S)^2}{kVA}\,\Omega$.

Thus, $\quad Z_{(pu)S} = \dfrac{Z_S}{Z_{bS}} = \dfrac{Z_S \times kVA}{1000(kV_S)^2}$ $\hspace{2cm}$ (2.4.2)

Dividing Eq. (2.4.1) by Eq. (2.4.2) and simplifying leads to

$$\frac{Z_{(pu)P}}{Z_{(pu)S}} = \frac{Z_P}{Z_S (kV_P/kV_S)^2} \qquad (2.4.3)$$

Recalling that the denominator in Eq. (2.4.3) represents the relationship to refer the secondary impedance to that of the primary side, that is, $Z_P = Z_S (kV_P/kV_S)^2 \, \Omega$. Thus, it can be seen that per unit impedance is independent of the two sides of a transformer.

Example 2.5 The power system shown in Fig. 2.10 has the following specifications:

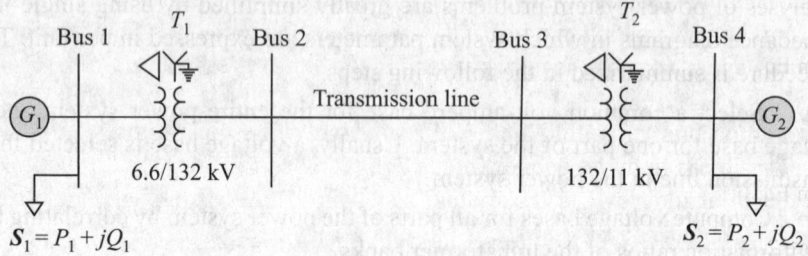

Fig. 2.10 Single-line diagram of a power system for Example 2.5

Generator G_1: 20 MVA, 6.6 kV, $X_{G1} = 0.10$ pu

Generator G_2: 25 MVA, 11 kV, $X_{G2} = 0.20$ pu

Transformer T_1: 25 MVA, 6.6/132 kV, $X_1 = 0.08$ pu

Transformer T_2: 30 MVA, 11/132 kV, $X_2 = 0.10$ pu

Transmission line: Line-to-line voltage = 132 kV,

impedance $Z = (30 + j \, 120) \, \Omega$

Load: $S_1 = 10$ MVA at 0.8 pf lagging and $S_2 = 25$ MVA at 0.9 pf leading.

Assuming MVA$_{base}$ = 50 for the system, calculate the pu values of generators, transformers, transmission line, and load. Draw a single-line diagram and show the pu values of the system components.

Solution The pu values of the equipment are on the name plate ratings as base quantities. Using Eq. (2.71), the pu values can be converted to the system MVA$_{base}$ and kV$_{base}$ as follows.

For generator G_1: MVA$_1$ = 20, kV$_1$ = 6.6, and $X_1 = 0.1$ pu

$\qquad$ MVA$_2$ = MVA$_{base}$ = 50, and kV$_2$ = 6.6

Using Eq. (2.71), X_{G1} on system base values = $(0.1) \times \dfrac{(6.6)^2 \times (50)}{(6.6)^2 \times (20)} = 0.25$ pu

For generator G_2: MVA$_1$ = 25, kV$_1$ = 11, and $X_1 = 0.2$ pu
MVA$_2$ = 50 and kV$_2$ = 11

Using Eq. (2.71), X_{G2} on system base values $(0.2) \times \dfrac{(11)^2 \times (50)}{(11)^2 \times (25)} = 0.4$ pu

For transformer T_1: MVA$_1$ = 25, kV$_1$ = 132, and $X_1 = 0.08$ pu

$\qquad$ MVA$_2$ = 50, and kV$_2$ = 132

Using Eq. (2.71), X_1 on system base values = $(0.08) \times \dfrac{(132)^2 \times (50)}{(132)^2 \times (25)} = 0.16$ pu

For transformer T_2: $MVA_1 = 30$, $kV_1 = 132$, and $X_2 = 0.08$ pu

$$MVA_2 = 50, \text{ and } kV_2 = 132$$

Using Eq. (2.71), X_2 on system base values $= (0.10) \times \dfrac{(132)^2 \times (50)}{(132)^2 \times (30)} = 0.167$ pu

For transmission line, $MVA_{base} = 50$ and $kV_{base} = 132$

Using Eq. (2.58), $Z_{pu} = \dfrac{(30 + j120) \times 50}{(132)^2} = (0.086 + j0.344)$

$$\text{Load } S_1 = 10 \times (0.8 + j0.6) = (8 + j6) \text{ MVA}$$

From Eq. (2.50), $S_{1pu} = \dfrac{(8 + j6)}{50} = (0.16 + j0.12)$

$$\text{Load } S_2 = 25 \times (0.9 - j0.436) = (22.5 - j10.897)$$

From Eq. (2.50), $S_{2pu} = \dfrac{(22.5 - j10.897)}{50} = (0.45 - j0.218)$

Figure 2.11 shows the pu diagram for a power system on a system base of 50 MVA.

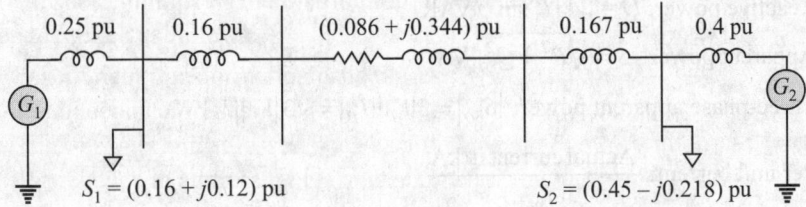

Fig. 2.11 Per unit diagram for the system given in Example 2.5 on a 50 MVA base

SUMMARY

- Modern-day power systems are invariably three-phase systems.
- In addition to delivering steady power for the same rating, the volume of conductor required for a three-phase system is less than that required for a single-phase system.
- In the steady-state operation, three-phase networks carry a balanced load and can be analysed by representing them as per-phase equivalent circuit consisting of a single line and the neutral as return.
- Standardized symbols are used to represent various components such as generators, transformers, and so on. Voltage and current variables are represented as phasor quantities. Alternating power is represented as a complex number whose real part represents active and imaginary part represents reactive power.
- The units for complex, real, and imaginary components of power are volt-amperes (VA), watts (W), and volt-amperes reactive (VAR), respectively. Since all the three components of power are of economic importance, supply companies encourage industrial consumers to improve their power factor at their terminals.
- Phase shift operator 'a' is a convenient tool to analyse power systems.
- Per unit normalization is a process of expressing electrical quantities to selected base values. Amongst several advantages of the use of per unit quantities, the most important is that analyses of power systems are simplified.

SIGNIFICANT FORMULAE

- Volume of conductor of a single-phase, two-wire system:

$$V_{1c} = a_{1p} \times l = \frac{4P^2}{V_m^2 \cos^2 \varphi} \times \frac{\rho l^2}{P_{1p}} \text{ m}^3$$

- Volume of conductor of a three-phase, three-wire system:

$$V_{3c} = 3 \times a_{3p} \times l = \frac{2P^2}{V_m^2 \cos^2 \varphi} \times \frac{\rho l^2}{P_{3p}} \text{ m}^3$$

- Ratio $\dfrac{V_{3c}}{V_{1c}} = 0.5$

- Voltage in rectangular form: $V = |V| (\cos \theta_1 + j \sin \theta_1)$ V
- Current in rectangular form: $I = |I| (\cos \theta_2 + j \sin \theta_2)$ A
- Instantaneous power: $p(t) = |V||I| \cos \varphi [1 + \cos 2(\omega t + \theta_2)]$
$$- |V||I| \sin 2(\omega t + \theta_2) \sin \varphi \text{ W}$$
- Real power: $P = |V||I| \cos \varphi$ W
- Reactive power: $Q = |V||I| \sin \varphi$ VAR
- Apparent power: $S = \sqrt{P^2 + Q^2}$ VA
- Three-phase apparent power: $|S_{3\phi}| = 3|V_P||I_P| = \sqrt{3}|V_L||I_L|$ W

- Per unit current: $\dfrac{\text{Actual current in kA}}{I_{\text{base}}}$

- Per unit voltage: $\dfrac{\text{Actual current in kV}}{V_{\text{base}}}$

- Per unit impedance: $(Z \text{ in ohms}) \times \dfrac{\text{MVA}_{\text{base}-1\phi}}{(\text{kV}_{\text{base}-LN})}$ or $(Z \text{ in ohms}) \times \dfrac{\text{MVA}_{\text{base}-3\phi}}{(\text{kV}_{\text{base}-LL})}$

- Change of base: $Z_{\text{new}} = (Z_{\text{old}}) \times \dfrac{(\text{kV}_{\text{old}})^2 (\text{MVA}_{\text{new}})}{(\text{kV}_{\text{new}})^2 (\text{MVA}_{\text{old}})}$

EXERCISES

Review Questions

2.1   Explain phasor representation of electrical quantities.

2.2   Derive expressions to represent a load $P + jQ$ as constant impedance (i) in series and (ii) in parallel. Determine the per unit R and X values in the two cases. Assume the load voltage to be V.

2.3   Derive expressions for complex power in a single-phase ac circuit and distinguish between the various components. Draw the phasor diagram for the complex power and state the sign convention.

2.4   Discuss the significance of lagging and leading power and explain the need for power factor improvement.

2.5   Explain the significance of the phase shift operator 'a' in the analysis of three-phase circuits.

2.6   Derive an expression for instantaneous power in a three-phase circuit and show that the power output is constant.

2.7 Explain per unit normalization and derive per unit formulae for (i) power and (ii) impedance.

2.8 Derive expressions to convert per unit impedance from (i) one voltage base to another voltage base, (ii) one power base to another power base, and (iii) one voltage and power base to another voltage and power base.

2.9 (a) Outline the advantages of working with per unit quantities.

(b) Write down the steps necessary to convert system parameters into per unit values.

Numerical Problems

2.1 For the power system shown in Fig. 2.12, convert all quantities to pu values on a system base of 25 MVA. Assume a base voltage of 33 kV for the transmission line.

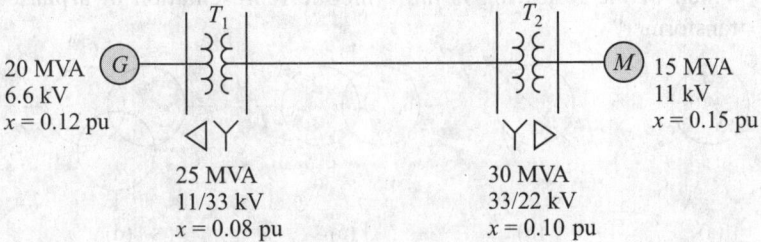

20 MVA (G)
6.6 kV
x = 0.12 pu

T_1

25 MVA
11/33 kV
x = 0.08 pu

T_2

(M) 15 MVA
11 kV
x = 0.15 pu

30 MVA
33/22 kV
x = 0.10 pu

Fig. 2.12

2.2 A resistor of 10 Ω, an inductor of 20 Ω and a capacitor of 25 Ω are connected in (i) series and (ii) in parallel across a voltage source of 1.1 kV. Determine the actual real, reactive and apparent power consumed in each case. If the system base is 500 kVA and 1.0 kV, compute the pu values of all the quantities.

2.3 Two generators are connected in parallel to a 6.6 kV bus. One of the generators has a rating of 20 MVA and a reactance of 15% while the second generator is rated at 15 MVA and has a reactance of 12%. Calculate the pu reactance on a 50 MVA and 6.6 kV base. What is the pu reactance of a single equivalent generator on 50 MVA and 6.6 kV base?

2.4 A three-phase 20 MVA, 11 kV generator has a reactance of 10% and is connected via a bank of three single-phase transformers to a transmission line whose series reactance is 100 Ω. The transmission line is supplying a load of 10 MVA at 10 kV and 0.8 power factor. The transformer bank at the generator end is connected in delta on the generator side and in star on the transmission line side and the transformer on the load side is star-star connected. Each of the single phase transformers in the bank of transformers is rated at 10 MVA, 11/110 kV with a reactance of 8%. Draw the pu circuit diagram for a system base of 10 MVA, 15 kV in the load circuit. Calculate the voltage at the generator terminals.

2.5 A three-winding transformer gave the following results when a short circuit test was performed on it:

Leakage reactance X_{ps} measured in primary with secondary short circuited and tertiary on open circuit = 0.25 Ω

Leakage reactance X_{pt} measured in primary with tertiary short circuited and secondary on open circuit = 0.10 Ω

Leakage reactance X_{st} measured in secondary with tertiary short circuited and primary on open circuit = 0.5 Ω

The windings of the transformer have the following ratings:

Primary: star connected, 11 kV, 20 MVA

Secondary: star connected, 6.6 kV, 15 MVA

Tertiary: delta connected 1.1 kV, 10 MVA.

On a base of 20 MVA, 11 kV, compute the reactances of the star connected equivalent circuit.

2.6 Determine (i) polar and (ii) rectangular forms of the following **a**-operator functions:

(a) $1 + \mathbf{a}^2$, (b) $1 - \mathbf{a}^2$, (c) $\mathbf{a} - 1$, (d) $\mathbf{a} - \mathbf{a}^2$, and (e) $\mathbf{a}^2 - \mathbf{a}$

Multiple Choice Objective Questions

2.1 Which of the following is the symbolic representation of a phase shift transformer?

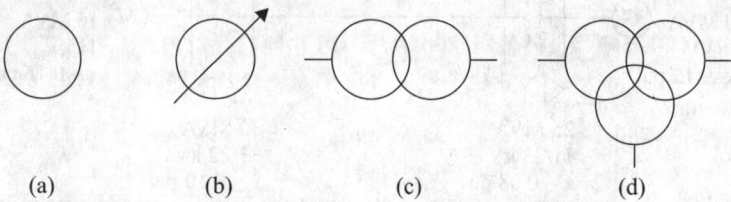

    (a)   (b)   (c)   (d)

2.2 Euler's identity relates

(a) an exponential function to a trigonometric function

(b) an exponential function to a polar function

(c) a polar function to a trigonometric function

(d) none of these

2.3 Which of the following applies to the instantaneous power in a single-phase ac circuit?

(a) Constant component   (b) Time varying component

(c) Frequency of time varying component is twice the voltage or current wave

(d) All of these

2.4 Which of the following is true for quadrature power component when the load is capacitive?

(a) It is a reactive power component.

(b) Power factor is leading.

(c) The average quadrature power is $|V|\,|I|\sin\varphi$.

(d) All of these.

2.5 As per the sign convention of power, which of the following signs will apply for leading power entering a node?

(a) Both P and Q are positive.   (b) P is negative and Q is positive.

(c) P is positive and Q is negative. (d) None of these.

2.6 Which of the following will be a result of a low power factor?

(a) High capital cost of electrical equipment

(b) Operational difficulties

(c) Low voltage at consumer terminals

(d) All of these

2.7 What is the general nature of the power factor of loads at consumer terminals?

(a) It is lagging.   (b) It is unity.

(c) It is leading.   (d) All of these.

2.8 Which of the following does not apply to the phase shift operator '*a*'?

    (a) It has unit magnitude.

    (b) It causes a shift of 120° in the clockwise direction.

    (c) It causes a shift of 120° in the anti-clockwise direction.

    (d) None of these.

2.9 The number of times power in a single-phase system passes through zero in one cycle is

    (a) once   (b) twice

    (c) thrice   (d) quadruple

2.10 Which of the following represents the volume of conductor required in a three-phase, three-wire system compared to a single-phase, two-wire system?

    (a) 0.40%   (b) 0.50%

    (c) 0.60%   (d) 0.70%

2.11 Which of the following expressions will return a zero value?

    (a) a (b) a^2

    (c) $1 + a$ (d) $1 + a + a^2$

2.12 Which of the following does not represent the apparent power of a three-phase system?

    (a) $\sqrt{3}|V_L|\,|I_L|\cos\varphi$ (b) $\sqrt{3}|V_L|\,|I_L|$

    (c) $3\,|V_P|\,|I_P|$ (d) None of these

2.13 Which of the following holds true?

    (a) Z_{pu} is directly proportional to base power.

    (b) Z_{pu} is inversely proportional to base power.

    (c) Z_{pu} is directly proportional to the square of base power.

    (d) Z_{pu} is inversely proportional to the square of base power.

2.14 If the per unit impedance of a transformer is Z_1 on voltage and power base of kV_1 and MVA_1, respectively, which of the following will represent Z_2 on a new voltage and power base of kV_2 and MVA_2, respectively?

    (a) $Z_1 \times \dfrac{kV_1^2 \times MVA_1}{kV_2^2 \times MVA_2}$ (b) $Z_1 \times \dfrac{kV_2^2 \times MVA_1}{kV_1^2 \times MVA_2}$

    (c) $Z_1 \times \dfrac{kV_1^2 \times MVA_2}{kV_2^2 \times MVA_1}$ (d) $Z_1 \times \dfrac{kV_1^2 \times MVA_2}{kV_2^2 \times MVA_1}$

2.15 Which of the following is not an advantage of per unit values?

    (a) Per unit values of electrical equipment having widely varying ranges lie within a narrow range while their ohmic values differ widely.

    (b) Per unit values of transformers are independent of the way they are connected.

    (c) Per unit values are independent of the type of system.

    (d) None of these.

Answers

2.1 (b)	2.2 (a)	2.3 (d)	2.4 (d)	2.5 (c)	2.6 (d)
2.7 (a)	2.8 (b)	2.9 (c)	2.10 (b)	2.11 (d)	2.12 (a)
2.13 (b)	2.14 (c)	2.15 (d)			

CHAPTER 3

Transmission Line Parameters

Learning Outcomes

A focussed study of this chapter will enable the reader to:
- Understand the purpose of a transmission line and visualize its general physical structure
- Identify the four line parameters (R, L, C, and G) which characterize a transmission line
- Know the relative importance of each parameter
- Identify the various types of materials employed as conductors and calculate their resistances
- Compute the inductance and capacitance for single-phase and three-phase circuits for different types of arrangements of conductors
- Calculate the effect of earth on the capacitance of a transmission line
- Become efficient in computing parameters of transmission lines through solving problems of various difficulty levels

OVERVIEW

Transmission lines constitute the backbone of power transmission systems. The availability of a well-developed, high-capacity system of transmission lines makes it technically and economically feasible to move large blocks of electric energy over long distances. Electric transmission of bulk power can be accomplished either by alternating current (ac) or direct current (dc), using aerial lines, or underground cables, or compressed gas insulated lines. The vast majority of the world's power lines is of the ac three-phase aerial (overhead) design with bare conductors that are surrounded by air serving as the insulating medium. Thus, the primary focus in this chapter is on overhead lines.

An electric transmission line is characterized by four parameters: series resistance, series inductance, shunt capacitance, and shunt conductance. The performance characteristics of a transmission line are dependent on these parameters. Inductance is the dominant series element due to its effect on power transmission capacity and voltage drop. Capacitance is the dominating shunt element and it represents a source of reactive power. Since the megavars generated are proportional to the square of the line voltage, the importance of the shunt capacitance parameter increases with an increase in the magnitude of the operating voltage. The series resistance and the shunt conductance are the least important parameters as their effect on the transmission capacity is relatively very less. However, the series resistance completely determines the real power transmission loss and its presence must be considered. The shunt conductance accounts for the resistive leakage current. The leakage current mainly flows along the insulator strings and ionized pathways in the air, and varies appreciably with changes in the weather, atmospheric humidity, pollution, and salt content. The effect of the shunt conductance under normal operating conditions is usually neglected. The remaining three parameters, however, are needed for the development of a transmission line model for use in power system studies.

3.1 CONDUCTORS

Although copper has a much higher conductivity, aluminium conductors are normally used for overhead transmission lines. An aluminium conductor is lighter, cheaper, and has a larger diameter than a copper conductor of the same resistance. Larger diameter results in a lower voltage gradient at the surface of the conductor and reduces undesirable effects called *corona*. Moreover, the supply of aluminium is abundant while that of copper is limited.

Stranded conductors provide flexibility when suspended from tower cross-arms by insulator strings. Stranded conductors are assembled by winding wires of small diameters in layers, with alternate layers wound in the opposite direction to prevent unwinding and to coincide the radius of outer layer with the inner radius of the next layer. Electrically, each strand in the conductor provides a parallel path for the flow of current. Thus, in this manner it is possible to build flexible conductors with large cross-sectional areas.

The advantage of having conductors with large diameters is that the electrical stress is reduced, thereby minimizing the occurrence of the corona effect on the line.

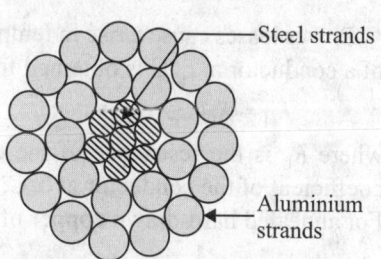

Fig. 3.1 Cross section of a 24/7 ACSR conductor

Aluminium-conductor steel-reinforced (ACSR) cables consisting of stranded steel wires for mechanical strength, surrounded by current-carrying layers of aluminium strands, are most commonly used in overhead transmission lines. Figure 3.1 shows the cross section of an ACSR cable with central core of seven steel strands and 24 aluminium strands in the two outer layers.

Apart from ACSR, other configurations of aluminium conductors that are used in transmission lines are all-aluminium conductors (AAC), all-aluminium-alloy conductors (AAAC), aluminium-conductor alloy-reinforced (ACAR), expanded ACSR, and aluminium clad steel conductor (Alumoweld). Transmission voltages above 220 kV, usually referred to as extra high voltage (EHV), and those at 765 kV and above, referred to as ultra high voltage (UHV), use bundled conductors with two, four, six, and eight conductors per bundle. In order to prevent corona in EHV lines, the diameter of the conductor is increased, thereby reducing electrical stress, by inserting a filler (paper or hessian) between the steel and aluminium layers. Such a configuration is called the expanded ACSR conductor.

Conductor manufacturers provide the characteristics of the stranded conductors such as number of strands, materials of the strands, diameter of a strand in mm, diameter of the complete conductor in mm, calculated resistance at 20°C in ohms per km, and approximate weight in kg per km. In addition, code words (bird and animal names) have been assigned to each conductor for easy reference.

3.2 RESISTANCE

Resistance causes power loss in a transmission line. Direct-current resistance R_{dc} is obtained from the formula

$$R_{dc} = \frac{\rho l}{A} \, \Omega \tag{3.1}$$

where ρ is the resistivity of the conductor material in Ω-m, l is the length of the conductor in m, and A is the cross-sectional area of the conductor in m². Table 3.1 shows the electrical properties of some common conductor materials.

Table 3.1 Properties of conductor materials

S. No.	Conductor material	Resistivity ρ at 20°C (Ω-m)	Temperature coefficient α at 0° C
1	Annealed copper	1.72×10^{-8}	234.5
2	Hard-drawn copper	1.77×10^{-8}	241.5
3	Aluminium	2.83×10^{-8}	228.1
4	Iron	10.00×10^{-8}	180.0
5	Silver	1.59×10^{-8}	243.0

Power losses cause a rise in temperature of the line conductors. Resistance R_{t1} of a conductor at t_1°C is obtained from the relation

$$R_{t1} = R_0(1 + \alpha_0 t_1) \tag{3.2}$$

where R_0 is the resistance of the conductor at 0°C and α_0 is the temperature coefficient of the conductor at 0°C.

For annealed hard-drawn copper of 100 % conductivity,

$$\alpha_0 = \frac{1}{234.5} = 0.00426$$

For hard-drawn copper of 97.3 % conductivity,

$$\alpha_0 = \frac{1}{241.5} = 0.00414$$

For hard-drawn aluminium of 61 % conductivity,

$$\alpha_0 = \frac{1}{228.1} = 0.00438$$

The resistance R_{t2} of a conductor at t_2°C can be found out from the expression

$$\frac{R_{t2}}{R_{t1}} = \frac{1/\alpha_0 + t_2}{1/\alpha_0 + t_1} \tag{3.3}$$

3.2.1 Effect of Non-uniform Distribution of Current

Distribution of direct current across the cross section of a conductor is uniform. When an alternating current flows through a conductor, distribution of current across its cross section is non-uniform. Non-uniform distribution of alternating current in a conductor is due to the following factors.

Skin effect   A detailed analytical explanation of the skin effect involves Bessel's functions which are beyond the scope of this text. However, the phenomenon of the alternating current to flow close to the surface of a conductor can be qualitatively explained by assuming a solid circular conductor to consist of elementary annular filaments of equal cross-sectional area. The current-carrying inner filaments produce flux which links the inner filaments only. On the other hand, the flux due to the current in the outer filaments enfolds both the inner and outer filaments. Thus, the flux linkages per ampere in the inner filaments are greater than in the outer filaments. Thus, the inductance and therefore the impedance of the inner filaments are higher than those of the outer filaments. Since the filaments are parallel, the alternating current has a tendency to flow close to the surface of the conductor. The factors which govern the skin effect are as follows:

- Frequency—higher the frequency, more the skin effect
- Diameter of the conductor—larger the diameter, higher the skin effect
- Shape of the conductor—more for solid conductors than stranded conductors
- Distance between conductors
- Resistivity of the material
- Permeability of the material
- Type of the material

At the power frequency, 50 Hz or 60 Hz, and for conductors with small diameters, the skin effect is negligible.

Proximity effect   Figure 3.2 shows a transmission line that has two conductors A and B of equal cross-sectional area. Line conductors A and B are divided into three sections possessing equal cross-sectional areas 1, 2, 3 and 1′, 2′, 3′, respectively.

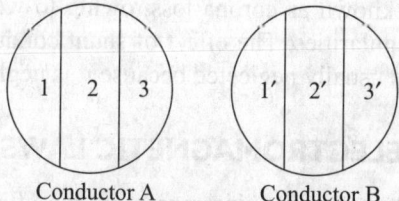

Conductor A   Conductor B

Fig. 3.2   Proximity effect

When current flows through Conductor A, the flux generated around it links with Conductor B and may be seen as forming three loops, 1 1′, 2 2′, and 3 3′, in parallel. Since the flux linking the loops progressively increases from loop 1 1′ to 3 3′, their inductance also increases accordingly. Thus, the density of the current flowing through the conductors is highest in loop 1 1′ (inner surface) and least in loop 3 3′ (outer surface), resulting in unequal distribution of the current in the conductor cross section. Proximity effect becomes more predominant as the distance between the conductors is reduced. Like the skin effect, the following factors govern the proximity effect.

- Frequency–higher the frequency, more the proximity effect
- Diameter of the conductor–larger the diameter, higher the proximity effect
- Distance between conductors
- Resistivity of the material
- Permeability of thc material

In overhead transmission lines, with normal spacing, the proximity effect is negligible. However, in underground cables, where the distance between the conductors is small, the proximity effect is of significance.

Both the skin effect and proximity effect lead to non-uniform current distribution in a conductor which results in an increase in the effective conductor resistance. As stated earlier, an analytical computation of the increase in resistance is complex. However, this computation is not required to be performed since manufacturers supply tables of electrical characteristics of their conductors. At power frequency, the ac resistance, R_{ac}, is approximately 2 per cent higher than the dc resistance, R_{dc}.

Spirality effect It depends on the type of conductor used, that is, solid, stranded, or hollow. Its effect is to increase the resistance as well as the internal reactance of stranded conductors. However, at power frequency, its effect is negligible and can be ignored in non-magnetic transmission line conductors.

3.3 LINE CONDUCTANCE

For the computation of shunt conductance G of a transmission line there is no reliable formula. The shunt conductance accounts for real power loss between conductors or between conductors and ground. For overhead lines, the real power loss is due to leakage current in the insulator and due to corona. The leakage current varies depending on the environmental conditions such as dirt, salt, and other contaminants, and the metereological conditions such as temperature and humidity around the conductor. When surface potential gradient of the conductor exceeds the dielectric strength of the surrounding air, ionization of the air space around the conductor takes place and this phenomenon is known as *corona*. Corona produces power loss, known as corona loss, owing to weather conditions and conductor surface irregularities. The effect of shunt conductance under normal operating conditions is usually neglected because it is negligible.

3.4 REVIEW OF ELECTROMAGNETIC LAWS

The inductance of a transmission line represents the effect of magnetic fields produced by the flow of current in the line conductors. Hence, a review of some basic laws of physics is presented here.

Electric current I amperes flowing through N number of turns of a conductor produces in its vicinity a magnetomotive force (mmf) $\mathcal{F} = NI$ ampere-turns (AT). By Ampere's circuital law, the magnetomotive force will equal the line integral of the magnetic field intensity H in AT/metre around a closed magnetic path. Thus,

$$NI = \mathcal{F} = \oint H dl = Hl \qquad (3.4)$$

or

$$H = \frac{NI}{l} = \frac{\mathcal{F}}{l} \text{ AT/m}$$

where l is the length of the mean flux path in metres.

The magnetic flux density B in Wb/m^2 in terms of the field intensity H is expressed as

$$B = \mu H \qquad (3.5)$$

where μ is the permeability constant of the medium. Then the total flux ϕ through a cross sectional area of A m^2 is obtained as

$$\phi = BA \text{ Wb} \tag{3.6}$$

and the flux linkages ψ in weber-turns (WbT) of N turns of the coil is given by

$$\psi = N\phi \tag{3.7}$$

Inductance L of a circuit is computed from flux linkage in weber-turns per unit current. If permeability μ is constant and the current flowing in the circuit is I amperes, then

$$L = \frac{\psi}{I} H \tag{3.8}$$

3.5 INDUCTANCE

The inductance of a transmission line depends on the size and type of conductors and the arrangement of the conductors. The arrangement of the phase conductors can be symmetrical with equal spacing between conductors of different phases, or unsymmetrical with unequal spacing between phase conductors. Each phase can have one conductor, or a number of conductors bundled together, and the transmission lines can have multicircuit arrangement on the same tower. The inductance of a transmission line would be different for different arrangements of the phase conductors.

In order to compute an accurate value for the inductance, both the flux inside the conductor as well as the flux external to the conductor must be considered.

3.5.1 Inductance of a Conductor due to Internal Flux

Cross section of a long cylindrical conductor with radius r metres and carrying a current I amperes is shown in Fig. 3.3. It is assumed that the return path for the current in the conductor is far away and does not affect the magnetic field of the conductor. Therefore, the magnetic flux lines can be assumed to be concentric with the conductor.

The field intensity H_x, at a distance x metres from the centre of the conductor, is constant over the circular path concentric with the conductor at x metres and is tangential to it. From Ampere's law, the line integration $\int H_x dl$ around the closed path is equal to the mmf $\mathcal{F}$.

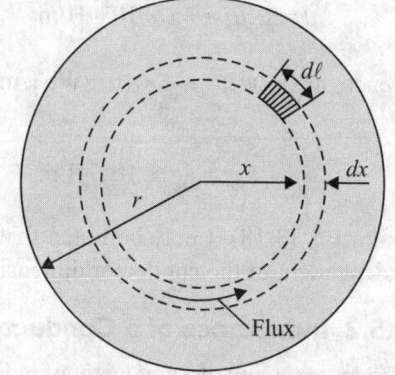

Fig. 3.3 Cross section of a current carrying cylindrical conductor

Thus,

$$\mathcal{F} = \int_0^{2\pi x} H_x dl = I_x \text{AT} \tag{3.9}$$

or $\qquad 2\pi x H_x = I_x \tag{3.10}$

where I_x is the enclosed current. If uniform current density across the conductor cross section is assumed, I_x can be written as

$$I_x = \frac{\pi x^2}{\pi r^2} I \tag{3.11}$$

Substitution of Eq. (3.11) in Eq. (3.10) leads to

$$H_x = \frac{x}{2\pi r^2} I \text{ AT/m} \tag{3.12}$$

Using Eq. (3.5), the flux density B_x, at x metres from the centre of the conductor, is given as

$$B_x = \mu H_x = \frac{\mu x I}{2\pi r^2} \text{ Wb/m}^2 \tag{3.13}$$

The flux $d\phi$, from Eq. (3.6), in a tubular element of thickness dx per metre of axial length, at x metres from the centre of the conductor is given by

$$d\phi = \frac{\mu x I}{2\pi r^2} dx \text{ Wb/m} \tag{3.14}$$

The flux linkage $d\psi$ per metre length of the conductor

$$d\psi = \frac{\pi x^2}{\pi r^2} d\phi = \frac{\mu x^3}{2\pi r^4} dx \text{ WbT/m} \tag{3.15}$$

Integrating Eq. (3.15), the total internal flux linkages ψ_{int} is obtained as

$$\psi_{int} = \int_0^r \frac{\mu I x^3}{2\pi r^4} = \frac{\mu I}{8\pi} \text{ WbT/m} \tag{3.16}$$

If relative permeability $\mu_r = 1$, then permeability μ which is equal to $\mu_0 \mu_r$ is given by

$$\mu = \mu_0 \mu_r = 4\pi \times 10^{-7} \text{ H/m}$$

Hence, $$\psi_{int} = \frac{I}{2} \times 10^{-7} \text{ WbT/m} \tag{3.17}$$

$$L_{int} = \frac{1}{2} \times 10^{-7} \text{ H/m} \tag{3.18}$$

From Eq. (3.18) it may be noted that the inductance due to the internal flux is independent of the conductor dimensions, that is, its radius.

3.5.2 Inductance of a Conductor due to External Flux

The cross section of a conductor of radius r metres and carrying a current I amperes is shown in Fig. 3.4.

Two points A_1 and A_2 external to the conductor surface and at distances D_1 and D_2, respectively, from the centre of the conductor are considered. It is assumed that the return path of the current is far removed from the conductor and does not affect the flux produced by the current flowing in the conductor. The magnetic flux paths are concentric circles around the conductor. All the flux lines between A_1 and A_2 lie within the concentric cylindrical surfaces passing through A_1 and A_2.

The magnetic field intensity at a distance x metres from the centre of the conductor is

$$H_x = \frac{I}{2\pi x} \text{ AT/m} \qquad (3.19)$$

The flux density B_x is then given by

$$B_x = \frac{\mu I}{2\pi x} \text{ Wb/m}^2 \qquad (3.20)$$

The flux $d\phi$ in the tubular element of thickness dx metres is

$$d\phi = \frac{\mu I}{2\pi x} dx \text{ Wb/m} \qquad (3.21)$$

Since the flux external to the conductor links the entire current only once, the flux linkages $d\psi$ are given by

$$d\psi = \frac{\mu I}{2\pi x} dx \text{ WbT/m} \qquad (3.22)$$

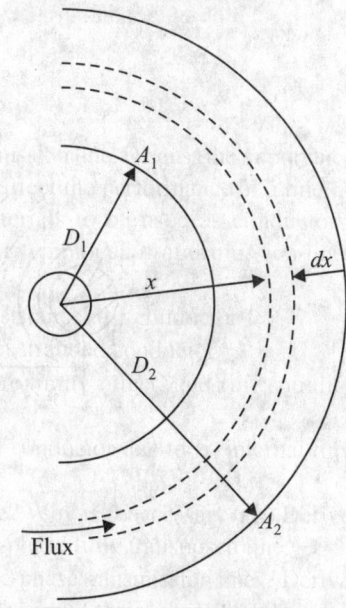

Fig. 3.4 External flux linkage due to a current-carrying conductor

The total flux ψ_{12} between A_1 and A_2 is

$$\psi_{12} = \int_{D_1}^{D_2} \frac{\mu I}{2\pi x} dx = \frac{\mu I}{2\pi} \ln \frac{D_2}{D_1} \text{ WbT/m} \qquad (3.23)$$

Since $\mu_r = 1$ and $\mu_0 = 4\pi \times 10^{-7}$, the total flux ψ_{12} between A_1 and A_2 is

$$\psi_{12} = 2 \times 10^{-7} \times I \ln \frac{D_2}{D_1} \text{ WbT/m} \qquad (3.24)$$

Therefore, the inductance due to the external flux linkages included between A_1 and A_2 is

$$L_{12} = 2 \times 10^{-7} \ln \frac{D_2}{D_1} \text{ H/m} \qquad (3.25)$$

3.5.3 Inductance of a Single-phase, Two-wire Line

A single-phase circuit of two parallel conductors of radii r_1 and r_2 metres, separated by a distance D metres, is shown in Fig. 3.5. One conductor is the return circuit for the other.

From Fig. 3.5, it is observed that the flux set up by the current in conductor 1 links it in the following ways:

(a) The external flux set up by current I in conductor 1 at a distance equal to or greater than $(D + r_2)$, that is, beyond conductor 2 from the centre of the conductor 1 does not link the circuit.

(b) The external flux between r_1 and $(D - r_2)$, that is, between conductors 1 and 2, links all the current I in conductor 1.

(c) The external flux over the surface of conductor 2, that is, between $(D - r_2)$ and $(D + r_2)$ links a fraction of the current varying from I to zero.

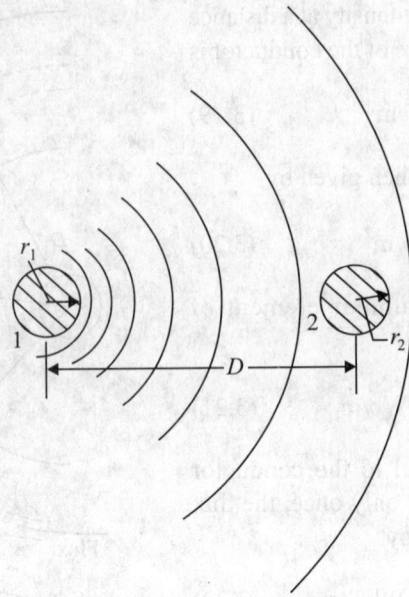

Fig. 3.5 Single-phase circuit

In transmission lines D is much greater than r_1 and r_2; hence, it can be assumed that D can be used instead of $(D - r_2)$ or $(D + r_2)$.

The inductance L_1 of the circuit due to current I in conductor 1 can be obtained by addition of inductance due to internal flux linkages determined by Eq. (3.18) and the inductance due to external flux linkages determined by Eq. (3.25) with r_1 replacing D_1 and D replacing D_2 to obtain

$$L_1 = 2 \times 10^{-7} \left(\frac{1}{4} + 2 \ln \frac{D}{r_1} \right) \text{ H/m} \tag{3.26}$$

Now, in Eq. (3.26) the term 1/4 may be written as $\ln e^{1/4}$.
Thus,

$$L_1 = 2 \times 10^{-7} \ln \frac{D}{r_1 e^{-1/4}} = 2 \times 10^{-7} \ln \frac{D}{r_1'} \text{ H/m} \tag{3.27}$$

where $r_1' = 0.7788 r_1$ is the geometric mean radius (GMR) or the self-geometric mean distance (self-GMD) of a solid round conductor. GMR represents an infinitesimally thin-walled conductor of radius r_1'. This mathematical convenience of replacing a conductor of radius r_1 by an equivalent thin-walled conductor of radius r_1' enables the designers to develop equations for line inductances without taking into consideration the field produced due to the internal flux.

The current in conductor 2 flows in a direction opposite to that of conductor 1, and the flux produced due to current in conductor 2, links the single-phase circuit in the same direction as that produced by conductor 1. Thus, the inductance due to current in conductor 2 is

$$L_2 = 2 \times 10^{-7} \ln \frac{D}{r_1'} \text{ H/m} \tag{3.28}$$

Hence, the total inductance of the circuit taking $r_1' = r_2' = r'$ can be written as,

$$L = L_1 + L_2 = 4 \times 10^{-7} \ln \frac{D}{r'} \text{ H/m} \tag{3.29}$$

$$L = 0.921 \log \frac{D}{r'} \text{ mH/km} \tag{3.29a}$$

3.5.4 Flux Linkages of One Conductor in a Group of Conductors

The transmission line conductors invariably employ stranded configuration, except when the size of the conductor is small. Hence, it would be desirable to develop more general expressions for flux linkages and inductance for transmission lines using stranded conductor configurations.

A group of n conductors, carrying phasor currents $I_1, I_2, \ldots, I_n$, whose sum equals zero is shown in Fig. 3.6. The distances of the conductors from a remote point P are shown in the figure as $D_{1P}, D_{2P}, \ldots, D_{iP}, D_{jP}, \ldots, D_{nP}$. The total flux linkages ψ_{iiP} of conductor i due to its current I_i including internal flux linkages but excluding all the flux linkages beyond point P is given by Eqs (3.17) and (3.24) as

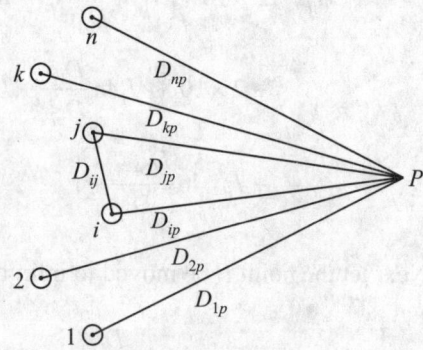

Fig. 3.6 Flux linkages of a conductor in a group of n conductors

$$\psi_{iiP} = 10^{-7} \left(\frac{I_i}{2} + 2I_i \ln \frac{D_{iP}}{r_i} \right) \text{ WbT/m} \tag{3.30}$$

or

$$\psi_{iiP} = 2 \times 10^{-7} \times I_i \ln \frac{D_{iP}}{r_i} \text{ WbT/m} \tag{3.31}$$

The flux linkages ψ_{ij} of conductor i due to current I_j in conductor j but excluding all the flux beyond point P is equal to the flux produced by I_j between point P and the conductor i, that is, within the limiting distances D_{jP} and D_{ij} from conductor j, is obtained by using Eq. (3.24) as

$$\psi_{ij} = 2 \times 10^{-7} I_j \ln \frac{D_{jP}}{D_{ij}} \text{ WbT/m} \tag{3.32}$$

where D_{ij} is the distance between the ith and jth conductor.

The total flux linkages ψ_{iP} of ith conductor due to all the conductors in the group but excluding flux beyond point P is

$$\psi_{iP} = 2 \times 10^{-7} \left(I_1 \ln \frac{D_{1P}}{D_{i1}} + I_2 \ln \frac{D_{2P}}{D_{i2}} + \cdots + I_i \ln \frac{D_{iP}}{r_i'} + \cdots + I_n \ln \frac{D_{nP}}{D_{in}} \right)$$

$$\tag{3.33}$$

Equation (3.33) may be written, by expanding the logarithmic terms, as

$$\psi_{iP} = 2 \times 10^{-7} \left(I_1 \ln \frac{1}{D_{i1}} + I_2 \ln \frac{1}{D_{i1}} + \cdots + I_i \ln \frac{1}{r_i'} + \cdots + I_1 \ln \frac{1}{D_{in}} \right)$$

$$+ 2 \times 10^{-7} \left(I_1 \ln D_{1P} + I_2 \ln D_{2P} + \cdots + I_i \ln D_{iP} \right.$$

$$\left. + \cdots + I_n \ln D_{nP} \right) \tag{3.34}$$

Now, $\quad I_1 + I_2 + \cdots + I_i + \cdots + I_{n-1} + I_n = 0$

Hence, $\quad I_n = - (I_1 + I_2 + \cdots + I_i + \cdots + I_{n-1})$

On substituting the value of I_n in Eq. (3.34), the following relation is obtained after simplification

$$\psi_{iP} = 2 \times 10^{-7} \left(I_1 \ln \frac{1}{D_{i1}} + I_2 \ln \frac{1}{D_{i1}} + \cdots + I_i \ln \frac{1}{r_i'} + \cdots + I_1 \ln \frac{1}{D_{in}} \right)$$

$$+ 2 \times 10^{-7} \left(I_1 \ln \frac{D_{1P}}{D_{nP}} + I_2 \ln \frac{D_{2P}}{D_{nP}} + \cdots + I_i \ln \frac{D_{iP}}{D_{nP}} + \cdots \right.$$

$$\left. + I_{n-1} \ln \frac{D_{(n-1)P}}{D_{nP}} \right) \tag{3.35}$$

Next let the point P be moved to infinity. The ratio of the distances $\dfrac{D_{1P}}{D_{nP}}$, $\dfrac{D_{2P}}{D_{nP}}$,

etc. approach unity and $\ln \dfrac{D_{1P}}{D_{nP}} \to 0$, $\ln \dfrac{D_{2P}}{D_{nP}} \to 0$, etc. Therefore, Eq. (3.35) reduces to

$$\psi_i = 2 \times 10^{-7} \left(I_1 \ln \frac{1}{D_{i1}} + I_2 \ln \frac{1}{D_{i2}} + \cdots + I_i \ln \frac{1}{r_i'} + \cdots + I_n \ln \frac{1}{D_{in}} \right) \tag{3.36}$$

3.5.5 Inductance of Composite Conductor Lines

A single-phase line composed of conductor X, having n identical parallel filaments, and conductor Y, which is the return circuit for the current in conductor X, having m identical parallel filaments, is shown in Fig. 3.7. Each filament of conductor X carries a current I/n, and each filament of conductor Y carries the return current $-I/m$ (negative sign indicates that the direction of I/m is opposite to that of I/n).

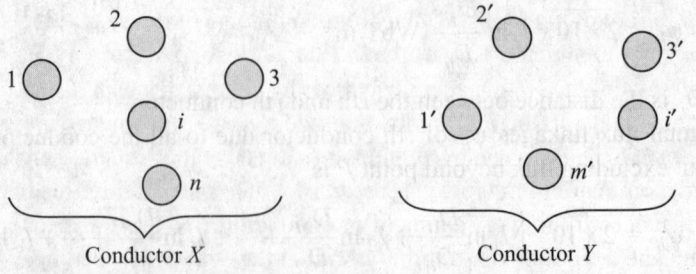

Fig. 3.7 Two composite conductors in a single-phase line

The flux linkages of the filament i of conductor X are obtained by using Eq. (3.36) as follows:

$$\psi_i = 2 \times 10^{-7} \frac{I}{n} \left(\ln \frac{1}{D_{i1}} + \ln \frac{1}{D_{i2}} + \cdots + \ln \frac{1}{r_i'} + \cdots + \ln \frac{1}{D_{in}} \right)$$

$$- 2 \times 10^{-7} \frac{I}{m} \left(\ln \frac{1}{D_{i1'}} + \ln \frac{1}{D_{i2'}} + \cdots + \ln \frac{1}{D_{im'}} \right) \tag{3.37}$$

$$\psi_i = 2 \times 10^{-7} I \ln \frac{(D_{i1'} \, D_{i2'} \, D_{i3'} \cdots D_{im'})^{1/m}}{(D_{i1} \, D_{i2} \, D_{i3} \cdots r_i' \cdots D_{in})^{1/n}} \text{ WbT/m} \tag{3.38}$$

where distances between the elements are designated by the letter D with appropriate subscripts.

If Eq. (3.38) is divided by the current (I/n), the inductance of filament i is obtained as

$$L_i = \frac{\psi_i}{(I/n)} = 2n \times 10^{-7} I \ln \frac{(D_{i1'} \, D_{i2'} \, D_{i3'} \cdots D_{im'})^{1/m}}{(D_{i1} \, D_{i2} \, D_{i3} \cdots r_i' \cdots D_{in})^{1/n}} \text{ H/m} \tag{3.39}$$

Hence, the average inductance of conductor X is obtained as

$$L_{av} = \frac{L_1 + L_2 + L_3 + \cdots + L_i + \cdots + L_n}{n} \tag{3.40}$$

The conductor X is composed of n identical electrically parallel elements. The inductance of conductor X, therefore, can be written as

$$L_X = \frac{L_{av}}{n} = \frac{L_1 + L_2 + \cdots + L_i + \cdots + L_n}{n^2} \tag{3.41}$$

Substituting the expression of each filament from Eq. (3.39) in Eq. (3.41), yields

$$L_X = 2 \times 10^{-7} \ln \frac{\left[\begin{matrix} (D_{11'} \, D_{12'} \, D_{13'} \cdots D_{1m'})(D_{21'} \, D_{22'} \, D_{23'} \cdots D_{2m'}) \cdots \\ (D_{i1'} \, D_{i2'} \, D_{i3'} \cdots D_{im'}) \cdots (D_{n1'} \, D_{n2'} \, D_{n3'} \cdots D_{nm'}) \end{matrix} \right]^{1/mn}}{\left[\begin{matrix} (D_{11} \, D_{12} \, D_{13} \cdots D_{1n})(D_{21} \, D_{22} \, D_{23} \cdots D_{2n}) \cdots \\ (D_{i1} \, D_{i2} \, D_{i3} \cdots D_{in}) \cdots (D_{n1} \, D_{n2} \, D_{n3} \cdots D_{nn}) \end{matrix} \right]^{1/n^2}} \text{ H/m}$$

$$\tag{3.42}$$

where r_1, r_2, and r_n have been replaced by D_{11}, D_{22}, and D_{nn} respectively to make the expression more symmetrical.

It may be noted that the numerator of the argument of the logarithm in Eq. (3.42) is the mnth root of mn terms, which are the products of the distances from all the n filaments of conductor X to all the m filaments of conductor Y. The mnth root of the product of the mn distances is called the geometric mean distance between conductor X and conductor Y and is denoted by D_m or geometric mean distance denoted by GMD. Between two conductors, this distance is called mutual GMD.

The denominator in Eq. (3.42) is the n^2 root of n^2 terms, which are the products of the distances from every filament in conductor X to itself and to every other filament. The n^2 root of these terms is called the self-GMD, denoted by D_{sX}, or geometric mean radius denoted by GMR.

In terms of D_m and D_{sX}, Eq. (3.42) becomes

$$L_X = 2 \times 10^{-7} \ln \frac{D_m}{D_{sX}} \text{ H/m} \tag{3.43}$$

$$= 0.4605 \log \frac{D_m}{D_{sX}} \text{ H/m} \tag{3.43a}$$

Similarly, the inductance of conductor Y can be determined as

$$L_Y = 2 \times 10^{-7} \ln \frac{D_m}{D_{sY}} \text{ H/m} \tag{3.44}$$

Thus, the inductance of the line is obtained as

$$L = L_X + L_Y$$

3.5.6 Inductance of Stranded Conductors

Stranded conductors are normally used for overhead transmission lines as discussed in Section 3.1. Inductance of such conductors can be found by the use of a procedure similar to that discussed in Section 3.5.5. The cross section of a seven-strand cable is shown in Fig. 3.8.

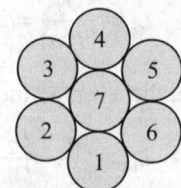

Fig. 3.8 Cross section of a seven-strand cable

The self-GMD of a seven-strand cable is the 49th root of 49 distances. If r is the radius of each strand then

$$D_s = \left[(r')^7 \left(D_{12}^2 \, D_{26}^2 \, D_{14} \, D_{17} \right)^6 (2r)^6 \right]^{1/49}$$

Substituting values of various distances,

$$D_s = \left[(0.7788r)^7 (2^2 r^2 \times 3 \times 2^2 r^2 \times 2^2 r \times 2r \times 2r)^6 \right]^{1/49} = 2.177r$$

For stranded conductors the distance between the sides of a line composed of one conductor per side is usually so great that the mutual GMD can be taken as equal to the centre-to-centre distance with negligible error.

3.6 INDUCTANCE OF THREE-PHASE LINES

The circuits employed for transmission of bulk power are invariably three-phase circuits. However, the methodology described in the previous sections to develop basic equations for the inductance of single-phase circuits can be conveniently applied to compute the inductance of three-phase transmission circuits.

3.6.1 Symmetrical Spacing

Arrangement of equilaterally spaced conductors in a three-phase circuit is shown in Fig. 3.9. Each conductor has a radius r metres and the spacing between the conductors is D metres. It is assumed that the neutral wire is not present, hence, $I_R + I_Y + I_B = 0$. As per Eq. (3.36), the flux linkages of the conductor 'a' can be written as

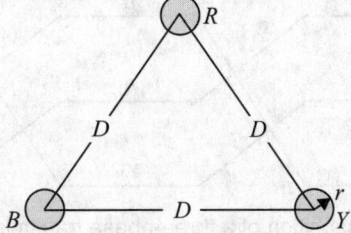

Fig. 3.9 Arrangement of conductors in a three-phase circuit (symmetrical spacing)

$$\psi_R = 2 \times 10^{-7} \left(I_R \ln \frac{1}{r'} + I_Y \frac{1}{D} + I_B \frac{1}{D} \right) \text{ WbT/m} \tag{3.45}$$

Since $I_R = -(I_Y + I_B)$, Eq. (3.45) becomes

$$\psi_R = 2 \times 10^{-7} \left(I_R \ln \frac{1}{r'} - I_R \frac{1}{D} \right) = 2 \times 10^{-7} I_R \ln \frac{D}{r'} \text{ WbT/m} \tag{3.46}$$

Hence,

$$L_R = \frac{\psi_R}{I_R} = 2 \times 10^{-7} \ln \frac{D}{r'} \text{ H/m} \tag{3.47}$$

$$L_R = 0.4605 \log \frac{D}{r'} \text{ mH/km} \tag{3.47a}$$

The inductances of conductors Y and B are the same as the inductance of the conductor R because of symmetry.

3.6.2 Unsymmetrical Spacing

A cross section of conductors of a three-phase line with unsymmetrical spacing is shown in Fig. 3.10.

Unsymmetrical spacing of the phase conductors of a three-phase line causes the flux linkages and inductance of each phase to be different and results in an unbalanced circuit. The three phases can be balanced by exchanging the positions of the conductors at regular intervals along the line so that each conductor occupies the original position of every other conductor over an equal distance. Such an exchange of conductor positions is called transposition. A complete transposition cycle is shown in Fig. 3.11, where conductors are designated as R, Y, and B, and the positions occupied are numbered 1, 2, and 3, respectively.

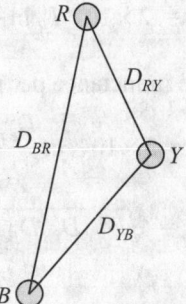

Fig.3.10 Arrangement of conductors in a three-phase circuit (unsymmetrical spacing)

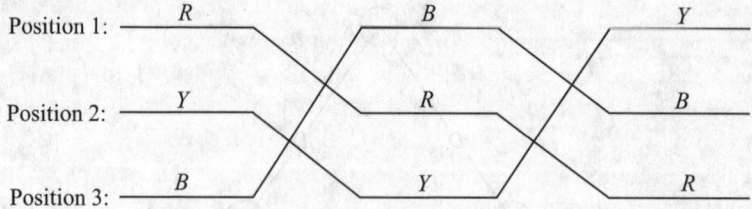

Fig. 3.11   Transposition of a three-phase transmission line circuit

The flux linkages of conductor R in position 1, when conductor Y is in position 2 and conductor B is in position 3 (see Fig. 3.10), may be obtained using Eq. (3.36) as follows:

$$\psi_{R1} = 2 \times 10^{-7} \left(I_R \ln \frac{1}{r_R'} + I_Y \ln \frac{1}{D_{RY}} + I_B \ln \frac{1}{D_{BR}} \right) \text{ WbT/m} \qquad (3.48)$$

With conductor R in position 2, when conductor Y is in position 3 and conductor B is in position 1, the flux linkages are

$$\psi_{R2} = 2 \times 10^{-7} \left(I_R \ln \frac{1}{r_R'} + I_Y \ln \frac{1}{D_{YB}} + I_B \ln \frac{1}{D_{RY}} \right) \text{ WbT/m} \qquad (3.49)$$

Finally, with conductor R in position 3, when conductor Y is in position 1 and conductor B is in position 2, the flux linkages are

$$\psi_{R3} = 2 \times 10^{-7} \left(I_R \ln \frac{1}{r_R'} + I_Y \ln \frac{1}{D_{BR}} + I_B \ln \frac{1}{D_{YB}} \right) \text{ WbT/m} \qquad (3.50)$$

The average value of the flux linkages of conductor R is

$$\psi_R = \frac{\psi_{R1} + \psi_{R2} + \psi_{R3}}{3}$$

$$= \frac{2 \times 10^{-7}}{3} \left(3 I_R \ln \frac{1}{r_R'} + 3 I_Y \ln \frac{1}{D_{RY} D_{YB} D_{BR'}} + 3 I_B \ln \frac{1}{D_{RY} D_{YB} D_{BR}} \right)$$

$$\qquad (3.51)$$

Since $\quad I_R = -(I_Y + I_B),$

$$\psi_R = 2 \times 10^{-7} I_R \ln \frac{D_{RY} D_{YB} D_{BR}}{r_R'} \text{ WbT/m} \qquad (3.52)$$

The average inductance per phase, therefore, is given by

$$L_R = 2 \times 10^{-7} \ln \frac{D_{eq}}{D_s} \text{ H/m} \qquad (3.53)$$

where $\quad D_{eq} = \sqrt[3]{D_{RY} D_{YB} D_{BR}}$ $\qquad\qquad\qquad\qquad (3.54)$

and $\quad D_s = r_R'$ $\qquad\qquad\qquad\qquad\qquad\qquad\qquad (3.54a)$

3.7 INDUCTANCE OF DOUBLE-CIRCUIT, THREE-PHASE LINES

It is a common practice to run two circuits of three-phase transmission lines on the same transmission tower so as to increase the reliability of power transmission.

It is also desirable to have low value of mutual GMD, D_m, and high value of self-GMD, D_s, so that the inductance of the parallel lines has a low value and power transfer over the lines increases. Thus, the individual conductors of a phase should be spaced as widely as possible while the distances between phases are kept as low as permissible.

The three sections of a commonly used transposition cycle of a double-circuit, three-phase transmission line, with vertical spacing, are shown in Fig. 3.12.

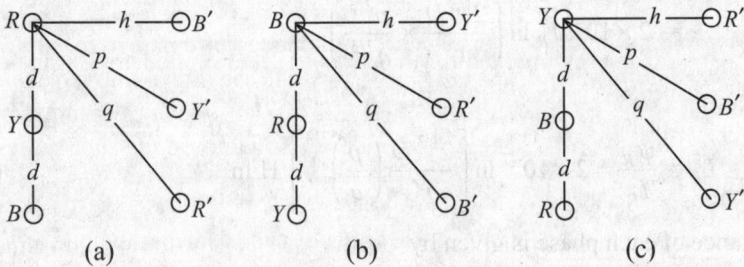

Fig. 3.12 Arrangement of conductors in a double-circuit three-phase transposed line (a) Phase R in position 1 (b) Phase R in position 2 (c) Phase R in position 3

The conductors R, Y, B belong to one circuit while conductors R', Y', B' belong to the other circuit. The various distances between conductors are shown in the figure. The flux linkages ψ_{R1}, ψ_{R2}, ψ_{R3} of phase R in position 1, position 2, and position 3 respectively are

$$\psi_{R1} = 2 \times 10^{-7} \left[I_R \left(\ln \frac{1}{r'} + \ln \frac{1}{q} \right) + I_Y \left(\ln \frac{1}{D} + \ln \frac{1}{p} \right) \right.$$

$$\left. + I_B \left(\ln \frac{1}{2D} + \ln \frac{1}{h} \right) \right] \tag{3.55}$$

$$\psi_{R2} = 2 \times 10^{-7} \left[I_R \left(\ln \frac{1}{r'} + \ln \frac{1}{h} \right) + I_Y \left(\ln \frac{1}{D} + \ln \frac{1}{p} \right) \right.$$

$$\left. + I_B \left(\ln \frac{1}{D} + \ln \frac{1}{p} \right) \right] \tag{3.56}$$

$$\psi_{R3} = 2 \times 10^{-7} \left[I_R \left(\ln \frac{1}{r'} + \ln \frac{1}{q} \right) + I_Y \left(\ln \frac{1}{2D} + \ln \frac{1}{h} \right) \right.$$

$$\left. + I_B \left(\ln \frac{1}{D} + \ln \frac{1}{p} \right) \right] \tag{3.57}$$

Hence, $\psi_R = \dfrac{\psi_{R1} + \psi_{R2} + \psi_{R3}}{3}$

$$= \frac{2 \times 10^{-7}}{3} \left[3I_R \ln \frac{1}{r'} + I_R \ln \frac{1}{q^2 h} + (I_Y + I_B) \ln \frac{1}{2D^3} \right.$$

$$\left. + (I_Y + I_B) \ln \frac{1}{p^2 h} \right] \tag{3.58}$$

Since $I_R + I_Y + I_B = 0$, therefore

$$\psi_R = \frac{2 \times 10^{-7}}{3} \left[3I_R \ln \frac{1}{r'} + I_R \ln \frac{1}{q^2 h} - I_R \ln \frac{1}{2D^3} - I_R \ln \frac{1}{p^2 h} \right]$$

$$= \frac{2 \times 10^{-7}}{3} \times 3I_R \left[\ln \left(\frac{2^{1/3} Dp^{2/3} h^{1/3}}{r' q^{2/3} h^{1/3}} \right) \right]$$

$$= 2 \times 10^{-7} I_R \ln \left(\frac{2^{1/3} D}{r'} \times \frac{p^{2/3}}{q^{2/3}} \right)$$

Thus,

$$L_R = \frac{\psi_R}{I_R} = 2 \times 10^{-7} \ln \left[\frac{2^{1/3} D}{r'} \left(\frac{p}{q} \right)^{2/3} \right] \text{ H/m} \tag{3.59}$$

Inductance of each phase is given by

$$L = \frac{L_R}{2} = 2 \times 10^{-7} \ln \left[2^{1/6} \left(\frac{D}{r'} \right)^{1/2} \left(\frac{p}{q} \right)^{2/3} \right] \text{ H/m} \tag{3.60}$$

3.8 COMPUTATION OF INDUCTANCE FOR BUNDLED CONDUCTORS

For EHV and UHV lines, bundled conductors having two or more conductors per phase in close proximity are used. The use of bundled conductors reduces corona loss and decreases the reactance of the transmission line. The reduction of reactance results from increased GMR of the bundle. The arrangements of bundles with two, three, and four conductors are shown in Fig. 3.13.

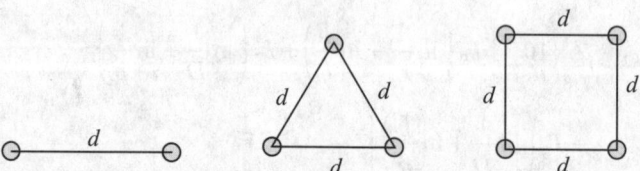

Fig. 3.13 Arrangement of bundled conductors

Let D_s^b indicate the GMR of a bundled conductor, D_s the GMR of the individual conductors, and d the distance between adjacent conductors of a bundled conductor (Fig. 3.13).

For a two-conductor bundle,

$$D_s^b = \left[(D_s \times d)^2 \right]^{1/4} = (D_s \times d)^{1/2} \tag{3.61}$$

For a three-conductor bundle,

$$D_s^b = \left[(D_s \times d \times d)^3 \right]^{1/9} = (D_s \times d \times d)^{1/3} \tag{3.62}$$

For a four-conductor bundle,

$$D_s^b = \left[(D_s \times d \times d \times \sqrt{2}d)^4 \right]^{1/16} = 1.09 (D_s \times d^3)^{1/4} \tag{3.63}$$

For computing inductance using Eq. (3.53), D_s^b replaces D_s of a single conductor. To compute D_{eq}, the distance from the centre of one bundle to the centre of another bundle is sufficiently accurate for D_{RY}, D_{YB}, and D_{BR}.

Example 3.1 Determine the self-GMD of the stranded conductor shown in Fig. 3.14. Assume that all the strands have the same radius r.

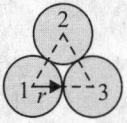

Solution There are nine distances and the self-GMD of the conductor will be the ninth root of the product of the nine distances. Thus,

$$\text{Self-GMD} = \left\{(r')^3 (D_{12}^2 D_{13}^2 D_{23}^2)\right\}^{1/9}$$

Fig. 3.14

$$D_{12} = D_{23} = D_{13} = 2r$$

Hence,

$$\text{Self-GMD} = \left\{(0.7788r)^3 (2r)^6\right\}^{1/9} = 1.4605r$$

Example 3.2 A transmission line is composed of ACSR conductors consisting of six aluminium strands and one steel strand. The diameter of the ACSR conductor is 6 cm, and each aluminium strand has a diameter of 2 cm. If the conductors are spaced at 120 cm in a horizontal plane, determine (i) the inductance per conductor, (ii) the loop inductance, and (iii) the loop reactance at 50 Hz. Neglect the effect of the steel strand.

Solution The cross section of the ACSR conductor is shown in Fig. 3.15.

From Fig. 3.15, the diameter of the steel strand $= (6 - 2 \times 2) = 2$ cm.

Since the effect of the steel strand is to be neglected, there are 36 distances and the self-GMD will be 1/36th root of the product of the distances. By referring to Fig. 3.15, the following distances are obtained:

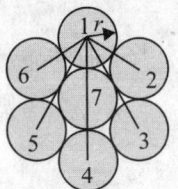

$$D_{12} = D_{16} = d = 2$$

Fig. 3.15

$$D_{13} = D_{15} = \sqrt{3}d = 2\sqrt{3}$$

$$D_{14} = 2d = 4$$

$$D_s = \left\{\left[(0.7788d)d^2 \left(\sqrt{3}d\right)^2 (2d)\right]^6\right\}^{\frac{1}{36}}$$

$$= \left\{\left[(0.7788 \times 2)(2)^2 \left(\sqrt{3} \times 2\right)^2 (2 \times 2)\right]^6\right\}^{\frac{1}{36}}$$

$$= 2.31 \text{ cm}$$

(i) Using Eq. (3.43) yields the value of inductance per conductor

$$L = 2 \times 10^{-7} \times \ln \frac{120}{2.31} = 0.79005 \text{ mH/km}$$

(ii) Loop inductance $= 2 \times L = 1.5801$ mH/km

(iii) Inductive reactance of the loop $= 2 \times \pi \times 50 \times 1.5801 \times 10^{-3} = 0.4964$ Ω/km

Example 3.3 A three-phase transmission system has the configuration shown in Fig. 3.16. If the neutral current is zero and the lines are not transposed, derive an expression for (i) flux linkages of the neutral conductor and (ii) voltage drop per unit length in each of the phases. Assume that all the four conductors have a radius of r.

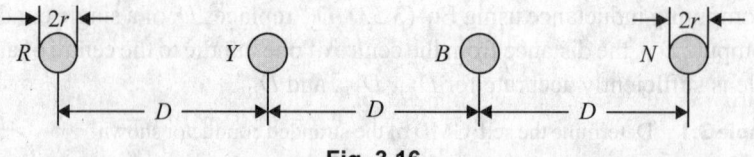

Fig. 3.16

If $D = 1.2$ m and $r = 0.75$ cm, determine (i) the voltage drop in the neutral and (ii) the voltage drop in each phase when the phase currents have the following values:

$$I_R = (25 - j30)A; \quad I_Y = (35 - j50)A; \quad I_B = (-60 + j80)A$$

Assume the system frequency is equal to 50 Hz. Take the length of the line = 10 km.

Solution From Fig. 3.16, the following distances can be written

$$D_{RN} = 3D, D_{YN} = 2D, \text{ and } D_{BN} = D.$$

(i) Since the neutral current is zero, the internal flux linkages are also zero. Hence Eq. (3.36) is modified and the expression for the flux linkages for the neutral conductor is written as

$$\psi_N = 2 \times 10^{-7} \left(I_R \ln \frac{1}{D_{RN}} + I_Y \ln \frac{1}{D_{YN}} + I_B \ln \frac{1}{D_{BN}} \right) \text{ WbT/m}$$

$$= 2 \times 10^{-7} \left(I_R \ln \frac{1}{3D} + I_Y \ln \frac{1}{2D} + I_B \ln \frac{1}{D} \right) \text{ WbT/m} \tag{3.3.1}$$

Voltage drop in the neutral is given by

$$\Delta V_N = j \times 2 \times \pi \times 50 \times 2 \times 10^{-7} \left(I_R \ln \frac{1}{3D} + I_Y \ln \frac{1}{2D} + I_B \ln \frac{1}{D} \right)$$

$$= j200\pi \times 10^{-7} \left(I_R \ln \frac{1}{3D} + I_Y \ln \frac{1}{2D} + I_B \ln \frac{1}{D} \right) \text{ V/m} \tag{3.3.2}$$

(ii) Again using Eq. (3.36) the expression for flux linkages for the R phase is written as

$$\psi_R = 2 \times 10^{-7} \left(I_R \ln \frac{1}{r'} + I_Y \ln \frac{1}{D} + I_B \ln \frac{1}{2D} \right) \text{ WbT/m}$$

Since $I_R + I_Y + I_B = 0$,

$$\psi_R = 2 \times 10^{-7} \left(I_R \ln \frac{2D}{r'} + I_Y \ln 2 \right) \text{ WbT/m}$$

Inductance of R phase can be written as

$$L_R = \frac{\psi_R}{I_R} = 2 \times 10^{-7} \left(\ln \frac{2D}{r'} + \frac{I_Y}{I_R} \ln 2 \right) \text{ H/m}$$

Hence, the voltage drop in phase R is given by

$$\Delta V_R = j\omega L_R I_R = j \times 2 \times \pi \times 50 \times 2 \times 10^{-7} I_R \left(\ln \frac{2D}{r'} + \frac{I_Y}{I_R} \ln 2 \right) \text{ V/m}$$

$$= j \times 200 \times \pi \times 10^{-7} \left(I_R \ln \frac{2D}{r'} + I_Y \ln 2 \right) \text{ V/m}$$

By following a similar process, expressions for voltage drops in phases Y and B are obtained as under

$$\Delta V_Y = j \times 200 \times \pi \times 10^{-7} \ln \frac{D}{r'} \text{ V/m}$$

$$\Delta V_B = j \times 200 \times \pi \times 10^{-7} \left(I_Y \ln 2 + I_B \ln \frac{2D}{r'} \right) \text{ V/m}$$

In matrix form the voltage drops per metre, in the three phases, can be expressed as

$$
\begin{bmatrix} \Delta V_R \\ \Delta V_Y \\ \Delta V_B \end{bmatrix} = j \times 200\pi \times 10^{-7}
\begin{bmatrix} \ln \dfrac{2D}{r'} & \ln 2 & 0 \\ 0 & \ln \dfrac{D}{r'} & 0 \\ 0 & \ln 2 & \ln \dfrac{2D}{r'} \end{bmatrix}
\begin{bmatrix} I_R \\ I_Y \\ I_B \end{bmatrix} \text{ V/m} \qquad (3.3.3)
$$

Utilizing the MATLAB facility, Eqs (3.3.1), (3.3.2), and (3.3.3) are solved by inputting the given data.

```
>> D=1.200;r=0.75*10^(-2);rd=0.7788*r;IR=25-i*30;IY=35-i*50;IB=-60+i*80;
                                                             %data input
>> fln=2*10^(-7)*(IR*log(1/(3*D))+IY*log(1/(2*D))+IB*log(1/(D)))
                                                             %data input
  fln = (-1.0345e-005 +1.3523e-005i) Wb-T/m          %flux linkages Ψ_N
>> DVN=i*2*pi*50*fln
DVN =DV_N = (-0.0042 - 0.0033i) V/m             %voltage induced in neutral
                                                              conductor

>> TDVN=DVN*10*1000
TDVN = (-42.4842 -32.5001i) V                %voltage induced in the neutral
                                                              conductor

>> abs(TDVN) =53.4898 V          %magnitude of the induced voltage in the
                                                %line in the neutral
>> LMAT=zeros(3);                       %initialises a 3 x 3 null matrix
>> LMAT(1,1)=log(2*D/rd);LMAT(1,2)=log(2);LMAT(2,2)=log(D/rd);
>> LMAT(3,2)=LMAT(1,2); LMAT(3,3)=LMAT(1,1);      %computes the non-zero
                                        %elements of the inductance matrix
>> LMAT
LMAT =                         %inductance matrix of the transmission system
    6.0183    0.6931    0
    0         5.3252    0
    0         0.6931    6.0183
>> I=[IR;IY;IB];                   %creates a column vector of currents
>> DVRYB=i*200*pi*10^(-7)*LMAT*I        %computes voltage drops per metre
DVRYB =
    (0.0135 + 0.0110) V/m                        %voltage drop in R phase
    (0.0167 + 0.0112) V/m                        %voltage drop in Y phase
    (-0.0281 - 0.0211) V/m                       %voltage drop in B phase
>> TDVRYB=DVRYB*10*1000
  TDVRYB =
  1.0eq + 002 *
  (1.3527 + 1.0982) V                   %voltage drop in R phase of the line.
  (1.6736 + 1.1715) V                   %voltage drop in Y phase of the line.
  (-2.8085 - 2.1173) V                  %voltage drop in B phase of the line.
  >> abs(TDVRYB)
  174.2408 V              %magnitude of voltage drop in R phase of the line
  204.2921 V              %magnitude of voltage drop in Y phase of the line
  351.7184 V              %magnitude of voltage drop in B phase of the line
```

Example 3.4 A three-phase, double-circuit, fully transposed line is shown in Fig. 3.17. The conductor diameter is 2 cm. Determine the inductance per phase of the line. What is the reactance per phase of the line if the power system frequency is equal to 50 Hz? Assume that the line is 30 km long.

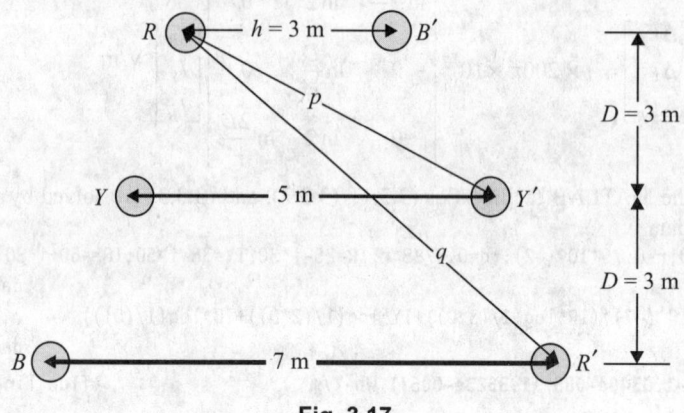

Fig. 3.17

Solution Self-GMR of the conductor = $0.7788 \times 1 = 0.7788$ cm
Referring to Fig. 3.17,

$$h = 300 \text{ cm}, p = 100\sqrt{3^2 + 4^2} = 100\sqrt{25} = 500 \text{ cm},$$

$$q = 100\sqrt{6^2 + 5^2} = 100\sqrt{61} = 781.02 \text{ cm}$$

Equation (3.60) for the inductance per phase of a double-circuit three-phase line is reproduced as follows

$$L = \frac{L_R}{2} = 2 \times 10^{-7} \ln \left[2^{1/6} \left(\frac{D}{r'}\right)^{1/2} \left(\frac{p}{q}\right)^{2/3} \right]$$

Substituting the various values yields

$$L = 2 \times 10^{-7} \ln \left[2^{1/6} \left(\frac{3}{0.7788}\right)^{1/2} \left(\frac{500}{781.02}\right)^{2/3} \right] \text{ H/m}$$

$$= 9.8501 \times 10^{-8} \text{ H/m}$$

Inductance of the line = $9.8501 \times 10^{-8} \times 300 \times 1000 = 29.6$ mH
Reactance of the line = $2\pi \times 50 \times 29.6 \times 10^{-3} = 9.2991 \ \Omega$

Example 3.5 Figure 3.18 shows the configuration of a transposed three-phase transmission line employing two bundle conductors per phase. The diameter of each conductor is 5 cm and the spacing between them is 50 cm, in each bundle. Calculate the inductance and reactance per phase of the line which is 200 km long.

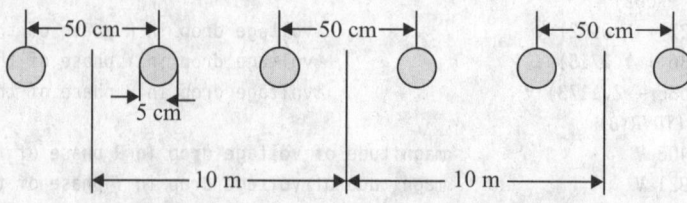

Fig. 3.18

Solution   Using Eq. (3.61) to calculate the self-GMR of the bundle gives

$$D_s^b = (D_s \times d)^{1/2} (0.7788 \times 2.5 \times 10^{-2} \times 0.5)^{1/2} = 0.0987 \text{ m}$$

$$D_{eq} = \sqrt[3]{D_{RY} D_{YB} D_{BR}} = \sqrt[3]{10 \times 10 \times 20} = 12.5992 \text{ m}$$

Equation (3.53) is utilized to obtain the inductance of the line

$$L = 2 \times 10^{-7} \ln \frac{D_{eq}}{D_s^b} = 2 \times 10^{-7} \ln \frac{12.5992}{0.0987}$$

$$= 9.6986 \times 10^{-7} \text{ H/m} = 0.96986 \text{ mH/km}$$

Inductance of the line = $0.96686 \times 200 = 0.1940$ H

Reactance of the line = $2\pi \times 50 \times 0.1940 = 60.9381$ Ω

3.9 REVIEW OF ELECTROSTATIC LAWS

A review of electrostatic laws, which form the basis for deriving expressions for the shunt capacitance of a transmission line, is presented here.

According to Gauss's law, the total charge q coulombs (C) over a closed surface of area A m^2 is expressed by

$$q = \int_A D \cdot da \tag{3.64}$$

where D is the uniform electric flux density in Coulomb per m^2 over the surface and da is an elementary strip of area perpendicular to the surface.

The electric field intensity E in V/m is written as

$$E = \frac{D}{\varepsilon} \tag{3.65}$$

where ε is the permittivity constant in F/m. For free space, permittivity constant is represented by ε_0 and is equal to 8.854×10^{-12} F/m.

The potential difference v_{ba} between two points a and b, in a field of electric intensity E in V/m, can be found by integrating E along any path joining the two points. Thus, from physics,

$$v_{ba} = v_b - v_a = -\int_a^b E dl = \int_b^a E dl \text{ V} \tag{3.66}$$

3.10 CAPACITANCE

The capacitance of a transmission line is due to the potential difference between conductors and is defined as the charge per unit of potential difference. When an alternating voltage is applied to a transmission line, the charge built up on the line conductors alternates with the frequency of the applied voltage. The alternate charging and discharging of the line causes a capacitive current to flow which is called the line charging current. Charging current flows in a transmission line, even when it is on no-load. The flow of charging current causes a voltage drop in the line which in turn affects the performance of the system.

Shunt capacitance is a source of reactive power. The effect of capacitance is negligible for power lines of lengths less than 80 km. For longer EHV transmission lines, the capacitance becomes increasingly important. In a high-voltage cable, close proximity of phase conductors results in a very large capacitance, of the order of 20 to 40 times that of an overhead transmission line. The reactive power generation for cables of even shorter lengths can then become a problem.

3.10.1 Electric Field of a Long Straight Conductor

A long straight cylindrical conductor with charge q coulombs per metre (C/m), isolated from other charges and far away from the ground, and placed in a uniform medium such as air, is shown in Fig. 3.19.

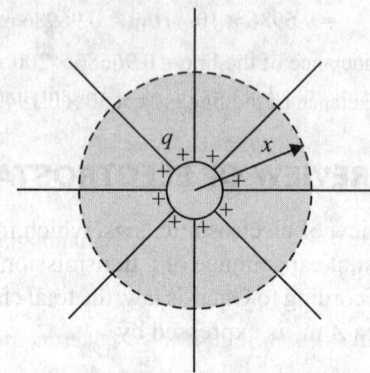

The electric flux produced is distributed uniformly around the periphery of the conductor. A cylindrical surface, x metres in radius, concentric with the conductor is an equipotential surface, and the electric flux density on the surface D_f is given by

Fig. 3.19 Electric flux lines on isolated conductor carrying charge of q C/m

$$D_f = \frac{q}{2\pi x} \text{ C/m}^2 \tag{3.67}$$

The electric field intensity at a distance x from the conductor is

$$E = \frac{q}{2\pi x \varepsilon} \text{ V/m} \tag{3.68}$$

where ε is the permittivity of the medium. In SI units, permittivity of free space ε_0 is 8.85×10^{-12} F/m. Relative permittivity for air is $\varepsilon = \varepsilon_r/\varepsilon_0 = 1$.

The potential difference between the two points in volts is numerically equal to the work done, in joules per coulomb (J/C), necessary to move one coulomb of charge between the two points.

The electric field intensity is a measure of the force on a charge in the field. Between two points, the line integral of the force in Newton (N) acting on q coulombs of positive charge is the work done in moving the charge from a point of lower potential to a point of higher potential and is numerically equal to the potential difference between the two points.

In Fig. 3.20 points X_1 and X_2 are located at distances D_1 and D_2 metres respectively from the centre of the conductor.

The simplest way to compute the voltage drop between two points is to compute the voltage, between the equipotential surfaces passing through X_1 and X_2, by integrating the field intensity over a radial path between the equipotential surfaces. Thus, the potential difference V_{12} between the points X_1 and X_2 is given by

$$V_{12} = \int_{D_1}^{D_2} E \, dx = \int_{D_1}^{D_2} \frac{q}{2\pi \varepsilon x} \, dx = \frac{q}{2\pi \varepsilon} \ln \frac{D_2}{D_1} \text{ V} \tag{3.69}$$

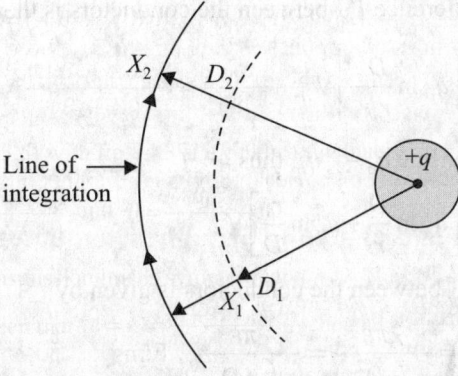

Fig. 3.20 Potential difference between points X_1 and X_2 due to a charge $+q$ coulomb

The potential difference between any two conductors in a group of n conductors shown in Fig. 3.21 can be obtained by adding the contributions of the individually charged conductors by using Eq. (3.69) repeatedly.

Thus, the potential difference between the conductors a and b is

$$V_{ab} = \frac{1}{2\pi\varepsilon}\left[q_a \ln \frac{D_{ab}}{r_a} + q_b \ln \frac{r_b}{D_{ba}} + q_c \ln \frac{D_{cb}}{D_{ca}} + \cdots + q_n \ln \frac{D_{nb}}{D_{na}} \right] \text{V} \quad (3.70)$$

It is observed from Eq. (3.70) that the potential difference between conductors a and b is the sum of the effect of the individually charged conductors. In writing Eq. (3.70), it has been assumed that the charge on the conductors is uniformly distributed over the surface and the length of the line. Such an assumption is realistic since the distances between the conductors, in power transmission lines, are much greater than the radii of the conductors.

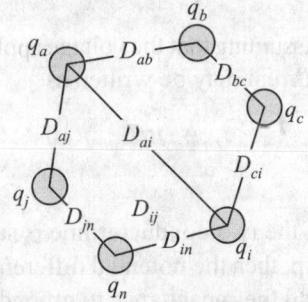

As stated earlier, both the voltages and charges on the conductors vary sinusoidally and are, therefore, phasor quantities.

Fig. 3.21 A group of n parallel charged conductors

3.10.2 Capacitance of a Two-conductor Line

A schematic arrangement of a two-conductor line connected to a single-phase ac supply is shown in Fig. 3.22. The radii of the conductors are r_a and r_b, and the charges on the conductors are q_a and q_b, respectively. The distance between the two conductors is D. It is assumed that the ground is far away from the conductors and the charge distribution over the surface of the conductors is uniform.

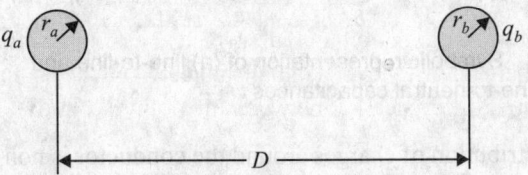

Fig. 3.22 A single phase two-conductor line

The potential difference V_{ab} between the conductors is then given by

$$V_{ab} = \frac{1}{2\pi\varepsilon}\left[q_a \ln\frac{D^2}{r_a r_b}\right] \text{ V} \tag{3.71}$$

For a two-conductor line $q_a = -q_b$, thus

$$V_{ab} = \frac{1}{2\pi\varepsilon}\left[q_a \ln\frac{D}{r} + q_b \ln\frac{r_b}{D}\right] \text{ V} \tag{3.72}$$

The capacitance C_{ab} between the conductors is given by

$$C_{ab} = \frac{q_a}{V_{ab}} = \frac{2\pi\varepsilon}{\ln\left(\dfrac{D^2}{r_a r_b}\right)} = \frac{\pi\varepsilon}{\ln\left(\dfrac{D}{\sqrt{r_a r_b}}\right)} \text{ F/m} \tag{3.73}$$

$$= \frac{0.01206}{\log\left(\dfrac{D}{\sqrt{r_a r_b}}\right)} \text{ } \mu\text{F/km} \tag{3.74}$$

In transmission lines, usually both the conductors have the same radius. Hence, substituting $r_a = r_b = r$,

$$C_{ab} = \frac{0.01206}{\log\left(\dfrac{D}{r}\right)} \text{ } \mu\text{F/km} \tag{3.75}$$

Assuming that the voltage applied to the two-conductor line is V volts, the charging current may be written as

$$I_{ch} = j\omega C_{ab} \times 10^{-6} \times V = j\omega\frac{0.01206}{\log\left(\dfrac{D}{r}\right)} \times 10^{-6} \times V \text{ A/km} \tag{3.76}$$

If the two-conductor line is supplied by a transformer having a grounded centre tap, then the potential difference between each conductor and the ground is $V_{ab}/2$ and the capacitance to ground, or neutral, is

$$C_n = C_{an} = C_{bn} = 2C_{ab} = \frac{0.02412}{\log\left(\dfrac{D}{r}\right)} \text{ } \mu\text{F/km} \tag{3.77}$$

The symbolic representation of line-to-line and line-to-neutral capacitances are shown in Fig. 3.23.

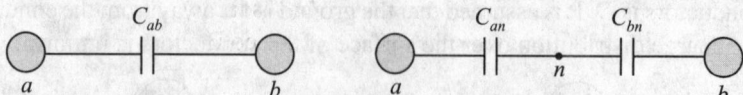

Fig. 3.23 Symbolic representation of (a) line-to-line and (b) line-to-neutral capacitances

When the distribution of charges around the conductor is non-uniform (which is often the case) due to the presence of charges other than those on the conductor, the computation of capacitance using Eq. (3.75) will introduce an error of only 0.01% for $(D/r) = 50$, which may be neglected.

For stranded conductors, the capacitance may be computed using Eq. (3.75) by substituting the outside radius of the conductor for r. Such an approximation does not introduce a significant error in the computation of the line capacitance.

3.10.3 Capacitance of a Three-phase Line with Equilateral Spacing

A three-phase line with the conductors equilaterally spaced is shown in Fig. 3.24. The three-phase conductors R, Y, and B have equal cross sections. Let the distance between the conductors be D, and the charges on the conductors be q_R, q_Y, and q_B respectively. Using Eq. (3.70), the voltages V_{RY} and V_{RB} can be written as

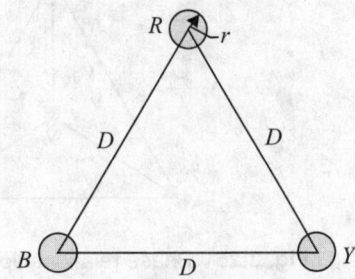

Fig. 3.24 Configuration of a three-phase line with conductors equilaterally spaced

$$V_{RY} = \frac{1}{2\pi\varepsilon}\left[q_R \ln\frac{D}{r} + q_Y \ln\frac{r}{D} + q_B \ln\frac{D}{D} \right] \text{V} \tag{3.78}$$

$$V_{RB} = \frac{1}{2\pi\varepsilon}\left[q_R \ln\frac{D}{r} + q_Y \ln\frac{D}{D} + q_B \ln\frac{r}{D} \right] \text{V} \tag{3.79}$$

Adding Eqs (3.78) and (3.79), gives

$$V_{RY} + V_{RB} = \frac{1}{2\pi\varepsilon}\left[2q_R \ln\frac{D}{r} + (q_Y + q_B)\ln\frac{r}{D} \right] \text{V} \tag{3.80}$$

If it is assumed that there are no charges in the neighbourhood of the conductors, then $(q_R + q_Y + q_B) = 0$, or $(q_Y + q_B) = -q_R$. Then, substituting $(q_Y + q_B) = -q_B$ in Eq. (3.80) yields

$$V_{RY} + V_{RB} = \frac{3q}{2\pi\varepsilon}\ln\frac{D}{r} \text{V} \tag{3.81}$$

Figure 3.25 shows the phase relationship between the voltages of a balanced three-phase ac system. From the figure, the relation between V_{RN}, the line-to-neutral voltage of phase R, and the line voltages V_{RY} and V_{RB} may be obtained as

$$V_{RY} = \sqrt{3}V_{RN}\angle 30° = \sqrt{3}V_{RN}(0.866 + j0.5) \tag{3.82}$$

$$V_{RB} = -V_{ca} = \sqrt{3}V_{RN}\angle -30° = \sqrt{3}V_{RN}(0.866 - j0.5) \tag{3.83}$$

Adding Eqs (3.82) and (3.83) gives

$$V_{RY} + V_{RB} = 3V_{RN} \tag{3.84}$$

Substituting Eq. (3.84) in Eq. (3.81) gives

$$V_{RN} = \frac{q_R}{2\pi\varepsilon}\ln\frac{D}{r} \text{V} \tag{3.85}$$

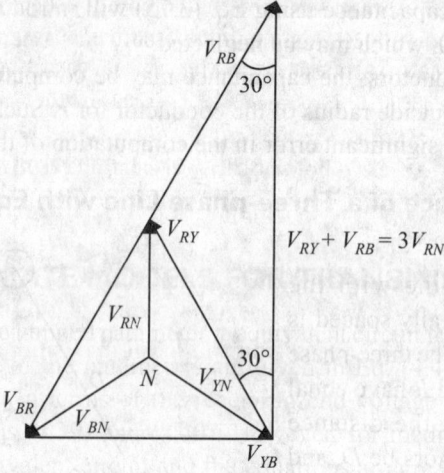

Fig. 3.25 Phase relationship between voltages of a balanced three-phase voltage system

The capacitance to neutral C_n, therefore, is

$$C_n = \frac{q_R}{V_{RN}} = \frac{2\pi\varepsilon}{\ln\dfrac{D}{r}} \text{ F/m} \tag{3.86}$$

$$C_n = \frac{0.02412}{\log\dfrac{D}{r}} \text{ } \mu\text{F/km} \tag{3.87}$$

The charging current per phase is obtained as

$$I_{ch} = j\omega C_n V_{RN} = j\omega \times \frac{2\pi\varepsilon}{\ln\dfrac{d}{r}} \times V_{RN} \text{ A/m} \tag{3.88}$$

3.10.4 Capacitance of a Three-phase Line with Unsymmetrical Spacing

Representation of a three-phase, fully transposed line with three identical conductors of radius r with unequal spacing is shown in Fig. 3.26.

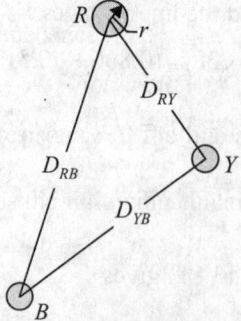

Fig. 3.26 Representation of a fully transposed three-phase line with unequal spacing

For the first section of the transposition cycle with phase R in position 1, Y in position 2, and B in position 3, the line voltage between the conductors of phases R and Y, V_{RY}, may be written as

$$V_{RY} = \frac{1}{2\pi\varepsilon}\left[q_R \ln\frac{D_{RY}}{r} + q_Y \ln\frac{r}{D_{RY}} + q_B \ln\frac{D_{YB}}{D_{BR}} \right] \text{V} \tag{3.89}$$

With phase R in position 2, Y in position 3, and B in position 1,

$$V_{RY} = \frac{1}{2\pi\varepsilon}\left[q_R \ln\frac{D_{YB}}{r} + q_Y \ln\frac{r}{D_{YB}} + q_B \ln\frac{D_{BR}}{D_{RY}} \right] \text{V} \tag{3.90}$$

With phase R in position 3, Y in position 1, and B in position 2,

$$V_{RY} = \frac{1}{2\pi\varepsilon}\left[q_R \ln\frac{D_{BR}}{r} + q_Y \ln\frac{r}{D_{BR}} + q_B \ln\frac{D_{RY}}{D_{YB}} \right] \text{V} \tag{3.91}$$

If the voltage drop along the line is neglected, V_{RY} is the same in each section of the transposition cycle. The charge on a conductor of a particular phase is different when its position changes with respect to other conductors during transposition. However, in reality the usual spacing between conductors of transmission lines is large, and sufficient accuracy is obtained by assuming that the charge per unit length on a conductor is the same in each part of the transposition cycle. Eqs (3.89), (3.90), and (3.91) have been written on the basis of the above assumption. However, with this assumption, the voltage between a pair of conductors of a transmission line becomes different for each part of the transposition cycle. Thus, an average value of voltage is used to calculate the capacitance. Using Eqs (3.89) to (3.91), the average voltage between conductors of phase R and Y is

$$V_{RY} = \frac{1}{6\pi\varepsilon}\left[q_R \ln\frac{D_{RY}D_{YB}D_{BR}}{r^3} + q_Y \ln\frac{D_{RY}D_{YB}D_{BR}}{r^3} + q_B \ln\frac{D_{RY}D_{YB}D_{BR}}{D_{RY}D_{YB}D_{BR}} \right]$$

$$= \frac{1}{2\pi\varepsilon}\left[q_R \ln\frac{D_{eq}}{r} + q_Y \ln\frac{r}{D_{eq}} \right] \text{V} \tag{3.92}$$

where

$$D_{eq} = \sqrt[3]{D_{RY}D_{YB}D_{BR}} \tag{3.93}$$

Similarly, the average voltage between conductors of phase R and B is

$$V_{RB} = \frac{1}{2\pi\varepsilon}\left[q_R \ln\frac{D_{eq}}{r} + q_B \ln\frac{r}{D_{eq}} \right] \text{V} \tag{3.94}$$

Adding Eqs (3.92) and (3.94), and applying Eq. (3.84) gives

$$V_{RY} + V_{RB} = 3V_{RN} = \frac{1}{2\pi\varepsilon}\left[2q_R \ln\frac{D_{eq}}{r} + (q_r + q_B)\ln\frac{r}{D_{eq}} \right] \text{V} \tag{3.95}$$

Since $q_R + q_Y + q_B = 0$, Eq. (3.95) takes the form

$$V_{RN} = \frac{3}{2\pi\varepsilon} q_R \ln\frac{D_{eq}}{r} \text{V} \tag{3.96}$$

Hence,

$$C_n = \frac{q_R}{V_{RN}} = \frac{2\pi\varepsilon}{\ln\dfrac{D_{eq}}{r}} \text{ F/m} \qquad (3.97)$$

$$= \frac{0.02412}{\ln\dfrac{D_{eq}}{r}} \text{ μF/km} \qquad (3.98)$$

Charging current may be written as

$$I_{ch} = j\omega C_n V_{RN} = j\omega \left(\frac{2\pi\varepsilon}{\ln\dfrac{D_{eq}}{r}}\right) \times V_{RN} \text{ A/m} \qquad (3.99)$$

3.11 EFFECT OF EARTH ON THE CAPACITANCE OF TRANSMISSION LINES

Earth is assumed to be a perfect conductor in the form of a horizontal equipotential surface of infinite extent. The electric field of charged conductors of a transmission line is altered by the presence of the equipotential surface of the earth. Hence, the effective capacitance between the conductors also changes.

The effect of the presence of ground can be accounted for by means of image charges. These imaginary charges are of the same magnitude as the actual physical charges on the conductors of the three-phase transmission line but carry opposite charges. These imaginary charges are placed as far below the ground level as the actual physical charges are placed above the ground. The electric flux between the overhead conductor and its image is perpendicular to the plane, which replaces the ground. This plane forms an equipotential surface. The electric flux above the plane is the same as it is when the ground is present instead of the image conductor.

3.11.1 Capacitance of a Single-phase Line

Figure 3.27 provides a schematic representation of a single-phase charged line in the presence of the ground. The line conductors a and b carry respective charges of q_a and q_b coulomb and the spacing between them is D metres.

The voltage between the conductors is obtained by applying Eq. (3.70) and is written as

$$V_{ab} = \frac{1}{2\pi\varepsilon}\left[q_a \ln\frac{D}{r} - q_b \ln\frac{r}{D} - q_a \ln\frac{\sqrt{4h^2 + D^2}}{2h} + q_b \ln\frac{2h}{\sqrt{4h^2 + D^2}}\right] \qquad (3.100)$$

In Eq. (3.100), r represents the radius of the conductors. Assume that $q_a = q_{a'} = q_b = q_{b'} = q$. On substitution and simplification Eq. (3.100) becomes

$$V_{ab} = \frac{q}{\pi\varepsilon}\ln\left[\frac{2hD}{r\sqrt{4h^2 + D^2}}\right] \qquad (3.101)$$

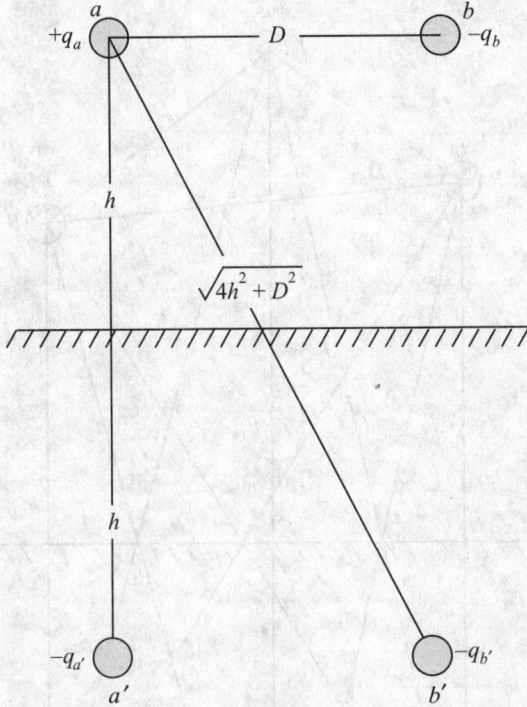

Fig. 3.27 Representation of a single-phase charged line and its images

Capacitance of the single-phase line, including the effect of earth, is given by

$$C_{ab} = \frac{\pi\varepsilon}{\ln\dfrac{D}{r\sqrt{\left\{1+\dfrac{D^2}{4h^2}\right\}}}} \text{ F/m} \tag{3.102}$$

Line-to-neutral capacitance of the line would be written as

$$C_{n} = \frac{2\pi\varepsilon}{\ln\dfrac{D}{r\sqrt{\left\{1+\dfrac{D^2}{4h^2}\right\}}}} \text{ F/m} \tag{3.103}$$

3.11.2 Capacitance of a Three-phase Line

A schematic representation of a three-phase charged transmission line with its images is shown in Fig. 3.28. It is assumed that the line is transposed and the three phase conductors R, Y, and B carry charges, and occupy positions 1, 2, and 3 respectively in the first section of the transposition cycle. The plane of the earth and below it the image conductors with charges $-q_R$, $-q_Y$ and $-q_B$ are also shown in the figure.

With the conductor of phase R in position 1, Y in position 2, and B in position 3, the voltage drop from the conductor of phase R to the conductor of phase Y may be determined by

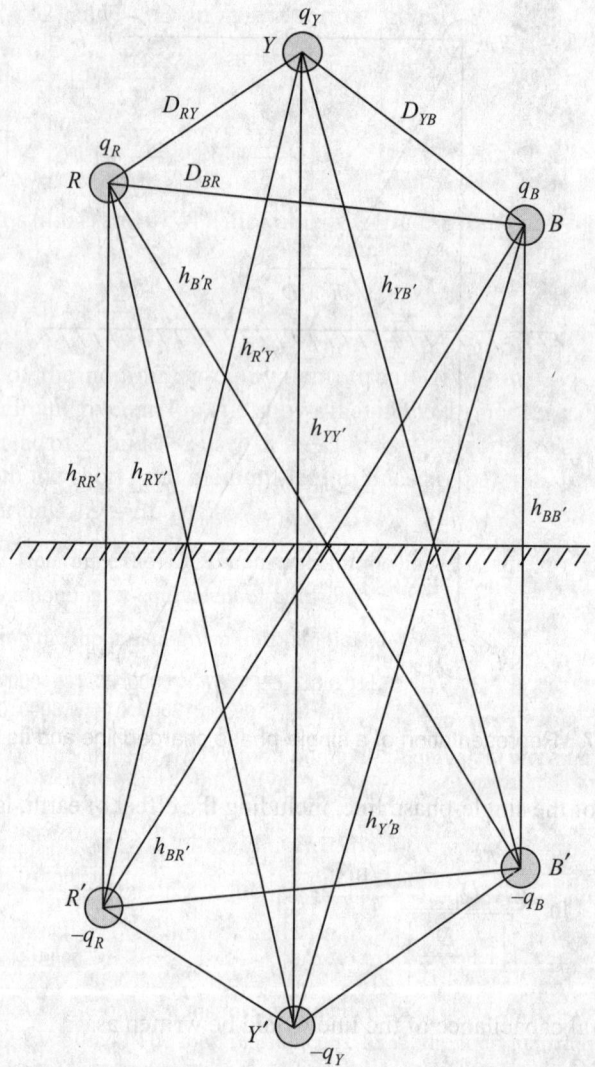

Fig. 3.28 Representation of a three-phase transmission line with its images

$$V_{RY} = \frac{1}{2\pi\varepsilon}\left[q_R\left(\ln\frac{D_{RY}}{r} - \ln\frac{h_{R'Y}}{h_{RR'}}\right) + q_Y\left(\ln\frac{r}{D_{RY}} - \ln\frac{h_{YY'}}{h_{R'Y}}\right)\right.$$

$$\left. + q_Y\left(\ln\frac{D_{YB}}{D_{BR}} - \ln\frac{h_{YB'}}{h_{BR'}}\right)\right] \qquad (3.104)$$

Similar equations may be written for the two other sections of the transposition cycle. Assuming charge per unit length of each conductor to be constant throughout the transposition cycle, the average value of V_{RY} for the entire line section is given by

$$V_{RY} = \frac{1}{2\pi\varepsilon}\left[q_R\left(\ln\frac{D_{eq}}{r} - \ln\frac{\sqrt[3]{h_{R'Y}h_{YB}h_{BR'}}}{\sqrt[3]{h_{RR'}h_{YY'}h_{BB'}}}\right)\right.$$

$$+ q_Y \left(\ln \frac{r}{D_{eq}} - \ln \frac{\sqrt[3]{h_{RR'} h_{YY'} h_{BB'}}}{\sqrt[3]{h_{R'Y} h_{YB'} h_{BR'}}} \right) \Bigg] \tag{3.105}$$

where $D_{eq} = \sqrt[3]{D_{RY} D_{YB} D_{BR}}$

The average value of V_{RB} can be found in a similar manner, and $3V_{RN}$ is obtained by adding the average values of V_{RY} and V_{RB}. Knowing that $q_R + q_Y + q_B = 0$, the capacitance to neutral may be obtained as

$$C_n = \frac{2\pi\varepsilon}{\ln\left(\dfrac{D_{eq}}{r}\right) - \ln\left(\dfrac{\sqrt[3]{h_{R'Y} h_{YB'} h_{BR'}}}{\sqrt[3]{h_{RR'} h_{YY'} h_{BB'}}}\right)} \ \text{F/m} \tag{3.106}$$

where $D_{eq} = \sqrt[3]{D_{RY} D_{YB} D_{BR}}$

Comparing Eq. (3.97) with Eq. (3.106) shows that the presence of earth increases the capacitance of a line. If the conductors are placed high above the ground, compared to the distances between the conductors themselves, then the effect of earth may be neglected.

3.12 COMPUTATION OF CAPACITANCE OF BUNDLED CONDUCTORS

Arrangement of a three-phase line employing bundled conductors is shown in Fig. 3.29.

The conductors of any one bundle are in parallel, and it can be assumed that the charge per bundle is divided equally between the conductors of the bundle. If charge on phase R is q_R, each of the sub-conductors R and R' has the charge $\dfrac{q_R}{2}$.

Then

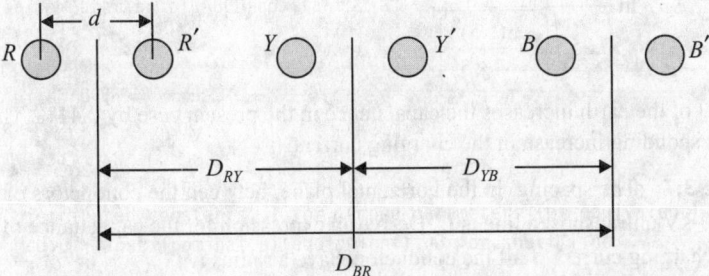

Fig. 3.29 Layout of a three-phase line employing bundled conductors

$$V_{RY} = \frac{1}{2\pi\varepsilon} \left[\frac{q_R}{2} \left(\ln \frac{D_{RY}}{r} + \ln \frac{D_{RY}}{d} \right) + \frac{q_Y}{2} \left(\ln \frac{r}{D_{RY}} + \ln \frac{d}{D_{RY}} \right) \right.$$

$$\left. + \frac{q_B}{2} \left(\ln \frac{D_{YB}}{D_{BR}} + \ln \frac{D_{YB}}{D_{BR}} \right) \right] \tag{3.107}$$

$$V_{RY} = \frac{1}{2\pi\varepsilon}\left[q_R\left(\ln\frac{D_{RY}}{\sqrt{rd}} \right) + q_Y\left(\ln\frac{\sqrt{rd}}{D_{RY}} \right) + q_B\left(\ln\frac{D_{YB}}{D_{BR}} \right) \right] \tag{3.108}$$

Depending on the line to be transposed, the capacitance may be computed in the usual manner, as described in earlier section, as follows

$$C_n = \frac{2\pi\varepsilon}{\ln\left(\dfrac{D_{eq}}{\sqrt{rd}} \right)} \quad \text{F/m to neutral} \tag{3.109}$$

where $D_{eq} = \sqrt[3]{D_{RY}D_{YB}D_{BR}}$

Example 3.6 A single-phase, 33-kV line, operating at 50 Hz, is spaced 1.5 m apart and the diameter of the conductors is 10 mm. Determine the capacitance between the conductors and the line charging current. Calculate the conductor capacitance to neutral and the charging current. How does the capacitance and charging current change if the effect of the earth is considered? Assume that the length of the line is equal to 5 km and the ground clearance is 5 m.

Solution Using Eq. (3.75), the capacitance between conductors is given by

$$C_{ab} = \frac{0.01206}{\log\left(\dfrac{1.5}{5\times10^{-3}} \right)} \times 5 \ \mu\text{F} = 0.0243 \ \mu\text{F}$$

From Eq. (3.77),

$$C_n = 2C_{ab} = 2 \times 0.0243 = 0.0486 \ \mu\text{F}$$

Line charging current is obtained using Eq. (3.76), that is,

$$I_{ch} = j \times 2 \times \pi \times 50 \times 0.0243 \times 10^{-6} \times 33 \times 10^3 \ \text{A} = j\, 0.2519 \ \text{A}$$

When the effect of the earth is considered, Eq. (3.102) is applicable. Thus,

$$C_{ab} = \frac{\pi \times 8.85 \times 10^{-12}}{\ln\dfrac{1.5}{0.05\sqrt{\left\{1+\dfrac{(1.5)^2}{4\times5^2}\right\}}}} \times 10^6 \times 5 \times 10^3 \ \mu\text{F} = 0.0244 \ \mu\text{F}$$

The effect of the earth increases the capacitance in the present case by 0.41%. There will be a corresponding increase in the charging current.

Example 3.7 The spacing, in the horizontal plane, between the conductors of a three-phase 110-kV untransposed line is D. Derive an expression for the capacitance of the line and line charging current if all the conductors have a radius r.

Solution The arrangement of the conductors is shown in Fig. 3.30.

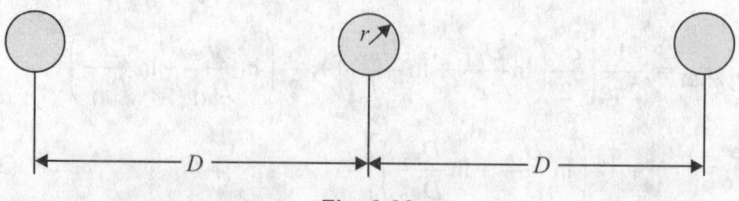

Fig. 3.30

Using Eq. (3.78) gives

$$V_{RY} = \frac{1}{2\pi\varepsilon}\left(q_R \ln\frac{D}{r} + q_Y \ln\frac{r}{D} + q_B \ln\frac{D}{2D}\right)$$

Putting $q_Y = -(q_R + q_B)$ in the preceding equation and simplifying leads to

$$V_{RY} = \frac{1}{2\pi\varepsilon}\left(q_R \ln\frac{D_2}{r_2} + q_B \ln\frac{D}{2r}\right) \tag{3.7.1}$$

Similarly, using Eq. (3.79), gives

$$V_{RB} = \frac{1}{2\pi\varepsilon}\left(q_R \ln\frac{2D}{r} + q_B \ln\frac{r}{2D} + q_Y \ln\frac{2D}{2D}\right)$$

or

$$q_B = \frac{2\pi\varepsilon V_{RB} - q_R \ln\dfrac{D}{2r}}{\ln\dfrac{r}{2D}}$$

Substituting for q_B in Eq. (3.7.1) and simplifying yields

$$q_R = \frac{2\pi\varepsilon\left(V_{RY}\ln\dfrac{r}{2D} - V_{RB}\ln\dfrac{D}{2r}\right)}{\left(\ln\dfrac{D^2}{r^2}\ln\dfrac{r}{2D} - \ln\dfrac{2D}{r}\ln\dfrac{D}{2r}\right)}$$

If V_{RY} is taken as the reference phasor, then $V_{RY} = V\angle 0^\circ$ and $V_{RB} = V\angle -60^\circ$. Thus,

$$q_R = \frac{2\pi\varepsilon V\left(\ln\dfrac{r}{2D}\angle 0^\circ \quad \ln\dfrac{D}{2r}\angle -60^\circ\right)}{\left(\ln\dfrac{D^2}{r^2}\ln\dfrac{r}{2D} - \ln\dfrac{2D}{r}\ln\dfrac{D}{2r}\right)}$$

Line capacitance

$$C_{RY} = \frac{q_R}{V} = \frac{2\pi\varepsilon\left(\ln\dfrac{r}{2D}\angle 0^\circ - \ln\dfrac{D}{2r}\angle -60^\circ\right)}{\left(\ln\dfrac{D^2}{r^2}\ln\dfrac{r}{2D} - \ln\dfrac{2D}{r}\ln\dfrac{D}{2r}\right)}\ \text{F/m}$$

Charging current

$$I_{ch} = j\omega\,\frac{2\pi\varepsilon V\left(\ln\dfrac{r}{2D}\angle 0^\circ - \ln\dfrac{D}{2r}\angle -60^\circ\right)}{\left(\ln\dfrac{D^2}{r^2}\ln\dfrac{r}{2D} - \ln\dfrac{2D}{r}\ln\dfrac{D}{2r}\right)} \times V\ \text{A/m}$$

By following a similar procedure, expressions for line capacitances C_{YB} and C_{BR} can also be obtained. It is left as a tutorial exercise for the readers to derive these expressions.

Example 3.8 In Example 3.7, taking $D = 6$ m and $r = 15$ mm, compute C_{RY} and line charging current for a 1-km long line. Take the operating frequency equal to 50 Hz.

Solution

$$C_{RY} = \frac{2\pi \times 8.85 \times 10^{-12} \left(\ln \dfrac{0.015}{2 \times 6} - \ln \dfrac{6}{2 \times 0.015} [0.5 - j0.866] \right) \times 10^6 \times 10^3}{\left(\ln \dfrac{6^2}{(0.015)^2} \ln \dfrac{0.015}{2 \times 6} - \ln \dfrac{2 \times 6}{0.015} \ln \dfrac{6}{2 \times 0.015} \right)} \ \mu F/km$$

$$= (0.0045 - j\,0.0022)\ \mu F/km$$

Line charging current

$$I_{ch} = j\,2 \times \pi \times 50 \times (0.0045 - j\,0.0022) \times 10^{-6} \times 110 \times 10^3$$

$$= (0.0763 + j\,0.1553)\ A$$

Example 3.9 The configuration of a three-phase 220-kV, 100-km long, fully-transposed, 50-Hz transmission line is given in Fig. 3.31. If the radius of the conductor is 20 mm, compute the capacitance to neutral of the line and the line charging current by considering the effect of the earth.

Solution Figure 3.32 provides the configuration of the line with its images. Referring to Fig. 3.32, the following distances can be obtained:

$$h_{R'Y} = \sqrt{\left(12 + 2\sqrt{3}\right)^2 + 2^2} = 15.93\ m; \quad h_{YB'} = \sqrt{12^2 + 4^2} = 12.65\ m;$$

$$h_{BR'} = 15.93\ m$$

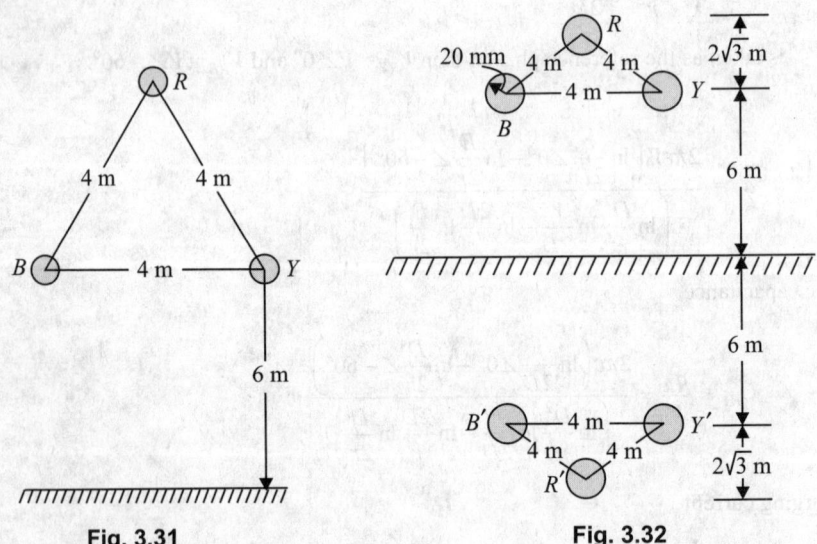

Fig. 3.31   **Fig. 3.32**

$$h_{RR'} = 12 + 2\sqrt{3} + 2\sqrt{3} = 18.93\ m;\ h_{yy'} = 12\ m;\ h_{BB'} = 12\ m$$

$$D_{eq} = \sqrt[3]{4 \times 4 \times 4} = 4\,m$$

Substituting these values in Eq. (3.106) yields

$$C_n = \frac{2\pi \times 8.854 \times 10^{-12}}{\ln\left(\dfrac{4}{0.02}\right) - \ln\left(\dfrac{\sqrt[3]{15.93 \times 12.65 \times 15.93}}{\sqrt[3]{18.93 \times 12 \times 12}}\right)}$$

$$= 1.0609 \times 10^{-11}\ F/m = 0.0106\ \mu F/km$$

Line capacitance to neutral = $0.0106 \times 100 = 1.06$ μF

Line charging current = $j\,2\pi \times 50 \times 1.06 \times 10^{-6} \times \dfrac{220}{\sqrt{3}} \times 10^3 = 42.2978$ A

Example 3.10 The arrangements of conductors, of equal radii, of a 150-km, 220-kV, 50-Hz, double circuit line is shown in Fig. 3.33.

If the diameter of the conductor is equal to 4 cm, determine (i) line capacitance to neutral, (ii) line charging current per phase and (iii) charging current per conductor. The transmission lines of both the circuits are fully transposed.

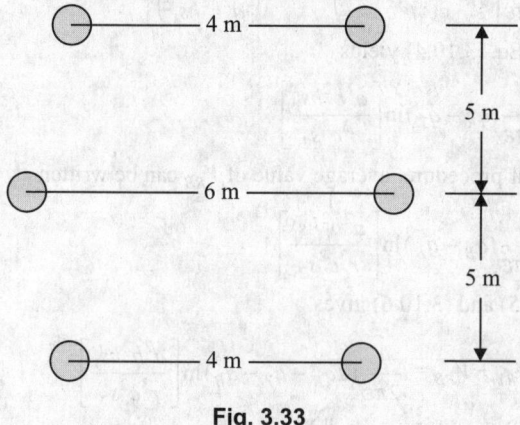

Fig. 3.33

Solution Figure 3.34 shows the cross section of the transmission line with the position of the phase conductors in the three sections of the transposition indicated.

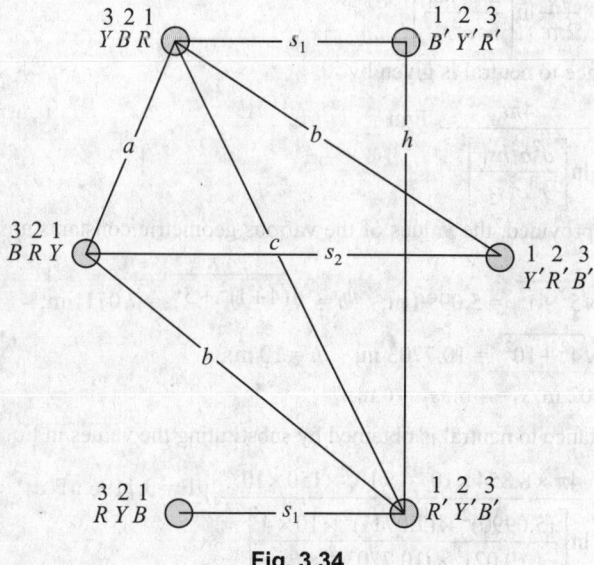

Fig. 3.34

Using Eq. (3.70), the voltage V_{RY} due to the charge built up in section 1 of the line is given by

$$V_{RY} = \frac{1}{2\pi\varepsilon}\left[q_R\left(\ln\frac{a}{r} + \ln\frac{b}{c} \right) + q_Y\left(\ln\frac{r}{a} + \ln\frac{s_2}{b} \right) + q_B\left(\ln\frac{a}{h} + \ln\frac{b}{s_1} \right) \right] \qquad (3.10.1)$$

Similarly, the voltages V_{RY} due to the charge built up in sections 2 and 3 of the line can be respectively written as

$$V_{RY} = \frac{1}{2\pi\varepsilon}\left[q_R\left(\ln\frac{a}{r}+\ln\frac{b}{s_2}\right)+q_Y\left(\ln\frac{r}{a}+\ln\frac{c}{b}\right)+q_B\left(\ln\frac{h}{a}+\ln\frac{s_1}{b}\right)\right] \qquad (3.10.2)$$

$$V_{RY} = \frac{1}{2\pi\varepsilon}\left[q_R\left(\ln\frac{h}{r}+\ln\frac{s_1}{c}\right)+q_Y\left(\ln\frac{r}{h}+\ln\frac{c}{s_1}\right)+q_B\left(\ln\frac{a}{a}+\ln\frac{b}{b}\right)\right] \qquad (3.10.3)$$

Average $V_{RY} = \dfrac{1}{6\pi\varepsilon}\left[q_R\ln\left(\dfrac{a^2b^2hs_1}{r^3c^2s_2}\right)+q_Y\ln\left(\dfrac{r^3c^2s_2}{a^2b^2hs_1}\right)\right] \qquad (3.10.4)$

On simplification Eq. (3.10.4) yields,

Average $V_{RY} = \dfrac{1}{2\pi\varepsilon}(q_R-q_Y)\ln\left[\dfrac{a^2b^2hs_1}{r^3c^2s_2}\right]^{1/3} \qquad (3.10.5)$

Following a similar procedure, average value of V_{RB} can be written as

Average $V_{RB} = \dfrac{1}{2\pi\varepsilon}(q_B-q_B)\ln\left[\dfrac{a^2b^2hs_1}{r^3c^2s_2}\right]^{1/3} \qquad (3.10.6)$

Adding Eqs (3.10.5) and (3.10.6) gives

$$3V_{RN} = V_{RY}+V_{RB} = \frac{1}{2\pi\varepsilon}(2q_R-q_Y-q_B)\ln\left[\frac{a^2b^2hs_1}{r^3c^2s_2}\right]^{1/3}$$

$$= \frac{3q_R}{2\pi\varepsilon}\ln\left[\frac{a^2b^2hs_1}{r^3c^2s_2}\right]^{1/3}$$

or $\qquad V_{RN} = \dfrac{q_R}{2\pi\varepsilon}\ln\left[\dfrac{a^2b^2hs_1}{r^3c^2s_2}\right]^{1/3}$

Thus, capacitance to neutral is given by

$$C_n = \frac{4\pi\varepsilon}{\ln\left[\dfrac{a^2b^2hs_1}{r^3c^2s_2}\right]^{1/3}} \quad \text{F/m} \qquad (3.10.7)$$

From the data provided, the values of the various geometric constants are calculated as follows:

$$a = \sqrt{5^2+1^2} = 5.0990 \text{ m}; \quad b = \sqrt{(4+1)^2+5^2} = 7.0711 \text{ m};$$

$$c = \sqrt{4^2+10^2} = 10.7703 \text{ m}; \quad h = 10 \text{ m};$$

$$r = 0.02 \text{ m}; \ s_1 = 4 \text{ m}; \ s_2 = 6 \text{ m};$$

(i) Line capacitance to neutral is obtained by substituting the values in Eq. (3.10.7).

$$C_n = \frac{4\pi\times8.854\times10^{-12}\times10^6\times150\times10^3}{\ln\left[\dfrac{(5.0990)^2\times(7.0711)^2\times10\times4}{(0.02)^3\times(10.7703)^2\times6}\right]^{1/3}} \ \mu\text{F} = 3.1196 \ \mu\text{F}$$

(ii) Line charging current $= 2\times\pi\times50\times3.1196\times10^{-6}\times\dfrac{220\times10^3}{\sqrt{3}} = 124.4832 \text{ A}$

(iii) Charging current per conductor $= \dfrac{124.4832}{2} = 62.2416 \text{ A}$

Example 3.11 The arrangement of conductors in a three-phase transmission line using bundled conductors is shown in Fig. 3.35.

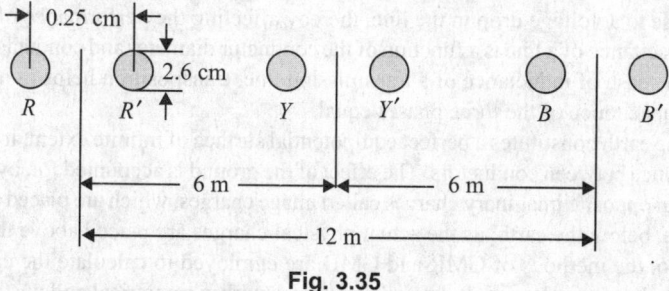

Fig. 3.35

Assuming the line to be fully transposed, calculate capacitance to neutral in F/km.

Solution Using Eq. (3.109), capacitance to neutral is obtained. However, first D_{eq} should be computed.

$$D_{eq} = \sqrt[3]{6 \times 6 \times 12} = 7.5443 \text{ m}$$

$$C_n = \frac{2 \times \pi \times 8.854 \times 10^{-12} \times 10^3}{\ln\left(\dfrac{7.5443}{\sqrt{0.013 \times 0.25}}\right)} = 1.1387 \times 10^{-8} \text{ F/km}$$

SUMMARY

- Electrical power transmission lines are characterized by four constants, namely series resistance, series inductance, shunt conductance, and shunt capacitance.
- Stranded aluminium conductors, instead of copper, are used in overhead transmission lines since the larger diameters reduce voltage gradients and leakage due to corona. Stranding makes the conductors flexible. ACSR conductors are most commonly employed.
- Since alternating current flows in the conductors, the tendency of the current to flow at the surface of the conductor is called the skin effect.
- The phenomenon of inductance in a transmission line is due to the magnetic field produced by the current flowing through the conductors and is a function of the conductor diameter and conductor spacing.
- The methods of geometric mean radius (GMR) and geometric mean distance (GMD) are employed to calculate the inductance of three-phase lines using stranded conductors.
- If the conductors are symmetrically spaced, all the conductors in the three phases have equal inductances. However, if the spacing of conductors is unsymmetrical, due to unequal flux linkages, the inductance of each phase is different and produces an unbalanced circuit.
- In order to make the inductances of the three phases equal, the position of each phase is changed, at equal intervals, so that each conductor occupies the original position of every other conductor. This process of exchanging conductor positions is called transposition of a line.
- The methods of GMR and GMD are also used to compute inductances of three-phase circuits using bundled conductors and three-phase double-circuit fully transposed transmission lines.
- Shunt conductance is a simulation of the leakage current over the surface of insulators. Since the magnitude of the leakage current is very small, it is usually neglected.

- Capacitance of a transmission line is defined as the charge per unit potential difference.
- A continuous exchange of charge between the line conductors at power frequency causes a shunt capacitive current to flow and is called the line charging current which in turn gives rise to a voltage drop in the line, thereby affecting the performance of the line.
- The capacitance of a line is a function of the conductor diameter and conductor spacing.
- As in the case of inductance of a transmission line, transposition helps in making the shunt capacitance of the three phases equal.
- Since the earth constitutes a perfect equipotential surface of infinite extent, it affects the capacitance between conductors. The effect of the ground is accounted for, by assuming equal but opposite imaginary charges called image charges, which are placed at an equal distance, below the earth, as the actual physical charges are placed above the earth.
- Here too, the methods of GMR and GMD are employed to calculate the capacitance of three-phase, single-circuit, transmission lines with symmetrical and unsymmetrical spacing (including bundled conductors) and double-circuit fully transposed transmission lines.

SIGNIFICANT FORMULAE

- Resistance R_{t2} of the conductor at $t_2°C$ $\quad \dfrac{R_{t2}}{R_{t1}} = \dfrac{1/\alpha_0 + t_2}{1/\alpha_0 + t_1}$

No. of phases	Inductance	Capacitance C	Effect of ground
Single	$0.921 \log \dfrac{D}{r'}$ mH/km	$\dfrac{0.01206}{\log\left(\dfrac{D}{r}\right)}$ μF/km	No effect
Three, symmetrical	$0.4605 \log \dfrac{D}{r'}$ mH/km	$\dfrac{0.02412}{\log\left(\dfrac{D}{r}\right)}$ μF/km	$\dfrac{2\pi\varepsilon}{\ln\left(\dfrac{D_{eq}}{r}\right) - \ln\left(\dfrac{\sqrt[3]{h_{R'Y}h_{YB'}h_{BR'}}}{\sqrt[3]{h_{RR'}h_{YY'}h_{BB'}}}\right)}$ F/m
Unsymmetrical	$2\times10^{-7} \ln \dfrac{D_{eq}}{D_s}$ H/m	$\dfrac{0.02412}{\ln\left(\dfrac{D_{eq}}{r}\right)}$ μF/km	$\dfrac{2\pi\varepsilon}{\ln\left(\dfrac{D_{eq}}{r}\right) - \ln\left(\dfrac{\sqrt[3]{h_{R'Y}h_{YB'}h_{BR'}}}{\sqrt[3]{h_{RR'}h_{YY'}h_{BB'}}}\right)}$ F/m
Double-circuit three-phase	$2\times10^{-7} \ln\left\{2^{1/6}\left(\dfrac{D}{r'}\right)^{1/2} \times \left(\dfrac{p}{q}\right)^{2/3}\right\}$ H/m	—	—
Bundled conductors	$2\times10^{-7} \ln \dfrac{D_{eq}}{D_s^b}$ H/m	$\dfrac{2\pi\varepsilon}{\ln\left(\dfrac{D_{eq}}{\sqrt{rd}}\right)}$ F/m	—

- Three-phase circuits
(i)   two-conductor bundle $D_s^b = (D_s \times d)^{1/2}$
(ii) three-conductor bundle $D_s^b = (D_s \times d \times d)^{1/3}$
(iii) four-conductor bundle $D_s^b = 1.09\,(D_s \times d^3)^{1/4}$
(iv) $D_{eq} = \sqrt[3]{D_{RY}D_{YB}D_{BR}}$

EXERCISES

Review Questions

3.1 List the parameters that characterize a transmission line. Discuss the importance of each of these parameters and how they affect the performance of a line.

3.2 Write a note on the requirements of materials to be used as conductors. Enumerate the different types of conductors employed in transmission lines and explain their properties.

3.3 Explain the effect of temperature on the resistance of a conductor.

3.4 Derive an expression for the self GMD of a stranded conductor.

3.5 Write clear notes on (i) skin effect, (ii) proximity effect, and (iii) spirality effect.

3.6 Derive expressions for the inductance of a conductor due to (i) internal flux and (ii) external flux.

3.7 What is transposition of a transmission line? Why is it necessary? (b) Derive an expression for the inductance of a three-phase fully transposed line.

3.8 What is the advantage of double-circuit three-phase transmission lines? Derive an expression for the inductance of a double-circuit three-phase transmission line. Assume the line to be fully transposed.

3.9 Based on first principle derive an expression for the potential difference between two conductors in a group of conductors.

3.10 Derive expressions for (i) capacitance and (ii) charging current for a single-phase two-conductor transmission line.

3.11 Derive expressions for (i) capacitance and (ii) charging current for a three-phase line with symmetrical spacing.

3.12 Derive expressions for (i) capacitance and (ii) charging current for a three-phase transmission line with unsymmetrical spacing.

3.13 A single conductor has the same current carrying capacity as that of a three-bundle conductor, with horizontal spacing, put together. Between the single conductor and the three bundle conductor arrangement, the performance of which line will be better and why? Explain.

3.14 Explain the effect of earth on the capacitance of a (i) single-phase, and (ii) three-phase line.

Numerical Problems

3.1 A conductor has a resistance of R_1 Ω at t_1°C and is made of copper with a resistance–temperature coefficient α_0 referred to 0°C. Find an expression for the resistance R_2 of the conductor at temperature t_2°C.

3.2 Determine the length l and diameter d of a cylinder of copper in terms of the volume v, resistivity ρ, and the resistance between opposite ends R.

3.3 Explain what is meant by the temperature coefficient of resistance of a material. A copper rod, 0.6 m long and 4 mm in diameter, has a resistance of 825 μΩ at 20°C. Calculate the resistivity of copper at that temperature. If the rod is drawn out into a wire having a uniform diameter of 0.8 mm, calculate the resistance of the wire when its temperature is 60°C. Assume the resistivity to be unchanged and the temperature coefficient of resistance of copper to be 0.00426 per °C.

3.4 A coil of insulated copper wire has a resistance of 160 Ω at 20°C. When the coil is connected to a 240 V supply, the current after several hours is 1.35 A. Calculate the average temperature rise throughout the coil assuming the temperature coefficient of resistance of copper at 20°C to be 0.0039 per °C.

3.5 A seven-strand aluminium conductor is made of identical wires, each of 1.5 mm diameter. If the length of the stranded conductor is 5 km, determine its resistance. Assume a 5% increase in the length of wires due to stranding and take resistivity of aluminium equal to 2.8×10^{-8} Ω-m at 20°C.

3.6 The maximum permissible loss in a 50-km, three-phase transmission line, carrying a current of 125 A per phase, should not exceed 225 kW. Calculate the diameter of the conductor, assuming the resistivity of the conductor equal to 2.8×10^{-8} Ω-m at 20°C.

3.7 Show that the inductance per metre of a coaxial cable which has a thin-walled outer conductor of radius r_1 and a solid inner conductor of radius r_2 is given by

$$L = \frac{\mu_0}{2\pi} \left(\frac{1}{4} + \ln \frac{r_1}{r_2} \right) \text{ H/m, where } \mu_0 \text{ represents the permeability of free space.}$$

Assume that the thickness of the outer conductor is negligible and the current density distribution is uniform over the conductor. If the outer diameter of the conductor is 1.0 cm and the inner conductor diameter is 0.5 cm, compute the inductance of such a 2-km-long co-axial cable.

3.8 A single-phase, 20-km-long transmission line is constituted of solid aluminium conductors spaced 1 m apart and each having a diameter 1.2 cm. Determine the equivalent diameter of a thin-walled, hollow conductor having the same inductance as the line with solid conductors.

3.9 Determine the self-GMD of the configurations of different types of bundled conductors shown in Fig. 3.36. Assume that the mean radius of a conductor is r.

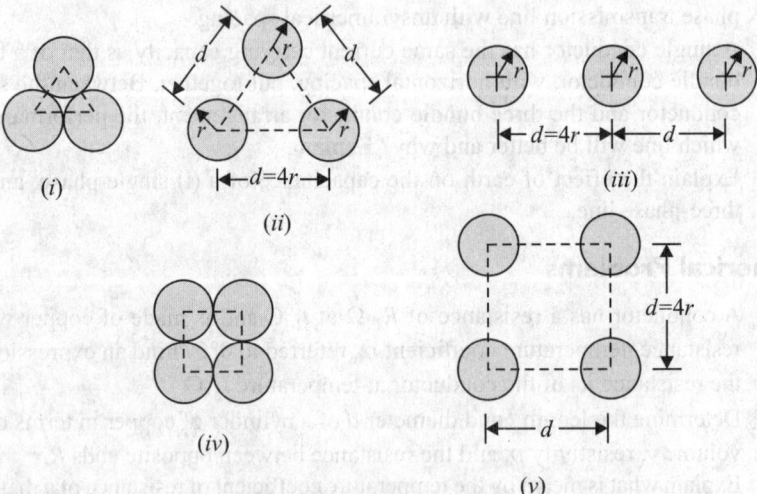

Fig. 3.36 Different arrangements of bundled conductors

3.10 The reactance of a 50-Hz, single-phase, 100-km-long transmission line is 80 Ω. Compute the (i) conductor diameter and (ii) geometric mean radius if the conductor spacing is 1 m.

3.11 A single-phase, double-circuit line is arranged as shown in Fig. 3.37. Derive an expression for the inductance per metre of each conductor. If $r = 0.5$ cm, $H = 1.2$ m, $k = 0.25$, and $f = 50$ Hz, compute the reactance of such a 100-km-long line.

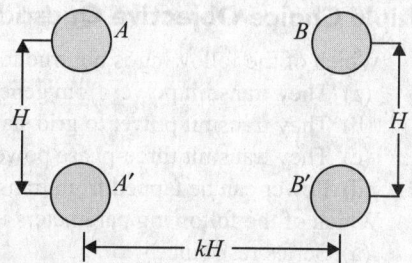

Fig. 3.37 Arrangement of conductors of a single-phase double-circuit line

3.12 The conductors of a 50-Hz, fully transposed line are horizontally arranged with an inter-conductor spacing of 5 m. If the capacitance to neutral per km of such a circuit is 0.010 μF, what will be the capacitance to neutral per km when the conductors are arranged at the corners of an equilateral triangle of sides 5 m? What is the ratio of charging currents per phase in the two cases?

3.13 A three-phase, 400-kV, 50-Hz, fully transposed circuit is constituted of bundle conductors and is arranged as shown in Fig. 3.38. If the outside diameter of each conductor is 2.0 cm, calculate the series resistance, series reactance, and shunt admittance per km when the three conductors share equal currents and equal charge magnitudes. Take the conductor temperature equal to 70°. Assume the following data: resistance at 20°C = 0.22 Ω per km and temperature coefficient of conductor = 0.0043 per °C. Calculate the series resistance, reactance, and shunt admittance if the length of the line is 200 km.

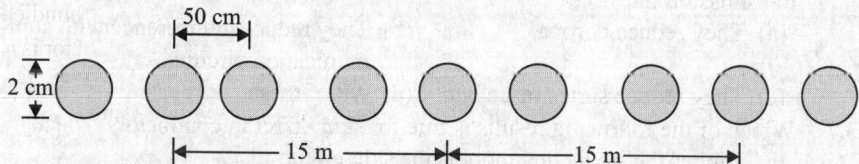

Fig. 3.38 Arrangement of bundled conductors of a three-phase line

3.14 A transmission tower is carrying a single-circuit, three-phase power line (untransposed) and a telephone line. The arrangement of conductors is as shown in Fig. 3.39. Derive an expression for the voltage induced in the telephone circuit due to the currents in the power line. For the arrangement shown in the figure, compute the voltage induced per km length in the telephone line when the power line carries a balanced current of 250 A.

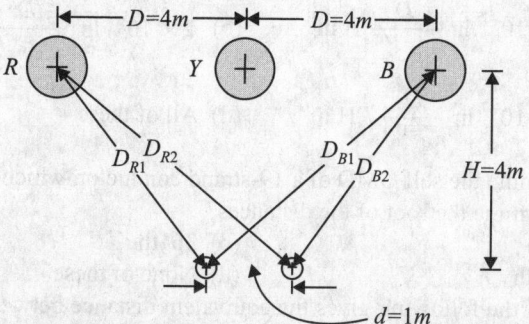

Fig. 3.39 Arrangement of conductors for power and telephone lines

Multiple Choice Objective Questions

3.1   Which of the following is not true for transmission lines?
   (a) They transmit power from generating stations to grid stations.
   (b) They transmit power to grid stations.
   (c) They transmit three-phase power.
   (d) Power can be tapped from transmission lines.

3.2   Which of the following parameters is of least importance?
   (a) Series resistance   (b) Series inductance
   (c) Shunt capacitance   (d) Shunt conductance

3.3   Which of the following is not an advantage when copper is used as a transmission line conductor?
   (a) Low specific resistivity
   (b) High tensile strength
   (c) High voltage gradient due to smaller diameter
   (d) Low wind loads due to smaller surface area

3.4   Which of the following is abbreviated as ACSR?
   (a) Aluminium conductor steel reinforced
   (b) Aluminium core strength reinforced
   (c) Aluminium centre steel reinforced
   (d) Aluminium conductor steel recovered

3.5   Bundled conductors are not used for which of the following operating voltages?
   (a) 132 kV   (b) 220 kV
   (c) 400 kV   (d) 765 and above

3.6   Which of the following is an advantage of using bundled conductors in transmission lines?
   (a) They reduce corona.   (b) They reduce interference with com-munication circuits.
   (c) They reduce series reactance.   (d) All of these.

3.7   Which of the following results is true for skin effect in conductors?
   (a) Higher the frequency more is the skin effect.
   (b) Larger the diameter higher is the skin effect.
   (c) Solid conductors have a higher skin effect than stranded conductors.
   (d) All of these.

3.8   The internal inductance of a conductor is represented by
   (a) 0.5×10^{-7} H/m   (b) $0.5 \times 10^{-7} \times r$ H/m
   (c) $0.5 \times 10^{-7} \times I$ H/m   (d) $0.5 \times 10^{-7} \times I \times r$ H/m

3.9   Which of the following is not the inductance of a single-phase, two-conductor circuit?

   (a) $2 \times 10^{-7} \ln \dfrac{D}{r_i e^{-1/4}}$ H/m   (b) $2 \times 10^{-7} \ln \dfrac{D}{r' e^{-1/4}}$ H/m

   (c) $2 \times 10^{-7} \ln \dfrac{D}{r_i e^{-1/4}} I$ H/m   (d) All of these

3.10 To determine the self GMD of a 19-strand conductor, which of the following will determine the root of the distances?
   (a) 19th   (b) 361th
   (c) 400th   (d) None of these

3.11 Which of the following gives the equivalent distance between conductors of a three-phase circuit?

(a) $D_{eq} = \sqrt[3]{D_{RY}D_{YB}D_{BR}}$ (b) $D_{eq} = \sqrt{D_{RY}D_{YB}D_{BR}}$

(c) $D_{eq} = D_{RY}D_{YB}D_{BR}$ (d) None of these

3.12 Each strand, in a 19-strand conductor, is of equal diameter and has an inductance of L H/m. The total inductance of the stranded conductor is represented by

     (a) 19L   (b) L/19

     (c) L/361   (d) None of these

3.13 Which of the following statements regarding bundled conductors is true?

     (a) They increase both corona loss and reactance.

     (b) They increase corona loss and decreases reactance.

     (c) They decrease corona loss and increases reactance.

     (d) They decrease both corona loss and reactance.

3.14 A 1 km-long, single-phase line employs conductors of 1 cm radius. The line-to-neutral capacitance is given by

     (a) $2 \times \pi \times 10^{-7}$ F/m   (b) $2 \times \pi^2 \times 10^{-7}$ F/m

     (c) $10^{-9}/36\pi$ F/m   (d) $10^{-9}/72$ F/m

3.15 Which of the following is a reason to justify the assumption that the charge on conductors is uniformly distributed?

     (a) Transmission voltage is high.

     (b) Conductors are stranded.

     (c) The distance between conductors is large compared to conductor radii.

     (d) All of these.

3.16 The condition $q_R + q_Y + q_B = 0$ is not applicable to which of the following circuits?

     (a) Two-conductor single-phase ac circuit

     (b) Three-phase ac circuit with symmetrical spacing

     (c) Three-phase ac circuit with unsymmetrical spacing

     (d) All of these

3.17 When the effect of earth on the capacitance of a transmission line is considered, the capacitance

     (a) remains unchanged   (b) reduces

     (c) increases   (d) None of these

3.18 The effect of earth on the capacitance of a line can be neglected when

     (a) conductor diameter is small compared to its height above ground.

     (b) bundled conductors are used.

     (c) height of conductors above ground is large compared to the spacing between conductors.

     (d) All of these.

Answers

3.1 (d)	**3.2** (d)	**3.3** (c)	**3.4** (a)	**3.5** (a)	**3.6** (d)
3.7 (d)	**3.8** (a)	**3.9** (c)	**3.10** (b)	**3.11** (a)	**3.12** (b)
3.13 (d)	**3.14** (d)	**3.15** (c)	**3.16** (a)	**3.17** (c)	**3.18** (c)

CHAPTER 4

Transmission Line Model and Performance

Learning Outcomes

A concentrated study of this chapter will enable the reader to:
- Grasp the functions of a transmission line and understand the meaning of power transmission and regulation
- Classify transmission lines into short, medium, and long lines based on the length of the line
- Develop long transmission line model and determine the nature of voltage and current at any point on the line
- Understand how voltage and current behave like travelling waves and power is transmitted over the lines at the speed of light
- Develop and analyse two-port network representation, lumped parametric equivalent circuit, and approximate models of transmission lines
- Understand the significance of surge impedance loading and Ferranti effect
- Determine complex power transfer, steady-state stability limit, in terms of the parameters and voltages at the two ends of a transmission line
- Comprehend the necessity of reactive line compensation
- Understand the phenomenon of transmission line transients and therefrom be able to explain voltage build-up at a transition point by drawing its Bewley lattice diagram for various types of loads

OVERVIEW

In Chapter 3, distributed parameters like resistance, inductance, and capacitance of power transmission lines, on per-phase basis, were derived and obtained per metre length of the line. It appears to be reasonable because if one is interested in the performance of a specific line of given length, then these parameters have to be multiplied by the actual length of the line in order to obtain the total parameter values of the line. This can be done for short lines. However, for longer lines the accuracy of this procedure is impaired, because the distributed effect of the parameters is neglected. The long line theory incorporates this distribution effect in the exact transmission line model. Short and medium line approximations of the transmission line model are obtained by lumping of distributed parameters. These models are used for the computation of voltages, currents, and power at the sending and receiving ends of a transmission line and the study of the performance of the line.

The ultra-fast transient overvoltages which occur in a power system are either of external origin, for example, lightning discharge, or generated internally by switching operations. Lightning is always a potential hazard to power system equipment, but switching operations can also cause equipment damage. These transients essentially involve only the transmission lines, and the associated voltage and current behave like travelling waves and power is transmitted over the lines at the speed of light.

4.1 REPRESENTATION OF TRANSMISSION LINES

The transmission line parameters—resistance, inductance, and capacitance—were derived as per unit length values in Chapter 3. These parameters are not lumped but are uniformly distributed along the length of the line, and the general equations relating voltage and current of a transmission line are formulated accordingly. However, the use of lumped parameters gives good accuracy for short lines of lengths less than 80 km and medium lines of lengths roughly between 80 km and 240 km.

For short length lines, the value of shunt capacitance is very small and it can be neglected without loss of accuracy. For short lines, only the series resistance and series inductance of the total length of the line are considered.

A medium length line can be represented well by the series resistance and series inductance of the total length of the line as lumped parameters and half the total capacitance of the line lumped at each end of the line.

Normally transmission lines are operated with balanced three-phase loads. Even when the lines are not transposed, the dissymmetry in line parameters is very small and the phases may be considered to be balanced. As the transmission line is considered to be in the sinusoidal steady state, the phasors and impedances are used in the analysis. In order to differentiate the series impedance/shunt admittance per unit length and the total series impedance/shunt admittance of the line, the following terminology is used:

$z = r + j\omega L$ = series impedance per unit length per phase

$y = G + j\omega C$ = shunt admittance per unit length per phase to neutral

l = length of line

$Z = zl$ = total series impedance per phase

$Y = yl$ = total shunt admittance per phase to neutral

For power transmission lines, the shunt conductance G is usually neglected.

4.2 LONG TRANSMISSION LINE MODEL

Figure 4.1 shows one phase and the neutral connection of a three-phase line. An elemental strip of length Δx, of a transmission line, at a distance x from the receiving end is shown in the figure. The length of the transmission line is l metres and z and y are respectively the series impedance and shunt admittance per unit length of the line.

If $I(x)$ is the current flowing through the elemental length Δx, then by Ohm's law the voltage at $(x + \Delta x)$ from the sending end may be written as

$$V(x + \Delta x) = V(x) + z\Delta x I(x) \qquad (4.1)$$

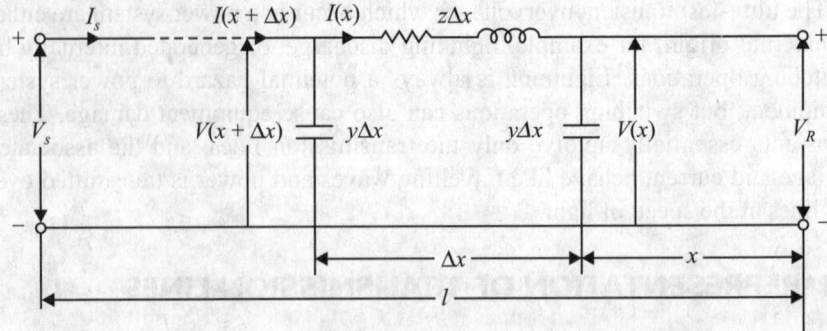

Fig. 4.1 Elemental length Δx of a transmission line with distributed parameters

Similarly, applying Kirchoff's current law to the junction of the elemental strip, the current at $(x + \Delta x)$ from the receiving end may be obtained as

$$I(x + \Delta x) = I(x) + y\Delta x V(x + \Delta x) \tag{4.2}$$

Equations (4.1) and (4.2) may be rewritten as

$$\frac{V(x + \Delta x) - V(x)}{\Delta x} = zI(x) \tag{4.3}$$

$$\frac{I(x + \Delta x) - I(x)}{\Delta x} = yV(x) \tag{4.4}$$

Taking the limit as $\Delta x \to 0$, Eqs (4.3) and (4.4) can be modified as

$$\frac{dV(x)}{dx} = zI(x) \tag{4.5}$$

$$\frac{dI(x)}{dx} = yV(x) \tag{4.6}$$

Equations (4.5) and (4.6) are two linear first-order homogeneous differential equations with two unknowns $V(x)$ and $I(x)$. Differentiating Eq. (4.5) with respect to x and substituting for $\frac{dI(x)}{dx}$ from Eq. (4.6), $I(x)$ may be eliminated as follows:

$$\frac{d^2V(x)}{dx^2} = z\frac{dI(x)}{dx} = yz\, V(x)$$

or

$$\frac{d^2V(x)}{dx^2} - yz\, V(x) = 0$$

or

$$\frac{d^2V(x)}{dx^2} - \gamma^2 V(x) = 0 \tag{4.7}$$

where $\gamma = \sqrt{yz}$ = the propagation constant $\tag{4.8}$

Equation (4.7) is a linear, second-order, homogeneous differential equation with one unknown, $V(x)$.

Similarly, differentiating Eq. (4.6) with respect to x and substituting for $\frac{dV(x)}{dx}$ from Eq. (4.5), $V(x)$ may be eliminated as follows:

$$\frac{d^2 I(x)}{dx^2} = y \frac{dV(x)}{dx} = yzI(x)$$

or $\quad \dfrac{d^2 I(x)}{dx^2} - yzI(x) = 0$

or $\quad \dfrac{d^2 I(x)}{dx^2} - \gamma^2 I(x) = 0 \qquad\qquad\qquad\qquad (4.9)$

Significance of propagation constant Substituting the defined values for series impedance per unit length per phase of $z = r + j\omega L$ and shunt admittance per unit length per phase to neutral of $y = G + j\omega C$ in Eq. (4.8), the expression for the propagation constant becomes

$$\gamma = \sqrt{zy} = \sqrt{(r + j\omega L)(G + j\omega C)} = \alpha + j\beta$$

It may be noted from the above equation that for finite values of series resistance and shunt conductance, the propagation constant γ is a complex number. The unit of γ is per metre or $(\text{metre})^{-1}$. The real part α is called the 'attenuation constant' and the imaginary component β is called the 'phase constant'.

However, in a lossless line both resistance and conductance are zero. Thus, $z = j\omega L$ and $y = j\omega C$, and the propagation constant γ becomes a purely imaginary quantity as shown below

$$\gamma = \sqrt{(j\omega L)(j\omega C)} = j\omega\sqrt{LC} = j\beta$$

From the above, it can be concluded that both Eqs (4.7) and (4.9) have complex solutions.

Substituting $p = d/dx$ in Eqs (4.7) and (4.9) yields the following:

$$\left(p^2 - \gamma^2\right) V(x) = 0; \qquad \left(p^2 - \gamma^2\right) I(x) = 0$$

From the above, it can be seen that the second-degree linear differential Eqs (4.7) and (4.9) will have the same form of solution. The characteristic equation is

$$\left(p^2 - \gamma^2\right) = 0$$

and the roots of the characteristic equation are

$$p_{1,2} = \pm\gamma$$

Thus, the solution of Eq. (4.7) is written as

$$V(x) = k_1 e^{\gamma x} + k_2 e^{-\gamma x} \qquad\qquad\qquad\qquad (4.10)$$

$$= (k_1 + k_2)\frac{e^{\gamma x} + e^{-\gamma x}}{2} + (k_1 - k_2)\frac{e^{\gamma x} - e^{-\gamma x}}{2}$$

$$= K_1 \cosh\gamma x + K_2 \sinh\gamma x \qquad\qquad\qquad (4.11)$$

where $K_1 = k_1 + k_2$ and $K_2 = k_1 - k_2$.

Similarly the general solution of Eq. (4.9) may be written as

$$I(x) = K_3\cosh\gamma x + K_4\sinh\gamma x \qquad\qquad\qquad (4.12)$$

In Eqs (4.11) and (4.12), K_1, K_2, K_3, and K_4 are constants and are determined from the boundary conditions.

To find the constants K_1 and K_2 the boundary conditions at the receiving end of the line can be applied; namely when $x = 0$, $V(0) = V_R$, and $I(0) = I_R$. Then, substitution of these conditions in Eq. (4.11) gives

$$K_1 = V_R \tag{4.13}$$

Further, at $x = 0$, Eq. (4.5) may be written as

$$\frac{dV(0)}{dx} = zI(0) = zI_R \tag{4.14}$$

Differentiating Eq. (4.11), with respect to x, gives

$$\frac{dV(x)}{dx} = -K_1\gamma\sinh\gamma x + K_2\gamma\cosh\gamma x \tag{4.15}$$

At $x = 0$, Eq. (4.15) becomes

$$\frac{dV(0)}{dx} = K_2\gamma \tag{4.16}$$

From Eqs (4.14) and (4.16), using $\gamma = \sqrt{yz}$, the following result is obtained

$$K_2 = \frac{z}{\gamma} I_R = \frac{z}{\sqrt{zy}} I_R = \sqrt{\frac{z}{y}} I_R = Z_c I_R \tag{4.17}$$

Substituting the values of the constants K_1 and K_2 in Eq. (4.11) yields

$$V(x) = V_R \cosh\gamma x + Z_c I_R \sinh\gamma x \tag{4.18}$$

where Z_c is the characteristic impedance or surge impedance of the line and is given by

$$Z_c = \sqrt{\frac{z}{y}} \tag{4.19}$$

For a lossless line, $r = 0$, $z = j\omega L$, and $y = j\omega C$, the characteristic impedance is then given by

$$Z_c = \sqrt{\frac{j\omega L}{j\omega C}} = \sqrt{\frac{L}{C}} \ \Omega \tag{4.20}$$

In a similar manner by applying the boundary condition at $x = 0$ to Eq. (4.12), the values of the constants K_3 and K_4 are obtained as

$$K_3 = I_R \text{ and } K_4 = \frac{V_R}{Z_c} \tag{4.21}$$

and Eq. (4.12) can be written as

$$I(x) = I_R \cosh\gamma x + \frac{V_R}{Z_c} \sinh\gamma x \tag{4.22}$$

Equations (4.18) and (4.22) provide expressions for per-phase voltage and current at a point x from the receiving end. Substituting $x = l$ in these two equations, the sending end voltage and current can be computed as

$$V_S = V_R \cosh\gamma l + Z_c I_R \sinh\gamma l = \cosh\gamma l \ V_R + Z_c \sinh\gamma l \ I_R \tag{4.23}$$

$$I_S = I_R \cosh\gamma l + \frac{V_R}{Z_c} \sinh\gamma l = \frac{1}{Z_c} \sinh\gamma l \ V_R + \cosh\gamma l \ I_R \tag{4.24}$$

In matrix form, Eqs (4.23) and (4.24) can be written as

$$\begin{bmatrix} V_S \\ I_S \end{bmatrix} = \begin{bmatrix} \cosh \gamma l & Z_c \sinh \gamma l \\ \dfrac{1}{Z_c} \sinh \gamma l & \cosh \gamma l \end{bmatrix} \begin{bmatrix} V_R \\ I_R \end{bmatrix} \tag{4.25}$$

Equation (4.25) can be used for calculating the sending end per-phase voltage and current in terms of the receiving end per-phase voltage and current. By following a similar procedure, it is possible to derive inverse expressions for computing receiving end per-phase voltage and current in terms of the sending end per-phase voltage and current. However, since supply of complex power and voltage is required to be maintained at the receiving end, from a power controller's perspective, it is important to know what quantum of complex power and voltage should be maintained at the sending end so that the load and voltage demand is met at the receiving end.

Example 4.1   A 132-kV, three-phase, 50-Hz transmission line has the following parameters: $r = 0.11 \ \Omega/\text{km}$ $L = 1.5 \ \text{mH/km}$ $C = 0.01 \ \mu\text{F/km}$
The transmission line is 150 km long and is supplying a lagging load of 50 MW (0.8 power factor) at 125 kV. Determine (i) sending end voltage, (ii) sending end current and efficiency of the line. Neglect shunt conductance of the line.

Solution $z = 0.11 + j\,(2\pi \times 50) \times 1.5 \times 10^{-3} = (0.11 + j0.471) \ \Omega/\text{km}$

$$= 0.4837 \ \angle 76.86° \ \Omega/\text{km}$$

$$y = j\,(2\pi \times 50) \times 0.01 \times 10^{-6} = 3.14 \times 10^{-6} \ \angle 90° \ \text{S/km}$$

$$Z_c = \sqrt{\frac{z}{y}} = \sqrt{\frac{(0.11 + j0.471)}{j3.14 \times 10^{-6}}} = 389.9 - j44.925 = 392.4748 \ \angle -6.57° \ \Omega$$

$$\gamma = \sqrt{yz} = \sqrt{j3.14 \times 10^{-6} \times (0.11 + j0.471)} = 0.0001 + j0.0012$$

$$= 0.0012 \ \angle 83.43°$$

$$\gamma l = 150\,(0.0001 + j0.0012) = 0.0150 + j0.1800 = 0.1806 \angle 85.24°$$

Receiving end voltage $V_R = \dfrac{125{,}000}{\sqrt{3}} \ \angle 0° = 72128.8 \angle 0° \text{V}$

Receiving end current $I_R = \dfrac{50 \times 10^3}{\sqrt{3} \times 125 \times 0.8} \ \angle -36.87° = 288.68 \ \angle -36.87°$

$$= 230.94 - j173.21 \ \text{A}$$

Now, $e^{\gamma l} = e^{(\alpha l + j\beta l)} = e^{\alpha l} e^{j\beta l} = e^{\alpha l} \angle \beta l$

$$e^{-\gamma l} = e^{-(\alpha l + j\beta l)} = e^{-\alpha l} e^{-j\beta l} = e^{-\alpha l} \angle -\beta l$$

$$\sinh \gamma l = \frac{1}{2}(e^{\alpha l} \angle \beta l - e^{-\alpha l} \angle -\beta l) = \frac{1}{2}(e^{0.015} \angle 0.18° - e^{-0.015} \angle -0.18°)$$

$$= \frac{1}{2}(1.0151 \angle 0.18° - 0.9851 \angle -0.18°)$$

$$= \frac{1}{2}[1.0151(0.9838 + j0.1790) - 0.9851(0.9838 - j017.90)]$$

$$= 0.0148 + j0.179 = 0.1796 \angle 85.27°$$

$$\cosh \gamma l = \frac{1}{2}(e^{\alpha l} \angle \beta l + e^{-\alpha l} \angle - \beta l) = \frac{1}{2}(e^{0.015} \angle 0.18° + e^{-0.015} \angle -0.18°)$$

$$= \frac{1}{2}(1.0151 \angle 0.18° + 0.9851 \angle -0.18°)$$

$$= \frac{1}{2}[1.0151(0.9838 + j0.1790) + 0.9851(0.9838 - j017.90)]$$

$$= 0.9838 + j0.0027 = 0.9838 \angle 0.22°$$

Substituting in Eqs (4.23) and (4.24) yields

$$V_S = 72128.8 \angle 0° \times 0.9838 \angle 0.22° + 392.4748 \angle -6.57° \times 0.1796 \angle 85.27° \times$$
$$288.68 \angle -36.87°$$

$$= 72128.8 \times (0.9838 + j0.0027) + 20348.6127 \angle -41.83°$$

$$= 86162 + j13766.6 = 87254.85 \angle 9.08° \text{ V}$$

$$I_S = \frac{72128.8 \times 0.1796 \angle 85.27°}{392.4748 \angle -6.57} + 0.9838 \angle 0.22° \times 288.68 \angle -36.87°$$

$$= 33.025 \angle 91.84 + 284 \angle -36.6°$$

$$= 226.9397 - j136.32 = 264.735 \angle -30.99°\text{A}$$

Sending end power $S_S = V_S I_S^* = 87254.85 \angle 9.08° \times 264.735 \angle 30.99°$

$$= 23.0994 \times 10^6 \angle 40.07° = (17.677 + j14.87) \times 10^6$$

Sending end real power per-phase $P_S = 17.677$ MW

Efficiency $\eta = \dfrac{50}{3 \times 17.677} \times 100 = 94.286\%$

4.3 NATURE OF VOLTAGE AND CURRENT WAVES

Equation (4.10) provides an expression for the rms or effective value of the voltage at a point which is at a distance x from the receiving end. Substituting $\gamma = \alpha + j\beta$ in the equation, the expression for voltage becomes

$$V(x) = k_1 e^{(\alpha + j\beta)x} + k_2 e^{-(\alpha + j\beta)x}$$

$$= k_1 e^{\alpha x} e^{j\beta x} + k_2 e^{-\alpha x} e^{-j\beta x} \tag{4.26}$$

It can be observed from Eq. (4.26) that the first term $k_1 e^{\alpha x} e^{j\beta x}$ increases in magnitude and advances in phase as the distance x from the receiving end increases. Conversely, as progress along the line from the sending end towards the receiving end is considered, the term decreases in magnitude and is retarded in phase. This is the characteristic of a travelling wave. This wave is similar to the behaviour of a wave in water, which varies in magnitude with time at any point, while its phase is retarded and its maximum value diminishes with distance from the origin. The first term is called the incident voltage wave.

The second term $k_2 e^{-\alpha x} e^{-j\beta x}$ decreases in magnitude and is retarded in phase as progress from the receiving end to the sending end is considered. This term is called the reflected voltage wave. At any point along the line, the voltage is the sum of two components—the incident voltage and the reflected voltage at that point.

Since the equation for current [Eq. (4.12)] is similar to the equation for voltage, current may be considered to be composed of incident and reflected currents. The instantaneous value of the voltage as a function of time t and distance x is obtained by transforming Eq. (4.26) from phasor domain to time domain by multiplying it with $e^{j\omega t}$ and taking the real part. The instantaneous voltage as a function of t and x thus becomes

$$v(t, x) = \sqrt{2}\,\mathrm{Re}\left[k_1 e^{\alpha x} e^{j(\omega t + \beta x)}\right] + \sqrt{2}\,\mathrm{Re}\left[k_2 e^{-\alpha x} e^{j(\omega t - \beta x)}\right] \qquad (4.27)$$

or $\qquad v(t, x) = v_1(t, x) + v_2(t, x) \qquad\qquad\qquad\qquad\qquad (4.28)$

where

$$v_1(t, x) = \sqrt{2}\,\mathrm{Re}\left[k_1 e^{\alpha x} e^{j(\omega t + \beta x)}\right] = \sqrt{2}k_1 e^{\alpha x}\cos(\omega t + \beta x) \qquad (4.29)$$

$$v_2(t, x) = \sqrt{2}\,\mathrm{Re}\left[k_2 e^{-\alpha x} e^{j(\omega t - \beta x)}\right] = \sqrt{2}k_2 e^{-\alpha x}\cos(\omega t - \beta x) \qquad (4.30)$$

and Re means the real part of the expression within the parenthesis.

Considering Eq. (4.30), neglecting α for a moment, it may be noted that for a fixed value of x, $v_2(t, x)$ varies sinusoidally with respect to time; and, for a fixed time t, $v_2(t, x)$ varies sinusoidally with respect to x. If the instantaneous value of voltage wave $v_2(t, x)$ is observed on the transmission line, with increasing time t at points x, which increase as per the relation

$$\omega t - \beta x = \text{constant} \qquad\qquad\qquad\qquad\qquad\qquad (4.31)$$

then, $v_2(t, x)$ is constant. This means that with varying time t and increasing x, the observation is monitored for a fixed point on the voltage wave $v_2(t, x)$, which is travelling from the receiving end of the line to the sending end. The velocity $\mathcal{V}$ of the wave $v_2(t, x)$ may be obtained by differentiating Eq. (4.31) with respect to t as follows:

$$\mathcal{V} = \frac{dx}{dt} = \frac{\omega}{\beta} = \frac{2\pi f}{\beta} = \frac{2\pi f}{\mathrm{Im}\left(\sqrt{zy}\right)} \qquad\qquad\qquad (4.32)$$

where f is the frequency in Hertz, and Im means the imaginary part of the quantity within the parenthesis.

Now time for a phase change of 2π radians is $1/f$ seconds. During this time the wave travels a distance equal to the wavelength λ. The wavelength λ or distance x on the wave which results in a phase shift of 2π radians is

$$\beta\lambda = 2\pi$$

or $\qquad \lambda = \dfrac{2\pi}{\beta} \qquad\qquad\qquad\qquad\qquad\qquad\qquad (4.33)$

For lossless lines, since resistance $R = 0$, conductance $G = 0$, $z = j\omega L$, $y = j\omega C$, $\gamma = \sqrt{zy} = j\beta = j\omega\sqrt{LC}$, and attenuation constant $\alpha = 0$, thus the phase constant becomes

$$\beta = \omega\sqrt{LC} \qquad\qquad\qquad\qquad\qquad\qquad\qquad (4.34)$$

Substituting Eq. (4.34) in Eqs (4.32) and (4.33), the velocity of propagation $\mathcal{V}$ and the wavelength λ can be written as

$$\mathcal{V} = \frac{1}{\sqrt{LC}} \qquad\qquad\qquad\qquad\qquad\qquad\qquad (4.35)$$

$$\lambda = \frac{1}{f\sqrt{LC}} \tag{4.36}$$

In Chapter 3, the expressions for inductance per unit length L and capacitance per unit length C of a transmission line were derived and given by Eqs (3.53) and (3.97) respectively. When the internal flux linkage of a conductor is neglected then $DS \approx r$. Substitution of Eqs (3.53) and (3.97) in Eqs (4.35) and (4.36), respectively gives

$$\mathcal{V} \approx \frac{1}{\sqrt{\mu_0 \varepsilon_0}} \tag{4.37}$$

$$\lambda \approx \frac{1}{f\sqrt{\mu_0 \varepsilon_0}} \tag{4.38}$$

Substituting values of $\mu_0 = 4\pi \times 10^{-7}$ and $\varepsilon_0 = 8.85 \times 10^{-12}$, the velocity of propagation of current and voltage waves along a lossless line is equal to 3×10^8 m/s, which is the velocity of light. For a power frequency of 50 Hz, the wavelength is

$$\lambda = \frac{3 \times 10^8}{50} = 6000 \text{ km}$$

In practice the length of transmission lines is only a fraction of the wavelength at 50 Hz.

The phenomenon of travelling waves is of significance when transmission lines are subjected to electromagnetic transients, which occur due to lightning strokes and switching. To study electromagnetic transients, it is required to study the mathematical simulation of transmission lines which lead to partial differential equations. The solutions of these equations are intricate, and hence will be treated later in this chapter in Section 4.12.

4.4 TWO-PORT NETWORK REPRESENTATIONS OF TRANSMISSION LINES

It is convenient to represent a transmission line by a two-port network. Figure 4.2 shows a two-port network representation of a transmission line. In the figure, the sending end per-phase voltage and current are respectively denoted by V_S and I_S. Similarly the receiving end per-phase voltage and current are represented by V_R and I_R, respectively. A, B, C, and D are complex numbers which depend on the transmission line parameters Z and Y.

The sending end voltage and current are related to the receiving end voltage and current through the $ABCD$ constants of the two-port network as follows

$$V_S = AV_R + BI_R \text{ V} \tag{4.39}$$

$$I_S = CV_R + DI_R \text{ A} \tag{4.40}$$

In order that the sending voltage and current, respectively, have the units of volts and amperes in Eqs (4.39) and (4.40), A and D should be dimensionless while B should have the unit of ohm and C the unit of siemen. In matrix form, Eqs (4.39) and (4.40) can be written as

$$\begin{bmatrix} V_S \\ I_S \end{bmatrix} = \begin{bmatrix} A & B \\ C & D \end{bmatrix} \begin{bmatrix} V_R \\ I_R \end{bmatrix} = [T] \begin{bmatrix} V_R \\ I_R \end{bmatrix} \tag{4.41}$$

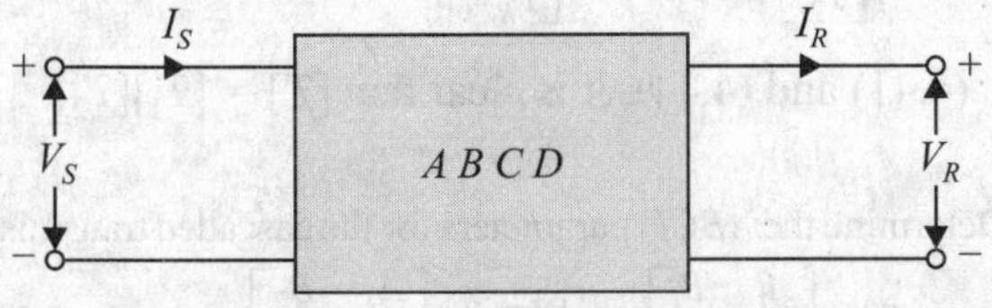

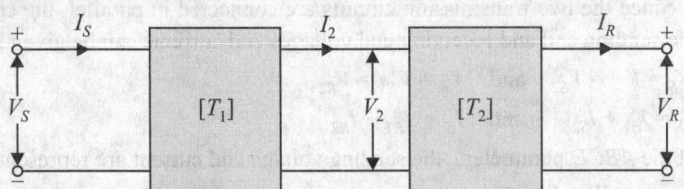

Fig. 4.2 Two-port network representation of a transmission line

Comparing Eq. (4.25) with Eq. (4.41), *ABCD* parameters may be obtained as

$A = D = \cosh \gamma l$ per unit

$B = Z_c \sinh \gamma l \ \Omega$ $\qquad\qquad\qquad\qquad\qquad\qquad\qquad$ (4.42)

$C = \dfrac{1}{Z_c} \sinh \gamma l$ S

From the foregoing, it may be observed that the terms A, B, C, and D will be complex if γ is complex. The relation $AD - BC = \cosh^2 \gamma l - \sinh^2 \gamma l = 1$. As per network theory, *ABCD* parameters apply to linear, time-invariant, passive, bilateral two-port networks and the relation, $AD - BC = 1$, always holds good.

A, B, C, and D are called transmission parameters and are represented by a matrix $[T]$ called the transmission matrix. It may be noted that A and D are dimensionless, B has the unit of ohm, and C has the unit of siemen. Since det $|T| = AD - BC = 1$, the inverse of matrix $[T]$ exists and is given by

$$[T]^{-1} = \begin{bmatrix} D & -B \\ -C & A \end{bmatrix} \tag{4.43}$$

Thus, the receiving end voltage and current can be expressed in terms of sending end voltage and current as

$$\begin{bmatrix} V_R \\ I_R \end{bmatrix} = [T]^{-1} \begin{bmatrix} V_S \\ I_S \end{bmatrix} = \begin{bmatrix} D & -B \\ -C & A \end{bmatrix} \begin{bmatrix} V_S \\ I_S \end{bmatrix}$$

The advantage of transmission matrix description is that the $[T]$ matrix for cascaded two-port networks is the product of individual $[T]$ matrices. Figure 4.3 shows two transmission matrices $[T_1]$ and $[T_2]$ connected in series.

Fig. 4.3 Cascading of transmission matrices

By inspection, the relation between V_2 and I_2 can be expressed as

$$\begin{bmatrix} V_2 \\ I_2 \end{bmatrix} = [T_2] \begin{bmatrix} V_R \\ I_R \end{bmatrix}$$

Similarly, V_S and I_S in terms of V_2 and I_2 is equal to

$$\begin{bmatrix} V_S \\ I_S \end{bmatrix} = [T_1] \begin{bmatrix} V_2 \\ I_2 \end{bmatrix} = [T_1][T_2] \begin{bmatrix} V_R \\ I_R \end{bmatrix} \tag{4.44}$$

Comparing Eqs (4.41) and (4.44), it is clear that $[T] = [T_1][T_2]$.

Example 4.2 Determine the *ABCD* parameters for the cascaded transmission lines shown in Fig. 4.3. Assume $[T_1] = \begin{bmatrix} A_1 & B_1 \\ C_1 & D_1 \end{bmatrix}$ and $[T_2] = \begin{bmatrix} A_2 & B_2 \\ C_2 & D_2 \end{bmatrix}$.

Solution It was shown in the previous section that $[T] = [T_1][T_2]$. Thus, the matrix of the cascaded transmission lines is expressed as

$$[T] = \begin{bmatrix} A_1 & B_1 \\ C_1 & D_1 \end{bmatrix} \times \begin{bmatrix} A_2 & B_2 \\ C_2 & D_2 \end{bmatrix} = \begin{bmatrix} (A_1A_2 + B_1C_2) & (A_1B_2 + B_1D_2) \\ (C_1A_2 + D_1C_2) & (C_1B_2 + D_1D_2) \end{bmatrix}$$

Hence, the parameters of the cascaded transmission lines are:

$$A = A_1A_2 + B_1C_2 \quad B = A_1B_2 + B_1D_2 \quad C = C_1A_2 + D_1C_2 \quad D = C_1B_2 + D_1D_2$$

Example 4.3 Two transmission lines are represented by two-port networks and are connected in parallel as shown in Fig. 4.4. The *ABCD* parameters of each line are shown in the figure. Compute the *ABCD* parameters of the composite line.

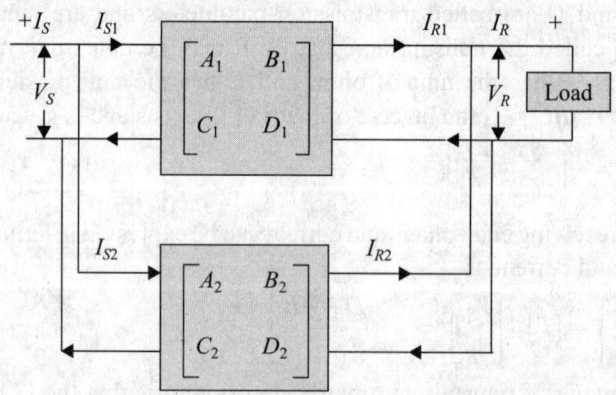

Fig. 4.4

Solution Since the two transmission circuits are connected in parallel, the correlation between the sending end and receiving end voltages and currents can be given by

$$V_S = V_{S1} = V_{S2} \quad \text{and} \quad V_R = V_{R1} = V_{R2}$$
$$I_S = I_{S1} + I_{S2} \quad \text{and} \quad I_R = I_{R1} + I_{R2}$$

In terms of the *ABCD* parameters, the sending voltage and current are represented as

$$V_S = A_1 V_R + B_1 I_{R1} \tag{4.3.1}$$
$$I_{S1} = C_1 V_R + D_1 I_{R1} \tag{4.3.2}$$
$$V_S = A_2 V_R + B_2 I_{R2} \tag{4.3.3}$$
$$I_{S2} = C_2 V_R + D_2 I_{R2} \tag{4.3.4}$$

Multiplying Eqs (4.3.1) and (4.3.3) by B_2 and B_1 respectively and adding the resulting equations leads to

$$(B_2 + B_1)V_S = (A_1B_2 + A_2B_1)V_R + B_1B_2(I_{R1} + I_{R2})$$

or
$$V_S = \frac{(A_1B_2 + A_2B_1)}{(B_2 + B_1)}V_R + \frac{B_1B_2}{(B_2 + B_1)}I_R \tag{4.3.5}$$

By comparing Eq. (4.3.5) with Eq. (4.39) it can be seen that

$$A = \frac{(A_1B_2 + A_2B_1)}{(B_2 + B_1)} \quad \text{and} \quad B = \frac{B_1B_2}{(B_2 + B_1)}$$

Remembering the transmission line is symmetrical $A = D$. Thus by comparing Eq. (4.3.5) with Eq. (4.41) it is seen that $A = D = \dfrac{(A_1B_2 + A_2B_1)}{(B_2 + B_1)}$ and $B = \dfrac{B_1B_2}{B_2 + B_1}$.

Next, the value of the parameter C is determined by making use of the relation $AD - BC = 1$. Substituting the values of A, B, and D in the relationship $AD - BC = 1$ results in

$$\frac{(A_1B_2 + A_2B_1)^2}{(B_2 + B_1)^2} - C \times \frac{B_1B_2}{B_1 + B_2} = 1$$

Simplification of the above expression leads to

$$C = \frac{(A_1B_2 + A_2B_1 + B_2 + B_1) \times (A_1B_2 + A_2B_1 - B_2 - B_1)}{(B_2 + B_1)B_1B_2} \tag{4.3.6}$$

Addition of Eqs (4.3.2) and (4.3.4) yields

$$I_S = (C_1 + C_2)V_R + D_2I_{R2} + D_1I_{R1} = (C_1 + C_2)V_R + D_2I_R + (D_1 - D_2)I_{R1}$$

Substituting I_{R1} from Eq. (4.3.1) in the above expression gives

$$I_S = (C_1 + C_2)V_R + D_2I_R + \frac{(D_1 - D_2)}{B_1} \times (V_S - A_1V_R)$$

Next, the value of V_S is substituted in the above expression as below.

$$I_S = (C_1 + C_2)V_R + D_2I_R + \frac{(D_1 - D_2)}{B_1} \times \left(\frac{[A_1B_2 + A_2B_1]}{B_2 + B_1}V_R + \frac{B_1B_2}{B_2 + B_1}I_R \right) - \frac{(D_1 - D_2)A_1}{B_1}V_R$$

Combining like terms yields

$$I_S = \left\{ (C_1 + C_2) + \frac{(D_1 - D_2)}{B_1} \times \left(\frac{[A_1B_2 + A_2B_1]}{B_2 + B_1} - A_1 \right) \right\} V_R$$
$$+ \left(D_2 + \frac{(D_1 - D_2)}{B_1} \times \frac{B_1B_2}{B_1 + B_2} \right) I_R$$

$$= \left\{ (C_1 + C_2) + \frac{(D_1 - D_2)(A_2 - A_1)}{B_2 + B_1} \right\} V_R + \frac{(B_1D_2 + B_2D_1)}{B_1 + B_2} I_R \tag{4.3.7}$$

It may be noted that Eq. (4.3.7) has the same form as given in Eq. (4.40). Hence, the parameter C of the composite transmission line is given by

$$C = (C_1 + C_2) + \frac{(D_1 - D_2)(A_2 - A_1)}{B_2 + B_1}$$

Further, it may also be seen from Eq. (4.3.7) that $D = \dfrac{(B_1D_2 + B_2D_1)}{B_1 + B_2} = \dfrac{(B_1A_2 + B_2A_1)}{B_1 + B_2} = A$ as shown earlier.

4.5 LUMPED PARAMETRIC π EQUIVALENT CIRCUIT OF TRANSMISSION LINES

In power system analysis, equivalent circuits of different components such as generators, transformers, and transmission lines are used. It is possible to find an equivalent π circuit model of a transmission line that can replace *ABCD* parameters of the line. In this section, the π equivalent circuit of a long transmission line will be derived.

Figure 4.5 shows the π equivalent circuit of a long transmission line in which parameters are lumped. It is intended to determine the parameters Z' and Y' so that the circuit given in Fig. 4.5 has the same *ABCD* parameters as those given in Eq. (4.42).

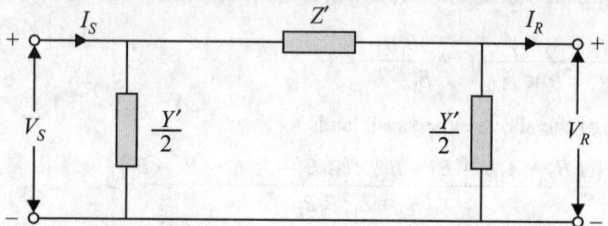

Fig. 4.5 π-equivalent circuit of a transmission line with lumped parameters

The parameters Z and Y represent the total series impedance per phase and shunt admittance per phase to neutral respectively of the line.

$$Z = zl = (r + j\omega L)l = rl + j\omega Ll$$

$$Y = yl = (G + j\omega C)l = Gl + j\omega Cl$$

In the preceding expressions, variables z, y, l, r, L, G, and C have the same meaning as stated in Section 4.1.

By applying Kirchoff's voltage and current laws to the circuit given in Fig. 4.5, the sending end voltage and current equations can be expressed, by inspection, as follows:

$$V_S = V_R + \left(I_R + \frac{Y'}{2}V_R\right)Z' = \left(1 + \frac{Y'Z'}{2}\right)V_R + Z'I_R \tag{4.45}$$

and $\quad I_S = \left(I_R + \frac{Y'}{2}V_R\right) + \frac{Y'}{2}V_S = \left(I_R + \frac{Y'}{2}V_R\right) + \frac{Y'}{2}\left\{\left(1 + \frac{Y'Z'}{2}\right)V_R + Z'I_R\right\}$

or $\quad I_S = Y'\left(1 + \frac{Y'Z'}{4}\right)V_R + \left(1 + \frac{Y'Z'}{2}\right)I_R \tag{4.46}$

Comparing Eqs (4.45) and (4.46) with matrix Eq. (4.41), the values of *ABCD* parameters can now be written as

$$A = D\left(1 + \frac{Y'Z'}{2}\right); B = Z' ; C = Y'\left(1 + \frac{Y'Z'}{4}\right) \tag{4.47}$$

Equating the value of B in Eq. (4.42) with that in Eq. (4.47) gives

$$Z_c \sinh \gamma l = Z' \tag{4.48}$$

Substituting $Z_c = \sqrt{\dfrac{z}{y}}$ in Eq. (4.48) and multiplying the denominator and numerator by γl, yields

$$Z' = \sqrt{\frac{z}{y}} \gamma l \frac{\sinh{(\gamma l)}}{\gamma l}$$

Putting $\gamma = \sqrt{zy}$ in the numerator and simplifying results leads to

$$Z' = zl \frac{\sinh{(\gamma l)}}{\gamma l} = Z \frac{\sinh{(\gamma l)}}{\gamma l} \tag{4.49}$$

Equation (4.49) gives the exact total series impedance of a transmission line. It is to be noted that the series impedance is a product of three factors, namely per unit impedance $(R + j\omega L)$, the length of the line l, and a factor $[(\sinh{(\gamma l)})/(\gamma l)]$. The factor $[(\sinh{(\gamma l)})/(\gamma l)]$ is called the correction factor. However, for transmission lines the absolute value $|\gamma| \ll 1$, so that $[(\sinh{(\gamma l)})/(\gamma l)]$ can be assumed to be approximately equal to unity. Hence, for practical purposes $Z' = Z = zl$.

Similarly, equating the value of parameter A in Eqs (4.42) and (4.47)

$$1 + \frac{Y'Z'}{2} = \cosh{(\gamma l)} \quad \text{or} \quad \frac{Y'}{2} = \frac{\cosh{(\gamma l)} - 1}{Z'}$$

Substituting the value of Z' from Eq. (4.48)

$$\frac{Y'}{2} = \frac{\cosh{(\gamma l)} - 1}{Z_c \sinh{(\gamma l)}} = \frac{1}{Z_c} \tanh\left(\frac{\gamma l}{2}\right) \tag{4.50}$$

$$= \sqrt{\frac{y}{z}} \tanh\left(\frac{\gamma l}{2}\right) = \frac{\gamma l}{2} \times \frac{\tanh\left(\dfrac{\gamma l}{2}\right)}{\left(\dfrac{l}{2} \sqrt{yz}\right)} = \frac{Y}{2} \times \frac{\tanh\left(\dfrac{\gamma l}{2}\right)}{\left(\dfrac{\gamma l}{2}\right)}$$

Observe that the shunt admittance is a product of three factors, that is, per unit admittance $y = (G + j\omega C)$, length l of the line, and a correction factor $[\tanh{(\gamma l/2)}]/(\gamma l/2)$, which is approximately equal to unity.

Thus, alternate expressions for Z_c and γl may be written as

$$Z_c = \sqrt{\frac{z}{y}} = \sqrt{\frac{zl}{yl}} = \sqrt{\frac{Z}{Y}} \tag{4.51}$$

$$\gamma l = \left(\sqrt{zy}\right) l = \sqrt{zlyl} = \sqrt{ZY} \tag{4.52}$$

Example 4.4 A 400-kV, three-phase fully transposed line of length 250 km has the following line constants:

$$z = (0.032 + j0.30) \; \Omega \text{ per km and } y = j3.5 \times 10^{-6} \text{ S per km}$$

Determine the *ABCD* constants of the line assuming distributed parameters.

Solution From the given data, the following constitutes the input data to the MATLAB command window

```
>> z = 0.032+0.30i; y = 3.5i*10^ (-6); l = 250;
Characteristic impedance
>> zc=sqrtm(z/y) = 293.18 - j 15.592 = 293.5993 ∠-3.0443° Ω
```

g = sqrtm(y*z) = 0.0001 + j 0.0010 = 0.0010 $\angle$86.9557°
$\gg$ A=cosh(g*l) = 0.9674 + j 0.0035 = 0.9674 $\angle$0.2050°
$\gg$ B=zc*sinh(g*l) = 7.8259 + j 74.1916 = 74.6032 $\angle$83.9786°Ω
$\gg$ C=sinh(g*l)/zc = -1.0141e-006 + j 8.6546e-004 = 0.000865 $\angle$90.0671° S
$\gg$ D = A

The computation of ABCD constants is verified as follows
A*D-B*C = 1.0000

4.6 APPROXIMATE MODELS FOR TRANSMISSION LINES

Figure 4.5 shows the lumped parameter π-equivalent circuit for a long transmission line and the values of the parameters are given in Eq. (4.47). The sending end voltage and current in terms of the receiving end voltage and current may be computed using Eqs (4.39) and (4.40). However, for medium-length lines and short lines, the equivalent circuit and the equations are greatly simplified. Since transmission lines operate at the power frequency, it was shown in Section 4.5 that the magnitude of $\gamma l \ll 1$, the series impedance and the shunt admittance values for the π-equivalent circuit can be computed without the use of the correction factors, and Z' can be replaced by Z and Y' can be replaced by Y. For short-length lines, Y may be neglected, being small, and the circuit may be further simplified.

Thus, depending upon the length and operating voltage, further approximations can be made in the simulation of a transmission line, without significantly affecting the accuracy of the results. These are discussed in the following sub-sections.

4.6.1 Short-length Line Simulation

In transmission lines, which are less than 80 km in length and operates at voltages below 66 kV, the shunt admittance can be ignored. The equivalent circuit of such a line is shown in Fig. 4.6.

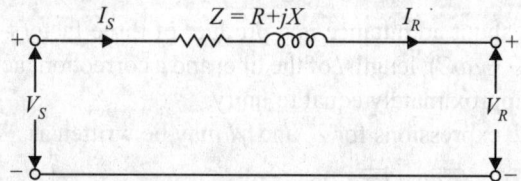

Fig. 4.6 Equivalent circuit of a short transmission line

In the figure, the series impedance

$$Z = zl = (r + j\omega L)l = rl + j\omega Ll = R + jX \tag{4.53}$$

where $R = rl$ = total series resistance of the line and $X = \omega Ll$ = total inductive reactance of the line.

Inspection of the figure yields the following equations:

$$V_S = V_R + ZI_R = V_R + (R + jX)I_R \tag{4.54}$$

$$I_S = I_R = 0 \times V_R + I_R \tag{4.55}$$

In matrix form, Eqs (4.54) and (4.55) may be written as

$$\begin{bmatrix} V_S \\ I_S \end{bmatrix} = \begin{bmatrix} 1 & Z \\ 0 & 1 \end{bmatrix} \begin{bmatrix} V_R \\ I_R \end{bmatrix} \qquad (4.56)$$

A comparison of Eq. (4.56) with Eq. (4.41) gives

$$A = D = 1; \; B = Z = (R + jX); \; C = 0$$

At no-load, $I_R = 0$. From Eq. (4.54) no-load voltage $V_{R(NL)}$ is obtained as

$$V_{R(NL)} = \frac{V_S}{A}$$

Voltage regulation of a line is defined as the ratio of the change in receiving end voltage from no-load to full-load to the receiving end full-load voltage. Mathematically, it is expressed as

$$\% \text{ Regulation} = \frac{\left|V_{R(NL)}\right| - \left|V_{R(FL)}\right|}{\left|V_{R(FL)}\right|} \qquad (4.57)$$

where $\left|V_{R(FL)}\right|$ is the receiving end voltage at full-load.

Voltage regulation of a line is a measure of the voltage drop of the line and depends on the magnitude and power factor of the load current. The effect of varying power factor of the load current is shown in Fig. 4.7. For lagging power factor and unity power factor loads, the voltage regulation has positive value, and for leading power factors, the regulation has negative values. In order to maintain specified voltage at the consumers' terminals, sending end voltage has to be regulated, and permissible voltage regulation of extra high voltage lines is specified at ± 5%.

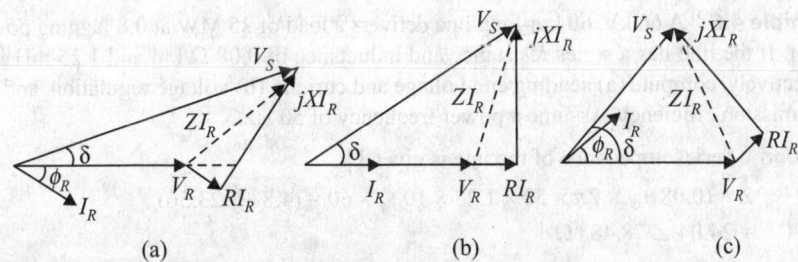

Fig. 4.7 Effect of varying power factor of the load on regulation of the line (a) lagging power factor, (b) unity power factor, and (c) leading power factor

Once the sending voltage V_S and the sending end current I_S are known, the sending end per-phase complex power S_S is given by

$$S_S = V_S I_S^* \qquad (4.58)$$

Sending end real power

$$P_S = \text{Re} \, (V_S I_S^*)$$

Line loss $= P_S - P_R \qquad (4.59)$

Transmission line efficiency

$$\eta = \frac{P_R}{P_S} = \frac{P_S - P_S + P_R}{P_S} = \left(1 - \frac{P_S - P_R}{P_S}\right) = \left(1 - \frac{\text{line loss}}{P_S}\right) \qquad (4.60)$$

Example 4.5 A 33-kV, three-phase transmission line supplies a load of 1000 kW at a lagging power factor of 0.8. If the per-phase resistance and reactance of the line is 20 Ω and 50 Ω, respectively, what will be the required sending end voltage if 33 kV is to be maintained at the receiving end? Also compute (i) sending end real power, (ii) line loss, (iii) efficiency, and (iv) per cent regulation.

Solution Power factor angle $\varphi = \cos^{-1} 0.8 = -36.87°$

Receiving end current $I_R = \dfrac{1000}{\sqrt{3} \times 33 \times 0.8} = 21.8693 \angle -36.87°$ A

If the receiving end per-phase voltage is taken as the reference phasor,

$$V_R = \frac{33}{\sqrt{3}} = 19.0526 \angle 0° \text{ kV}$$

$$I_R = 21.8693 \times (0.8 - j0.6) = (17.4954 - j13.1216) \text{ A}$$

Sending end voltage $V_S = 19.0526 + \dfrac{(20 + j50) \times (17.4954 - j13.1216)}{1000}$

$$= (20.059 + j0.6123) \text{ kV}$$
$$= 20.068 \angle 1.7486° \text{ kV}$$

Sending end line voltage $= (20.068 \angle 1.7846°) \times \sqrt{3} = 34.759 \angle 1.7486°$ kV

Per cent regulation $= \dfrac{34.759 - 33}{33} \times 100 = 5.33$

Line loss $= |I_R^2| \times R = (21.8693)^2 \times 20$ W $= 9.5653$ kW

Sending end power $P_S = 1000 + 9.5653 = 1009.5653$ kW

Efficiency $\eta = \dfrac{1000}{1009.5653} \times 100 = 99.05\%$

Example 4.6 A 66 kV, 60 km-long line delivers a load of 25 MW at 0.8 lagging power factor. If the line has a series resistance and inductance of 0.08 Ω/km and 1.25 mH/km, respectively, compute (a) sending end voltage and current, (b) voltage regulation, and (c) transmission efficiency. Assume a power frequency of 50 Hz.

Solution Series impedance of the line is given by

$$Z = (0.08 + j \times 2\pi \times 50 \times 1.25 \times 10^{-3}) \times 60 = (4.8 + j\,23.56)$$
$$= 24.04 \angle 78.48° \ \Omega$$

Taking the receiving end voltage as the reference: $V_R = \dfrac{66}{\sqrt{3}} = 38.11$ kV/phase

The receiving end current

$$I_R = \frac{25000}{\sqrt{3} \times 66 \times 0.8} \cos^{-1}(-0.8) = \frac{25000}{\sqrt{3} \times 66 \times 0.8}(0.8 - j0.6)$$
$$= (218.69 - j164.02) = 273.37 \angle -36.87° A$$

Since the length of the line is 60 km, it is classified as a short line. Hence the *ABCD* parameters of the line are: $A = D = 1$, $Z = 24.31 \angle 57.30°\Omega$, and $C = 0$. Writing the transmission line equation in the matrix form of Eq. (4.41) yields

$$\begin{bmatrix} V_S \\ I_S \end{bmatrix} = \begin{bmatrix} 1 & (4.8 + j23.56) \\ 0 & 1 \end{bmatrix} \begin{bmatrix} (38.11 + j0) \times 10^3 \\ 218.69 + (-j164.02) \end{bmatrix}$$

$$= \begin{bmatrix} (43.02 + j4.37) \times 10^3 \\ 218.69 - j164.02 \end{bmatrix} = \begin{bmatrix} 43.24 \angle 5.79° \\ 273.37 \angle -36.87° \end{bmatrix} \qquad (4.6.1)$$

(a) From Eq. (4.6.1), it can be seen that the sending end voltage

$$V_S = \sqrt{3} \times 43.24 \angle 5.79° = 74.89 \angle 5.79° \text{ kV}$$

Similarly, the sending end current $I_S = 273.37 \angle{-36.87°}$ which is the same as the receiving end current, thereby verifying the computations.

(b) Voltage regulation as per Eq. (4.57) is equal to $\dfrac{(43.24 - 38.11)}{38.11} \times 100 = 13.46\%$

(c) Referring to the phasor diagram in 4.7(a), phase angle between the sending end voltage and sending end current is given by

$$\varphi_S = (\varphi_R + \delta) = 36.87° + 5.79° = 42.66°.$$

Thus, using phase quantities, the sending end power is computed as follows

$$P_S = 3 \times V_S \times I_S \times \cos\varphi_S = \frac{3 \times 43.24 \times 273.37 \times \cos(42.66°)}{1000} = 26.08 \text{ MW}$$

Therefore, transmission efficiency $\eta = \dfrac{25}{26.08} \times 100 = 95.86\%$.

4.6.2 Medium-length Line Simulation

For lines of length greater than 80 km and less than 250 km, the shunt capacitance must be considered as the line charging current becomes appreciable with increase in line length. Such lines are categorized as medium-length lines. It is common to lump the total shunt capacitance and locate half at each end of the line. Such a circuit, called a nominal π circuit, is shown in Fig. 4.8. In the figure, Z is the total series impedance of the line and Y is the total shunt admittance given by

$$Y = (G + j\omega C)l \text{ S}$$

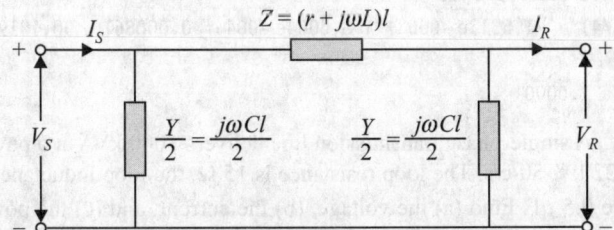

Fig. 4.8 Nominal π-equivalent circuit of a medium-length line

The shunt conductance G is negligible and can be ignored. Therefore,

$$Y = j\omega Cl \text{ S} \tag{4.61}$$

An inspection of the equivalent circuit in Fig. 4.8 yields the following equations for sending end voltage and current

$$V_S = V_R + \left(I_R + \frac{Y}{2}V_R\right)Z = \left(1 + \frac{YZ}{2}\right)V_R + ZI_R \tag{4.62}$$

and $\quad I_S = \left(I_R + \dfrac{Y}{2}V_R\right) + \dfrac{Y}{2}V_S = \left(I_R + \dfrac{Y}{2}V_R\right) + \dfrac{Y}{2}\left\{\left(1 + \dfrac{YZ}{2}\right)V_R + ZI_R\right\}$

or $\quad I_S = Y\left(1 + \dfrac{YZ}{4}\right)V_R + \left(1 + \dfrac{YZ}{2}\right)I_R \tag{4.63}$

Writing Eqs (4.62) and (4.63) in matrix format, gives

$$
\begin{bmatrix} V_S \\ I_S \end{bmatrix} = \begin{bmatrix} \left(1+\dfrac{YZ}{2}\right) & Z \\ Y\left(1+\dfrac{YZ}{4}\right) & \left(1+\dfrac{YZ}{2}\right) \end{bmatrix} \begin{bmatrix} V_R \\ I_R \end{bmatrix}
\tag{4.64}
$$

Comparing Eqs (4.64) and (4.41), the values of *ABCD* parameters can be obtained as

$$
A = D = \left(1+\frac{YZ}{2}\right); \ B = Z; \ C = Y\left(1+\frac{YZ}{4}\right)
\tag{4.65}
$$

Comparison of the nominal π-equivalent circuit given in Fig. 4.8 with the π-equivalent circuit given in Fig. 4.5 shows that in the former case Z and $Y/2$ have been used instead of Z' and $Y'/2$ used in the latter case. Since the equivalent circuit models of both the short- and medium-length lines are passive, linear, bilateral, and time invariant, $AD - BC = 1$.

Example 4.7   Determine the *ABCD* constants for the line in Example 4.4, when the line is simulated by a nominal π-equivalent circuit.

Solution   Using the input data given in Example 4.4,

```
>> Z=z*l = 8.0000 + j 75.0000 = 75.4255 ∠83.9115°
>> Y=y*l = j 0.000875 = 0.000875 ∠90°
```

Using Eq. (4.65), the values of *ABCD* for a nominal π-equivalent circuit are obtained as follows

```
>> A=1+Y*Z/2 = 0.9672 + j 0.0035 = 0.9672 ∠0.2073°
>> B=Z = 75.4255 ∠83.9115° Ω
>> C=Y*(1+Y*Z/4) = −1.5313e−006 + j 8.6064e−004 = 0.000861 ∠90.1019°S
>> D=A
>> A*D − B*C = 1.0000
```

Example 4.8   A single-phase transmission line delivers 1000 kVA at a power factor of 0.71 lagging, 22 kV, 50-c/s. The loop resistance is 15 Ω, the loop inductance 0.2 H, and the capacitance 0.5 μF. Find (a) the voltage, (b) the current, and (c) the power factor at the sending end. Use nominal-π method. (d) If the sending end voltage be maintained unaltered, to what value will the receiving end voltage rise on no-load?

Solution   As described in Section 4.6.2, $Z = zl = (r + j\omega L)l = R + jX_L$ and $Y = yl = (g + j\omega C)l = G + jX_C$, therefore

```
>> Z=15+2*pi*50*0.2i = (15.0000 + 62.8319i) Ω
>> Y=2*pi*50*0.5i*(10^−6) = 0 + 1.5708e−004i S
```

Using Eq. (4.65),

```
>> A=(1+Y*Z/2) = 0.9951 + 0.0012i
>> B=Z=(15.0000 + 62.8319i) Ω
>> C=Y*(1+Y*Z/4) = (−9.2528e−008 + 1.5669e−004i) S
>> D=A = 0.9951 + 0.0012i
```

Receiving end current is

```
>> IR=1000/(22) = 45.4545 A
```

In complex form, receiving end current is

```
>> IR=IR*(0.71-0.7042i) = (32.2727-32.0091i) A
>> VR=22000*(1+0i) = 22000 V
```

By using Eq. (4.64), the sending voltage and current are computed as follows

```
>> VS=[A B;C D]*[VR;IR]
VS = 1.0e+004 * [2.4387 + 0.1574i
                 0.0032 - 0.0028i]
>> abs(VS) = 1.0e+004 * [2.4437
                         0.0043]
>> angle(VS) in degrees = [3.6919
                          -41.4227]
```

From the foregoing, it is observed that the sending end voltage and current respectively are 24.437 kV and 43 A. The power factor at the sending end is $\cos(3.69° + 41.42°) = 0.71$. When the load current at the receiving end is zero, Eq. (4.39) gets modified as

$$V_S = AV_R$$

Therefore, receiving end voltage at no-load with the sending end voltage maintained at 24.437 kV is obtained as

$$V_R = 24.437/\text{abs}(A) = 24.5582 \text{ kV}$$

4.7 MATLAB FUNCTION

The function trmlnper is a generalized MATLAB program for computing the performance of a three-phase transmission line. The function computes the *ABCD* constants, from the line data, and provides the option of modelling with (a) distributed parameters, (b) lumped parameters as (i) π-equivalent circuit, (ii) nominal π-equivalent, and (iii) short line.

```
function [A,B,C,D,Vs] = trmlnper(r,L,g,CC,l,Vr,Pr,pf,pftype);
% Programme to compute the performance of transmission lines
% Developed by the authors
% Part-I: Computation of ABCD constants
Vs=zeros(2,1);
z=r+i*2*pi*50*L;
y=g+i*2*pi*50*CC;
gamma=sqrtm(z*y);
Zc=sqrtm(z/y);
type=input ('type of line');
% 0 for distributed constants, 1 for lumped parameters (pi equivalent)
% 2 for medium lines (nominal pi equivalent), 3 for short line
if type == 0;
    A=cosh(gamma*l);
    B=Zc*sinh(gamma*l);
    C=sinh(gamma*l)/Zc;
    D=A;
else
end
if type == 1;
    Zd=(z*l*sinh(gamma*l))/(gamma*l);
    Yd=(y*l*tanh(qamma*l/2))/(gamma*l/2);
```

```
    A=(1+(Yd*Zd/2));
    B=Zd;
    C=Yd*(1+(Yd*Zd/4));
    D=A;
else
end
if type == 2;
    Z=z*l;
    Y=y*l;
    A=(1+(Y*Z/2));
    B=Z;
    C=Y*(1+(Y*Z/4));
    D=A;
else
end
if type == 3;
    Z=z*l;
    A=1;
    B=Z;
    C=0;
    D=A;
else
end
Ir=Pr/(sqrtm(3)*Vr*pf);
Vr=Vr/sqrtm(3);
phi=acos(pf);
if pftype == 0;
    phi=-phi;
else
end
Ir=Ir*(cos(phi)+i*sin(phi));
%Part-II Computation of sending end voltage and current
[Vs]=[A B;C D]*[Vr;Ir]
VS=abs(Vs(1))
deltaVs=angle(Vs(1))*180/pi
IS=abs(Vs(2))
deltaIs=angle(Vs(2))*180/pi
% Part-II Computation of sending end power, line loss, efficiency
% and regulation
Ps=(Vs(1)*(Vs(2))')
PS=real(Ps)*3
lnlos=PS-Pr
effy=(1-lnlos/PS)*100
reg=(VS-abs(Vr))/(abs(Vr))
VS=sqrtm(3)*VS
```

Details of the input to the function trmlnper are as follows:

r	Per-phase series resistance of the line in Ω per unit length
L	Per-phase inductance of the line in H per unit length
g	Per-phase shunt conductance of the line in mho per unit length
CC	Per-phase shunt capacitance of the line in F per unit length
l	Length of the line
Vr	Line-to-line voltage in kV at the receiving end
Pr	Three-phase receiving end power in MW

pf   Power factor angle of the load in radians at the receiving end

pftype Type of power factor, that is, lagging or leading. For lagging power factor, pftype = 0

During execution, the function trmlnper asks for input in respect of type of line.

Type of line = 0 when the line is modelled with distributed parameters

                = 1 for lumped parameters when the line is modelled as a π-equivalent circuit

                = 2 for lumped parameters when the line is modelled as a nominal π-equivalent circuit

                = 3 when the line is modelled as a short line

Example 4.9 The constants of a three-phase, 130-km transmission line are as follows: resistance of the line = 0.0781 Ω/km/conductor, line-to-neutral inductance = 0.746 mH/km, line-to-neutral shunt capacitance = 0.00995 μF/km. Assume shunt conductance between line and neutral is equal to zero. Use the MATLAB function trmlnper to compute sending end voltage, current, power, line loss, efficiency, and regulation of the line when it is modelled as (i) π-equivalent circuit and (ii) nominal π-equivalent circuit. The line is delivering 24 MW at 0.8 lagging power factor.

Solution

Input data: ≫ r=0.0781;L=0.746*(10^-3);g=0;CC=0.00995*(10^-6); Vr=66;l=130; Pr=24; pf=0.8;pftype=0;

≫ [A,B,C,D,Vs] = trmlnper(r,L,g,CC,l,Vr,Pr,pf,pftype);

(i) When the line is modelled as a π-equivalent circuit

type of line1

Output:

Sending end per-phase voltage, Vs = 44.7808 + j4.8711 = 45.0449 ∠6.2080° kV

Sending end line-to-line voltage = 78.0201 kV

Sending end current, I_S = 0.2090 − j0.1406 = 0.2519 ∠−33.9349° kA

Three-phase sending end power = 26.0176 MW

Total line loss = 3 × 2.0176 = 6.0528 MW

Efficiency = 92.2451%

Regulation of the line = 0.1821 = 18.21%

(ii) When the line is modelled as a nominal π-equivalent circuit

type of line2

Output:

Sending end per-phase voltage, V_S = 44.7981 + j4.8764 = 45.0628 ∠6.2123° kV

Sending end line-to-line voltage = 78.0510 kV

Sending end current, I_S = 0.2090 − j0.1406 = 0.2519 ∠−33.9383° kA

Three-phase sending end power = 26.0253 MW

Total line loss = 3 × 2.0253 = 6.0759 MW

Efficiency = 92.2180%

Regulation of the line = 0.1826 = 18.26%

4.8 SURGE IMPEDANCE LOADING OF LINES

Surge impedance loading (SIL) is the power delivered at rated voltage by a lossless line to a load resistance whose value is equal to the surge impedance $Z_c = \sqrt{L/C}$. For a lossless line, Z_c is purely resistive. Figure 4.9 shows a lossless line of length *l* terminated in a resistance equal to surge impedance.

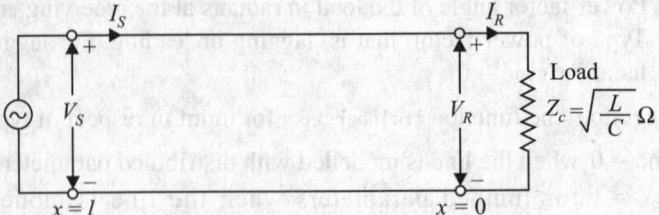

Fig. 4.9 Surge impedance loading of a lossless line

Recalling that for a lossless line $\gamma = j\beta$, using Eq. (4.42), the *ABCD* constants are written as

$$A = D = \cosh(j\beta l) = \frac{e^{j\beta l} + e^{-j\beta l}}{2} = \cos(\beta l)$$

$$B = Z_c \sinh(j\beta l) = jZ_c \sin(\beta l)\ \Omega$$

$$C = \frac{1}{Z_c}\sinh(j\beta l) = j\frac{1}{Z_c}\sin(\beta l)\ S$$

From Eq. (4.18), the voltage at a distance x from the receiving end is written as

$$V(x) = \cos(\beta x)V_R + jZ_c\sin(\beta x)I_R \qquad (4.66)$$

$$= \cos(\beta x)V_R + jZ_c\sin(\beta x)\left(\frac{V_R}{Z_c}\right)$$

$$= [\cos(\beta x) + j\sin(\beta x)]V_R$$

$$= e^{j\beta x}\,V_R = V_R\,\angle\beta x \qquad (4.67)$$

$$\boxed{|V(x)| = |V_R|} \qquad (4.68)$$

Equation (4.68) shows that in a lossless line under surge impedance loading, the magnitude of voltage at any point x is constant and is equal to the sending end value, but has a different phase position.

Similarly, from Eq. (4.22)

$$I(x) = j\frac{1}{Z_C}\sin(\beta x)V_R + \cos(\beta x)\left(\frac{V_R}{Z_C}\right)$$

$$= [\cos(\beta x) + j\sin(\beta x)]\left(\frac{V_R}{Z_C}\right)$$

$$= (e^{j\beta x})\frac{V_R}{Z_c} = I_R\,\angle\beta x \qquad (4.69)$$

$$\boxed{|I(x)| = |I_R|} \qquad (4.70)$$

Equation (4.70) shows that in a lossless line under surge impedance loading, the magnitude of current at any point along the line is constant and is equal to the sending end value.

The complex power $S(x)$ at any point x along the line may be obtained using Eqs (4.67) and (4.69) as

$$S(x) = P(x) + jQ(x) = V(x) I(x)^*$$

$$= \left(e^{j\beta x} V_R\right) \left(e^{j\beta x} \frac{V_R}{Z_c}\right)^* = \frac{|V_R|^2}{Z_c} \tag{4.71}$$

From Eq. (4.71), it is observed that the real power flow along a lossless line at SIL remains constant from the sending end to the receiving end. Since Z_c is resistive, there is no reactive power flow in the line, and $Q_S = Q_R = 0$. This indicates that under SIL, the reactive loss in the line inductance $\omega L |I_R|^2$ is exactly offset by the reactive power supplied by the line charging capacitance $\omega C |V_R|^2$, that is, $\omega L |I_R|^2 = \omega C |V_R|^2$. From this relation, the surge impedance may be obtained as $Z_c = V_R/I_R = \sqrt{L/C}$, which verifies the relation given in Eq. (4.20).

At the rated line voltage, the surge impedance loading of the line is given by

$$\text{SIL} = \frac{V_{R(\text{rated})}^2}{Z_c} \tag{4.72}$$

where rated voltage is used for a single-phase line and rated line-to-line voltage is used for a three-phase line. The surge impedance loading of the line in itself is not a measure of the maximum power that can be delivered over the line.

From Eq. (4.68), note that the magnitude of the voltage at the sending end is equal to the magnitude of the voltage at the receiving end. Thus, there is no voltage drop and the voltage profile along the line is flat. However, in power engineering practice, the load on lines can vary significantly and is different from SIL. It may only be a fraction of SIL when the line is lightly loaded, and it may be several times SIL under heavy load, depending on the length of the line and reactive compensation. For loads much higher than SIL, shunt capacitors may be needed to minimize voltage drop along the line. On the other hand, for light load conditions, that is, loads much lesser than SIL, shunt inductors may be needed to compensate the line charging current.

4.8.1 Voltage Profile

When a transmission line does not carry a load equal to SIL, the voltage profile along the line is no longer horizontal. Figure 4.10 shows a plot of the voltage

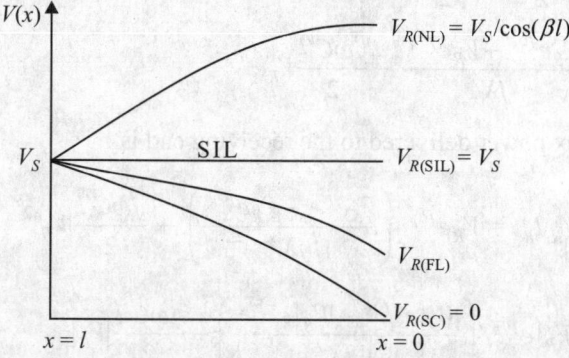

Fig. 4.10 Voltage profile along a transmission line for various loading conditions

profile for a transmission line of length 1500 km (quarter wavelength), for different loading conditions. The voltage at the sending end is maintained constant for all loading conditions.

For a lossless line, the voltage at any point x of the line is given by Eq. (4.66) as $V(x) = \cos(\beta x)V_R + jZ_c \sin(\beta x)I_R$. The following cases may be considered:

(i) When the line carries no load, that is, $I_{R(NL)} = 0$, voltage at any point x from the receiving end is $V_{(NL)}(x) = \cos(\beta x)\ V_{R(NL)}$, where the subscript (NL) stands for no-load values. Then, sending end voltage $V_{S(NL)} = \cos(\beta l)\ V_{R(NL)}$, and the no-load voltage increases from $V_{S(NL)}$ at the sending end to $V_{R(NL)}$ at the receiving end, where $x = 0$.

Ferranti effect The phenomenon of the receiving end voltage being greater than the sending end voltage when a long transmission line carries no load or, is lightly loaded, is called *Ferranti effect*. It occurs due to the line carrying a considerable amount of charging current because of the line capacitance to ground. It may be noted from the relation $V_{R(NL)} = V_{S(NL)}/\cos(\beta l)$ that the receiving end voltage will always be higher than the sending end voltage. Ferranti effect is more predominant in long HV transmission lines and cables.

(ii) When the line carries a load equal to the surge impedance Z_c, the voltage profile at the receiving end $V_{R(SIL)}$ is flat as discussed earlier.

(iii) When there is a short circuit at the receiving end, then the receiving end voltage $V_{R(SC)} = 0$, where the subscript (SC) stands for short-circuit values. Using Eq. (4.66), the voltage at the sending end $V_{S(SC)} = Z_c \sin(\beta l)\ I_{R(SC)}$, and the voltage decreases from this value at the sending end to $V_{R(SC)} = 0$ at the receiving end.

(iv) For full-load, the voltage profile $V_{R(FL)}$ will depend on the magnitude of load current and the power factor. The voltage profile will lie above the short-circuit voltage profile.

4.8.2 Steady-state Stability Limit

The real power delivered by a lossless line can be derived using Fig. 4.5. The sending end and receiving end bus voltages are assumed to be $V_S = |V_S|\angle\delta_S$ and $V_R = |V_R|\angle\delta_R$. The voltage magnitudes $|V_S|$ and $|V_R|$ are held constant. From Fig. 4.5, the receiving end current I_R is obtained as

$$I_R = \frac{V_S - V_R}{Z'} - \frac{Y'}{2}V_R$$

$$= \frac{V_S e^{j\delta_S} - V_R e^{j\delta_R}}{jX'} - \frac{j\omega C'l}{2}V_R e^{j\delta_R} \qquad (4.73)$$

The complex power delivered to the receiving end is

$$S_R = V_R I_R^* = V_R e^{j\delta_R}\left(\frac{V_S e^{j\delta_S} - V_R e^{j\delta_R}}{jX'}\right)^* + \frac{j\omega C'l}{2}|V_R|^2$$

$$= |V_R|e^{j\delta_R}\left(\frac{|V_S|e^{-j\delta_S} - |V_R|e^{-j\delta_R}}{-jX'}\right) + \frac{j\omega C'l}{2}|V_R|^2$$

$$= \frac{j|V_R| \ |V_S|\cos(\delta_R - \delta_S) - |V_R| \ |V_S|\sin(\delta_R - \delta_S) - j|V_R|^2}{X'}$$

$$+ \frac{j\omega C'l}{2}|V_R|^2 \tag{4.74}$$

$$= \frac{j|V_R||V_S|\cos\delta + |V_R||V_S|\sin\delta + j|V_R|^2}{X'} + \frac{j\omega C'l}{2}|V_R|^2$$

where $\delta = \delta_S - \delta_R$ is the power angle.

As the line is lossless, sending end real power P_S is equal to the receiving end real power P_R, that is, $P_R = P_S$. Equating the real part of Eq. (4.74), the real power delivered at the receiving end is obtained as

$$P_R = P_S = \mathrm{Re}\ (S_R) = \frac{|V_R| \ |V_S|}{X'}\sin\delta \tag{4.75}$$

The voltage magnitudes being constant, P_R increases with increase of power angle δ. The line delivers maximum power $P_{\max}$ when $\delta = 90°$, and is given by

$$P_{\max} = \frac{|V_S| \ |V_R|}{X'} \tag{4.76}$$

In Section 4.2, the parameters of a nominal π-equivalent circuit for a lossless line were defined and are given here for ready reference: $Z_c = \sqrt{L/C}$ is real and $\gamma = j\beta$ is a purely imaginary number. The values of Z' and Y' can be calculated as

$$Z' = jZ_c\sin\gamma l = jZ_c\sin\beta l = jX'$$

Substituting for X' in Eq. (4.75) gives

$$P_R = P_S = \frac{|V_R| \ |V_S|}{Z_c\sin\beta l}\sin\delta \tag{4.77}$$

Expressing $|V_S|$ and $|V_R|$ in per unit of rated line voltage $|V_{\text{rated}}|$ gives

$$P_R = P_S = \left(\frac{|V_R|}{|V_{\text{rated}}|}\right)\left(\frac{|V_S|}{|V_{\text{rated}}|}\right)\frac{|V_{\text{rated}}|^2}{Z_c}\frac{\sin\delta}{\sin(\beta l)} \tag{4.78}$$

Using Eqs (4.72) and (4.33) in Eq. (4.78), yields

$$P_S = P_R = V_{R(\text{pu})} \times V_{S(\text{pu})} \times (\text{SIL})\frac{\sin\delta}{\sin\left(\dfrac{2\pi l}{\lambda}\right)}\ \text{W} \tag{4.79}$$

Equations (4.76) to (4.79) indicate that steady-state stability limit increases with an increase in the magnitude of the line voltage. If the line voltage is doubled, the stability limit is increased four times. However, steady-state stability limit decreases with an increase in length. For $\delta = 90°$, the theoretical steady-state stability limit is given by

$$P_{\max} = \frac{V_{S(\text{pu})} \times V_{R(\text{pu})} \times (\text{SIL})}{\sin\left(\dfrac{2\pi l}{\lambda}\right)} \tag{4.80}$$

Example 4.10 A three-phase, 400-kV, 50-Hz transmission line has a series inductive reactance of 0.30 Ω/km and a shunt admittance of 3.75×10^{-6} S/km. If the line is 300 km long, determine its (i) surge impedance Z_c, (ii) propagation constant γ, (iii) *ABCD* constants, (iv) wavelength, and (v) SIL.

Solution   Inductance $L = \dfrac{0.30}{2 \times \pi \times 50} = 9.5493 \times 10^{-4}$ H/km

Shunt capacitance $C = \dfrac{3.75 \times 10^{-6}}{2 \times \pi \times 50} = 1.1937 \times 10^{-8}$ F/km

(i) Surge impedance $Z_c = \sqrt{\dfrac{9.5493 \times 10^{-4}}{1.1937 \times 10^{-8}}} = 282.8427\ \Omega$

Phase constant $\beta = 2 \times \pi \times 50 \sqrt{\left(9.5493 \times 10^{-4}\right) \times \left(1.1937 \times 10^{-8}\right)}$

$= 0.0011$ rad/km

(ii) Propagation constant $= j\beta = j0.0011$ per km

(iii) *ABCD* constants

$A = \cos(0.0011 \times 300) = 0.9498$

$D = A = 0.9498$

$B = j282.8427 \times \sin(0.0011 \times 300) = j88.4889\ \Omega$

$C = j\sin(0.0011 \times 300)/282.8427 = -j0.0011$ S

Verification: $AD - BC = 1.0000$

(iv) Wave length $\lambda = \dfrac{3 \times 10^8}{50} = 6000000$ metres $= 6000$ km

(v) From Eq. (4.72), SIL $= \dfrac{V_{R(\text{rated})}^2}{Z_c}$

$= \dfrac{(400)^2}{282.8427} = 565.6854$ MW

Example 4.11 A three-phase, loss-free, 600-km-long transmission line has a series inductive reactance and shunt capacitance of 0.35 Ω/km/phase and 4.2 S/km/phase, respectively. Plot the voltage profile of the line when it carries loads equal to (i) 25%, (ii) 50%, (iii) 100%, (iv) 125%, and (v) 150% of the SIL. Assume the receiving end voltage is equal to 1.0 pu in all cases.

Solution   The MATLAB script file voltage profile is used to compute the voltage profile.

```
%Plotting a 'voltage profile'
>> L=0.35/(2*pi*50)              % computes the inductance of the line
L =
0.0011
>> C=4.2*(10^-6)/(2*pi*50)       %computes the shunt capacitance of the line
C =
1.3369e-008
>> Zc=sqrtm(L/C)                 %computes the surge impedance of the line
Zc =
288.6751
>> beta=2*pi*50*sqrtm(L*C)       %computes the phase constant β of the line
beta =
0.0012
```

```
>> Vr=1.0;X=1.0;                          %sets the receiving end voltage = 1.0 p. u.
                                          %X represents the load as a percentage of SIL
>> step=600/20;                           %voltage is computed at 20 points. Step
                                          %fixes the increment in x which represents
                                          %the distance in km from the receiving
                                          %end.
>> x=600:-step:0;
>> y=(cos(beta*x)+i*sin(beta*x)/X)*Vr;    %expression for computing the
                                          %voltage at a point x from the
                                          %receiving end.Eq. (4.66) is
```

%modified by substituting $I_R = \dfrac{V_R}{XZ_c}$

```
>> plot(x,abs(y),'k')                     %plots the distance along the x-
                                          %axis and absolute value of the
                                          %voltage at point x km from the
                                          %receiving end in black solid line

hold on                                   %activates the hold for overlay
                                          %plots
>> X=0.25;                                %fixes the receiving end load at
                                          %25% of SIL

>> y=(cos(beta*x)+i*sin(beta*x)/X)*Vr;
>> plot(x,abs(y),'k-')                    %plots the distance along the x-
                                          %axis and absolute value of the
                                          %voltage at point x km from the
                                          %receiving end in black solid line

>> X=0.5;
>> y=(cos(beta*x)+i*sin(beta*x)/X)*Vr;
>> plot(x,abs(y),'k+')                    %plots the distance along the
                                          %x-axis and absolute value of the
                                          %voltage at point x km from the
                                          %receiving end in black with
                                          %a + sign.
>> X=0.75;
>> y=(cos(beta*x)+i*sin(beta*x)/X)*Vr;
>> plot(x,abs(y),'k.')                    %plots the distance along the
                                          %x-axis and absolute value of the
                                          %voltage at point x km from the
                                          %receiving end in black with a .
                                          %(point).

>> X=1.25;
>> y=(cos(beta*x)+i*sin(beta*x)/X)*Vr;
>> plot(x,abs(y),'k*')                    %plots the distance along the
                                          %x-axis and absolute value of the
                                          %voltage at point x km from the
                                          %receiving end in black with a *.

>> X=1.5;
>> y=(cos(beta*x)+i*sin(beta*x)/X)*Vr;
>> plot(x,abs(y),'kdiamond')              %plots the distance along the
                                          %x-axis and absolute value of the
                                          %voltage at point x km from the
```

```
                                       %receiving   end   in   black   with   a
                                       %diamond.
>> Xlabel('Distance from receiving     %labels the x-axis as
end in km.')                           %Distance   from   receiving   end
                                       %in km.
>> Ylabel('Sending end voltage         %labels the y-axis as Sending
in p. u.')                             %end voltage in p. u.
>> title('VOLTAGE PROFILE OF A THREE-PHASE TRANSMISSION LINE')
                                       %provides the title VOLTAGE
                                       %PROFILE OF A THREE-PHASE
                                       %TRANSMISSION LINE
>> hold off                            %de-activates the hold for
                                       %overlay of plots
>>
```

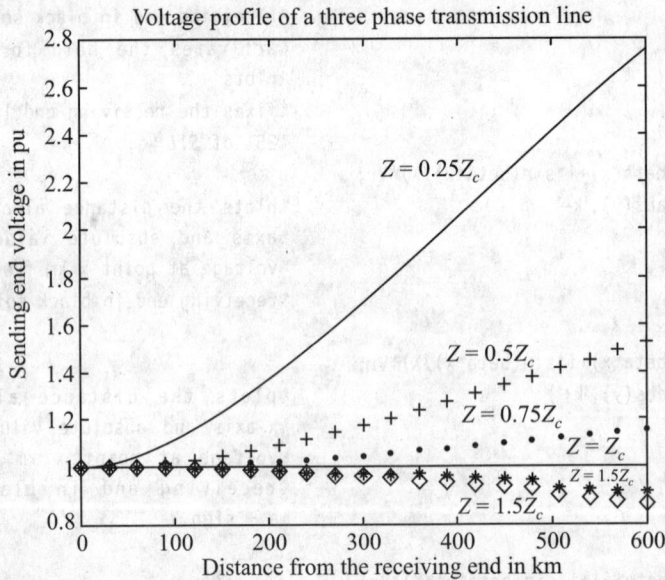

Fig. 4.11 Voltage profile of a transmission line for varying load

4.9 COMPLEX POWER FLOW THROUGH A TRANSMISSION LINE

The main objective of transmission lines is to transfer power from the generating stations to the load centre for consumption. In this section, equations will be developed for complex power transfer, in terms of the parameters and voltages at the two ends of a transmission line, using the short, medium, and long line models.

4.9.1 Complex Power Transmission in Short-length Line

Figure 4.12(a) represents a per-phase circuit of a short transmission line. The line being short is represented by the per-phase series impedance $Z = R + j\omega L$, where R represents the resistance in ohm and L is the inductance of the line in henry. Let the voltages at the sending end and receiving end, respectively, be $V_S = |V_S| \angle \delta_S$

$= |V_S| e^{j\delta_S}$ and $V_R = |V_R| \angle \delta R = |V_R| e^{j\delta_R}$. In polar form, the transmission line impedance $Z = |Z| \angle \theta = |Z| e^{j\theta}$, where θ is the impedance angle $= \tan^{-1}(\omega L/R)$. It is assumed that the two ends of the transmission line have generators connected and the voltage magnitudes at the two ends of the line V_S and V_R are maintained constant. In the analysis, the constant voltages are replaced by voltage sources and the corresponding single-phase circuit model of the short-length line is as shown in Fig. 4.12(b).

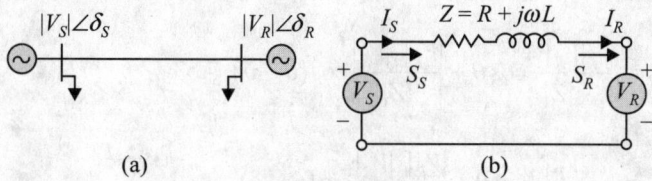

(a)   (b)

Fig. 4.12 (a) Per-phase circuit diagram for a short-length line
(b) Equivalent circuit for a short-length line

Complex power flow S_S at the sending end can be written as

$$S_S = V_S I_S^* = |V_S| \angle \delta_S \left(\frac{|V_S| \angle \delta_S - |V_R| \angle \delta_R}{|Z| \angle \theta} \right)^*$$

$$= |V_S| \angle \delta_S \left(\frac{|V_S| \angle -\delta_S - |V_R| \angle -\delta_R}{|Z| \angle -\theta} \right)$$

$$= \frac{|V_S|^2}{|Z| \angle -\theta} - \frac{|V_S| \, |V_R| \angle (\delta_S - \delta_R)}{|Z| \angle -\theta} \tag{4.81}$$

Substituting $\delta_S - \delta_R = \delta$ in Eq. (4.81) gives

$$S_S = \frac{|V_S|^2}{|Z|} e^{j\theta} - \frac{|V_S| \, |V_R|}{|Z|} e^{j\theta} e^{j\delta} \tag{4.82}$$

$$= \frac{|V_S|^2}{|Z|} (\cos\theta + j\sin\theta) - \frac{|V_S| \, |V_R|}{|Z|} \{\cos(\theta+\delta) + j\sin(\theta+\delta)\} \tag{4.83}$$

Equating the real and imaginary parts in Eq. (4.83) yields

$$P_S = \frac{|V_S|^2}{|Z|} \cos\theta - \frac{|V_S| \, |V_R|}{|Z|} \cos(\theta+\delta) \tag{4.84}$$

$$Q_S = \frac{|V_S|^2}{|Z|} \sin\theta - \frac{|V_S| \, |V_R|}{|Z|} \sin(\theta+\delta) \tag{4.85}$$

where P_S and Q_S are the active and reactive power flows, respectively, at the sending end.

Similarly, an expression analogous to Eq. (4.82) can be developed for complex power flow S_R at the receiving end and is given below:

$$S_R = -\frac{|V_R|^2}{|Z|}e^{j\theta} + \frac{|V_R|\,|V_S|}{|Z|}e^{j\theta}e^{-j\delta} \tag{4.86}$$

$$S_R = P_R + jQ_R$$

$$= -\frac{|V_R|^2}{|Z|}(\cos\theta + j\sin\theta) + \frac{|V_R|\,|V_S|}{|Z|}\{\cos(\theta-\delta) + j\sin(\theta-\delta)\} \tag{4.87}$$

Equating real and imaginary parts in Eq. (4.87) gives

$$P_R = -\frac{|V_R|^2}{|Z|}\cos\theta + \frac{|V_R|\,|V_S|}{|Z|}\cos(\theta-\delta) \tag{4.88}$$

$$Q_R = -\frac{|V_R|^2}{|Z|}\sin\theta + \frac{|V_R|\,|V_S|}{|Z|}\sin(\theta-\delta) \tag{4.89}$$

It is noted that for a given transmission line $|Z|$ is fixed and $\theta \approx 90°$. Hence, active and reactive powers at the two ends depend on $|V_S|$, $|V_R|$ and the power angle δ. Equations (4.84) and (4.88) indicate that the flow of real power is predominantly controlled by the power angle δ, while Eqs (4.85) and (4.89) show that the reactive power flow is dependent largely on the magnitudes of the voltages at the two ends.

For a lossless line, that is, for $R = 0$, $Z = j\omega L = jX\,\Omega$, and $\theta = \pi/2$; Eqs (4.84), (4.85), (4.88), and (4.89) take the following forms

$$P_S = P_R = \frac{|V_S||V_R|}{X}\sin\delta \tag{4.90}$$

$$Q_S = \frac{|V_S|^2}{X} - \frac{|V_S||V_R|}{X}\cos\delta \tag{4.91}$$

$$Q_R = -\frac{|V_R|^2}{X} + \frac{|V_S||V_R|}{X}\cos\delta \tag{4.92}$$

It can be observed from Eq. (4.90) that the maximum power transferred P_{max} across a line is given by

$$P_{max} = \frac{|V_S|\,|V_R|}{X} \tag{4.93}$$

Thus, $P_S = P_R = P_{max}\sin\delta \tag{4.94}$

Sending End and Receiving End Power Circles

The variation in complex power transferred with a change in the power angle δ can be plotted on a complex plane and yields useful information for a quick assessment of the performance of a line.

In the steady-state operation of power systems, voltage magnitudes $|V_S|$ and $|V_R|$ are maintained within specified values and, therefore, can be assumed as constant. However, the power angle δ undergoes considerable changes as the real power transfer varies. Hence, since the parameters of the line, $|Z|$ and θ, are fixed and the voltage magnitudes at the sending end and the receiving end are controlled and assumed to remain constant, Eqs (4.82) and (4.86) become functions of the power angle δ and may be written as

$$S_S = C_S - Be^{j\delta} \tag{4.95}$$

$$S_R = C_R + Be^{-j\delta} \tag{4.96}$$

where

$$C_S = \frac{|V_S|^2}{|Z|}e^{j\theta}$$

$$C_R = -\frac{|V_R|^2}{Z}e^{j\theta} \tag{4.97}$$

$$B = \frac{|V_S|\,|V_R|}{|Z|}e^{j\theta}$$

On a complex plane, if real power is represented along the x-axis and reactive power along the y-axis, it is observed that as δ is varied, the loci of both S_S and S_R are circles on the complex plane, whose radii are determined by B. The circle for S_S is called the sending end power circle whose centre is given by C_S. Similarly, the centre of the receiving end power circle traced by S_R is determined by C_R. Figure 4.13 shows the sending end and receiving end power circles on a complex plane.

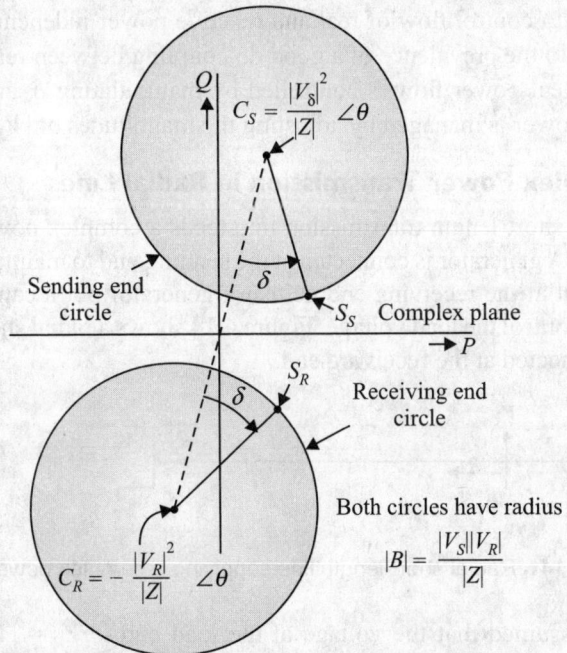

Fig. 4.13 Sending end and receiving end power circle diagrams for a short-length line

By studying the power circle diagrams given in Fig. 4.13, the following observations can be made.

(a) If the power angle δ is zero, C_S, C_R, and B lie on a straight line.

(b) For $V_S \neq V_R$, the circles do not intersect.

(c) The power received is equal to the power at sending end minus the transmission line losses.

(d) From the sending end power circle diagram and Eq. (4.84), it is observed that the maximum power, which can be transmitted, is limited by the angle ($\theta + \delta$). In fact power sent is maximum when ($\theta + \delta$) = 180°.

(e) Similarly, from the receiving end power circle diagram and Eq. (4.88), it is observed that the maximum power which can be received occurs when ($\theta - \delta$) = 0, that is, when $\theta = \delta$.

(f) Real power cannot be transmitted beyond the maximum limit for a transmission line, by making $\delta > (180° - \theta)$, since synchronism between the generators at the two ends of the transmission line is lost and effective transfer of real power from the sending end to the receiving end is disrupted. The phenomenon of losing synchronism will be discussed in Chapter 11.

(g) Equation (4.90), however, suggests a method for increasing the transmission capability of a line. The maximum limit of power transmission can be increased by decreasing the inductive reactance X of the line. This technique is called compensation and will be discussed in a subsequent section.

(h) In the operation of high-voltage power systems in practice, $|V_S| \approx |V_R|$, and transmission lines have a high X/R ratio, that is, $\theta \approx 90°$, and the power angle δ is of the order of 10°. Such types of operating conditions enable power engineers to control flow of real and reactive power independently of each other due to the prevalence of a good de-coupling between real and reactive powers. Real power flow is controlled by manipulating δ, and the flow of reactive power is managed by adjusting the magnitudes of $|V_S|$ and $|V_R|$.

4.9.2 Complex Power Transmission in Radial Lines

In this case, a short-length transmission line feeds a complex power load at the receiving end. A generator is connected at the sending end to maintain the voltage at that end, but at the receiving end neither a generator nor a capacitor bank is connected to control the load voltage. Figure 4.14 shows a radial short-length line with load connected at the receiving end.

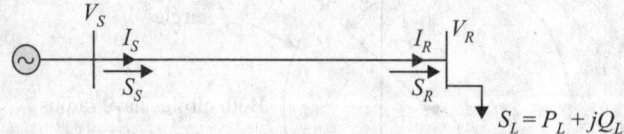

Fig. 4.14   Radial short-length line supplying a complex power load

Let it be assumed that the voltage at the load end is $V_R = |V_R| \angle 0°$, and the load draws a current I_R at a fixed lagging power factor angle φ, thus $I_R = |I_R| \angle -\varphi = |I_R| e^{-j\varphi}$. Then complex power S_R at the receiving end is equal to complex load power S_L. The complex load power may be expressed as

$$S_L = P_L + j\,Q_L = S_R = V_R I_R^* = |V_R||I_R| e^{j\varphi}$$

$$= |V_R||I_R| (\cos\varphi + j\sin\varphi)$$

$$= |V_R||I_R|\cos\varphi\left(1 + j\frac{\sin\varphi}{\cos\varphi}\right) = P_L\left(1 + j\tan\varphi\right) \tag{4.98}$$

where P_L and Q_L are active power and reactive power components of the load. In Eq. (4.98), $P_L\tan\varphi = Q_L$. If resistance of the line is neglected, then $Z = X$, and using Eqs (4.90) and (4.92), the real and reactive components of load can be expressed as

$$P_R = P_L = \frac{|V_S|\,|V_R|}{X}\sin\delta \tag{4.99}$$

$$Q_L = P_L\tan\varphi = Q_R = -\frac{|V_R|^2}{X} + \frac{|V_R|\,|V_S|}{X}\cos\delta$$

or

$$P_L\tan\varphi + \frac{|V_R|^2}{X} = \frac{|V_R|\,|V_S|}{X}\cos\delta \tag{4.100}$$

or

$$\left(P_L\tan\varphi + \frac{|V_R|^2}{X}\right)^2 = \left(\frac{|V_R|\,|V_S|}{X}\right)^2\cos^2\delta$$

$$= \left(\frac{|V_R|\,|V_S|}{X}\right)^2\left(1 - \sin^2\delta\right) \tag{4.101}$$

Substituting for $\sin^2\delta$ from Eq. (4.99) and also noting that $P_L = P_R$, Eq. (4.101) after simplification becomes

$$\left(P_L\tan\varphi + \frac{|V_R|^2}{X}\right)^2 = \left(\frac{|V_R|\,|V_S|}{X}\right)^2 - P_L^2$$

On re-arranging the above expression to solve for V_R

$$|V_R|^2 = \frac{|V_S|^2}{2} - P_L X\tan\varphi \pm \sqrt{\frac{|V_S|^4}{4} - P_L X\left(P_L X + |V_S|^2\tan\varphi\right)} \tag{4.102}$$

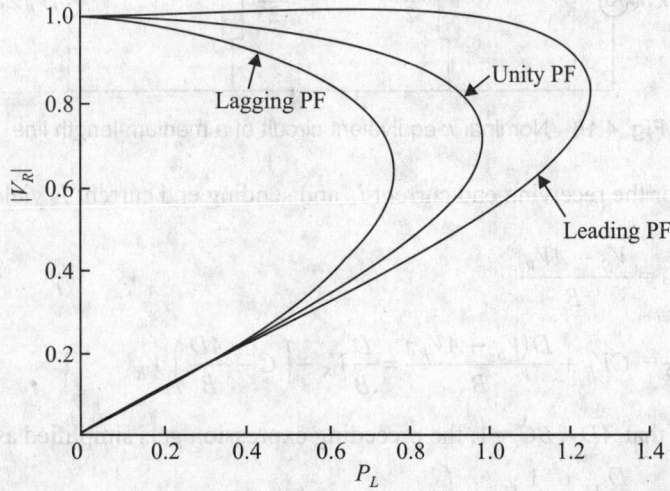

Fig. 4.15 Plot of load voltage magnitude versus varying load

Equation (4.102) provides an expression for computing the magnitude of the receiving end voltage or the load voltage, for a varying load at a given power factor and for a fixed magnitude of sending end voltage and line reactance X. Figure 4.15 shows a plot of the magnitude of the voltage V_R versus the load P_L, for a fixed magnitude of the sending end voltage and line reactance X, when the power factor changes from lagging to unity to leading.

A study of the curves in Fig. 4.15 shows the following.

(a) The feeder has a maximum capability to supply a load.
(b) As can be observed from Eq. (4.102) for each load there are two magnitudes of the load voltage at which it can be theoretically supplied. However, the lower value of the load voltage magnitude gives rise to unstable operation and may even lead to a voltage collapse.
(c) Control of load voltage is better as the power factor changes from lagging to leading. Thus, there is a distinct advantage in supplying reactive power at the load end, through capacitors or synchronous condensers, compared to supplying the same from a remote end.

4.9.3 Complex Power Transmission in Medium or Long Lines

In Section 4.9.1, complex power flow in a short-length line was discussed. For analysis of medium and long lines, the π-equivalent circuit model shown in Fig. 4.16 will be used. It is assumed that the voltage magnitudes at the two ends of the line are constant, and V_S and V_R represent voltages at the sending end and receiving end respectively.

Equations (4.39) and (4.40) are repeated here for ready reference

$$V_S = AV_R + BI_R; \quad I_S = CV_R + DI_R$$

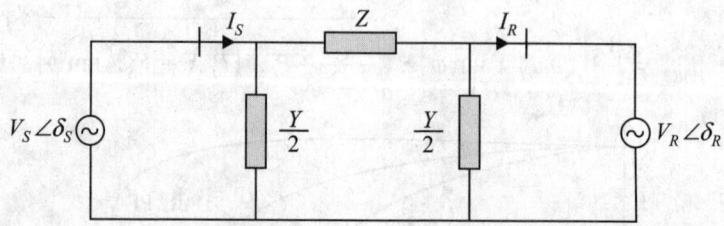

Fig. 4.16 Nominal π-equivalent circuit of a medium-length line

Solving for the receiving end current I_R and sending end current I_S yields

$$I_R = \frac{V_S - AV_R}{B}$$

$$I_S = CV_R + \frac{D(V_S - AV_R)}{B} = \frac{D}{B}V_S + \left(C - \frac{AD}{B}\right)V_R$$

Recalling that $AD - BC = 1$, the preceding expression gets simplified as follows

$$I_S = \frac{D}{B}V_S - \frac{1}{B}V_R \qquad (4.103)$$

For medium and long lines represented by nominal π-equivalent circuit, $A = D$. Thus, Eq. (4.103) can be written as

$$I_S = \frac{A}{B}V_S - \frac{1}{B}V_R \tag{4.104}$$

Sending end complex power is given by

$$S_S = V_S I_S^* = V_S \left(\frac{A}{B}V_S - \frac{1}{B}V_R\right)^* = \frac{A^*}{B^*}V_S^2 - \frac{1}{B^*}V_S V_R^* \tag{4.105}$$

For the sake of understanding the concept of transmission of complex power, the nominal π-equivalent circuit shown in Fig. 4.16 is employed. For a nominal π-equivalent circuit, $A = (1 + YZ/2)$ and $B = Z$. Substituting the values of A and B and putting $V_S = |V_S| \angle\delta_S$, $V_R = |V_R| \angle\delta_R$, $Z = |Z| \angle\theta$, and $Y = |Y| \angle\phi$, in Eq. (4.105) yields on simplification the following equation:

$$S_S = \frac{|Y|\,|V_S|^2}{2}e^{-j\phi} + \frac{|V_S|^2}{|Z|}e^{j\theta} - \frac{|V_S|\,|V_R|}{|Z|}e^{j(\delta_S - \delta_R + \theta)}$$

$$= \frac{|Y|\,|V_S|^2}{2}e^{-j\phi} + \frac{|V_S|^2}{|Z|}e^{j\theta} - \frac{|V_S|\,|V_R|}{|Z|}e^{j\theta}e^{j\delta} \tag{4.106}$$

where $\delta = \delta_S - \delta_R$. Similarly an expression for the received complex power S_R can be derived and is given by

$$S_R = -\frac{Y^*}{2}|V_R|^2 + \frac{1}{Z^*}|V_R|^2 - \frac{|V_S|\,|V_R|}{Z^*}e^{-j\delta}$$

$$= -\frac{|Y|\,|V_R|^2}{2}e^{-j\phi} + \frac{|V_R|^2}{|Z|}e^{j\theta} - \frac{|V_S|\,|V_R|}{|Z|}e^{j\theta}e^{-j\delta} \tag{4.107}$$

In the steady-state operation of power systems, if the voltage magnitudes $|V_S|$ and $|V_R|$ are assumed as constant, the power angle δ undergoes considerable changes as the real power transfer varies. Since the parameters of the line, $|Z|$, $|Y|$, θ, and ϕ, are fixed, Eqs (4.106) and (4.107) become functions of the power angle δ and may be written as

$$S_S = C_S - Be^{j\delta} \tag{4.108}$$

$$S_R = C_R + Be^{-j\delta} \tag{4.109}$$

where $\quad C_S = \dfrac{|Y||V_S|^2}{2}e^{-j\phi} + \dfrac{|V_S|^2}{|Z|}e^{j\theta}$

$$C_R = -\frac{|Y||V_R|^2}{2}e^{-j\phi} + \frac{|V_R|^2}{|Z|}e^{j\theta} \tag{4.110}$$

$$B = \frac{|V_S||V_R|}{|Z|}e^{j\theta}$$

On a complex plane, with real power along the x-axis and reactive power along the y-axis, it may be seen that as δ is varied the loci of both S_S and S_R are circles on the complex plane, whose radii are determined by B. The sending end power circle has a centre given by C_S, and the centre of the receiving end power circle traced by S_R is determined by C_R. Figure 4.17 shows the sending end and receiving end power circles on a complex plane.

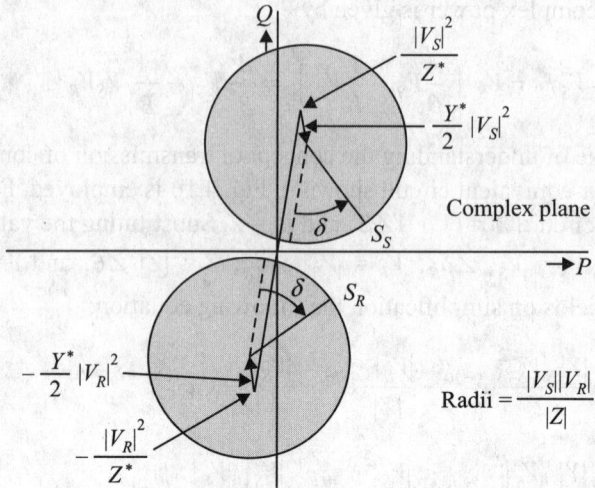

Fig. 4.17 Power circle diagram of a nominal π-equivalent circuit

It may be noted that the values of C_S and C_R in Eq. (4.97) for a short-length line differ from those in Eq. (4.110) for a medium-length/long line by an additional constant term $\left(|Y||V_R|^2/2\right)e^{-j\phi}$. This term represents the power consumed in $Y/2$ and has a relatively low magnitude. The centres of the sending end and receiving end power circles for medium-length/long line, when compared with respect to those or short-length line, respectively get shifted by $\left[\left(|Y||V_R|^2/2\right)e^{-j\phi}\right]$ and $\left[-\left(|Y||V_R|^2/2\right)e^{-j\phi}\right]$. It may further be noted that since Y is predominantly reactive, (i) the shift in the positions of the centres is nearly vertical and (ii) the real power transferred remains unchanged. Generally, the conclusions drawn in the case of a short line model are also applicable in the present case.

Example 4.12 Figure 4.18 shows a three-phase, 33-kV line feeding a per-phase load of 10 MW. (i) If the impedance of the line is $Z = j20\,\Omega$, determine the load angle and the reactive power to be supplied by the capacitive source connected at the load end to maintain a line voltage of 33 kV at the load. (ii) If the capacitive source is removed, what is the maximum real power load which can be supplied? What will be the power angle and the voltage to supply the load?

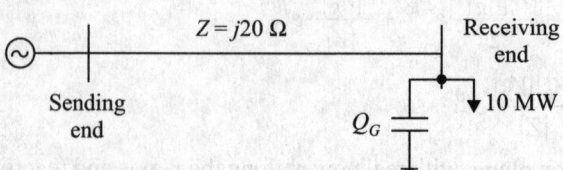

Fig. 4.18

Solution   Per-phase sending and receiving end voltage $V_S = V_R = \dfrac{33}{\sqrt{3}} = 19.0526$ kV

(i) Using Eq. (4.93),

$$P_{max} = \frac{19.0526 \times 19.0526}{20} = 18.15 \text{ MW}$$

Power angle is given by

$$\delta = \sin^{-1}\left(\frac{10}{18.15}\right) = 33.4332°$$

Use Eq. (4.92) to compute the reactive power to be transmitted from the sending to the receiving end

$$Q_R = -\frac{(19.0526)^2}{20} + \frac{19.0526 \times 19.0526}{20} \cos(33.4332°)$$

$$= -3.0031 \text{ MVAR}$$

If the reactive power source is removed, the only power transmitted along the line is the real power. Suppose the power transmitted is P at a power angle δ and the magnitude of the voltage is V_R. From Eq. (4.90),

$$\frac{|V_S||V_R|}{X} \sin\delta = P$$

or

$$\left(\frac{|V_S||V_R|}{X}\right)^2 \sin^2\delta = P^2 \qquad (4.12.1)$$

Since no reactive power is transmitted, Eq. (4.92) is written as

$$-\frac{|V_R|^2}{X} + \frac{|V_R||V_S|}{X}\cos\delta = 0$$

or

$$\left(\frac{|V_2||V_2|}{X}\right)^2 \cos^2\delta = \left(\frac{|V_2|}{X}\right)^2 \qquad (4.12.2)$$

Adding of (4.12.1) and (4.12.2) and simplifying yields

$$V_R^4 - V_S^2 V_R^2 + X^2 P^2 = 0$$

Put $v = V_R^2$ and the above equation gets converted into a quadratic equation

$$v^2 - (19.0526)^2 v + 400 \, P^2 = 0$$

$$v = \frac{(19.0526)^2 \pm \sqrt{(19.0526)^4 - 4 \times 400 \times P^2}}{2}$$

A solution for V_R can be obtained if v has a real solution. Hence for maximum real power

$$(19.0526)^4 - 4 \times 400 \times P^2 = 0$$

$$P = 9.075 \text{ MW}$$

$$v = \frac{(19.0526)^2}{2} = 181.5008$$

$$V_R = \sqrt{181.5008} = 13.4722 \text{ kV}$$

$$\delta = \sin^{-1}\left(\frac{9.075 \times 20}{19.0526 \times 13.4722}\right) = 44.9998° = 45°$$

Example 4.13   The sending end voltage V_S and receiving end voltage V_R of a line whose impedance is $0.1 \angle 80°$ Ω are respectively 1.02 and 0.9 pu. Draw the sending end and receiving end power circle diagrams and determine: (i) sending end P_{max}, (ii) power angle

δ_{12} at which P_{max} is transferred from sending end, (iii) receiving end P_{max}, (iv) power angle δ at which P_{max} is transferred from receiving end, and (v) real power line loss at $\delta = 15°$.

Solution   Equations (4.95) and (4.96) for sending end and receiving end power circles from Section 4.9.1 are reproduced here for easy reference

$$S_S = C_S - Be^{j\delta}; \quad S_R = C_R + Be^{-j\delta}$$

where $C_S = \dfrac{|V_S|^2}{|Z|}e^{j\theta}$, $C_R = -\dfrac{|V_R|^2}{Z}e^{j\theta}$, and $B = \dfrac{|V_S|\,|V_R|}{|Z|}e^{j\theta}$

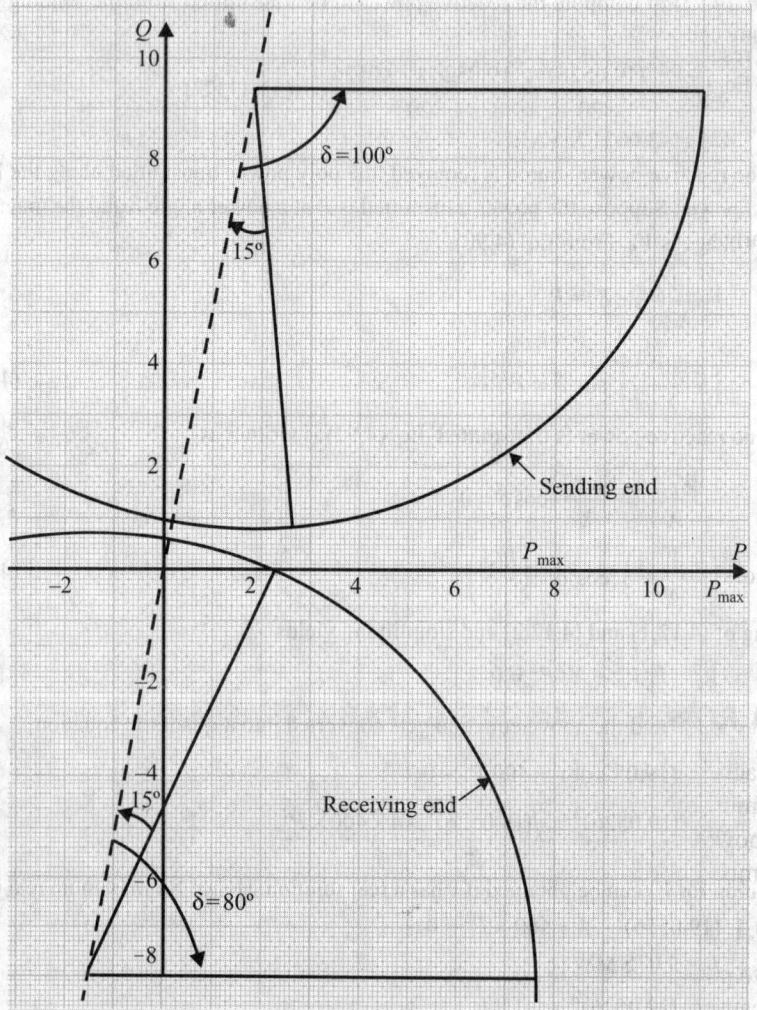

Fig. 4.19   Power circle diagram of Example 4.13

The coordinates on a complex plane for the sending end power circle are

$$C_S = \frac{(1.02)^2}{0.1}e^{j80°} = 1.8066 + j10.2459$$

The coordinates on a complex plane for the receiving end power circle are

$$C_R = -\frac{(0.9)^2}{0.1}e^{j80°} = -1.4066 - j7.9769$$

The radius $B = \dfrac{(1.02) \times (0.9)}{0.1} = 9.18$

The sending end and receiving end power circles are drawn on a complex plane as shown in Fig. 4.19, wherein x-axis represents the real power and y-axis represents the reactive power.

From the sending end circle diagram, $P_{max} = 11$ pu and $\delta = 100°$

Similarly, from the receiving end circle diagram, $P_{max} = 7.75$ pu and $\delta = 80°$

For $\delta = 15°$, sending end power $P_S = 2.65$ pu and receiving power $P_R = 2.50$ pu

Line losses $= 2.65 - 2.50 = 0.15$ pu

These results can be easily verified by using Eqs (4.84) and (4.86).

Using Eq. (4.84) yields,

$$\text{Sending end } P_{max} = \frac{(1.02)^2}{0.1} \cos 80° + \frac{1.02 \times 0.9}{0.1} = 10.9886 \text{ pu at } \delta$$

$$\delta = 180° - 80° = 100°$$

Similarly, using Eq. (4.86) gives,

$$\text{Receiving end } P_{max} = \frac{(0.9)^2}{0.1} \cos 80° - \frac{0.9 \times 1.02}{0.1} = 7.7734 \text{ pu at } \delta = 80°.$$

$$\text{Sending end power} = \frac{(1.02)^2}{0.1} \cos 80° - \frac{1.02 \times 0.9}{0.1} \cos(80° + 15°)$$

$$= 2.6067 \text{ pu at } \delta = 15°$$

$$\text{Receiving end power} = \frac{(0.9)^2}{0.1} \cos 80° - \frac{0.9 \times 1.02}{0.1} \cos(80° - 15°)$$

$$= 2.4731 \text{ pu at } \delta = 15°$$

Line loss at $\delta = 15°$ is given by

Sending power at $\delta = 15°$ – Receiving end power at $\delta = 15°$

$$= 2.6067 - 2.4731 = 0.1336 \text{ pu}$$

4.10 POWER TRANSFER CAPABILITY OF LINES

In practice power lines are not operated to deliver their theoretically derived maximum power, which is based on rated terminal voltages at the ends of a line and power angular displacement $\delta = 90°$. The power transfer capability of a transmission line is governed by its thermal loading limit and the stability limit.

4.10.1 Effect of Temperature on Power Transfer Capacity

Due to power loss in the conductors of a transmission line, heat is generated and this causes a temperature rise. The power loss reduces the efficiency of power transmission while the rise in temperature causes stretching of the conductors. The latter results in increase of sag of the conductors between the towers thereby reducing ground clearance, which could be hazardous. At higher temperatures, irreversible stretching of the conductors (which occurs around 100°C) is likely to take place. Safe limit to prevent irreversible stretching imposes a limit on the maximum value of current a line can carry. This thermal limit depends on line design, that is, conductor cross section, conductor geometry, and spacing of conductors between towers, etc., and environmental conditions, that is, ambient temperature, wind velocity, etc. Bundling of conductors is also useful as it provides increased surface area for heat dissipation. The thermal limit is specified by the current carrying capacity of the conductor and is available in the manufacturers' data.

Further, design of transmission lines is primarily based on the voltage at which the power is to be transmitted. The selection of transmission voltage level, in turn, is governed largely by the quantum of power and the distance over which it is to be transmitted. The conductor size, geometry, spacing between phases, insulation, etc. are decided based on the voltage at which power is to be transmitted. Further, statutory requirements to supply power at the specified voltage at the consumers' terminals, within a tolerance limit, impose that the transmission voltage is maintained at or around the rated voltage. Thus, the capability of a transmission line to transfer power is limited by the transmission voltage level and the maximum current carrying capacity. Apart from the above, the maximum limit of power transmission of a line may be restricted by the thermal limitations of the terminal equipment, such as transformers.

4.10.2 Effect of Stability on Power Transfer Capacity

In Section 4.9.1 it was shown that for a short-length, lossless line with voltage support at both ends, the power transferred was represented by Eq. (4.90). The maximum power transferred is given by $P_{max} = |V_S||V_R|/X$, and the power angle $\delta = 90°$. In practice, however, the power angle δ should be strictly limited to maximum values around 30° to 45°. This is because the generator and transformer reactances, which when added to the line, will result in a larger value of δ for a given value of power delivered. In this case, the stability limit is reduced to less than the ultimate transmission capability of the line. Thus, for short lines, the power transfer capability is set by the thermal limits rather than the stability limit.

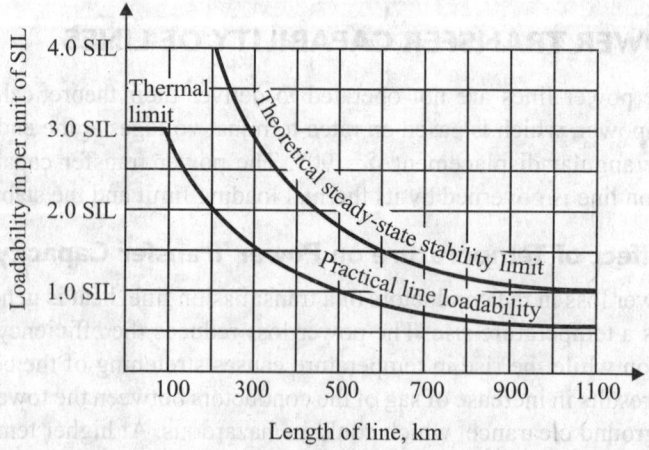

Fig. 4.20 Steady-state stability limit versus length of a transmission line

In Section 4.8.2, for a lossless long line, the maximum power transfer is given by $P_{max} = |V_S||V_R|/X'$ [Eq. (4.76)], when the sending and receiving voltage magnitudes are maintained constant. P_{max} is called the theoretical steady-state stability limit of the line. Equation (4.80) gives the theoretical steady-state stability limit. For $V_{Spu} = V_{Rpu} = 1$, $\lambda = 6000$ km, and $\delta = 45°$, the stability limit is plotted using Eq. (4.80). For comparison, a typical thermal limit is also shown. It may be noted from Fig. 4.20 that for short-length lines, the thermal limit governs, whereas for long lines the stability limit is important for determining power transfer capability of transmission lines.

Example 4.14 For the transmission line in Example 4.10, compute the receiving end voltage when the line (i) is on open circuit, (ii) is terminated by its surge impedance, and (iii) carries a load which is 60% of the surge impedance. Rated voltage is applied at the sending end.

Solution From Example 4.10, the following data is available:

Surge impedance $Z_c = 282.8427\ \Omega$

Phase constant $\beta = 0.0011$ rad/km

Per-phase sending end voltage $V_S = \dfrac{400}{\sqrt{3}} = 230.9401$ kV

(i) When the line is on no-load, $I_R = 0$,

$$V_S = \cos(\beta l)V_R.$$

Hence, $V_R = \dfrac{230.9401}{\cos(0.0011 \times 300)} = 243.1459$ kV

(ii) When the line is terminated by its surge impedance, the receiving end voltage is equal to the sending end voltage. However, Eq. (4.67) can be used to verify the statement.

$$V_S = V_R e^{j\beta l}$$

or $V_R = \dfrac{230.9401}{e^{(j \wedge 0.0011 \wedge 300)}} = 219.35 - j72.251$

Magnitude of sending end voltage = 230.9401 kV

(iii) When the line carries a load equal to $0.6Z_c$, from Eq. (4.67),

$$V_R = \dfrac{230.9401}{\left[\cos(0.0011 \times 300) + j\dfrac{\sin(0.0011 \times 300)}{0.6}\right]}$$

$$= 186.84 - j102.57$$

Absolute value of receiving end voltage = 213.1396 kV

4.11 REACTIVE LINE COMPENSATION

Referring to the voltage profiles of a long transmission line under different load conditions, it is to be observed that at light loads which are markedly less than SIL, the voltage at the receiving end is more than the sending end voltage. Conversely, when the loads are several times more than the SIL the voltage at the receiving end plunges downwards. Of course, when the load is equal to SIL, there is no flow of reactive power and the voltage is constant along the line. From the foregoing, it is understood that when the line is lightly loaded and also when the line carries a heavy load, the voltage has to be controlled at both the ends of the line.

4.11.1 Inductive Shunt Reactor Compensation

Under light load or open line conditions, shunt reactor compensation is employed. Figure 4.21 shows an equivalent circuit with an inductive shunt reactor having a reactance of jX_{Lsh} connected at the receiving end.

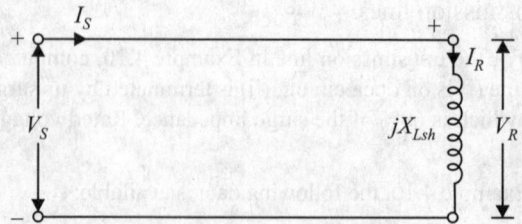

Fig. 4.21   Shunt reactive compensation of a long line

If V_R is the receiving end voltage, then the receiving end current is given by

$$I_R = \frac{V_R}{jX_{Lsh}} \tag{4.111}$$

Substituting for I_R in Eq. (4.66) yields

$$V_S = \cos(\beta l)V_R + jZ_C \sin(\beta l)\frac{V_R}{jX_{Lsh}} = \left[\cos(\beta l) + \frac{Z_C \sin(\beta l)}{X_{Lsh}}\right]V_R$$

Simplification leads to

$$X_{Lsh} = \frac{Z_C \sin(\beta l)}{\left(\dfrac{V_S}{V_R} - \cos(\beta l)\right)} \tag{4.112}$$

Equation (4.112) provides the value of the inductive reactance to be connected at the receiving end. It may also be noted that since no real power is flowing along the line, the sending end and receiving end voltages are in phase. Sending end current for such a long line can be expressed as

$$I_S = \left[j\frac{1}{Z_C}\sin(\beta l)\right]V_R + \cos(\beta l)I_R$$

Substituting for V_R from Eq. (4.111) yields

$$I_S = \left[-\frac{X_{Lsh}}{Z_C}\sin(\beta l) + \cos(\beta l)\right]I_R \tag{4.113}$$

If the value of X_{Lsh} is substituted by assuming $V_S/V_R = 1.0$, it is observed that $I_S = I_R$.

4.11.2 Shunt Capacitor Compensation

When the power factor at the receiving end load is low, then either the magnitude of the sending end voltage is required to be increased, or reactive power is necessary to be supplied at the load bus, in order to maintain the voltage at the consumers' terminals. The latter technique for improving the power factor has been found to be more effective. Shunt capacitors are, therefore, connected directly across the load bus or in the tertiary winding of the transformer to reduce voltage drop and line losses. Equations (4.88) and (4.89) which express real and reactive receiving end power for a nominal π-equivalent circuit can be used to compute the reactive volt-ampere (VAR) rating of the shunt capacitive reactor X_{Csh} required to be connected for a given load.

4.11.3 Capacitive Series Reactor Compensation

Capacitive reactors connected in series with a long line help in raising both the steady state and transient stability limits, in addition to minimizing the voltage dip at the consumers' terminals during heavy loads. Series capacitive compensation also assists in improving the economic loading of a line.

Figure 4.22 shows series and shunt capacitors connected in a long power transmission line.

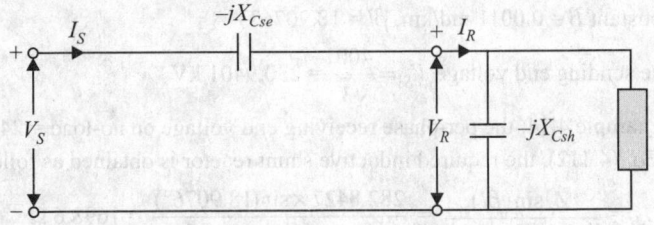

Fig. 4.22 Shunt and series compensation of a long line

The per-phase active power transferred over a loss-free long line for the equivalent circuit shown in Fig. 4.22 can be written as

$$P = \frac{|V_S|\,|V_R|}{(X' - X_{Cse})}\sin\delta \qquad (4.114)$$

where X_{Cse} is the reactance of the capacitor connected in series with the line. The ratio X_{Cse}/X' is defined as the compensation. The ratio varies from 0.25 to 0.70.

From Eq. (4.114) it may be noted that the theoretical capability of the line to transmit power has been enhanced. Experience has shown that by incurring an additional cost of series capacitors (which is a fraction of the cost of a new line), the transmission capability of extra high voltage lines can be more than doubled. Another advantage of the use of series capacitors is that their reactive volt-ampere rating varies in concert with the line loading.

The disadvantage of series compensation is that under short-circuit conditions, the series capacitors have to be protected against the resultant high currents. Special protective devices, which bypass the short-circuit current, are employed for this purpose. Another disadvantage is that when a disturbance occurs in the system, a resonant circuit, which resonates at a frequency below the normal synchronous frequency, is set up. The sub-synchronous frequency f_r is given by

$$f_r = f_s\sqrt{\frac{1}{L'C_{se}}} \qquad (4.115)$$

where L' is the inductance in henry of the π-equivalent circuit, C_{se} is the series capacitance in farads, and f_s is the synchronous frequency in hertz. The sub-synchronous resonance (SSR), as it is called, is likely to cause damage to the turbo-generator, if it occurs.

Example 4.15 The sending end rated voltage of the three-phase transmission line in Example 4.10 is 400 kV.
(i) Compute the reactance and rating of an inductive shunt reactor to be connected at the receiving end, if the receiving end voltage is to be maintained at 400 kV on no-load.
(ii) A three-phase load of 800 MVA at 0.8 power factor lagging is connected at the receiving end of the line. It is required to maintain a voltage of 400 kV at the receiving end when the line is supplying the load. Compute the MVAR and capacitance of a shunt capacitor for the purpose.
(iii) Compute the sending end voltage and regulation if a series capacitor connected in the middle of the line provides 50% compensation.

Solution   From Example 4.10, the following data is available;

Surge impedance $Z_c = 282.8427 \ \Omega$

Phase constant $\beta = 0.0011$ rad/km, $\beta l = 18.9076°$

Per-phase sending end voltage $V_S = \dfrac{400}{\sqrt{3}} = 230.9401$ kV

(i) From Example 4.14, the per-phase receiving end voltage on no-load = 243.1459 kV
Using Eq. (4.112), the required inductive shunt reactor is obtained as follows

$$XL_{sh} = \frac{Z_c \sin(\beta l)}{\left(\dfrac{V_S}{V_R} - \cos(\beta l)\right)} = \frac{282.8427 \times \sin(18.9076°)}{[1 - \cos(18.9076°)]} = 1698.6 \ \Omega$$

Shunt reactor rating $= 3 \times \dfrac{230.9401 \times 230.9401}{1698.6} = 94.1945$ MVAR

(ii) Using Eq. (4.75), the power angle for supplying real power = $800 \times 0.8 = 640$ MW at the receiving end is given by

$$\frac{400 \times 400}{91.6532} \sin \delta_{12} = 640 \text{ or } \delta_{12} = \sin^{-1}\left(\frac{640 \times 91.6532}{400 \times 400}\right) = 21.5069°$$

In Eq. (4.105), if $A = \cos (\beta l)$ and $B = jX'$, which are the values of the constants, are substituted and the equation is simplified and separated into real and imaginary parts, then the expression for sending end reactive power is obtained as follows

$$Q_{12} = \frac{V_1^2}{X'} \cos(\beta l) - \frac{V_1 V_2}{X'} \cos(\delta_{12})$$

It was also shown that for a lossless line $X' = Z_C \sin \beta l = 91.6532 \ \Omega$
Substituting this value in the above expression, yields the value for the sending end reactive power as follows:

$$Q_{12} = \frac{(230.9401)^2}{91.6532} \cos(18.9076°) - \frac{230.9401 \times 230.9401}{91.6532} \cos(21.5069°)$$

$$= 9.1178 \text{ MVAR/phase}$$

Three-phase reactive power of the load = $800 \times 0.6 = 480$ MVAR

Three-phase rating of the shunt capacitor = $j(3 \times 9.1178) - j480 = -j452.6466$ MVAR

and capacitive reactanc $X_C = \dfrac{(400)^2}{j452.6466} = -j353.4766 \ \Omega$

Magnitude of shunt capacitance $= C = \dfrac{10^6}{2 \times \pi \times 50 \times 353.4766} = 9.0051 \ \mu F$

(iii) For a line to be loss free, equivalent reactance X' has been computed as

$X' = Z_C \sin\beta l = 91.6532 \ \Omega$

Capacitive reactance to be connected in series for 50% line compensation

$X_{Cse} = 0.5 \times 91.6532 \ \Omega = 45.8266 \ \Omega$

Receiving end current $I_R = \dfrac{800}{\sqrt{3} \times 400 \times 0.8} = 1.4434 \ kA$

The constants of the π-equivalent circuit with line compensation are

$Z' = j(X' - X_{Cse}) = j(91.6532 - 45.8266) = j45.8266 \ \Omega$

$Y' = j\dfrac{2}{Z_c} \tan\left(\dfrac{\beta l}{2}\right) = j\dfrac{2}{282.8427} \tan\left(\dfrac{18.9076}{2}\right) = j0.0012 \ S$

$A = \left(1 + \dfrac{Y'Z'}{2}\right) = \left(1 + \dfrac{j0.0012 \times j45.8266}{2}\right) = 0.9730$

and $B = Z' = j\ 45.8266 \ \Omega$

Assume that the per-phase receiving end voltage V_R is the reference vector. Thus,

$I_R = 1.4434(0.8 - j0.6) = (1.1547 - j0.8660) \ kA$

$V_S = 0.9519 \times 230.9401 + j45.8266 \times (1.1547 - j0.8660)$

$= (259.52 + j52.917) kV$

$= 264.8596 \ \angle 11.5248° \ kV$

Sending end line voltage $= \sqrt{3} \times 264.8596 \angle 11.5248° = 458.7503 \ \angle 11.5248° \ kV$

Regulation % $= \dfrac{458.7503 - 400}{400} \times 100 = 14.6876\%$

4.12 TRANSMISSION LINE TRANSIENTS

The ultra-fast transient overvoltages occur on a power system either originated externally by atmospheric lightning discharges on the exposed transmission lines or generated internally by the abrupt but normal network changes resulting from regular switching operations. These transients, termed *surge phenomena*, are entirely electric in nature and involve, essentially, only the transmission lines. Physically, a disturbance of this type causes an electromagnetic wave that travels with almost the speed of light along the lines, giving rise to reflected waves at the line terminations. The surge phenomena associated with these waves take place, therefore, during the first few milliseconds after their initiation. Due to the line losses, the travelling waves attenuate fast and die out after a few reflections. Also, the series inductances of transformer windings effectively block the disturbances, thereby preventing them from entering generator windings. However, due to the reinforcing action of several reflected waves, it is possible for voltages to build up to a level that could destroy the insulation of the high-voltage equipment.

Lightning strokes hitting either ground wires or power conductors cause an injection of current, and divides the current into half, in two directions. The crest value of current along the struck conductor varies widely because of the wide variation in the intensity of the strokes; typical values being 10 kA and above.

In a case where a power line receives a direct stroke, the damage to equipment at line terminals is caused by the voltages between the line and the ground resulting from the injected charges, which travel along the line as current. These voltages are typically above 10^6 volts. Strokes to the ground wires can also cause high-voltage surges on the power lines by electromagnetic induction.

Transient overvoltages due to the surges provide a basis for selection of equipment insulation levels and surge-protection devices. At voltages up to about 220 kV, the insulation level of the lines and equipment is determined from the point of view of lightning protection. For voltages above 220 kV but less than 700 kV, switching operations as well as lightning are potentially damaging to insulation. At voltages above 700 kV, the level of insulation is decided mainly by the magnitude of switching surges.

Overhead lines can be protected from direct strokes of lightning, in most cases, by one or more wires at ground potential strung at the highest point above the power-line conductors. These protecting wires, called ground wires, or shield wires, are connected to ground through the transmission towers supporting the line. The ground wires, rather than the power line, receive the lightning strokes in most cases.

Fastest circuit breakers that operate within a few cycles (1 cycle = 20 ms) are too slow to protect against lightning or switching surges. Lightning surges can rise to peak levels within a few microseconds and switching surges within a few hundred microseconds, which is fast enough to destroy insulation before a circuit breaker could open. Surge arresters are used to protect equipment insulation against transient overvoltages. These devices limit voltage to a ceiling level and absorb the energy from lightning and switching surges.

4.12.1 Travelling Waves

The study of transmission-line surges, regardless of their origin, is very complex. For lightning surges on transmission lines, the study of a lossless line is considered. This simplification enables understanding of some of the phenomena without involving complicated theory.

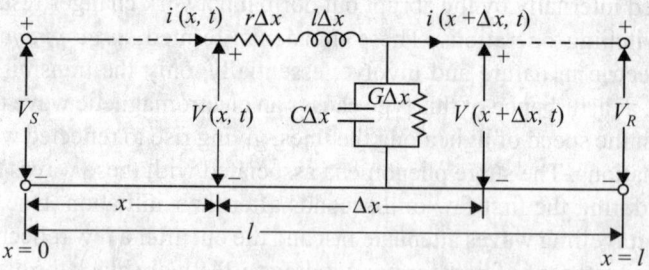

Fig. 4.23 Long line with distributed parameters

One phase and the neutral return of a transmission line section of length Δx metres is shown in Fig. 4.23. If the line has a series inductance L H/m, series resistance $r\ \Omega$/m, and a line-to-line capacitance C F/m, then the line section has a series inductance $L\Delta x$ H, shunt capacitance $C\Delta x$ F, and shunt conductance G S/m as shown. In Section 4.2, the direction of line position x was selected to be from the receiving end ($x = 0$) to the sending end ($x = l$); this selection was unimportant, since the variable x was subsequently eliminated when relating the steady-state sending end quantities V_S and I_S to the receiving end quantities V_R and I_R. Here, however, voltages and current waveforms travelling along the line are of interest. Therefore, for the present analysis, the direction of increasing x is selected to be from the sending end ($x = 0$) towards the receiving end ($x = l$).

Writing a KVL and KCL equation for the circuit in Fig. 4.23

$$v(x + \Delta x, t) - v(x, t) = -r\Delta x \, i\,(x, t) - L\Delta x \, \frac{\partial i(x,t)}{\partial t} \tag{4.116}$$

$$i\,(x + \Delta x, t) - i\,(x, t) = -G\Delta x \, v(x, t) - C\Delta x \, \frac{\partial i(x,t)}{\partial t} \tag{4.117}$$

Dividing Eqs (4.116) and (4.117) by Δx and taking the limit as $\Delta x \to 0$ yields

$$\frac{\partial v(x, t)}{\partial x} = -ri(x, t) - L\frac{\partial i(x, t)}{\partial t} \tag{4.118}$$

$$\frac{\partial i(x, t)}{\partial x} = -Gv(x, t) - C\frac{\partial v(x, t)}{\partial t} \tag{4.119}$$

The partial derivatives are used here because $v(x, t)$ and $i(x, t)$ are differentiated with respect to both position x and time t. Also, the negative signs in Eqs (4.118) and (4.119) are due to the reference direction for x. For example, with a positive value of $\partial i/\partial t$ in Fig. 4.23, $v(x, t)$ decreases as x increases.

Taking the Laplace transform of (4.118) and (4.119) gives

$$\frac{dV(x,s)}{dx} = -(r + sL)I(x,s) = -z(s)I(x,s) \tag{4.120}$$

$$\frac{dI(x,s)}{dx} = -(G + sC)V(x,s) = -y(s)V(x,s) \tag{4.121}$$

where zero initial conditions are assumed. $V(x, s)$ and $I(x, s)$ are the Laplace transforms of $v(x, t)$ and $i(x, t)$. Also, since the derivatives are now with respect to only one variable x, ordinary, rather than partial derivatives are used.

Now differentiating Eq. (4.120) with respect to x and substituting Eq. (4.121) leads to

$$\frac{d^2V(x,s)}{dx^2} = -z(s)\frac{dI(x,s)}{dx} = y(s)z(s)V(x,s)$$

$$= \gamma^2(s)V(x, s) \tag{4.122}$$

Similarly, differentiating Eq. (4.121) with respect to x and substituting Eq. (4.120) yields

$$\frac{d^2I(x,s)}{dx^2} = -y(s)\frac{dV(x,s)}{dx} = y(s)z(s)I(x,s)$$

$$= \gamma^2(s)I(x, s) \tag{4.123}$$

where $\quad \gamma(s) = \sqrt{(r + sL)(G + sC)} \tag{4.124}$

Equation (4.122) is a linear, second-order, homogeneous differential equation. By inspection, its solution is obtained as

$$V(x, s) = V^f(s) e^{-\gamma(s)x} + V^b(s) e^{\gamma(s)x} \tag{4.125}$$

$$I(x, s) = I^f(s) e^{-\gamma(s)x} + I^b(s) e^{\gamma(s)x} \tag{4.126}$$

where $V^f(s)$, $V^b(s)$, $I^f(s)$, and $I^b(s)$ are constants that can be evaluated from the boundary conditions at the sending and receiving ends of the line. The superscripts f and b refer to waves travelling in the positive x and negative x directions, which will be explained later in the section.

In general, it is impossible to obtain a closed form expression for $v(x, t)$ and $i(x, t)$, which are inverse Laplace transforms of Eqs (4.125) and (4.126). However for a lossless line in which $r = 0$, and $G = 0$, the inverse transform can be obtained as follows. Rewriting Eq. (4.124) gives

$$\gamma(s) = s\sqrt{LC} = s/\mathcal{V} \tag{4.127}$$

where $\quad \mathcal{V} = \dfrac{1}{\sqrt{LC}}$ m/s $\tag{4.128}$

Thus, Eqs (4.125) and (4.126) become

$$V(x, s) = V^f(s)e^{-sx/\mathcal{V}} + V^b(s)e^{sx/\mathcal{V}} \tag{4.129}$$

$$I(x, s) = I^f(s)e^{-sx/\mathcal{V}} + I^b(s)e^{sx/\mathcal{V}} \tag{4.130}$$

Taking the inverse Laplace transform of Eqs (4.129) and (4.130), and recalling the time shift property, $\mathcal{L}[f(t - \tau)] = F(s)e^{-s\tau}$, yields

$$v(x, t) = v^f\left(t - \frac{x}{\mathcal{V}}\right) + v^b\left(t + \frac{x}{\mathcal{V}}\right) \tag{4.131}$$

$$i(x, t) = i^f\left(t - \frac{x}{\mathcal{V}}\right) + i^b\left(t + \frac{x}{\mathcal{V}}\right) \tag{4.132}$$

where the functions $v^f(t - x/\mathcal{V})$, $v^b(t + x/\mathcal{V})$, $i^f(t - x/\mathcal{V})$, and $i^b(t + x/\mathcal{V})$, can be evaluated from the boundary conditions.

The physical significance of the solutions given by Eqs (4.131) and (4.132) may be explained as follows.

Any function of the form

$$f\left(t \pm \frac{x}{\mathcal{V}}\right) \text{ or } f(x \pm t\mathcal{V}) \tag{4.133}$$

represents a rigid distribution, or a travelling wave, because for any value of t a corresponding value of x can be found such that $f(x \pm t\mathcal{V})$ has a constant value, and therefore defines a fixed point on $f(x \pm t\mathcal{V})$. Corresponding values of x and t, which define the same point on a wave, are given by

$$t - \frac{x}{\mathcal{V}} = \lambda_2 \text{ for the forward wave} \tag{4.134a}$$

$$t + \frac{x}{\mathcal{V}} = \lambda_1 \text{ for the backward wave} \tag{4.134b}$$

where λ_1 and λ_2 are constants.

Travelling wave on a lossless line of the form is shown in Fig. 4.24. Since the argument $(x - t\mathcal{V})$ will not change if t and x are increased by the amounts Δt and

$\mathcal{V}\Delta t$, respectively, it is clear that the wave of the form represents a wave travelling in positive x direction with a velocity $\mathcal{V} = \Delta x/\Delta t$.

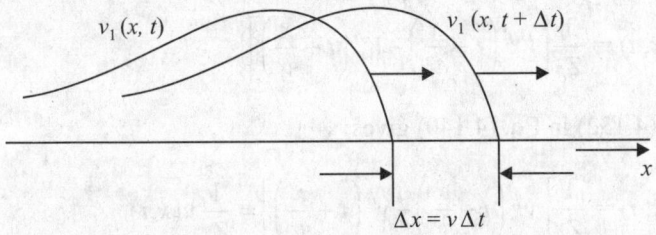

$$\Delta x = v \Delta t$$

Fig. 4.24 Travelling voltage wave

The velocities of propagation are found by differentiating Eqs (4.134a) and (4.134b).
Thus

$$\frac{dx}{dt} = \mathcal{V} = \frac{1}{\sqrt{LC}} \quad \text{for the forward wave} \tag{4.135a}$$

$$\frac{dx}{dt} = -\mathcal{V} = -\frac{1}{\sqrt{LC}} \quad \text{for the backward wave} \tag{4.135b}$$

Hence, it may be seen that the voltage and current distribution of Eqs (4.131) and (4.132) are propagated as travelling waves, and may consist of forward waves v^f, i^f moving in the direction of positive x and backward waves v^b, i^b moving in the direction of negative x, both waves having the same velocity $\mathcal{V} = 1/\sqrt{LC}$. As mentioned earlier in Section 4.3, the velocity of propagation of voltage and current waves for an overhead line is approximately equal to 3×10^8 m/s, the speed of light in free space. For underground cables, the speed of propagation is much lower than for overhead lines.

The terms $I^f(s)$ and $I^b(s)$ may be evaluated by using Eqs (4.129) and (4.130) in Eq. (4.121), neglecting shunt conductance,

$$\frac{s}{\mathcal{V}}\left[-I^f(s)e^{-sx/\mathcal{V}} + I^b(s)e^{sx/\mathcal{V}}\right] = -sC\left[V^f(s)e^{-sx/\mathcal{V}} + V^b(s)e^{sx/\mathcal{V}}\right] \tag{4.136a}$$

Equating the coefficients of $e^{-sx/\mathcal{V}}$ on both sides of the above equation gives

$$I^f(s) = \mathcal{V}CV^f(s) = \frac{C}{\sqrt{LC}}V^f(s) = \frac{V^f(s)}{\sqrt{\frac{L}{C}}} = \frac{V^f(s)}{Z_c} \tag{4.136b}$$

where

$$Z_c = \sqrt{\frac{L}{C}} = \text{surge impedance of the lossless line in ohms} \tag{4.137}$$

Similarly equating the coefficients of $e^{sx/\mathcal{V}}$ in Eq. (4.136a)

$$I^b(s) = \frac{-V^b(s)}{Z_c} \tag{4.138}$$

Thus Eqs (4.130) and (4.132) may be written as

$$I(x, s) = \frac{1}{Z_c}\left[V^f(s)e^{-sx/\mathcal{V}} - V^b(s)e^{sx/\mathcal{V}} \right]$$

(4.139)

$$i(x, t) = \frac{1}{Z_c}\left[V^f\left(t - \frac{x}{\mathcal{V}}\right) - V^b\left(t + \frac{x}{\mathcal{V}}\right) \right]$$

(4.140)

Using Eq. (4.132) in Eq. (4.140) gives

$$i(x, t) = \frac{1}{Z_c}\left[v^f\left(t - \frac{x}{\mathcal{V}}\right) - v^b\left(t + \frac{x}{\mathcal{V}}\right) \right] = \frac{1}{Z_c}v(x,t)$$

$$= \left[i^f\left(t - \frac{x}{\mathcal{V}}\right) + i^b\left(t + \frac{x}{\mathcal{V}}\right) \right]$$

From the preceding equation it may be noted that

$$\frac{v(x,t)}{i(x,t)} = \sqrt{\frac{L}{C}} = Z_c \text{ for a forward wave}$$

(4.141)

$$\frac{v(x,t)}{i(x,t)} = -\sqrt{\frac{L}{C}} = -Z_c \text{ for a backward wave}$$

(4.142)

That is, the ratio of voltage to current is a constant and is equal to the surge impedance. The voltage and current waves are replicas of each other. While the forward voltage v^f and forward current i^f have the same sign, the backward voltage v^b and backward current i^b are of opposite signs. This means that the current flows in the direction of propagation of a positive voltage wave.

4.12.2 Boundary Conditions for a Line of Finite Length

Figure 4.25 shows a lossless line of length l, with surge impedance $Z_c = \sqrt{L/C}$, terminated by an impedance $Z_R(s)$ at the receiving end. At the sending end, the line is represented by a source whose Thevenin equivalent voltage and impedance are $E(s)$ and $Z_S(s)$, respectively. $V(x, s)$ and $I(x, s)$ are the Laplace transforms of the voltage and current at a position x on the line and velocity $\mathcal{V} = 1/\sqrt{LC}$. It is assumed that the line is initially unenergized.

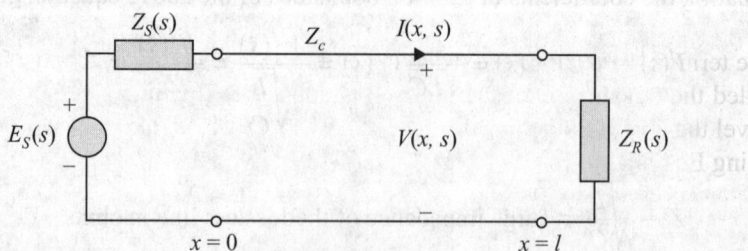

Fig. 4.25 Lossless line with source and load terminations

From Eqs (4.129) and (4.139), at $x = 0$,

$$V(0, s) = V^f(s) + V^b(s)$$

(4.143)

$$I(0, s) = \frac{1}{Z_c}\left[V^f(s) - V^b(s)\right] \tag{4.144}$$

From Fig. 4.25, the boundary condition at the sending end is

$$E(s) = V(0, s) + Z_S(s)I(0, s) \tag{4.145}$$

Substituting Eqs (4.143) and (4.144) in Eq. (4.145)

$$E(s) = V^f(s) + V^b(s) + Z_S(s) \times \frac{1}{Z_c}\left[V^f(s) - V^b(s)\right]$$

$$= \left(1 + \frac{Z_S(s)}{Z_c}\right)V^f(s) + \left(1 - \frac{Z_S(s)}{Z_c}\right)V^b(s) \tag{4.146}$$

From Fig. 4.25, the boundary condition at the receiving end is

$$V(l, s) = Z_R(s)I(l, s) \tag{4.147}$$

From Eqs (4.129) and (4.139), at $x = l$,

$$V(l, s) = V^f(s)e^{-sl/\mathcal{V}} + V^b(s)e^{sl/\mathcal{V}} \tag{4.148}$$

$$I(l, s) = \frac{1}{Z_c}[V^f(s)e^{-sl/\mathcal{V}} - V^b(s)e^{sl/\mathcal{V}}] \tag{4.149}$$

Substituting Eqs (4.148) and (4.149) in Eq. (4.147) and simplifying gives

$$V^f(s)e^{-sl/\mathcal{V}} + V^b(s)e^{sl/\mathcal{V}} = \frac{Z_R(s)}{Z_c}[V^f(s)e^{-sl/\mathcal{V}} + V^b(s)e^{sl/\mathcal{V}}]$$

or $$\left(1 - \frac{Z_R(s)}{Z_c}\right)e^{-sl/\mathcal{V}}V^f(s) + \left(1 + \frac{Z_R(s)}{Z_c}\right)e^{sl/\mathcal{V}}V^b(s) = 0$$

Solving for $V^b(s)$ leads to

$$V^b(s) = \alpha_R(s)V^f(s)e^{-2sl/\mathcal{V}} = \alpha_R(s)V^f(s)e^{-2s\tau} \tag{4.150}$$

where

$$\alpha_R(s) = \frac{\dfrac{Z_R(s)}{Z_c} - 1}{\dfrac{Z_R(s)}{Z_c} + 1} = \frac{Z_R(s) - Z_c}{Z_R(s) + Z_c} \quad \text{per unit} \tag{4.151}$$

The term $\alpha_R(s)$ is called the receiving end reflection coefficient, and $\tau = (l/\mathcal{V})$ is called the transit time of the line and is defined as the time taken by a wave to travel the length of the line.

Using Eqs (4.150) in Eqs (4.129) and (4.139) gives

$$V(x, s) = V^f(s)\left[e^{-sx/\mathcal{V}} + \alpha_R(s)e^{s[(x/\mathcal{V}) - 2\tau]}\right] \tag{4.152}$$

$$I(x, s) = \frac{V^f(s)}{Z_c}\left[e^{-sx/\mathcal{V}} - \alpha_R(s)e^{s[(x/\mathcal{V}) - 2\tau]}\right] \tag{4.153}$$

A study of Eq. (4.150) provides an interesting physical insight into the problem. The equation indicates that the backward travelling wave is a scaled version of the forward travelling wave, the scaling factor being $\alpha_R = [(Z_R(s) - Z_c)/(Z_R(s)$

+ Z_c)], and is delayed by $2(l - x)/\mathcal{V}$. Hence, the backward travelling wave may be visualized as a reflection of the forward travelling wave.

Substituting Eqs (4.152) and (4.153) in Eq. (4.145), which was developed for the receiving end, leads to

$$V^f(s) \left[1 + \alpha_R(s)\ e^{-2s\tau} \right] = E(s) - \frac{Z_s(s)}{Z_c} V^f(s)[1 - \alpha_R(s)e^{-2s\tau}]$$

(4.154)

Solving for $V^f(s)$ gives

$$V^f(s) \left[\left(\frac{Z_S(s)}{Z_c} + 1 \right) - \alpha_R(s)e^{-2s\tau} \left(\frac{Z_S(s)}{Z_c} - 1 \right) \right] = E(s)$$

or $\quad V^f(s) \left(\dfrac{Z_S(s)}{Z_c} + 1 \right) \left[1 - \alpha_R(s)\ \alpha_S(s)e^{-2s\tau} \right] = E(s)$

$\therefore \quad V^f(s) = E(s) \left[\dfrac{Z_S(s)}{Z_c} + 1 \right] \left[\dfrac{1}{1 - \alpha_R(s)\alpha_S(s)e^{-2s\tau}} \right]$ (4.155)

where

$$\alpha_S(s) = \frac{\dfrac{Z_S(s)}{Z_c} - 1}{\dfrac{Z_S(s)}{Z_c} + 1} = \frac{Z_S(s) - Z_c}{Z_S(s) + Z_c} \text{ per unit}$$

(4.156)

The term α_S is called the sending end reflection coefficient. By substituting Eq. (4.156) in Eqs (4.152) and (4.153), the complete solution is obtained as

$$V(x, s) = E_S(s) \left[\frac{Z_c}{Z_S(s) + Z_c} \right] \left[\frac{e^{-sx/\mathcal{V}} + \alpha_R(s)\ e^{s[(x/\mathcal{V}) - 2\tau]}}{1 - \alpha_R(s)\alpha_S(s)\ e^{-2s\tau}} \right] \quad (4.157)$$

$$I(x, s) = \left[\frac{E_S(s)}{Z_S(s) + Z_c} \right] \left[\frac{e^{-sx/\mathcal{V}} - \alpha_R(s)\ e^{s\ [(x/\mathcal{V}) - 2\tau]}}{1 - \alpha_R(s)\alpha_S(s)\ e^{-2s\tau}} \right] \quad (4.158)$$

The general equations (4.157) and (4.158) will be made use of to establish simplified procedures in connection with reflection of travelling waves.

4.12.3 Behaviour of a Wave at a Transition Point

When a travelling wave on a line reaches a transition point at which there is an abrupt change of circuit constants, such as an open-or short-circuited terminal, or a junction with another line, or a machine winding, then a part of the wave is reflected back on the line, and a part may pass on to the other sections of the circuit. At the transition point itself, the voltage (or current) may be anything from zero to double the magnitude of the wave, depending on the terminal characteristics. The incoming wave is called the incident wave, and the two waves to which it gives rise at the transition point are called the reflected and transmitted waves, respectively. Such waves are formed in accordance with Kirchhoff's laws, and they satisfy the differential equations of the line.

One phase and the neutral return of a transmission line section shown in Fig. 4.25, which is initially unenergized, connected to a step voltage source $Eu(t)$ at sending end are considered. The line has length l, characteristic impedance Z_c, and velocity of propagation V. The effect of different types of line terminations at the receiving end on the travelling waves will be discussed in the following sections.

Line Terminated by Surge Impedance of the Line

The receiving end is connected by Z_c, hence $Z_R = Z_c$. The Laplace transform of source voltage is $E(s) = E/s$. Let it be assumed that the source impedance $Z_S = 0$. Then, from Eqs (4.151) and (4.156)

$$\alpha_R(s) = \frac{Z_c - Z_c}{Z_c + Z_c} = 0$$

$$\alpha_S(s) = \frac{Z_S - Z_c}{Z_S + Z_c} = \frac{0 - Z_c}{0 + Z_c} = -1$$

Thus from Eqs (4.157) and (4.158),

$$V(s, x) = \left(\frac{E}{s}\right)\left(\frac{Z_c}{0 + Z_c}\right)\left(e^{-sx/\mathcal{V}}\right) = \left(\frac{E}{s}\right)\left(e^{-sx/\mathcal{V}}\right) \tag{4.159}$$

$$I(s, x) = \left(\frac{E}{s}\right)\left(\frac{1}{0 + Z_c}\right)\left(e^{-sx/\mathcal{V}}\right) = \left(\frac{E/Z_c}{s}\right)\left(e^{-sx/\mathcal{V}}\right) \tag{4.160}$$

Taking the inverse Laplace transform of Eqs (4.159) and (4.160) leads to

$$v(x, t) = Eu\left(t - \frac{x}{\mathcal{V}}\right) \tag{4.161}$$

$$i(x, t) = \frac{E}{Z_c}u\left(t - \frac{x}{\mathcal{V}}\right) \tag{4.162}$$

At $t = 0$, the step voltage E is applied at the sending end of the line that has a surge impedance Z_c. A forward travelling voltage wave $v^f = E$ associated with a current wave of value $i^f = E/Z_c$ is initiated, and these waves travel in the positive x direction. At the receiving end, the load connected being Z_c, the reflection coefficient $\alpha_R = 0$, no backward wave is initiated, and $v^b = 0$, and $i^b = 0$. The transmitted (or refracted) voltage waves v^t and current i^t to the load, in accordance with Kirchhoff's laws, is then $v^t = v^f = E$ and $i^t = i^f = E/Z_c$. Thus the forward wave brings energy to the point of termination, which is transmitted or refracted to the surge impedance load Z_c connected at that point, and the energy is dissipated in it. The line is energized to steady-state voltage E volts and current E/Z_c amperes. The voltage and current waves are plotted in Fig. 4.26 for $x = l/2$, and $x = l$. At the middle of the line $x = l/2$, hence,

$$v\left(\frac{l}{2}, t\right) = Eu\left(t - \frac{l}{2\mathcal{V}}\right) = Eu\left(t - \frac{\tau}{2}\right) \tag{4.163a}$$

$$i\left(\frac{l}{2}, t\right) = \frac{E}{Z_c}u\left(t - \frac{\tau}{2}\right) \tag{4.163b}$$

At the transition point $x = l$, hence,

$$v(l, t) = Eu\left(t - \frac{l}{\mathcal{V}}\right) = Eu(t - \tau) \qquad (4.164a)$$

$$i(l, t) = \frac{E}{Z_c}u(t - \tau) \qquad (4.164b)$$

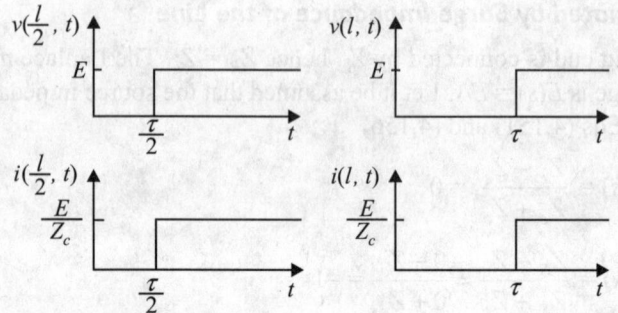

Fig. 4.26 Voltage and current waveforms for $Z_R = Z_C$ and $Z_S = 0$

Line Open at the Receiving End

The source voltage at the sending end is step input voltage $E(t) = E$, then $E(s) = E/s$. The step voltage source is matched at the sending end with the source impedance $Z_S(s) = Z_c$ and the receiving end is open, so $Z_R(s) = \infty$. Using Eqs (4.151) and (4.156) yields

$$\alpha_R(s) = \lim_{Z_R \to \infty}\left(\frac{Z_R - Z_c}{Z_R + Z_c}\right) = 1$$

$$\alpha_S(s) = \frac{Z_S - Z_c}{Z_S + Z_c} = \frac{1-1}{1+1} = 0$$

From Eqs (4.157) and (4.158),

$$V(s, x) = \left(\frac{E}{s}\right)\left(\frac{Z_c}{Z_c + Z_c}\right)\left[e^{-sx/\mathcal{V}} + e^{s\,[(x/\mathcal{V})-2\tau]}\right]$$

or $\qquad V(s, x) = \frac{E}{s}\left(\frac{1}{2}\right)\left[e^{-sx/\mathcal{V}} + e^{s\,[(x/\mathcal{V})-2\tau]}\right] \qquad (4.165)$

$$I(s, x) = \frac{E}{s}\left(\frac{1}{2Z_c}\right)\left[e^{-sx/\mathcal{V}} - e^{s\,[(x/\mathcal{V})-2\tau]}\right] \qquad (4.166)$$

Taking Laplace inverse of Eqs (4.165) and (4.166) gives

$$v(x, t) = \frac{E}{2}u\left(t - \frac{x}{\mathcal{V}}\right) + \frac{E}{2}u\left(t + \frac{x}{\mathcal{V}} - 2\tau\right) \qquad (4.167)$$

$$i(x, t) = \frac{E}{2Z_c}u\left(t - \frac{x}{\mathcal{V}}\right) - \frac{E}{2Z_c}u\left(t + \frac{x}{\mathcal{V}} - 2\tau\right) \qquad (4.168)$$

At $x = l/2$,

$$v\left(\frac{l}{2}, t\right) = \frac{E}{2} u\left(t - \frac{\tau}{2}\right) + \frac{E}{2} u\left(t - \frac{3\tau}{2}\right) \tag{4.169a}$$

$$i\left(\frac{l}{2}, t\right) = \frac{E}{2Z_c} u\left(t - \frac{\tau}{2}\right) - \frac{E}{2Z_c} u\left(t - \frac{3\tau}{2}\right) \tag{4.169b}$$

At $x = l$,

$$v(l, t) = \frac{E}{2} u(t - \tau) + \frac{E}{2} u(t - \tau) = Eu(t - \tau) \tag{4.170a}$$

$$i(l, t) = \frac{E}{2Z_c} u(t - \tau) - \frac{E}{2Z_c} u(t - \tau) = 0 \tag{4.170b}$$

These travelling waves at the middle of the line are plotted in Fig. 4.27. At $t = 0$, the step voltage source of E volts is impressed on the source impedance $Z_S = Z_c$ in series with the line surge impedance Z_c. Due to the division of voltage, the sending-end voltage at $t = 0$ is $E/2$, and a forward travelling step voltage wave $v^f = E/2$ volts, and a step current wave $i^f = E/(2Z_c)$ amperes are initiated at the sending end. These waves arrive at the sending end of the line at $t = \tau$. With $\alpha_R(s)$ $= 1$, the backward travelling voltage wave $v^b = v^f = E/2$, and the total voltage at the receiving end at $t = \tau$ is $v^b + v^f = E$. The backward travelling current wave $i^b = -i^f = E/(2Z_c)$ is also initiated at the receiving end at time $t = \tau$, and the total current at the receiving end at $t = \tau$ is $i^b + i^f = 0$. These backward voltage and current waves travel towards the sending end and are superimposed on the waves travelling in the forward direction. No additional forward or backward travelling waves are initiated as at the sending end $\alpha_S(s) = 0$. In steady-state, the line, which is open at the receiving end, is energized at E volts with zero current.

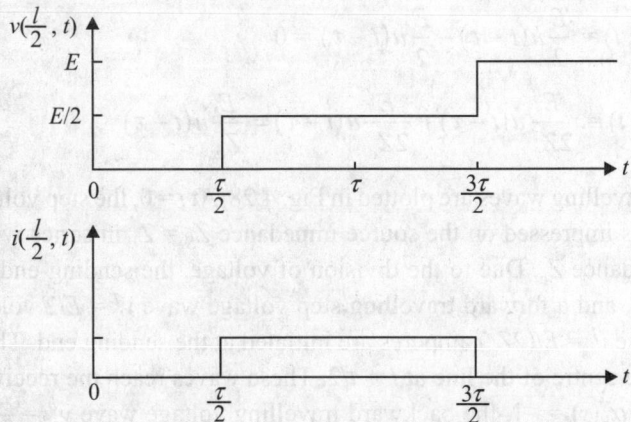

Fig. 4.27 Voltage and current waveforms at the middle
of the line for $Z_R = \alpha$, $Z_S = Z_C$

Line Short-circuited at the Receiving End

The source voltage at the sending end is step input voltage $E(s) = E/s$. The step voltage source is matched at the sending end with the source impedance $Z_S(s) = Z_c$ and the receiving end is open, so $Z_R(s) = 0$. The use of Eqs (4.151) and (4.156) gives

$$\alpha_R(s) = \frac{0 - Z_c}{0 + Z_c} = -1$$

$$\alpha_S(s) = \alpha_S(s) = \frac{Z_c - Z_c}{Z_c + Z_c} = 0$$

From Eqs (4.157) and (4.158),

$$V(x, s) = \frac{E}{s}\left(\frac{1}{2}\right)\left[e^{-sx/\mathcal{V}} - e^{s[(x/\mathcal{V}) - 2\tau]}\right] \tag{4.171}$$

$$I(x, s) = \frac{E}{s}\frac{1}{2Z_c}\left[e^{-sx/\mathcal{V}} + e^{s[(x/\mathcal{V}) - 2\tau]}\right] \tag{4.172}$$

Taking the inverse Laplace transform of Eqs (4.171) and (4.172)

$$v(x, t) = \frac{E}{2}u\left(t - \frac{x}{\mathcal{V}}\right) - \frac{E}{2}u\left(t + \frac{x}{\mathcal{V}} - 2\tau\right) \tag{4.173}$$

$$i(x, t) = \frac{E}{2Z_c}u\left(t - \frac{x}{\mathcal{V}}\right) + \frac{E}{2Z_c}u\left(t + \frac{x}{\mathcal{V}} - 2\tau\right) \tag{4.174}$$

At $x = l/2$,

$$v\left(\frac{l}{2}, t\right) = \frac{E}{2}u\left(t - \frac{\tau}{2}\right) - \frac{E}{2}u\left(t - \frac{3\tau}{2}\right) \tag{4.175a}$$

$$i\left(\frac{l}{2}, t\right) = \frac{E}{2Z_c}u\left(t - \frac{\tau}{2}\right) + \frac{E}{2Z_c}u\left(t - \frac{3\tau}{2}\right) \tag{4.175b}$$

At $x = l$,

$$v(l, t) = \frac{E}{2}u(t - \tau) - \frac{E}{2}u(t - \tau) = 0 \tag{4.176a}$$

$$i(l, t) = \frac{E}{2Z_c}u(t - \tau) + \frac{E}{2Z_c}u(t - \tau) = \frac{E}{Z_c}u(t - \tau) \tag{4.176b}$$

These travelling waves are plotted in Fig. 4.28. At $t = 0$, the step voltage source of E volts is impressed on the source impedance $Z_S = Z_c$ in series with the line surge impedance Z_c. Due to the division of voltage, the sending-end voltage at $t = 0$ is $E/2$, and a forward travelling step voltage wave $v^f = E/2$ volts and step current wave $i^f = E/(2Z_c)$ amperes are initiated at the sending end. These waves arrive at the centre of the line at $t = \tau/2$. These waves reach the receiving end at $t = \tau$. With $\alpha_R(s) = -1$, the backward travelling voltage wave $v^b = -v^f$, and the total voltage at the receiving end at $t = \tau$ is $v^b + (-v^f) = 0$. At $t = \tau$, the backward travelling current wave $i^b = i^f$ is initiated at the receiving end and at time $t = \tau$, the total current at the receiving end is $i^b + i^f = E/Z_c$. These waves arrive at the centre of the line at $t = 3\tau/2$ and are superimposed on the forward travelling waves. No additional forward or backward travelling waves are initiated at the sending end as $\alpha_S(s) = 0$. In steady-state, voltage at receiving end is zero and current is equal to $i^f + i^b = E/Z_c$.

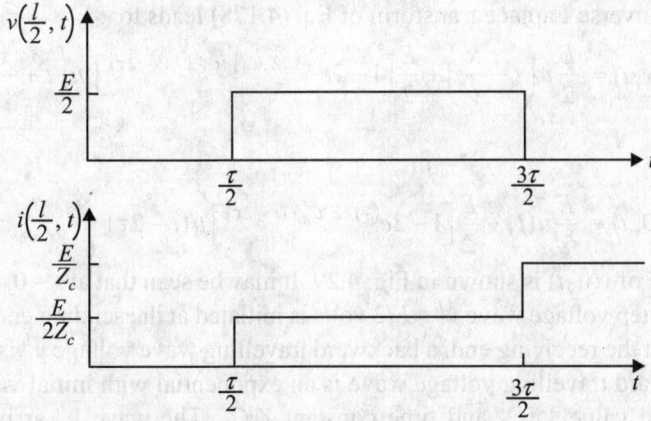

Fig. 4.28 Voltage and current waveforms at the middle of the line for $Z_R = 0$, $Z_S = Z_c$

Line Terminated by Capacitive Load

The receiving end of the line, whose surge impedance is Z_c, is terminated by a capacitor of capacitance C_R farads, which is initially unenergized. At the sending end, source voltage is unit step voltage $e_S(t) = Eu(t)$ and source impedance $Z_S = Z_c$. Then $E_S(s) = E/s$ and $Z_R = 1/(sC_R)$. Substitution in Eq. (4.151) gives

$$\alpha_R(s) = \frac{\dfrac{1}{sC_R} - Z_c}{\dfrac{1}{sC_R} + Z_c} = \frac{1 - sC_R Z_c}{1 + sC_R Z_c} = \frac{-s + \dfrac{1}{Z_c C_R}}{s + \dfrac{1}{Z_c C_R}}$$

$$\alpha_S(s) = \frac{Z_c - Z_c}{Z_c + Z_c} = 0$$

From Eq. (4.157),

$$V(x, s) = \frac{E}{s}\left(\frac{1}{2}\right)\left[e^{-sx/v} + \left(\frac{-s + \dfrac{1}{Z_c C_R}}{s + \dfrac{1}{Z_c C_R}}\right) e^{s[(x/v) - 2\tau]} \right]$$

$$= E\left(\frac{1}{2}\right)\left[\frac{e^{-sx/v}}{s} + \left(\frac{-s + \dfrac{1}{Z_c C_R}}{s\left(s + \dfrac{1}{Z_c C_R}\right)}\right) e^{s[(x/v) - 2\tau]} \right] \tag{4.177}$$

Taking partial fraction expansion of the second term of Eq. (4.177) yields

$$V(x, s) = \frac{E}{2}\left[\frac{e^{-sx/v}}{s} + \left(\frac{1}{s} - \frac{2}{s + \dfrac{1}{Z_c C_R}}\right)\left[e^{s[(x/v) - 2\tau]}\right] \right] \tag{4.178}$$

Taking inverse Laplace transform of Eq. (4.178) leads to

$$v(x, t) = \frac{E}{2}u\left(t - \frac{x}{\mathcal{V}}\right) + \frac{E}{2}\left[1 - 2e^{(-1/Z_cC_R)(t + x/\mathcal{V} - 2\tau)}\right]u\left(t + \frac{x}{\mathcal{V}} - 2\tau\right)$$

(4.179)

At $x = 0$,

$$v(0, t) = \frac{E}{2}u(t) + \frac{E}{2}\left[1 - 2e^{(-1/Z_cC_R)(t - 2\tau)}\right]u(t - 2\tau)$$

(4.180)

The plot of $v(0, t)$ is shown in Fig. 4.29. It may be seen that at $t = 0$, a forward travelling step voltage wave $v^f = E/2$ volts is initiated at the sending end. At $t = \tau$, v^f arrives at the receiving end, a backward travelling wave voltage v^b is initiated. The backward travelling voltage wave is an exponential with initial value $-E/2$, steady-state value $+E/2$, and time constant Z_cC_R. The wave v^b arrives at the sending end at $t = 2\tau$, where it is superimposed on the forward travelling wave. No additional waves are initiated, since the source impedance is matched to the line and $\alpha_S = 0$. In steady-state, the line and the capacitor at the receiving end are energized at E volts with zero current.

The capacitor at the receiving end can also be viewed as a short circuit at the instant $t = \tau$, when v^f arrives at the receiving end. For a short circuit at the receiving end, $\alpha_R = -1$, and therefore wave front of $v^b = -v^f = -E/2$. However, in steady-state, the capacitor is an open circuit, for which $\alpha_R = +1$, and the steady-state backward travelling voltage wave equals the forward travelling voltage wave.

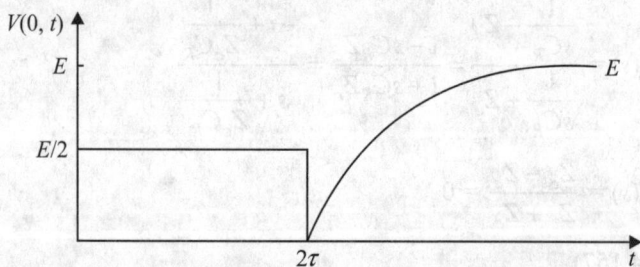

Fig. 4.29 Voltage waveform at sending end with line
terminated with C, $Z_S = Z_c$

General Transition Points

Figure 4.30 shows a general transition point normally met with on transmission systems, which consist of a junction at which there is a shunt impedance to ground $Z_g(s)$; and n transmission lines of surge impedances $Z_{c1}, ..., Z_{cn}$ joined through series impedances $Z_1(s), ..., Z_n(s)$, respectively. Let the total impedance looking beyond the transition point be taken as $Z_0(s)$. When an incident wave v^f approaching along the line having surge impedance Z_{c1} reaches the junction point, it will give rise to a wave v_1^b reflected back on line Z_{c1}; transmitted waves $v_1^t, ..., v_n^t$ on lines Z_{c2} to Z_{cn}, respectively; a potential v at the junction; and a potential v_0 at the transition point. Let the transition point be taken as the origin of coordinates, and distance along the line away from the transition point be counted as negative, so that the approaching incident wave is travelling in the positive direction. All lines will be taken as lossless lines, so that, in accordance with Eqs (4.141) and

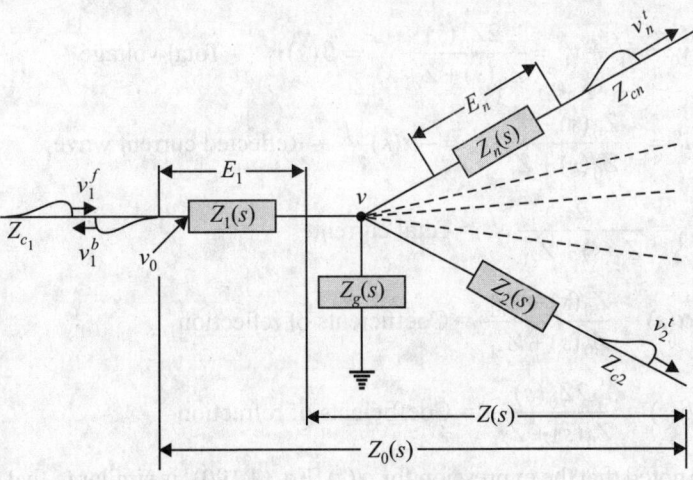

Fig. 4.30 General transition point

(4.142), the voltage and current incident (forward) waves are related by the surge impedance of the line, as follows:

$$\frac{v_1^f}{i_1^f} = Z_{c1} \tag{4.181}$$

The reflected (backward) waves are related by the negative of the surge impedance by

$$\frac{v_1^b}{i_1^b} = -Z_{c1} \tag{4.182}$$

Transmitted waves on outgoing lines are given by

$$\frac{v_k^t}{i_k^t} = Z_{ck} \quad \text{for } k = 2, 3, ..., n \tag{4.183}$$

Quantities, such as v^f and i^f will denote incident waves; v^b and i^b denote reflected waves; and v^t and i^t denote transmitted waves. The total current and voltage at the transition point are

$$i_0 = i_1^f + i_1^b \tag{4.184}$$

$$v_0 = v_1^f + v_1^b = Z_0(s)i_0 \tag{4.185}$$

Substituting Eqs (4.181) and (4.182) in Eqs (4.184) and (4.185) and solving for the various quantities gives

$$v_1^b = v_0 - v_1^f = Z_0(s)\left(i_1^f + i_1^b\right) - v_1^f = Z_0(s)\left(\frac{v_1^f}{Z_{c1}} - \frac{v_1^b}{Z_{c1}}\right) - v_1^f$$

or $\quad \left[\dfrac{Z_0(s)}{Z_{c1}} + 1\right] v_1^b = \left[\dfrac{Z_0(s)}{Z_{c1}} - 1\right] v_1^f$

Hence, $\quad v_1^b = \dfrac{Z_0(s) - Z_{c1}}{Z_0(s) + Z_{c1}} v_1^f = \alpha(s)v_1^f = $ Reflected voltage wave $\tag{4.186}$

$$v_0 = v_1^f + v_1^b = \frac{2Z_0(s)}{Z_0(s) + Z_{c1}} v_1^f = \beta(s) v_1^f = \text{Total voltage} \qquad (4.187)$$

$$i_1^b = -\frac{Z_0(s) - Z_{c1}}{Z_0(s) + Z_{c1}} i_1^f = -\alpha(s) \, i_1^f = \text{Reflected current wave} \qquad (4.188)$$

$$i_0 = \frac{2}{Z_0(s) + Z_{c1}} v_1^f = \text{Total current} \qquad (4.189)$$

where $\quad \alpha(s) = \dfrac{Z_0(s) - Z_{c1}}{Z_0(s) + Z_{c1}} = \text{Coefficients of reflection} \qquad (4.190)$

$$\beta(s) = \frac{2Z_0(s)}{Z_0(s) + Z_{c1}} = \text{Coefficients of refraction} \qquad (4.191)$$

It may be noted that the expression for $\alpha(s)$ [Eq. (4.190)] is similar to that of $\alpha_R(s)$ [Eq. (4.151)] and $\alpha_S(s)$ [Eq. (4.156)].

Now the total impedance $Z_0(s)$ consists of $Z_1(s)$ in series with the other impedances in parallel, and the impedance looking into any outgoing line k is $Z_k(s) + z_{ck}$. Hence,

$$Z_0(s) = Z_1(s) + Z(s) = Z_1(s) + \cfrac{1}{\dfrac{1}{Z_g(s)} + \sum_{k=2}^{n} \dfrac{1}{Z_k(s) + Z_{ck}}} \qquad (4.192)$$

Eliminating $Z_0(s)$ from Eqs (4.187) and (4.189) yields

$$v_0 = 2v_1^f - Z_{c1} i_0 \qquad (4.193)$$

The potential at the junction, using Eq. (4.189) is given by

$$v = Z(s) \, i_0 = \frac{2Z(s)}{Z_0(s) + Z_{c1}} v_1^f \qquad (4.194)$$

and the current through the shunt impedance then is given by

$$i_g = \frac{v}{Z_g(s)} = \frac{2Z(s)}{Z_0(s) + Z_{c1}} \times \frac{1}{Z_g(s)} v_1^f \qquad (4.195)$$

The current and voltage waves transmitted to any line k (where $n \geq k \geq 2$), by Eqs (4.194) and (4.182), are

$$i_k^t = \frac{v}{Z_k(s) + Z_{ck}} = \frac{2Z(s)}{Z_0(s) + Z_{c1}} \frac{1}{Z_k(s) + Z_{ck}} v_1^f \qquad (4.196)$$

$$i_k^t = Z_{ck} i_k^t = \frac{2Z(s)}{Z_0(s) + Z_{c1}} \frac{Z_{ck}}{Z_k(s) + Z_{ck}} v_1^f \qquad (4.197)$$

The potential drops across the lumped series impedances, using Eqs (4.189) and (4.196), are

$$V_1 = Z_1(s) i_0 = \frac{2Z_1(s)}{Z_0(s) + Z_{c1}} v_1^f \qquad (4.198)$$

$$Vk = Z_k(s) i_k^t = \frac{2Z(s)}{Z_0(s) + Z_{c1}} \frac{Z_k(s)}{Z_k(s) + Z_{ck}} v_1^f \qquad (4.199)$$

Thus, if v_1^f is given at the transition point as a function of time and all the impedances are known, then the other voltages and currents are determined by solving the above differential equations, that is, Eqs (4.198) and (4.199). In particular, if v_1^f is a rectangular wave with an infinite tail, it may be taken as a unit step function and the solution can be obtained by Laplace inverse transformation.

4.12.4 Bewley Lattice Diagram

In the study of transmission line transients, it is sometimes necessary to consider the successive reflections of travelling waves, such as in the theory of ground wires, the effect of short lengths of cables, and the process of charging or discharging a line. But it is exceedingly difficult to keep track of the multiplicity of these successive reflections. A lattice diagram, or a time-space diagram, devised by L.V. Bewley in 1930, shows at a glance the position and direction of motion of every incident, reflected, and refracted wave on the system at every instant of time. Even the effects of attenuation can be entered on the lattice.

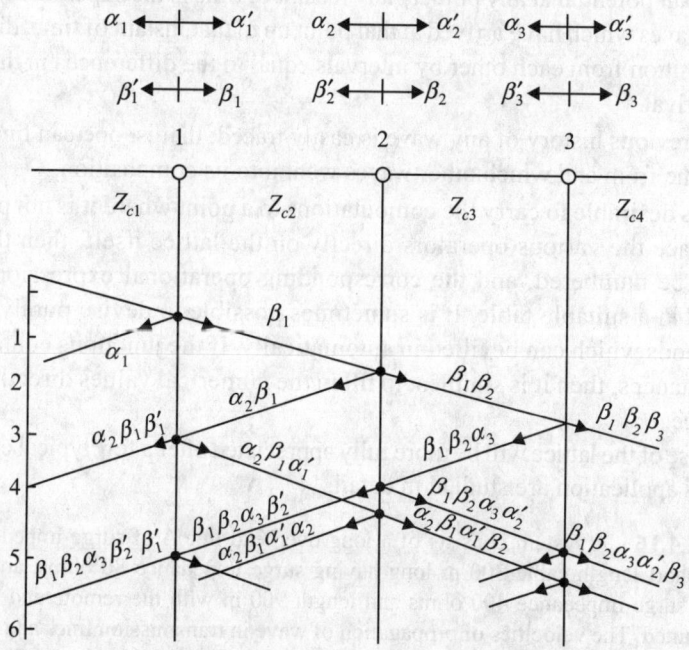

Fig. 4.31 Bewley lattice diagram

The principle of the reflection lattice is illustrated for three junctions, 1, 2, and 3, placed along a line as shown Fig. 4.31. The circuits between junctions may be either overhead lines or cables, having different surge impedances, different velocities of wave propagation, and attenuation factors. The layout of the junctions is drawn to scale at intervals equal to the times of passage of the wave instead of the actual lengths between junctions. A suitable vertical time scale in units of τ is drawn at the left of the lattice. The diagonal lines represent travelling waves. Laying off the junctions at intervals equal to the times of wave passage has the advantage that all the diagonals have the same slope, and the time scale is applicable to every branch. At the top of the lattice, on the junctions, indicators with the reflection and refraction operators are marked. These indicators are shown as double-headed arrows and are labelled as follows:

α = reflection operator for waves approaching from the left

α' = reflection operator for waves approaching from the right

β = refraction operator for waves approaching from the left

β' = refraction operators for waves approaching from the right

Now, starting at the origin of the initial incident wave at the upper left-hand corner of the lattice, the reflected and refracted waves at each junction are obtained by applying the reflection and refraction operators at that junction to the incident waves arriving there from both the left and the right; and this procedure is continued until the lattice is completed. It will be observed that :

(a) all waves travel downhill.

(b) the position of any wave at any time is given by the timescale at the left of the lattice.

(c) the total potential at any point at any instant of time is the superposition of all the waves which have arrived at that point up to that instant of time, displaced in position from each other by intervals equal to the differences in their time of arrival.

(d) the previous history of any wave is easily traced; that is, one can find where it came from and which other waves went into its composition.

(e) if it is desirable to carry the computations to a point where it is not practical to place the various operators directly on the lattice itself, then the arms may be numbered, and the corresponding operational expressions tabulated in a suitable table. It is sometimes possible to devise purely tabular methods which can be filled in automatically. If the junctions contain only resistances, then it is simplest to fill in the numerical values directly on the lattice.

(f) the use of the lattice will be more fully appreciated after a few typical examples of its application are studied in detail.

Example 4.16 A system consists of a long overhead line A of surge impedance 400 ohms, a short length cable 300 m long having surge impedance 80 ohms, and a short line B of surge impedance 400 ohms and length 900 m with the remote end of line B open-circuited. The velocities of propagation of wave in transmission lines and cable are respectively 300 m/μs and 150 m/μs. Determine by means of Bewley lattice diagram the pu voltage at the junction of the long line and the cable at $t = 10$ μs, if the surge originates at the remote end of the line A.

Solution The system under consideration is shown in Fig. 4.32. The distances between the junctions are drawn to scale as per the time of travel of a wave in between the junctions.

Time taken to travel the length of cable = 300/150 = 2 μs

Time taken to travel the length of line B = 900/300 = 3 μs

The reflection and refraction coefficients are

$$\alpha_1 = -\alpha_2 = -\alpha'_1 = \alpha'_2 = \frac{80-400}{400+80} = -\frac{2}{3} = -0.667$$

$$\beta_1 = \beta'_2 = \frac{2 \times 80}{400+80} = \frac{1}{3} = 0.333$$

$$\beta_2 = \beta_1' = \frac{2 \times 400}{400 + 80} = \frac{5}{3} = 1.667$$

Since line A is a long line, the reflection at the remote end will be neglected. The remote end of line B is open-circuited giving the value $\alpha = 1$, and $\beta = 0$ at that end.

The Bewley lattice diagram is shown in Fig. 4.32. From the diagram the magnitude of the voltage at $t = 10$ μs at the junction 1 is

$$v_1 = (1.00 - 0.667 + 0.37 + 0.163) = 0.866 \text{ pu}$$

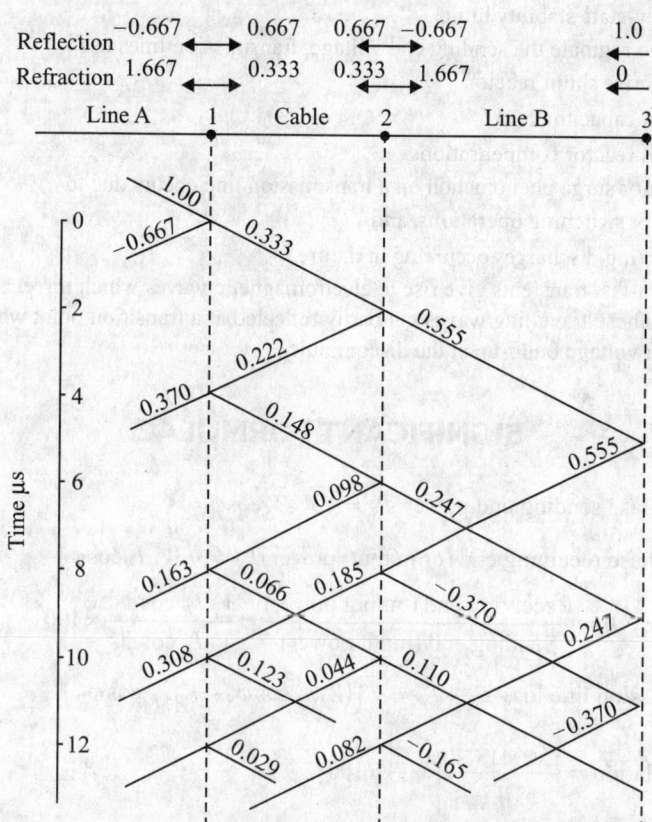

Fig. 4.32 Bewley lattice diagram for Example 4.16

SUMMARY

- Power transmission lines are employed to transmit bulk electrical energy from generating (sending) end to utilization (receiving) end.
- In the steady state, a transmission line is studied for its ability to (i) transmit power efficiently (that is minimum line losses) and (ii) maintain voltage at receiving end, from no load to full load, within the stipulated ±5 per cent for high-voltage lines.
- Based on the length, a transmission line is classified into (i) short line, (ii) medium line, and (iii) long line.
- The performance of a transmission line in steady-state operation is studied by using two-port network representations and developing the lumped $ABCD$ constants of π-equivalent models of short, medium, and long lines.

- Surge impedance loading is used to compute the voltage profile of transmission lines and explain the Ferranti effect.
- Equations in terms of the parameters and sending and receiving end voltages of a line are employed for drawing sending and receiving end power circle diagrams which help in determining the line losses from the power at the sending and receiving ends.
- The power transfer capability of a transmission line is based on
  (i) rise in temperature of the line which is governed by line losses and ambient temperature, and
  (ii) steady-state stability limit.
- In order to regulate the sending end voltage, transmission lines use
  (i)   inductive shunt reactor,
  (ii) shunt capacitor, and
  (iii) series reactor compensations.
- Transient or surge phenomenon on a transmission line occurs due to
  (i) abrupt switching operations, and
  (ii) lightning discharges occurring in nature.
- Such ultra-fast transients give rise to electromagnetic waves which travel at the speed of light. These travelling waves get partly reflected at a transition point which in turn leads to a voltage build-up at the discontinuity.

SIGNIFICANT FORMULAE

- Three-phase sending end power $P_S = \sqrt{3}V_S I_S \cos\varphi_S$

- Three-phase receiving end (or output) power $P_R = \sqrt{3}V_R I_R \cos\varphi_R$

- Efficiency $\eta = \dfrac{\text{Receiving end (output power)}}{\text{Sending end (input power)}} = \dfrac{V_R I_R \cos\varphi_R}{V_S I_S \cos\varphi_S} \times 100$

- Transmission line loss $P_S - P_R = \sqrt{3}\left(V_S I_S \cos\varphi_S - V_R I_R \cos\varphi_R\right)$

- % Regulation $= \dfrac{\left|V_{R(NL)}\right| - \left|V_{R(FL)}\right|}{\left|V_{R(FL)}\right|} \times 100$

- Propagation constant $\gamma = \sqrt{zy} = \alpha + j\beta$

- Surge impedance $Z_C = \sqrt{\dfrac{L}{C}}\ \Omega$

- *ABCD* parameters for different types of lines

Type of line	A = D	B in Ω	C in S
Long	$\cosh(\gamma l)$	$Z_C \sinh(\gamma l)$	$\dfrac{\sinh(\gamma l)}{Z_C}$
π-equivalent	$\left(1 + \dfrac{Y'Z'}{2}\right)$	Z'	$\left(1 + \dfrac{Y'Z'}{4}\right)$
Short	1	Z	0
Medium	$\left(1 + \dfrac{YZ}{2}\right)$	Z	$\left(1 + \dfrac{YZ}{4}\right)$

$$\text{SIL} = \frac{V_{R(\text{rated})}^2}{Z_C}$$

Sending end power $P_S = \dfrac{|V_S|^2}{|Z|}\cos\theta - \dfrac{|V_S|\,|V_R|}{|Z|}\cos(\theta+\delta)$

$$Q_S = \frac{|V_S|^2}{|Z|}\sin\theta - \frac{|V_S|\,|V_R|}{|Z|}\sin(\theta+\delta)$$

Receiving end power $P_R = -\dfrac{|V_R|^2}{|Z|}\cos\theta + \dfrac{|V_R|\,|V_S|}{|Z|}\cos(\theta-\delta)$

$$Q_R = -\frac{|V_R|^2}{|Z|}\sin\theta + \frac{|V_R|\,|V_S|}{|Z|}\sin(\theta-\delta)$$

For a lossless line $P_{\max} = \dfrac{|V_S|\,|V_R|}{X}$

Power circle equations $S_S = \mathbf{C}_S - Be^{j\delta}$

$$S_R = \mathbf{C}_R + Be^{-j\delta}$$

where $\quad C_S = \dfrac{|V_S|^2}{|Z|}e^{j\theta}$

$$C_R = -\frac{|V_R|^2}{Z}e^{j\theta}$$

$$B = \frac{|V_S|\,|V_R|}{|Z|}e^{j\theta}$$

Travelling waves: $v(x,t) = v^f\left(t - \dfrac{x}{\mathcal{V}}\right) + v^b\left(t + \dfrac{x}{\mathcal{V}}\right)$

$$i(x,t) = i^f\left(t - \frac{x}{\mathcal{V}}\right) + i^b\left(t + \frac{x}{\mathcal{V}}\right)$$

EXERCISES

Review Questions

4.1   Derive expressions for receiving end voltage and current in terms of the sending end voltage and current and distributed parameters of a transmission line.

4.2   Derive an expression for the driving point and transfer impedance of a radial transmission line which is terminated at the receiving end by its characteristic impedance. Also determine the (i) voltage and current gains, (ii) complex and real power gains, and (iii) efficiency of the line.

4.3   Determine (i) propagation constant, (ii) characteristic impedance, (iii) attenuation constant, and (iv) phase constant of a lossless transmission line.

4.4   Derive expressions for sending end voltage and current for a lossless transmission line.

4.5   Show that for a transmission line of infinite length there is no reflected wave. Assume that $\alpha > 0$.

4.6 For a transmission line having distributed parameters, show that $\det[T] = 1$ and prove that $[T]^{-1} = \begin{bmatrix} D & -B \\ -C & A \end{bmatrix}$.

4.7 Prove that $\dfrac{Y}{2} = \dfrac{yl}{2}\dfrac{\tanh\left(\dfrac{\gamma l}{2}\right)}{\left(\dfrac{\lambda l}{2}\right)}$ when a transmission line is represented by a π-equivalent circuit with lumped parameters.

4.8 Derive expressions for the *ABCD* constants for a lossless long transmission line. Assume distributed parameters for the line.

4.9 A medium-length power transmission line is represented as a nominal π-equivalent circuit with lumped parameters. The total series impedance of the line is $Z\ \Omega$ and the total shunt capacitance is $Y = j\omega C$ S. Derive equations for the sending end voltage and current and therefrom determine the *ABCD* constants of the line. Prove $AD - BC = 1$.

4.10 Derive an expression for the receiving end power in terms of the *ABCD* constants of a long line represented by the π-equivalent circuit. Assume that the line has losses. Show that the maximum receiving end power is less when compared to a lossless line.

4.11 Prove that the voltage at the mid-point of a long line terminated at the receiving end by an inductive shunt reactor X_{Lsh} is a maximum and is given by $V_R/\cos(\beta l/2)$. Take the length of the line equal to l and $V_S/V_R = 1.0$. Also show that the current at mid-point is zero. Assume that the line is loss free.

4.12 A single-phase, lossless line is connected to a step voltage source $Eu(t)$ matched at the sending end, that is, $Z_S = Z_c$. The receiving end is connected to an inductance L. Derive the expression of voltage $v(x, t)$ at the sending end and plot $v(x, t)$ versus time.

Numerical Problems

4.1 Can the MATLAB function `trmlnper` be used to compute performance of lossless transmission lines? Explain. Write a generalized MATLAB function for computing the sending end variables and regulation for a loss-free transmission line. The function should also provide surge impedance, phase constant, velocity of propagation, and wavelength of the line as output.

A loss free 300-km, 400-kV, 50-Hz transmission line has an inductance $L = 0.95$ mH/km/phase and a shunt capacitance of 0.15 µF/km/phase. Use the developed MATLAB function to compute sending end (i) voltage, (ii) current, (iii) complex power, and (iv) percentage regulation when the load supplied at the receiving end is 650 MW at 0.8 lagging power factor. The receiving end voltage is 400 kV. What are the values of Z_c, β, λ, and the velocity of propagation?

Solve this problem by executing the developed MATLAB function and verify the answers by hand calculation.

4.2 A 132-kV, three-phase transmission line has a resistance of 20 Ω/phase and a reactance of 100 Ω/phase. A synchronous condenser automatically controls the receiving end voltage at 132 kV for all loads. Calculate the rating of the synchronous condenser and the power factor of the load if the KVAR rating of the synchronous condenser remains constant for zero and 50 MW loads. Assume the supply voltage is equal to 145 kV.

4.3 A three-phase, 33-kV transmission line supplies a load of 30 MW, at a lagging factor of 0.8, at 33 kV. If the per-phase resistance and reactance of the line is 7.5 Ω and 15 Ω, respectively, calculate the sending end voltage and the voltage angle. The voltage at the sending and receiving ends is maintained at 33 kV by a shunt capacitor at the receiving end. Determine the three-phase rating of the capacitor when the line is supplying the above load. Compute the maximum transmission capacity of the line.

4.4 Write a MATLAB function for studying the performance of lossless, shunt, and series compensated lines, for varying load conditions, when they are represented by π-equivalent circuits.

4.5 A system consists of a long line of surge impedance 400 ohms, a cable of length 300 m, surge impedance 50 ohms, a line of length 300 m, surge impedance 400 ohms, a cable of length 300 m, surge impedance 50 ohms and a long line of surge impedance 400 ohms. The velocity of propagation of wave is 300 m/μs in line and 150 m/μs in cable. A step wave of 100 kV travels along one of the long lines. Draw the Bewley lattice diagram and plot voltage versus time at the junction of long line and cable for 10 μs.

Multiple Choice Objective Questions

4.1 Which of the following is not true for transmission lines?
   (a) Three-phase system is used for transmission of power.
   (b) Single-phase representation is sufficient for analysis.
   (c) Bulk power is transmitted.
   (d) None of these.

4.2 Which of the following represents the efficiency of a line?
   (a) P_R/P_S (b) P_S/P_R
   (c) $(P_R - P_S)/P_S$ (d) $(P_R - P_S)/P_R$

4.3 Voltage regulation of a transmission can be
   (a) positive   (b) negative
   (c) both positive and negative   (d) None of these

4.4 The stipulated voltage regulation of a transmission line is
   (a) ±2.5%   (b) ±5%
   (c) ±7.5%   (d) ±10%

4.5 In which of the following cases will the voltage regulation of a line be zero?
   (a) Unity power factor   (b) Lagging power factor
   (c) Leading power factor   (d) All of these

4.6 For a loss-free line which of the following represents the characteristic impedance of a long transmission line?
   (a) $1/\sqrt{LC}$ (b) $\sqrt{LC}$
   (c) LC (d) $\sqrt{L/C}$

4.7 A 110-kV, three-phase transmission line has a per unit capacitance of 0.015 μF and an inductance of 3.0 mH. Which of the following is the magnitude of the characteristic impedance of the line?
   (a) 350 Ω   (b) 400 Ω
   (c) 450 Ω   (d) 500 Ω

4.8 A long transmission is on no-load. Which of the following is true?
    (a) $V_S > V_R$ (b) $V_S = V_R$
    (c) $V_S < V_R$ (d) None of these

4.9 Two transmission lines having parameters A_1, B_1, C_1, D_1 and A_2, B_2, C_2, D_2 are connected in parallel. Which of the following correctly represents the parameter A of the combination?

    (a) $\dfrac{(A_1 B_2 + A_2 B_1)}{(B_2 + B_1)}$ (b) $\dfrac{(A_2 B_1 + A_1 B_2)}{(B_2 + B_1)}$

    (c) $\dfrac{(B_2 + B_1)}{(A_1 B_2 + A_2 B_1)}$ (d) $\dfrac{(B_2 + B_1)}{(A_2 B_1 + A_1 B_2)}$

4.10 Which of the following represents B parameter for a long line?
    (a) $\cosh(\gamma l)$ (b) $Z_C \sinh(\gamma l)$
    (c) $\sinh(\gamma l)/Z_C$ (d) None of these

4.11 Which of the following is true?
    (a) $AC - DB = 1$ (b) $AD - BC = 1$
    (c) $AB - DC = 1$ (d) All of these

4.12 For which of the following conditions is $V_R > V_S$?
    (a) $Z = 1.5 Z_C$ (b) $Z = 1.25 Z_C$
    (c) $X = Z_C$ (d) $Z = 0.25 Z_C$

4.13 The parameter C of a transmission line is zero. Which of the following lines is represented?
    (a) π-equivalent   (b) Long
    (c) Medium   (d) Short

4.14 What is the effect of $V_S < V_R$ in a long transmission line called?
    (a) Induction effect   (b) Transformer effect
    (c) Ferranti effect   (d) None of these

4.15 Which of the following is not the radius of a power circle?

    (a) $\dfrac{|V_S||V_R|}{Z} e^{j\theta}$ (b) $\dfrac{|V_S|^2}{Z} e^{j\theta}$

    (c) $\dfrac{|V_R|^2}{Z} e^{j\theta}$ (d) All of these

4.16 Which of the following determines the power transmission capability of a line?
    (a) Conductor temperature rise   (b) Transmission voltage
    (c) Steady-state stability limit   (d) All of these

4.17 Which of the following is a reason to resort to shunt compensation in an extra high voltage line?
    (a) Improve stability   (b) Restrict fault currents
    (c) Improve voltage profile   (d) None of these

4.18 Which of the following is a reason to resort to series compensation in an extra high voltage line?
    (a) Improve stability   (b) Restrict fault currents
    (c) Improve voltage profile   (d) None of these

4.19 Which of the following is applicable to complex power transmission in medium/long lines?
    (a) There is a vertical shift in positions of centres of sending and receiving end power circles.

(b) Real power transmitted remains unchanged.

(c) Power consumed in shunt capacitance is relatively low.

(d) All of these.

4.20 Which of the following is related to the speed of transients set up in transmission lines?

(a) Slow (b) Medium

(c) Fast (d) Ultra-fast

4.21 Which of the following is used for protecting transmission lines against direct strokes of lightning?

(a) Ground wires (b) Surge arrestors

(c) Circuit breakers (d) All of these

4.22 Which of the following represents the reflection coefficient $\alpha_R(s)$ of a travelling wave?

(a) $\dfrac{Z_R(S)+Z_C}{Z_R(S)-Z_C}$ (b) $\dfrac{Z_R(S)-Z_C}{Z_R(S)+Z_C}$

(c) $\dfrac{Z_R(S)Z_S(S)+Z_C}{Z_R(S)Z_S(S)-Z_C}$ (d) $\dfrac{Z_R(S)Z_S(S)-Z_C}{Z_R(S)Z_S(S)+Z_C}$

Answers

4.1 (d)	4.2 (a)	4.3 (c)	4.4 (b)	4.5 (a)	4.6 (d)
4.7 (c)	4.8 (a)	4.9 (a)	4.10 (b)	4.11 (b)	4.12 (d)
4.13 (d)	4.14 (c)	4.15 (a)	4.16 (d)	4.17 (c)	4.18 (a)
4.19 (d)	4.20 (d)	4.21 (a)	4.22 (b)		

CHAPTER 5

Simulation of Power System Components

Learning Outcomes

An intense study of this chapter will enable the reader to
- Enumerate various types of power system components at the transmission level
- Develop mathematical models of synchronous machines (salient and non-salient pole types) for simulating their behaviour in the steady state and transient state
- Starting with an ideal transformer, develop equivalent circuits of practical three-phase transformers and autotransformers, and also understand the working of tap-changing and regulating transformers
- Categorize different types of consumer loads and simulate their mathematical models
- Easily work with per unit quantities for analysing power system networks

OVERVIEW

An electrical power system, at the transmission level, consists of synchronous generators for generating electrical energy, transformers for stepping up and down voltages, high-voltage transmission lines for transferring bulk power, compensating devices, and loads. In order to analyse a power system, it is essential to develop mathematical models of all its components.

The mathematical simulation of transmission lines was described in Chapter 4. There are many books on the subject of ac machinery which provide an adequate analysis of generators, motors, and transformers. From the point of view of facilitating power system analysis, simple equivalent circuit models for synchronous generators, transformers, and loads are essential. For readers who have already studied the subject, this chapter would be a helpful review.

For analysing power systems in which there are many voltage levels, it is convenient to use the per unit system. The application of per unit quantities to calculations for the system is also discussed.

5.1 SYNCHRONOUS MACHINES

For the generation of electrical energy, three-phase synchronous machines operating as generators are employed by the utilities, the world over. A synchronous generator, also known as an alternator, is a rotary energy conversion device driven by steam turbines, hydro turbines, or gas turbines. It consists of a rotor, which carries the dc field winding for producing the flux, and a stator, which carries the three-phase armature winding, displaced both in time and space by 120°. A balanced three-phase voltage is developed across the armature terminals, when the rotor is driven by a prime mover at its synchronous speed

$$N_S = \left(\frac{120f}{P}\right) \text{rpm} \tag{5.1}$$

where P is the number of poles of the alternator and f is the frequency in Hz.

Due to the low voltage rating and the consequent low power consumption (0.2–3% of the machine rating), the dc winding is placed on the rotor. The rotor is also equipped with one or more short-circuited windings known as damper windings. The dc winding is excited, through brushes and slip rings, by a dc generator (referred to as exciter) mounted on the rotor shaft of the synchronous generator. Modern-day synchronous generators, however, employ brushless excitation, which use ac generators with rotating rectifiers. The insulation requirement of a dc winding, which operates at 400 V dc, is not very stringent and the weight of copper winding (due to low power consumption) is less. Therefore, the rotor weight and inertia are both low. This aspect makes it possible to design high-speed synchronous generators.

The practice of locating the armature winding on the stator makes the design of high-voltage synchronous generators feasible, since it is easy to insulate the static windings. This arrangement has made it possible to design synchronous generators with operating voltages up to 33 kV.

Depending upon the type of rotor, two types of synchronous generators are designed, namely (i) non-salient (cylindrical or round) rotor, and (ii) salient rotor. The former type of rotors have uniform air gap and distributed winding. Generators with round rotors essentially have high speeds (1500 rpm, 3000 rpm) and are used with steam turbines. The non-salient pole machines constitute nearly 70% of the large category of synchronous generators and have a power rating which ranges between 150 MVA to 1500 MVA. The salient-pole machines, on the other hand, have a non-uniform air gap and the excitation windings are concentrated on the poles. Such machines have low speeds (500–1000 rpm), large number of poles, large diameters, and short axial lengths. Hydro generators have salient pole construction.

5.1.1 Generated emf in a Synchronous Machine

An elementary two-pole, three-phase synchronous generator is shown in Fig. 5.1. The stator contains three coils AA_1, BB_1, and CC_1, shown as concentrated full-pitched coils, with axes of the coils displaced from each other by 120° electrical. The coils may be considered to represent distributed windings producing sinusoidal mmf waves concentrated on the magnetic axes of their respective phases. It is assumed that the field winding of the rotor, when excited by a dc source, produces a sinusoidal flux distribution at its surface along the air gap of the machine.

The flux density may be expressed as

$$B = B_m \cos\theta \quad \text{Wb/m}^2 \tag{5.2}$$

where B_m is the maximum value of flux density at the centre of the pole and θ is the angle in electrical radians measured from the rotor pole axis.

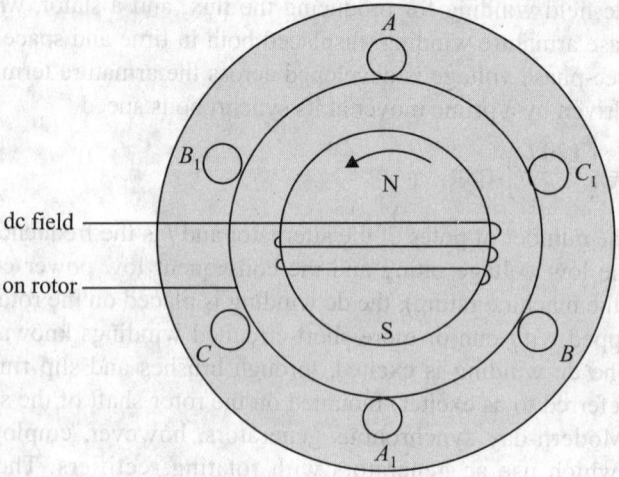

Fig. 5.1   Three-phase, two-pole synchronous generator

Let l be the length of the conductor or the axial stator length and r the mean radius at the air gap, then the air gap flux per pole ϕ_f is the integral of the flux density over the pole area. Thus, for a two-pole machine,

$$\phi_f = \int_{-\pi/2}^{\pi/2} B_m \cos\theta \, lr \, d\theta = 2B_m lr \text{ Wb} \tag{5.3}$$

For a P-pole machine,

$$\phi_f = \frac{2}{P} \times 2 B_m lr = \frac{4}{P} B_m lr \text{ Wb} \tag{5.4}$$

since the pole area is $2/P$ times that of a two-pole machine of the same length and diameter.

When the rotor magnetic axis and the coil axis coincide, the stator coil links flux ϕ_f. As the rotor is rotated at a constant angular velocity ω, the flux linking T turns of the stator coil at any time t is

$$\psi = T\phi_f \cos\omega t \text{ Wb-turns} \tag{5.5}$$

where $t = 0$, when the rotor magnetic axis and the coil axis coincide. By Faraday's law, emf e induced in the stator coil may be obtained as

$$e = -\frac{d\psi}{dt} = \omega T\phi_f \sin\omega t \tag{5.6}$$

$$= E_m \sin\omega t \text{ V} \tag{5.7}$$

The induced emf varies sinusoidally with time with a peak value $E_m = \omega T\phi_f$, and the rms value of the induced phase emf is given by

$$E = \frac{\omega T\phi_f}{\sqrt{2}} = \frac{2\pi f T\phi_f}{\sqrt{2}} = 4.44 f T\phi_f \text{ V} \tag{5.8}$$

where f is the frequency in hertz. In an actual ac machine, the stator coils of each phase are distributed in a number of slots, and the emfs induced in different slots are not in phase; hence the phasor sum of the induced emfs is less than their algebraic sum. Thus the rms value of the induced phase emf is to be modified as

$$E = 4.44 K_w f T\phi_f \text{ V} \tag{5.9}$$

where K_w is called the winding factor. For three-phase windings, K_w is about 0.85 to 0.95.

5.1.2 Armature Reaction in a Synchronous Machine

If a balanced three-phase load is connected to a three-phase generator, balanced currents will flow in the phases of the armature winding. These armature currents produce a magnetic field flux of constant amplitude rotating at a uniform speed around the circumference of the air gap. This fact may be demonstrated mathematically by considering a three-phase, two-pole, cylindrical-rotor synchronous machine having one slot, per pole, per phase as shown in Fig. 5.1.

Let the three phases of an ac machine carry balanced alternating currents. If I_m is the peak value of the currents, then the current in the three phases may be expressed as

$$i_a = I_m \sin \omega t \text{ A}$$

$$i_b = I_m \sin \left(\omega t - \frac{2\pi}{3} \right) \text{ A}$$

$$i_c = I_m \sin \left(\omega t - \frac{4\pi}{3} \right) \text{ A}$$

If the magnetic circuit is unsaturated and B_m is the maximum flux density due to maximum current in any phase, then the flux densities for the three phases may be expressed as

$$B_a = B_m \sin \omega t \text{ Wb/m}^2$$

$$B_b = B_m \sin \left(\omega t - \frac{2\pi}{3} \right) \text{ Wb/m}^2$$

$$B_c = B_m \sin \left(\omega t - \frac{4\pi}{3} \right) \text{ Wb/m}^2$$

At any point θ radians (electrical) in space, (θ being measured from the point of zero flux density for phase a), the flux density distribution due to the three phases a, b, and c can be mathematically represented as

$$B_1 = B_m \sin \omega t \sin \theta \text{ Wb/m}^2$$

$$B_2 = B_m \sin \left(\omega t - \frac{2\pi}{3} \right) \sin \left(\theta - \frac{2\pi}{3} \right) \text{ Wb/m}^2$$

$$B_3 = B_m \sin \left(\omega t - \frac{4\pi}{3} \right) \sin \left(\theta - \frac{4\pi}{3} \right) \text{ Wb/m}^2$$

The resultant flux density distribution at θ radians (electrical) is the sum of the flux density distributions due to current in the three phases. Therefore,

$$B_1 + B_2 + B_3 = B_m \left[\sin \omega t \sin \theta + \sin \left(\omega t - \frac{2\pi}{3} \right) \sin \left(\theta - \frac{2\pi}{3} \right) \right.$$

$$\left. + \sin \left(\omega t - \frac{4\pi}{3} \right) \sin \left(\theta - \frac{4\pi}{3} \right) \right]$$

$$= 1.5 B_m (\sin \omega t \sin \theta + \cos \omega t \cos \theta)$$

$$= 1.5 B_m \cos(\omega t - \theta) \text{ Wb/m}^2 \tag{5.10}$$

From Eq. (5.10) it can be observed that when $\theta = \omega t = 2\pi ft$, the resultant flux density distribution has a maximum value of $1.5B_m$. For a value of $t = 1/f$, the time duration of one cycle is $\theta = 2\pi$ electrical radians, which means the position of the peak value of the resultant flux rotates through two pole pitches in one cycle. It may thus be concluded that the armature reaction flux rotates around the armature at the angular velocity ω equal to the angular velocity of the rotor. Therefore, this flux is stationary with respect to the flux produced by the dc field winding of the rotor. The net flux produced in the air gap between the stator and the rotor is the resultant of the field flux and the armature reaction flux.

5.1.3 The Circuit Model of a Synchronous Machine

Under no-load operation, the stator winding current is zero. The field winding flux ϕ_f induces an emf per phase E volts, the rms value of which is given by Eq. (5.9). It follows directly from Eqs (5.5) and (5.7) that the emf phasor E, called the excitation voltage, lags the flux ϕ_f by 90° as shown in Fig. 5.2, and this voltage always appears at the armature terminals at no-load with the field winding excited. At no-load, the terminal voltage phasor $V_t = E$.

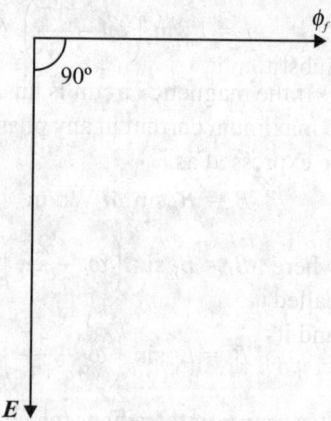

If a balanced three-phase load is connected to the armature terminals, balanced three-phase currents will flow in the phases of the stator armature winding with phase rms current I_a A at a terminal voltage per-phase of V_t volts. The stator currents produce armature reaction flux ϕ_{ar} per pole that rotates in the same direction as the rotor field flux ϕ_f. The resultant air gap flux ϕ_r is the phasor sum of ϕ_{ar} and ϕ_f. The flux ϕ_r generates the voltage E_r volts in the armature phase winding.

Fig. 5.2 Phasor diagram on no-load of a synchronous generator

The phasor diagram for the synchronous generator is shown in Fig. 5.3. If magnetic saturation is neglected, then the phasor voltages E, E_{ar}, E_r are

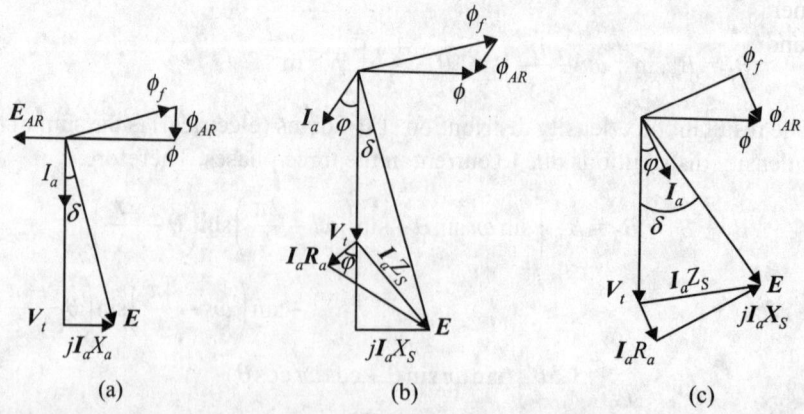

(a)   (b)   (c)

Fig. 5.3 Phasor diagram for (a) lagging pf; (b) unity pf; and (c) leading pf

proportional to the flux phasors ϕ_f, ϕ_{ar}, and ϕ_r respectively that generate them; and further, the voltage phasors lag the flux phasors by 90°. It may be noted that the phasor E_{ar} lags the armature current phasor I_a by 90° and the magnitude of E_{ar} is determined by ϕ_{ar}, which in turn is proportional to the magnitude of I_a. So an inductive reactance X_{ar} can be specified such that

$$E_{ar} = -jI_a X_{ar} \text{ V} \tag{5.11}$$

Then the phasor voltage E_r may be expressed as

$$E_r = E + E_{ar} = (E - jI_a X_{ar}) \text{ V} \tag{5.12}$$

A small part of the flux lines produced by the armature reaction, which do not cross the air gap, is called leakage flux, and X_L denotes leakage reactance due to this flux. If the synchronous generator is loaded then voltage generated by the resultant flux exceeds the terminal voltage V_t only by the voltage drops in the armature resistance $R_a \Omega$ and the armature leakage reactance $X_L \Omega$. Then

$$V_t = (E_r - I_a R_a - jI_a X_L) \text{ V} \tag{5.13}$$

Substituting for E_r from Eq. (5.12) into Eq. (5.13)

$$V_t = E - jI_a X_{ar} - I_a R_a - jI_a X_L$$

$$= E - I_a R_a - jI_a (X_{ar} + X_L) = E - I_a R_a - jI_a X_S \tag{5.14a}$$

$$= E - I_a (R_a + jX_S) = (E - I_a Z_S) \text{ V} \tag{5.14b}$$

where $X_S = (X_{ar} + X_L) \Omega$ is called the synchronous reactance and $Z_S = R_a + jX_S$ is called the synchronous impedance. The magnitude of synchronous impedance Z_S and its angle θ is given by

$$Z_S = \sqrt{R_a^2 + X_S^2} \ \Omega \tag{5.15a}$$

$$\theta = \tan^{-1} \frac{X_S}{R_a} \tag{5.15b}$$

The equivalent circuit of a non-salient pole synchronous generator follows directly from Eq. (5.14b). E is considered the source voltage and Z_S is treated as internal source impedance. Figure 5.4 shows the resulting equivalent circuit. Generally armature resistance R_a is much smaller than the synchronous reactance X_S and is often neglected.

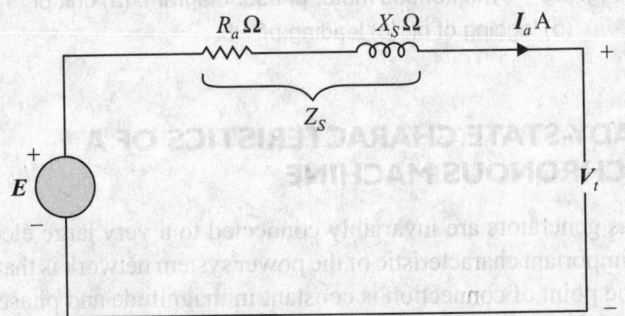

Fig. 5.4 Equivalent circuit of a synchronous generator

The phasor diagrams for a cylindrical rotor synchronous generator with the terminal voltage V_t as the reference, and load current I_a with lagging, unity, and leading power factors are shown in Fig. 5.3(a), 5.3(b), and 5.3(c) respectively. The angle φ is the phase angle (lag or lead) of the armature current I_a with respect to the terminal voltage V_t. The angle δ, called load or torque angle, is the phase displacement between the generated voltage E and the terminal voltage V_t. The generated voltage E leads the terminal voltage, a condition necessary for generator action to take place.

The principles that have been discussed could be extended to a synchronous motor. The equivalent circuit for a synchronous machine operating as a motor is identical to that of a synchronous generator with the direction of I_a reversed. The phasor diagrams may be derived from those of the synchronous motor by making two changes in Eq. (5.14b). The terms E and V_t are interchanged in that V_t is now the source voltage applied to the machine and E is a counter emf generated in the machine. The voltage equation for the machine operating as a motor becomes

$$V_t = (E + I_a Z_S) \text{ V} \tag{5.16}$$

The angle δ between E and V_t, which was positive in case of generator because of the driving action of the prime mover, is now negative. The angle δ is known as load (torque) angle. It may be noted that E is associated with the rotor field and V_t is associated with the resultant field. At no load, the angle δ is zero implying that the axes of the field poles and the resultant field are in time phase. Based on these two changes, the phasor diagrams of the synchronous machine operating as a motor for unity pf, lagging pf, and leading pf currents are shown in Figs 5.5(a), (b) and (c) respectively.

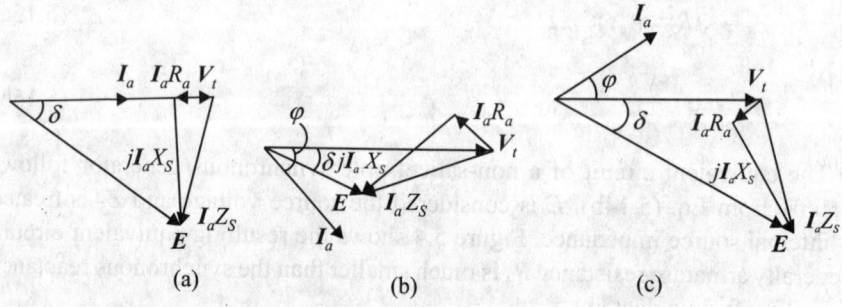

Fig. 5.5 Synchronous motor phasor diagram: (a) unit pf; (b) lagging of pf; (c) leading pf

5.2 STEADY-STATE CHARACTERISTICS OF A SYNCHRONOUS MACHINE

Synchronous generators are invariably connected to a very large electric power system. An important characteristic of the power system network is that the system voltage at the point of connection is constant in magnitude and phase angle, and the system frequency is constant. Such a point in a power system is termed as *infinite bus*. Thus, when a generator is connected to an infinite bus, the terminal voltage V_t of the generator will not be altered by any changes in the excitation of the generator, and no frequency change will occur, regardless of any change in the prime mover input of the synchronous machine. Thus, infinite bus is an ideal voltage source and has infinite inertia.

Hence a synchronous machine, when connected to an infinite bus, has its speed and terminal voltage fixed. However, the field current and the mechanical torque on the shaft are the two controllable variables of the synchronous machine. The variation of the field excitation current I_f, when applied to a synchronous machine, operating as a generator or a motor, supply or absorb a variable amount of reactive power. Because the synchronous machine runs at constant speed, the real power can only be controlled through the variation of torque imposed on the shaft of the machine by either the prime mover in the case of a generator or the mechanical load in the case of a motor.

5.2.1 Reactive Power Control

A synchronous machine connected to an infinite bus is shown in Fig. 5.6. It is assumed that a round-rotor generator with negligible R_a is delivering constant active power P_G to the infinite bus and the corresponding current is I_a that lags the terminal voltage V_t by an angle φ, and power or load angle δ exists between V_t and the generated emf E of the machine.

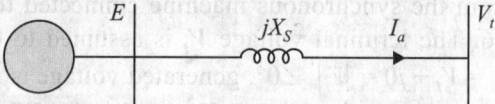

Fig. 5.6   Synchronous machine connected to infinite bus

The complex power delivered to the system by the generator is given in per unit by

$$S_G = P_G + jQ_G = V_t I_a^* = |V_t||I_a| \ (\cos\varphi + j\sin\varphi) \tag{5.17}$$

Then,

$$P_G = |V_t||I_a|\cos\varphi \tag{5.18}$$

$$Q_G = |V_t||I_a|\sin\varphi \tag{5.19}$$

Now in Eq. (5.18), $|V_t|$ is constant, being infinite bus voltage, and as power P_G delivered to the infinite bus system is constant, it is clear that $|I_a| \cos\varphi$, the projection of phasor I_a on V_t must remain constant. Under these conditions, as the dc field current I_f is varied, the generated emf E varies proportionally maintaining $|I_a|\cos\varphi$ always constant. Thus, the tip of phasor I_a must fall on a vertical line as shown in Fig. 5.7. The phasor diagrams for three different values of armature currents are shown in the figure. Using Eq. (5.14b) for lagging power factor, current I_{a1} results in E_1. For unity power factor, armature current has minimum value I_{a2} which results in E_2. Similarly for leading power factor current I_{a3} the emf phasor is E_3. Thus it can be seen that while maintaining constant active power output of the generator, the generation of reactive power can be controlled by means of the rotor excitation. The variation in $|I_a|$ as a function of the field current I_f resembles the letter V and is often referred to as V-curve of a synchronous machine.

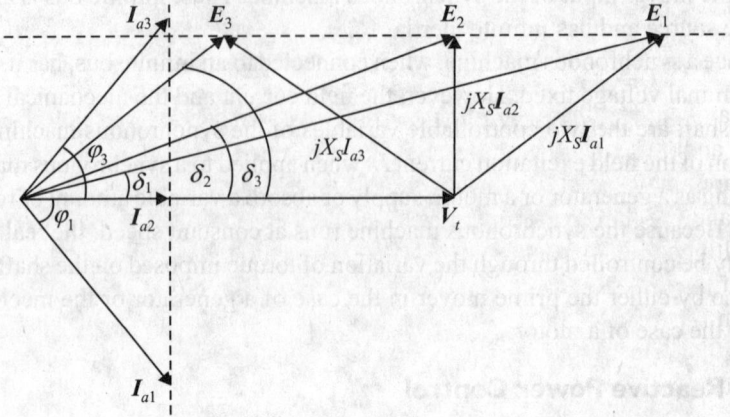

Fig. 5.7 Variation of field excitation current at constant power

5.2.2 Active Power Control

For a round rotor synchronous generator, the expression for complex power can be written from the synchronous machine connected to an infinite bus shown in Fig. 5.6. The terminal voltage V_t is assumed to be the reference phasor, that is, $V_t = V_t + j0 = |V_t| \angle 0°$, generated voltage is $E = |E| \angle \delta$, and the synchronous impedance $Z_S = (R_a + jX_S) = |Z_S| \angle \theta$. When V_t, E, and Z_S are in per unit then,

$$I_a = \frac{|E| \angle \delta - |V_t| \angle 0°}{|Z_S| \angle \theta} \quad \text{and} \quad I_a^* = \frac{|E| \angle - \delta - |V_t|}{|Z_S| \angle - \theta}$$

Then the complex power delivered to the system at the terminals of the generator in per unit is

$$S_G = (P_G + jQ_G) = V_t I_a^* = |V_t| \frac{|E| \angle - \delta - |V_t|}{|Z_S| \angle - \theta}$$

$$= \frac{|E||V_t|}{|Z_S|} [\cos(\theta - \delta) + j \sin(\theta - \delta)] - \frac{|V_t|^2}{|Z_S|} (\cos \theta + j \sin \theta) \qquad (5.20)$$

From Eq. (5.20),

$$P_G = \frac{|E||V_t|}{|Z_S|} \cos(\theta - \delta) - \frac{|V_t|^2}{|Z_S|} \cos \theta \qquad (5.21a)$$

$$Q_G = \frac{|E||V_t|}{|Z_S|} \sin(\theta - \delta) - \frac{|V_t|^2}{|Z_S|} \sin \theta \qquad (5.21b)$$

Equations (5.21a) and (5.21b) provide expressions for the real and reactive power generated by a round rotor synchronous generator. If R_a is neglected, then $Z_S = jX_S$. Equations (5.21a) and (5.21b), respectively, take the following form

$$P_G = \frac{|E||V_t|}{X_S} \sin \delta \qquad (5.22a)$$

$$Q_G = \frac{|E||V_t|}{X_S} \cos \delta - \frac{|V_t|^2}{X_s} \qquad (5.22b)$$

It may be noted that when V_t and E are in volts and Z_S is in ohms, then Eqs (5.22a) and (5.22b) will give P_G and Q_G in per-phase quantities, provided that V_t and E are line-to-neutral voltages. In case both V_t and E are line-to-line values, then P and Q will provide three-phase values. The per unit P and Q of Eq. (5.22) are multiplied by three-phase base MVA or base MVA per-phase depending on whether three-phase power or power per-phase is required.

From Eq. (5.22a), it may be seen that when $|V_t|$ and $|E|$ are constant, then by varying the prime mover input to the generator, power angle δ is changed and consequently the power P varies.

5.3 SYNCHRONOUS CONDENSER

A synchronous condenser is an unloaded synchronous machine. When connected to an infinite bus, it can provide a very convenient and continuous control of reactive power by adjusting its field current.

The active power output of a synchronous machine connected to an infinite bus is given by Eq. (5.22a). When the power angle δ is positive, the machine functions as a generator and delivers power to the bus, and when δ is negative, it draws power from the bus and functions as a synchronous motor. When $\delta = 0$, the machine neither delivers nor receives power from the bus.

When a synchronous generator operates at no load, that is, $\delta = 0$, Eq. (5.22b) takes the following form

$$Q_G = \frac{|E||V_t|}{X_S} - \frac{|V_t|^2}{X_S} = \frac{|V_t|(|E| - |V_t|)}{X_S} \qquad (5.23)$$

Thus, if the rotor field excitation is changed, then the induced emf $|E|$ changes. From Eq. (5.23) it is seen that by increasing the field excitation, the synchronous machine supplies reactive power to the bus, and therefore behaves like a capacitor bank. In large sizes, synchronous condensers are cheaper than capacitor banks.

Example 5.1 A 250-kVA, 3300-V, star-connected, three-phase synchronous generator has resistance and synchronous reactance per-phase of 0.25 Ω and 3.5 Ω respectively. Calculate the voltage regulation at full load 0.8 power factor lagging.

Solution Output power in VA $= \sqrt{3} V_L I_L$

Then $I_L = \dfrac{250 \times 10^3}{\sqrt{3} \times 3300} = 43.74$ A

For a star-connected machine, line current is equal to phase current. Hence,

Phase current $I_a = 43.74$ A

Phase voltage $V_t = \dfrac{V_L}{\sqrt{3}} = \dfrac{3300}{\sqrt{3}} = 1905.26$ V

Let V_t be the reference phasor. Then $V_t = 1905.26\angle 0°$ V

$$R_a = 0.25 \ \Omega; \ X_S = 3.5 \ \Omega, \text{ then } Z_S = 0.25 + j3.5 = 3.51\angle 85.91° \ \Omega$$

$$\cos\varphi = 0.8; \text{ then } \varphi = \cos^{-1}0.8 = 36.87°; \sin\varphi = 0.6$$

Then $I_a = 43.74 \ \angle -36.87°$A

Using Eq. (5.14b), $E = V_t + I_a Z_S$

$$= (1905.26 + j0) + 43.74 \times 3.51\angle(85.91 - 36.87)°$$

$$= 1905.26 + 153.53\angle 49.04° = 2009.06\angle 3.31 \text{ V}$$

$\therefore$ Percentage voltage regulation $= \dfrac{2009.06 - 1905.26}{1905.26} \times 100 = 5.45\%$

Example 5.2   A 33-kV, three-phase, 45-MVA synchronous generator is connected to a 50-Hz infinite bus. The generator is supplying rated power at 0.8 power factor lagging. If the synchronous armature has negligible resistance and reactance per phase of the generator is 10 Ω, determine (i) the excitation voltage per-phase and the load angle, (ii) the armature current, load angle, and power factor when the generator output is reduced to 30 MW by reducing the turbine input, and (iii) the maximum power that the generator can deliver without losing synchronism, armature current, and the power factor. In parts (i) and (ii) assume constant excitation. Determine the minimum excitation required for the generator to deliver rated output without losing synchronism.

Solution   Rated power output $= 45 \times 0.8 = 36$ MW

Armature current $I_a = \dfrac{36 \times 1000}{\sqrt{3} \times 33 \times 0.8} = 787.2958$ A

Terminal voltage per-phase $V_t = \dfrac{33}{\sqrt{3}} = 19.0526$ kV

Put $V_t = 19.0526\angle 0°$ kV. Thus, $I_a = 787.2958 \ (0.8 - j0.6)$ A

From Eq. (5.12) with $R_a = 0$,

(i) $E = 19052.6 + j10 \times 787.2958 \ (0.8 - j0.6) = 23776 + j6298.4$

$$= 24.596\angle 14.8369° \text{ kV per phase}$$

(ii) From Eq. (5.22a)

$$\delta = \sin^{-1}\left(\frac{(30/3) \times 10}{24.596 \times 19.0526}\right) = 12.3213°$$

$$I_a = \frac{(24.596\angle 12.3213° - 19.0526\angle 0°) \times 1000}{j10} = 524.86 - j497.69$$

$$= 723.3068 \ \angle -43.4776° \text{ A}$$

Power factor of the load $= \cos(-43.4776°) = 0.7256$ lagging

(iii) For maximum power $\delta = 90°$ and Eq. (5.22a) takes the form

$$P_{\max} = \frac{3 \times 24.596 \times 19.0526}{10} = 140.5853 \text{ MW}$$

Armature current $I_a = \dfrac{(24.596\angle 90° - 19.0526\angle 0°) \times 1000}{j10} = 2459.6 + j1905.3$

$$= 3111.2 \ \angle 37.7621° \text{ A}$$

Power factor $= \cos \ (37.7621°) = 0.7903$ leading

The upper limit of the load angle at which power can be supplied without losing synchronism is 90°. Since, at minimum excitation, power factor is equal to unity, the power output $= (36/3) = 12$ MW/phase

Then, $E = \dfrac{12 \times 10}{19.0526} = 6.2984$ kV

$$I_a = \frac{(6.2984\angle 90° - 19.0526\angle 0°) \times 1000}{j10} = 629.84 + j1905.26$$

$$= 2006.7\angle 71.7072° \text{ A}$$

Power factor at minimum excitation = cos(71.7072°) = 0.3139 (leading).

Example 5.3   For the synchronous generator in Example 5.2, plot the V-curves when it supplies power at (i) 25%, (ii) 50%, (iii) 75%, and (iv) 100% rated power. Assume that the power factor angle ϕ varies from –70° to 70° and $E = 1500I_f$.

Solution   The MATLAB function vcurves given below is an executable program for plotting the V-curves for varying power factor. The function is an interactive program. For example at the command keyboard the control is returned to the keyboard. At this point, the user inserts commands through the keyboard for plotting of the curves (shown in Fig. 5.8).

```
function [Ecom, Emag] = vcurves(Prtd,Xs,Vt,phi,K);
%Program for plotting v-curves of a synchronous machine
%Developed by the authors.
index=1.0;
if index == 1;
X = input ('percentage load');
Pg=X*Prtd;
Vt=Vt/sqrtm(3);
pfang=phi*pi/180;
x=pfang:-0.005:-pfang;
Pg=Pg*ones(1,length(x));
Ia=Pg./(3*abs(Vt)*cos(x));
Iacom=Ia.*(cos(x)+i*sin(x));
Ecom=Vt+i*Xs*Iacom;
Emag=abs(Ecom);
Ifld=Emag*1000/K;
keyboard;
else
end
```

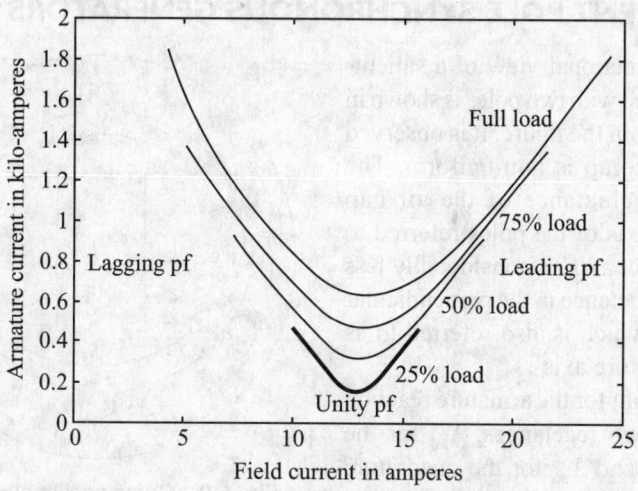

Fig. 5.8   Plot of V-curves of a synchronous machine

Description of the input symbols for the function vcurves is as follows:

Prtd = 36 MW   Rated three-phase power output of the synchronous machine

Vt = 33 kV   Rated line-to-line terminal voltage

Xs = 10 Ω   Synchronous reactance per phase of the machine

phi = – 70° to 70°   Range of power factor angle for plotting the V-curves

K = 1500   Proportionality constant for computing the excitation voltage

```
≫ Prtd=36;Xs=10;Vt=33;phi=70;K=1500;
≫ [Ecom,Emag] = vcurves(Prtd,Xs,Vt,phi,K);
percentage load1.0
K≫ plot(Ifld,Ia,'k')     %plots field current vs. armature in black
K≫ hold on               %invokes hold on command for plotting more curves
K≫ return                %returns command to the function for execution
≫ [Ecom,Emag] = vcurves(Prtd,Xs,Vt,phi,K);
percentage load0.75
K≫ plot(Ifld,Ia,'k-')    %plots field current vs. armature in
                         %black solid line
K≫ return                %returns command to the function for execution
≫ [Ecom,Emag] = vcurves(Prtd,Xs,Vt,phi,K);
percentage load0.5
K≫ plot(Ifld,Ia,'k-.')   %plots field current vs. armature in
                         %black dash-dot line
K≫ return                %returns command to the function for execution
≫ [Ecom,Emag] = vcurves(Prtd,Xs,Vt,phi,K);
percentage load0.25
K≫ plot(Ifld,Ia,'k.')    %plots field current vs. armature in
                         %black point line
K≫ index=0;              %set index to zero to show no more v-curves
                         %are to be plotted
K≫ hold off              %turns off the hold on command
K≫ return                %returns command to the function for execution
≫ xlabel('Field current in amperes')              %labels the x-axis
≫ ylabel('Armature current in kilo-amperes')      %labels the y-axis
≫ title('PLOT OF V-CURVES OF A SYNCHRONOUS MACHINE')
                                       %adds a title to the graph
≫                        %execution of the function 'vcurves, is terminated
```

5.4 SALIENT-POLE SYNCHRONOUS GENERATORS

The cross-sectional view of a salient-pole machine with two poles is shown in Fig. 5.9. From the figure, it is observed that the air gap is non-uniform. The magnetic reluctance of the air gap along the axis of the poles referred to as the 'direct axis' is considerably less than the reluctance in the perpendicular direction, which is also referred to as the 'quadrature axis'.

To account for the armature reaction there are two reactances, X_{ad} for the direct axis and X_{aq} for the quadrature axis, and $X_{ad} > X_{aq}$. The reactance of the

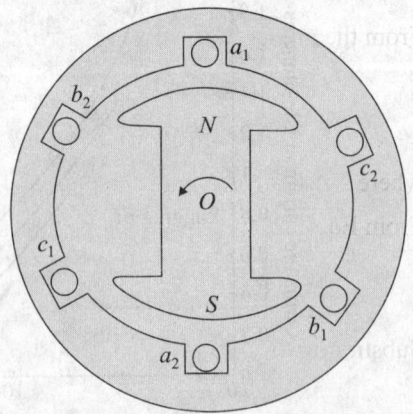

Fig. 5.9 Cross-sectional view of a salient-pole machine with two poles

generator along the direct axis $X_d = X_L + X_{ad}$, and the reactance in the quadrature axis $X_q = X_L + X_{aq}$, where X_L is the leakage reactance. The armature voltage drops due to the direct axis and quadrature axis reactances are accounted for by resolving the armature current I_a into quadrature axis component I_q in time phase, and direct axis component I_d in time quadrature with the excitation voltage E. The phasor diagram for a salient pole generator, in which the armature resistance R_a is neglected, is shown in Fig. 5.10. From the phasor diagram

$$E = V_t + jI_dX_d + jI_qX_q \tag{5.24}$$

Unfortunately, there is no simple equivalent circuit to represent a salient pole synchronous generator.

If the terminal voltage phasor V_t is resolved along and perpendicular to the excitation voltage phasor E, in the phasor diagram shown in Fig. 5.10, the following equations result

$$|E| - I_dX_d = |V_t| \cos\delta \tag{5.25a}$$

$$I_qX_q = |V_t| \sin\delta \tag{5.25b}$$

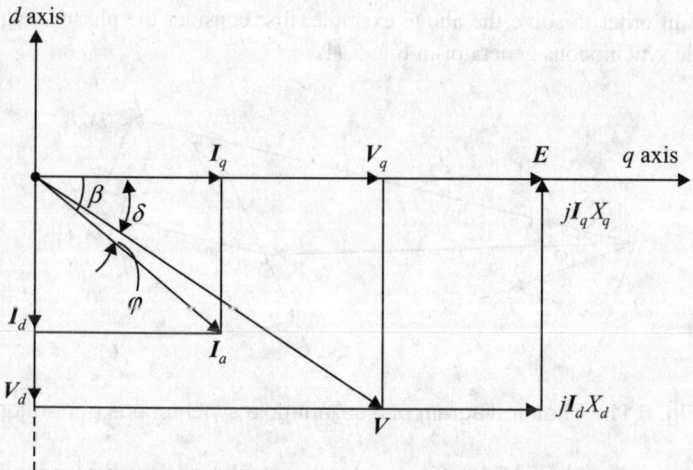

Fig. 5.10   Phasor diagram of a salient-pole generator

From the phasor diagram (Fig. 5.10), the following relations are obtained

$$I_d = |I_a| \sin\beta \tag{5.26a}$$

$$I_q = |I_a| \cos\beta \tag{5.26b}$$

where $\beta = \delta + \varphi$ \hfill (5.27)

From Eq. (5.27), the following expressions may be written

$$\cos\varphi = \cos(\beta - \delta) = \cos\beta\cos\delta + \sin\beta\sin\delta \tag{5.28a}$$

$$\sin\varphi = \sin(\beta - \delta) = \sin\beta\cos\delta + \cos\beta\sin\delta \tag{5.28b}$$

Substituting from Eq. (5.26) in Eqs (5.28) gives

$$|I_a| \cos\varphi = I_q\cos\delta + I_d\sin\delta \tag{5.29a}$$

$$|I_a| \sin\varphi = I_d\cos\delta - I_q\sin\delta \tag{5.29b}$$

Substituting the above expressions in Eq. (5.20) and using Eq. (5.22), the following results are obtained for P_G per-phase and Q_G per-phase

$$P_G = \frac{|E||V_t|}{X_d}\sin\delta + \frac{|V_t|^2}{2}\left(\frac{1}{X_q} - \frac{1}{X_d}\right)\sin 2\delta \tag{5.30a}$$

$$Q_G = \frac{|E||V_t|}{X_d}\cos\delta - |V_t|^2\left(\frac{\cos^2\delta}{X_d} + \frac{\sin^2\delta}{X_q}\right) \tag{5.30b}$$

It is observed from Eq. (5.30a) that the expression for real power contains a sin 2δ component called the reluctance torque or saliency torque. However, $(|E||V_T|/X_d)$ sin δ is the dominating component.

Example 5.4 A three-phase salient-pole synchronous generator has a terminal voltage of 1.0 pu. If the generator armature supplies a current of 0.8 pu at a lagging power factor of 0.8, determine the excitation voltage and the load angle. Also calculate the real and reactive power generated. Plot the variation in real power generated versus the load angle. Neglect the armature resistance and assume $X_d = 1.1$ pu and $X_q = 0.8$ pu.

Solution In order to solve the above example, first consider the phasor diagram of a salient-pole synchronous generator in Fig. 5.11.

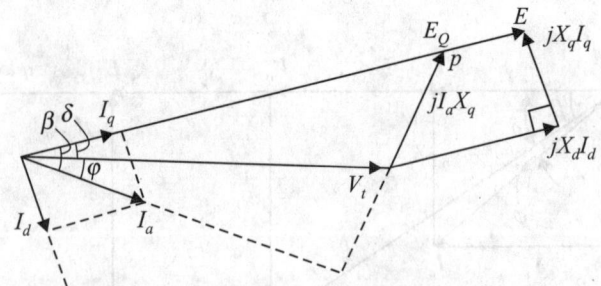

Fig. 5.11 Phasor diagram of a salient-pole synchronous generator

On comparison with Fig. 5.10, it is seen that the two phasor diagrams are similar except that a phasor jX_qI_a has been added to Fig. 5.11. It will suffice to show that the addition of the phasor jX_qI_a is correct if it can be proved that the point E_Q lies on and is in the direction of E. From Fig. 5.11, it can be seen that

$$p = V_t + jI_aX_q = V_t + jX_q(I_d + I_q) \tag{5.31}$$

From Fig. 5.11, it is also observed that

$$E = V_t + jI_d\,X_d + jI_qX_q \tag{5.32}$$

Subtracting Eq. (5.31) from Eq. (5.32) leads to

$$E - p = j(X_d - X_q)I_d \tag{5.33}$$

It may be observed that jI_d is parallel to E, thus $E - p$ is also parallel to E. Hence 'p' is collinear with E. Using Eq. (5.28),

$$\varphi = \cos^{-1}(0.8) = 36.87°$$

$$p = 1.0 + j0.8 \times 0.8 \angle{-36.87°} = 1.3840 + j0.5120 = 1.4757 \angle 20.3015°$$

From Fig. (5.11), it is observed that the power angle $\delta = 20.3015°$ and

$$|I_q| = |0.8| \cos (36.87 + 20.30) = 0.4337 \text{ pu}$$
$$|I_d| = |0.8| \sin (36.87 + 20.30) = 0.6772 \text{ pu}$$
$$I_q = 0.4337 \angle 20.3015° \text{ pu},$$
$$I_d = 0.6722 \angle (20.3015° - 90°) = 0.6722 \angle -69.6985° \text{ pu}$$

Using Eq. (5.32) gives

$$E = 1.0 + j\, 1.1 \times 0.6722 \angle -69.6985° + j\, 0.8 \times 0.4337 \times \angle 20.3015°$$
$$= 1.5731 + j\, 0.5820 = 1.6773 \angle 20.3015° \text{ pu}$$

Using Eq. (5.30a) gives

$$P_G = \frac{1.6773 \times 1.0}{1.1} \sin \delta + \frac{(1.0)^2}{2}\left(\frac{1}{0.8} - \frac{1}{1.1}\right) \sin 2\delta$$

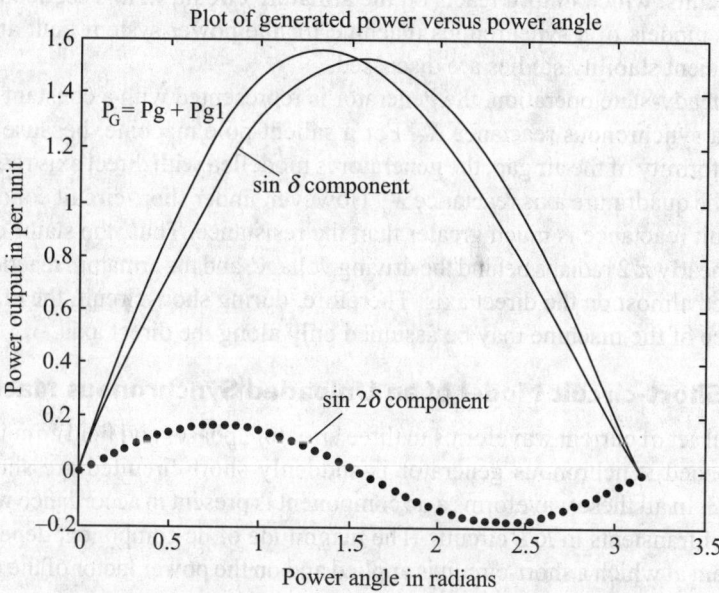

Fig. 5.12 Power angle characteristic of salient-pole machine

The following script file plots the variation of generated power versus the power angle.

```
% Plot of generator output versus power angle
>> step=2*pi/100;              % computes the increment in δ in radians
>> delta=0:step:pi;            % range of delta from 0 to π radians
>> Pg=(1.6773*1.0/1.1)         % computes the sin δ component of P_G
        *sin(delta);
>> plot(delta,Pg,'k-')         % plots the sin δ component of P_G against δ
>> hold on                     % Invokes the hold on key for over lay plots
>> Pg1=0.5*(1/0.8-1/1.1)       % computes the sin 2δ component of P_G
        *sin(2*delta);
>> plot(delta,Pg1,'k.')        % plots the sin 2δ component of P_G against δ
>> PG=Pg+Pg1;                  % computes the total generated power
>> plot(delta,PG,'k')          % plots total power output versus δ
>> xlabel('Power angle         % labels the x-axis
        in radians');
```

```
>> ylabel('Power output        % labels the y-axis
          in per unit');
>> title('Plot of Generated    % adds the title to the plot
Power versus Power Angle');
>> hold off                     % turns off the hold-on key
```

5.5 CIRCUIT MODEL OF A SYNCHRONOUS MACHINE FOR TRANSIENT STUDIES

The equivalent circuit of a synchronous machine developed in earlier sections is suitable for describing its steady-state performance. However under transient conditions, such as a short-circuit at the generator terminals, the flux linkages with the rotor circuits change with time. This causes a transient current to flow in the rotor circuits, which in turn reacts on the armature circuit. In this section simple network models of a synchronous machine for the power system fault analysis and transient stability studies are discussed.

For steady-state operation, the generator is represented with a constant emf E behind a synchronous reactance X_S. For a salient-pole machine, because of the non-uniformity of the air gap, the generator is modelled with direct axis reactance X_d and the quadrature axis reactance X_q. However, under short-circuit conditions, the circuit reactance is much greater than the resistance. Thus, the stator current lags by nearly $\pi/2$ radians behind the driving voltage, and the armature reaction flux is centred almost on the direct axis. Therefore, during short-circuit, the effective reactance of the machine may be assumed only along the direct axis.

5.5.1 Short-circuit Model of an Unloaded Synchronous Machine

A typical set of current waveforms in three armature phases and field circuit when an unloaded synchronous generator is suddenly short-circuited are shown in Fig. 5.13. In all these waveforms a dc component is present in accordance with the theory of transients in R-L circuits. The magnitude of dc component depends on the instant at which a short-circuit is applied and on the power factor of the circuit.

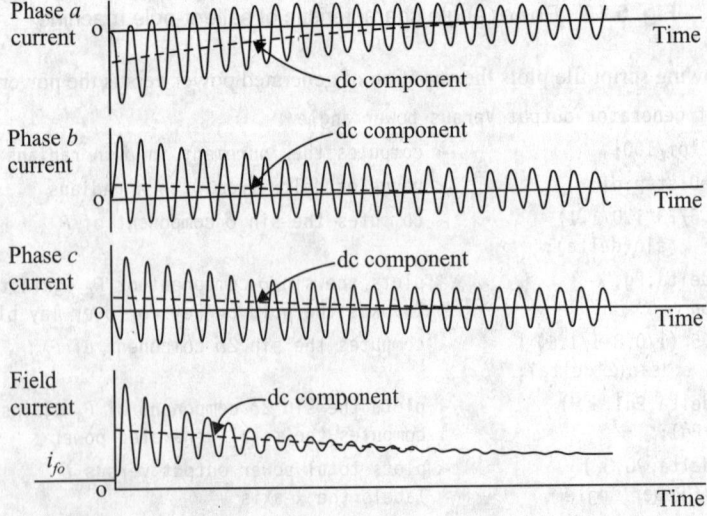

Fig. 5.13   Balanced three-phase short-circuit waveforms of a generator

If the dc component of current is eliminated from the current in each phase, the resulting plot of each phase current versus time is shown in Fig. 5.14. The plot of phase current in Fig. 5.14 shows that the ac component of the armature current decays from a very high initial value to the steady state value. This is because the machine reactance changes due to the effect of the armature reaction. At the instant prior to the short-circuit, there is some flux on the direct axis linking both the stator and rotor, due only to rotor flux if the machine is on open-circuit, or due to the resultant of rotor flux and stator flux if some stator current is flowing. When there is a sudden increase of stator current on short-circuit resulting in an increase of armature reaction mmf, the main air gap flux cannot change to a new value instantaneously, as it is linked with low resistance circuits consisting of: (i) rotor field winding which is a closed circuit, and (ii) damper windings which consist of short-circuited turns of copper strip, set in poles to dampen oscillatory tendencies. The air gap flux remains unchanged initially, and the stator armature currents are large, and can flow only through the creation of opposing currents in the rotor field and damper windings, which is essentially a transformer action. Since the stator flux is unable at first to establish any armature reaction, the reactance of armature reaction X_{ar} is negligible, and the initial machine reactance is very small, approximately equal to the leakage reactance. The eddy currents in the damper circuit decay rapidly owing to the high resistance, and armature current commences to fall. After this the current in the rotor field circuit decays, and the armature reaction mmf is gradually established. The generated emf and stator current fall until the steady-state condition on short-circuit is reached. Here, the full armature reaction effect is operational, and the machine is represented by the direct axis synchronous reactance.

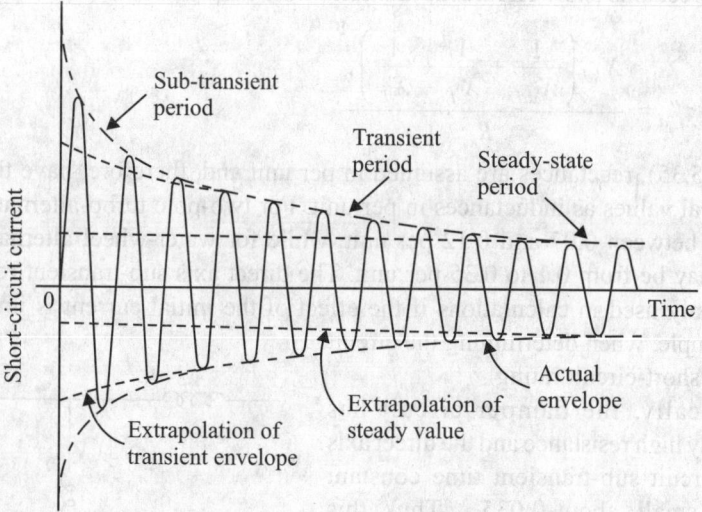

Fig. 5.14   Symmetric trace of armature short-circuit current of a generator

For purely qualitative purposes, a useful picture can be obtained by thinking of the field and damper windings as the secondaries of a transformer, whose primary is the armature winding. During normal steady-state condition, there is no transformer action between stator and rotor windings of the synchronous machine, as the resultant field flux produced by both the stator and rotor revolve with the same synchronous speed. This is similar to a transformer with open-circuited secondaries. For this condition, its primary may be described by the synchronous reactance X_d. During disturbance, the rotor speed is no longer the same as that of the revolving field produced by stator windings resulting in the transformer action. Thus, field and damper circuits resemble more nearly as short-circuited secondaries. The equivalent circuit for this condition, referred to the stator side, is shown in Fig. 5.15. Ignoring winding resistances, the equivalent reactance of Fig. 5.15, known as the direct axis sub-transient reactance, is

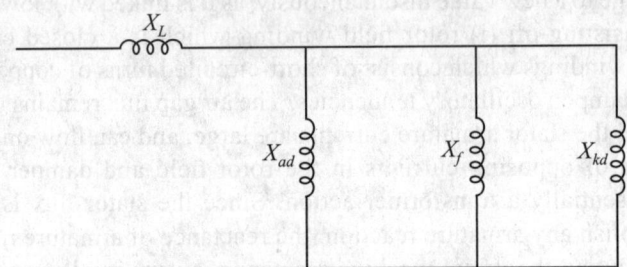

Fig. 5.15   Equivalent circuit for sub-transient period

$$X''_d = X_L + \left(\frac{1}{X_{ad}} + \frac{1}{X_f} + \frac{1}{X_{kd}} \right)^{-1} \tag{5.34}$$

If the damper winding resistance R_k is inserted in Fig. 5.15, and the Thevenin's inductance seen at the terminals of R_k is obtained, the circuit time constant, known as the direct axis short-circuit sub-transient time constant, becomes

$$\tau''_d = \frac{X_{kd} \left(\dfrac{1}{X_{ad}} + \dfrac{1}{X_f} + \dfrac{1}{X_L} \right)^{-1}}{R_k} \tag{5.35}$$

In Eq. (5.35), reactances are assumed in per unit and, therefore, have the same numerical values as inductances in per unit. For two-pole turbo-alternators, X''_d may be between 0.07 and 0.12 per unit, while for water-wheel alternators the range may be from 0.1 to 0.35 per unit. The direct axis sub-transient reactance X''_d is only used in calculations if the effect of the initial current is important, for example, when determining the circuit breaker short-circuit rating.

Typically, the damper circuit has relatively high resistance and the direct axis short-circuit sub-transient time constant is very small, about 0.035 s. Thus, this component of current decays quickly. It is then permissible to ignore the branch of the equivalent circuit, which takes account of the damper windings, and the equivalent circuit reduces to that of Fig. 5.16.

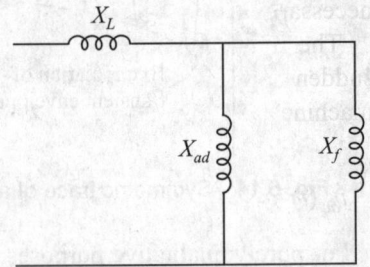

Fig. 5.16   Equivalent circuit for transient period

Ignoring winding resistances, the equivalent reactance of Fig. 5.16, known as the direct axis short-circuit transient reactance, is

$$X'_d = X_L + \left(\frac{1}{X_{ad}} + \frac{1}{X_f}\right)^{-1} \tag{5.36}$$

If the field winding resistance R_f is inserted in Fig. 5.16, and the Thevenin's inductance seen at the terminals of R_f is obtained, the circuit time constant, known as the direct axis short-circuit transient time constant, becomes

$$\tau'_d = \frac{X_f + \left(\dfrac{1}{X_L} + \dfrac{1}{X_{ad}}\right)^{-1}}{R_f} \tag{5.37}$$

The direct axis transient short-circuit reactance X'_d may be in the range of 0.10 to 0.25 per unit. The short-circuit transient time constant T'_d is usually of the order of 1 to 2 seconds.

The field winding time constant which characterizes the decay of transients with the armature open-circuited is called the *direct axis open-circuit transient time constant*. This is given by

$$\tau'_{do} = \frac{X_f}{R_f} \tag{5.38}$$

Typical values of the direct axis open-circuit transient time constant are about 5 seconds.

Finally, when the disturbance is altogether over, there will be no hunting of the rotor, and hence there will not be any transformer action between the stator and the rotor, and the circuit reduces to that of Fig. 5.17. The equivalent reactance becomes the direct axis synchronous reactance, given by

$$X_d = X_L + X_{ad} \tag{5.39}$$

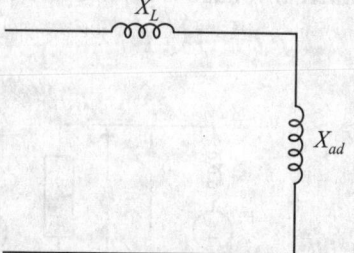

Fig. 5.17 Equivalent circuit for steady state

Similar equivalent circuits are obtained for reactances along the quadrature axis. These reactances X''_q, X'_q, and X_q may be considered for cases when the circuit resistance results in a power factor appreciably above zero, and the armature reaction is not necessarily totally on the direct axis.

The fundamental-frequency component of armature current following the sudden application of a short-circuit to the armature of an initially unloaded machine can be expressed as

$$i_{ac}(t) = \sqrt{2}E_0\left[\left(\frac{1}{X''_d} + \frac{1}{X'_d}\right)e^{-t/\tau''} + \left(\frac{1}{X'_d} + \frac{1}{X_d}\right)e^{-t/\tau'_d} + \frac{1}{X_d}\right]\sin(\omega t + \delta)$$

$$(5.40)$$

It should be recalled that in the derivation of the above results, resistance was neglected except in consideration of the time constant. In addition, in the above treatment the dc and the second harmonic components corresponding to the decay of the trapped armature flux have been neglected. It should also be emphasized that the representation of the short-circuited paths of the damper windings and the solid iron rotor by a single equivalent damper circuit is an approximation to the actual situation. However, this approximation has been found to be quite valid in many cases. The synchronous machine reactances and time constants are provided by the manufacturers.

5.5.2 Short-circuit Model of a Loaded Synchronous Machine

The equivalent circuit of a generator that has a balanced three-phase load Z_L per phase is shown in Fig. 5.18(a). The current flowing through the load before the fault occurs is I_L and the terminal voltage of the generator is V_t. As discussed in Section 5.1.3 the equivalent circuit of the synchronous generator is its no-load voltage E in series with its synchronous reactance X_S (resistance of the armature R_a neglected). It may be noted that for a round rotor machine $X_S = x_d$. If a three-phase short-circuit, simulated by closing of switch S, occurs at point F in the system, then a short-circuit from F to neutral in the equivalent circuit does not satisfy the conditions for calculating sub-transient current since the reactance of the generator must be X''_d. Similarly, for computing transient current the reactance X_d' of the generator must be used. The desired result may be achieved by using the equivalent circuit shown in Fig. 5.18(b). In this circuit sub-transient voltage E'' behind X_d'' supplies a steady load current I_L when switch S is open, and supplies the short-circuit current when switch S is closed. The sub-transient internal voltage is given by

$$E'' = V_t + j\,I_L X''_d \tag{5.41}$$

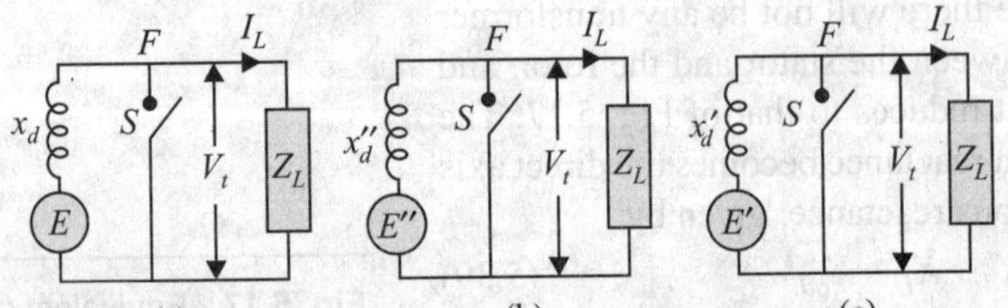

Fig. 5.18 Modification of equivalent circuit to allow for initial load current

Similarly transient current I' is supplied through X'_d and the driving voltage is transient voltage E' behind X'_d and is given by

$$E' = V_t + j\,I_L X'_d \tag{5.42}$$

The corresponding equivalent circuit is shown in Fig. 5.18(c).

5.5.3 Synchronous Machine Model for Stability Studies

A synchronous machine is represented in transient stability studies by its classical model, where saliency is neglected, and the machine is represented by constant internal voltage E' behind transient reactance in series with the direct-axis transient reactance X'_d of the machine.

5.6 POWER TRANSFORMERS

Transformers are a vital constituent of an electrical power system since they step up the voltage, at the generator end, to a level suitable for transmission of bulk power over long distances. Similarly, at the receiving end the transformers step down the voltage to levels desirable for utilization of energy. Between the generating point and the consumers' terminals, the voltage levels of electrical energy may undergo several (four to five) transformations.

5.6.1 The Ideal Transformer

An ideal transformer means that the permeability μ of the core is infinite, and all the magnetic flux created by the primary winding links with the secondary winding, core loss is zero, and the windings have zero resistance.

The schematic representation of an ideal two-winding transformer is shown in Fig. 5.19. The primary winding when connected to an ac source v_1 induces emf e_1 in the primary winding, and is given by

$$v_1 = e_1 = N_1 \frac{d\phi}{dt} \text{ V} \tag{5.43}$$

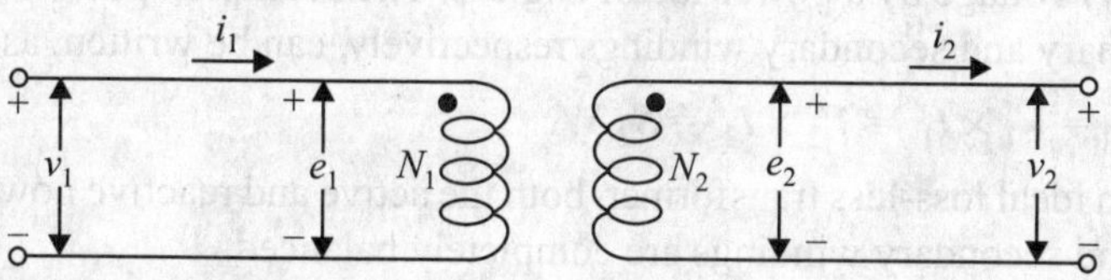

Fig. 5.19 Schematic representation of an ideal two-winding transformer

The whole of the flux ϕ links the secondary winding as well. Hence, the emf e_2 induced in the secondary winding is in phase with e_1; and with the secondary open-circuited, the following equation holds good

$$v_2 = e_2 = N_2 \frac{d\phi}{dt} \text{ V} \tag{5.44}$$

Dividing Eq. (5.43) by Eq. (5.44), the ratio $v_1 : v_2$ is obtained as

$$\frac{v_1}{v_2} = \frac{e_1}{e_2} = \frac{N_1}{N_2} \tag{5.45}$$

The relationship between the instantaneous voltages of Eq. (5.45) can be expressed alternatively in terms of the effective or rms values as

$$\frac{V_1}{V_2} = \frac{E_1}{E_2} = \frac{N_1}{N_2} = a \tag{5.46}$$

where a is called the turns ratio. If the instantaneous flux is $\phi = \phi_m \sin\omega t$ Wb, then from Eq. (5.43) the induced voltage is

$$e_1 = N_1 \frac{d\phi}{dt} = \omega N_1 \phi_m \cos\omega t \text{ V} \tag{5.47}$$

where ϕ_m is the maximum value of the flux and $\omega = 2\pi f$ rad./sec, the frequency being f hertz. It may be noted that the induced emf leads the flux by 90°. The rms value of the induced emf is

$$E_1 = V_1 = \frac{2\pi}{\sqrt{2}} f N_1 \phi_m = \sqrt{2} \pi f N_1 \phi_m$$

$$= 4.44 f N_1 \phi_m \text{ V} \tag{5.48}$$

When a load is connected across the secondary winding rms current I_2 A flows through it. In the ideal transformer, the core is assumed to have zero reluctance and there is an exact mmf balance between the primary and the secondary. The net mmf acting on the core therefore is zero, since for an ideal transformer it is assumed that the exciting current is zero. If I_2 A, which flows through N_2 turns of the winding, sets up an mmf that will oppose the core flux ϕ, then

$$I_1 N_1 = I_2 N_2 \tag{5.49}$$

From Eqs (5.46) and (5.49)

$$\frac{E_1}{E_2} = \frac{I_2}{I_1} = \frac{N_1}{N_2} = a \tag{5.50}$$

In order to show the phase relationship between primary and secondary, voltages and currents, let it be assumed that the load (secondary) current lags the load (secondary) voltage by a power factor angle φ. Then complex powers S_1 and S_2, in the primary and secondary windings respectively, can be written, as

$$S_1 = V_1 \times I_1^* = V_2 \times I_2^* = S_2 \text{ VA} \tag{5.51}$$

Thus, in an ideal loss-less transformer, both the active and reactive powers, in the primary and secondary windings are completely balanced.

The equivalent circuit of an ideal transformer is shown in Fig. 5.20. The secondary voltage and secondary current can be referred to the primary side, that is

$$V_1 = V_2' = aV_2 \; ; \; I_1 = I_2' = \frac{1}{a} I_2 \tag{5.52}$$

where V_2' and I_2' respectively denote secondary voltage and current referred to the primary. Similarly, primary voltage and current referred to the secondary side are

$$V_2 = V_1' = \frac{1}{a} V_1 \; ; \; I_2 = I_1' = aI_1 \tag{5.53}$$

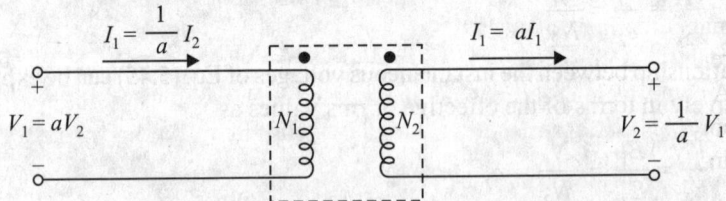

Fig. 5.20 Equivalent circuit of an ideal transformer

Figure 5.21(a) shows an ideal transformer, the secondary of which is connected across a load of impedance Z_L and it draws a current of I_2 amperes.

$$Z_L = \frac{V_2}{I_2} \tag{5.54}$$

Substituting the values of V_2 and I_2 from Eq. (5.53), Eq. (5.54) becomes

$$Z_L = \frac{V_1/a}{aI_1} \;;\; \text{or} \; \frac{V_1}{I_1} = a^2 Z_L = Z'_L \tag{5.55}$$

Z'_L in Eq. (5.55) is the equivalent load impedance transformed to the primary side. Figure 5.21(b) shows the load impedance connected after transformation to the primary of the ideal transformer. Figure 5.21(c) is the final reduced network as looked at from the source of supply. It may be noted that so far as the performance of the ideal transformer is concerned, the three circuits shown in Fig. 5.21 are the same. It is also clear that, if needed, the voltage and currents may also be referred to the secondary side.

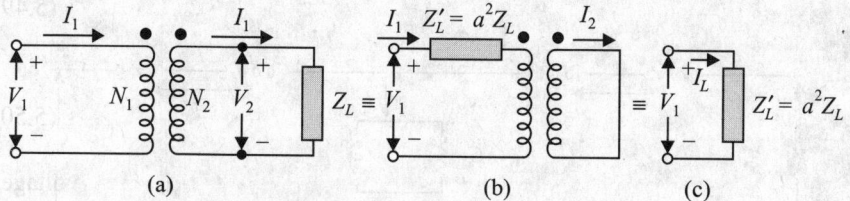

(a)   (b)   (c)

Fig. 5.21   Transformation of load impedance

5.6.2 The Equivalent Circuit of a Practical Transformer

In a practical two-winding transformer core permeability is finite, and some of the magnetic flux linking the primary winding does not link the secondary winding. This flux is proportional to primary current and causes a voltage drop that is accounted for by an inductive reactance x_{l1} called leakage reactance, which is added in series with the primary winding of an ideal transformer. For similar reasons a leakage reactance x_{l2} is added in series with the secondary winding. The winding resistances r_1 and r_2 are also accounted for by connecting them in series with their respective windings.

A practical transformer has finite reluctance, and when the secondary current I_2 A is zero, the primary current has a finite value I_0 A, called the no-load current. This current has a component I_m A, known as the magnetizing current, which sets up the core flux, and is in phase with the flux. Since flux lags the induced voltage E_1 by 90°, I_m also lags E_1 by 90°. In the equivalent circuit, I_m is taken into account by the magnetizing reactance jX_m.

In a real transformer losses occur in the core due to hysteresis and eddy-current losses. A component of current I_c A in phase with E_1 accounts for the energy loss in the core, and is represented in the equivalent circuit by the resistance R_c as shown in Fig. 5.22.

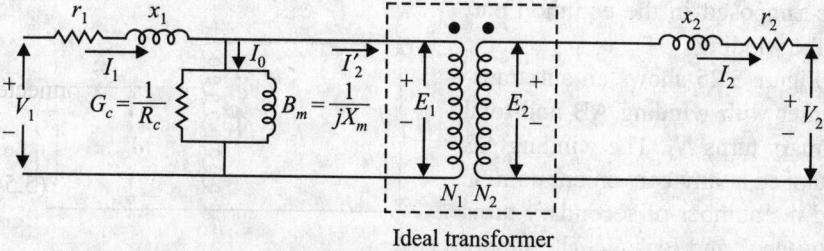

Ideal transformer

Fig. 5.22   Complete equivalent circuit of a transformer

The equivalent circuit is usually drawn with the ideal transformer not shown and all voltages, currents, and impedances are referred to either the low- or high-voltage side of the transformer as shown in Fig. 5.23. In the given figure, the referred values are indicated with primes to distinguish them from the actual values given in Fig. 5.22. Very often the magnetizing branch is neglected, because magnetizing current is too small compared to the usual load currents, in which case the transformer is represented by equivalent series impedance $Z_{eq} = R_{eq} + jX_{eq}$, as shown in Fig. 5.24(a). The equivalent resistance R_{eq} and the equivalent reactance X_{eq} are given by

$$R_{eq} = r_1 + r'_2 = r_1 + \frac{r_2}{a^2}; \text{ and } X_{eq} = x_1 + x'_2 = x_1 + \frac{x_2}{a^2} \qquad (5.56)$$

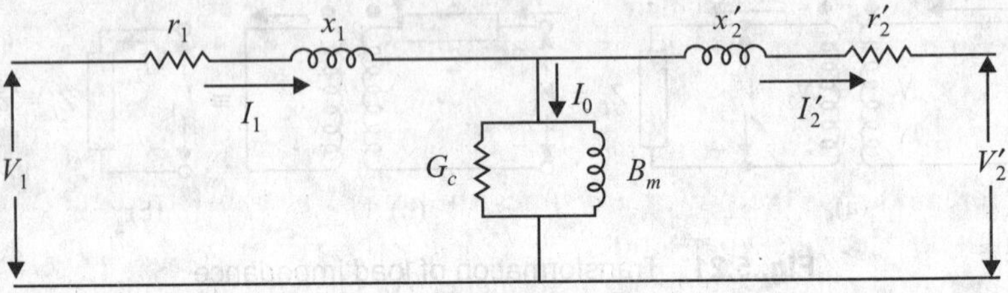

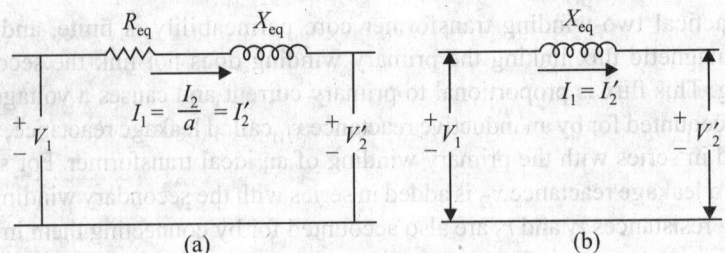

Fig. 5.23 Equivalent circuit of a transformer with the ideal transformer left out

(a)

(b)

Fig. 5.24 Approximate equivalent circuits

For large power transformers, the equivalent resistance R_{eq} is very small compared with the equivalent reactance X_{eq} and can be neglected, and the equivalent circuit reduces to that shown in Fig. 5.24(b).

5.7 AUTOTRANSFORMERS

An autotransformer is a transformer having a part of its winding common to the primary and secondary circuits. The input and output are electrically connected as well as coupled by a mutual flux, and the input and output currents are superimposed in the common part of the winding.

Figure 5.25 shows an autotrans-former with winding AB and total primary turns N_1. The winding AB is tapped at any convenient point C and the number of secondary turns between C and B is N_2. The supply voltage V_1 volts is applied across

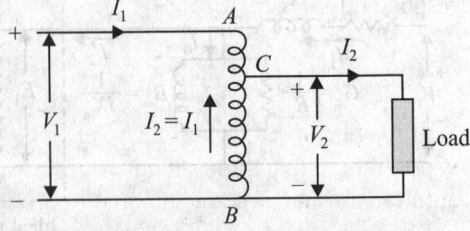

Fig. 5.25 An autotransformer

AB and the load, being connected across CB, has a voltage V_2 volts. The primary input current is I_1 A and the load current is I_2 A. Neglecting voltage drop due to leakage impedances,

$$\frac{V_1}{V_2} = \frac{N_1}{N_2} = \frac{E_1}{E_2} = a \qquad (5.57)$$

where E_1 and E_2 are the induced emfs and a is the turns ratio. The mmfs in the two windings counterbalance each other. Neglecting the exciting current, the following holds good,

$$N_2(I_2 - I_1) = (N_1 - N_2)I_1 \text{ AT} \quad \text{or} \quad N_1 I_1 = N_2 I_2$$

or

$$\frac{I_2}{I_1} = \frac{N_1}{N_2} = a \qquad (5.58)$$

Power transferred inductively or by transformer action is given by

$$V_2(I_2 - I_1) = \frac{a-1}{a} \, V_2 I_2 \text{ VA}$$

The remainder of the power $1/a$ times the output is transferred conductively through the electrical connection.

A conventional two-winding transformer can be changed into an autotransformer by connecting the primary and the secondary windings in series as shown in Fig. 5.26. With respect to a two-winding transformer with the same input–output specifications, there is a saving of copper in case of an autotransformer. In addition, the I^2R loss in an autotransformer is lower and the efficiency higher than in the two-winding transformer. (However, the main disadvantage of an autotransformer is that electrical isolation between the primary and secondary windings is lost.)

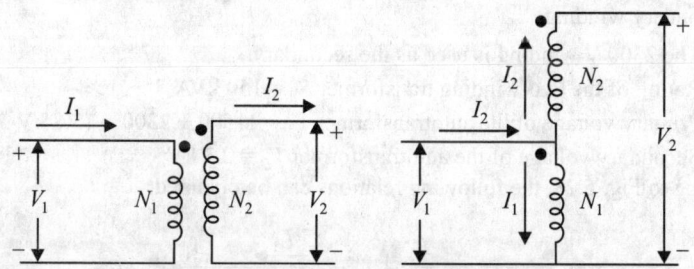

Fig. 5.26 (a) Conventional two-winding transformer
(b) Two-winding transformer connected as autotransformer

Another advantage of autotransformers is that they provide a higher power rating in comparison to two-winding transformers. Assume S_T and S_{AT} are the apparent power ratings of a two-winding transformer and an autotransformer respectively. Then,

$$\frac{S_{AT}}{S_T} = \frac{(V_1 + V_2)I_1}{V_1 I_1} = 1 + \frac{N_2}{N_1} = 1 + \frac{1}{a} \qquad (5.59)$$

It may be noted from Eq. (5.59) that a higher output from an autotransformer is possible with a higher number of turns (N_2) in the common winding. Thus an autotransformer, when compared to a two-winding transformer of equal rating, has a lower internal impedance, higher efficiency, and is smaller in size.

5.7.1 Equivalent Circuit of an Autotransformer

From an inspection of the connection diagram given in Fig. 5.26, the equivalent circuit of the autotransformer can be deduced and is shown in Fig. 5.27. In the equivalent circuit in Fig. 5.27, R_{eq} Ω and X_{eq} Ω represent the equivalent series resistance and leakage reactance respectively of the autotransformer, with the series resistance and leakage reactance of N_2 turns referred to the N_1 turns side.

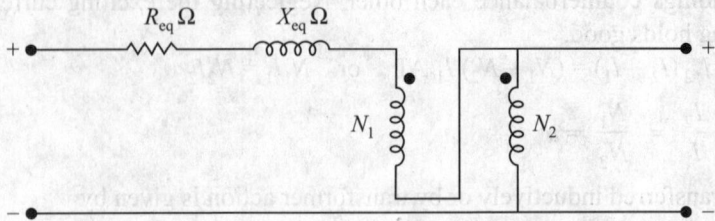

Fig. 5.27 Equivalent circuit of an autotransformer

Single-phase autotransformers can be connected for Y–Y three-phase operation, or a three-phase unit can be built. Three-phase autotransformers are sometimes used to connect two transmission lines operating at different voltage levels.

Example 5.5 A 11500/2300-V transformer is rated at 150 kVA as a two-winding transformer. If the two windings are connected in series to form an autotransformer, what will be the voltage ratio and output?

Solution   The two windings of a two-winding transformer can be connected in series to form an autotransformer with either winding being used as a secondary. Therefore, the voltage ratio and output of the autotransformer will depend on the winding which is used as the secondary winding.

Case I: The 2300 V winding is used as the secondary.

Rating of the two-winding transformer S_T = 150 kVA

Primary voltage of the autotransformer V_1 = 11500 + 2300 = 13.8 kV

Secondary voltage of the autotransformer V_2 = 2.3 kV

By referring to Fig. 5.28, the following relations can be obtained:

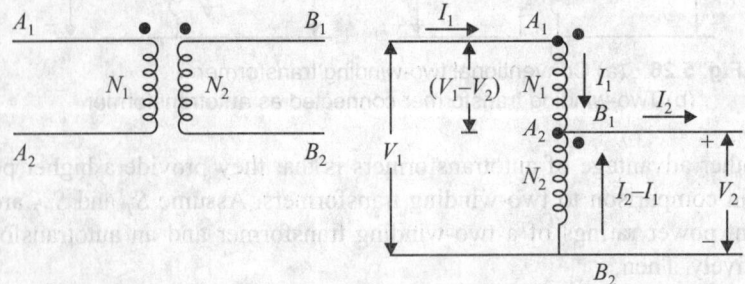

Fig. 5.28 Autotransformer connection of a two-winding transformer

Two-winding transformer voltage ratio $a = \dfrac{V_1 - V_2}{V_2} = \dfrac{N_1}{N_2} = \dfrac{11.5}{2.3} = 5$

The autotransformer voltage ratio $a' = \dfrac{V_1}{V_2} = \dfrac{V_1 - V_2 + V_2}{V_2} = a + 1 = 6$

Turns ratio $a = \dfrac{13.8}{2.3} = 6$

Rating of the transformer $S_T = (V_1 - V_2)I_1 = (I_2 - I_1)V_2$ (5.5.1)

Rating of the autotransformer $S_{AT} = V_1 I_1 = V_2 I_2$ (5.5.2)

But $\dfrac{I_2 - I_1}{I_1} = \dfrac{N_1}{N_2} = a$ (5.5.3)

Then $I_1 = \left(\dfrac{1}{1+a}\right) I_2$ (5.5.4)

Substituting Eq. (5.5.4) in Eq. (5.5.1)

$$S_T = V_2 \left(\frac{V_1}{V_2} - 1\right)\left(\frac{1}{1+a}\right) I_2 = (1 + a - 1)\left(\frac{1}{1+a}\right) V_2 I_2$$

$$= \left(\frac{a}{1+a}\right) S_{AT} \qquad\qquad (5.5.5)$$

Using Eq. (5.5.5), $S_{AT} = \left(\dfrac{1+a}{a}\right) \times 150 = 180\ \text{kVA}$

Case II: The 11500 V winding is used as the secondary

$V_1 = 13.8\ \text{kV};\ V_2 = 11.5\ \text{kV}$

Voltage ratio $a' = \dfrac{13.8}{11.5} = 1.2$

Voltage ratio $a = \dfrac{13.8 - 11.5}{11.5} = \dfrac{2.3}{11.5} = 0.2$

Using Eq. (5.5.5), $S_{AT} = \left(\dfrac{1+a}{a}\right) \times 150 = 900\ \text{kVA}$

5.8 THREE-PHASE TRANSFORMERS

A three-phase transformer is made up of a bank of three single-phase transformers. Alternatively it may be made of three primary and three secondary windings on a common three-limbed magnetic core and contained in a single tank. A single-unit three-phase transformer requires less floor space, weighs less, has higher efficiency, and costs less by about 15% than a bank of three single-phase transformers. However, the disadvantage of a three-phase transformer is that in case of a fault in one phase of a winding the transformer is required to be replaced.

The bank of three transformers or the three windings can be connected on the primary and secondary sides in four ways, namely delta–delta, star–star, delta–star, star–delta as shown in Fig. 5.29. In all the connections of this figure (i) windings to the left are primaries, and those on the right are secondaries, and (ii) windings as shown in parallel are placed on the same core and their primary and secondary voltages are in phase. When the primary side is connected to a balanced three-phase line-to-line supply voltage V and line currents I, the resulting currents and voltages are also indicated in Fig. 5.29. In all these connections the ratio of primary to secondary turns $a = N_1/N_2$ and ideal transformers are assumed. It may be noted that for fixed line-to-line voltages, the total kVA rating of each transformer is one-third the kVA rating of the bank, regardless of the connections

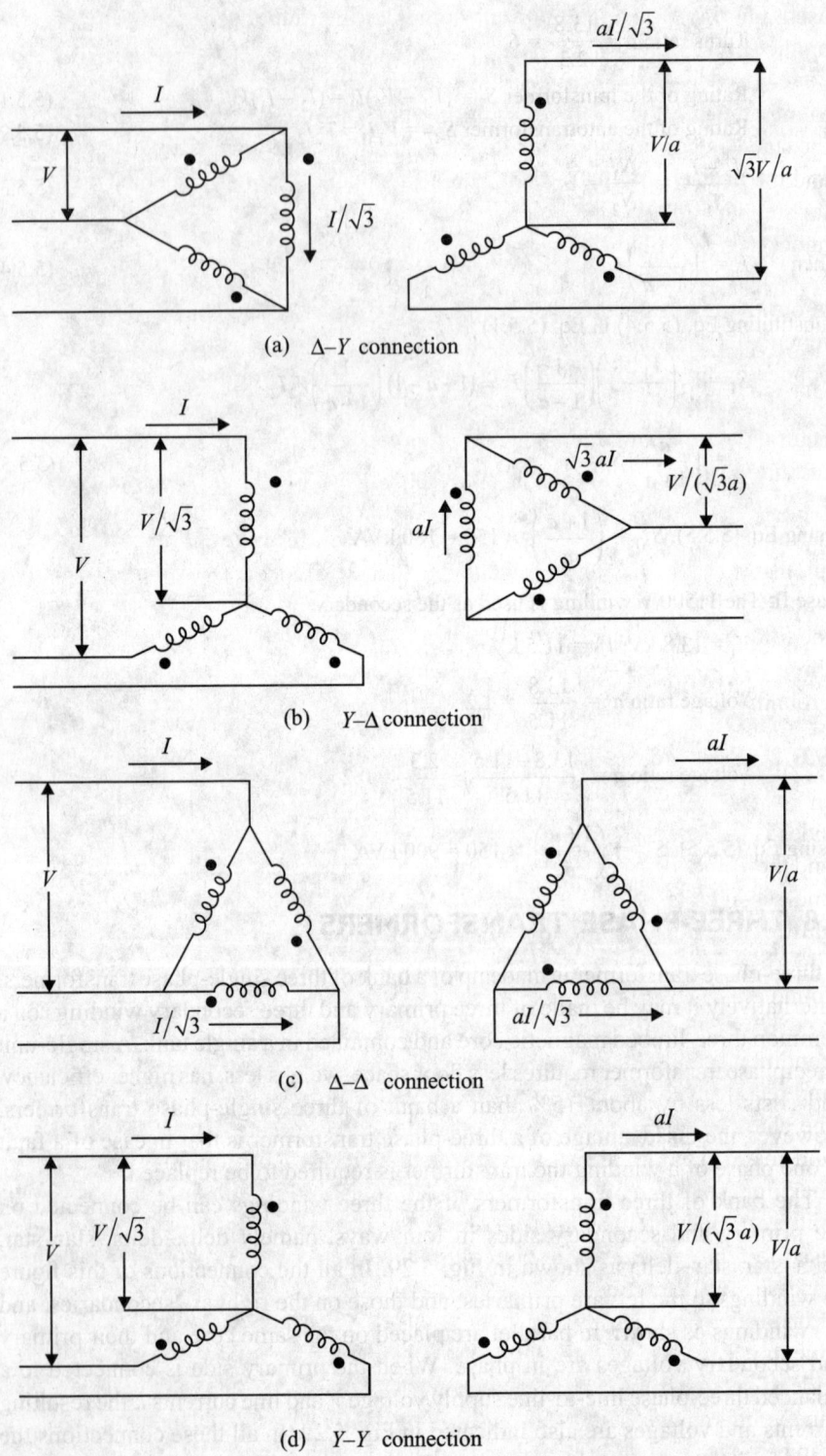

(a) Δ–Y connection

(b) Y–Δ connection

(c) Δ–Δ connection

(d) Y–Y connection

Fig. 5.29 Common three-phase transformer connections

used, but the voltage and current ratings of the individual transformers depend on the connections.

A star–delta (Y–Δ) connection is used for stepping down from high voltage to medium and low voltages with the star point connected to the ground. Since the star side carries the high voltage, insulation requirements are also less severe and the neutral point is earthed. Similarly a delta–star (Δ–Y) connection is used for stepping up to a high voltage. Both Y–Δ and Δ–Y provide stable connections under unbalanced load conditions.

A delta–delta (Δ–Δ) connection has the advantage that even when one phase is out of service for maintenance and repair, the other two phases continue to function with a reduced rating of 58% of the original capacity. This type of a connection is called a V-connection. A Δ–Δ connection provides a path for the third harmonics, thereby eliminating the associated problems. However, since no neutral point is available, the winding insulation has to be designed to withstand full line-to-line voltage.

A star–star (Y–Y) connection is rarely used due to the problems of third harmonics and unbalanced operation. To eliminate the problem of third harmonics, another set of delta-connected windings (called the *tertiary winding*) is employed. The *tertiary winding* is placed on the transformer core and provides a path for the third harmonic current. A transformer with a *tertiary winding* is also called a three-winding transformer. A star–star connection provides the facility of neutral grounding on both sides and decreased insulation level, and hence lowers the cost.

5.8.1 The Per-phase Model of Three-phase Transformer

In Y–Y and Δ–Δ connections the ratio of line voltages on primary and secondary sides is the same as the ratio of phase voltages on primary and secondary sides; and there is no phase shift between the corresponding line voltages on the primary and secondary sides. However, in Y–Δ and Δ–Y connections there is a phase shift of 30° between primary and secondary line-to-line voltages. Referring to Fig. 5.29, the ratio of the line voltage magnitude for Y–Δ transformer can be found out to be $\sqrt{3}\,a$.

Circuit computations involving three-phase transformer banks under balanced conditions can be made by dealing with only one of the transformers or phases and recognizing that conditions are the same in the other two phases except for the phase displacements associated with a three-phase system. As the core loss and magnetization current for power transformers are less than one % of the maximum ratings, the shunt branch of the equivalent circuit is neglected, and only resistance and leakage reactance referred to primary or secondary side are used to represent equivalent circuit of the transformer. It is usually convenient to carry out the computations on a per-phase-Y line-to-neutral basis, since transformer impedances can then be added directly in series with transmission line impedances. The impedances of transmission lines can be referred from one side of the transformer bank to the other by the use of square of the ideal line-to-line voltage ratio of the bank. In dealing with Y–Δ or Δ–Y banks, all quantities can be referred to the Y-connected side. In dealing with Δ–Δ banks in series with transmission lines, it is convenient to replace the Δ-connected impedances of the transformers by equivalent Y-connected impedances. It can be shown that a balanced Δ-connected circuit of Z_Δ Ω/phase is equivalent to balanced Y-connected circuit of Z_Y Ω/phase if $Z_Y = Z_\Delta/3$.

5.9 THREE-WINDING TRANSFORMERS

A three-winding transformer has three windings—primary, secondary, and tertiary. Since the three windings have different voltages and voltampere ratings, such types of transformers find applications in power systems in which two independent loads having different voltage ratings have to be supplied from the same source. The tertiary winding is used to supply auxiliary power in a sub-station or to feed a local distribution system. Three-winding transformers are also used to interconnect systems operating at different transmission voltages. Static capacitors or synchronous condensers may be connected to the tertiary windings for power factor correction or voltage regulation. Sometimes delta connected tertiary windings are put on three-phase banks to provide a circuit for the third harmonics of the exciting current.

A schematic diagram of a single-phase three-winding transformer is shown in Fig. 5.30.

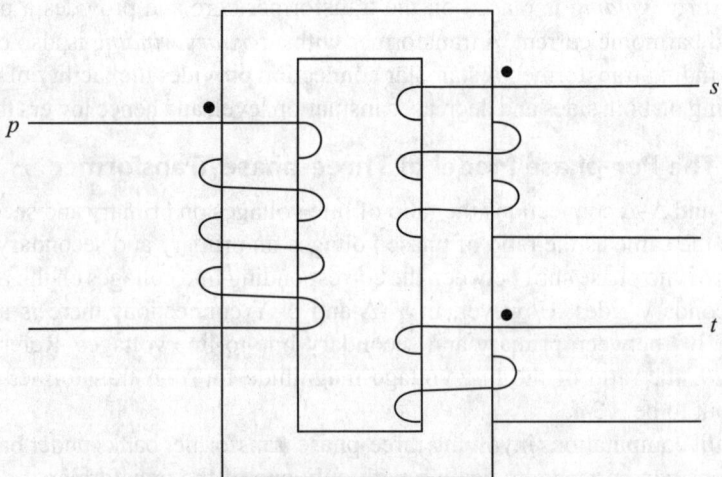

Fig. 5.30 Schematic of a three-winding transformer

Based on the principles of transformer theory and operation, the following equations can be written.

$$\frac{E_p}{E_s} = \frac{N_p}{N_s}, \frac{E_S}{E_t} = \frac{N_S}{N_t}, \text{ and } \frac{E_t}{E_p} = \frac{N_t}{N_p}$$

or
$$\frac{E_p}{N_p} = \frac{E_s}{N_s} = \frac{E_t}{N_t} \tag{5.60}$$

The subscripts *p*, *s*, and *t* are used to designate primary, secondary, and tertiary windings respectively. The equivalent circuit of a three-winding transformer can be developed by the standard short-circuit test. The following parameters can be measured experimentally:

Z_{ps}: the leakage impedance between primary and secondary; secondary is short-circuited and tertiary is kept on open-circuit.

Z_{pt} : the leakage impedance between primary and tertiary; tertiary is short-circuited and secondary is kept on open-circuit.

Z'_{st} : the leakage impedance between secondary and tertiary; tertiary is short-circuited and primary is kept on open-circuit.

The impedance Z'_{st} when referred to primary side gives

$$Z_{st} = \left(\frac{N_p}{N_s}\right)^2 Z'_{st} \tag{5.61}$$

Let the equivalent circuit for the transformer be sought in the form as in Fig. 5.31 with Z_p, Z_s, and Z_t, being referred to the primary side, where Z_p, Z_s, and Z_t are respectively the impedances of the primary, secondary, and tertiary windings, referred to the primary side. Then,

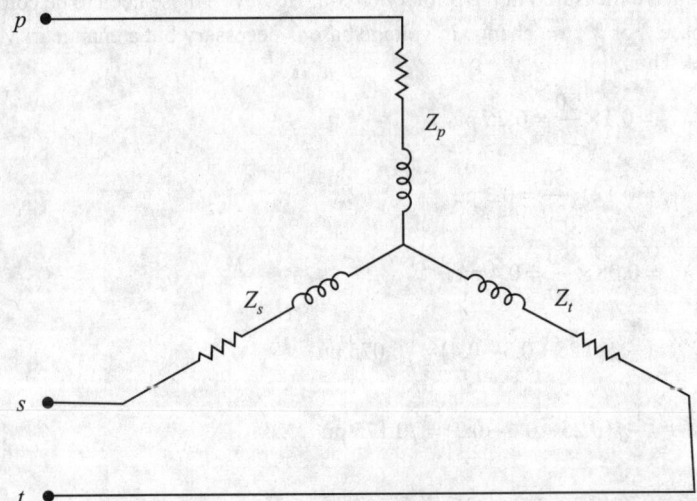

Fig. 5.31 Equivalent circuit of a three-winding transformer

$$Z_p + Z_s = Z_{ps} \tag{5.62}$$

$$Z_p + Z_t = Z_{pt} \tag{5.63}$$

$$Z_s + Z_t = Z_{st} \tag{5.64}$$

Solving Eqs (5.62), (5.63), and (5.64) gives

$$Z_p = \frac{1}{2}\left(Z_{ps} + Z_{pt} - Z_{st}\right) \tag{5.65}$$

$$Z_s = \frac{1}{2}\left(Z_{ps} + Z_{st} - Z_{pt}\right) \tag{5.66}$$

$$Z_t = \frac{1}{2}\left(Z_{pt} + Z_{st} - Z_{ps}\right) \tag{5.67}$$

In the equivalent circuit shown in Fig. 5.31, the shunt branch representing the magnetizing current and core loss components has been left out.

Example 5.6 A three-phase, three-winding transformer has primary and secondary windings connected in star and tertiary winding in delta. The primary, secondary, and tertiary windings have the following ratings respectively: 66 kV, 20 MVA; 11 kV, 10 MVA; and 6.6 kV, 5 MVA. The leakage reactances of the three-windings and their bases are given as follows:

$$X_{ps} = 0.1 \text{ on } 66 \text{ kV, } 20 \text{ MVA}$$

$$X_{pt} = 0.12 \text{ on } 66 \text{ kV, } 20 \text{ MVA}$$

$$X_{st} = 0.08 \text{ on } 11 \text{ kV, } 10 \text{ MVA}$$

Selecting 50 MVA and 66 kV as base values in the primary, calculate the pu leakage reactances of the equivalent star-connected circuit.

Solution As specified, base values in primary circuit are 50 MVA and 66 kV. Appropriate base values for the equivalent circuit in the three circuits are; primary circuit: 50 MVA, 66 kV; secondary circuit: 50 MVA, 11 kV; tertiary circuit: 50 MVA, 6.6 kV.

From the given data, it is observed that X_{ps} and X_{pt} are on the appropriate voltage base, since these were measured in the primary circuit. However, these need to be converted to 50 MVA base. For X_{st}, no change in voltage base is necessary but a change in MVA base is essential. Thus,

$$X_{ps} = 0.1 \times \frac{50}{20} = 0.25 \text{ pu}$$

$$X_{pt} = 0.12 \times \frac{50}{20} = 0.3 \text{ pu}$$

$$X_{st} = 0.08 \times \frac{50}{10} = 0.4 \text{ pu}$$

Then, $$X_p = \frac{1}{2} j(0.25 + 0.3 - 0.4) = j0.075 \text{ pu}$$

$$X_s = \frac{1}{2} j(0.25 + 0.4 - 0.3) = j0.175 \text{ pu}$$

$$X_s = \frac{1}{2} j(0.3 + 0.4 - 0.25) = j0.225 \text{ pu}$$

5.10 REGULATING TRANSFORMERS

It is mandatory that an electric energy supply authority delivers power at the specified voltage, within a prescribed tolerance, at the consumers' terminals. From a power systems operation engineer's perspective, therefore, it is essential that the voltage drop due to a continuously varying real and reactive power load is compensated. Transformers play the important role of managing (i) the real power flow by controlling the phase angle and (ii) the reactive power flow by controlling the voltage. Regulating transformers are used to control the voltage magnitude and phase angle in a power system and are of two types—(i) tap-changing transformers and (ii) regulating transformers or boosters.

5.10.1 Tap-changing Transformers

Practically all power transformers have taps in one or more windings for changing the turns ratio. This method is most popular for controlling voltage at all levels.

Changing tap settings affects the voltage magnitude as the distribution of reactive power is altered.

Transformers with Fixed Tap Setting

A transformer with off-nominal turns ratio can be represented by its impedance, or admittance, connected in series with an ideal transformer as shown in Fig. 5.32(a). An equivalent π-circuit, suitable for power flow studies, as shown in Fig. 5.32(b) may be obtained.

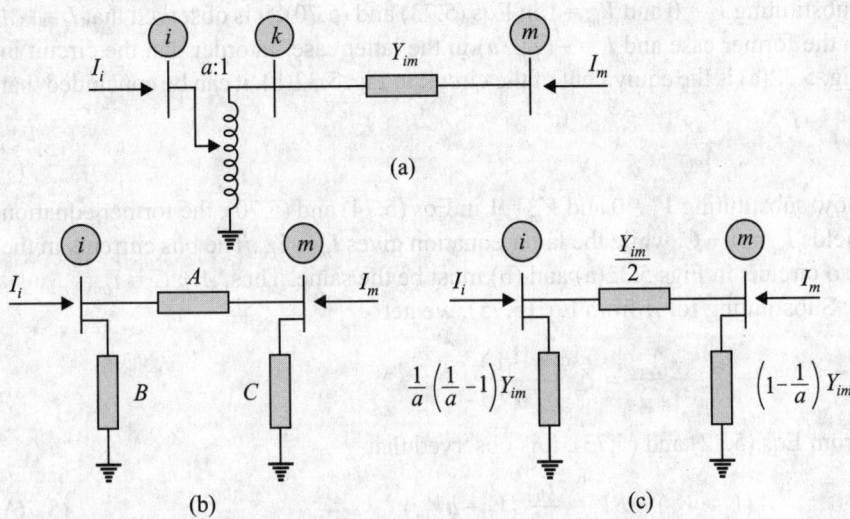

Fig. 5.32 Tap-changing transformer: (a) Equivalent circuit; (b) equivalent p-circuit; (c) parameters expressed in terms of admittance and off-nominal turns ratios

From Fig. 5.32(a), the current I_i may be determined as

$$I_i = \frac{I_{km}}{a} \tag{5.68}$$

where I_{km} is the current flowing from k to m and a is the turns ratio of an ideal transformer. The line current from k to m can also be expressed as

$$I_{km} = Y_{km}(V_k - V_m)$$

Substituting for I_{km} in Eq. (5.68) gives

$$I_i = \frac{Y_{km}}{a}(V_k - V_m) \tag{5.69}$$

Remembering that $\dfrac{V_i}{a} = V_k$, Eq. (5.69) can be written as

$$I_i = \frac{Y_{km}}{a}\left(\frac{V_i}{a} - V_m\right) = \frac{Y_{km}}{a^2}(V_i - aV_m) \tag{5.70}$$

Similarly I_m can be expressed as

$$I_m = Y_{im}(V_m - V_k) \tag{5.71}$$

$$= Y_{im}\left(V_m - \frac{V_i}{a}\right) = \frac{Y_{im}}{a}(aV_m - V_i) \tag{5.72}$$

It is required to establish expressions to determine the A, B, and C constants of the equivalent π-circuit in terms of the parameters of the off-nominal turns ratio transformer.

From Fig. 5.32(b), the terminal currents flowing into the bus i and bus m can be written as

$$I_i = A(V_i - V_m) + BV_i \tag{5.73}$$

$$I_m = A(V_m - V_i) + CV_m \tag{5.74}$$

Substituting $V_i = 0$ and $V_m = 1$ in Eqs (5.73) and (5.70), it is observed that $I_i = -A$ in the former case and $I_i = -(Y_{im}/a)$ in the latter case. In order that the circuit in Fig. 5.32(a) is the equivalent of the circuit in Fig. 5.32(b), it can be concluded that

$$A = \frac{Y_{im}}{a} \tag{5.75}$$

Now substituting $V_i = 0$ and $V_m = 1$ in Eqs (5.74) and (5.70), the former equation yields $I_m = A + C$, while the latter equation gives $I_m = Y_{im}$. The bus currents in the two circuits in Figs 5.32(a) and (b) must be the same. Thus, $A + C = Y_{im}$.

Substituting for A from Eq. (5.75), we get

$$C = Y_{im} - \frac{Y_{im}}{a} = Y_{im}\left(\frac{a-1}{a}\right)$$

From Eqs (5.72) and (5.73), it is observed that

$$A (V_i - V_m) + BV_i = \frac{Y_{im}}{a^2}(V_i - aV_m) \tag{5.76}$$

Substituting A from Eq. (5.75) in Eq. (5.76) and simplifying, B is obtained as

$$B = \frac{Y_{im}}{a}\left(\frac{1}{a}-1\right) \tag{5.77}$$

The π-equivalent circuit of off-nominal turns ratio transformer is shown in Fig. 5.32(c).

Transformers with Tap-changing under Load (TCUL)

In order to maintain the voltage at specified values, under varying loads in a power system, tap-changing under load (TCUL) transformers are used. The voltage at a specified bus is maintained by varying the tap settings and thereby changing the turns ratio of the TCUL transformer at that bus. A small change Δa is affected in changing the turns ratio. The change in tap setting is in percentage per step and is supplied by the manufacturer.

5.10.2 Regulating Transformers

Regulating transformers or boosters are used to change the voltage magnitude and phase angle at a certain point in the system by a small amount. These transformers add a small component of voltage, typically less than 0.1 pu to the line or phase voltages.

Regulating Transformers for Voltage Magnitude Control

There are a number of different schemes by means of which it is possible to obtain magnitude control of the added voltage ΔV. Figure 5.33(a) shows one common arrangement. From a three-phase autotransformer, called exciting transformer, an adjustable voltage is picked off and applied to the primary of the series transformer, which is placed at the point where the voltage ΔV is to be added. The voltage picked up by means of the tap changer T_a is in phase with V_a. Therefore, the series transformer in phase a must add a voltage ΔV_a to V_a which is in phase with V_a itself, illustrated by the phasor diagram in Fig. 5.33(b). The magnitude $|\Delta V|$ is obviously dependent upon the tap changer position. Since the voltages are in phase, a booster of this type is called an in-phase booster. If the polarity of the voltage ΔV_a is reversed, then the output voltage is less than the input voltage.

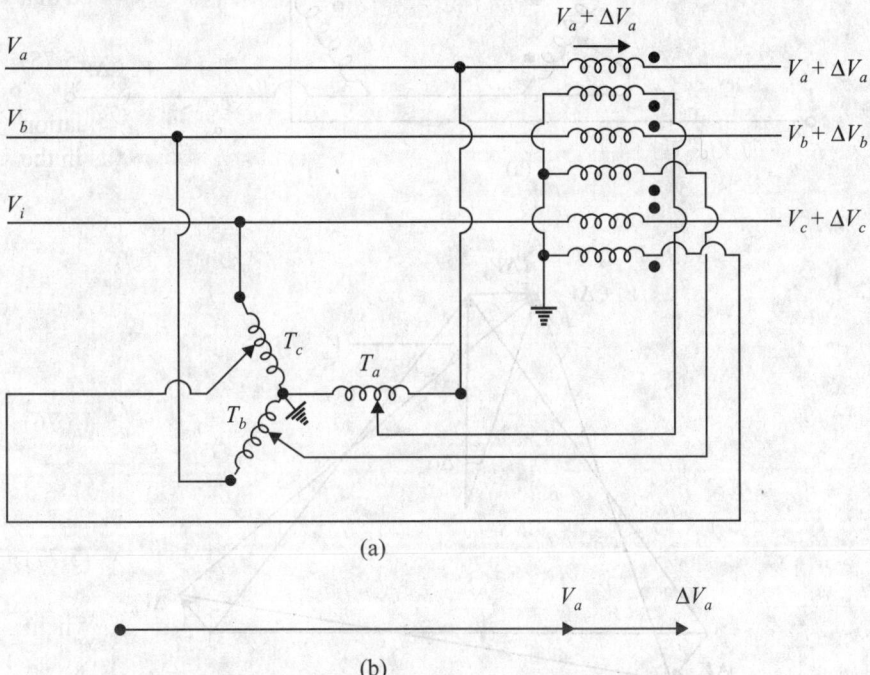

(a)

(b)

Fig. 5.33 Regulating transformer for voltage magnitude control

Regulating Transformers for Voltage Phase Angle Control

A possible arrangement to accomplish this type of control is shown in Fig. 5.34(a). If the added voltage ΔV has a phase of $\pm 90°$ relative to the system voltage V, then the effect of the added voltage is to change the phase angle of the system voltage, as shown in the phasor diagram of Fig. 5.34(b). The phase angle adjustment $\Delta \alpha$ is approximately proportional to the magnitude of the added regulator voltage ΔV; smaller the adjustments, better is the approximation. It is also easily confirmed from the phasor diagram that for small angular adjustments, the voltage magnitude practically remains unaffected.

The derivation of equivalent circuit of a phase-shifting transformer for phase angle control is somewhat more complex. However simulation of phase-shifting transformers for power system analysis is described in detail in Section 7.9.

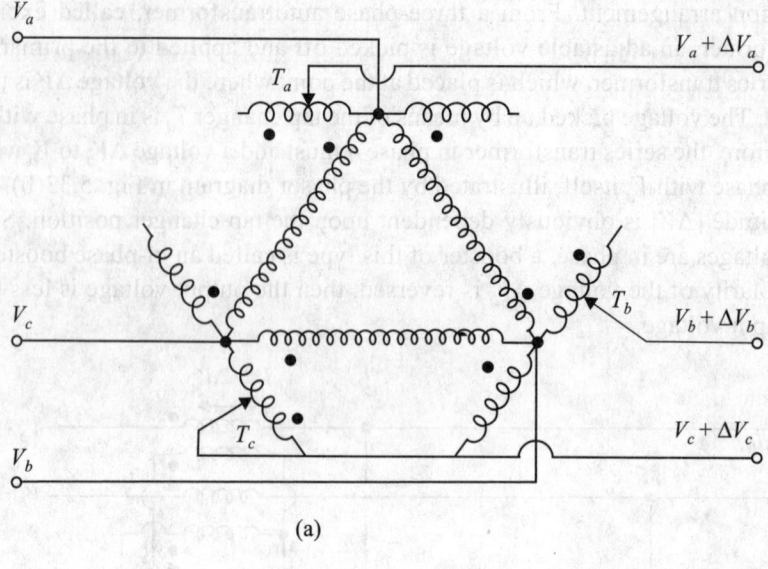

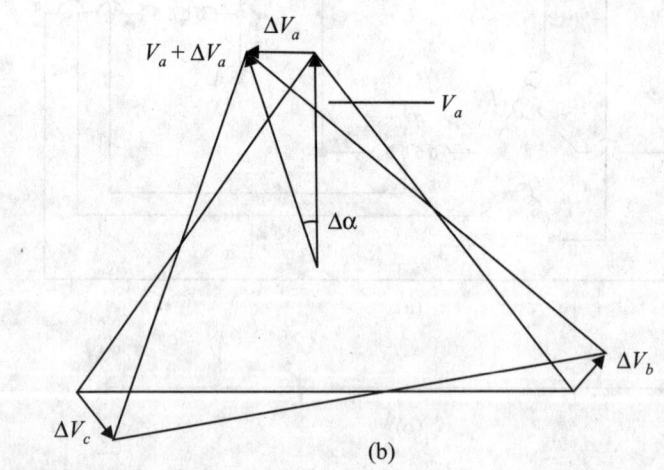

Fig. 5.34 Regulating transformer for voltage phase angle control

5.10.3 Conversion of Transformer Parameters into pu Quantities

The ohmic values of resistance and leakage reactance of a transformer depend on whether they are measured on the high-voltage or low-voltage side of the transformer. If they are expressed in per unit, the base voltampere is the voltampere rating of the transformer, and the base voltage is understood to be the voltage rating of the low-voltage winding if the ohmic values of resistance and leakage reactance are referred to the low-voltage side of the transformer. Likewise, if the ohmic values are referred to the high-voltage side of the transformer then the base voltage is taken to be the voltage rating of the high-voltage winding. The per unit impedance of a transformer is the same regardless of whether it is determined from ohmic values referred to the high-voltage or low-voltage sides of the transformers.

For circuits connected to each other through a transformer, the proper selection of voltage bases is to be made. The voltage bases for the circuits connected through the transformer must have the same ratio as the turns ratio of the transformer windings.

When magnetizing current is neglected, the transformer is represented completely by its impedance $(R + jX)$ in per unit. No per unit voltage transformation occurs when this system is used, and the current will also have the same per unit value on both sides of the transformer.

Example 5.7 A 5-MVA, three-phase, 11/220 kV, delta-star transformer has a leakage reactance of 4.5 Ω referred to the low-voltage side. Calculate the pu leakage reactance when $MVA_{base} = 50$ and $kV_{base} = 11$. If $kV_{base} = 132$, what is the pu leakage reactance? What is the ratio of the pu leakage reactances and what conclusion can be drawn from the results?

Solution Using Eq. (2.47), if $kV_{base} = 11$,

$$Z_{base} = \frac{(11)^2}{5} = 24.2 \ \Omega$$

$$Z_{pu} = \frac{4.5}{24.2} = 0.186 \qquad (5.7.1)$$

Transformation ratio $a = \dfrac{11}{220} = \dfrac{1}{20}$

Leakage reactance referred to the star side $= \dfrac{4.5}{a^2} = 4.5 \times 400 = 1800 \ \Omega$

If $kV_{base} = 220$, $Z_{base} = \dfrac{(220)^2}{5} = 9680 \ \Omega$

$$Z_{pu} = \frac{1800}{9680} = 0.186 \text{ which is the same as (5.7.1)}.$$

The value of the pu leakage reactance (or impedance) is independent of the low or high-voltage windings, if the base voltages are so selected that they have the same transformation ratio as that of the rated line-to-line transformer voltages.

Example 5.8 Each of the three identical single-phase transformers has the following nameplate rating: Capacity 5 MVA, primary/secondary voltage: 11 kV/66 kV, leakage reactance $X_1 = 0.08$ pu, and mutual reactance $X_m = 75$ pu.
Calculate (a) actual reactances of the single-phase transformers referred to (i) primary and (ii) secondary; (b) the pu reactances when connected as (i) delta–delta, (ii) delta–star, (iii) star–delta, and (iv) star-star.

Solution Assume $MVA_{base} = 5$
(a) (i) If primary voltage is selected as base voltage, i.e., $kV_{base} = 11$, then

$$Z_{1Base} = \frac{(11)^2}{5} = 24.2 \ \Omega$$

$X_1 = 0.08 \times 24.2 = 1.936 \ \Omega$, and $X_m = 75 \times 24.2 = 1815 \ \Omega$

(ii) If secondary is selected as base voltage, that is, $kV_{base} = 66$

$$Z_{2base} = \frac{(66)^2}{5} = 871.2 \ \Omega$$

$X_1 = 0.08 \times 871.2 = 696.96 \ \Omega$, and $X_m = 75 \times 871.2 = 65340 \ \Omega$.

(b) For three-phase connections of three identical single-phase transformer units, base values selected are the line-to-line voltage and three-phase voltampere rating. The pu values calculated are same as that calculated for single-phase transformer with base values selected as phase voltage and voltampere rating of the single-phase transformer. The pu values are independent of the side to which they are referred. Therefore, the pu values of X_1 and X_m remain unchanged for all types of three-phase transformer connections.

5.11 REPRESENTATION OF LOADS

The variation of real and reactive power taken by a load with the variation of voltage is important while representing the load for power flow and stability studies. For system studies usually the load on a substation is to be considered, and this load is of a composite nature consisting of industrial, commercial, and domestic consumers. At a typical substation bus, load may consist of the following:

Induction motors 50–70%

Heating and lighting 20–30%

Synchronous motors 5–10%

Transmission losses 10–12%

Although it would be accurate to consider the P-V and Q-V characteristics of each of these loads for simulation, the analytic treatment would be very complicated. For analytic purposes there are mainly three ways of representing the load.

Constant power representation Here both the specified active power P and reactive power Q are assumed constant. This representation is used in load flow studies.

Constant current representation In this scheme current I is computed as

$$I = \frac{P - jQ}{V^*} = |I| \angle (\theta - \varphi) \tag{5.78}$$

where $V = |V| \angle \theta$ and $\varphi = \tan^{-1}(Q/P)$ is the power factor angle. The magnitude of I is held constant.

Constant impedance representation This representation is mostly used in stability studies. If P and Q at a bus are known and assumed to remain constant, then the constant impedance Z connected between the bus and the ground represents the load and it is computed as

$$Z = \frac{V}{I} = \frac{|V|^2}{P - jQ} \tag{5.79}$$

or the load admittance is given as

$$Y = \frac{I}{V} = \frac{P - jQ}{|V|^2} \tag{5.80}$$

Combination of all the three representations Hence,

$$P = a + b|V| + c|V|^2$$
$$Q = d + e|V| + f|V|^2$$

where $a, b, c, d, e,$ and f are constants which separate the load into constant power, constant current, and constant impedance components. The constants a and d

represent a part of the load as a combination of constant real and reactive power, constants b and e represent a part of the load as constant current load at constant power factor, and constants c and f represent a part of the load as static impedance.

There is no restriction on the magnitudes of the constants; they may be zero, positive or negative. While emphasizing the importance of accurate representation of system loads, it is pointed out that determination of the load characteristics of an electric power utility is quite involved and requires accurate data for computing the constants.

Example 5.9 A 50-MVA, 11-kV, three-phase generator has a sub-transient reactance of 10%. The generator connected to the power system shown in Fig. 5.35 has the following specifications:

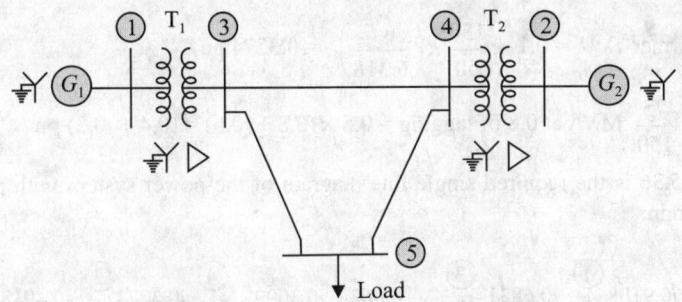

Fig. 5.35 Single line diagram for Example 5.9

Generator G_1: 50-MVA, 11-kV, three-phase generator, $X'' = 0.10$ pu
Generator G_2: 40-MVA, 6.6-kV, three-phase generator, $X'' = 0.12$ pu
Transformer T_1: Three-phase transformer, 100 MVA, 11/220 kV, $X = 0.15$ pu
Transformer T_2: Three single-phase units each rated 50 MVA, 6.6/132 kV,

$X = 0.10$ pu
Load: 75 MVA at 0.8 pf lagging
Transmission line: $Z_{34} = 30 + j150$ Ω; $Z_{35} = 20 + j40$ Ω;

$Z_{45} = 25 + j60$ Ω
Assuming a voltampere base of 150 MVA, compute the pu values for the power system. Draw a single line diagram of the power system and show the pu values of the system parameters.

Solution The three-phase rating of transformer $T_2 = 50 \times 3 = 150$ MVA
and its line-to-line voltage ratio $= \sqrt{3} \times 132/6.6 = 230.36/6.6$ kV
Assume a voltage base of 230 kV for the transmission line part of the power system.

For transmission line part of the system: $Z_{\text{base}} = \dfrac{(220)^2}{100} = 484$ Ω

$$Z_{34} = \frac{(30 + j150)}{484} = (0.062 + j0.3099) \text{ pu}$$

$$Z_{35} = \frac{(20 + j40)}{484} = (0.0413 + j0.0826) \text{ pu}$$

$$Z_{45} = \frac{(25 + j60)}{484} = (0.0516 + j0.1239) \text{ pu}$$

Base kV in G_1 circuit of the system $= 220 \times \dfrac{11}{220} = 11$ kV

Base kV in G_2 circuit of the system $= 220 \times \dfrac{6.6}{230} = 6.313$ kV

Generator G_1: $X = 0.1 \times \dfrac{150}{50} \times \left(\dfrac{11}{6.313}\right)^2 = 0.9108$ pu

Transformer T_1: $X = 0.15 \times \dfrac{150}{100} \times \left(\dfrac{11}{6.313}\right)^2 = 0.6831$ pu

Generator G_2: $X = 0.12 \times \dfrac{150}{40} \times \left(\dfrac{6.6}{6.313}\right)^2 = 0.4918$ pu

Transformer T_2: $X = 0.1 \times \dfrac{150}{50} \times \left(\dfrac{6.6}{6.318}\right)^2 = 0.3274$ pu

Load $= \dfrac{75}{150}$ MVA at 0.8 pf lagging $= 0.5 \times (0.8 + j0.6) = (0.4 + j0.3)$ pu

Figure 5.36 is the required single line diagram of the power system with pu values shown therein.

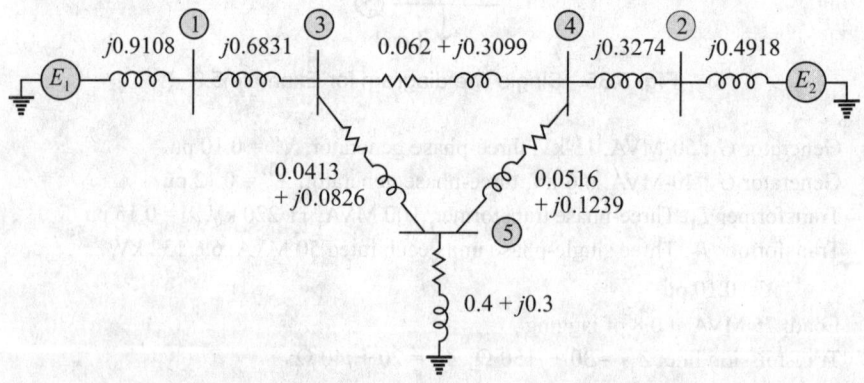

Fig. 5.36 Reactances diagram with values in pu for Example 5.10

SUMMARY

- A synchronous generator, also known as an alternator, is a rotary converting device which is run at synchronous speed by a prime mover for generating electric energy. Its stator carries the three-phase armature windings which are uniformly displaced by 120° (electrical), both in space and time.
- The practice of placing the armature windings on the stator has made it feasible to design alternators with operating voltages up to 33 kV. Due to a low operating voltage of 400 V, the DC field winding is placed on the rotor.
- Depending on the type of rotor, an alternator may be classified as (i) non-salient pole (round or cylindrical rotor) or (ii) salient pole.
- Cylindrical rotor alternators have high speeds (1500–3000 rpm), whereas salient pole machines have a power rating of 150–1500 MVA.
- The voltage generated by an alternator is proportional to the resultant flux, which is the phasor sum of field and armature fluxes.

- In the steady state, a synchronous generator is modelled by a constant generated voltage source E behind synchronous impedance Z_S.
- In the steady-state operation, an alternator when connected to an infinite bus can be used for managing both the reactive and real power flows by controlling the field current and power angle respectively.
- A synchronous condenser is an unloaded synchronous machine operating at leading power factor.
- Under sub-transient conditions of operation, an alternator is represented by its sub-transient voltage E'' behind sub-transient reactance X''_d. Similarly, when an alternator is in the transient stage it is represented by its transient voltage E' behind transient reactance X'_d.
- Transformers operate on the principle of electro-magnetic induction.
- In power systems, step-up transformers are employed to raise voltage at the generation end and step-down transformers at the distribution and consumer ends to lower the voltage.
- Three-phase transformers are three single-phase transformers or three winding transformers and can be connected in star or delta.
- Autotransformers have a part of their winding common to both primary and secondary windings.
- Taps on transformers are used to control the voltage levels by changing the turns ratio and regulating transformers (or boosters) function to vary both the voltage magnitude and phase angle, by small amounts at any point in a system.
- Loads in a power system are represented as (i) constant power, (ii) constant current, (iii) constant impedance, and (iv) a combination of all three.

SIGNIFICANT FORMULAE

- Synchronous speed $N_S = \left(\dfrac{120f}{P}\right)$

- Voltage equation of the alternator when it operates as a generator $V_t = (E - I_a Z_S)$

- Voltage equation of the alternator when it operates as a motor $V_t = (E + I_a Z_S)$

- Real power generated by a round rotor machine $P_G = \dfrac{|E||V_t|}{X_S}\sin\delta$

- Reactive power generated by a round rotor machine $Q_G = \dfrac{|E||V_t|}{X_S}\cos\delta - \dfrac{|V_t|^2}{X_S}$

- Generated voltage of a salient-pole machine $E = V_t + j\,I_d\,X_d + j\,I_q\,X_q$

- Real power generated by a salient-pole machine

$$P_G = \frac{|E||V_t|}{X_d}\sin\delta + \frac{|V_t|^2}{2}\left(\frac{1}{X_q} - \frac{1}{X_d}\right)\sin 2\delta$$

- Reactive power generated by a salient-pole machine

$$Q_G = \frac{|E||V_t|}{X_d}\cos\delta - |V_t|^2\left(\frac{\cos^2\delta}{X_d} + \frac{\sin^2\delta}{X_q}\right)$$

- Sub-transient internal voltage $E'' = V_t + jI_L X''_d$
- Transient internal voltage $E' = V_t + jI_L X'_d$

- Transformation ratio $\dfrac{V_1}{V_2} = \dfrac{E_1}{E_2} = \dfrac{N_1}{N_2} = a$
- Induced voltage $E_1 = 4.44\, f N_1\, \Phi_m\, \mathrm{V}$
- Simulation of a three-winding transformer in per unit:

$$\text{Primary impedance } Z_p = \frac{1}{2}\left(Z_{ps} + Z_{pt} - Z_{st}\right)$$

$$\text{Secondary impedance } Z_s = \frac{1}{2}\left(Z_{ps} + Z_{st} - Z_{pt}\right)$$

$$\text{Tertiary impedance } Z_t = \frac{1}{2}\left(Z_{pt} + Z_{st} - Z_{ps}\right)$$

- Load representation:

  As constant load; both P and Q are maintained constant.

  As constant current load; $I = \dfrac{P - jQ}{V^*} = |I| \angle(\theta - \varphi)$

  As constant impedance $Z = \dfrac{V}{I} = \dfrac{|V|^2}{(P - jQ)}$

  Combination of all three $P = a + b\,|V| + c\,|V|^2$ and $Q = d + e\,|V| + f\,|V|^2$

EXERCISES

Review Questions

5.1 (a) Explain with sketches the constructional features of a synchronous machine.

    (b) Enumerate the relative merits of a stationary armature winding and a rotating magnetic field.

5.2 Prove that for a salient-pole machine, having zero armature resistance, the rotor angle at which maximum power output occurs is less than $\pi/2$.

5.3 Develop the equivalent circuit of a three-winding transformer on a per unit basis.

Numerical Problems

5.1 For the one line diagram shown in Fig. 5.37, calculate the voltages at buses 1 and 2 in (i) per unit and (ii) kV if the loads at buses 2 and 3 are respectively given by $40\angle-64.5°$ and $80\angle10°$ MVA. (iii) What is the actual real and reactive power supplied by the generator? The voltage at bus 3 is maintained equal to 400 kV. Assume the base power and voltage to be 100 MVA and 400 kV respectively for the system. The reactance, in per unit, of the lines is shown in the figure.

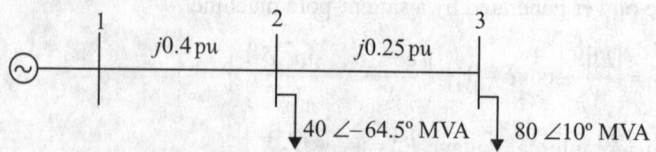

Fig. 5.37 Single line diagram for Problem 5.1

5.2 A synchronous generator is delivering 0.8 per unit real power at 0.8 power factor lagging to an infinite bus whose voltage is held equal to 1.0 per unit. Determine the generator voltage and phase difference between the bus voltage and generator voltage assuming (i) a round rotor machine having an armature reactance of $X = 0.8$ per unit and (ii) a salient-pole machine whose reactances are given by $X_d = 0.8$ and $X_q = 0.48$ per unit. In each case, draw the phasor diagram. Assume zero armature resistance for both the round rotor and salient-pole machines.

5.3 Write a MATLAB function to plot the real and reactive power output versus rotor angle in Numerical Problem 5.2. Determine the rotor angles at maximum real power output for the round rotor and salient-pole machines. Explain why the rotor angle is not the same for the two machines.

5.4 A round rotor generator is delivering 0.6 pu real power at 0.8 power factor lagging to an infinite bus whose voltage is $1.0\angle0°$ pu. The generator has a synchronous reactance of $j0.6$ pu and negligible armature resistance. Calculate (i) the power angle δ and excitation voltage E, (ii) the maximum power the generator can deliver without losing synchronism, (iii) the armature current and power factor if the excitation is increased by 20% with the power delivered to the infinite bus remaining constant, and (iv) the armature current and power factor when the mechanical torque is increased to raise the power delivered by 25% with the magnitude of the excitation voltage E held constant at the value in (i).

5.5 A synchronous generator is supplying a load of $(1.7 - j1.0)$ per unit via a transmission line as shown in Fig. 5.38. If the reactance of the line $X_{12} = j0.08$ per unit and its line-charging admittance $Y/2 = j4.0$ per unit, then determine the following: (i) the voltage magnitude of bus 1 if the voltage at bus 2 is to be maintained at 1.05 per unit, (ii) the real and reactive power supplied by the generator, (iii) the phase difference between the voltage at bus 1 and bus 2, and (iv) the magnitude of the internal voltage of the generator if its synchronous reactance is 0.22 per unit.

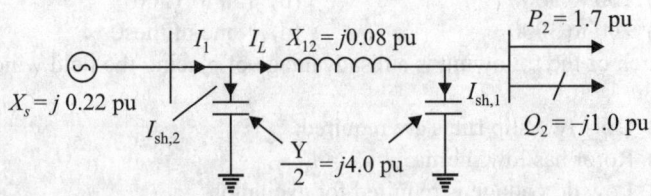

Fig. 5.38 Single-line diagram for Problem 5.5

5.6 A generator, having a synchronous reactance of 0.75 per unit is supplying 0.6 real power at 0.8 power factor lagging to a remote bus via a transmission line of reactance $j0.2$ per unit. If the remote bus voltage is maintained at $1.0\angle0°$ per unit, determine (i) the reactive power supplied by the generator and the load angle, (ii) the maximum power which can be delivered, and (iii) the magnitude of the generator terminal voltage when maximum power is delivered. (iv) What is the reactive power of the load?

5.7 A 50-Hz, 50-kVA, 220/1100-volts, two-winding transformer is supplying rated load at 0.8 lagging power factor at 95% efficiency. It is now required to connect the transformer as a step-down autotransformer to be used in a distribution system. (i) If the autotransformer winding is connected to obtain a voltage ratio of 1320/1100 volts, calculate its kVA rating and (ii) determine its efficiency if the autotransformer delivers load at 0.8 power factor lagging, as per its kVA rating in (i). Assume an ideal transformer.

5.8   A three-phase, 220-kV transmission line is connected by two transformers at the sending and receiving ends as shown in Fig. 5.39. If the impedance of the line at 220 kV is $(20 + j50)$ Ω and it supplies a load of 200 MVA at 0.8 lagging power factor, calculate the tap settings of the two transformers to maintain a voltage of 11 kV at the load end.

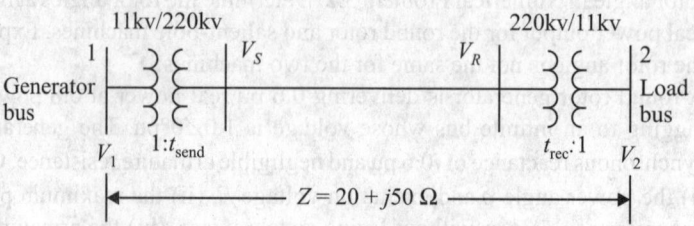

Fig. 5.39   Single-line diagram for Problem 5.8

Multiple Choice Objective Questions

5.1   Which of the following represents the synchronous speed of an 8-poles, 50-Hz synchronous generator?
   (a) 750 rpm   (b) 800 rpm
   (c) 900 rpm   (d) 950

5.2   What number of poles will generate a frequency of 100 Hz if the generator is running at 2000 rpm?
   (a) 4   (b) 6
   (c) 8   (d) 10

5.3   Which of the following represents the MVA range of round rotor synchronous generators?
   (a) 100 to 500   (b) 150 to 1500
   (c) 200 to 3000   (d) None of these

5.4   Which of the following is a disadvantage of placing the field winding on the rotor?
   (a) Only two slip rings are required.
   (b) Rotor has low inertia.
   (c) Low dc voltage is required for excitation.
   (d) None of these.

5.5   Which of the following represents the RMS value of the generated voltage in a three-phase ac machine?
   (a) $2.22 \, \phi f Z$ (b) $4.44 \, \phi f T$
   (c) $4.44 \phi \left(\dfrac{P N_S}{120} \right) T$ (d) All of these

5.6   Which of the following represents the terminal voltage of a synchronous motor?
   (a) $V_t = E + I_a Z_S$ (b) $V_t = E - I_a Z_S$
   (c) $E = V_t + I_a Z_S$ (d) None of these

5.7   Which of the following represents the apparent power of a synchronous generator?

(a) $P_G + jQ_G$
(b) $V_t I_a^*$
(c) $|V_t||I_a| (\cos \varphi + j \sin \varphi)$
(d) All of these

5.8 Which of the following controls the reactive power flow?
(a) Change in field excitation
(b) Change in power factor
(c) Change in load angle
(d) All of these

5.9 Which of the following does not apply when a synchronous machine is operating as a condenser?
(a) It is on no load.
(b) It is supplying maximum load.
(c) Power factor is unity.
(d) Field excitation is minimum.

5.10 Which of the following represents the generated voltage in a salient-pole synchronous generator?
(a) $E = V_t + jI_d (X_d + X_q)$
(b) $E = V_t + jI_q (X_d + X_q)$
(c) $E = V_t + jI_q (X_d + X_q)$
(d) $E = V_t + jI_d X_d + jI_q X_q$

5.11 Which of the following represents real power output of a salient pole synchronous generator?

(a) $P_G = \dfrac{|E||V_t|}{X_d} \sin 2\delta + \dfrac{|V_t|^2}{2} \left(\dfrac{1}{X_q} - \dfrac{1}{X_d} \right) \sin \delta$

(b) $P_G = \dfrac{|E||V_t|}{X_d} \cos 2\delta + \dfrac{|V_t|^2}{2} \left(\dfrac{1}{X_q} - \dfrac{1}{X_d} \right) \cos \delta$

(c) $P_G = \dfrac{|E||V_t|}{X_d} \sin \delta + \dfrac{|V_t|^2}{2} \left(\dfrac{1}{X_q} - \dfrac{1}{X_d} \right) \sin 2\delta$

(d) $P_G = \dfrac{|E||V_t|}{X_d} \cos \delta + \dfrac{|V_t|^2}{2} \left(\dfrac{1}{X_q} - \dfrac{1}{X_d} \right) \sin 2\delta$

5.12 In a short-circuit condition of a salient-pole synchronous generator, the armature reaction flux is centred on
(a) quadrature axis
(b) direct axis
(c) both quadrature and direct axes
(d) None of these

5.13 Which of the following represents the equivalent load impedance transformed to the primary side?
(a) Z_L/a
(b) Z_L/a^2
(c) $a^2 Z_L$
(d) $a Z_L$

5.14 Which of the following represents the component of power transferred inductively in an autotransformer?
(a) $1/a$
(b) a
(c) $\dfrac{a-1}{a}$
(d) $\dfrac{a}{a-1}$

5.15 If N_2 is the number of turns in the common winding of an autotransformer, which of the following relation will lead to a higher output?
(a) $N_2 < N_1$
(b) $N_2 = N_1$
(c) $N_2 > N_1$
(d) All of these

5.16   Which of the following represents the per unit value of the secondary winding in a three-winding three-phase transformer?

(a) $\frac{1}{2}\left(Z_{pt}+Z_{st}-Z_{ps}\right)$

(b) $\frac{1}{2}\left(Z_{ps}+Z_{pt}-Z_{st}\right)$

(c) $\frac{1}{2}\left(Z_{ps}+Z_{st}-Z_{pt}\right)$

(d) None of these

5.17   Which of the following is the value of B in a π-equivalent representation of a fixed tap-setting transformer?

(a) $\dfrac{Y_{im}}{a}$

(b) $\dfrac{Y_{im}}{a}\left(\dfrac{1}{a}-1\right)$

(c) $Y_{im}\left(\dfrac{a-1}{a}\right)$

(d) None of these

5.18   Which of the following is a function of a regulating transformer?
(a) Change the voltage magnitude   (b) Change the phase angle
(c) Works as a booster   (d) All of these

5.19   In which of the following manner loads can be simulated?
(a) Constant power only
(b) Constant current only
(c) Constant impedance only
(d) A proportionate combination of all three

Answers

5.1 (c)	5.2 (b)	5.3 (b)	5.4 (d)	5.5 (d)	5.6 (a)
5.7 (d)	5.8 (a)	5.9 (b)	5.10 (d)	5.11 (c)	5.12 (b)
5.13 (c)	5.14 (a)	5.15 (c)	5.16 (c)	5.17 (b)	5.18 (d)
5.19 (d)					

CHAPTER 6

Formulation of Network Matrices

Learning Outcomes

A directed study of this chapter will enable the reader to
- Understand the basic principles of graph theory and be able to distinguish between terms such as graph, path, tree, co-tree, and so on
- Form and manipulate bus admittance and impedance matrices, based on an understanding of incidence and primitive networks, so as to reflect changes in networks
- Simulate mathematical models of integrated power system networks from an understanding of the characteristics of the individual components and be able to understand how to handle mutually coupled elements
- Take advantage of techniques such as triangular factorization for network reduction and solutions
- Become proficient in writing MATLAB based applications to analyse complex integrated power system networks

OVERVIEW

Analyses of present-day integrated networks require the simulation of electrical properties of different components that constitute a power system. The formulation of a suitable mathematical model must describe the characteristics of individual network components as well as the relations that govern the interconnection of these elements. Matrices, as a mathematical tool, have been found to be a very convenient method of representing such types of large interconnected networks for digital computer solutions. The basic graph theory, incidence matrices, and primitive networks are commonly used for representation of networks and formulate bus admittance matrix Y_{bus} and bus impedance matrix, Z_{bus}, of the network. Modifications of Y_{bus} and Z_{bus} of the network can be done to reflect changes in the network. Treatment of mutual coupling between elements can also be taken into account in the formulation of network matrices. The network equations in admittance and impedance frame of reference can be solved by direct and indirect methods. Because the network matrices are very large, sparsity techniques are used to enhance the speed of computation in solving power system problems.

6.1 BASIC GRAPH THEORY

The formulation of network matrices is the first step in the analyses of electrical power systems. To assemble a network matrix, the knowledge of the physical layout of the electrical power system, that is, how the various system components, such as generators, transformers, and transmission lines, are connected is required. Graph theory is a very handy tool in describing the physical structure of a power system. Some terms commonly used in graph theory are defined as follows.

Element   An individual component such as a generator, transformer, transmission line, or load, which irrespective of its characteristics, is represented in the single line diagram of a power system by single line segment is called an element.

Node   The terminals of an element are known as nodes.

Graph   A graph shows the physical interconnections of the elements of a network.

Sub-graph   A sub-graph is any subset of elements of the graph.

Path   A path is a sub-graph of connected elements with no more than two elements connected to any one node.

Connected graph   If there is a path between every pair of nodes then the graph is said to be a connected graph.

Oriented graph   When all the elements of a connected graph are assigned directions, the graph is called an oriented graph.

Tree   A tree is defined as a connected sub-graph constituting all the nodes but containing no closed loops.

Branch   A branch is an element of a tree.

Links   Those elements of the connected graph that are not included in the tree are called links.

Co-tree   A co-tree is a sub-graph that is constituted of links. Co-tree is a complement of a tree.

The addition of a link to a tree gives rise to one closed path called a loop. Every addition of a link gives rise to one or more closed loops. A single line representation of a four-bus electrical power system network is shown in Fig. 6.1(a), and the corresponding oriented connected graph is given in Fig. 6.1(b). Figure 6.2 shows a tree and the corresponding co-tree of the connected graph given in Fig. 6.1(b). In the figure, branches and links have been shown in solid and dashed lines respectively.

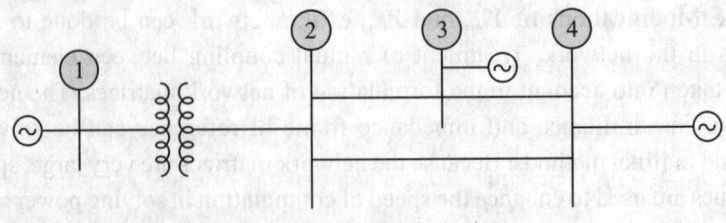

(a)

(Contd)

(*Contd*)

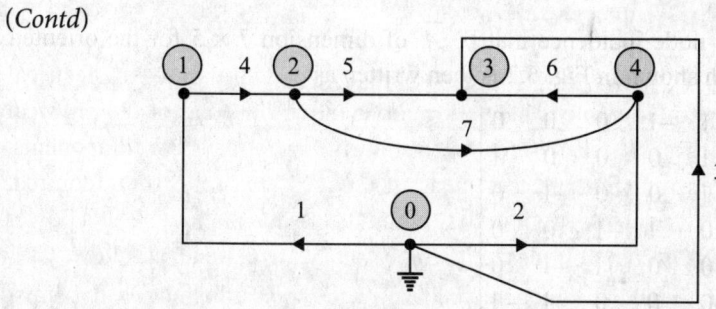

(b)

Fig. 6.1 (a) Single-line diagram of a four-bus electrical power system and (b) oriented connected graph of the network

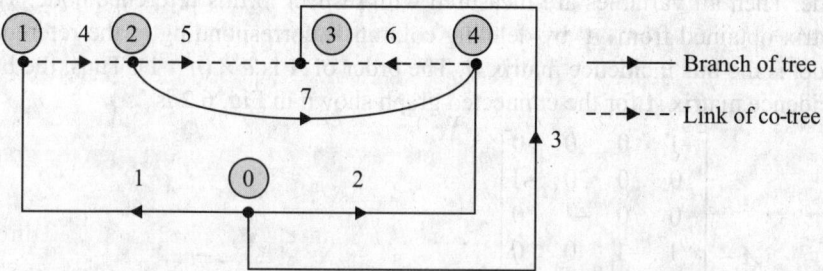

Fig. 6.2 Tree and co-tree of the connected graph shown in Fig. 6.1(b)

If in a connected graph the number of elements is e, the number of nodes including the reference node is n, then the number of branches b is given by

$$b = n - 1$$

and the number of links is given by

$$l = e - b = e - n + 1$$

In Fig. 6.2, $n = 5$ and $e = 7$, hence $b = 5 - 1 = 4$ and $l = 7 - 5 + 1 = 3$.

6.2 INCIDENCE MATRICES

The physical structure of a connected graph can be represented with the help of incidence matrices. The element-node incidence matrix and element-bus incidence matrix provide information pertaining to the network connections and will be presented in the following sections.

6.2.1 Element-node Incidence Matrix

As the name suggests, element-node incidence matrix $\overline{A}$ indicates how an element is connected to a node in an oriented connected graph. The dimension of the matrix $\overline{A}$ is $e \times n$, and the elements a_{km} of the matrix are defined as follows:

$a_{km} = 1$ if the element k is incident on and oriented away from node m

$a_{km} = -1$ if the element k is incident on and oriented towards node m

$a_{km} = 0$ if the element k is not incident on node m

The element-node incidence matrix $\overline{A}$ of dimension 7×5 for the oriented connected graph shown in Fig. 6.2 is then written as

$$\overline{A} = \begin{bmatrix} 1 & -1 & 0 & 0 & 0 \\ 1 & 0 & 0 & 0 & -1 \\ 1 & 0 & 0 & -1 & 0 \\ 0 & 1 & -1 & 0 & 0 \\ 0 & 0 & 1 & -1 & 0 \\ 0 & 0 & 0 & 1 & -1 \\ 0 & 0 & 1 & 0 & -1 \end{bmatrix}$$

6.2.2 Element-bus Incidence Matrix

In a power system network, the ground node (node 0) is selected as the reference node. Then all variables are measured with respect to this reference node. The matrix obtained from $\overline{A}$ by deleting column 1, corresponding to the reference node, is the bus incidence matrix A. The order of A is $e \times (n-1)$. Thus, the bus incidence matrix A for the connected graph shown in Fig. 6.2 is

$$A = \begin{bmatrix} -1 & 0 & 0 & 0 \\ 0 & 0 & 0 & -1 \\ 0 & 0 & -1 & 0 \\ 1 & -1 & 0 & 0 \\ 0 & 1 & -1 & 0 \\ 0 & 0 & 1 & -1 \\ 0 & 1 & 0 & -1 \end{bmatrix}$$

6.3 PRIMITIVE NETWORKS

A power network is essentially an interconnection of two-terminal components. When a set of components is not connected, it constitutes a primitive network. The two-terminal components can be represented in impedance form and in admittance form as shown in Fig. 6.3.

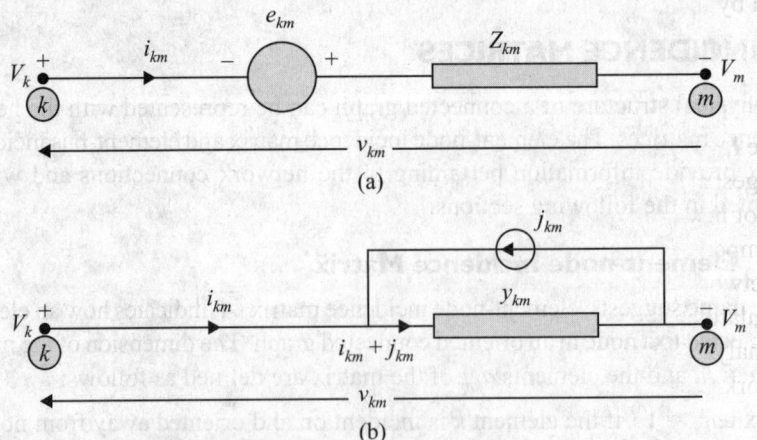

Fig. 6.3 Primitive network representation in (a) impedance form and (b) admittance form

The performance equation of a primitive network can be expressed with the help of voltage and current variables using either impedance or admittance form. Generally, the equation in impedance form is written as

$$v_{km} + e_{km} = z_{km}i_{km} \tag{6.1}$$

where v_{km} is the voltage across element k-m, e_{km} is the source voltage connected in series with the element k-m, i_{km} is the current through the element k-m and z_{km} is the self-impedance of the element. Similarly, the performance equations in admittance form are

$$i_{km} + j_{km} = y_{km}v_{km} \tag{6.2}$$

where j_{km} is the source current connected in parallel with the element k-m and y_{km} is the self-admittance of the element.

Based on Eqs (6.1) and (6.2), the performance equations for the primitive impedance and admittance networks, shown in Fig. 6.3, can be respectively written as

$$\boldsymbol{v} + \boldsymbol{e} = \boldsymbol{z}\boldsymbol{i} \tag{6.3a}$$

$$\boldsymbol{i} + \boldsymbol{j} = \boldsymbol{y}\boldsymbol{v} \tag{6.3b}$$

where $\boldsymbol{e}$ and $\boldsymbol{j}$ are specified voltage and current vectors respectively, and $\boldsymbol{v}$ and $\boldsymbol{i}$ are the element voltage and current vectors respectively. The matrices $\boldsymbol{z}$ and $\boldsymbol{y}$ are diagonal matrices of the primitive networks whose diagonal elements represent the self-impedance and self-admittance, respectively, when there is no mutual coupling between the elements. A non-zero, off-diagonal element in either $\boldsymbol{z}$ matrix or $\boldsymbol{y}$ matrix gives the mutual impedance or mutual admittance, as the case may be, between the two elements. The primitive admittance matrix $\boldsymbol{y}$ is the inverse of primitive impedance matrix $\boldsymbol{z}$.

6.4 FORMATION OF NETWORK MATRICES

A network is made up of an interconnected set of primitive elements. The general form of the network performance equation in admittance and impedance form is given by the performance equation in admittance form as

$$\boldsymbol{I}_{\text{bus}} = \boldsymbol{Y}_{\text{bus}}\boldsymbol{V}_{\text{bus}} \tag{6.4}$$

$$\boldsymbol{V}_{\text{bus}} = \boldsymbol{Z}_{\text{bus}}\boldsymbol{I}_{\text{bus}} \tag{6.5}$$

where $\boldsymbol{I}_{\text{bus}}$ is a current vector of injected currents and $\boldsymbol{V}_{\text{bus}}$ is a voltage vector of bus voltages. The bus voltages are specified with respect to a reference node, usually, but not necessarily the ground. $\boldsymbol{Y}_{\text{bus}}$ is the bus admittance matrix and $\boldsymbol{Z}_{\text{bus}}$ is the bus impedance matrix for the network.

Network equations can be formulated systematically in a variety of forms. The formulation of network in the form of bus admittance matrix is extensively used in the analyses of power systems. The bus admittance matrix of the interconnected network can be obtained by singular transformation of primitive admittance matrix using bus incidence matrix A. An alternate simple method of forming bus admittance matrix by inspection is widely used. This method uses the node voltage equations of the power system network.

6.4.1 Formation of Network Matrices by Singular Transformation

The bus admittance matrix of the interconnected network can be obtained by using the bus incidence matrix A to relate the variables and parameters of the primitive network to the bus quantities of the interconnected network.

The performance equation of the primitive network in admittance form, given by Eq. (6.3b), when pre-multiplied by the transpose of the bus incidence matrix A gives

$$A^t i + A^t j = A^t y v \tag{6.6}$$

In Eq. (6.6), by Kirchhoff's current law, the term $A^t i = 0$, since the product $A^t i$ represents a vector in which each element gives the sum of currents flowing through the elements terminating at a bus. Equation (6.6), therefore, simplifies to

$$A^t j = A^t y v \tag{6.7}$$

Now, the term $A^t j$ represents a vector in which each element is the algebraic sum of source currents injected into each bus and thus equals the vector of injected bus currents. Therefore, the bus current vector I_{bus} is given by

$$I_{\text{bus}} = A^t j = A^t y v \tag{6.8}$$

If the bus voltage vector is represented by V_{bus}, the complex power S in the network is given by

$$S = (I^*_{\text{bus}})^t V_{\text{bus}}$$

Similarly, the total power in the primitive network is given by $(j^*)^t v$. Since power is invariant,

$$S = (I^*_{\text{bus}})^t V_{\text{bus}} = (j^*)^t v \tag{6.9}$$

From Eq. (6.8), taking conjugate on both sides gives $(I^*_{\text{bus}})^t = (j^*)^t A^*$. Substitution of $(I^*_{\text{bus}})^t$ in Eq. (6.9) yields

$$(j^*)^t A^* V_{\text{bus}} = (j^*)^t v \tag{6.10}$$

Since A is real, $A^* = A$. Hence Eq. (6.10) may be written as

$$(j^*)^t A V_{\text{bus}} = (j^*)^t v \tag{6.11}$$

Equation (6.11) holds for all values of source currents j. Therefore, it is observed that

$$A V_{\text{bus}} = v \tag{6.12}$$

Substituting for v from Eq. (6.12) in Eq. (6.8) gives

$$I_{\text{bus}} = A^t y A V_{\text{bus}} \tag{6.13}$$

The network performance equation is given by Eq. (6.4) as $I_{\text{bus}} = Y_{\text{bus}} V_{\text{bus}}$. Comparing Eqs (6.4) and (6.13), it follows that

$$Y_{\text{bus}} = A^t y A \tag{6.14}$$

As bus incidence matrix A is singular, therefore, $A^t y A$ is a singular transformation of primitive matrix y.

The general form of the network performance equation in impedance form is given by Eq. (6.5) as $V_{\text{bus}} = Z_{\text{bus}} I_{\text{bus}}$. The Z_{bus} matrix for the network can be obtained from

$$Z_{\text{bus}} = Y_{\text{bus}}^{-1} = (A^t y A)^{-1} \tag{6.15}$$

Effect of Mutual Coupling

In the absence of mutual coupling between the elements, both the z and y matrices are diagonal matrices. As already stated in Section 6.3, the effect of mutual coupling in a power system network between elements is reflected by the corresponding off-diagonal terms of the primitive z or y matrices being non-zero. This will be illustrated by considering a sample three-bus system shown in Fig. 6.4(a). The elements 4 and 5 are mutually coupled. These lines are assumed to be short lines and are represented by equivalent series impedances of the respective lines. The resulting system graph is shown in Fig. 6.4(b). The ground node is represented by 0.

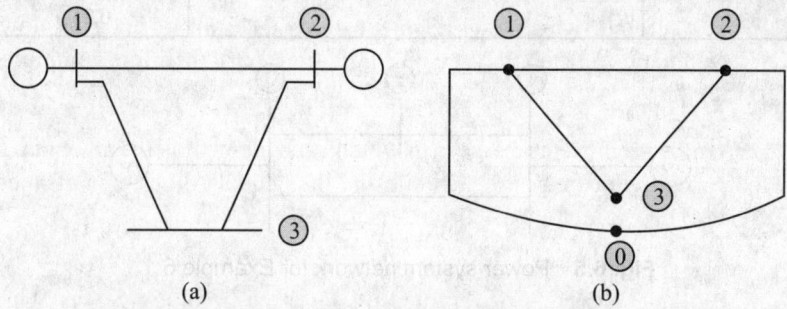

(a) (b)

Fig. 6.4 (a) A sample three-bus system and
(b) Connected graph of the three-bus system

The primitive impedance matrix z for the system is given by

$$z = \begin{bmatrix} z_{11} & & & & \\ & z_{22} & & & \\ & & z_{33} & & \\ & & & z_{44} & z_{45} \\ & & & z_{45} & z_{55} \end{bmatrix} \qquad (6.16)$$

where z_{ii} represents the self-impedance of element-i ($i = 1, 2, \ldots, 5$); and z_{45} is the mutual impedance between lines 1-3 and 2-3.

The primitive admittance matrix y of the system is the inverse of z and is given by

$$y = \begin{bmatrix} z_{11}^{-1} & & & & \\ & z_{22}^{-1} & & & \\ & & z_{33}^{-1} & & \\ & & & \dfrac{z_{55}}{\Delta} & \dfrac{-z_{45}}{\Delta} \\ & & & \dfrac{-z_{45}}{\Delta} & \dfrac{z_{44}}{\Delta} \end{bmatrix} \qquad (6.17)$$

where $\Delta = z_{44}z_{55} - z_{45}^2$ $\qquad (6.18)$

Example 6.1 For the power system network shown in Fig. 6.5, the primitive element data is as follows:

Element number	Bus number		Primitive impedance
	From	To	
1	1	0	0.05
2	3	0	0.10
3	1	2	0.50
4	2	3	0.40
5	1	3	0.25

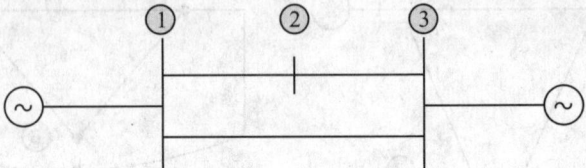

Fig. 6.5 Power system network for Example 6.1

(a) Compute the Y_{bus} matrix assuming zero mutual coupling between the elements.
(b) Compute the Y_{bus} matrix by taking a mutual coupling of 0.2 between elements 4 and 5.

Solution The oriented connected graph of the power system network graph is shown in Fig. 6.6. For the assumed orientation in the connected graph, the bus incidence matrix [A] is of order 5×3 and is given by

$$A = \begin{bmatrix} -1 & 0 & 0 \\ 0 & 0 & -1 \\ 1 & -1 & 0 \\ 0 & -1 & 1 \\ -1 & 0 & 1 \end{bmatrix}$$

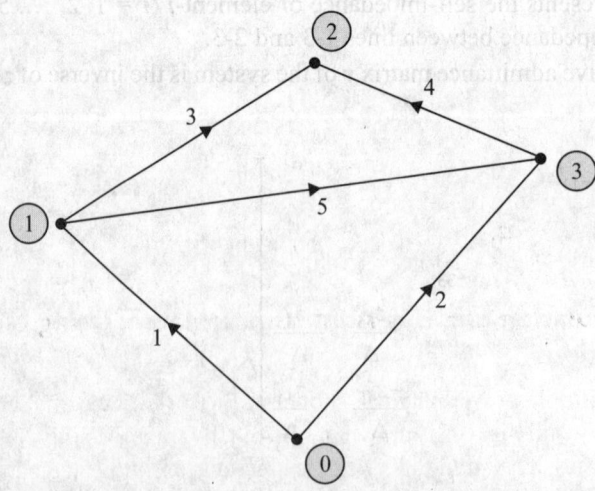

Fig. 6.6 Oriented connected graph of the network shown in Fig. 6.5

$$A^t = \begin{bmatrix} -1 & 0 & 1 & 0 & -1 \\ 0 & 0 & -1 & -1 & 0 \\ 0 & -1 & 0 & 1 & 1 \end{bmatrix}$$

$$z = \begin{bmatrix} 0.05 & 0 & 0 & 0 & 0 \\ 0 & 0.10 & 0 & 0 & 0 \\ 0 & 0 & 0.50 & 0 & 0 \\ 0 & 0 & 0 & 0.40 & 0 \\ 0 & 0 & 0 & 0 & 0.25 \end{bmatrix}$$

$$y = z^{-1} = \begin{bmatrix} 20.0 & 0 & 0 & 0 & 0 \\ 0 & 10.0 & 0 & 0 & 0 \\ 0 & 0 & 2.0 & 0 & 0 \\ 0 & 0 & 0 & 2.5 & 0 \\ 0 & 0 & 0 & 0 & 4.0 \end{bmatrix}$$

(a) $$Y_{bus} = A^t y A = \begin{bmatrix} 24.5000 & -2.0000 & -2.5000 \\ -2.0000 & 4.5000 & -2.5000 \\ -2.5000 & -2.5000 & 15.0000 \end{bmatrix}$$

(b) In this case

$$z = \begin{bmatrix} 0.05 & 0 & 0 & 0 & 0 \\ 0 & 0.10 & 0 & 0 & 0 \\ 0 & 0 & 0.50 & 0 & 0 \\ 0 & 0 & 0 & 0.40 & 0.20 \\ 0 & 0 & 0 & 0.20 & 0.25 \end{bmatrix}$$

$$y = z^{-1} = \begin{bmatrix} 20.0 & 0 & 0 & 0 & 0 \\ 0 & 10.0 & 0 & 0 & 0 \\ 0 & 0 & 2.0 & 0 & 0 \\ 0 & 0 & 0 & 4.17 & -3.33 \\ 0 & 0 & 0 & -3.33 & 6.67 \end{bmatrix}$$

$$Y_{bus} = A^t y A = \begin{bmatrix} 28.6667 & -5.3333 & -3.3333 \\ -5.3333 & 6.1667 & -0.8333 \\ -3.3333 & -0.8333 & 14.1667 \end{bmatrix}$$

$$Z_{bus} = Y_{bus}^{-1} = \begin{bmatrix} 0.0448 & 0.0405 & 0.0129 \\ 0.0414 & 0.2009 & 0.0216 \\ 0.0193 & 0.0271 & 0.0767 \end{bmatrix}$$

6.4.2 Formulation of a Bus Admittance Matrix Using Nodal Equations

Consider the simple power network shown in Fig. 6.7. The equivalent circuit of the system shown in Fig. 6.7 is drawn in Fig. 6.8, by transforming voltage sources to current sources. The admittances of the elements are indicated on the diagram. The ground is chosen as the reference.

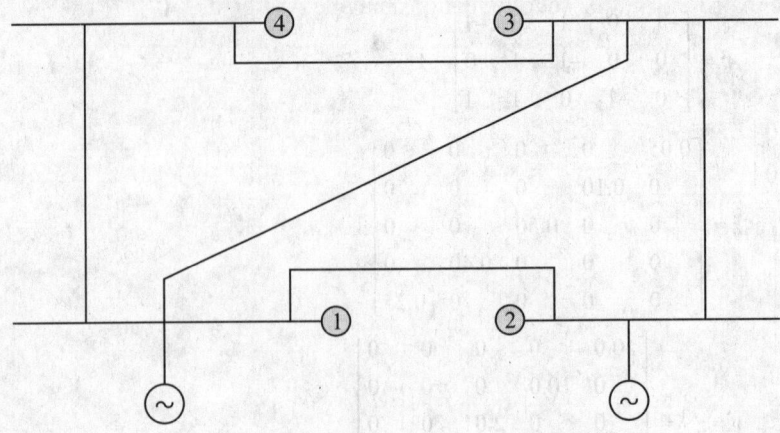

Fig. 6.7 A power system network

Applying Kirchhoff's current law KCL to all the nodes gives

$$I_1 = y_{10}V_1 + y_{12}(V_1 - V_2) + y_{13}(V_1 - V_3) + y_{14}(V_1 - V_4)$$

$$I_2 = y_{20}V_1 + y_{12}(V_2 - V_1) + y_{23}(V_2 - V_3)$$

$$0 = y_{23}(V_3 - V_2) + y_{13}(V_3 - V_1) + y_{34}(V_3 - V_4)$$

$$0 = y_{14}(V_4 - V_1) + y_{34}(V_4 - V_3)$$

By rearranging the equations, we get

$$I_1 = (y_{10} + y_{12} + y_{13} + y_{14})V_1 - y_{12}V_2 - y_{13}V_3 - y_{14}V_4$$

$$I_2 = -y_{12}V_1 + (y_{20} + y_{12} + y_{23})V_2 - y_{23}V_3$$

$$0 = -y_{13}V_1 - y_{23}V_2 + (y_{13} + y_{23} + y_{34})V_3 - y_{34}V_4$$

$$0 = -y_{14}V_1 - y_{34}V_3 + (y_{14} + y_{34})V_4$$

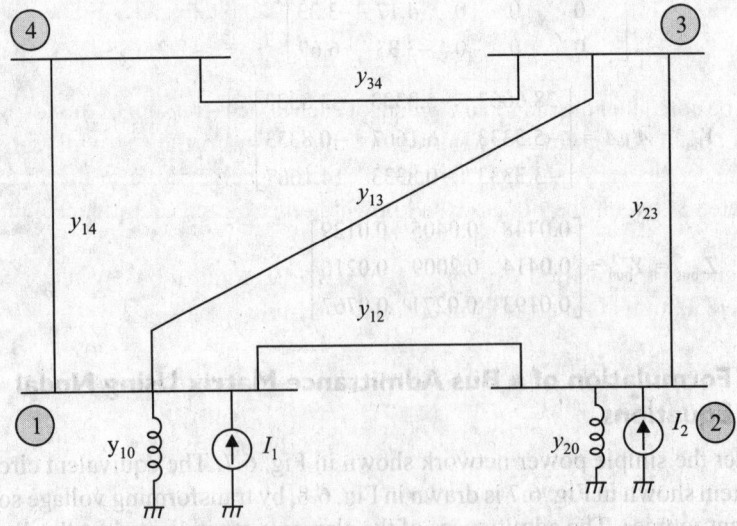

Fig. 6.8 An equivalent circuit of the system shown in Fig. 6.7

In the matrix form, this set of equations may be written as

$$
\begin{bmatrix} I_1 \\ I_2 \\ 0 \\ 0 \end{bmatrix} =
$$

$$
\begin{bmatrix}
y_{10}+y_{12}+y_{13}+y_{14} & -y_{12} & -y_{13} & -y_{14} \\
-y_{12} & y_{20}+y_{12}+y_{23} & -y_{23} & 0 \\
-y_{13} & -y_{23} & y_{13}+y_{23}+y_{34} & -y_{34} \\
-y_{14} & 0 & -y_{34} & y_{14}+y_{34}
\end{bmatrix}
\begin{bmatrix} V_1 \\ V_2 \\ V_3 \\ V_4 \end{bmatrix}
$$

$$
= \begin{bmatrix}
Y_{11} & Y_{12} & Y_{13} & Y_{14} \\
Y_{21} & Y_{22} & Y_{23} & Y_{24} \\
Y_{31} & Y_{32} & Y_{33} & Y_{34} \\
Y_{41} & Y_{42} & Y_{43} & Y_{44}
\end{bmatrix}
\begin{bmatrix} V_1 \\ V_2 \\ V_3 \\ V_4 \end{bmatrix} \tag{6.19}
$$

or $\qquad I_{\text{bus}} = Y_{\text{bus}} V_{\text{bus}}$ $\qquad\qquad\qquad\qquad\qquad$ (6.20)

where

$$Y_{11} = y_{10} + y_{12} + y_{13} + y_{14}$$
$$Y_{22} = y_{20} + y_{12} + y_{23}$$
$$Y_{33} = y_{13} + y_{23} + y_{34}$$
$$Y_{44} = y_{14} + y_{34}$$
$$Y_{12} = Y_{21} = -y_{12}$$
$$Y_{13} = Y_{31} = -y_{13}$$
$$Y_{14} = Y_{41} = -y_{14}$$
$$Y_{23} = Y_{32} = -y_{23}$$
$$Y_{24} = Y_{42} = 0$$
$$Y_{34} = Y_{43} = -y_{34}$$

It may be noted that the diagonal element of each node is the sum of the admittances connected to the node. It is known as the self-admittance or driving-point admittance. The off-diagonal element is equal to the negative of the admittance connected between the nodes. It is known as the mutual admittance or transfer admittance.

Extending the relation expressed by Eq. (6.19) to an *n*-bus system, the node voltage equations in matrix form are

$$
\begin{bmatrix} I_1 \\ I_2 \\ \vdots \\ I_i \\ \vdots \\ I_n \end{bmatrix} =
\begin{bmatrix}
Y_{11} & Y_{12} & \cdots & Y_{1i} & \cdots & Y_{1n} \\
Y_{21} & Y_{22} & \cdots & Y_{2i} & \cdots & Y_{2n} \\
\vdots & \vdots & & \vdots & & \vdots \\
Y_{i1} & Y_{i2} & \cdots & Y_{ii} & \cdots & Y_{in} \\
\vdots & \vdots & & \vdots & & \vdots \\
Y_{n1} & Y_{n2} & \cdots & Y_{ni} & \cdots & Y_{nn}
\end{bmatrix}
\begin{bmatrix} V_1 \\ V_2 \\ \vdots \\ V_i \\ \vdots \\ V_n \end{bmatrix} \tag{6.21}
$$

Equation (6.21) is also written mathematically as

$$I_i = \sum_{j=1}^{j=n} Y_{ij}V_j \qquad \text{for } i = 1, 2, 3, \ldots, n \qquad (6.22)$$

Equation (6.21) is the network performance equation and is the same as $I_{bus} = Y_{bus}V_{bus}$ as given in Eq. (6.4). The diagonal term Y_{ii} and off-diagonal term Y_{ij} of the Y_{bus} are given by

$$Y_{ii} = \sum_{j=0}^{n} y_{ij} \qquad j \neq i \qquad (6.23)$$

$$Y_{ij} = Y_{ji} = -y_{ij} \qquad (6.24)$$

Y_{bus} can also be obtained by the singular transformation of the primitive admittance matrix y and the result will be the same.

Building the Y_{bus} Matrix

Since power system networks are composed of linear bilateral components, Y_{bus} is a square symmetric matrix. If the elements connected to a bus are not mutually coupled to other elements then the formulation of a Y_{bus} matrix of a network is a two-step procedure as follows.

Step 1   To obtain the diagonal term Y_{ii}, add the admittances of all the branches (including those connected between the node and ground) to ith node.

Step 2   To obtain the off-diagonal term Y_{ij}, connected between node i and node j, take the negative of the sum of all the primitive admittances of the branches connected between nodes i and j. Since the matrix is symmetric, $Y_{ij} = Y_{ji}$.

Example 6.2   For the system network shown in Fig. 6.7, draw its oriented connected graph and compute Y_{bus}. Assume zero mutual coupling. The primitive admittances of the elements are same as shown in the diagram.

Solution   The oriented connected graph of the power system network graph is shown in Fig. 6.9.

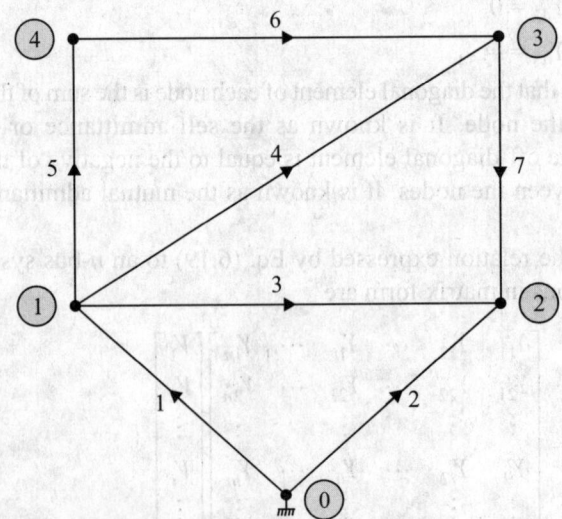

Fig. 6.9   Oriented connected graph of the network shown in Fig. 6.7

For the assumed orientation in the connected graph, the bus incidence matrix A is of order 7×4 and is given by

$$A = \begin{bmatrix} -1 & 0 & 0 & 0 \\ 0 & -1 & 0 & 0 \\ 1 & -1 & 0 & 0 \\ 1 & 0 & -1 & 0 \\ 1 & 0 & 0 & -1 \\ 0 & 0 & -1 & 1 \\ 0 & -1 & 1 & 0 \end{bmatrix}$$

$$A^t = \begin{bmatrix} -1 & 0 & 1 & 1 & 1 & 0 & 0 \\ 0 & -1 & -1 & 0 & 0 & 0 & -1 \\ 0 & 0 & 0 & -1 & 0 & -1 & 1 \\ 0 & 0 & 0 & 0 & -1 & 1 & 0 \end{bmatrix}$$

and the primitive element matrix is

$$y = \begin{bmatrix} y_{10} & 0 & 0 & 0 & 0 & 0 & 0 \\ 0 & y_{20} & 0 & 0 & 0 & 0 & 0 \\ 0 & 0 & y_{12} & 0 & 0 & 0 & 0 \\ 0 & 0 & 0 & y_{13} & 0 & 0 & 0 \\ 0 & 0 & 0 & 0 & y_{14} & 0 & 0 \\ 0 & 0 & 0 & 0 & 0 & y_{34} & 0 \\ 0 & 0 & 0 & 0 & 0 & 0 & y_{23} \end{bmatrix}$$

Using Eq. (6.14),

$$Y_{bus} = A^t y A$$

$$= \begin{bmatrix} y_{10} + y_{12} + y_{13} + y_{14} & -y_{12} & -y_{13} & -y_{14} \\ -y_{12} & y_{20} + y_{12} + y_{23} & -y_{23} & 0 \\ -y_{13} & -y_{23} & y_{13} + y_{23} + y_{34} & -y_{34} \\ -y_{14} & 0 & -y_{34} & y_{14} + y_{34} \end{bmatrix}$$

It may be noted that the Y_{bus} obtained here is the same as that obtained by using nodal equations in Section 6.4.2.

Example 6.3   For the power system shown in Fig. 6.10, build the Y_{bus} matrix. The branch impedances of the lines are as follows:

Line 1-2: $(10 + j40)\ \Omega$; Line 1-4: $(15 + j50)\ \Omega$; Line 2-3: $(5 + j25)\ \Omega$;

Line 2-4: $(15 + j20)\ \Omega$; Line 3-4: $(10 + j30)\ \Omega$

Assume that an impedance of $(20 + j40)$ is connected between node 4 and the ground.

Solution $y_{12} = 1/(10 + j40) = 0.0059 - j0.0235$ S

$y_{14} = 1/(15 + j50) = 0.0055 - j0.0183$ S

$y_{23} = 1/(5 + j25) = 0.0077 - j0.0385$ S

$y_{24} = 1/(15 + j20) = 0.0240 - j0.0320$ S

$y_{34} = 1/(10 + j30) = 0.0100 - j0.0300$ S

$y_{40} = 1/(20 + j40) = 0.0100 - j0.0200$ S

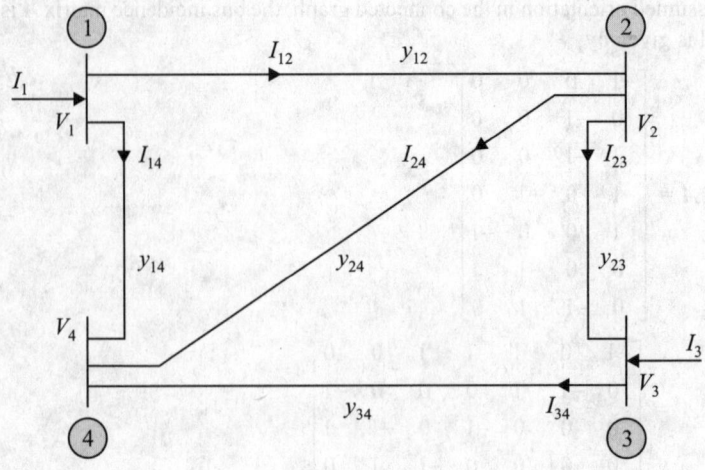

Fig. 6.10 The power system network for Example 6.3

$Y_{11} = y_{12} + y_{14} = 0.0059 - j0.0235 + 0.0055 - j0.0183$
$\quad\quad = 0.104 - j0.0418$ S

$Y_{22} = y_{12} + y_{23} + y_{24} = 0.0059 - j0.0235 + 0.0077 - j0.0385$
$\quad\quad + 0.0240 - j0.0320 = 0.0376 - j0.094$ S

$Y_{33} = y_{23} + y_{34} = 0.0077 - j0.0385 + 0.0100 - j0.0300$
$\quad\quad = 0.0177 - j0.0685$ S

$Y_{44} = y_{14} + y_{24} + y_{34} + y_{40} = 0.0055 - j0.0183 + 0.0240 - j0.0320$
$\quad\quad + 0.0100 - j0.0300 + 0.0100 - j0.0200$
$\quad\quad = 0.0495 - j0.1003$ S

$Y_{12} = Y_{21} = -(0.0059 - j0.0235)$ S; $\quad\quad\quad\quad Y_{13} = Y_{31} = 0;$

$Y_{14} = Y_{41} = -(0.0055 - j0.0183)$ S;

$Y_{23} = Y_{32} = -(0.0077 - j0.0385)$ S;

$Y_{24} = Y_{42} = -(0.0240 - j0.0320)$ S;

$Y_{34} = Y_{43} = -(0.0100 - j0.0300)$ S

Having computed the elements of Y_{bus}, the admittance matrix can be written as

$$\begin{bmatrix} 0.104 - j0.0418 & -0.0059 + j0.0235 & 0 & -0.0055 + j0.0183 \\ -0.0059 + j0.0235 & 0.0376 - j0.094 & -0.0077 + j0.0385 & -0.0240 + j0.0320 \\ 0 & -0.0077 + j0.0385 & 0.0177 - j0.0685 & -0.0100 + j0.0300 \\ -0.0055 + j0.0183 & -0.0240 + j0.0320 & -0.0100 + j0.0300 & 0.0495 + j0.1003 \end{bmatrix}$$

6.4.3 MATLAB Program for Building Y_{bus} Matrix

From the foregoing, it is evident that, though feasible, building a Y_{bus} matrix by inspection is very time-consuming. Hence, using computers for building Y_{bus} matrices is essential. The following is an executable MATLAB program for building and modifying the Y_{bus} matrix from the line impedance data.

```
function[Ybus]=admitmat1(loc,elemz,col,nbus,locs);
% Program for building Ybus
```

```
% Developed by the authors
Ybus = zeros (nbus);
locs = locs - 1;
nbus = nbus - 1;
I = 1;
k = 1;
for II = 1:locs;
   add = loc (II+1) - loc (II);
   for kk = 1:add;
      J = col (k);
   Ybus(I,I) = Ybus(I,I) + 1/elemz(k);
   if J == 0;
      disp ('branch between node and ground')
      k=k+1;
      else
      Ybus(J,J) = Ybus(J,J) + 1/elemz(k);
      Ybus(I,J) = Ybus (I,J)-1/elemz(k);
      Ybus(J,I) = Ybus (I,J);
      k = k + 1;
      end
   end
   I = I + 1;
end
Ybus
% Program for modifying Ybus
disp('Do you wish to modify Ybus?')
ans = input('enter 1 if YES, 0 if NO');
if ans == 0
   return
else
   keyboard
   for n = 1:noc
      keyboard
   Ybus(I,I) = Ybus(I,I) - 1/elemz;
   if J == 0;
      disp ('branch between node and ground')
      else
      Ybus(J,J) = Ybus(J,J) - 1/elemz;
      Ybus(I,J) = Ybus (I,J)+1/elemz;
      Ybus(J,I) = Ybus (I,J);
      end
   end
   Ybus
   end
```

The program is executed through a function file. The program is saved in a file named admitnat1.m and the output gives the elements of Y_{bus}. The various input parameters are as follows:

loc   a row vector which stores the start of each row of Y_{bus}
elemz   a row vector which stores the primitive impedances of the lines
col   a row vector which provides the column number of Y_{bus}
nbus   number of buses in the system
locs   the size of the loc vector

For the network shown in Fig. 6.10, the input vectors would look as shown here.

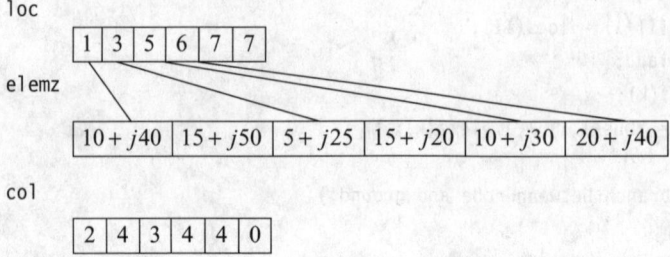

It can be observed that the first element is connected between node 1 and 2 and its primitive impedance is entered at elemz (1). There are two elements [loc (2) − loc (1) = 3 − 1 = 2] connected to node 1 and the nodes to which these elements are connected are 2 and 4. Similarly, data for node two begins from the third element in elemz [loc (2) = 3], and here also two elements [loc (3) − loc (2) = 5 − 3 = 2] are connected to the node. The two elements are connected between nodes 2 and 3 [col (3)] and 2 and 4 [col (4)]. The entry col(6) = 0 shows that the element is connected between the node and the reference ground node.

After assembling Y_{bus}, the program questions whether any modifications are required to Y_{bus}. If the answer is yes, the control is passed to the keyboard and data is input with regard to the number of changes, primitive impedance of the element, and the nodes between which it has to be connected. Thus, depending on whether Y_{bus} is to be modified or not, the program returns one or two Y_{bus} matrices.

The Y_{bus} matrices obtained with the data from the network shown in Fig. 6.10 are given below. The second modified Y_{bus} is when line 2-4 is removed.

Y_{bus} =

```
0.0114 − j0.0419     −0.0059 + j0.0235           0            −0.0055 + j0.0183
−0.0059 + j0.0235     0.0376 − j0.0940    −0.0077 + j0.0385    −0.0240 + j0.0320
        0            −0.0077 + j0.0385     0.0177 − j0.0685    −0.0100 + j0.0300
−0.0055 + j0.0183    −0.0240 + j0.0320    −0.0100 + j0.0300     0.0419 − j0.0803
```

```
Do you wish to modify Ybus?
enter 1 if YES, 0 if N01
K» noc=1;
K» return
K» I=2;J=4;elemz=15+20i;
K» return
```

Y_{bus} =

```
 0.0114 − j0.0419     −0.0059 + j0.0235           0            −0.0055 + j0.0183
−0.0059 + j0.0235      0.0136 − j0.0620    −0.0077 + j0.0385           0
        0            −0.0077 + j0.0385     0.0177 − j0.0685    −0.0100 + j0.0300
−0.0055 + j0.0183            0             −0.0100 + j0.0300     0.0179 − j0.048
```

6.4.4 Sparsity in Y_{bus} Matrix

An element in Y_{bus} has a non-zero value, if the two buses are directly connected through a transmission line or a transformer. The non-existence of a direct connection between two buses means that the off-diagonal transfer admittances of the Y_{bus} matrix will be zero. Normally in a large interconnected power system, each bus may be connected with a few neighbouring buses. The non-existence of a direct connection between many buses means that the vast majority of a real world Y_{bus} matrix will be zero. Thus, for a large power system, Y_{bus} matrix is highly sparse. In large systems sparsity may be as high as 97%. It may, however, be noted that whereas Y_{bus} is a highly sparse matrix, its inverse Z_{bus} is always full.

Further, the Y_{bus} matrix of a power system is symmetric; therefore, it suffices if only the diagonal self-admittances and the elements of upper or lower triangular half of the Y_{bus} matrix are stored. Thus, in a Y_{bus} matrix of order n, the number of elements required to be stored is $n(n + 1)/2$.

The storage requirement of a Y_{bus} matrix, for a large system having buses of the order of a few thousand, increases tremendously when it is stored in the conventional two-dimensional arrays. Further, the computational time also increases since several arithmetic operations are required to be performed on zero elements. In order to save both the storage requirement and computation time, it is essential to take advantage of the sparsely populated Y_{bus} matrix.

The non-zero elements of a sparse matrix are stored in a single dimensional array with two tables; one of which gives the column number and the other gives the starting position of the next row. The storage of only non-zero elements can be best explained by an example. Figure 6.11 shows a sparse matrix of a seven-bus hypothetical network. Only the non-zero elements are indicated in the matrix.

y_{11}		y_{13}				
	y_{22}	y_{23}			y_{26}	
y_{31}	y_{32}	y_{33}	y_{34}			
		y_{43}	y_{44}			
				y_{55}		y_{57}
	y_{62}				y_{66}	
				y_{75}		y_{77}

Fig. 6.11 Y_{bus} matrix of a seven-bus power system

The storage of the Y_{bus} matrix, along with the two single dimension tables, is shown in Fig. 6.12.

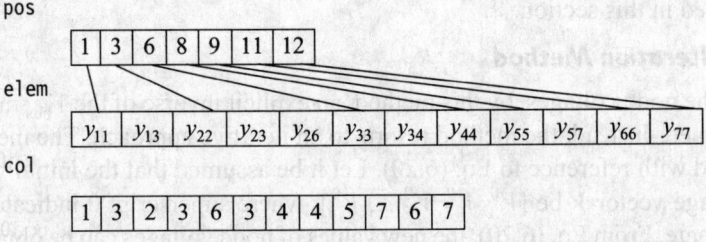

Fig. 6.12   Storage of the non-zero elements of Y_{bus} and position and column tables

The non-zero elements of the Y_{bus} matrix are stored in a single dimensional table elem and the position of the diagonal element of each row is shown in the pos table. Thus, the diagonal element of row two is in the third position, while the third row starts from the sixth position. It is also to be noted that the difference between any two values in the pos table gives the number of elements in a row. For example, the difference between the third and fourth positions in the pos table [pos (4) – pos (3) = 8 – 6 = 2] shows that the number of elements in the third row is two. It is also to be noted that the length of the elem and col tables is equal to the number of non-zero elements of Y_{bus}, when only the upper triangle along with the diagonal elements are stored.

6.5 SOLUTION OF NODAL EQUATIONS

The Y_{bus} of an electrical power system remains unchanged when the configuration and the system parameters remain the same, while the operating conditions of the system may undergo a change. A change in the operating conditions is simulated by a change in the external voltage or current sources connected to the nodes. Thus, if node voltages with reference to the ground are specified, the node currents can be obtained by multiplying Y_{bus} by the node voltage column vector V_{bus} as per Eq. (6.20). On the other hand, if the node currents are known, the node voltages can be computed by the product of Y_{bus}^{-1} and the node current column vector I_{bus}.

For the users of MATLAB programming, the two computations can be conveniently performed by the inbuilt functions. Thus, $I_{bus} = Y_{bus}V_{bus}$ and $V_{bus} = inv[Y_{bus}]I_{bus}$. The techniques for solving the set of node voltage equations, given by Eq. 6.20, may be categorized as indirect methods and direct methods.

6.5.1 Indirect Methods

The indirect methods, also known as *iterative* methods, assume a set of initial values for the variables whose solution is to be determined. The initial values of the variables are used to start the process of computation. The new values of the variables obtained are then used to compute more accurate values. Thus, through the process of successive approximations, the unknown variables are computed. The iterative methods provide a systematic approach to finding approximate solutions with reasonable accuracy. The number of iterations required or the rate of convergence depends on the characteristics of the Y_{bus} matrix, that is, the actual power system network. Thus, depending on the characteristics of the system admittance matrix, approximate successive values may converge quickly or slowly or may not converge at all. A judicious choice of the initial values also affects the rate of convergence. The two most commonly used iterative methods are described in this section.

Gaussian Iteration Method

To obtain the node voltages, by this method, an explicit inverse of the Y_{bus} matrix is not obtained. Instead, the method resorts to an iterative approach. The method is explained with reference to Eq. (6.20). Let it be assumed that the initial value of the voltage vector V be $[V_1^0 \ V_2^0 \ V_3^0 \dots V_n^0]$, where superscript 0 indicates an initial estimate. From Eq. (6.20), the new values of node voltages can be obtained by operating on each row individually, that is,

$$V_1^1 = \frac{I_1}{Y_{11}} - \left(\frac{Y_{12}}{Y_{11}} V_2^0 + \frac{Y_{13}}{Y_{11}} V_3^0 + \ldots + \frac{Y_{1n}}{Y_{11}} V_n^0 \right) \tag{6.25a}$$

$$V_2^1 = \frac{I_2}{Y_{22}} - \left(\frac{Y_{21}}{Y_{22}} V_1^0 + \frac{Y_{23}}{Y_{22}} V_3^0 + \ldots + \frac{Y_{2n}}{Y_{22}} V_n^0 \right) \tag{6.25b}$$

$$\vdots$$

$$V_n^1 = \frac{I_n}{Y_{nn}} - \left(\frac{Y_{n1}}{Y_{nn}} V_1^0 + \frac{Y_{n2}}{Y_{nn}} V_2^0 + \ldots + \frac{Y_{nm-1}}{Y_{nn}} V_{nm-1}^0 \right) \tag{6.25n}$$

Equations (6.25) may be written in a general mathematical form as

$$V_i^{k+1} = \frac{1}{Y_{ii}} \left\{ I_i - \sum_{j=1,\, j\neq i}^{n} Y_{ij} V_j^k \right\} \qquad i = 1, 2, 3, \ldots, n \tag{6.26}$$

where k is the iteration count. At the end of each iteration, the node voltages are checked whether the tolerance condition is met, that is,

$$\left| V_i^{k+1} - V_i^k \right| \leq \text{tolerance limit} \qquad i = 1, 2, 3, \ldots, n \tag{6.27}$$

Generally, the method operates as follows.

Step 1   An initial estimate of all the node voltages $[V_1^0 \ V_2^0 \ V_3^0 \ldots V_n^0]$ is made.

Step 2   One line at a time of Eq. (6.26) is operated upon to obtain the new value of the node voltage.

Step 3   Step 2 is repeated for all the nodes.

Step 4   The new computed values of the node voltages are compared with the initial estimated values assumed in step 1. If the new values of the node voltages are within a certain acceptable value (tolerance limit), the solution of Eq. (6.26) has been obtained.

Step 5   If the new values of the node voltages are not within the specified tolerance limit (normally taken as 0.0001), the new values of node voltages are substituted for the estimated (or previous) values of node voltages.

Step 6   Steps 2 to 5 are repeated till two consecutive values of the node voltages fall within the specified tolerance limit.

Gauss–Seidel Iteration Method

This method is similar to the Gauss iteration method, since here too each linear algebraic equation is processed row wise. The method varies from the previous method in the sense that the latest value of the node voltage is immediately substituted in the processing of subsequent rows. For example, as soon as the voltage of node 1 in the first iteration is obtained by processing Eq. (6.26) for $i = 1$, the new value of the voltage of node 1 is substituted in Eqs (6.25b) to (6.25n). Similarly, when voltage of node 2 is computed, the updated voltage values of nodes 1 and 2 are substituted for calculating the voltages of node 3, and so on till computation of all the nodes is complete. Mathematically, this substitution can be expressed as follows:

$$V_i^{k+1} = \frac{1}{Y_{ii}} \left\{ I_i - \sum_{j=1}^{i-1} Y_{ij} V_j^{k+1} - \sum_{j=i+1}^{n} Y_{ij} V_j^k \right\} \qquad i = 1, 2, 3, \ldots, n \tag{6.28}$$

Acceleration Factor

Often the convergence of two consecutive values within the specified tolerance limit can be achieved with the use of an acceleration factor at the end of each iteration. If α is the acceleration factor, the accelerated voltage is obtained by

$$V_i^{k+1} = V_i^k + \alpha\left(V_i^{k+1} - V_i^k\right) \tag{6.29}$$

The accelerated values then replace the calculated values of the voltages. The selection of both the starting voltage values and the acceleration factor is based on conjecture and is often governed by experience.

6.5.2 Direct Methods

Direct methods require the determination of the inverse of the Y_{bus} matrix of order n. The solutions by direct methods are obtained in definite arithmetic operations and are accurate, except for the round-off errors. The number of arithmetic operations is determined by the size of the system, which governs the number of linear algebraic equations and methods employed. However, in power systems, since Y_{bus} is symmetric, the number of arithmetic operations gets reduced.

An inverse of Y_{bus} converts a sparse matrix into a full matrix. Therefore, from the perspective of the direct solution of the node equations, matrix inversion is very inefficient. The method of factorization of the Y_{bus} matrix for computing direct solutions of the node equations is more efficient and is described in the next section.

Triangular Factorization of a Matrix

The triangulation of matrices or the triangular decomposition of matrices to solve simultaneous linear equations is the most widely used method for manipulating coefficient matrices. These methods factorize the coefficient matrix into their triangular form. The LU method of factorization is discussed here.

The LU method of factorization consists of expressing the coefficient matrix A as the product of two factor matrices, such that

$$A = LU \tag{6.30}$$

where L is the lower triangular matrix and U is the upper triangular matrix, which has unity elements on its diagonal. For a third-order matrix A, these factor matrices are

$$A = \begin{bmatrix} a_{11} & a_{12} & a_{13} \\ a_{21} & a_{22} & a_{23} \\ a_{31} & a_{32} & a_{33} \end{bmatrix} = LU = \begin{bmatrix} l_{11} & & \\ l_{21} & l_{22} & \\ l_{31} & l_{32} & l_{33} \end{bmatrix} \begin{bmatrix} 1 & u_{12} & u_{13} \\ & 1 & u_{23} \\ & & 1 \end{bmatrix} \tag{6.31}$$

Multiplying L and U and equating the elements of this matrix with the corresponding elements of A gives:

$$
\begin{array}{lll}
a_{11} = l_{11} & a_{12} = l_{11}\,u_{12} & a_{13} = l_{11}\,u_{13} \\
a_{21} = l_{21} & a_{22} = l_{21}\,u_{12} + l_{22} & a_{23} = l_{21}\,u_{13} + l_{22}\,u_{23} \\
a_{31} = l_{31} & a_{32} = l_{31}\,u_{12} + l_{32} & a_{33} = l_{31}\,u_{13} + l_{32}\,u_{23} + l_{33}
\end{array}
\tag{6.32}
$$

Rearranging the terms in Eq. (6.32) yields

$$l_{11} = a_{11} \qquad u_{12} = a_{12}/l_{11} \qquad u_{13} = a_{13}/l_{11}$$

$$l_{21} = a_{21} \qquad l_{22} = a_{22} - l_{21}u_{12} \qquad u_{23} = \frac{1}{l_{22}}(a_{23} - l_{21}u_{13}) \qquad (6.33)$$

$$l_{31} = a_{31} \qquad l_{32} = a_{32} - l_{31}u_{12} \qquad l_{33} = a_{33} - l_{31}u_{13} - l_{32}u_{23}$$

From these equations, it is evident that there are unique values of l and u which give the original elements of A.

Algorithm for Factorization of Y_{bus}

The procedure for calculating the table of factors is explained for a four-bus system whose nodal equations are given by

$$
\begin{bmatrix}
y_{11} & y_{12} & y_{13} & y_{14} \\
y_{22} & y_{22} & y_{23} & y_{24} \\
y_{31} & y_{32} & y_{33} & y_{34} \\
y_{41} & y_{42} & y_{43} & y_{44}
\end{bmatrix}
\begin{bmatrix}
V_1 \\
V_2 \\
V_3 \\
V_4
\end{bmatrix}
=
\begin{bmatrix}
I_1 \\
I_2 \\
I_3 \\
I_4
\end{bmatrix}
\qquad (6.34)
$$

The algorithm performs Gaussian elimination by rows since it is more convenient to work with rows for developing a computer program. The elements of L and U are stored in the same space as Y_{bus} since it can be assumed that the diagonal element of U is unity.

$$\begin{bmatrix} l_{11} & u_{12} & u_{13} & u_{14} \end{bmatrix} \qquad l_{11} = y_{11}, \ u_{1j} = \frac{1}{l_{11}}(y_{1j}) \ j = 2, 3, 4 \qquad (6.35)$$

$$
\begin{bmatrix}
l_{11} & u_{12} & u_{13} & u_{14} \\
l_{21} & u_{22} & u_{23} & u_{24}
\end{bmatrix}
\qquad
\begin{aligned}
& l_{21} = y_{21}, \ l_{22} = y_{22} - l_{21}u_{12} \qquad (6.36) \\
& u_{2j} = \frac{1}{l_{22}}(y_{2j} - l_{21}u_{1j}) \ j = 3, 4
\end{aligned}
$$

$$
\begin{bmatrix}
l_{11} & u_{12} & u_{13} & u_{14} \\
l_{21} & u_{22} & u_{23} & u_{24} \\
l_{31} & l_{32} & l_{33} & u_{34}
\end{bmatrix}
\qquad
\begin{aligned}
& l_{31} = y_{31} \\
& l_{32} = y_{32} - l_{31}u_{12} \\
& l_{33} = y_{33} - l_{13}u_{31} - l_{32}u_{23} \qquad (6.37) \\
& u_{34} = \frac{1}{l_{33}}(y_{34} - l_{31}u_{14} - l_{32}u_{24})
\end{aligned}
$$

$$
\begin{bmatrix}
l_{11} & u_{12} & u_{13} & u_{14} \\
l_{21} & u_{22} & u_{23} & u_{24} \\
l_{31} & l_{32} & l_{33} & u_{34} \\
l_{41} & l_{42} & l_{43} & l_{44}
\end{bmatrix}
\qquad
\begin{aligned}
& l_{41} = y_{41} \\
& l_{42} = y_{42} - l_{41}u_{12} \qquad (6.38) \\
& l_{43} = y_{43} - l_{41}u_{13} - l_{42}u_{23} \\
& l_{44} = y_{44} - l_{41}u_{14} - l_{42}u_{24} - l_{43}u_{34}
\end{aligned}
$$

The final array, usually referred to as the table of factors, is shown in Eq. (6.39)

$$
\begin{bmatrix}
l_{11} & u_{12} & u_{13} & u_{14} \\
l_{22} & u_{22} & u_{23} & u_{24} \\
l_{31} & l_{32} & l_{33} & u_{34} \\
l_{41} & l_{42} & l_{43} & l_{44}
\end{bmatrix}
\qquad (6.39)
$$

It may be noted that the diagonal elements of the upper triangular matrix, $u_{ii} = 1$, for $i = 1, 2, \ldots, n$, is not shown in Eq. (6.39).

MATLAB Program for Computing Table of Factors for Y_{bus}

The MATLAB program named tof computes the table of factors for the Y_{bus} matrix of order n. The input is the Y_{bus} matrix and the number of buses nbus.

```
function [Ybus]=tof(Ybus,nbus);
%Program for computing table of factors
%Developed by the authors
for k = 1:nbus;
   if k ==1;
      for j = 2:nbus;
         Ybus(k,j)=Ybus(k,j)/Ybus(k,k);
      end
   else
      for j = 2:nbus;
         if j <=k;
            for m = 1:j-1;
            Ybus(k,j)=Ybus(k,j)-Ybus(k,m)*Ybus(m,j);
         end
      else
         for m = 1:k-1;
            Ybus(k,j)=Ybus(k,j)-Ybus(k,m)*Ybus(m,j);
         end
         Ybus(k,j)=Ybus(k,j)/Ybus(k,k);
      end
   end
 end
end
```

Example 6.4 For the matrix given below, compute the table of factors. Verify the table of factors using the procedure employed in Section 6.5.2 (triangular factorization).

$$\begin{bmatrix} 4 & 3 & 6 \\ 2 & 8 & 5 \\ 1 & 5 & 9 \end{bmatrix}$$

Solution The table of factors obtained by using MATLAB program tof are as follows:

```
≫ Ybus=[4 3 6;2 8 5;1 5 9];nbus=3;
≫ [Ybus]=tof(Ybus,nbus)
Ybus =
 4.0000 0.7500 1.5000
 2.0000 6.5000 0.3077
 1.0000 5.2500 6.1923
```

Following the procedure, outlined in Section 6.5.2 (triangular factorization), the elements of the table of factors are computed as follows:

$$l_{11} = 4$$

$$u_{12} = \frac{1}{4}(3) = 0.75$$

$$u_{13} = \frac{1}{4}(6) = 1.5$$

$$l_{21} = 2$$

$$l_{22} = 8 - 2 \times (0.75) = 6.5$$

$$u_{23} = \frac{1}{6.5}(5 - 2 \times 1.5) = 0.31$$

$$l_{31} = 1$$

$$l_{32} = 5 - 1 \times 0.75 = 5.25$$

$$l_{21} = 2$$

$$l_{33} = 9 - 1 \times 1.5 - 5.25 \times 0.31 = 6.18$$

It is observed that the output from the MATLAB program tof is verified.

Solving Nodal Equations by Forward and Backward Substitution

If a set of simultaneous linear equations are written in matrix form as:

$$AX = b \tag{6.40}$$

where A is a coefficient matrix of order n, X is a column vector of n variables whose values are to be determined, and b is a column vector of n specified values. Then substituting Eq. (6.30) into Eq. (6.40) gives

$$LUX = b \tag{6.41}$$

Let $\qquad UX = Y \tag{6.42}$

then from Eqs (6.41) and (6.42),

$$LY = b \tag{6.43}$$

Writing explicitly Eqs (6.42) and (6.43) for a third-order problem gives

$$x_1 + u_{12} x_2 + u_{13} x_3 = y_1$$

$$x_2 + u_{23} x_3 = y_2 \tag{6.44}$$

$$x_3 = y_3$$

$$l_{11} y_1 = b_1$$

$$l_{21} y_1 + l_{22} y_2 = b_2 \tag{6.45}$$

$$l_{31} y_1 + l_{32} y_2 + l_{33} y_3 = b_3$$

Since L is a lower triangular matrix, Y can be found from L and b by forward substitution, and since U is an upper triangular matrix, the unknown vector X can be found from U and Y by backward substitution. For a third-order problem, this technique gives the values of Y and X as

$$y_1 = \frac{b_1}{l_{11}}$$

$$y_2 = \frac{l}{l_{22}}(b_2 - l_{21}y_1) \tag{6.46}$$

$$y_3 = \frac{1}{l_{33}}(b_3 - l_{31}y_1 - l_{32}y_2)$$

$$x_3 = y_3$$

$$x_2 = y_2 - u_{23}x_3 \tag{6.47}$$

$$x_1 = y_1 - u_{12}x_2 - u_{13}x_3$$

The nodal equations for a power system may be written in the matrix form as

$$I_{bus} = Y_{bus}V_{bus} \tag{6.48}$$

The bus admittance matrix Y_{bus} may be written as the product of the lower triangular matrix L and the upper triangular matrix U. Thus Eq. (6.48) becomes

$$I_{bus} = LUV_{bus} \tag{6.49}$$

The unknown node voltages vector V_{bus} in Eq. (6.49) is obtained in two steps, that is, by first solving for the vector V' such that

$$LV' = I \tag{6.50}$$

And then solving

$$UV_{bus} = V' \tag{6.51}$$

Equation (6.50) can be solved by forward substitution as follows:

$$V'_1 = \frac{I_1}{l_{11}} \tag{6.52a}$$

$$V'_k = \frac{1}{l_{kk}}\left[I_k - \sum_{j=1}^{k-1}l_{kj}V'_j\right] \qquad k = 2, 3, 4, ..., n \tag{6.52b}$$

while Eq. (6.51) can then be solved by backward substitution as follows:

$$V_n = \frac{V'_n}{u_{nn}} = V'_n \qquad \text{as } u_{nn} = 1 \tag{6.53a}$$

$$V_i = \frac{1}{u_{ii}}\left[V'_i - \sum_{j=i+1}^{n}u_{ij}V_j\right] \qquad i = n-1, n-2, ..., 1 \tag{6.53b}$$

where n is the order of the Y_{bus} matrix.

Using Eqs (6.52) and (6.53) together makes it possible to obtain the node voltages without computing the inverse of Y_{bus} matrix.

The use of the table of factors for the computation of the inverse of Y_{bus} matrix will now be illustrated for a four-bus power system.

Forward substitution

$$\begin{bmatrix} \frac{1}{l_{11}} & & & \\ & 1 & & \\ & & 1 & \\ & & & 1 \end{bmatrix}\begin{bmatrix} I_1 \\ I_2 \\ I_3 \\ I_4 \end{bmatrix} = \begin{bmatrix} V'_1 \\ I_2 \\ I_3 \\ I_4 \end{bmatrix}$$

which gives $V'_1 = \left(\dfrac{1}{l_{11}}\right)I_1$

$$\begin{bmatrix} 1 & & & \\ & \frac{1}{l_{22}} & & \\ & & 1 & \\ & & & 1 \end{bmatrix}\begin{bmatrix} 1 & & & \\ -l_{21} & 1 & & \\ & & 1 & \\ & & & 1 \end{bmatrix}\begin{bmatrix} V'_1 \\ I_2 \\ I_3 \\ I_4 \end{bmatrix} = \begin{bmatrix} V'_1 \\ V'_2 \\ I_3 \\ I_4 \end{bmatrix}$$

Then, $V'_2 = \left(\dfrac{1}{l_{22}}\right)(I_2 - l_{21}V'_1)$

$$
\begin{bmatrix} 1 & & & \\ & 1 & & \\ & & \dfrac{1}{l_{33}} & \\ & & & 1 \end{bmatrix}
\begin{bmatrix} 1 & & & \\ & 1 & & \\ -l_{31} & -l_{32} & 1 & \\ & & & 1 \end{bmatrix}
\begin{bmatrix} V'_1 \\ V'_2 \\ I_3 \\ I_4 \end{bmatrix}
=
\begin{bmatrix} V'_1 \\ V'_2 \\ V'_3 \\ I_4 \end{bmatrix}
$$

Hence, $V'_3 = \left(\dfrac{1}{l_{33}}\right)(I_3 - l_{31}V'_1 - l_{32}V'_2)$

$$
\begin{bmatrix} 1 & & & \\ & 1 & & \\ & & 1 & \\ & & & \dfrac{1}{l_{44}} \end{bmatrix}
\begin{bmatrix} 1 & & & \\ & 1 & & \\ & & 1 & \\ -l_{41} & -l_{42} & -l_{43} & 1 \end{bmatrix}
\begin{bmatrix} V'_1 \\ V'_2 \\ V'_3 \\ I_4 \end{bmatrix}
=
\begin{bmatrix} V'_1 \\ V'_2 \\ V'_3 \\ V'_4 \end{bmatrix}
$$

and $\qquad V'_4 = \left(\dfrac{1}{l_{44}}\right)(I_4 - l_{41}V'_1 - l_{42}V'_2 - l_{43}V'_3)$

It is to be noted that $V_4 = V'_4$. Starting with V_4, backward substitution is performed to obtain the node voltages.

The preceding pictorial description of the forward substitution method helps in writing general mathematical formulae for computing of the voltage vector V' and the formulae so developed will be the same as that given in Eqs (6.52a) and (6.52b).

Backward substitution

$$
\begin{bmatrix} 1 & & & \\ & 1 & & \\ & & 1 & -u_{34} \\ & & & 1 \end{bmatrix}
\begin{bmatrix} V'_1 \\ V'_2 \\ V'_3 \\ V'_4 \end{bmatrix}
=
\begin{bmatrix} V'_1 \\ V'_2 \\ V'_3 \\ V_4 \end{bmatrix}
$$

which gives $V_3 = V'_3 - u_{34}V_4$

$$
\begin{bmatrix} 1 & & & \\ & 1 & -u_{23} & -u_{24} \\ & & 1 & \\ & & & 1 \end{bmatrix}
\begin{bmatrix} V'_1 \\ V'_2 \\ V_3 \\ V_4 \end{bmatrix}
=
\begin{bmatrix} V'_1 \\ V_2 \\ V_3 \\ V_4 \end{bmatrix}
$$

then $\qquad V_2 = V'_2 - u_{23}V_3 - u_{24}V_4$

$$
\begin{bmatrix} 1 & -u_{12} & -u_{13} & -u_{14} \\ & 1 & & \\ & & 1 & \\ & & & 1 \end{bmatrix}
\begin{bmatrix} V'_1 \\ V_2 \\ V_3 \\ V_4 \end{bmatrix}
=
\begin{bmatrix} V_1 \\ V_2 \\ V_3 \\ V_4 \end{bmatrix}
$$

and $\qquad V_1 = V'_1 - u_{12}V_2 - u_{13}V_3 - u_{14}V_4$

Similar to the forward substitution method, general computer adaptable mathematical formulae can be written using the backward substitution method and they will be same as that given in Eqs (6.53a) and (6.53b).

Effect of Symmetry

The admittance matrix Y_{bus} of an electrical power system network is symmetrical about the major diagonal. Consequently, both the storage requirement and the number of arithmetic operations in computing the table of factors reduce by approximately 50%. The number of operations in solving Eq. (6.50) by forward substitution and Eq. (6.51) by backward substitution, however, is not affected by symmetry.

6.5.3 MATLAB Program for Solving Node Equations

The MATLAB program named fbsub is given here. It performs the forward and backward substitutions to solve Eq. (6.4) to obtain the node voltages from the given node currents. The program is an extension of the MATLAB program tof.

```
function[Ybus,I]=fbsub(Ybus,nbus,I);
%Program for forward and backward substitution
%Developed by the authors
for k = 1:nbus;
if   k ==1;
     for j = 2:nbus;
         Ybus(k,j)=Ybus(k,j)/Ybus(k,k);
     end
   else
     for j = 2:nbus;
        if j <=k;
           for m = 1:j-1;
               Ybus(k,j)=Ybus(k,j)-Ybus(k,m)*Ybus(m,j);
           end
         else
           for m = 1:k-1;
               Ybus(k,j)=Ybus(k,j)-Ybus(k,m)*Ybus(m,j);
           end
           Ybus(k,j)=Ybus(k,j)/Ybus(k,k);
        end
     end
   end
end
for k = 1:nbus;
  if k ==1;
     I(k)=I(k)/Ybus(k,k);

else
for j = 1:k-1;
     I(k)=I(k)-Ybus(k,j)*I(j);
   end
   I(k)=I(k)/Ybus(k,k);
end
end
```

```
for k = nbus:-1:1;
   if k == nbus;
      disp ('node voltages')
      else
      for j = nbus:-1:k+1;
         I(k)=I(k)-Ybus(k,j)*I(j);
      end
   end
end
```

Example 6.5   For the matrix in Example 6.4, assume $I = [1\ 1\ 1]^t$. Solve for V by using the MATLAB program fbsub. Verify the answer by actually performing backward and forward substitutions.

Solution

```
≫ Ybus=[4 3 6;2 8 5;1 5 9];nbus=3;I=[1;1;1];
≫ [Ybus,I]=fbsub(Ybus,nbus,I)
node voltages
Ybus =
   4.0000 0.7500 1.5000
   2.0000 6.5000 0.3077
   1.0000 5.2500 6.1923
V =
   0.1056
   0.0559
   0.0683
```

Use Eq. (6.52) to perform forward substitution,

$$V'_1 = \frac{1}{4} \times 1 = 0.25$$

$$V'_2 = \frac{1}{6.5} \times (1 - 2 \times 0.25) = 0.0769$$

$$V'_3 = \frac{1}{6.1923} \times (1 - 1 \times 0.25 - 4.25 \times 0.0769) = 0.06834$$

Using Eq. (6.53) to perform backward substitution,

$$V_3 = V'_3 = 0.06834$$

$$V_2 = (0.0769 - 0.3077 \times 0.06834) = 0.0559$$

$$V_1 = (0.25 - 0.75 \times 0.0559 - 1.5 \times 0.06834) = 0.1056$$

Again it is observed that the output from fbsub is verified.

6.6 NETWORK REDUCTION

At times, computation of voltages at all nodes (or buses) is of no immediate interest. In stability studies of power systems, non-generator buses or buses which are neither to be faulted nor required for metering purposes can be eliminated from the network. In such cases, variables associated with these buses are not required and, therefore, these can be eliminated from the node equations.

Equation (6.54) expresses the node equations for a hypothetical power system consisting of five buses.

$$\begin{bmatrix} Y_{11} & Y_{12} & 0 & Y_{14} & 0 \\ Y_{21} & Y_{22} & Y_{23} & 0 & Y_{25} \\ 0 & Y_{32} & Y_{33} & 0 & Y_{35} \\ \hline Y_{41} & 0 & 0 & Y_{44} & Y_{45} \\ 0 & Y_{52} & Y_{53} & Y_{54} & Y_{55} \end{bmatrix} \begin{bmatrix} V_1 \\ V_2 \\ V_3 \\ V_4 \\ V_5 \end{bmatrix} = \begin{bmatrix} I_1 \\ I_2 \\ I_3 \\ I_4 \\ I_5 \end{bmatrix} \tag{6.54}$$

Assume that buses 4 and 5 are required to be eliminated. The Y_{bus} matrix and the voltage and current vectors in Eq. (6.54) are partitioned such that they are conformable to matrix multiplication.

Let

$$Y'_{11} = \begin{bmatrix} Y_{11} & Y_{12} & 0 \\ Y_{21} & Y_{22} & Y_{23} \\ 0 & Y_{32} & Y_{33} \end{bmatrix} \qquad Y'_{12} = \begin{bmatrix} Y_{14} & 0 \\ 0 & Y_{25} \\ 0 & Y_{35} \end{bmatrix}$$

$$Y'_{21} = \begin{bmatrix} Y_{41} & 0 & 0 \\ 0 & Y_{52} & Y_{53} \end{bmatrix} \qquad Y'_{22} = \begin{bmatrix} Y_{44} & Y_{45} \\ Y_{54} & Y_{55} \end{bmatrix}$$

$$V'_1 = \begin{bmatrix} V_1 \\ V_2 \\ V_3 \end{bmatrix} \qquad V'_2 = \begin{bmatrix} V_4 \\ V_5 \end{bmatrix} \qquad I'_1 = \begin{bmatrix} I_1 \\ I_2 \\ I_3 \end{bmatrix} \qquad I'_2 = \begin{bmatrix} I_4 \\ I_5 \end{bmatrix}$$

After making the above substitutions, Eq. (6.54) takes the following form

$$\begin{bmatrix} Y'_{11} & Y'_{12} \\ Y'_{21} & Y'_{22} \end{bmatrix} \begin{bmatrix} V'_1 \\ V'_2 \end{bmatrix} = \begin{bmatrix} I'_1 \\ I'_2 \end{bmatrix} \tag{6.55}$$

From the compound matrix form of Eq. (6.55), the following equations are obtained

$$Y'_{11}V'_1 + Y'_{12}V'_2 = I'_1 \tag{6.56}$$

$$Y'_{21}V'_1 + Y'_{22}V'_2 = I'_2 \tag{6.57}$$

Equation (6.57) is used to obtain the expression for V'_2 as

$$V'_2 = Y'_{22}{}^{-1}(I'_2 - V'_{21}V'_1) \tag{6.58}$$

and substituting for V'_2 in Eq. (6.56) gives

$$Y'_{11}V'_1 + Y'_{12}\{Y'_{22}{}^{-1}(I'_2 - Y'_{21}V'_1)\} = I'_1$$

Simplification of the above expression results in the following

$$(Y'_{11} - Y'_{12}Y'_{22}{}^{-1}Y'_{21})V'_1 = I'_1 - Y'_{12}Y'_{22}{}^{-1}I'_2 \tag{6.59}$$

Substituting

$$Y''_{11} = (Y'_{11} - Y'_{12}Y'_{22}{}^{-1}Y'_{21}) \tag{6.60}$$

and $\qquad I''_1 = I'_1 - Y'_{12}Y'_{22}{}^{-1}I'_2 \tag{6.61}$

in Eq. (6.59) gives

$$Y''_{11}V'_1 = I''_1$$

or $\qquad V'_1 = Y''_{11}{}^{-1}I''_1 \tag{6.62}$

It may be observed from Eq. (6.60) that Y''_{11} represents the reduced admittance matrix (also called Kron reduction). Buses 4 and 5 have also been eliminated and the order of the equation is equal to the number of buses retained in the system. Further, the second term in Eq. (6.61) gives the distributed currents from the eliminated buses into the retained buses. If, however, the loads at the eliminated buses have been represented as static loads, then $I'_2 = 0$. Hence, Eq. (6.62) reduces to

$$V'_1 = Y''^{-1}_{11} I'_1$$

The method described here is similar to Gaussian elimination. An alternate approach is to perform elimination of buses one row and column at a time. The algorithm for the elimination of buses, for a power system having n buses, is mathematically written as

$$Y^{new}_{ij} = Y_{ij} - \frac{Y_{ik} Y_{kj}}{Y_{kk}} \qquad i, j = 1, 2, 3, \ldots, n \quad j \neq k \qquad (6.63)$$

In Eq. (6.63), bus k is being eliminated and Y^{new}_{ij} represents the modified elements of the reduced Y_{bus} matrix of order $(n-1) \times (n-1)$.

It may be observed that when a bus k is eliminated, all branches connected to the bus are redistributed. The distribution is such that when a bus k is eliminated, for each pair of buses (i, j) which are connected through the bus k, one of the following occurs:

(i)   If there is a direct connection available between i and j, it gets modified.
(ii) If there is no direct connection available between i and j, a new branch linking bus i with bus j is created.

This can be easily explained by referring to the Y_{bus} matrix of the five-bus system given in Eq. (6.54). It is desired to eliminate bus 5. It is also observed that buses 2, 3, and 4 are connected to bus 5. Hence, when bus 5 is eliminated, new branches are created between buses 2 and 4 and buses 3 and 4. On the other hand, if bus 3 is eliminated, no new branch is created but the branch between buses 2 and 5 is modified. Clearly, the time required for elimination of buses depends on the number of branches added and the number of branches modified, between pairs of buses adjacent to the bus being eliminated. It is also noted that network reduction reduces sparsity.

6.7 NUMBERING SCHEMES TO CONSERVE SPARSITY

From the foregoing, it is evident that both triangular factorization (LU decomposition) of Y_{bus} and network reduction lead to the creation of non-zero terms. In other words, the number of non-zero terms added to Y_{bus} during triangular factorization or network reduction depends on the sequence in which the rows are processed. Three numbering schemes, presented by Tinney, help in conserving sparsity in the Y_{bus} matrix. Essentially the schemes re-number (or re-order) the system nodes in a defined manner so that the number of added non-zero terms is minimum.

Tinney's schemes for near optimal ordering are as follows:

Major diagonal banding scheme   Number the rows of the matrix according to the number of off-diagonal terms before elimination. The row with the least number of terms is eliminated first. In this scheme, the elements form a band around the major diagonal as shown here with non-zero elements shown by ×.

$$\begin{bmatrix} \times & \times & 0 & 0 & 0 & 0 \\ \times & \times & 0 & 0 & 0 & 0 \\ 0 & 0 & \times & \times & 0 & 0 \\ 0 & 0 & \times & \times & \times & \times \\ 0 & 0 & 0 & \times & \times & \times \\ 0 & 0 & 0 & \times & \times & \times \end{bmatrix}$$

Tinney's numbering scheme I   Number the rows such that at each step of the process the next row to be operated is the one with the fewest non-zero terms. If more than one row meets this criterion, select any one. This scheme requires a simulation of the effects on the accumulation of non-zero terms of the elimination process. The input data is a list by rows of the column numbers of the non-zero off-diagonal terms.

Tinney's numbering scheme II   Number the rows so that at each step of the process the next row to be operated is the one that will introduce the fewest non-zero terms. If more than one row meets this criterion, select any one. This scheme involves a trial simulation of every feasible alternative of the elimination process at each step. Input information is the same as for scheme I.

6.8 FORMULATION OF BUS IMPEDANCE MATRIX

The method of representation of power systems in the impedance form for analyses under fault conditions is more convenient since the solution of voltage equations can be obtained directly and requires no iteration. Before embarking on a description of the algorithm for the formulation of the impedance matrix, it would be prudent to describe how the node voltages and node currents are related.

6.8.1 Node Voltages and Currents

A general n-bus linear passive network is shown in Fig. 6.13. Assume that the bus impedance matrix (superscript n signifies the number of buses) is known for the network. The general form of the network performance equation in impedance form is

$$V_{bus}^n = Z_{bus}^n \, I_{bus}^n$$

where V_{bus}^n is an $n \times 1$ vector of bus voltages with respect to the reference node

and I_{bus}^n is an $n \times 1$ vector of injected bus currents. The node voltage equations, to describe the network performance in expanded form, can be written as

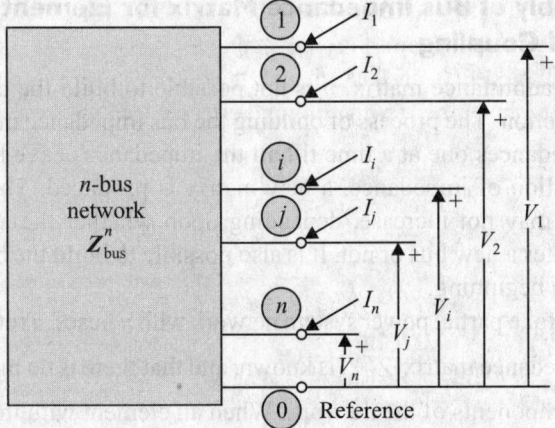

Fig. 6.13 General *n*-bus linear passive network with ground as reference

$$
\begin{bmatrix} V_1 \\ V_2 \\ \cdots \\ V_j \\ \cdots \\ V_n \end{bmatrix} = \begin{bmatrix} Z_{11} & Z_{12} & \cdots & Z_{1j} & \cdots & Z_{1n} \\ Z_{21} & Z_{22} & \cdots & Z_{2j} & \cdots & Z_{2n} \\ \cdots & \cdots & \cdots & \cdots & \cdots & \cdots \\ Z_{j1} & Z_{j2} & \cdots & Z_{jj} & \cdots & Z_{jn} \\ \cdots & \cdots & \cdots & \cdots & \cdots & \cdots \\ Z_{n1} & Z_{n2} & \cdots & Z_{nj} & \cdots & Z_{nn} \end{bmatrix} \begin{bmatrix} I_1 \\ I_2 \\ \cdots \\ I_j \\ \cdots \\ I_n \end{bmatrix}
\qquad (6.64)
$$

If a unit current is injected into the *j*th bus, with all other bus currents being zero, and the open-circuit voltage at each bus is measured, then

$$
Z_{j1} = \frac{V_1}{I_j}
$$

$$
Z_{j2} = \frac{V_2}{I_j}
$$

$$
\vdots \qquad \vdots
$$

$$
Z_{j2} = \frac{V_j}{I_j}
\qquad (6.65)
$$

$$
\vdots \qquad \vdots
$$

$$
Z_{jn} = \frac{V_n}{I_j}
$$

Since all impedances are defined with all the nodes open circuited except one (*j*th node), the impedance Z_{jj} is called the open-circuit, driving-point impedance and Z_{ij} is called open-circuit transfer impedance for $i \neq j$.

Since the network is made up of linear passive impedances, $Z_{ji} = Z_{ij}$. Thus, from Eq. (6.65), it is evident that by injecting a unit current at each bus, in turn, and keeping all the other bus ports on open circuit, it is possible to determine the elements of the bus impedance matrix.

6.8.2 Assembly of Bus Impedance Matrix for Elements without Mutual Coupling

Unlike the bus admittance matrix, it is not possible to build the bus impedance matrix by inspection. The process of building the bus impedance matrix proceeds by adding impedances one at a time till all the impedances have been included. With each addition of impedance, a new matrix is produced. The order of the matrix may or may not increase, depending upon whether the addition of the impedance creates a new bus or not. It is also possible to build the bus impedance matrix from the beginning.

Assume that for a partial power system network with n buses, a reference node 0, and the bus impedance matrix Z_{bus}^n is known, and that there is no mutual coupling between the components of the network. When an element with impedance Z_b is added to the partial network, four types of modifications are possible:

1. Addition of a tree branch with impedance Z_b from a new bus to the reference bus
2. Addition of a tree branch with impedance Z_b from a new bus to an existing bus
3. Adding a co-tree link with impedance Z_b between two existing buses
4. Adding a co-tree link with impedance Z_b between an existing bus to the reference bus

Type I Modification

A new branch with impedance Z_b is added to the n-bus network by connecting it to the reference, as shown in Fig. 6.14. A new bus $(n + 1)$ is created and it is desired to obtain the new bus impedance matrix Z_{bus}^{n+1}. If current injected at bus $(n + 1)$ is $I_{n+1} = 1$, then the voltage at bus $(n + 1)$ is

$$V_{n+1} = Z_{n+1,\,n+1}I_{n+1} = Z_b I_{n+1} \tag{6.66}$$

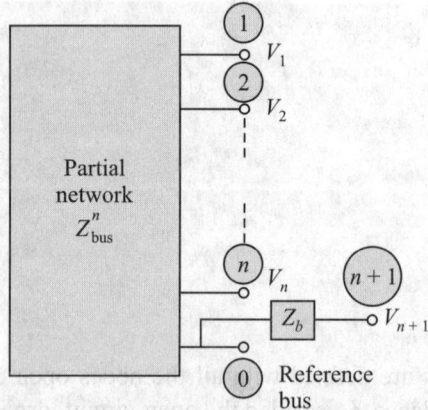

Fig. 6.14 Addition of impedance Z_b to reference bus

As there is no mutual impedance between the new bus and the previous partial network, there cannot be any voltages induced at the other buses due to injected current at bus $(n + 1)$. Thus,

$$V_j = Z_{n+1,j} I_{n+1} = Z_{j,n+1} I_{n+1} = 0 \quad \text{for } j = 1, 2, \ldots; n \tag{6.67}$$

Thus, $\quad Z_{n+1,j} = Z_{j,n+1} = 0 \tag{6.68}$

$$Z_{n+1,n+1} = Z_b \tag{6.69}$$

The new bus impedance matrix equation for addition of a tree branch without mutual coupling to the reference may be written as follows:

$$
V_{bus}^{n+1} =
\begin{bmatrix} V_1 \\ \vdots \\ V_n \\ \hline V_{n+1} \end{bmatrix}
=
\begin{bmatrix}
Z_{11} & & Z_{1n} & | & 0 \\
\vdots & & \vdots & | & 0 \\
& & & | & \vdots \\
Z_{n1} & & Z_{nm} & | & 0 \\
\hline
0 & 0 & \cdots & 0 & | & Z_b
\end{bmatrix}
I_{bus}^{n+1}
\tag{6.70}
$$

Thus, addition of a new branch having impedance Z_b between the reference and the new bus $(n + 1)$ augments the original bus impedance matrix Z_{bus}^n by a row and a column. In the augmented matrix of order $(n + 1) \times (n + 1)$, only the diagonal element $Z_{n+1,n+1} = Z_b$ and all the off-diagonal elements of the new row and column are zero.

Type 2 Modification

A new branch with impedance Z_b, and without mutual coupling, is connected to an existing bus p of the n-bus network creating a new bus $(n + 1)$ as shown in Fig. 6.15, and it is desired to obtain the new bus impedance matrix Z_{bus}^{n+1}.

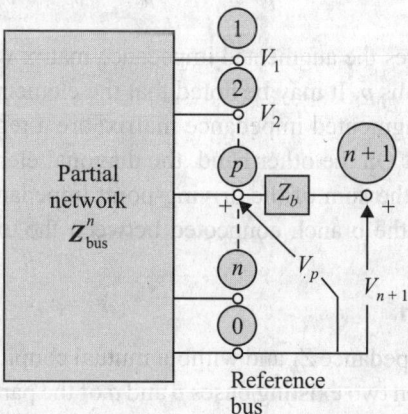

Fig. 6.15 The addition of impedance Z_b to an existing bus p and new bus $(n + 1)$

Assume that a unit current is injected into the bus $(n + 1)$ with respect to the reference and all the other buses are on open circuit. Then the voltage at this bus is

$$V_{n+1,n+1} = Z_{n+1,n+1} I_{n+1} = (Z_{pp} + Z_b) I_{n+1} \tag{6.71}$$

Thus, the driving-point impedance at the bus $(n + 1)$ is the driving-point impedance at bus p plus the new impedance Z_b.

$$Z_{n+1,n+1} = Z_{pp} + Z_b \tag{6.72}$$

As the unit current is injected into bus p also, the voltages at all the buses, except bus p, and hence the transfer impedances are the same as those for bus p.

$$Z_{k, n+1} = Z_{n+1, k} = Z_{kp} = Z_{pk} \quad k = 1, 2, ..., p, ..., n; \quad k \neq n+1 \quad (6.73)$$

Hence, the relation between the bus voltages and bus currents for the $(n + 1)$-bus system is given by

$$V_{bus}^{n+1} = \begin{bmatrix} V_1 \\ V_2 \\ \vdots \\ V_p \\ \vdots \\ V_n \\ \overline{V_{n+1}} \end{bmatrix} = Z_{bus}^{n+1} \, I_{bus}^{n+1}$$

$$= \begin{bmatrix} Z_{11} & \cdots & Z_{1p} & \cdots & Z_{1n} & \vdots & Z_{p1} \\ Z_{21} & & \vdots & & \vdots & \vdots & \vdots \\ \vdots & & \vdots & & \vdots & \vdots & \\ Z_{p1} & & Z_{pp} & & Z_{pm} & \vdots & Z_{pp} \\ \vdots & & \vdots & & \vdots & \vdots & \\ Z_{n1} & \cdots & Z_{np} & \cdots & Z_{nm} & \vdots & Z_{pn} \\ \hline Z_{p1} & & Z_{pp} & & Z_{pn} & \vdots & Z_{pp} + Z_b \end{bmatrix} \begin{bmatrix} I_1 \\ I_2 \\ \vdots \\ I_p \\ \vdots \\ I_n \\ \overline{I_{n+1}} \end{bmatrix} \quad (6.74)$$

Equation (6.74) gives the augmented impedance matrix with a new bus $(n + 1)$ added to an existing bus p. It may be noted that the elements of the $(n +1)$th row and column, in the augmented impedance matrix, are a reproduction of the row elements of the bus p. On the other hand, the diagonal element $Z_{n+1, n+1}$ of the augmented matrix is the sum of the driving-point impedance of bus p, Z_{pp}, and the impedance Z_b of the branch connected between the existing bus p and the new bus $(n + 1)$.

Type 3 Modification

A co-tree link with impedance Z_b and without mutual coupling is connected to the n-bus network between two existing buses p and q of the partial network as shown in Fig. 6.16. With the introduction of the element Z_b, no new bus is formed and due to the loop formed by the link the driving-point impedances at buses p and q become parallel current paths. From Fig. 6.16, it is observed that the current I_b flowing through the impedance Z_b is given by

$$I_b = \frac{V_p - V_q}{Z_b} = Y_b(V_p - V_q) \quad (6.75)$$

and it can also be observed the current injections in the partial network get modified, which in turn modify the bus voltages.

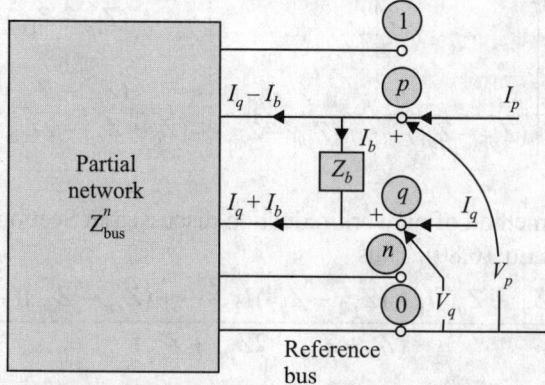

Fig. 6.16 Addition of a new impedance Z_b between two existing buses p and q

Due to the modified current injections at the buses p and q in the original network, the bus voltages are expressed as follows

$$V_1 = Z_{11}I_1 + \cdots + Z_{1p}(I_p - I_b) + Z_{1q}(I_q + I_b) + \cdots + Z_{1n}I_n$$

$$\vdots \qquad\qquad \vdots \qquad\qquad\qquad \vdots$$

$$V_p = Z_{p1}I_2 + \cdots + Z_{pp}(I_p - I_b) + Z_{pq}(I_q + I_b) + \cdots + Z_{pn}I_n$$

$$V_q = Z_{q1}I_2 + \cdots + Z_{qp}(I_p - I_b) + Z_{qq}(I_q + I_b) + \cdots + Z_{qn}I_n \qquad (6.76)$$

$$\vdots \qquad\qquad \vdots \qquad\qquad\qquad \vdots$$

$$V_n = Z_{n1}I_2 + \cdots + Z_{np}(I_p - I_b) + Z_{nq}(I_q + I_b) + \cdots + Z_{nn}I_n$$

On rearranging, Eq. (6.76) yields

$$V_1 = Z_{11}I_1 + \cdots + Z_{1p}I_p + Z_{1q}I_q + \cdots + Z_{1n}I_n - (Z_{1p} - Z_{1q})I_b$$

$$\vdots \qquad\qquad \vdots \qquad\qquad\qquad \vdots$$

$$V_p = Z_{p1}I_1 + \cdots + Z_{pp}I_p + Z_{pq}I_q + \cdots + Z_{pn}I_n - (Z_{pp} - Z_{pq})I_b$$

$$V_q = Z_{q1}I_1 + \cdots + Z_{pq}I_p + Z_{qq}I_q + \cdots + Z_{qn}I_n - (Z_{qp} - Z_{qq})I_b$$

$$\vdots \qquad\qquad \vdots \qquad\qquad\qquad \vdots$$

$$V_n = Z_{n1}I_1 + \cdots + Z_{np}I_p + Z_{nq}I_q + \cdots + Z_{nn}I_n - (Z_{np} - Z_{nq})I_b \qquad (6.77)$$

The voltages for buses p and q, expressed by Eq. (6.76), may be written as

$$V_p - V_q - Z_b I_b = 0 \qquad (6.78)$$

By using the relations for V_p and V_q from Eq. (6.77), substituting in Eq. (6.78), and rearranging, the following relation is obtained

$$(Z_{p2} - Z_{q1})I_1 + (Z_{p2} - Z_{q2})I_2 + \cdots + (Z_{pn} - Z_{qn})I_n$$
$$- (Z_b + Z_{pp} - 2Z_{pq} + Z_{qq})I_b = 0 \qquad (6.79)$$

Equations (6.77) and (6.79) may be combined to write the voltage–current relations with the augmented $(n + 1)$ matrix.

$$
\begin{bmatrix} V_{\text{bus}}^n \\ \hline 0 \end{bmatrix} = \begin{bmatrix} Z_{\text{bus}}^n & \vdots & \begin{matrix} (Z_{1p} - Z_{1p}) \\ \vdots \\ (Z_{np} - Z_{nq}) \end{matrix} \\ \hline (Z_{p1} - Z_{q1}) \cdots (Z_{pn} - Z_{qn}) & \vdots & (Z_b + Z_{pp} - 2Z_{pq} + Z_{qq}) \end{bmatrix} \begin{bmatrix} I_{\text{bus}}^n \\ \hline I_b \end{bmatrix}
$$

$$(6.80)$$

Employing the method of network reduction discussed in Section 6.6, I_b can be eliminated from Eq. (6.80). Thus,

$$
I_b = \frac{\left[(Z_{p1} - Z_{q1})I_1 + (Z_{p2} - Z_{q2})I_2 + \cdots + (Z_{pn} - Z_{qn})I_n \right]}{-(Z_b + Z_{pp} - 2Z_{pq} + Z_{qq})}
$$

$$(6.81)$$

With the elimination of I_b from Eq. (6.81), the elements of the modified Z_{bus}^{n+1} matrix can be written as

$$
Z_{\text{bus}}^{n+1} = Z_{\text{bus}}^n - \left[\frac{-1}{(Z_b + Z_{pp}^n - 2Z_{pq}^n + Z_{qq}^n)} \right] \times \begin{bmatrix} Z_{1p}^n - Z_{1q}^n \\ Z_{2p}^n - Z_{2q}^n \\ \vdots \\ Z_{np}^n - Z_{nq}^n \end{bmatrix}
$$

$$
\left[\left(Z_{p1}^n - Z_{q1}^n \right) \left(Z_{p2}^n - Z_{q2}^n \right) \cdots \left(Z_{pn}^n - Z_{qn}^n \right) \right]
$$

$$(6.82)$$

Employing Eq. (6.82), a computer adaptable algorithm for adding impedance Z_b to existing buses p and q can be written as

$$
Z_{\text{bus}}^{n+1} = Z_{\text{bus}}^n - yZZ^t
$$

$$(6.83)$$

where

$$
y = -\frac{1}{(Z_b + Z_{pp}^n - 2Z_{pq}^n + Z_{qq}^n)}
$$

$$(6.84a)$$

and

$$
Z = \begin{bmatrix} Z_{1p}^n - Z_{1q}^n \\ Z_{2p}^n - Z_{2q}^n \\ \vdots \\ Z_{np}^n - Z_{nq}^n \end{bmatrix}
$$

$$(6.84b)$$

It can be observed from Eq. (6.84b) that Z is a column vector constituted of the difference between the elements of columns p and q in the Z_{bus}^n matrix.

From the foregoing it is clear that when an impedance is added between two existing buses, the order of the original Z_{bus}^n remains unchanged; however, every driving point and transfer impedance in the Z_{bus}^n matrix may change. From Eq. (6.83) it is also noted that the product of ZZ^t is a square matrix of order $n \times n$. However, if symmetry in Z_{bus}^n is taken advantage of, then only upper or lower triangular part of Z_{bus}^n may be computed.

As can be observed, the algorithm described earlier assembles the bus impedance matrix element by element. The same algorithm can be employed to modify the bus impedance matrix to reflect changes in a network. The procedure is similar to adding an element except that negative impedance is to be used for the element being removed from the network.

Type 4 Modification

A new branch with impedance Z_b and without mutual coupling is connected to the n-bus network between an existing bus p and the reference bus of the partial network as shown in Fig. 6.17.

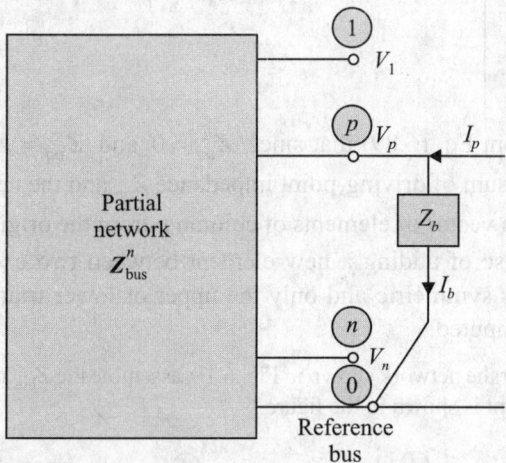

Fig. 6.17 Addition of impedance Z_b between existing bus p and the reference

In the Type 3 modification where impedance Z_b is added between two existing network buses p and q, if the q bus is made the reference bus, then this becomes type 4 modification. Then

$$Z_{qq}^n - 0 \text{ and } Z_{jq}^n - Z_{qj}^n - 0 \quad \text{for } j = 1, 2, ..., n \tag{6.85}$$

Hence, Eq. (6.80) is modified as

$$
\begin{bmatrix} V_{\text{bus}}^n \\ \hline 0 \end{bmatrix} =
\begin{bmatrix} Z_{\text{bus}}^n & \begin{matrix} Z_{1p} \\ \vdots \\ Z_{np} \end{matrix} \\ \hline Z_{p1} \quad \cdots \quad \cdots \quad \cdots \quad Z_{pn} & -(Z_b + Z_{pp}) \end{bmatrix}
\begin{bmatrix} I_{\text{bus}}^n \\ \hline I_b \end{bmatrix}
\tag{6.86}
$$

Eliminating I_b from Eq. (6.86), the elements of Z_{bus}^{n+1} can be written as

$$Z_{\text{bus}}^{n+1} = Z_{\text{bus}}^n - \left[\frac{-1}{(Z_b + Z_{pp}^n)} \right] \begin{bmatrix} Z_{1p}^n \\ Z_{2p}^n \\ \vdots \\ Z_{np}^n \end{bmatrix} \begin{bmatrix} Z_{p1}^n & Z_{p2}^n & \cdots & Z_{pn}^n \end{bmatrix} \tag{6.87}$$

From Eq. (6.87), the algorithm for computer application can be written as

$$Z_{\text{bus}}^{n+1} = Z_{\text{bus}}^n - y_1 Z Z^t \tag{6.88}$$

where $\quad y_1 = \dfrac{1}{\left(Z_b + Z_{pp}^n\right)}$ (6.89a)

and $\quad Z = \begin{bmatrix} Z_{1p}^n \\ Z_{2p}^n \\ \vdots \\ Z_{np}^n \end{bmatrix}$ (6.89b)

It is observed from Eq. (6.89) that since $Z_{qq}^n = 0$ and $Z_{pq}^n = Z_{qp}^n = 0$, y_1 is the reciprocal of the sum of driving-point impedance Z_{pp} and the new impedance Z_b and Z is a column vector of elements of column p from the original Z_{bus}^n matrix. Similar to the case of adding a new element between two existing buses, the product of ZZ' is symmetric and only the upper or lower triangular portion of Z_{bus}^n may be computed.

Example 6.6 For the network shown in Fig. 6.18, assemble the Z_{bus} matrix. The imped-ance of each element is shown in the figure.

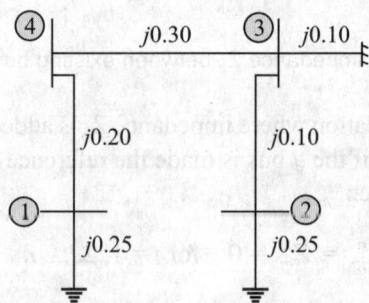

Fig. 6.18 The network for Example 6.6

Solution Begin to assemble Z_{bus} by selecting buses which are connected to the reference bus. In this case buses 1 and 2 are connected, in turn to the reference bus.

Commence assembling Z_{bus} by connecting bus 1 to the reference bus. Using Eq. (6.70) and taking $Z_b = j0.25$. Now,

$$Z_{bus}^1 = V_1^1 = j0.25$$

Similarly, when bus 2 is connected to the reference, Z_{bus}^1 gets augmented by a row and a column as follows:

$$Z_{bus}^2 = \begin{bmatrix} V_1^2 \\ V_2^2 \end{bmatrix} = j\begin{bmatrix} 0.25 & 0 \\ 0 & 0.25 \end{bmatrix}$$

Next connect existing bus 2 with new bus 3. In Eq. (6.72), use values $p = 2$, $q = 3$, and $Z_b = j0.10$. Thus,

$$Z_{33} = Z_{22} + Z_b = j0.25 + j0.10 = j0.35, \quad Z_{13} = Z_{31} = 0,$$

and $\quad Z_{23} = Z_{32} = j0.25$

$$[Z_{bus}^2] = \begin{bmatrix} V_1^3 \\ V_2^3 \\ V_3^3 \end{bmatrix} = j\begin{bmatrix} 0.25 & 0 & 0 \\ 0 & 0.25 & 0.25 \\ 0 & 0.25 & 0.35 \end{bmatrix}$$

The addition of bus 4 proceeds in the same manner as bus 3. Here $p = 1$ and $q = 4$.

$$Z_{44} = Z_{11} + Z_b = j0.25 + j0.20 = j0.45, \ Z_{14} = Z_{41} = 0, \ Z_{24} = Z_{42} = 0,$$

and $\quad Z_{34} = Z_{43} = j0.25$

$$[Z_{bus}^4] = \begin{bmatrix} V_1^4 \\ V_2^4 \\ V_3^4 \\ V_4^4 \end{bmatrix} = j \begin{bmatrix} 0.2500 & 0 & 0 & 0.2500 \\ 0 & 0.2500 & 0.2500 & 0 \\ 0 & 0.2500 & 0.3500 & 0 \\ 0.2500 & 0 & 0 & 0.4500 \end{bmatrix}$$

$Z_b = j0.3$ is now to be added between bus 3 and bus 4. Equations (6.82) to (6.84) apply; in this case $p = 3$ and $q = 4$.

From Eqs (6.84a) and (6.84b)

$$y = \frac{1}{Z_b \times Z_{33}^4 - 2 \times Z_{34}^4 + Z_{44}^4} = \frac{1}{j0.3 + j0.35 - 2 \times j0.0 + j0.45}$$

$$= -j0.9091$$

$$Z = \begin{bmatrix} Z_{13}^4 - Z_{14}^4 \\ Z_{23}^4 - Z_{24}^4 \\ Z_{33}^4 - Z_{34}^4 \\ Z_{43}^4 - Z_{44}^4 \end{bmatrix} = j \begin{bmatrix} -0.2500 \\ 0.2500 \\ 0.3500 \\ -0.4500 \end{bmatrix}$$

$$Z_{bus}^5 = Z_{bus}^4 - (-y) Z Z^t = j \begin{bmatrix} 0.2500 & 0 & 0 & 0.2500 \\ 0 & 0.2500 & 0.2500 & 0 \\ 0 & 0.2500 & 0.3500 & 0 \\ 0.2500 & 0 & 0 & 0.4500 \end{bmatrix}$$

$$-(j0.9091) \begin{bmatrix} -0.25 \\ 0.25 \\ 0.35 \\ -0.45 \end{bmatrix} \begin{bmatrix} -0.25 & 0.25 & 0.35 & -0.45 \end{bmatrix}$$

$$= j \begin{bmatrix} 0.1932 & 0.0568 & 0.0795 & 0.1477 \\ 0.0568 & 0.1932 & 0.1705 & 0.1023 \\ 0.0795 & 0.1705 & 0.2386 & 0.1432 \\ 0.1477 & 0.1023 & 0.1432 & 0.2659 \end{bmatrix}$$

For the impedance to be connected between bus 3 and the reference bus, $p = 0$ and $q = 3$. Therefore, from Eqs (6.89a) and (6.89b), y_1 and Z, respectively can be obtained as follows

$$y_1 = \frac{1}{Z_b + Z_{33}^5} = \frac{1}{j0.10 + j0.2386} = j2.9533$$

$$Z = \begin{bmatrix} -Z_{13}^5 \\ -Z_{23}^5 \\ -Z_{33}^5 \\ -Z_{43}^5 \end{bmatrix} = j \begin{bmatrix} -0.0795 \\ -0.1705 \\ -0.2386 \\ -0.1432 \end{bmatrix}$$

$$Z_{bus}^6 = Z_{bus}^5 - (-y_1) Z Z^t = j \begin{bmatrix} 0.1745 & 0.0168 & 0.0235 & 0.1141 \\ 0.0168 & 0.1074 & 0.0503 & 0.0302 \\ 0.0235 & 0.0503 & 0.0705 & 0.0423 \\ 0.1141 & 0.0302 & 0.0423 & 0.2054 \end{bmatrix}$$

6.9 MATLAB PROGRAM FOR ASSEMBLY OF Z_{bus} MATRIX

The MATLAB program function [Zbus] = zeebus (nelemnt,ind) assembles the impedance matrix $[Z_{bus}]$ from a partially assembled $[Z_{bus}]$ or from the very beginning. The input parameters are nelemnt for the total number of elements in the network and ind is an indicator. ind = 1 if partial $[Z_{bus}]$ is available and ind = 0 if $[Z_{bus}]$ is to be assembled from the beginning. The program also requires four parameters, for each element, to be input from the keyboard, that is, bus numbers (p and q), impedance of the element, and the type of bus. The type of bus = 1 when the element is connected between reference and a new bus, type of bus = 2 when the element is connected between an existing bus and a new bus, type of bus = 3 when the element is connected between two existing buses, and type of bus = 4 when an element is connected between an existing bus and the reference.

```
function [Zbus]=zeebus(nelemnt,ind,node);
if ind == 1;
   Zbus=input('partial matrix Zbus');
   ind=0;
else
end
if ind == 0;
   for l=1:nelemnt
   p=input('bus number p');
   q=input('bus number q');
   zb=input('impedance');
   type=input('type of bus');
   if type == 1;
     for k=1:q
        if k == q;
           Zbus(k,k)=zb;
        else
           Zbus(k,q)=0;
        end
     end
end
if type == 2;
   for k=1:q
      if k == q;
         Zbus(q,q)=zb+Zbus(p,p);
      else
         if k == p;
            Zbus(p,q)=Zbus(p,p);
            Zbus(q,p)=Zbus(p,q);
         else
            Zbus(k,q)=0;
            Zbus(q,k)=0;
         end
      end
   end
end
```

```
if type == 3;
    y=1/(zb+Zbus(p,p)-2*Zbus(p,q)+Zbus(q,q));
    for k=1:node
        X(k,1)=Zbus(k,p)-Zbus(k,q);
        Xt(1,k)=(k.1);
    end
    Zbus=Zbus-(-y)*X*X',
end
if type ==4;
    y=1/(zb+Zbus(q,q));
    for k=1:node
    X(k,1)=-Zbus(k,q);
    Xt(1,k)=(k.1);
end
    Zbus=Zbus-(-y)*X*X'
    end
end
else
end
```

Example 6.7 Use the MATLAB program function [Zbus]=zeebus(nelemnt,ind) to assemble Z_{bus} matrix for the network shown in Fig. 6.19. Impedances of the elements are shown in the figure.

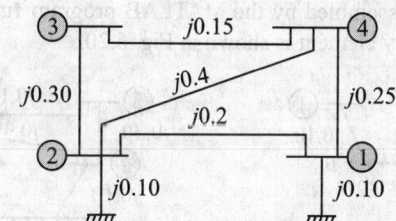

Fig. 6.19 The network for Example 6.7

Solution

```
≫ nelemnt=7;ind=0;
≫ [Zbus]=zeebus(nelemnt,ind)
bus number p0
bus number q1
impedance0.1i
type of bus1                                    See Fig. 6.20 (a)
bus number p0
bus number q2
impedance0.1i
type of bus1                                    See Fig. 6.20 (b)
bus number p2
bus number q3
impedance0.3i
type of bus2                                    See Fig. 6.20 (c)
```

```
bus number p2
bus number q4
impedance0.4i
type of bus2                                    See Fig. 6.20 (d)
bus number p3
bus number q4
impedance0.15i
type of bus3                                    See Fig. 6.20 (e)
bus number p1
bus number q2
impedance0.2i
type of bus3                                    See Fig. 6.20 (f)
bus number p1
bus number q4
impedance0.25i
type of bus3                                    See Fig. 6.20 (g)
  Zbus =
    0 + 0.0701i   0 + 0.0299i   0 + 0.0387i   0 + 0.0457i
    0 + 0.0299i   0 + 0.0701i   0 + 0.0613i   0 + 0.0543i
    0 + 0.0387i   0 + 0.0613i   0 + 0.1841i   0 + 0.1073i
    0 + 0.0457i   0 + 0.0543i   0 + 0.1073i   0 + 0.1492i
```

How the Z_{bus} matrix is assembled by the MATLAB program function [Zbus]=zeebus (nelemnt,ind), element by element is shown in Fig. 6.20.

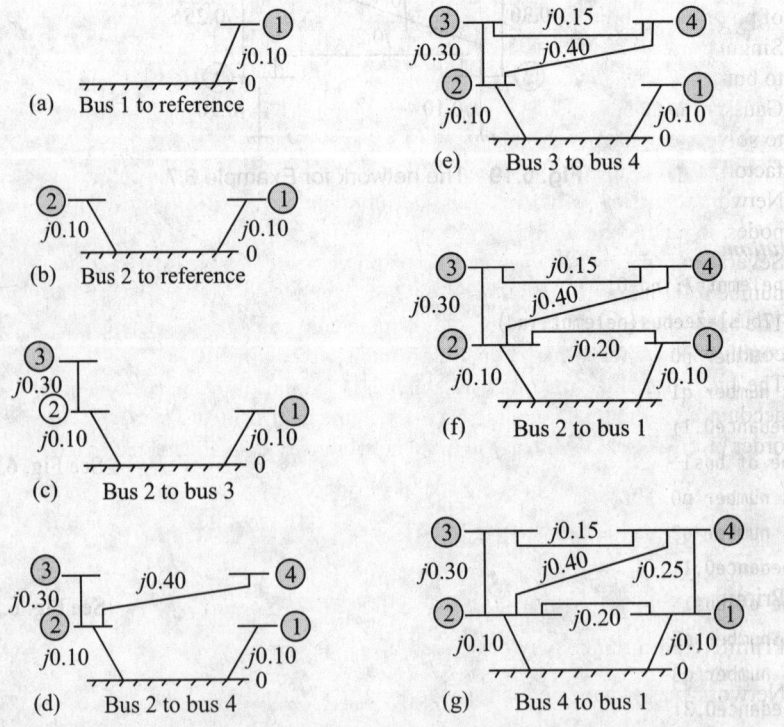

(a)   Bus 1 to reference

(b)   Bus 2 to reference

(c)   Bus 2 to bus 3

(d)   Bus 2 to bus 4

(e)   Bus 3 to bus 4

(f)   Bus 2 to bus 1

(g)   Bus 4 to bus 1

Fig. 6.20   Step-by-step assembly of Z_{bus}

Example 6.8 Determine Z_{bus} when the element between buses 1 and 2 in Fig. 6.19 is removed.

Solution

```
≫ nelemnt=1;ind=1;
≫ [Zbus]=zeebus(nelemnt,ind)
partial matrix Zbus[0.0701 0.0299 0.0387 0.0457;0.0299 0.0701 0.0613 0.0543;0.0387
0.0613 0.1841 0.1073;0.0457 0.0543 0.1073 0.1492]*i
bus number p1
bus number q2
impedance−0.2i
type of bus3
Zbus =
```

0 + 0.0836i	0 + 0.0164i	0 + 0.0311i	0 + 0.0428i
0 + 0.0164i	0 + 0.0836i	0 + 0.0689i	0 + 0.0572i
0 + 0.0311i	0 + 0.0689i	0 + 0.1884i	0 + 0.1089i
0 + 0.0428i	0 + 0.0572i	0 + 0.1089i	0 + 0.1498i

SUMMARY

- The principle of graph theory by defining a node, graph, link, branch, and so on is utilized to draw circuit diagrams of interconnected power networks, which in turn are employed to formulate element-node and element-bus incidence matrices.
- A set of unconnected, two-terminal power system components, in either impedance or admittance representations, constitute a primitive network.
- Using element-node or element-bus incidence matrix and primitive networks, matrices of power system networks, with or without mutual coupling are formulated.
- Singular transformation is utilized to build and transform bus admittance matrix $[Y_{bus}]$ to bus impedance matrix $[Z_{bus}]$. Y_{bus} matrices have non-zero values and are sparse.
- Gaussian and Gauss–Seidel iteration methods are two indirect methods which are applied to solve the nodal equation $[I_{bus}] = [Y_{bus}] [V_{bus}]$. Under the direct method, triangular factorization of the $[Y_{bus}]$ is employed to solve the nodal equations.
- Network reduction, which requires node elimination, is utilized when voltages at a few nodes, in a network, are of interest.
- Several numbering schemes can be employed to conserve sparsity and reduce the number of arithmetic operations.
- The bus impedance form of analyses is more convenient to solve networks under fault condition since the voltage equations can be solved directly.
- The bus impedance matrix is assembled by adding impedances one at a time and accounting for mutual coupling between elements. During assembly of the matrix the order of the matrix will increase if the new branch being added creates a new bus.

SIGNIFICANT FORMULAE

- Primitive impedance network equation: $v + e = z\,i$
- Primitive admittance network equation: $i + j = yv$
- Network performance equations: $I_{bus} = Y_{bus}V_{bus}$ and $V_{bus} = Z_{bus}\,I_{bus}$
- Bus admittance matrix: $Y_{bus} = A^t yA$

- Bus impedance matrix: $Z_{\text{bus}} = Y_{\text{bus}}^{-1} = (A^t y A)^{-1}$

- Node voltage equations for n-bus system: $I_i = \sum\limits_{j=1}^{j=n} Y_{ij} V_j$ for $i = 1, 2, 3, \ldots, n$

- Gaussian iteration method: $V_i^{k+1} = \dfrac{1}{Y_{ii}} \left\{ I_i - \sum\limits_{j=1, j \neq i}^{n} Y_{ij} V_j^k \right\}$ $i = 1, 2, 3, \ldots, n$

- Condition for convergence: $\left| V_i^{k+1} - V_i^k \right| \leq$ tolerance limit $i = 1, 2, 3, \ldots, n$

- Gauss–Seidel iteration method: $V_i^{k+1} = \dfrac{1}{Y_{ii}} \left\{ I_i - \sum\limits_{j=1}^{i-1} Y_{ij} V_j^{k+1} - \sum\limits_{j=i+1}^{n} Y_{ij} V_j^k \right\}$

  $i = 1, 2, 3, \ldots, n$

- Forward substitution for solving nodal equations: $V_k' = \dfrac{1}{l_{kk}} \left[I_k - \sum\limits_{j=1}^{k-1} l_{kj} V_j' \right]$

  $k = 2, 3, 4, \ldots, n$

- Backward substitution for solving nodal equations: $V_i = \dfrac{1}{u_{ii}} \left[V_i' - \sum\limits_{j=i+1}^{n} u_{ij} V_j \right]$

  $i = n - 1, n - 2 \ldots, 1$

- Network reduction one node at a time: $Y_{ij}^{\text{new}} = Y_{ij} - \dfrac{Y_{ik} Y_{kj}}{Y_{kk}}$ i, j

  $= 1, 2, 3, \ldots, n$ $j \neq k$

- Performance equations in impedance form: $V_{\text{bus}}^n = Z_{\text{bus}}^n I_{\text{bus}}^n$

EXERCISES

Numerical Problems

6.1 For the power system network shown in Fig. 6.21, draw an oriented connected graph and clearly show the branches and links. Also determine the number of links and branches. Assume ground as the reference.

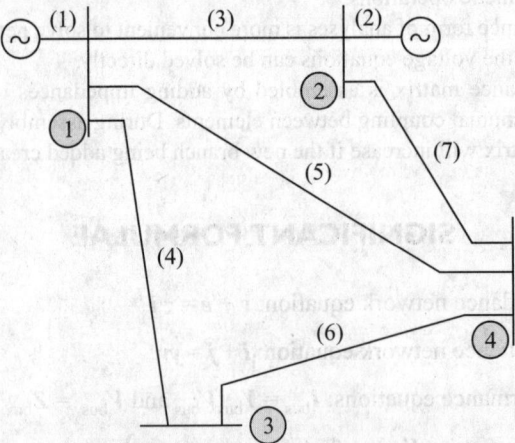

Fig. 6.21 Power system network for Problem 6.1

6.2 Formulate the node element and bus incidence matrices with ground as reference for the network in Fig. 6.21.

6.3 Repeat Problem 6.2 with node 1 as reference.

6.4 The following table provides the impedances for the network in Fig. 6.21:

Element number	Bus number		Primitive impedance
	From	To	
1	1	0	0.08
2	2	0	0.10
3	1	2	0.5
4	1	3	0.4
5	1	4	0.25
6	3	4	0.20
7	2	4	0.25

Compute the impedance and admittance matrices for the network by singular transformations. Assume zero mutual coupling and ground as reference.

6.5 With node 4 as reference, form $\hat{A}$ and A matrices for the network shown in Fig. 6.22. The reactance data for the network elements is as follows:

Element number	Bus number		Primitive reactance
	From	To	
1	1	2	$j0.06$
2	2	3	$j0.10$
3	2	3	$j0.10$
4	3	4	$j0.2$
5	4	5	$j0.25$
6	1	5	$j0.15$

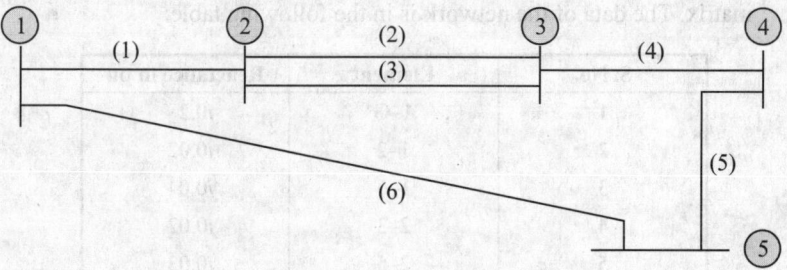

Fig. 6.22 Power system network for Problem 6.5

Assume a mutual coupling equal to $j0.1$ between elements 2 and 3. Compute Z_{bus} and Y_{bus} by singular transformations. Verify the result by non-singular transformations.

6.6 For the power system shown in Fig. 6.23, write the admittance matrix Y_{bus} by inspection. All values shown in the figure are in per unit.

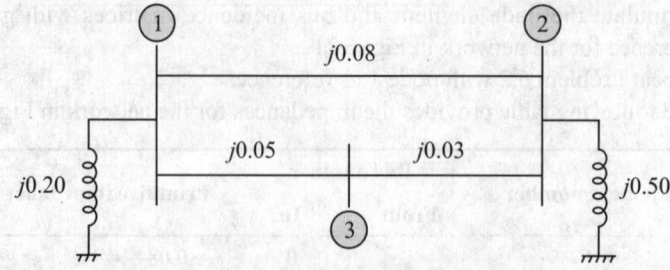

Fig. 6.23 Power system network for Problem 6.6

6.7 Verify the Y_{bus} matrix obtained in Problem 6.6 by executing the MATLAB program admitmat1.

6.8 The following is the per unit bus admittance matrix of a hypothetical three-bus network.

$$Y_{bus} = \begin{bmatrix} 2.0000 & -0.8000 & -0.2000 \\ -0.8000 & 2.6000 & -0.6000 \\ -0.2000 & -0.6000 & 1.5000 \end{bmatrix}$$

Develop the node equations in admittance form to represent the network. Use the Gauss iterative method to solve the node equations for two iterations. Assume initial voltages of all the buses to be equal to 1.0 pu. Take the current injections at buses 1, 2, and 3 equal to 1.1, 1.2, and 0.75 pu, respectively.

6.9 Repeat Problem 6.8 to compute the bus voltages by Gauss–Seidel method. Discuss the importance of acceleration factor and its effect on convergence.

6.10 Develop a MATLAB program for an n-bus system to solve the node voltage equations by the Gauss–Seidel iterative technique. Assume that the Y_{bus} matrix is obtained from the MATLAB admitmat1 program. The program should be interactive and should accept the input of initial voltages, acceleration factor, etc. from the keyboard.

6.11 Test the MATLAB program by executing Problem 6.9.

6.12 Figure 6.24 shows an electrical power system network. Build the Y_{bus} network matrix. The data of the network is in the following table:

S. No.	Element	Reactance in pu
1	1–G	$j0.2$
2	1–2	$j0.02$
3	1–6	$j0.04$
4	2–3	$j0.02$
5	2–5	$j0.03$
6	3–4	$j0.05$
7	4–5	$j0.025$
8	4–6	$j0.05$
9	5–6	$j0.02$

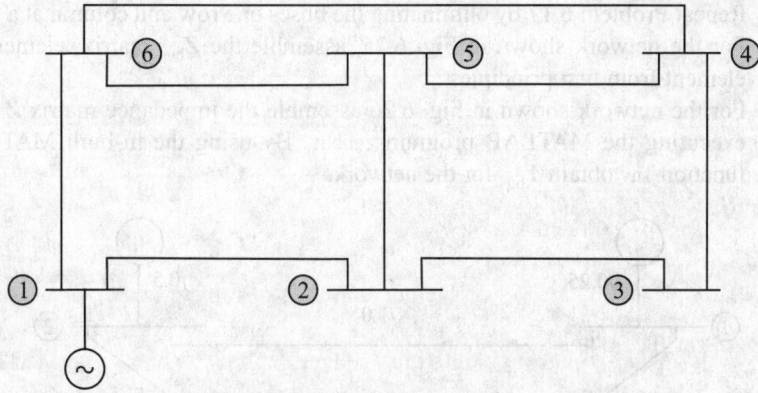

Fig. 6.24 Power system network for Problem 6.12

6.13 How is the Y_{bus} modified if the lines between buses 2-5 and 4-6 are removed and a new line having a reactance $j0.02$ pu is connected between buses 3-6?

6.14 For the system shown in Fig. 6.25, assemble the Y_{bus} matrix and compute the table of factors manually. The data is given in the following table:

S. No.	Element	Reactance in pu
1	1–2	$j0.025$
2	1–4	$j0.04$
3	2–G	$j0.02$
4	2–3	$j0.05$
5	3–G	$j0.01$
6	3–4	$j0.025$

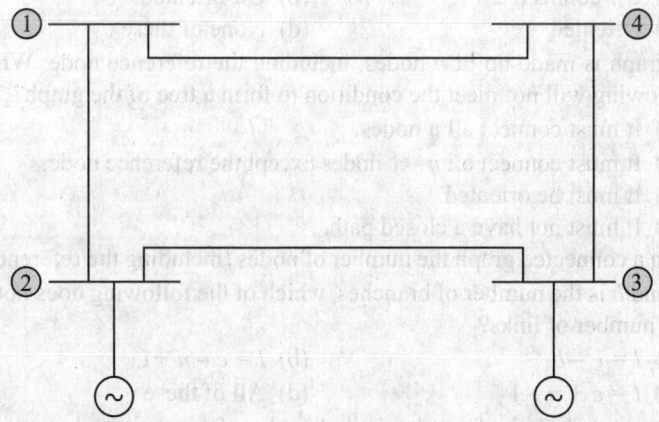

Fig. 6.25 Power system network for Problem 6.14

6.15 For the network of Problem 6.14, assume that the bus currents are $I_1 = I_2 = I_3 = I_4 = 1.0$ pu. Compute the bus voltages by the method of forward and backward substitution.

6.16 Verify the result in Problem 6.15 by executing the MATLAB program fbsub.

6.17 In the network shown in Fig. 6.24, bus numbers 5 and 6 are to be eliminated. By the method of matrix partitioning, obtain the reduced Y_{bus} matrix of the four-bus power system.

6.18 Repeat Problem 6.17 by eliminating the buses one row and column at a time.

6.19 For the network shown in Fig. 6.25, assemble the Z_{bus} matrix, element by element from first principles.

6.20 For the network shown in Fig. 6.26 assemble the impedance matrix Z_{bus} by executing the MATLAB program zeebus. By using the in-built MATLAB function inv obtain Y_{bus} for the network.

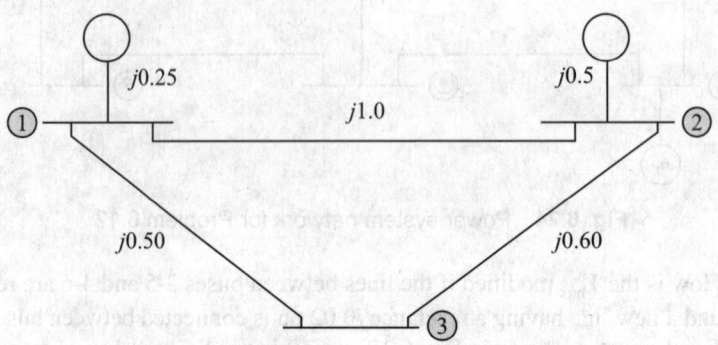

Fig. 6.26 Network for Problem 6.20

6.21 What is the change in the Z_{bus} matrix in Problem 6.20 when the branch connecting bus 1 and bus 2 is removed? Verify the answer by the inverse of Y_{bus} matrix.

Multiple Choice Objective Questions

6.1 Which of the following will define a graph when it is marked to show directions from one node to another node?
  (a) Un-connected   (b) Un-oriented
  (c) Oriented   (d) None of these

6.2 A graph is made up of n nodes, including the reference node. Which of the following will not meet the condition to form a tree of the graph?
  (a) It must connect all n nodes.
  (b) It must connect all $n - 1$ nodes except the reference node.
  (c) It must be oriented.
  (d) It must not have a closed path.

6.3 If in a connected graph the number of nodes (including the reference node) is n, and b is the number of branches, which of the following does not represent the number of links?
  (a) $l = e - b$ (b) $l = e - n + 1$
  (c) $l = e + n - 1$ (d) All of these

6.4 A connected graph of n nodes, including the reference, is made up of e elements and l links. The order of the bus incidence matrix A obtained from the node element matrix A will be
  (a) $e \times (n - 1)$ (b) $e \times n$
  (c) $e \times (n + 1)$ (d) None of these

6.5 Which of the following do not apply to a primitive network?
  (a) It has two terminals.   (b) It is unattached.
  (c) It can be either in impedance or admittance form.
  (d) None of these.

6.6 Which of the following represents the transformation for determining the $\mathbf{Y}_{bus}$ of a network?

(a) $A^t yA$

(b) $A y^t A$

(c) $Ay A^t$

(d) $A^t yA^t$

6.7 Which of the following is not a transformation for obtaining Z_{bus}?

(a) $[A^t y^{-1} A]$

(b) $[A^t y^{-1} A^{-1}]$

(c) $[A^t yA]^{-1}$

(d) None of these

6.8 In a primitive impedance network, two elements m and n are mutually coupled. Which of the following elements will reflect the effect of mutual coupling?

(a) $z\,(m,m)$

(b) $z\,(n,n)$

(c) Either of $z\,(m,\,n)$ or $z(n,m)$

(d) Both $z\,(m,\,n)$ and $z(n,m)$

6.9 While formulating the bus admittance matrix of a power system, which of the following is not applied?

(a) Self-admittance is the sum of all the elements connected to the node.

(b) Mutual admittance between two nodes is the admittance of the element.

(c) Mutual admittance between two nodes is the negative of the admittance of the element.

(d) None of these.

6.10 Which of the following applies to sparse Y_{bus} matrices?

(a) Absence of an element between two nodes means the off-diagonal element is 0.

(b) Sparsity in Y_{bus} matrices of integrated large power systems is high.

(c) The number of elements in a Y_{bus} matrix of order n is $n(n+1)/2$.

(d) All of these

6.11 Which of the following is not applied to solve nodal equations by Gauss–Seidel method?

(a) Iteration

(b) An initial estimate of bus voltages

(c) LU factorization

(d) Acceleration factor

6.12 Which of the following represents the relationship for solving $I_{bus} = Y_{bus} V_{bus}$?

(a) LV

(b) UV

(c) LUV

(d) All of these

6.13 When a bus is eliminated in network reduction which of the following happens?

(a) Elimination takes place one row and column at a time.

(b) The order of the bus matrix reduces by the number of buses eliminated.

(c) All branches connected to the bus are redistributed.

(d) All of these.

6.14 Which of the following is the purpose of numbering schemes?

(a) To make the matrix symmetrical

(b) To conserve sparsity

(c) To increase the number of mathematical operations

(d) All of these

6.15 Which of the following cannot be feasible while assembling Z_{bus} matrix of a power system?

(a) Assemble the matrix by inspection.

(b) Z_{bus} can be assembled ab initio.

(c) Assembly of Z_{bus} proceeds one impedance at a time.

(d) When an impedance is added to the partial Z_{bus} the number of modifications is four.

6.16 In which of the following will the order of the matrix increase during the formulation of Z_{bus}?
  (a) Addition of a branch between two existing buses
  (b) Addition of a new branch, without mutual coupling, between an existing bus and a reference bus
  (c) Addition of a new branch, without mutual coupling, between an existing bus and a new bus
  (d) All of these

Answers

6.1 (c)	**6.2** (b)	**6.3** (c)	**6.4** (a)	**6.5** (d)	**6.6** (a)
6.7 (a)	**6.8** (d)	**6.9** (b)	**6.10** (d)	**6.11** (c)	**6.12** (c)
6.13 (d)	**6.14** (b)	**6.15** (b)	**6.16** (c)		

CHAPTER 7

Power Flow Studies

Learning Outcomes

A concentrated study of this chapter will enable the reader to
- Understand the importance of performing load flow studies from the perspective of designing and planning power system operation in steady state
- Envisage the power flow problem based on an understanding of the system requirements such as thermal and stability loading limits of transmission lines, and the need to maintain voltages and scheduled power exchanges over tie lines with neighbouring systems within prescribed limits
- Formulate power flow equations, both in the admittance and impedance bus frame of reference
- Become adept to solving the non-linear algebraic power flow equations by applying the two most commonly accepted numerical methods: Gauss–Seidel and Newton–Raphson
- Comprehend decoupled method and fast decoupled method that are derived from Newton–Raphson method for faster solution of power flow study
- Develop algorithms and write MATLAB programs for power flow solutions by iterative techniques such as Gauss–Seidel and Newton–Raphson
- Familiarize with various types of control devices and their simulation in power flow studies

OVERVIEW

A three-phase electrical power system under normal steady-state conditions of operation is expected to deliver the demanded real and reactive power to the consumers' terminals at the rated voltage and frequency, within the specified limits of tolerance. To obtain a complete description of the behaviour of the power system, it is essential to know the voltages at various node points or buses and the power flowing through the elements of the system. Power flow studies provide the required information regarding bus voltages and power flowing through transmission lines, transformers, and other elements of a power system for a specified load demand subject to the regulating capabilities of generators, condensers, tap-changing transformers, phase shifting transformers as well as specified net interchange of power with adjoining power systems. To keep pace with increasing load growth, new generating facilities are planned and added periodically by electric utilities. Addition of a new generating unit causes some transmission facilities to become overloaded during peak load periods. Numerous power flow studies are carried out which help in critically assessing alternate plans for system expansion to meet the ever-increasing load demand.

Power flow studies also help the planning and operation engineers to meet contingency situations such as loss of a large generating unit or a major line outage due to thermal overloading of the line. Power flow studies also help to determine the best size and the most favourable location for the power capacitors for improving the power factor as well as the voltage profile of a power system. Some of these are considered to be offline executions of a power flow computation, which are usually performed months or years ahead of the actual situation.

Additionally, online power flow studies are periodically executed for monitoring and controlling a power system. The real time results of a power flow computation may be used to determine the reactive compensation needed to establish bus voltages. It may also be used to establish the incremental transmission line losses associated with changing the output of a generator that is useful in deciding optimal generation allocations to the generating stations so that the cost of generation is minimal.

A power flow study is the steady-state analysis of an interconnected power system during normal operating conditions. The system is assumed to be operating under balanced conditions and hence can be represented by a single-phase network. The power system network may contain hundreds of nodes, connected by branches and links with impedances specified in per unit on a common MVA base. The transmission lines, transformers, and shunt or series reactances of the power system are modelled by their linear equivalent circuits, as discussed in Chapters 4 and 5. However, generation and load demand at the buses has non-linear electrical characteristics. The power demand at a bus is closely modelled by constant real and reactive power, so that if terminal voltages increase, the current demand decreases, and vice versa. This type of load is suitably described as a fixed power demand at a bus. The generating plants normally operate at a regulated voltage level and fixed real power injection. The voltage phase angles between generators on the system are not known. If a generator's reactive power output is within acceptable limits, it is permitted to vary so as to match the system demand. Thus, the mathematical formulation of the power flow problem, also known as the load flow problem, results in a system of non-linear algebraic equations. These equations can be established using the bus frame of reference. The coefficients of the equations depend on the selection of the independent variables, that is, voltages or currents. Thus, either the bus admittance or the bus impedance matrices can be used. The solution of algebraic equations describing the power system is based on iterative techniques because of their non-linearity. The solution must satisfy Kirchhoff's laws. The solution must also satisfy the constraints such as the tap setting range of underload tap-changing transformers and specified power interchange with adjoining power systems.

7.1 THE POWER FLOW PROBLEM

In the power flow analysis, the load powers are assumed as known constants. A given set of loads on the buses can be served from a given set of generators in an infinite number of power flow configurations. Power flow analysis concerns itself not only with the actual physical mechanism which controls the power flow in the network meshes, but also with how a best or optimum flow configuration from among the myriad of possibilities is selected. Some important aspects of power flow analysis are as follows.

(a) The sum of real power injected at the generating buses must equal, at each instant of time, the sum of total system load demand plus system losses. To achieve optimum economic operation, this total generated power must be scheduled between the generators in a unique ratio, and the individual generator outputs must be closely maintained at the predetermined set points. As the load demand slowly changes throughout the day, therefore, these set points change slowly with time. Thus load-flow results for a certain hour of the day may be quite different from the next hour.

(b) The power transfer capability of a transmission line is limited to the thermal loading limit and the stability limit. It must be ascertained that the transmission lines do not operate too close to their stability or thermal limits.

(c) It is necessary to keep the voltage levels of certain buses within close tolerances. This can be achieved by proper scheduling of reactive powers.

(d) The power system must fulfil contractual scheduled interchange of power to neighbouring systems, if any, via its tie lines.

(e) Power flow analyses are very important in the planning stages of new networks or additions to existing ones.

The power flow or load flow problem may be split into the following sub-problems:

(a) the formulation of a mathematical model that describes the relationships between voltages and powers in the interconnected system,

(b) specification of the power and voltage constraints that must apply to the various buses of the network, and

(c) the computation of voltage magnitude and phase angle of each node or bus in a power system under balanced three-phase, steady-state conditions. As an offshoot of this calculation, the real and reactive power flows in transmission lines, transformers, as well as the equipment and system losses can be computed.

7.2 FORMULATION OF THE POWER FLOW EQUATIONS

The general form of network performance equations (as described in Section 6.4) for an *n*-bus system in the bus frame of reference, in admittance and impedance forms, are given by

$$I_{bus} = Y_{bus} V_{bus} \tag{7.1}$$

$$V_{bus} = Z_{bus} I_{bus} \tag{7.2}$$

where I_{bus} is the bus current vector, whose elements are injected bus currents, V_{bus} is the bus voltage vector, whose elements are bus voltages, Y_{bus} is the bus admittance matrix, and Z_{bus} is the bus impedance matrix for the network.

The bus voltages are specified with respect to a reference node, which is usually, but not necessarily, the ground. When ground is taken as the reference, the effect of shunt elements such as line charging, static capacitors, and shunt elements of transformer equivalent circuits are included in the elements of Z_{bus} or Y_{bus}, as the case maybe. In case one of the buses of a power system is taken as the reference bus, then all voltages are measured with respect to the reference bus and the shunt elements are treated as current sources at the buses.

Nodal equations are extensively employed in admittance form as Y_{bus} can be formed easily and the changes in the system parameters can be easily incorporated. An additional benefit of Y_{bus} is owing to its symmetry and sparsity. Both these features can be exploited to save on computer storage requirements and computer operation time.

Equation (7.1) is a vector equation, in the bus frame of reference in admittance form, consisting of n scalar equations for an n-bus system. In general, injected current at any bus i takes the form

$$I_i = Y_{i1}V_1 + Y_{i2}V_2 + \dots + Y_{in}V_n \qquad \text{for } i = 1, 2, \dots, n \qquad (7.3a)$$

$$= \sum_{m=1}^{n} Y_{im}V_m \qquad \text{for } i = 1, 2, \dots, n \qquad (7.3b)$$

However, in a power system, the injected generator complex power S_{Gi} and tapped load complex power S_{Li} at bus i are specified. The complex bus power S_i is defined as the difference between generator and load complex power at the bus. Then,

$$S_i = S_{Gi} - S_{Li} \qquad (7.4a)$$

or $\qquad P_i + jQ_i = (P_{Gi} - P_{Li}) + j(Q_{Gi} - Q_{Li}) \qquad (7.4b)$

Equating real and imaginary parts of Eq. (7.4b) gives

$$P_i = P_{Gi} - P_{Li} \qquad (7.5a)$$

$$Q_i = Q_{Gi} - Q_{Li} \qquad (7.5b)$$

where P_i and Q_i are the injected active and reactive power respectively at bus i, P_{Gi} and Q_{Gi} are the generated active and reactive power respectively at bus i, and P_{Li} and Q_{Li} are the active and reactive power load respectively at bus i.

Since the information about the net injected complex power S_i for a power system is available while information about bus currents I_i is not available, therefore it would be prudent to replace I_i in Eq. (7.3) by the bus power. The complex power at bus i is

$$S_i = (P_i + jQ_i) = V_i I_i^* \qquad \text{for } i = 1, 2, \dots, n \qquad (7.6)$$

or $\qquad I_i = \dfrac{S_i^*}{V_i^*} = \dfrac{P_i - jQ_i}{V_i^*} \qquad \text{for } i = 1, 2, \dots, n \qquad (7.7)$

Substituting the value of I_i from Eq. (7.7) in Eq. (7.3a) yields

$$\frac{P_i - jQ_i}{V_i^*} = Y_{i1}V_1 + Y_{i2}V_2 + \dots + Y_{in}V_n$$

$$\text{for } i = 1, 2, \dots, n \qquad (7.8)$$

or $\qquad S_i^* = P_i - jQ_i = Y_{i1}V_1V_i^* + Y_{i2}V_2V_i^* \dots + Y_{in}V_nV_i^*$

$$\text{for } i = 1, 2, \dots, n \qquad (7.9)$$

Equation (7.9) may be written in a compact form as

$$P_i - jQ_i = \sum_{m=1}^{n} Y_{im}V_mV_i^* \qquad \text{for } i = 1, 2, \dots, n \qquad (7.10)$$

Equation (7.10) constitutes the general form for static power flow equations (SPFE). It may be noted that n power flow equations [Eq. (7.9)] are complex.

Let the bus voltages and admittance be written in rectangular and polar form as follows:

$$V_i = e_i + jf_i = |V_i| \angle \delta_i \tag{7.11}$$

and

$$Y_{im} = G_{im} + jB_{im} = |Y_{im}| \angle \theta_{im} \tag{7.12}$$

where

$$|V_i| = \sqrt{e_i^2 + f_i^2}; \quad \delta_i = \tan^{-1} \frac{f_i}{e_i} \tag{7.13}$$

and

$$|Y_{im}| = \sqrt{G_{im}^2 + B_{im}^2}; \quad \theta_{im} = \tan^{-1} \frac{B_{im}}{G_{im}} \tag{7.14}$$

Substituting for V_i and Y_{im} in Eq. (7.10) yields

$$P_i - jQ_i = \sum_{m=1}^{n} (G_{im} + jB_{im})(e_m + jf_m)(e_i - jf_i)$$

$$\text{for } i = 1, 2, \dots, n \tag{7.15}$$

Equating the real and imaginary parts in Eq. (7.15) leads to

$$P_i = e_i \sum_{m=1}^{n} (G_{im}e_m - B_{im}f_m) + f_1 \sum_{m=1}^{n} (G_{im}f_m + B_{im}e_m)$$

$$\text{for } i = 1, 2, \dots, n \tag{7.16}$$

$$Q_i = e_i \sum_{m=1}^{n} (G_{im}f_m + B_{im}e_m) - f_i \sum_{m=1}^{n} (G_{im}e_m - B_{im}f_m)$$

$$\text{for } i = 1, 2, \dots, n \tag{7.17}$$

If the polar form of bus voltage V_i and admittance Y_{im} are substituted in Eq. (7.10), it takes the following form

$$S_i^* = (P_i - jQ_i) = |V_i|e^{-j\delta_i} \sum_{m=1}^{n} |Y_{im}| e^{j\theta_{im}} |V_m|e^{j\delta_m}$$

$$\text{for } i = 1, 2, \dots, n \tag{7.18}$$

Once again equating the real and imaginary parts in Eq. (7.18),

$$P_i = |V_i|\sum_{k=1}^{n} |Y_{im}||V_m|\cos(\theta_{im} - \delta_i + \delta_m) \quad \text{for } i = 1, 2, \dots, n \tag{7.19}$$

$$Q_i = -|V_i|\sum_{m=1}^{n} |Y_{im}||V_m|\sin(\theta_{im} - \delta_i + \delta_m) \quad \text{for } i = 1, 2, \dots, n \tag{7.20}$$

Equations (7.19) and (7.20) constitute $2n$ real power flow equations for the n-bus system.

7.3 COMPUTATIONAL ASPECTS OF POWER FLOW PROBLEM

The starting point of a power flow problem is a single-line diagram of the power system, from which input data for the computer solution is obtained. The input data consists of transmission line data, transformer data, and the bus generation and load data.

For solving a power flow problem, iterative methods using both Y_{bus} and Z_{bus} are available in theory. The bus admittance matrix Y_{bus} and bus impedance matrix Z_{bus} can be constructed from the transmission line and transformer input data.

A typical bus of a power system network is shown in Fig. 7.1. Each bus i is associated with four variables: voltage magnitude $|V_i|$, phase angle δ_i, net injected real power P_i, and net injected reactive power Q_i supplied to the bus. Thus for an n-bus system there are a total number of $4n$ variables. Equations (7.19) and (7.20) can be solved for $2n$ variables, provided the other $2n$ variables are specified as input data. Practical considerations allow a power system engineer to fix a priori two variables at each bus. The other two are unknown variables, and may then be computed by solving $2n$ power flow equations given by Eqs (7.19) and (7.20). A priori fixing of $2n$ variables for an n-bus system is achieved by classification of buses which will be discussed in the next section.

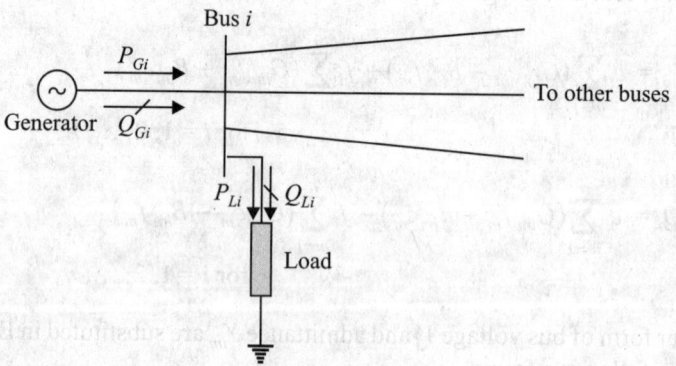

Fig. 7.1 A typical bus of a power system

7.3.1 Categorization of Buses

Each bus i of the system buses is categorized into one of the three types:

Load bus At this bus, a priori, the total injected complex power is specified. This means that active and reactive power load demands are known and active and reactive power generation are specified. The magnitude $|V_i|$ and phase angle δ_i of such a bus i are unknown. This bus is also called P-Q bus. Thus, at the ith load bus

$$S_i^{sp} = P_i^{sp} + jQ_i^{sp}$$

$$= S_{Gi}^{sp} - S_{Li}^{sp} = \left(P_{Gi}^{sp} - P_{Li}^{sp}\right) + j\left(Q_{Gi}^{sp} - Q_{Li}^{sp}\right) \tag{7.21}$$

where the subscripts Gi and Li refer to generation and load respectively at the load bus i, while the superscript sp stands for specified value.

Voltage-controlled bus At this bus, the active and reactive power load demands are known and active power generation and voltage magnitude are specified. Thus, injected active power is specified. The voltage magnitude is maintained constant by injection of reactive power. The maximum and minimum limits on the value of reactive power are also specified. For such a bus i, the phase angle of the voltage δ_i and the reactive power Q_i are to be determined. Generator buses are the voltage-controlled buses in the system. This bus is also called P-V bus. Thus, at such a P-V bus,

$$P_i^{sp} = P_{Gi}^{sp} - P_{Li}^{sp} \tag{7.22}$$

$$|V_i|^{sp} = \sqrt{e_i^2 + f_i^2} \tag{7.23}$$

Slack bus One bus, known as the slack or swing bus, is taken as the reference where a priori complex power load demand is known, while the voltage magnitude and its phase angle are specified, the phase angle usually being set equal to 0. Therefore, no computations are needed for this bus. Usually for power flow solution one of the system buses is selected as the slack bus. The concept of a slack bus is necessary because in the system the real power loss or I^2R losses are not known in advance, hence it is not possible to specify the injected real power at all the buses. It is customary to designate one of the voltage-controlled buses, generally having the largest generation, as the slack bus. The solution of power flow equations will render P_{Gs} and Q_{Gs} (and thus P_s and Q_s) for this bus, where subscript s stands for slack bus.

7.3.2 Methods of Power Flow Solution

The power flow equations are non-linear algebraic equations for which explicit solutions are not possible. These equations can be solved by iterative techniques. Historically, the first algorithm for power flow problem, employing the Gauss–Siedel method, using Y_{bus} was reported by Ward and Hale in 1956. The Gauss–Siedel method has minimum storage requirement but has slower convergence, and in some studies, fails to converge. Even to this day this method is still used when the system size is not very large. In 1961, Van-Ness and Griffins suggested Newton's method using Gaussian elimination that has the advantage of superior convergence characteristics. Sato and Tinney (1963) introduced the concept of optimally ordered elimination for the solution of large, sparse systems and showed such methods to be very efficient. Later in 1967, Tinney and Hart showed that by the use of optimally ordered Gaussian elimination and special programming techniques, both the storage requirement and computing speed are drastically reduced by Newton's method. For systems having 200 buses or more this method is inevitably used. Newton–Raphson method with quadratic convergence characteristics has become popular for general purpose power flow studies for most electric utilities.

In 1963, Brown et al. developed a method which employed the system bus impedance matrix Z_{bus} for power flow solution. This method of power flow solution has excellent convergence characteristics, is not sensitive to initial values of the voltage profile, and can process negative impedance (series compensation of lines). However, this method suffers from the disadvantage that the bus impedance matrix Z_{bus} is a full matrix and requires very large computer memory. This barrier was overcome by applying diakoptics. However, this barrier has now become insignificant as the present-day computers have very large storage capacity.

In 1972, B. Stott suggested approximations in Newton–Raphson method by decoupling MW-frequency and MVAR-voltage loops in the power system. A simplified version of *decoupled method* known as the *fast decoupled method*, with faster convergence characteristics, was suggested by B. Stott and O. Alsac in 1974. Subsequently, more work on the fast decoupled method, for faster convergence, was reported. In this chapter, only the principles underlying these methods will be discussed.

7.3.3 Iterative Computation of PFE

Numerical solutions of the power flow equations (PFE), which are simultaneous non-linear algebraic equations, necessitate the method of systematic initial guessing of the unknown variables. In power-flow studies, the solution is approached in the following manner:

1. An initial solution, $V_i^{(0)}$, is guessed.
2. This solution is used in conjunction with Eq. (7.10) to compute a new and better first solution, $V_i^{(1)}$.
3. The first solution is used for finding a second one, and so on.

This repetitive process of converging on the solution is referred to as an iterative method. The different methods of power flow solutions utilize somewhat different schemes in computing the new estimates. The term algorithm means a list of computer instructions specifying the sequence of operations used. The quality of the algorithm must be judged by the speed of convergence. Generally, an increase in the speed of convergence can be obtained by paying a price in terms of algorithm complexity.

Bus Mismatches and Convergence Criteria

The two popular convergence criteria are (i) voltage change magnitude criteria and (ii) bus mismatch power criteria.

Voltage change magnitude criteria   In iterative processes of a power-flow computation, one would stop the computations when all bus voltages in the $(k+1)$th iteration are not different from those in the kth iteration by more than a specified value of ε, the convergence tolerance limit. Typical magnitudes of ε range from 10^{-2} to 10^{-4} pu. The computation is stopped when the difference in magnitudes of change in voltage in two subsequent iterations of all the buses is less than ε, that is,

$$\left| V_i^{(k+1)} - V_i^k \right| < \varepsilon \qquad \text{for } i = 1, 2, \ldots, n \text{ and } i \neq s \qquad (7.24a)$$

and, in the presence of voltage-controlled buses, the following tolerance condition is met

$$\left| V_i^{(k+1)} - V_i^k \right| \leq \varepsilon \qquad \text{for all load buses} \qquad (7.24b)$$

where s indicates the slack bus number.

Bus mismatch power criteria   At bus i, the complex power mismatch ΔS_i is the difference between the scheduled and calculated values of complex power at the bus. In the kth iteration, the complex power mismatch may be written using Eqs (7.6) and (7.21) as

$$\Delta S_i^k = S_i^{\text{sp}} - S_i^k = P_i^{\text{sp}} + jQ_i^{\text{sp}} - \left(P_i^k + jQ_i^k\right) \tag{7.25}$$

Separating the real and imaginary parts in Eq. (7.25), the active and reactive power mismatches may be written as

$$\Delta P_i^k = P_i^{\text{sp}} - P_i^k \tag{7.26}$$

$$\Delta Q_i^k = Q_i^{\text{sp}} - Q_i^k \tag{7.27}$$

The most common convergence criteria used in practice are

$$\left|\Delta P_i^k\right| \leq \varepsilon \text{ for all } P\text{-}Q \text{ and } P\text{-}V \text{ buses} \tag{7.28a}$$

$$\left|\Delta Q_i^k\right| \leq \varepsilon \text{ for all } P\text{-}Q \tag{7.28b}$$

Typical magnitudes of ε range from 0.01 to 10 MW or MVAR as the case may be.

7.3.4 Computational Procedure for Different Bus Types

The computations to be performed at a certain bus i will depend upon the bus type as follows:

Slack bus For this bus both $|V_s|$ and $\angle\delta_s$ are known. Thus, no computations are needed, and at every bus scan, computation for the bus s (slack bus) can be skipped.

Load bus For this type of bus both $|V_i|$ and $\angle\delta_i$ are unknown. Thus computations must be performed which yields upgraded values for both voltage magnitude and phase angle.

Voltage-controlled bus Here the magnitude $|V_i|$ is known but $\Delta\delta_i$ must be computed for the upgraded value. Also, as Q_{Gi} is unknown for this type of bus, Q_{Gi} is computed in each iteration cycle in order to check that it lies within the capability range of the reactive generation source at the bus in question. Should it fall outside the range then, from that iteration onwards, Q_{Gi} is assumed to be equal to the limit value. With Q_i thus fixed, the bus in effect becomes a load bus. The formula for computing Q_{Gi} is obtained from Eq. (7.17) in rectangular coordinates as follows

$$Q_{Gi} = Q_{Li} + Q_i = Q_{Li} + e_i \sum_{k=1}^{n} (B_{ik}e_k + G_{ik}f_k) - f_i \sum_{k=1}^{n} (G_{ik}e_k - B_{ik}f_k) \tag{7.29a}$$

In polar coordinates, Q_{Gi} is obtained from Eq. (7.20) as

$$Q_{Gi} = Q_{Li} + Q_i = Q_{Li} - |V_i| \sum_{k=1}^{n} |Y_{ik}||V_k| \sin(\theta_{ik} - \delta_i + \delta_k) \tag{7.29b}$$

7.3.5 Computation of Slack Bus Power

Since all the voltages are known, the injected power at bus s, S_s, may be computed using Eq. (7.10) as

$$P_s - jQ_s = \sum_{k=1}^{n} Y_{ik} V_k V_s^* \tag{7.30}$$

Thus P_{Gs} and Q_{Gs} which are unknowns at this bus can be determined from

$$P_{Gs} = P_s + P_{Ls} \tag{7.31}$$

$$Q_{Gs} - Q_s + Q_{Ls} \tag{7.32}$$

7.3.6 Computation of Line Flows

The computation of the power flows on the various transmission lines of the network is the last step in power flow analysis. Figure 7.2 shows the equivalent π-model of the line connecting the two buses i and j of the system network. The line can be represented by the total series admittance Y and $Y_{sh}/2$ connected to bus i and bus m, where Y_{sh} is the total shunt admittance of the line. The line current I_{im}, measured at bus i and defined positive in the direction i to m, may be computed from Fig. 7.2 as

$$I_{im} = (V_i - V_m)Y + V_i Y_{sh}/2 \tag{7.33}$$

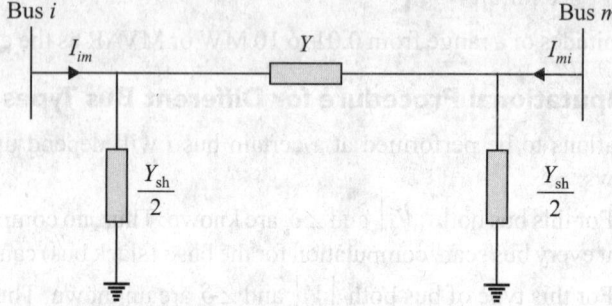

Bus i Bus m

Fig. 7.2   Transmission line model for line flow computations

Then the line powers S_{im} and S_{mi} measured at buses i and m respectively and both defined positive into the line are now

$$S_{im} = P_{im} + jQ_{im} = V_i I_{im}^* = V_i(V_i^* - V_m^*)Y^* + |V_i|^2 Y_{sh}^*/2 \tag{7.34}$$

$$S_{mi} = P_{mi} + jQ_{mi} = V_m I_{mi}^* = V_m(V_m^* - V_i^*)Y^* + |V_m|^2 Y_{sh}^*/2 \tag{7.35}$$

The algebraic sum of Eqs (7.34) and (7.35) determines the losses in the line. This completes the power flow study.

7.4 GAUSS–SEIDEL ITERATIVE TECHNIQUE

(G–S) This is one of the simplest iterative methods known. It is suitable for power flow study of small power systems where program simplicity is more important than computing costs.

The solutions of the bus voltage equations, both in the Y_{bus} and Z_{bus} frame of reference, with the indirect methods, namely Gauss iterative method and Gauss–Seidel iterative technique, have been discussed in Section 6.5.1, and with the direct method such as Gauss elimination technique has been discussed in Section 6.5.2. In this section, simulation of the load buses and voltage-controlled buses in the algorithms for solving the bus voltage equations by Gauss–Seidel method is explained.

7.4.1 Gauss–Seidel Method Applied to Load Buses

For an n-bus system, Eq. (6.28) of Section 6.5.1 (Gauss–Seidel iterative method) gives the computation scheme for calculating the bus voltages, which is reproduced below for ease of reference

$$V_i^{k+1} = \frac{1}{Y_{ii}} \left\{ I_i^k - \sum_{j=1}^{i-1} Y_{ij} V_j^{k+1} - \sum_{j=i+1}^{n} Y_{ij} V_j^k \right\}$$

$$\text{for } i = 1, 2, 3, ..., n \text{ and } i \neq s \qquad (7.36)$$

where the superscript k is the iteration count.

For each load bus i, the bus current I_i^k in Eq. (7.36) is computed from the complex power S_i injected into the bus, which can be determined from the complex power of the generator S_{Gi} and complex power S_{Li} of the load of the bus i, available as system data. If no generator is connected to the bus i, then $S_{Gi} = 0$. Hence $S_i = S_{Gi} - S_{Li} = -S_{Li}$. The bus current I_i^k can be computed as

$$I_i^k = \frac{S_i^*}{\left(V_i^k\right)^*} = \frac{P_i - jQ_i}{\left(V_i^k\right)^*} \qquad (7.37)$$

Substituting the value of I_i^k from Eq. (7.37) into Eq. (7.36) yields

$$V_i^{k+1} = \frac{1}{Y_{ii}} \left\{ \frac{P_i - jQ_i}{\left(V_i^k\right)^*} - \sum_{m=1}^{i-1} Y_{im} V_m^{k+1} - \sum_{m=i+1}^{n} Y_{im} V_m^k \right\}$$

$$i = 1, 2, 3, ..., n \text{ and } i \neq s \qquad (7.38)$$

At the end of the iteration, the node voltages are checked to ascertain whether the tolerance condition is met, that is,

$$\left| V_i^{k+1} - V_i^k \right| \leq \varepsilon \qquad i = 1, 2, 3, ..., n \text{ and } i \neq s$$

7.4.2 Gauss–Seidel Method Applied to Voltage-controlled Buses

In case of a voltage-controlled bus, real power and voltage magnitude are known, while reactive power Q_i is not known. With the help of Eq. (7.17) or Eq. (7.20), Q_i^{k+1} is computed by using the bus voltage estimates at the kth iteration, and in rectangular coordinates the reactive power is computed as

$$Q_i^{k+1} = +e_i^k \sum_{m=1}^{n} \left(B_{im} e_m^k + G_{im} f_m^k \right) - f_i^k \sum_{m=1}^{n} \left(G_{im} e_m^k - B_{im} f_m^k \right) \qquad (7.39a)$$

while in polar form the reactive power is computed as

$$Q_i^{k+1} = -\left| V_i^k \right| \sum_{m=1}^{n} \left| Y_{im} \right| \left| V_m^k \right| \sin \left(\theta_{im} - \delta_i + \delta_m \right) \qquad (7.39b)$$

On the right hand side of Eqs 7.39(a) and (b) the latest available values of the bus voltages are used.

Since the voltage at the bus i is to be maintained at $|V_i|^{\text{sp}}$, being a voltage-controlled bus, only the imaginary part of the bus voltage V_i^{k+1} is retained, while the real part is adjusted to satisfy this condition

$$\left(e_i^{k+1} \right)^2 + \left(f_i^{k+1} \right)^2 = \left(|V_i|^{\text{sp}} \right)^2 . \qquad (7.40)$$

or $\qquad e_i^{k+1} = \sqrt{\left(|V_i|^{\text{sp}} \right)^2 - \left(f_i^{k+1} \right)^2} \qquad (7.41)$

where e_i^{k+1} and f_i^{k+1} are the real and imaginary components of the voltage V_i^{k+1}.

At a voltage-controlled bus i, usually the maximum reactive power limit $Q_{i,\,max}$ and the minimum reactive power limit $Q_{i,\,min}$ are specified. Therefore, if the computed reactive power Q_i^{k+1} falls within the specified values of $Q_{i,\,max}$ and $Q_{i,\,min}$, that is, $Q_{i,\,max} \geq Q_i^{k+1} \geq Q_{i,\,min}$ then the voltage magnitude of bus i can be maintained constant. If this condition is violated, that is, either $Q_i^{k+1} < Q_{i,\,min}$ or $Q_i^{k+1} > Q_{i,\,max}$, then the bus voltage cannot be maintained constant, and the voltage-controlled bus i is treated like a P-Q bus or a load bus. If $Q_i^{k+1} > Q_{i,\,max}$, then Q_i^{k+1} is set equal to $Q_{i,\,max}$, and if $Q_i^{k+1} < Q_{i,\,min}$, then Q_i^{k+1} is set equal to $Q_{i,\,min}$. In all these cases, V_i^{k+1} is computed from Eq. (7.38) using this value of Q_i^{k+1}. There is no adjustment to be made for e_i^{k+1} in this case. If in subsequent computation Q_i^{k+1}, does fall within the available reactive power at that bus, then the bus is switched back to a P-V bus.

The new voltage estimates so obtained immediately take the place of the previous voltage values and the process is continued till all the bus voltages are calculated. At the end of the iteration, the bus voltages for all load buses are checked whether the tolerance condition is met, that is,

$$\left| V_i^{k+1} - V_i^k \right| \leq \varepsilon \text{ for all load buses}$$

If these voltage values are within the specified tolerance limit ε, for each bus, a solution to the bus voltage Eq. (7.39) is obtained and the process of iteration is stopped. On the other hand, if even any one of the bus voltage values, between two successive iterations, is not within the specified tolerance limit ε, the entire process of iteration is repeated for all the bus voltages, till all the bus voltage values have converged to within the specified tolerance limit.

7.4.3 Acceleration of Convergence

The rate of convergence of Gauss–Siedel method can be increased by applying acceleration factor as already discussed under Section 6.5.1 (Accelaration factor). After every iteration a correction factor is applied to each load bus voltage as follows:

$$\Delta V_i^{k+1} = \alpha \left(V_i^{k+1} - V_i^k \right) \tag{7.42}$$

and V_i^{k+1} becomes

$$V_i^{k+1} = V_i^k + \Delta V_i^{k+1} \tag{7.43}$$

with α being the acceleration factor. Its value depends upon the system. For typical systems, the range of 1.3 to 1.7 is found to be satisfactory.

Example 7.1   For the power system network shown in Fig. 7.3, compute the bus voltages using the Gauss–Seidel iteration method. Line reactance and loads are shown in the figure. Bus 1 is the slack bus and buses 2 and 3 are the load and voltage-control buses, respectively. Assume tolerance equal to 0.00001.

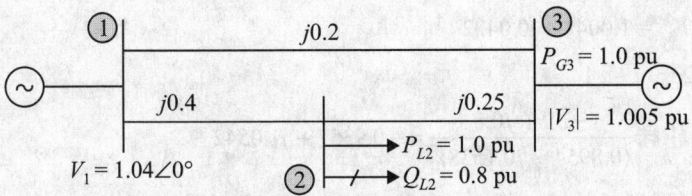

Fig. 7.3 Three-bus system for Example 7.1

Solution The Y_{bus} matrix of the system can be assembled by inspection and is given as

$$Y_{bus} = j \begin{bmatrix} -7.5 & 2.5 & 5.0 \\ 2.5 & -6.5 & 4.0 \\ 5.0 & 4.0 & -9.0 \end{bmatrix}$$

Iteration 1:

Assume an initial estimate of voltage equal to $(1.0 + j0.0)$ on the load bus 2. Using Eq. (7.37),

$$I_2^0 = \frac{-1.0 + j0.8}{1.0 - j0} = -1.0 + j0.8$$

From Eq. (7.36),

$$V_2^1 = \frac{I_2^0 - (Y_{21}V_1 + Y_{23}V_3^0)}{Y_{22}}$$

$$= \frac{-1.0 + j0.8 - (j2.5 \times 1.04 + j4.0 \times 1.005)}{(-j6.5)}$$

$$= 0.8954 - j0.1538$$

Since bus 3 is a voltage-controlled bus, real power P and magnitude of the bus voltage are specified. Using Eq. (7.18), reactive power Q_3 at the bus is computed as follows

$$Q_3^1 = \text{Im}(V_3^0 \times I_3^*) = \text{Im}\left[1.005 \times \left(Y_{31}V_1^0 + Y_{32}V_2^1 + Y_{33}V_3^0\right)^*\right]$$

$$= \text{Im}[(1.005 + j0) \times \{j5.0 \times (1.04 + j0)$$

$$+ j4.0 \times (0.8954 - j0.1538) + (-j9.0) \times (1.005 + j0)\}^*]$$

$$= 0.2647$$

$$I_3^0 = \frac{S_3^*}{\left(V_3^0\right)^*} = \frac{(1.0 - j0.2647)}{\left(V_3^0\right)^*} = \frac{(1.0 - j0.2647)}{(1.005 - j0)}$$

$$= 0.9950 - j0.2635$$

$$V_3^1 = \frac{\left[I_3^* - \left(Y_{31}V_1^1 + Y_{32}V_2^1\right)\right]}{Y_{33}}$$

$$= \frac{(0.9950 - j0.2635) - \left[j5.0 \times (1.04 + j0) + j4.0(0.8954 - j0.1538)\right]}{(-j9.0)}$$

$$= 1.0050 + j0.0422$$

Since the voltage magnitude at bus 3 is held constant, the real part of V_3^1 is modified as per Eq. (7.41)

$$e_3^1 = \sqrt{(1.005)^2 - (0.0422)^2} = 1.0041$$

Hence, $V_3^1 = 1.0041 + j0.0422$

Iteration 2:

$$I_2^1 = \frac{-1.0 + j0.8}{(0.8954 - j0.1538)^*} = -0.9357 + j1.0542$$

From Eq. (7.36)

$$V_2^2 = \frac{I_2^1 - (Y_{21}V_1 + Y_{23}V_3^1)}{Y_{22}}$$

$$= \frac{-1.0 + j0.8 - [j2.5 \times 1.04 + j4.0 \times (1.0041 + j0.0422)]}{(-j6.5)}$$

$$= 0.8557 - j0.1180$$

$$Q_3^2 = \text{Im}(V_3^1 I_3^*)$$

$$= \text{Im}\left[(1.0041 + j0.0422) \times (Y_{31}V_1 + Y_{32}V_2^2 + Y_{33}V_3^1)^*\right]$$

$$= \text{Im}\,[(1.0041 + j0.0422) \times \{j0.5 \times (1.04 + j0) + j4.0$$
$$\times (0.8557 - j0.1180) + (-j9.0) \times (1.0041 + j0.0422)\}]$$

$$= -0.38$$

$$I_3^1 = \frac{1 + j0.38}{V_3^{1*}} = \frac{1 + j0.38}{1.0041 - j0.0422} = 0.9783 + j0.4197$$

$$V_3^2 = \frac{I_3^1 - (Y_{31}V_1 + Y_{32}V_2^2)}{Y_{33}}$$

$$= \frac{(0.9783 + j0.4197) - [j0.5(1.04 + j0) + j4.0(0.8557 - j0.1180)]}{-j9.0}$$

$$= 0.9114 + j0.0563$$

In order to keep the voltage magnitude at bus 3 constant at 1.005, the real part of V_3^2 is re-computed as follows:

$$\text{Re}\left(V_3^2\right) = \sqrt{(1.005)^2 - (0.0563)^2} = 1.0034$$

Hence,

$$V_3^2 = 1.0034 + j0.0563$$

The process of iteration is repeated till the voltages converge to within the specified tolerance which in the present case is 0.00001. The voltages at buses 2 and 3 are as follows:

Iteration	V_1	V_2	V_3
1	$1.0400 + j0.0000$	$0.8954 - j0.1538$	$1.0041 + j0.0422$
2	$1.0400 + j0.0000$	$0.8557 - j0.1180$	$1.0034 + j0.0563$
3	$1.0400 + j0.0000$	$0.8519 - j0.1200$	$1.0032 + j0.0602$
4	$1.0400 + j0.0000$	$0.8507 - j0.1201$	$1.0032 + j0.0602$
5	$1.0400 + j0.0000$	$0.8505 - j0.1202$	$1.0032 + j0.0602$

7.4.4 MATLAB Program for Power Flow Studies by Gauss–Seidel Method Using Y_{bus}

The MATLAB program function pfsg for computing the bus voltages by the Gauss–Seidel iteration method is as follows.

```
function [Ybus,VN]=pfsg(nbus,nlns,ld,pfbd,accrucy);
%Program for performing power flow studies
%Part-I: Assembly of Ybus from line data
Ybus = zeros(nbus);
VP(nbus,1)=0;
VN(nbus,1)=0;
for k=1:nlns;
  m=ld(k,1);
  n=ld(k,2);
  Ybus(m,m)=Ybus(m,m)+1/ld(k,3)+ld(k,4);
  Ybus(n,n)=Ybus(n,n)+1/ld(k,3)+ld(k,4);
  Ybus(m,n)=-1/ld(k,3);
  Ybus(n,m)=Ybus(m,n);
end
Ybus
itr=0;
count=0;
%Gauss–Seidel iteration method
VP(1)=input('slack bus voltage');
for k=2:nbus;
  if pfbd(k,1) == 0;
    VP(k)=VP(1);
  else
    VP(k)=pfbd(k,3);
  end
end
while count <= nbus-1;
  sum=0;
  itr=itr+1;
  for k=2:nbus;
    if pfbd(k,1) == 0;
      I(k)=(pfbd(k,2)-pfbd(k,4)+i*(pfbd(k,3)-pfbd(k,5)))/(VP(k))';
    else
      sum=0;
      for m=1:nbus;
        sum=sum+Ybus(k,m)*VP(m);
      end
      q(k)=imag(VP(k)*(sum)');
      I(k)=((pfbd(k,2)+i*q(k))/(VP(k)))';
end
sum=0;
```

```
for m=1:nbus;
    if m ~= k;
        sum=sum+Ybus(k,m)*VP(m);
    else
    end
end
VN(k)=(I(k)-sum)/Ybus(k,k);
if abs(real(VN(k))-real(VP(k))) <= accrucy;
    count=count+0.5;
else
end
if abs(imag(VN(k))-imag(VP(k))) <= accrucy;
    count=count+0.5;
else
end
if pfbd(k,1) == 0;
VP(k)= VN(k);
        else
            z=imag(VN(k));
            zz=(pfbd(k,3)^2-(z^2))^0.5;
            VN(k)=zz+i*z;
            VP(k)=VN(k);
        end
    end
end
fprintf('iteration  voltage  voltage\n')
for k=1:nbus;
    fprintf('%6.2f\',itr);
    fprintf('%13.4f\',real(VP(k)));
    fprintf('%15.4f\n',imag(VP(k)));
end
```

The inputs to the program function pfsg are

nbus	Number of buses in the power system
nlns	Number of lines
ld	Line data. The line data is in the form of a matrix whose dimensions are (nlns × 4). The first two columns in the ld matrix are the bus numbers between which the line is connected, the third column shows the complex impedance of the line and the fourth column provides one half of the line susceptance. The complex impedance and susceptance pertaining to the lines are shown in pu.
pfbd	pfbd matrix includes data related to the loads on each bus and the type of buses, that is, slack, load, or voltage-controlled bus. The dimensions of the matrix are (nbus × 5). Column 1 shows the type of bus, that is, 1 for slack bus, 0 for load bus, and 2 for a voltage-controlled bus. For a slack bus, columns 2 to 4 are zero. For a load bus, columns 2 and 3 indicate real and reactive generated power while columns 4 and 5 show real and reactive load. Lagging reactive load is shown as

negative. For a voltage-controlled bus, column 2 gives the real power, column 3 shows the specified voltage magnitude, and columns 4 and 5 indicate the specified maximum and minimum Q, respectively.

accrucy   Tolerance limit

7.5 GAUSS ELIMINATION (TRIANGULAR FACTORIZATION) METHOD

The method of triangular factorization (LU decomposition) of the Y_{bus} matrix and obtaining the bus (node) voltage solution of the equations by forward and backward substitution has been described in Section 6.5.2 under Solving nodal equations by forward and backward substitution. For obtaining a bus-by-bus solution of the power flow equations, represented by Eq. (7.3), the following algorithm can be used.

Based on the initial estimates of the bus voltages, the initial magnitudes of the injected bus currents are estimated as follows:

(i) When the shunt elements to ground have been included in the self admittance of the Y_{bus} matrix

$$I_i^k = \frac{(P_i + jQ_i)^*}{(V_i^k)^*} \qquad i = 2, 3, 4, \dots, n \qquad (7.44a)$$

(ii) When the shunt elements to ground have not been including in the self-admittance of the Y_{bus} matrix

$$I_i^k = \frac{(P_i + jQ_i)^*}{(V_i^k)^*} - Y_i(V_i^k)^* \qquad i = 2, 3, 4, \dots, n \qquad (7.44b)$$

The bus-by-bus forward substitution is performed as follows

$$V_i^{k+1} = \frac{i}{l_{ii}}\left(I_i^k - \sum_{m=2}^{i-1} l_{im}V_m^{k+1} - \sum_{m=i+1}^{n} l_{im}V_m^k \right)$$
$$i = 2, 3, 4, \dots, n \qquad (7.45)$$

It may be noted that in describing the algorithm for forward substitution, bus 1 has been taken as the slack bus. Further, it may also be observed that when performing the forward substitution, latest value of the bus voltage is used, thereby, implementing the Gauss–Seidel iteration method.

The algorithm for performing the backward substitution follows the following sequence.

$$V_n = V_n^{k+1} \qquad (7.46a)$$

$$V_i = V_i - \sum_{m=i+1}^{n} u_{im}V_m \qquad i = n-1, n-2, n-3, \dots, 2 \qquad (7.46b)$$

Once the bus voltage equations are solved by performing forward substitution [Eq. (7.45)] and backward substitution [Eqs (7.46)], the test for convergence is applied and the process of iteration is halted when all the bus voltages are within the specified tolerance ε.

In the Gauss elimination method also, the treatment of load and voltage-controlled buses is the same as for the Gauss–Seidel iteration method.

For bus voltage values, the range of tolerance limit ε within 0.00001 to 0.00005, which leads to acceptable power mismatch, is found to provide satisfactory accuracy. In actual power flow studies, mismatch in real bus power ΔP and reactive bus power ΔQ is specified as tolerance for indicating convergence of the bus voltages. The tolerance limits both for real and reactive bus powers are specified as 0.01 pu.

7.6 POWER FLOW SOLUTION USING Z_{bus} MATRIX

Power flow programs usually always employ the Y_{bus} matrix. In the method using Y_{bus}, it was possible to ignore the equation for slack bus as the voltage at that bus was specified. In the formulation of power flow problem using Z_{bus}, the slack bus currents appear in all the equations which is not known a priori since the slack bus power is not specified.

7.6.1 Power Flow Equations in Z_{bus} Frame of Reference

In the formulation of the power flow equations in the Z_{bus} frame of reference, Z_{bus} is assembled taking the slack bus as the reference bus. Consequently all bus loads and shunt elements, such as line charging capacitance and π-equivalent circuit representation of transformers are converted into current sources. As such there are no shunt connections between a bus and the ground. Figure 7.4 shows an n-bus system, where the slack bus is the reference bus.

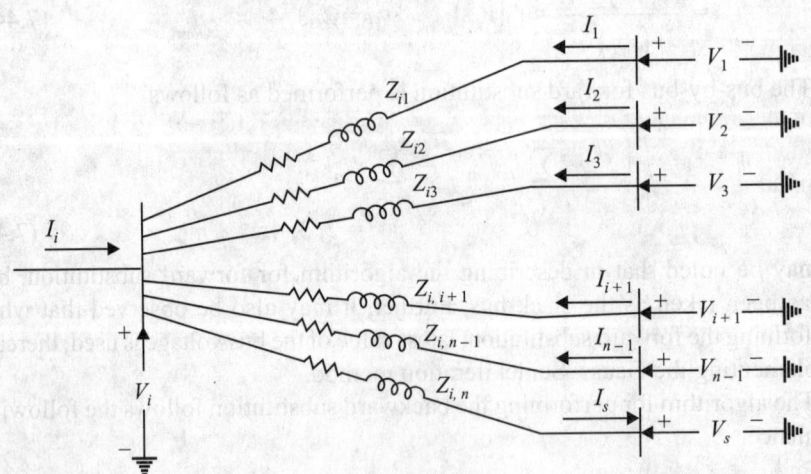

Fig. 7.4 Z_{bus} representation of an n-bus power system with the slack bus as reference bus

It may be noted that since the slack bus is the reference bus, each bus has a driving point impedance with respect to the reference bus and mutual impedances as shown in the figure.

The general form of bus voltage Eq. (7.2), with the slack bus, as the reference bus, gets modified as under

$$V_{\text{bus}} - V_s = Z_{\text{bus}} I_{\text{bus}} \tag{7.47}$$

where V_s is a column vector of $(n-1)$ slack bus voltage values.

For an n-bus power system, with the slack bus as the reference bus, there would be $(n-1)$-bus voltage equations and Eq. (7.47) is written as

$$\begin{bmatrix} V_1 - V_s \\ V_2 - V_s \\ \vdots \\ V_{n-1} - V_s \end{bmatrix} = Z_{\text{bus}} I_{\text{bus}} \tag{7.48}$$

It is to be observed that since the bus impedance Z_{bus} matrix is formulated with the slack bus as the reference bus, there will be no connection between any bus and the ground.

From the assumed directions of flow of currents in Fig. 7.4, it may be noted that the slack bus current is the negative of the sum of injected currents in all $(n-1)$ buses, that is,

$$I_s = -\sum_{i=1}^{n-1} I_i \tag{7.49}$$

The injected bus currents can be computed from the injected bus powers. Therefore, using Eq. (7.7)

$$I_i = \frac{(P_i + jQ_i)^*}{(V_i)^*} \tag{7.50}$$

Equations (7.48) and (7.50) are the bus voltage equations in the Z_{bus} frame of reference and can be used to perform power flow studies when there are no shunt elements connected between the bus and the ground. In order to reflect the effect of shunt elements, when present between a bus i and the ground, Eq. (7.50) for computing the injected currents gets modified as under

$$I_i = \frac{(P_i + jQ_i)^*}{(V_i)^*} - Y_i E_i \tag{7.51}$$

In the latter case, when shunt elements have to be accounted for, Eq. (7.48) is used in conjunction with Eq. (7.51) for performing power flow studies.

Solutions to the power flow equations for performing power flow studies of electrical power systems can be obtained with the help of appropriate numerical techniques.

In the Z_{bus} frame of reference, the bus voltage equations represented by Eq. (7.48) are employed to obtain power flow solutions of systems by the Gauss–Seidel iterative method. The voltage at any bus i, in a system having n buses, at the $(k+1)$th iteration is obtained from Eq. (7.48) as under

$$\left[V_i^{k+1} - V_s \right] = \sum_{m=1}^{i-1} Z_{im} I_i^{k+1} + \sum_{m=i+1}^{n-1} Z_{im} I_i^k \tag{7.52}$$

where

$$I_i^k = \frac{(P_i + jQ_i)^*}{(V_i^k)^*}$$

when there are no shunt elements. When shunt elements are present

$$I_i^k = \frac{(P_i + jQ_i)^*}{(V_i^k)^*} - Y_i (V_i^k)^*$$

7.6.2 Inclusion of Voltage-controlled Buses

The method to include voltage-controlled buses in the computation scheme is similar to that in Gauss–Siedel method using Y_{bus}. While computing bus current, Q_i^k must replace Q_i^{sp}. Bus voltages are computed using this value of Q_i^k, then the real and imaginary parts of the bus voltage are adjusted so that the magnitude satisfies the voltage magnitude constraint while the voltage angle is kept unchanged. The limits of reactive power Q are taken into account by changing a bus from P-V to P-Q bus and vice versa, as the situation arises.

Example 7.2   Figure 7.5 shows a hypothetical power system in which bus 3 is the slack bus. With the slack bus as the reference, assemble the Z_{bus} matrix. Use the bus voltage equations in the Z_{bus} frame of reference to compute voltages at bus 2 and 3. The load on each bus is shown in the figure. Assume slack bus voltage equal to $(1.05 + j0)$ pu and tolerance equal to 0.00001.

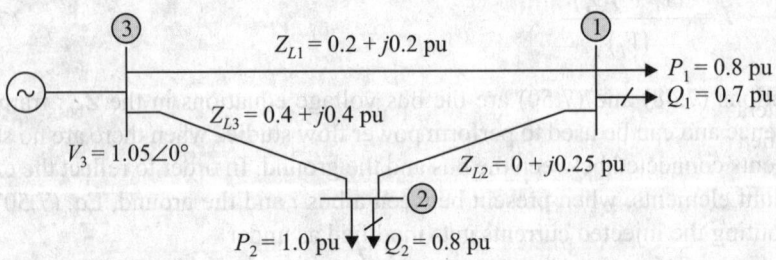

Fig. 7.5   Power system for Example 7.2

Solution   The Z_{bus} matrix for the system is

$$\begin{bmatrix} 0.1372 + j0.1557 & 0.1256 + j0.0887 \\ 0.1256 + j0.0887 & 0.1487 + j0.2226 \end{bmatrix}$$

Assume that the initial voltages at bus 2 and 3 are each equal to the slack bus voltage, that is, $V_1^0 = V_2^0 = (1.05 + j0)$ pu.

Iteration 1: Bus currents are obtained by employing Eq. (7.50)

$$I_1^1 = \frac{-0.8 - j0.7}{1.05 - j0} = -0.7619 - j0.6667$$

$$I_2^1 = \frac{-1.0 - j0.8}{1.05 - j0} = -0.9524 - j0.7619$$

Equation (7.52) is used to calculate the new estimates of voltages at bus 1 and 2 as follows:

$$\begin{bmatrix} V_1^1 - 1.05 \\ V_2^1 - 1.05 \end{bmatrix}$$

$$= \begin{bmatrix} 0.1372 + j0.1557 & 0.1256 + j0.0887 \\ 0.1256 + j0.0887 & 0.1487 + j0.226 \end{bmatrix} \begin{bmatrix} -0.7619 - j0.6667 \\ -0.9524 - j0.7619 \end{bmatrix}$$

$$V_1^1 = 0.99717 - j0.39023$$

$$V_2^1 = 1.04138 - j0.47667$$

Iteration 2: The bus currents are now computed with the new estimates of voltages at bus 1 and 2. Thus,

$$I_1^1 = \frac{-0.8 - j0.7}{0.99717 - j0.39023} = -0.9340 - j0.3365$$

$$I_2^1 = \frac{-1.0 - j0.8}{1.04138 - j0.47667} = -1.0846 - j0.2717$$

$$\begin{bmatrix} V_1^2 - 1.05 \\ V_2^2 - 1.05 \end{bmatrix}$$

$$= \begin{bmatrix} 0.1372 + j0.1557 & 0.1256 + j0.0887 \\ 0.1256 + j0.0887 & 0.1487 + j0.2226 \end{bmatrix} \begin{bmatrix} -0.9340 - j0.3365 \\ -1.0846 - j0.2717 \end{bmatrix}$$

$$V_1^2 = 0.86208 - j0.32187$$

$$V_2^1 = 0.86168 - j0.40669$$

The iterative process is repeated till two consecutive voltages at each bus are within the specified tolerance limit. The path of convergence of the voltages at bus 1 and 2 is shown here.

Iteration	Bus	Voltage
3	1	$0.82869 - j0.37924$
3	2	$0.82525 - j0.48446$
3	3	$1.05000 + j0.00000$
4	1	$0.79913 - j0.36127$
4	2	$0.78485 - j0.46439$
4	3	$1.05000 + j0.00000$
5	1	$0.78833 - j0.37680$
5	2	$0.77241 - j0.48584$
5	3	$1.05000 + j0.00000$
-	-	–
-	-	–
-	-	–

(*Contd*)

(Contd)

Iteration	Bus	Voltage
19	1	$0.76859 - j0.37563$
19	2	$0.74634 - j0.48642$
19	3	$1.05000 + j0.00000$
20	1	$0.76858 - j0.37563$
20	2	$0.74633 - j0.48641$
20	3	$1.05000 + j0.00000$

7.6.3 MATLAB Program for Power Flow Studies by Gauss Method using Z_{bus}

The MATLAB program function pfsz computes the bus voltages by the Gauss iterative method in the Z_{bus} frame of reference. The Z_{bus} matrix for the system employs the MATLAB program function zeebus described in Section 6.9.

```
function [VP,VN]=pfsz(Zbus,nbus,pfbd,accrucy);
%Program for performing power flow studies using Zbus
%Developed by the authors
%Part-I: output from the function zeebus gives Zbus
VP=zeros(nbus,1);
VN=zeros(nbus,1);
I=zeros(nbus,1);
itr=0;
count=0;
%Part-II Gauss iteration method
VP(1)=input('slack bus voltage');
for k=2:nbus;
   VP(k)=VP(1);
end
while count <= nbus-1;
   itr=itr+1
   for k=2:nbus;
     I(k)=(pfbd(k,2)-pfbd(k,4)+i*(pfbd(k,3)-pfbd(k,5)))/(VP(k))';
   end
   for k=2:nbus;
     sum=0;
     for m=2:nbus;
     sum=sum+Zbus(k-1,m-1)*I(m);
     end
     VN(k)=sum+VP(1);
     if abs(real(VN(k))-real(VP(k))) < accrucy;
        count=count+0.5;
     else
     end
```

```
    if abs(imag(VN(k))-imag(VP(k))) < accrucy;
        count=count+0.5;
    else
    end
    VP(k)=VN(k);
  end
  fprintf(' itr  Bus  Voltage  Voltage\n');
  for k=1:nbus;
    fprintf('%4i    \',itr);
    fprintf('%4i    \',k);
    fprintf('%15.5f\',real(VP(k)));
    fprintf('%15.5f\n',imag(VP(k)));
  end
end
```

The inputs [nbus, pfbd, accrucy] used for the program function pfsz are similar to the inputs used for the program function pfsg.

7.7 NEWTON–RAPHSON (N–R) METHOD

The Newton–Raphson method is the most widely used method for solving simultaneous, non-linear algebraic equations. Newton–Raphson method is an iterative procedure based on an initial estimate of the unknown variables and the use of Taylor's series expansion. Before extending the method to a system of non-linear algebraic equations, it is helpful to illustrate the solution of a non-linear equation in one variable and then generalize it for the *n*-dimensional case.

7.7.1 Newton–Raphson Method for Single-dimensional Case

Let the single-dimensional, non-linear equation be expressed by

$$f(x) = y \tag{7.53}$$

Let the initial guess be x^0, and Δx^0 is a small deviation from the correct solution.

Then $f(x^0 + \Delta x^0) = y$ (7.54)

The left hand side of Eq. (7.54) by Taylor's series expansion around operating point x^0 gives

$$f\left(x^0\right) + \Delta x^0 \, f'\left(x^0\right) + \frac{\left(\Delta x^0\right)^2}{2!} f''\left(x^0\right) + \cdots = y \tag{7.55}$$

where f' and f'' are the first and second derivatives respectively of f with respect to x. As the error Δx^0 is very small, the higher order terms can be neglected, retaining only the linear terms, Eq. (7.55) becomes

$$\Delta x^0 f'\left(x^0\right) = y - f\left(x^0\right) = \Delta y^0 \tag{7.56}$$

where $\Delta y^0 \approx y - f(x^0)$ (7.57)

Then $$\Delta x^0 = \frac{\Delta y^0}{f'\left(x^0\right)} = \frac{y - f\left(x^0\right)}{f'\left(x^0\right)} \tag{7.58}$$

Then an improved estimate x^1 is obtained by adding Δx^0 to the initial estimate x^0. Thus

$$x^1 = x^0 + \Delta x^0 = x^0 + \frac{\Delta y^0}{f'(x^0)} = x^0 + \frac{y - f(x^0)}{f'(x^0)} \tag{7.59}$$

Next $f(x)$ is expanded around x^1 and an improved estimate x^2 is obtained in a similar manner, and so on. Generalizing for kth iteration gives

$$\Delta y^k = y - f(x^k) \tag{7.60}$$

$$\Delta x^k = \frac{\Delta y^k}{f'(x^k)} = \frac{y - f(x^k)}{f'(x^k)} \tag{7.61}$$

$$x^{k+1} = x^k + \Delta x^k = x^k + \frac{y - f(x^k)}{f'(x^k)} \tag{7.62}$$

The iterative process is continued till the function $f(x)$ converges, within a specified tolerance, to the real root.

Example 7.3 The resistive network shown in Fig. 7.6 is supplying a load of 0.5 pu over a line with resistance 0.4 pu. If bus 1 is assumed to be the slack bus having a voltage of 1.0 pu, using the N–R iterative method, determine (i) voltage at load bus 2, (ii) current in line 1-2, (iii) the slack bus power, and (iv) power loss in the line.

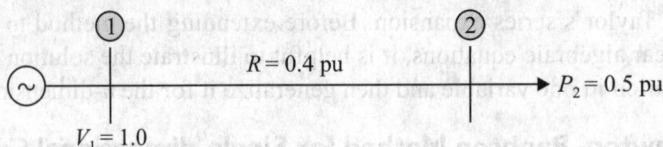

$$V_1 = 1.0$$

Fig. 7.6 Network for Example 7.3

Solution The power at bus 2 in terms of the bus voltages and the line resistance can be written as

$$V_2 \frac{(V_1 - V_2)}{r} = \frac{(V_1 V_2 - V_2^2)}{0.4} = 0.5$$

$$V_1 V_2 - V_2^2 = 0.2$$

The single-dimension, non-linear function in V, therefore, is written as

$$f(V) = V_2^2 - V_1 V_2 + 0.2 = 0$$

$$f'(V) = \frac{df(V)}{dV_2} = 2V_2 - V_1$$

The recursive or iterative formulae represented by Eq. (7.61) in this case are given by

$$\Delta V_2^k = -\frac{f(V^k)}{\dfrac{df(V^k)}{dV_2}} = -\frac{f(V)}{f'(V)}$$

$$V_2^{k+1} = V_2^k + \Delta V_2^k$$

(i) Taking $V_1 = 1.0$ and using the iterative formulae, V_2 is obtained as follows:

k	V_2^k	$f(V)$	$\dfrac{df(V)}{dV_2}$	ΔV_2^k	$V_2^{k+1} = V_2^k + \Delta V_2^k$
0	0.9	0.1100	0.8000	− 0.1375	0.7625
1	0.7625	0.0189	0.5250	− 0.0360	0.7265
2	0.7265	0.0013	0.4530	− 0.0029	0.7236
3	0.7236	8.1966×10^{-6}	0.4473	-1.8327×10^{-5}	0.7236

(ii) Line current $= \dfrac{V_1 - V_2}{r} = \dfrac{1.0 - 0.7236}{0.4} = 0.691 \, \text{pu}$

(iii) Slack bus power $= V_1 \times$ line current $= 0.691 \, \text{pu}$

(iv) Power loss in the line $= 0.691 - 0.50 = 0.191 \, \text{pu}$

7.7.2 Newton–Raphson Method for an *n*-dimensional Case

The single-dimensional, non-linear case can be extended to an *n*-dimensional, non-linear case. Let the non-linear equations be expressed in matrix form by

$$F(X) = Y \tag{7.63}$$

where $F(X)$ is a non-linear analytical function of the dependent *n*-vector X. The system of n equations expressed by Eq. (7.63) is of the form

$$f_i(x_1, x_2, \ldots, x_n) = y_i, \qquad i = 1, 2, \ldots, n \tag{7.64}$$

Let the initial estimate of the solution vector be

$$x_1^0, x_2^0, \ldots, x_n^0$$

and let it be assumed that the corrections $\Delta x_1, \Delta x_2, \ldots, \Delta x_n$, are required for $x_1^0, x_2^0, \ldots, x_n^0$, respectively, so that Eq. (7.58) is solved, that is,

$$f_i(x_1 + \Delta x_1, x_2 + \Delta x_2, \ldots, x_n + \Delta x_n) = y_i, \qquad i = 1, 2, \ldots, n \tag{7.65}$$

Expanding the left hand side of Eq. (7.65) by Taylor's series, about the initial estimates, and neglecting higher order terms, yields

$$f_i(x_1 + \Delta x_1, x_2 + \Delta x_2, \ldots, x_n + \Delta x_n)$$

$$= f_i\left(x_1^0, x_2^0, \ldots, x_n^0\right) + \Delta x_1^0 \left(\frac{\partial f_i}{\partial x_1}\right)^0 + \Delta x_2^0 \left(\frac{\partial f_i}{\partial x_2}\right)^0 + \ldots + \Delta x_i^0 \left(\frac{\partial f_i}{\partial x_n}\right)^0 = y_i$$

$$i = 1, 2, \ldots, n \tag{7.66}$$

In matrix form, Eq. (7.66) may be written as

$$
\begin{bmatrix}
y_1 - f_1\left(x_1^0, x_2^0, \ldots, x_n^0\right) \\[2ex]
y_2 - f_2\left(x_1^0, x_2^0, \ldots, x_n^0\right) \\[2ex]
\vdots \\[2ex]
y_n - f_n\left(x_1^0, x_2^0, \ldots, x_n^0\right)
\end{bmatrix}
=
\begin{bmatrix}
\left(\dfrac{\partial f_1}{\partial x_1}\right)^0 & \left(\dfrac{\partial f_1}{\partial x_2}\right)^0 & \cdots & \left(\dfrac{\partial f_1}{\partial x_n}\right)^0 \\[2ex]
\left(\dfrac{\partial f_2}{\partial x_1}\right)^0 & \left(\dfrac{\partial f_2}{\partial x_2}\right)^0 & \cdots & \left(\dfrac{\partial f_2}{\partial x_n}\right)^0 \\[2ex]
\vdots & \vdots & \cdots & \vdots \\[2ex]
\left(\dfrac{\partial f_n}{\partial x_1}\right)^0 & \left(\dfrac{\partial f_n}{\partial x_2}\right)^0 & \cdots & \left(\dfrac{\partial f_n}{\partial x_n}\right)^0
\end{bmatrix}
\begin{bmatrix}
\Delta x_1^0 \\[2ex]
\Delta x_2^0 \\[2ex]
\vdots \\[2ex]
\Delta x_n^0
\end{bmatrix}
$$

$$\tag{7.67}$$

In matrix form, Eq. (7.67) can also be written as

$$\Delta Y^0 = J^0 \, \Delta X^0 \tag{7.68}$$

where J^0 is the Jacobian for the functions f_i and ΔX^0 is the change vector Δx_i. The elements of the matrices ΔY^0 and J^0 are evaluated by substituting the current values of x_i's. Then the elements of column vector ΔX^0 can be obtained from

$$\Delta X^0 = [J^0]^{-1} \, \Delta Y^0 \tag{7.69}$$

and the new values of x_i's are computed from

$$x_i^1 = x_i^0 + \Delta x_i^0 \tag{7.70}$$

The process is repeated until two successive values of each x_i differ only by a specified tolerance. Then, similar to Eq. (7.62) for one-dimensional case, the Newton–Raphson iterates for an n-dimensional case are defined as

$$\Delta Y^k = J^k \, \Delta X^k \tag{7.71}$$

Then
$$\Delta X^k = [J^k]^{-1} \Delta Y^k \tag{7.72}$$

and
$$X^{k+1} = X^k + [J^k]^{-1} \Delta Y^k \tag{7.73}$$

where

$$\Delta Y^k = \begin{bmatrix} y_1 - f_1\left(x_1^k, x_2^k, \ldots, x_n^k\right) \\ y_2 - f_2\left(x_1^k, x_2^k, \ldots, x_n^k\right) \\ \vdots \\ y_n - f_n\left(x_1^k, x_2^k, \ldots, x_n^k\right) \end{bmatrix} \tag{7.74}$$

and

$$[J^k] = \begin{bmatrix} \left(\dfrac{\partial f_1}{\partial x_1}\right)^k & \left(\dfrac{\partial f_1}{\partial x_2}\right)^k & \cdots & \left(\dfrac{\partial f_1}{\partial x_n}\right)^k \\[2mm] \left(\dfrac{\partial f_2}{\partial x_1}\right)^k & \left(\dfrac{\partial f_2}{\partial x_2}\right)^k & \cdots & \left(\dfrac{\partial f_2}{\partial x_n}\right)^k \\[2mm] \vdots & \vdots & \cdots & \vdots \\[2mm] \left(\dfrac{\partial f_n}{\partial x_1}\right)^k & \left(\dfrac{\partial f_n}{\partial x_2}\right)^k & \cdots & \left(\dfrac{\partial f_n}{\partial x_n}\right)^k \end{bmatrix} \tag{7.75}$$

ΔX^k is a column vector of n independent variables and is equal to

$$\begin{bmatrix} \Delta x_1^k \\ \Delta x_2^k \\ \vdots \\ \Delta x_n^k \end{bmatrix} \tag{7.76}$$

It may be observed from Eq. (7.72) that the determination of ΔX^k requires the calculation of the inverse of the Jacobian matrix $[J^k]$, which not only entails cumbersome computations but is also time-consuming. However, MATLAB provides the inverse matrix operator (\) which facilitates the computation of inverse of the Jacobian matrix $[J^k]$. Thus,

$$\Delta X^k = [J^k] \backslash \Delta Y^k \qquad (7.77)$$

Hence, X^{k+1} can be determined using Eq. (7.73).

Equations (7.71), (7.72), and (7.73) provide the recursive formulae to solve a set of n non-linear equations by the N–R iterative method.

The Newton–Raphson method has a unique feature in possessing quadratic convergence characteristics. It has a drawback in that the initial guess must be close to the solution. However, for power systems this is not a serious drawback as an initial guess can always be made from operating experience.

Example 7.4 Figure 7.7 shows a three-bus resistive network. By employing the N–R iterative method in the Y_{bus} frame of reference compute (i) bus voltages, (ii) line currents, (iii) slack bus power, and (iv) total losses in the system. Assume bus 1 as the slack bus with a voltage of 1.0 pu. The parameters of the system in pu are shown in the figure.

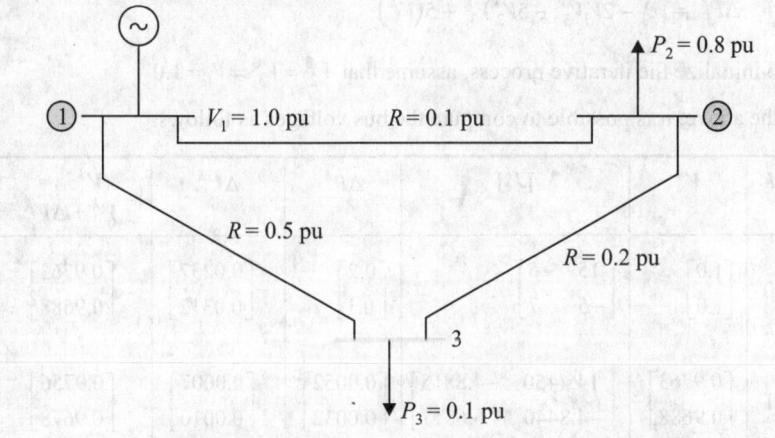

Fig. 7.7 Three-bus network for Example 7.4

Solution The Y_{bus} matrix is assembled by inspection of the system and is given as follows:

$$Y_{bus} = \begin{bmatrix} 12 & -10 & -2 \\ -10 & 15 & -5 \\ -2 & -5 & 7 \end{bmatrix}$$

In terms of the admittances and the bus voltages, the power at bus 2 and bus 3, respectively is as follows:

$$P_2 = -10V_1V_2 + 15V_2^2 - 5V_3V_2$$

$$P_3 = -2V_1V_3 - 5V_2V_3 + 5V_3^2$$

The Jacobian matrix $[J^k]$ is written as

$$[J^k] = \begin{bmatrix} \left(\dfrac{\partial P_2}{\partial V_2}\right)^k & \left(\dfrac{\partial P_2}{\partial V_3}\right)^k \\ \left(\dfrac{\partial P_3}{\partial V_2}\right)^k & \left(\dfrac{\partial P_3}{\partial V_3}\right)^k \end{bmatrix}$$

The elements of the Jacobian matrix are obtained by differentiating equations representing bus powers P_2 and P_3, that is,

$$\left(\frac{\partial P_2}{\partial V_2}\right)^k = -10V_1 + 30V_1^k - 5V_3^k$$

$$\left(\frac{\partial P_2}{\partial V_3}\right)^k = -5V_2^k$$

$$\left(\frac{\partial P_3}{\partial V_2}\right)^k = -5V_3^k$$

$$\left(\frac{\partial P_3}{\partial V_3}\right)^k = -2V_1 - 5V_2^k + 14V_3^k$$

$$\Delta P_2^k = P_2 - 10V_1V_2^k + 15\left(V_2^k\right)^2 - 5V_3V_2^k$$

$$\Delta P_3^k = P_3 - 2V_1V_3^k - 5V_2^kV_3^k + 5\left(V_3^k\right)^k$$

(i) To initialize the iterative process, assume that $V_2^0 = V_3^0 = V_1 = 1.0$

From the above, it is possible to compute the bus voltages as follows:

k	V^k	$[J^k]$	ΔP^k	ΔV^k	$V^{k+1} = V^k + \Delta V^k$
0	$\begin{bmatrix} 1.0 \\ 1.0 \end{bmatrix}$	$\begin{bmatrix} 15 & -5 \\ -5 & 7 \end{bmatrix}$	$\begin{bmatrix} 0.2 \\ 0.1 \end{bmatrix}$	$-\begin{bmatrix} 0.0237 \\ 0.0312 \end{bmatrix}$	$\begin{bmatrix} 0.9763 \\ 0.9688 \end{bmatrix}$
1	$\begin{bmatrix} 0.9763 \\ 0.9688 \end{bmatrix}$	$\begin{bmatrix} 14.4450 & -4.8815 \\ -4.8440 & 6.6817 \end{bmatrix}$	$\begin{bmatrix} 0.0052 \\ 0.0032 \end{bmatrix}$	$\begin{bmatrix} 0.0007 \\ 0.0010 \end{bmatrix}$	$\begin{bmatrix} 0.9756 \\ 0.9678 \end{bmatrix}$
2	$\begin{bmatrix} 0.9756 \\ 0.9678 \end{bmatrix}$	$\begin{bmatrix} 14.4291 & -4.8815 \\ -4.8391 & 6.6714 \end{bmatrix}$	$\begin{bmatrix} 0.0000 \\ 0.0000 \end{bmatrix}$	$\begin{bmatrix} 0.0000 \\ 0.0000 \end{bmatrix}$	$\begin{bmatrix} 0.9756 \\ 0.9678 \end{bmatrix}$

(ii) Current in line 1-2 = $\dfrac{1.0 - 0.9756}{0.1} = 0.2440$ pu

Current in line 1-3 = $\dfrac{1.0 - 0.9678}{0.5} = 0.0644$ pu

Current in line 2-3 = $\dfrac{0.9756 - 0.9678}{0.2} = 0.0390$ pu

(iii) Slack bus power = $1.0 \times (0.2440 + 0.0644) = 0.3084$ pu

(iv) Losses in the system = (Slack bus power) – (System load)

$$= 0.3084 - (0.2 + 0.1) = 0.0084 \text{ pu}$$

7.8 POWER FLOW SOLUTION BY NEWTON–RAPHSON METHOD

The formulation of the power flow problem by Newton–Raphson method can be done using the expressions for active and reactive powers, expressed either in polar coordinates or in rectangular coordinates.

7.8.1 Polar Version

The complex power at a bus i of an electrical power system in terms of the bus voltages and the system admittances is given by Eq. (7.6). On the other hand, the real power P_i and reactive power Q_i at the bus in the polar form can be expressed respectively by Eqs (7.19) and (7.20). Equations (7.19) and (7.20) have been rewritten here for ready reference.

$$P_i = |V_i| \sum_{k=1}^{n} |Y_{im}| |V_m| \cos(\theta_{im} - \delta_i + \delta_m)$$

$$\text{for } i = 1, 2, \ldots, n$$

$$Q_i = -|V_i| \sum_{m=1}^{n} |Y_{im}| |V_m| \sin(\theta_{im} - \delta_i + \delta_m)$$

$$\text{for } i = 1, 2, \ldots, n$$

These two equations can be expressed as

$$P_i - |V_i| \sum_{k=1}^{n} |Y_{im}| |V_m| \cos(\theta_{im} - \delta_i + \delta_m) = 0$$

$$\text{for } i = 1, 2, \ldots, n \qquad (7.78)$$

$$Q_i - \left[-|V_i| \sum_{m=1}^{n} |Y_{im}| |V_m| \sin(\theta_{im} - \delta_i + \delta_m) \right] = 0$$

$$\text{for } i = 1, 2, \ldots, n \qquad (7.79)$$

Of the total number of n buses, let the number of P-Q buses be m_1, P-V buses be m_2, and let there be one slack bus, so that $n = m_1 + m_2 + 1$.

It may be observed that each bus is associated with two non-linear equations, one for real power and the other for reactive power. Power flow study of the system involves solution of these equations for voltage magnitude $|V|$ and voltage phase angle δ. Thus, for a power system constituted of n buses, as the voltage magnitude and voltage phase angle for slack bus is specified, $2(n-1)$ non-linear equations have to be solved. The N–R iterative method is a handy technique for achieving this objective.

The basic power flow problem here is to find m_1 unknown bus voltage magnitudes $|V|$ at the P-Q buses, and $m_1 + m_2$ unknown bus voltage angles δ at the P-Q buses and P-V buses. Let X be the vector of all unknown $|V|$ and δ, and Y be the vector of all specified variables. The dimension of X is $2m_1 + m_2$ and that of Y is $2m_1 + 2m_2 + 2$ (equal to $2n$). Then, if bus s is specified as the slack bus, then

$$X = \begin{bmatrix} \delta \\ |V| \end{bmatrix} \text{ on each P-Q bus} \\ \delta \} \text{ on each P-V bus} \end{bmatrix}$$

$$\text{and} \quad Y = \begin{bmatrix} \begin{Bmatrix} V_s \\ \delta_s \end{Bmatrix} & \text{on slack bus} \\[2ex] \begin{Bmatrix} P_i^{sp} \\ Q_i^{sp} \end{Bmatrix} & \text{on each } P\text{-}Q \text{ bus} \\[2ex] \begin{Bmatrix} P_i^{sp} \\ |V_i|^{sp} \end{Bmatrix} & \text{on each } P\text{-}Q \text{ bus} \end{bmatrix}$$

Now the selection is made from the set of Eqs (7.78) and (7.79), such that the number of equations equal to the number of unknowns in X to form the non-linear power flow equations $F(X, Y) = 0$. Thus

$$F(X, Y) = \begin{bmatrix} \text{Eq. (7.78) for each } P\text{-}Q \text{ bus and } P\text{-}V \text{ bus with } P_i = P_i^{sp} \\ \text{Eq. (7.79) for each } P\text{-}Q \text{ bus with } Q_i = Q_i^{sp} \end{bmatrix} = 0$$

(7.80)

Thus there are $2m_1 + m_2$ equations and this is equal to the number of unknowns in X. Equation (7.80) may be written in the form

$$\begin{bmatrix} \Delta P \\ \Delta Q \end{bmatrix} = 0 \tag{7.81}$$

where

$$\Delta P_i = P_i^{sp} - |V_i| \sum_{k=1}^{n} |Y_{im}||V_m| \cos(\theta_{im} - \delta_i + \delta_m)$$

$$\text{for } i = 1, 2, \ldots, n; \quad i \neq s \tag{7.82a}$$

$$\Delta Q_i = Q_i^{sp} - \left[-|V_i| \sum_{m=1}^{n} |Y_{im}||V_m| \sin(\theta_{im} - \delta_i + \delta_m) \right]$$

$$\text{for } i = 1, 2, \ldots, n; \quad i \neq s; \ i \neq P\text{-}V \text{ bus} \tag{7.82b}$$

Then, similar to Eq. (7.71) for the n-dimensional case, the Newton–Raphson iterates for the power flow studies take the following form:

$$\begin{bmatrix} \Delta P^k \\ \hline \Delta Q^k \end{bmatrix} = \begin{bmatrix} J_1^k & J_2^k \\ \hline J_3^k & J_4^k \end{bmatrix} \begin{bmatrix} \Delta \delta^k \\ \hline \Delta |V|^k \end{bmatrix} = [J^k] \begin{bmatrix} \Delta \delta^k \\ \hline \Delta |V|^k \end{bmatrix} \tag{7.83}$$

where $\Delta \delta$ is the sub-vector of incremental angles at P-Q buses and P-V buses, $\Delta|V|$ is the sub-vector of incremental voltage magnitude at P-Q buses, J is the Jacobian matrix of partial derivatives; J_1 is the sub-matrix of the partial derivative of P's, given by Eq. (7.19), with respect to relevant δ s, J_2 is the sub-matrix of the partial derivative of P's, given by Eq. (7.19), with respect to relevant V's, J_3 is the sub-matrix of the partial derivative of Q's, given by Eq. (7.20), with respect to relevant δ s, and J_4 is the sub-matrix of the partial derivative of Q's, given by Eq. (7.20), with respect to relevant V's.

Computation for Systems having P-Q Buses

If bus 1 is specified as slack bus and all the other $(n-1)$ buses are P-Q buses, then Eq. (7.83) may be written in the expanded form as follows:

$$
\begin{bmatrix} \Delta P_2^k \\ \vdots \\ \Delta P_n^k \\ -- \\ \Delta Q_2^k \\ \vdots \\ \Delta Q_n^k \end{bmatrix} =
\begin{bmatrix}
\left(\dfrac{\partial P_2}{\partial \delta_2}\right)^k & \cdots & \left(\dfrac{\partial P_2}{\partial \delta_n}\right)^k & \left(\dfrac{\partial P_2}{\partial |V_2|}\right)^k & \cdots & \left(\dfrac{\partial P_2}{\partial |V_n|}\right)^k \\
\vdots & \ddots & \vdots & \vdots & \ddots & \vdots \\
\left(\dfrac{\partial P_n}{\partial \delta_2}\right)^k & \cdots & \left(\dfrac{\partial P_n}{\partial \delta_n}\right)^k & \left(\dfrac{\partial P_2}{\partial |V_2|}\right)^k & \cdots & \left(\dfrac{\partial P_n}{\partial |V_n|}\right)^k \\
\left(\dfrac{\partial Q_2}{\partial \delta_2}\right)^k & \cdots & \left(\dfrac{\partial Q_2}{\partial \delta_n}\right)^k & \left(\dfrac{\partial Q_2}{\partial |V_2|}\right)^k & \cdots & \left(\dfrac{\partial Q_2}{\partial |V_n|}\right)^k \\
\vdots & \ddots & \vdots & \vdots & \ddots & \vdots \\
\left(\dfrac{\partial Q_n}{\partial \delta_2}\right)^k & \cdots & \left(\dfrac{\partial Q_n}{\partial \delta_n}\right)^k & \left(\dfrac{\partial Q_2}{\partial |V_2|}\right)^k & \cdots & \left(\dfrac{\partial Q_n}{\partial |V_n|}\right)^k
\end{bmatrix}
\begin{bmatrix} \Delta \delta_2^k \\ \vdots \\ \Delta \delta_n^k \\ -- \\ \Delta |V_2|^k \\ \vdots \\ \Delta |V_n|^k \end{bmatrix}
$$

(7.84)

Thus,

$$
\begin{bmatrix} \Delta P^k \\ -- \\ \Delta Q^k \end{bmatrix} =
\begin{bmatrix} \Delta P_2^k \\ \vdots \\ \Delta P_n^k \\ -- \\ \Delta Q_2^k \\ \vdots \\ \Delta Q_n^k \end{bmatrix}
\text{ and }
\begin{bmatrix} \Delta \delta^k \\ -- \\ \Delta |V|^k \end{bmatrix} =
\begin{bmatrix} \Delta \delta_2^k \\ \vdots \\ \Delta \delta_n^k \\ -- \\ \Delta |V_2|^k \\ \vdots \\ \Delta |V_n|^k \end{bmatrix}
$$

(7.85)

The terms ΔP_i^k and ΔQ_i^k, known as the power residues, can be calculated as the difference between the specified and computed values. These are given by

$$\Delta P_i^k = P_i^{\text{sp}} - P_i^k \qquad \text{for } i = 2, 3, \ldots, n \qquad (7.86)$$

$$\Delta Q_i^k = Q_i^{\text{sp}} - Q_i^k \qquad \text{for } i = 2, 3, \ldots, n \qquad (7.87)$$

In order to compute the voltage angles δ and voltage magnitudes $|V|$, the inverse of the Jacobian matrix $[J]$ has to be computed. Hence, in the iterative form, Eq. (7.83) at the kth iteration will be written as

$$
\begin{bmatrix} \Delta \delta^k \\ -- \\ \Delta |V|^k \end{bmatrix} =
\begin{bmatrix} J_1^k & J_2^k \\ J_3^k & J_4^k \end{bmatrix}^{-1}
\begin{bmatrix} \Delta P^k \\ -- \\ \Delta Q^k \end{bmatrix}
$$

(7.88)

and

$$\begin{bmatrix} \delta^{k+1} \\ -- \\ |V|^{k+1} \end{bmatrix} = \begin{bmatrix} \delta^{k} \\ -- \\ |V|^{k} \end{bmatrix} + \begin{bmatrix} \Delta\delta^{k} \\ -- \\ \Delta|V|^{k} \end{bmatrix} \tag{7.89}$$

Equations (7.88) and (7.89) provide the mathematical simulation for performing power flow studies, by iteration, of electrical power systems.

It may be noted that in order to use the recursive formulae represented by Eqs (7.88) and (7.89), it is necessary to compute elements of the Jacobian matrix $[J]$, that is, elements of sub-matrices $\left[J_1^k\right]$, $\left[J_2^k\right]$, $\left[J_3^k\right]$, and $\left[J_4^k\right]$. The diagonal and off-diagonal elements of these sub-matrices are as follows.

The diagonal and off-diagonal elements of $[J_1]$ are

$$\frac{\partial P_i}{\partial \delta_i} = \sum_{m\neq i} |V_i||V_m||Y_{im}|\sin(\theta_{im} - \delta_i + \delta_m) \tag{7.90}$$

$$\frac{\partial P_i}{\partial \delta_m} = -|V_i||V_m||Y_{im}|\sin(\theta_{im} - \delta_i + \delta_m) \quad \text{for } m \neq i \tag{7.91}$$

The diagonal and off-diagonal elements of $[J_2]$ are

$$\frac{\partial P_i}{\partial |V_i|} = 2|V_i||Y_{im}|\cos\theta_{ii} + \sum_{m\neq i} |V_m||Y_{im}|\cos(\theta_{im} - \delta_i + \delta_m) \tag{7.92}$$

$$\frac{\partial P_i}{\partial |V_m|} = |V_i||Y_{im}|\cos(\theta_{im} - \delta_i + \delta_m) \quad \text{for } m \neq i \tag{7.93}$$

The diagonal and off-diagonal elements of $[J_3]$ are

$$\frac{\partial Q_i}{\partial \delta_i} = \sum_{m\neq i} |V_i||V_m||Y_{im}|\cos(\theta_{im} - \delta_i + \delta_m) \tag{7.94}$$

$$\frac{\partial Q_i}{\partial \delta_m} = -|V_i||V_m||Y_{im}|\cos(\theta_{im} - \delta_i + \delta_m) \quad \text{for } m \neq i \tag{7.95}$$

The diagonal and off-diagonal elements of $[J_4]$ are

$$\frac{\partial Q_i}{\partial |V_i|} = -2|V_i||Y_{im}|\sin\theta_{ii} - \sum_{m\neq i} |V_m||Y_{im}|\sin(\theta_{im} - \delta_i + \delta_m) \tag{7.96}$$

$$\frac{\partial Q_i}{\partial |V_m|} = -|V_i||Y_{im}|\sin(\theta_{im} - \delta_i + \delta_m) \quad \text{for } m \neq i \tag{7.97}$$

Computation for Systems having Both P-Q and P-V Buses

As already stated in Section 7.3.1, the voltage magnitude and real power at a voltage-controlled bus in a power system are specified. Therefore, equations involving $\Delta|V|$ and ΔQ corresponding columns of the Jacobian matrix associated with the control buses are eliminated. In other words, if a power system of n buses contains n_2 buses specified as voltage-controlled buses, then the number of buses with real power constraints are equal to $(n - 1)$ and the number of buses having reactive power control is equal to $(n - n_2 - 1)$. Thus, the order of the $[J]$ matrix will be $2(n - 1) - n_2$. The order of the sub-matrices will be as follows:

$$[J_1] \qquad (n - 1) \times (n - 1)$$
$$[J_2] \qquad (n - 1) \times (n - 1 - n_2)$$
$$[J_3] \qquad (n - 1 - n_2) \times (n - 1)$$
$$[J_4] \qquad (n - 1 - n_2) \times (n - 1 - n_2)$$

Algorithm for Power Flow Solution by N–R Method

The procedure for the power flow solution of a system by Newton–Raphson method is as follows.

Step 1: Initialize the N–R iterative process by setting the iteration count $k = 0$ and set the voltage magnitudes $|V_i|^0$ equal to the slack bus voltage or equal to 1.0. Set the bus voltage angle δ_i^0 equal to zero for the P-Q or load buses. Set the voltage angles δ_i^0 equal to zero for the P-V or voltage-controlled buses.

Step 2: For the load buses, compute the real and reactive powers by using Eqs (7.19) and (7.20) respectively. Then compute power residuals using Eqs (7.86) and (7.87).

For voltage-controlled buses, real power at the buses is calculated by using Eq. (7.19). Then compute $\left[\Delta P_i^k\right]$ using Eq. (7.86).

Step 3: Compute the elements of Jacobian matrix by computing the sub-matrices $\left[J_1^k\right], \left[J_2^k\right], \left[J_3^k\right],$ and $\left[J_4^k\right]$, using Eqs (7.90) to (7.97).

Step 4: Solve Eq. (7.84), by either forward and backward substitution (Gauss elimination) or the MATLAB function (\\) for computing the inverse of a matrix, to obtain $\Delta \delta_i^k$ and ΔV_i^k.

Step 5: Compute the new estimates of bus voltage magnitudes and voltage angles by using Eq. (7.89).

Step 6: Apply the following test for convergence.

$$\left|\Delta P_i^k\right| = \left|P_i^{sp} - P_i^k\right| \leq \text{tolerance}$$

$$\left|\Delta Q_i^k\right| = \left|Q_i^{sp} - Q_i^k\right| \leq \text{tolerance}$$

The power mismatch at each bus is used to specify the tolerance which is usually of the order of 0.01 pu for real and reactive powers. If the tolerance condition for each bus is satisfied, the solution of the power flow equations has been obtained, if not, put $k = k + 1$, and go to *Step* 2.

Example 7.5 Use the Newton–Raphson iterative technique to solve Example 7.1.

Solution The Y_{bus} matrix in polar form of the system is calculated and given as follows:

$$Y_{bus} = \begin{bmatrix} 7.5000 & 2.5000 & 5.0000 \\ 2.5000 & 6.5000 & 4.0000 \\ 5.0000 & 4.0000 & 9.0000 \end{bmatrix}$$

and

$$[\theta] = \begin{bmatrix} -1.5708 & 1.5708 & 1.5708 \\ 1.5708 & -1.5708 & 1.5708 \\ 1.5708 & 1.5708 & -1.5708 \end{bmatrix}$$

Step 1: Assume $V_2^0 = 1.04$ pu and $\delta_2^0 = 0$ radians; for the voltage-controlled bus, take

$\delta_3^0 = 0$ radians.

Step 2: Employing Eqs (7.19) and (7.20), the real and reactive powers for bus 2 and the real power for bus 3 are computed as follows

$$P_2 = 1.04 \times [2.5 \times 1.04 \times \cos(-1.5708) + 6.5 \times 1.04 \times \cos(1.5708)$$
$$+ 4.0 \times 1.005 \times \cos(-1.5708)] = 0$$

Similarly, $P_3 = 0$

$$Q_2 = 1.04 \times [2.5 \times 1.04 \times \sin(-1.5708) + 6.5 \times 1.04 \times \sin(1.5708)$$
$$+ 4.0 \times 1.005 \times \sin(-1.5708)] = 0.1456$$

$$\Delta P_2 = P_2^{sp} - P_2 = -1.0 - 0 = -1.0$$

$$\Delta P_3 = P_3^{sp} - P_3 = 1.0 - 0 = 1.0$$

$$\Delta Q_3 = Q_3^{sp} - Q_3 = -0.8 - 0.1456 = -0.9456$$

Use Eqs (7.90) to (7.97) to compute the elements of $[J_1]$, $[J_2]$, $[J_3]$, and $[J_4]$. Thus,

$$\frac{\partial P_2}{\partial \delta_2} = 1.0 \times 1.04 \times 2.5 \times \sin(-1.5708) + 1.0 \times 4.0 \times \sin(-1.5708) = 6.8848$$

$$\frac{\partial P_2}{\partial \delta_3} = 1.0 \times 1.005 \times 4.0 \times \sin(-1.5708) = -4.1808$$

$$\frac{\partial P_3}{\partial \delta_2} = 1.0 \times 1.005 \times 4.0 \times \sin(-1.5708) = -4.1808$$

$$\frac{\partial P_3}{\partial \delta_3} = 1.005 \times 1.04 \times 5.0 \times \sin(-1.5708) + 1.0 \times 4.0 \times \sin(-1.5708)$$

$$= 9.4068$$

$$[J_1] = \begin{bmatrix} 6.8848 & -4.1808 \\ -4.1808 & 9.4608 \end{bmatrix}$$

$$\frac{\partial P_2}{\partial V_2} = 2 \times 1.0 \times 6.5 \times \cos(1.5708) + 1.04 \times 2.5 \times \cos(-1.5708)$$

$$+ 1.0 \times 4.0 \times \cos(-1.5708) = 0$$

$$\frac{\partial P_3}{\partial V_2} = 1.0 \times 4.0 \times \cos(-1.5708) = 0$$

$$[J_2] = \begin{bmatrix} 0 \\ 0 \end{bmatrix}$$

$$\frac{\partial Q_2}{\partial \delta_2} = 1.0 \times (1.04 \times 2.5 \cos(-1.5708) + 1.005 \times \cos(-1.5708)) = 0$$

$$\frac{\partial Q_2}{\partial \delta_3} = 1.0 \times 1.005 \times 4.0 \times \cos(-1.5708) = 0$$

$$[J_3] = [0 \quad 0]$$

$$\frac{\partial Q_2}{\partial V_2} = -2 \times 1.0 \times 6.5 \times \sin(1.5708) + 1.04 \times 2.5 \times \sin(-1.5708)$$
$$+1.005 \times 4.0 \times \sin(-1.5708) = 6.9$$

Therefore, the Jacobian matrix is formulated as follows:

$$[J] = \begin{bmatrix} 6.8848 & -4.1808 & 0.0000 \\ -4.1808 & 9.4068 & 0.0000 \\ 0.0000 & -0.0000 & 6.9000 \end{bmatrix}$$

Step 3: Using Eq. (7.72), the following relation is obtained

$$\begin{bmatrix} \Delta\delta_2 \\ \Delta\delta_3 \\ \Delta V_2 \end{bmatrix} = [J]^{-1} \begin{bmatrix} \Delta P_2 \\ \Delta P_3 \\ \Delta Q_2 \end{bmatrix}$$

The MATLAB facility (\) for computing the inverse of a matrix is taken advantage of to compute the changes in δ_1, δ_2, and V_2. Hence,

$$\begin{bmatrix} \Delta\delta_2 \\ \Delta\delta_3 \\ \Delta V_2 \end{bmatrix} = \begin{bmatrix} 6.8848 & -4.1808 & 0.0000 \\ -4.1808 & 9.4068 & 0.0000 \\ 0.0000 & -0.0000 & 6.9000 \end{bmatrix} \backslash \begin{bmatrix} -1.0000 \\ 1.0000 \\ -0.9456 \end{bmatrix} = \begin{bmatrix} -0.1105 \\ 0.0572 \\ -0.1370 \end{bmatrix}$$

Step 4: Using Eq. (7.70), the new δ_2, δ_3, and V_2 are estimated, that is,

$\delta_2 = 0 - 0.1105 = -0.1105$, $\delta_3 = 0 + 0.0572 = 0.0572$, and
$V_2 = 1.04 - 0.1370 = 0.9030$

Step 5: Clearly, the mismatch in real and reactive powers at buses 2 and 3, that is,

$$\text{abs} \begin{bmatrix} \Delta P_2 \\ \Delta P_3 \\ \Delta Q_2 \end{bmatrix} \gg 0.01$$

Hence, it is necessary to return to Step 2 and repeat Steps 3, 4, and 5 till the absolute values of the real and reactive powers is < or = 0.01.
The results of the subsequent iterations are as follows:

itr	V_2	δ_2	δ_3	ΔP_2	ΔP_3	ΔQ_2	$\Delta\delta_2$	$\Delta\delta_3$	ΔV_2
2	0.9030	−0.1105	0.0572	0.1352	0.0954	−0.1873	−0.0279	0.0026	−0.0410
3	0.8619	−0.1384	0.0598	−0.008	0.0053	−0.0122	−0.0021	0.0001	−0.0031
4	0.8589	−0.1405	0.0599	≪0.01	≪0.01	≪0.01	≪0.0001	≪0.0001	≪0.0001

The corresponding values of the [J] matrices are given here from a tutorial perspective:
Iteration [J]

2
$$\begin{bmatrix} 5.9123 & -3.5790 & -0.9578 \\ -3.5790 & 8.7964 & 0.6710 \\ -0.8648 & 0.6059 & 5.1907 \end{bmatrix}$$

3
$$\begin{bmatrix} 5.6165 & -3.3970 & -1.1506 \\ -3.3970 & 8.6137 & 0.7917 \\ -0.9917 & 0.6824 & 4.6884 \end{bmatrix}$$

4
$$\begin{bmatrix} 5.5945 & -3.3835 & -1.1643 \\ -3.3835 & 8.6001 & 0.8002 \\ -1.0000 & 0.6872 & 4.6511 \end{bmatrix}$$

7.8.2 Rectangular Version

Rectangular version of power flow solution by N–R method uses e_i and f_i, the real and imaginary part of the voltage respectively, as the variables. The real power P_i and reactive power Q_i at the bus in the rectangular form can be expressed respectively by Eqs (7.16) and (7.17). Equations (7.16) and (7.17) have been rewritten here for ready reference.

$$P_i = e_i \sum_{m=1}^{n} (G_{im}e_m - B_{im}f_m) + f_i \sum_{m=1}^{n} (G_{im}f_m + B_{im}e_m)$$

$$\text{for } i = 1, 2, \ldots, n \quad (7.16)$$

$$Q_i = e_i \sum_{m=1}^{n} (G_{im}f_m + B_{im}e_m) - f_i \sum_{m=1}^{n} (G_{im}e_m - B_{im}f_m)$$

$$\text{for } i = 1, 2, \ldots, n \quad (7.17)$$

Since at P-V buses, the voltage magnitude $|V_i|$ is constant but e_i and f_i can vary as $|V_i|^2 = e_i^2 + f_i^2$. Thus the number of unknowns is increased by n_2 (the number of P-V buses) compared to the polar version. For each of the P-V buses, the set of non-linear equations is

$$\left(|V_i|^{\text{sp}}\right)^2 - \left(e_i^2 + f_i^2\right) = 0 \quad (7.98)$$

The column vector X of unknown variables will have e_i and f_i for $i = 1, 2, \ldots, n$; and $i \neq s$. Thus, the total number of non-linear equations are $2(n_1 + n_2) = 2(n - 1)$. If bus 1 is slack bus, then these equations are

$$P_i - \left[e_i \sum_{m=1}^{n} (G_{im}e_m - B_{im}f_m) + f_i \sum_{m=1}^{n} (G_{im}f_m + B_{im}e_m) \right] = 0$$

$$\text{for } i = 2, 3, \ldots, n; \quad (7.99)$$

$$Q_i - \left[e_i \sum_{m=1}^{n} (G_{im}f_m + B_{im}e_m) - f_i \sum_{m=1}^{n} (G_{im}e_m - B_{im}f_m) \right] = 0$$

$$\text{for } i = 2, 3, \ldots, n_1 \quad (7.100)$$

$$|V_i|^2 - e_i^2 + f_i^2 = 0 \qquad \text{for } i = (n_1 + 1), (n_1 + 2), \ldots, (n_1 + n_2) \quad (7.101)$$

The linear equation in the iterative process is

$$\begin{bmatrix} \Delta P^k \\ \Delta Q^k \\ \Delta |V|^{2^k} \end{bmatrix} = \begin{bmatrix} J_1^k & J_2^k \\ J_3^k & J_4^k \\ J_5^k & J_6^k \end{bmatrix} \begin{bmatrix} \Delta e^k \\ \Delta f^k \end{bmatrix} \quad (7.102)$$

The diagonal and off-diagonal elements of these sub-matrices are given as follows. The diagonal and off-diagonal elements of $[J_1]$ are

$$\frac{\partial P_i}{\partial e_i} = \sum_{m=1}^{n} (G_{im}e_m - B_{im}f_m) + G_{ii}e_i + B_{ii}f_i \quad (7.103)$$

$$\frac{\partial P_i}{\partial e_m} = G_{im}e_i + B_{im}f_i \tag{7.104}$$

The diagonal and off-diagonal elements of $[J_2]$ are

$$\frac{\partial P_i}{\partial f_i} = \sum_{m=1}^{n} (G_{im}f_m + B_{im}e_m) - B_{ii}e_i + G_{ii}f_i \tag{7.105}$$

$$\frac{\partial P_i}{\partial e_m} = G_{im}f_i - B_{im}e_i \tag{7.106}$$

The diagonal and off-diagonal elements of $[J_3]$ are

$$\frac{\partial Q_i}{\partial e_i} = -\sum (B_{im}e_m + G_{im}f_m) - B_{ii}e_i + G_{ii}f_i \tag{7.107}$$

$$\frac{\partial Q_i}{\partial f_m} = G_{im}f_i - B_{im}e_i \tag{7.108}$$

The diagonal and off-diagonal elements of $[J_4]$ are

$$\frac{\partial Q_i}{\partial e_i} = \sum_{m=1}^{n} (G_{im}e_m - B_{ii}f_m) - G_{ii}e_i - B_{ii}f_i \tag{7.109}$$

$$\frac{\partial Q_i}{\partial e_m} = -G_{im}e_i - B_{im}f_i \tag{7.110}$$

The diagonal and off-diagonal elements of $[J_5]$ are

$$\frac{\partial |V_i|^2}{\partial e_i} = 2e_i \tag{7.111}$$

$$\frac{\partial |V_i|^2}{\partial e_m} = 0 \tag{7.112}$$

The diagonal and off-diagonal elements of $[J_6]$ are

$$\frac{\partial |V_i|^2}{\partial f_i} = 2f_i \tag{7.113}$$

$$\frac{\partial |V_i|^2}{\partial f_m} = 0 \tag{7.114}$$

After solving linear Eq. (7.100) for Δe and Δf, the corrections are applied to e and f and the computations are repeated till convergence is reached.

7.8.3 MATLAB Program for Power Flow Studies using N–R Method

The function pfsnr is a MATLAB program function for computing the bus voltages for power flow studies, by the Newton–Raphson technique. The input data is similar to the function pfsg described in Section 7.4.4. An additional input, mbus, to indicate the number of voltage-controlled buses in the system, is required to be made.

```
function [Ybus,theta] = pfsnrnew(nbus,nlns,ld,mbus,pfbd);
%Program for performing power flow studies by N-R method
%Developed by the authors
%Part-I: Assembly of Ybus from line data
Ybus = zeros(nbus);
J2=zeros(nbus-1,nbus-1-mbus);
J3=zeros(nbus-1-mbus,nbus-1);
J4=zeros(nbus-1-mbus,nbus-1-mbus);
J1=zeros(nbus-1,nbus-1);
delp=zeros(nbus-1,1);
if mbus == 0
   delq=zeros(nbus-1,1);
else
   delq=zeros(nbus-1-mbus,1);
end
pie=3.141592;
for k=1:nbus;
   delta(k,1)=0;
   if pfbd(k,1) == 0;
   Vmag(k,1)=1.0;
else
end
end
for k=1:nlns;
   m=ld(k,1);
   n=ld(k,2);
   Ybus(m,m)=Ybus(m,m)+1/ld(k,3)+ld(k,4);
   Ybus(n,n)=Ybus(n,n)+1/ld(k,3)+ld(k,4);
   Ybus(m,n)=-1/ld(k,3);
   Ybus(n,m)=Ybus(m,n);
end
Ybus
theta=angle(Ybus)
Ybus=abs(Ybus)
%Initialise the iterative process
Vmag(1)=input('slack bus voltage magnitude');
delta(1)=input('slack bus voltage angle');
for k=2:nbus;
   delta(k)=delta(1);
   if pfbd(k,1) == 0;
      Vmag(k)=Vmag(1);
   else
      Vmag(k)=pfbd(k,3);
   end
end
itr=0;
count=0;
while count < 2*(nbus-1)-mbus;
   itr=itr+1;
```

```
%Compute sub-matrices J1, J2, J3 and J4
for i=2:nbus;
    for m=2:nbus;
        sum=0;
        if m == i;
            for k=1:nbus;
                if k ~= m;
                    sum=sum+Ybus(i,k)*Vmag(k)*sin(theta(i,k)-
delta(i)+delta(k));
                else
                end
                J1(i-1,m-1)=Vmag(i)*sum;
            end
        else
            J1(i-1,m-1)=-Vmag(i)*Ybus(i,m)*Vmag(m)*sin(theta(i,m)-
delta(i)+delta(m));
        end
    end
end
I=0;
for i=2:nbus;
    if pfbd(i,1) == 0;
        I=I+1;J=0;
        for m=2:nbus;
            sum1=0;J=J+1;
            if i == m;
                for k=1:nbus;
                    if k ~= m;
                        sum1=sum1+Ybus(i,k)*Vmag(k)*cos(theta(i,k)-
delta(i)+delta(k));
                    else
                    end
                    J2(J,I)=2*Vmag(i)*Ybus(i,i)*cos(theta(i,i))+sum1;
                end
            else
                J2(J,I)=Vmag(i)*Ybus(i,m)*cos(theta(i,m)-
delta(i)+delta(m));
            end
        end
    else
    end
end
I=0;
for i=2:nbus;
    if pfbd(i,1) == 0;
        I=I+1;J=0;
        for m=2:nbus;
            Sum1=0;J=J+1;
            if i == m;
```

```
                for k=1:nbus;
                    if k ~= m;
                        sum1=sum1+Ybus(i,k)*Vmag(k)*cos(theta(i,k)-
delta(i)+delta(k));
                    else
                    end
                    J3(I,J)=Vmag(i)*sum1;
                end
            else
                J3(I,J)=-Vmag(i)*Ybus(i,m)*Vmag(m)*cos(theta(i,m)-
delta(i)+delta(m));
            end
        end
    else
    end
end
I=0;
for i=2:nbus;
    if pfbd(i,1) == 0;
        I=I+1;J=0;
        for m=2:nbus-mbus;
            Sum1=0;J=J+1;
            if i == m;
                for k=1:nbus;
                    if k ~= m;
                        sum1=sum1+Ybus(i,k)*Vmag(k)*sin(theta(i,k)-
delta(i)+delta(k));
                    else
                    end
                    J4(I,J)=-2*Vmag(i)*Ybus(i,i)*sin(theta(i,i))-sum1;
                end
            else
                J4(I,J)=-Vmag(i)*Ybus(i,m)*sin(theta(i,m)-
delta(i)+delta(m));
            end
        end
    else
    end
end
JJ=[J1 J2;J3 J4]
%compute delp & delq from scheduled bus powers
for k=2:nbus;
    sum=0;
    for m=1:nbus;
        sum=sum+Vmag(k)*Ybus(k,m)*Vmag(m)*cos(theta(k,m)-
delta(k)+delta(m));
    end
    delp(k-1)=(pfbd(k,2)-pfbd(k,4))-sum;
end
```

```
    for k=2:nbus-mbus;
        sum=0;
        if pfbd(k,1) == 0;
            for m=1:nbus;
                sum=sum-Vmag(k)*Ybus(k,m)*Vmag(m)*sin(theta(k,m)-
delta(k)+delta(m));
            end
            delq(k-1)=(pfbd(k,3)-pfbd(k,5))-sum;
        else
        end
    end
    ddpq=[delp;delq]
    ddv=JJ\ddpq
    for k=1:nbus-1;
        j=k+1;
        delta(j)=delta(j)+ddv(k);
        if pfbd(j,1) == 2;
            f=pfbd(j,3)*sin(delta(j));
            e=sqrtm((pfbd(j,3))^2-f^2);
            delta(j)=atan(f/e);
        else
        end
    end
    for k=1:nbus-1;
        j=k+1;
        if pfbd(j,1) == 0;
            Vmag(j)=Vmag(j)+ddv(k+nbus-1);
        else
        end
    end
    % Check power mismatch at buses;
    count=0;
    T1 = nbus-1
    if mbus==0
    T2 = nbus-1,
    else
    T2 = nbus-1-mbus,
    end
    for k=1:T1 + T2;
        if abs(ddpq(k)) <= 0.01;
            count=count+1
        else
        end
    end
end
pause
itr
Vmag
delta
delp
delq
end
```

Example 7.6 Compute the bus voltages for the power system of Example 7.5 by executing the MATLAB program function pfsnr. Show the input data.

Solution The input data and its format corresponding to the variables are depicted as follows

nbus	Number of buses in the system = 3
nlns	Number of lines in the system = 3
ld	Line data table [1 2 0.4i 0;1 3 0.2i 0;2 3 0.25i 0]
mbus	Number of voltage-controlled buses = 1
pfbd	Power flow bus data [1 0 0 0 0;0 0 0 1.0 0.8;2 1.0 1.005 0 0]

≫ [Ybus,theta] = pfsnr(nbus,nlns,ld,mbus,pfbd);

Slack bus voltage magnitude = 1.04
Slack bus voltage angle = 0

Iteration		Voltage at bus 2						
		V_{mag}	δ	V_{mag}	δ	P_2	P_3	Q_2
1	0.9030	−0.1105	1.0050	0.0572	1.0000	1.0000	−0.9456	
2	0.8619	−0.1384	1.0050	0.0598	−0.1352	0.0954	−0.1873	
3	0.8589	−0.1405	1.0050	0.0599	−0.0083	0.0053	−0.0122	
4	0.8588	−0.1405	1.0050	0.0599	1.0e-004	*(−0.4422)	0.2734	−0.6578)

7.8.4 Decoupled Power Flow Solution

In general, power transmission lines are mostly reactive with resistance value quite small, that is, the transmission lines have high X/R ratio. Further during the steady-state operation of electrical power systems, the difference in bus voltage angle between two adjacent buses is reasonably small, typically a few degrees. The effect of these factors on the sub-matrix $[J_2]$ will now be examined. The diagonal and the off-diagonal terms of the sub-matrix $[J_2]$ are given in Eqs (7.92) and (7.93), and are reproduced here for ready reference.

$$\frac{\partial P_i}{\partial |V_i|} = 2|V_i||Y_{im}|\cos\theta_{ii} + \sum_{m \neq i}|V_m||Y_{im}|\cos(\theta_{im} - \delta_i + \delta_m) \qquad (7.92)$$

$$\frac{\partial P_i}{\partial |V_m|} = |V_i||Y_{im}|\cos(\theta_{im} - \delta_i + \delta_m) \quad \text{for } m \neq i \qquad (7.93)$$

Since the transmission lines have a high X/R ratio, the admittance angle θ_{im} is close to 80°, while $(\delta_i - \delta_m)$ is only a few degrees (less than 10°). Thus, the diagonal and off-diagonal terms of $[J_2]$ are quite small. Based on a similar reasoning, the diagonal and off-terms of $[J_3]$ can also be taken as negligibly small.

On the other hand, from an examination of the off-diagonal, non-zero terms of $[J_1]$ and $[J_4]$, it is observed that these have significant values. Thus, it can be said that the flow of real power is strongly dependent on the bus voltage angle δ_i and is nearly independent of voltage magnitude $|V_i|$. Similarly the flow of reactive power is largely dependent on voltage magnitude $|V_i|$, whereas δ_i has no significant influence. The couplings between P-δ and Q-V components of the problem are relatively weak. Hence, the trend is to solve the P-δ and Q-V problems separately, that is, treat P-δ and Q-V problems as decoupled. Although the quadratic convergence characteristic suffers due to these approximations, but there are computational benefits such as saving of computer storage for the Jacobian matrix.

The decoupling characteristic of the Jacobian matrix enables the power engineers to simplify the solution of the power flow equations, without much loss of accuracy, by equating sub-matrices $[J_2]$ and $[J_3]$ to zero. The resulting linear equations become

$$[\Delta P] = [J_1] [\Delta \delta] \tag{7.115}$$

$$[\Delta Q] = [J_4] [\Delta |V|] \tag{7.116}$$

The diagonal and off-diagonal elements of $[J_1]$ are given by Eqs (7.90) and (7.91) respectively, and that of $[J_4]$ are given by Eqs (7.96) and (7.97).

The iterations for solution of power flow problem takes the form

$$\left[\Delta \delta^k\right] = \left[J_1^k\right]^{-1} \left[\Delta P^k\right] \tag{7.117}$$

$$\left[\Delta |V|^k\right] = \left[J_4^k\right]^{-1} \left[\Delta Q^k\right] \tag{7.118}$$

There are two ways of solving Eqs (7.92) and (7.93).

1. Simultaneous solution for $\Delta \delta$ and $\Delta |V|$.
2. Solution of $\Delta \delta$ first, and then use updated value of δ in Eq. (7.114) to solve for $\Delta |V|$.

7.8.5 Fast Decoupled Power Flow Solution

It is observed that at the end of the iteration the elements of the Jacobian matrix have to be re-computed. Considerable time can be saved if simplifications are introduced to avoid the re-computation of the Jacobian matrix at the end of the iteration. Such modifications were introduced by Stott and Alsac in 1974 leading to the fast decoupled power flow solution.

The diagonal elements of sub-matrix $[J_1]$ described by Eq. (7.90) may be rewritten by adding and subtracting the term $V_i^2 Y_{ii} \sin \theta_{ii}$ as

$$\frac{\partial P_i}{\partial \delta_i} = \sum_{m \neq i} |V_i||V_m||Y_{im}|\sin(\theta_m - \delta_i + \delta_m) + |V_i|^2 |Y_{ii}|\sin \theta_{ii}$$

$$-|V_i|^2 |Y_{ii}|\sin \theta_{ii}$$

$$= \sum |V_i||V_m||Y_{im}|\sin(\theta_{im} - \delta_i + \delta_m) - |V_i|^2 |Y_{ii}|\sin \theta_{ii}$$

Replacing the first term on the right side of the preceding equation by $-Q_i$, as given by Eq. (7.20), and substituting $|Y_{ii}| \sin \theta_{ii} = B_{ii}$ [from Eq. (7.12)], yields

$$\frac{\partial P_i}{\partial \delta_i} = -Q_i - |V_i|^2 B_{ii} \tag{7.119}$$

In practical power systems, $B_{ii} \gg Q_i$. Hence, Q_i in Eq. (7.119) can be neglected. Further simplification is obtained by assuming $|V_i|^2 \approx ||V_i|$. Therefore,

$$\frac{\partial P_i}{\partial \delta_i} = -|V_i| B_{ii} \tag{7.120}$$

As already stated, in the normal steady-state operation of a power system $\delta_i - \delta_m$ is very small. If it is assumed that $(\theta_{im} - \delta_i + \delta_m) \approx \theta_{im}$, the expression for the off-diagonal elements of $[J_1]$ described by Eq. (7.91) can be written as

$$\frac{\partial P_i}{\partial \delta_m} = -|V_i||V_m||Y_{im}|\sin\theta_{im} = -|V_i||V_m|\,B_{im} \qquad (7.121)$$

where B_{im} is the susceptance of the mutual admittance between bus i and bus m. A further simplification can be introduced to Eq. (7.121) by assuming $V_m \approx 1.0$, that is,

$$\frac{\partial P_i}{\partial \delta_m} = -|V_i|\,B_{im} \qquad (7.122)$$

Similarly, the expression for the diagonal terms of $[J_4]$ described by Eq. (7.96) is modified as follows:

$$\frac{\partial Q_i}{\partial |V_i|} = -2|V_i||Y_{im}|\sin\theta_{ii} - \sum_{m \neq i}|V_m||Y_{im}|\sin(\theta_{im} - \delta_i + \delta_m)$$

$$= -|V_i||Y_{im}|\sin\theta_{ii} - \sum_{m=1}^{n}|V_m||Y_{im}|\sin(\theta_{im} - \delta_i + \delta_m)$$

Replacing second term of the preceding equation with $-Q_i$, as given by Eq. (7.20), yields

$$\frac{\partial Q_i}{\partial |V_i|} = -|V_i||Y_{ii}|\sin\theta_{ii} + Q_i \qquad (7.123)$$

Based on the earlier assumption that $B_{ii} \gg Q_i$, Eq. (7.123) simplifies to

$$\frac{\partial Q_i}{\partial |V_i|} = -|V_i|\,B_{ii} \qquad (7.124)$$

Similarly, assuming $(\theta_{im} - \delta_i + \delta_m) \approx \theta_{im}$, the relation for the off-diagonal terms of $[J_4]$ gets modified as follows

$$\frac{\partial Q_i}{\partial |V_m|} = -|V_i|\,B_{im} \qquad (7.125)$$

The simplified Eqs (7.120), (7.122), (7.124), and (7.125) can be used to obtain fast power flow solutions of power systems instead of the decoupled equations represented by Eqs (7.115) and (7.116). Thus,

$$[\Delta P] = [|V_i|\,B'_{im}]\,[\Delta\delta] \qquad (7.126)$$

$$[\Delta Q] = [|V_i|\,B''_{im}]\,[\Delta|V|] \qquad (7.127)$$

where B'_{im} and B''_{im} are elements of $[-B]$ matrix.

Dividing Eqs (7.126) and (7.127) by $|V_i|$ gives

$$\left[\frac{\Delta P}{|V|}\right] = [B'][\Delta\delta] \qquad (7.128)$$

$$\left[\frac{\Delta Q}{|V|}\right] = [B''][\Delta|V|] \qquad (7.129)$$

Matrices [**B'**] and [**B''**] are the negative of the imaginary component of the Y_{bus} matrix or the negative susceptance of the system. For a system of n buses, containing n_2 voltage-controlled buses, the order of [**B'**] matrix in Eq. (7.128) is $(n-1) \times (n-1)$. The order of [**B''**] matrix in Eq. (7.129) is $(n-1-n_2) \times (n-1-n_2)$ since P and V are specified and Q is not specified for n_2 buses, so the corresponding rows and columns in [**B''**] are eliminated.

Thus in the fast decoupled power flow algorithm, the successive changes of the voltage magnitude and phase angle are given by

$$[\Delta\delta] = [\boldsymbol{B'}]^{-1}\left[\frac{\Delta\boldsymbol{P}}{|V|}\right] \qquad (7.130)$$

$$[\Delta|V|] = [\boldsymbol{B''}]^{-1}\left[\frac{\Delta\boldsymbol{Q}}{|V|}\right] \qquad (7.131)$$

An inspection of [**B**] matrix shows that it is constant and need not be computed at every iteration. The power flow solution provided by Eqs (7.130) and (7.131) takes more number of iterations to converge, compared to the N–R method. However, the time per iteration is considerably less. Thereby, overall time taken for a power flow solution is considerably less and the fast decoupled method finds favour for conducting online and contingency studies wherein a large number of outages are to be simulated.

7.8.6 DC Power Flow Solution

Full power flow allows for management of both active and reactive power flows. Recently, with the liberalization of electricity markets, active power and reactive power are treated as different products. Active power is a tradable commodity, whereas reactive power is rather regarded as an ancillary service that has to be provided by the system operator and its costs are socialized among all users of the system. Due to the separation of these products, methods considering only the active power flow have gained increasing interest.

DC power flow is a further simplification of the fast decoupled power flow solution. It considers only active power flows, neglecting voltage support, reactive power management, and transmission losses. However, there are situations when speed of a solution is very important and inaccuracy can be lived with. Such types of situations arise in contingency analysis, where it is necessary to quickly assess its techno-economic impact. The fact that DC power flow problem is linear and hence simple, it is very often used for techno-economic studies of power systems for assessing the influence of commercial energy exchanges on active power flows in the transmission network.

The DC load flow solution is a linearization of the non-linear AC load flow problem. It is based on the following assumptions:

- Since line resistance $r \ll X$, all series resistances are neglected.
- The voltage angle $\delta_{im} = (\delta_i - \delta_m)$ is small; therefore $\sin\delta_{im} \approx \delta_{im}$ and $\cos\delta_{im} \approx 1$.
- All bus voltages have magnitude 1 pu.
- All in-phase transformers are on nominal tap.
- All shunt network elements are neglected.

As a consequence of the above assumptions, the only variables are voltage angles δ_{im} and all active power injections P_i. Hence, Eq. (7.131) can be dropped. The simulation of the DC power flow problem, therefore, as obtained from Eq. (7.130) is given below.

$$[\Delta\delta] = [B']^{-1} [\Delta P] \qquad (7.132)$$

The advantages of the DC power flow solution method are:

- Storage requirement of the system matrix $[B']$ is about half that of the full AC problem.
- Inversion of the $[B']$ matrix is required once only.
- Solution of the problem requires only one iteration.

The above advantages appreciably simplify the algorithm of computer programming and reduce the computation time.

The disadvantage of the method is that only flow of real power lines can be ascertained. No hint of reactive or apparent power flows in lines can be obtained.

7.9 COMPARISON OF POWER FLOW SOLUTION METHODS

Gauss–Seidel and Newton–Raphson iterative techniques are the two most commonly employed methods for solving PFE. Table 7.1 presents a comparison of the two methods.

Table 7.1 Comparison of the characteristics of the Gauss–Seidel and Newton–Raphson Methods

Parameter	Gauss–Seidel method	Newton–Raphson method
Type of formulation	Linear	Non-linear
No. of equations to be solved	$(n-1)$	$2(n-1)$
Format of power flow equations	Rectangular or polar	Rectangular or polar, but the latter is preferred due to better reliability of results
Algorithm for computer simulation	Simple	Complex
Memory requirement	Low	High
Accuracy	Low	High
Convergence type	Linear	Quadratic
Convergence rate	Slow	Fast
Number of iterations	More	Less
Time per iteration	Less	More
Time for solution	More	Less

The fast decoupled and DC load flow methods are simplifications of the basic Newton–Raphson method. It would not be out of place to review the main features of the three methods. Table 7.2 provides a brief comparison of the three methods.

Table 7.2 Comparison of the Newton–Raphson, fast decoupled, and DC load flow methods

Parameter	N-R method	Fast decoupled method	DC method
Accuracy of results	Accurate	Accurate	Approximate
Speed of solution	Slow	Fast	Very fast
Reliability of results	Good	Not so good	Good

7.10 POWER FLOW CONTROL BY REGULATING THE OPERATING CONDITIONS

From a power engineer's perspective, information about the system voltages and line flows available as output from solution of power flow equations does not provide insight as to how real power flow or voltage levels are adjusted. It is essential to control the flow of power in the system to meet the demands of load changes due to daily and seasonal variations. Both real and reactive power flows are controlled to meet the variations in load. The different types of control devices available for regulating the flow of power are as follows:

(i)   Generators at the power stations

(ii) Series and shunt capacitors and reactors

(iii) Regulating transformers

The real power output of a generator is governed by the torque angle, which in turn is dependent on the turbine torque. On the other hand, the reactive power output is dependent on the magnitude of the generator emf which is controlled by the generator field excitation. Hence, the two operating conditions which can be regulated at a generating station for power control are the turbine torque for real power control and field excitation for reactive power control. However, it is to be noted that in a power system, the number of generator buses available are relatively less in number.

Series and shunt capacitors and reactors constitute a secondary source of reactive power for controlling the flow of reactive power, thereby, maintaining the voltage profile. The series and shunt capacitors and reactors are categorized as static voltage control (SVC) devices and are strategically located in the power system. The SVC devices are switched in or switched out to regulate the flow of reactive power, and thus, maintain the voltage profile in the system.

Regulating transformers, including tap-changing transformers, constitute another source for regulating the operating conditions to control the flow of active and reactive power. Simulation of the regulating transformers and their representation in power flow studies are discussed in detail in the next section.

7.10.1 Simulation of Regulating Transformers

The flow of real and reactive power along transmission lines in power systems is controlled by tap-setting transformers. Tap-setting transformers perform the important function of regulating the power flow in a system. Therefore, their mathematical simulations in power flow studies are critical.

Transformers with Fixed Tap Setting

The mathematical simulation of fixed tap-setting transformers in the form of π-equivalent circuit has been discussed earlier in Section 5.10.1. The parameters of π-equivalent circuit [Fig. 5.32(c)] modify the system bus admittance matrix Y_{bus}. If an off-nominal turns ratio transformer is connected between bus i and bus m in a power system, the self and mutual elements Y_{ii}, Y_{mm}, and Y_{im} (Y_{mi}) of the Y_{bus} matrix get modified as follows.

$$Y_{ii} = Y_{i1} + Y_{i2} + \ldots + \frac{Y_{im}}{a} + \ldots + Y_{in} + \frac{Y_{im}}{a}\left(\frac{1}{a} - 1\right)$$

or
$$Y_{ii} = Y_{i1} + Y_{i2} + \ldots + \frac{Y_{im}}{a^2} + \ldots + Y_{in} \qquad (7.133)$$

Similarly,

$$Y_{mm} = Y_{m1} + Y_{m2} + \ldots + \frac{Y_{mi}}{a} + \ldots + Y_{in} + Y_{mi}\left(\frac{a-1}{a}\right)$$

or
$$Y_{mm} = Y_{m1} + Y_{m2} + \ldots + Y_{mi} + \ldots + Y_{in} \qquad (7.134)$$

The mutual admittance term between bus i and m or m and i is given by

$$Y_{im} = Y_{mi} = -\frac{Y_{im}}{a} \qquad (7.135)$$

From the foregoing, the following is concluded.

- The self-admittance of one of the buses between which the transformer is connected has a term Y_{im}/a^2 instead of Y_{im}, Eq. (7.133).
- The self-admittance of the second bus remains unchanged, Eq. (7.134).
- The mutual admittance between the buses, which connect the transformer is $-(Y_{im}/a)$, Eq. (7.135).

Representation of off-nominal turns ratio transformers in the Z_{bus} matrix The π-equivalent circuit given in Fig. 5.32(c) is used to represent the off-nominal turns ratio transformers for power flow studies. The following is the approach to include the effect of such type of transformers in the Z_{bus} matrix.

(a) The elements of the Z_{bus} are computed with the impedance between bus i and bus m taken equal to (a/Y_{im}).

(b) If the effect of the shunt elements, when computing the elements of the Z_{bus} matrix, has not been included, the bus currents are given by

$$I_i = \frac{(P_i + jQ_i)^*}{V_i^*} - Y_i V_i - \frac{Y_{im}}{a}\left(\frac{1}{a} - 1\right)V_i \qquad (7.136)$$

$$I_i = \frac{(P_m + jQ_m)^*}{V_m^*} - Y_m V_m - Y_{im}\left(\frac{a-1}{a}\right)V_m \qquad (7.137)$$

where y_i and y_m are shunt admittance to ground at bus i and bus m, respectively.

Transformers with Tap Changing Under Load (TCUL)

In order to maintain the voltage at specified values, under varying loads in a power system, tap changing under load (TCUL) transformers are used. The voltage at a specified bus is maintained by varying the tap settings, that is, the turns ratio of the TCUL transformer at that bus.

In order to maintain the voltage at a bus i in a system, the effect of the change in the turns ratio is simulated in power flow studies in the following manner.

Representation of TCUL transformers in the Y_{bus} matrix The voltage at bus i, is checked at the end of each alternate iteration. If the voltage at the bus exceeds the specified value, that is,

$$\text{abs}\left(V_i^k\right) - \text{abs}(V_i) > \varepsilon$$

where ε is a tolerance in voltage magnitude and is usually of the order of 0.01, a small change Δa is affected in the turns ratio. The change in tap setting is in percentage per step and is supplied by the manufacturer.

If the transformer is connected between bus i and bus m, the self-admittance Y_{ii} and Y_{mm} elements and mutual admittance Y_{im} and Y_{mi} elements of the Y_{bus} matrix system are re-computed when a change in tap setting is affected during a power flow study. In the Gauss and Gauss–Seidel iterative methods, the affected elements of the Y_{bus} matrix are computed before the commencement of the $(k + 1)$th iteration.

Representation of TCUL transformers in the Z_{bus} matrix The elements of the Z_{bus} matrix are required to be modified to incorporate a change in the tap setting. A change in tap setting is similar to adding impedance between two existing buses, as discussed in Section 6.8.2 (Type 3 modification).

Figure 7.8 shows a TCUL transformer of impedance aZ_{im} connected between two buses i and m, where a is the transformation ratio at a particular tap setting. It is necessary to compute the impedance between i and m when the turns ratio changes to $(a + \Delta a)$ due to a change in the tap setting. Assume that impedance bZ_{im} is connected in parallel with aZ_{im} such that the parallel combination has an impedance of $(a + \Delta a)Z_{im}$. Therefore,

$$(a + \Delta a)Z_{im} = \frac{aZ_{im} \times bZ_{im}}{aZ_{im} + Z_{im}}$$

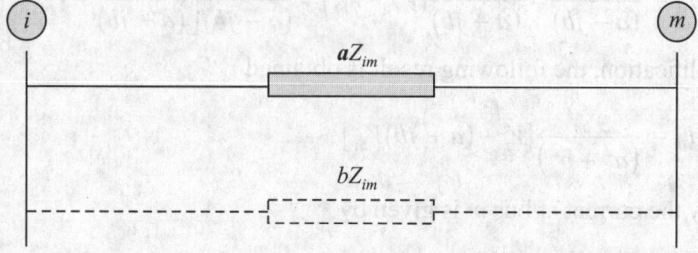

Fig. 7.8 Addition of an element to incorporate a change in tap setting

Simplification to obtain bZ_{im} gives

$$bZ_{im} = -\frac{a(a + \Delta a)}{\Delta a} = a\left(1 + \frac{a}{\Delta a}\right) \qquad (7.138)$$

The disadvantage of including TCUL transformers in the Z_{bus} matrix frame of reference is that for every change in tap setting, the Z_{bus} matrix has to be re-computed. The re-computation of the Z_{bus} matrix can be avoided by connecting the original impedance of the transformer in series between the two buses i and m and the shunt elements are varied to incorporate the changes in the tap setting.

Representation of Phase Shifting Transformers

Phase shifting transformers, in power systems are employed to control the flow of power. In power flow studies, a phase shifting transformer connected between buses i and m, is represented by admittance y_{im} in series with an ideal autotransformer having a complex turns ratio as shown in Fig. 7.9. Then

$$\frac{V_i}{V_j} = a + jb \tag{7.139}$$

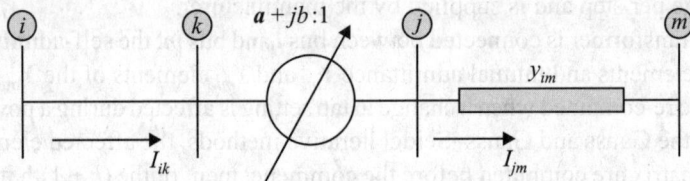

Fig. 7.9   Representation of a phase shifting transformer

Since the transformer is ideal,

$$V_i I_{ik}^* = V_j I_{jm}^*$$

Then
$$\frac{I_{ik}}{I_{jm}} = \frac{V_j^*}{V_i^*} = \frac{1}{a - jb} \tag{7.140}$$

From Fig. 7.9 it is observed that

$$I_{jm} = y_{im} (V_j - V_m)$$

Also, from Eqs (7.139) and (7.140)

$$I_{ik} = \frac{I_{jm}}{(a - jb)} = \frac{y_{im}}{(a - jb)}(V_j - V_m) = \frac{y_{im}}{(a - jb)}\left[\frac{V_i}{(a + jb)} - V_m\right]$$

On simplification, the following result is obtained

$$I_{ik} = \frac{y_{im}}{(a^2 + b^2)}[V_i - (a + jb)V_m] \tag{7.141}$$

Similarly the current at bus m is given by

$$I_{mj} = y_{im}(V_m - V_j) \tag{7.142}$$

Substituting for V_j from Eq. (7.139) in Eq. (7.142), and simplifying gives

$$I_{mj} = \frac{y_{im}}{(a + jb)}[(a + jb)V_m - V_i] \tag{7.143}$$

Modification of Y_{bus} to include phase shifting transformers Assume that a phase shifting transformer is connected between a bus i and bus m in a power system. The current at bus i is written as

$$I_i = I_{i1} + I_{i2} + \ldots + I_{ik} + \ldots + I_{in} \tag{7.144}$$

where $I_i = Y_{ii}V_i$

$$I_{i1} = Y_{i1}V_1$$

$$\vdots$$

$$I_{ik} = \frac{y_{im}}{(a^2 + b^2)}[V_i - (a + jb)V_m]$$

$$\vdots$$

$$I_{in} = Y_{in}V_n$$

In order to compute the self-admittance Y_{ii} of bus i, let $V_i = 1.0$ pu and short-circuit all other buses in the system. It is observed that

$$I_i = Y_{ii}$$

$$I_{i1} = Y_{i1}$$

$$I_{i2} = Y_{i2} \tag{7.145}$$

$$I_{ik} = \frac{y_{im}}{(a^2 + b^2)}$$

$$I_{in} = Y_{in}$$

Substituting Eq. (7.145) in Eq. (7.144)

$$Y_{ii} = Y_{i1} + Y_{i2} + \ldots + Y_{im} + \ldots + Y_{in} \tag{7.146}$$

where $Y_{im} = \dfrac{y_{im}}{(a^2 + b^2)}$

Thus, Eq. (7.150) provides the self-admittance of bus i when a phase shifting transformer is connected between bus i and bus m. From Fig. 7.9, it is observed that current from bus m to bus i is $-I_{im}$. Thus,

$$Y_{mi} = -I_{jm} = -y_{im}V_j$$

Substituting for V_i from Eq. (7.139) and putting $V_i = 1.0$ pu in the resultant expression, the mutual admittance Y_{mi} can be written as

$$Y_{mi} = -\frac{y_{im}}{a + jb}$$

The self-admittance of bus m is similarly obtained by putting $V_m = 1.0$ pu and short-circuiting all the other buses in the system.

$$Y_{mm} = Y_{m1} + Y_{m2} + \ldots + Y_{mi} + \ldots + Y_{mn}$$

where $Y_{mi} = Y_{mi}$

The mutual admittance is computed next. From Fig. 7.9, the current flowing from bus i to bus m is I_{ik}. Hence,

$$Y_{im} = I_{ik}$$

Knowing $I_{ik} = \dfrac{I_{jm}}{(a - jb)} = \dfrac{y_{im}}{(a - jb)}(V_j - V_m)$, hence,

$$Y_{im} = I_{ik} = \dfrac{y_{im}}{(a - jb)}(V_j - V_m)$$

Since $V_j = 0$, and $V_m = 1.0$ pu

$$Y_{im} = -\dfrac{y_{im}}{(a - jb)}$$

It may be noted from the expressions for mutual admittances that $Y_{im} \neq Y_{mi}$. This is to be expected since a phase shifting transformer is not a bilateral circuit.

For a given angular displacement and tap setting, the complex turns ratio is determined by the following expression

$$a + jb = a(\cos\theta + j\sin\theta)$$

where

$$a = \dfrac{\text{abs}(V_i)}{\text{abs}(V_j)}$$

If the phase shift from bus i to bus j is positive, that is, if the sign of θ is plus, then the voltage at the bus i leads the voltage at bus j.

SUMMARY

- Power flow studies of interconnected systems are undertaken in the steady state under normal operating conditions for the purpose of planning, operation, and control of power systems.
- The power flow equations are simulated in either admittance or bus frame of reference.
- Associated with each bus are four variables, namely $|V|$, $|\delta|$, P, and Q, leading to $4n$ variables in an n node system.
- Depending upon which of the variables specified, a bus is classified into (i) load bus (P and Q specified), (ii) voltage controlled bus (P and $|V|$ specified), and (iii) slack or swing bus ($|V|$ and δ specified).
- Iterative techniques are employed to solve power flow equations. In the Gauss–Seidel method, normally Y_{bus} frame of reference is used, while Newton–Raphson method employs the Z_{bus} frame of reference to formulate the PFE.
- In the iterative method, initial estimates of the unknown variables are made and new values are then computed. As soon as the new estimates of the variables at a bus are known, these replace the earlier values.
- At the end of an iteration, the new estimates are compared with the earlier estimates to determine whether the tolerance condition is reached. Once all the estimates of variables are within the specified tolerance, a solution of the power flow equations is achieved.
- Gauss elimination method employs triangular factorization to solve the power flow equations.
- Once all the $2n$ unknown variables are estimated, the load flows in the network are computed by applying the familiar Kirchoff's laws of electrical engineering.

SIGNIFICANT FORMULAE

- Complex power : $P_i - jQ_i = \sum\limits_{m=1}^{n} Y_{im} V_m V_i^*$ for $i = 1, 2, ..., n$

- Convergence criterion for voltage at load bus : $\left| V_i^{(k+1)} - V_i^k \right| < \varepsilon$ for $i = 1, 2, ... n$; and $i \neq s$

- Convergence criterion for voltage controlled bus : $\left| V_i^{(k+1)} - V_i^k \right| \leq \varepsilon$ for all load buses.

- Practical convergence criterion : $\left| \Delta P_i^k \right| \leq \varepsilon$ for all P-Q and P-V buses and $\left| \Delta Q_i^k \right| \leq \varepsilon$ for all P-Q buses.

- Slack bus power : $P_s - jQ_s = \sum\limits_{k=1}^{n} Y_{ik} V_k V_s^*$

- Line flow from bus i to m : $P_{im} + jQ_{im} = V_i (V_i^* - V_m^*) Y^* + |V_i|^2 Y_{sh}^* / 2$

- Line flow from bus m to i : $P_{im} + jQ_{im} = V_m (V_m^* - V_i^*) Y^* + |V_m|^2 Y_{sh}^* / 2$

- Gauss–Seidel method for load buses :

$$V_i^{k+1} = \frac{1}{Y_{ii}} \left\{ \frac{P_i - jQ_i}{\left(V_i^k \right)^*} - \sum\limits_{m=1}^{i-1} Y_{im} V_m^{k+1} - \sum\limits_{m>i}^{n} Y_{im} V_m^k \right\} \quad i = 2, 3, ..., n \text{ and } i \neq s$$

- Gauss–Seidel method for voltage controlled buses :

$$Q_i^{k+1} = V_i^k \sum\limits_{m=1}^{n} Y_{im} V_m^k \sin\left(\delta_i^k - \delta_m^k - \theta_{im} \right)$$

- Gauss elimination method:

  Forward substitution

$$V_i^{k+1} = \frac{1}{l_{ii}} \left(I_i^k - \sum\limits_{m=2}^{i-1} l_{im} V_m^{k+1} - \sum\limits_{m=i+1}^{n} l_{im} V_m^k \right) \text{ for } i = 2, 3, ..., n$$

  Backward substitution

$$V_n = V_n^{k+1}$$

$$V_i = V_i - \sum\limits_{m=i+1}^{n} u_{im} V_m \text{ for } i = n-1, n-2, ..., 2$$

- PFE in Z_{bus} frame of reference $\left[V_i^{k+1} - V_s \right] = \sum\limits_{m=1}^{i-1} Z_{im} I_i^{k+1} + \sum\limits_{m=i+1}^{n-1} Z_{im} I_i^k$

  where $I_i^k = \dfrac{(P_i + jQ_i)^*}{\left(V_i^k \right)^*}$ when no shunt elements are present and

$$I_i^k = \frac{(P_i + jQ_i)^*}{\left(V_i^k \right)^*} - Y_i \left(V_i^k \right)^* \text{ when shunt elements are present.}$$

- PFE for N-R method $\begin{bmatrix} \Delta\delta^k \\ \Delta|V|^k \end{bmatrix} = \begin{bmatrix} J_1^k & J_2^k \\ J_3^k & J_4^k \end{bmatrix}^{-1} \begin{bmatrix} \Delta P^k \\ \Delta Q^k \end{bmatrix}$

- Diagonal and off-diagonal elements of $[J_1]$:

$$\frac{\partial P_i}{\partial \delta_i} = \sum_{m \neq i} |V_i|\,|V_m|\,|Y_{im}|\sin(\theta_{im} - \delta_i + \delta_m)$$

$$\frac{\partial P_i}{\partial \delta_m} = -|V_i|\,|V_m|\,|Y_{im}|\sin(\theta_{im} - \delta_i + \delta_m) \text{ for } m \neq i$$

- Diagonal and off-diagonal elements of $[J_2]$:

$$\frac{\partial P_i}{\partial |V_i|} = 2|V_i|\,|Y_{im}|\cos\theta_{ii} + \sum_{m \neq i} |V_m|\,|Y_{im}|\cos(\theta_{im} - \delta_i + \delta_m)$$

$$\frac{\partial P_i}{\partial |V_m|} = |V_i|\,|Y_{im}|\cos(\theta_{im} - \delta_i + \delta_m) \text{ for } m \neq i$$

- Diagonal and off-diagonal elements of $[J_3]$:

$$\frac{\partial Q_i}{\partial \delta_i} = \sum_{m \neq i} |V_i|\,|V_m|\,|Y_{im}|\cos(\theta_{im} - \delta_i + \delta_m)$$

$$\frac{\partial Q_i}{\partial \delta_m} = -|V_i|\,|V_m|\,|Y_{im}|\cos(\theta_{im} - \delta_i + \delta_m) \text{ for } m \neq i$$

- Diagonal and off-diagonal elements of $[J_4]$:

$$\frac{\partial Q_i}{\partial |V_i|} = -2|V_i|\,|Y_{im}|\sin\theta_{ii} - \sum_{m \neq i} |V_m|\,|Y_{im}|\sin(\theta_{im} - \delta_i + \delta_m)$$

$$\frac{\partial Q_i}{\partial |V_m|} = -|V_i|\,|Y_{im}|\sin(\theta_{im} - \delta_i + \delta_m) \text{ for } m \neq i$$

- Rectangular version of PFE for N-R method: $\begin{bmatrix} \Delta P^k \\ \Delta Q^k \\ \Delta|V|^{2k} \end{bmatrix} = \begin{bmatrix} J_1^k & J_2^k \\ J_3^k & J_4^k \\ J_5^k & J_6^k \end{bmatrix} \begin{bmatrix} \Delta e^k \\ \Delta f^k \end{bmatrix}$

- Diagonal and off-diagonal elements of $[J_1]$:

$$\frac{\partial P_i}{\partial e_i} = \sum_{m=1}^{n} (G_{im}e_m - B_{im}f_m) + G_{ii}e_i + B_{ii}f_i$$

$$\frac{\partial P_i}{\partial e_m} = G_{im}e_i + B_{im}f_i$$

- Diagonal and off-diagonal elements of $[J_2]$:

$$\frac{\partial P_i}{\partial f_i} = \sum_{m=1}^{n} (G_{im}f_m + B_{im}e_m) - B_{ii}e_i + G_{ii}f_i$$

$$\frac{\partial P_i}{\partial f_m} = G_{im}f_i - B_{im}e_i$$

- Diagonal and off-diagonal elements of $[J_3]$:

$$\frac{\partial Q_i}{\partial e_i} = -\sum (B_{im}e_m + G_{im}f_m) - B_{ii}e_i + G_{ii}f_i$$

$$\frac{\partial Q_i}{\partial f_m} = G_{im}f_i - B_{im}e_i$$

- Diagonal and off-diagonal elements of $[J_4]$:

$$\frac{\partial Q_i}{\partial e_i} = \sum_{m=1}^{n}(G_{im}e_m - B_{im}f_m) - G_{ii}e_i - B_{ii}f_i$$

$$\frac{\partial Q_i}{\partial e_m} = -G_{im}e_i - B_{im}f_i$$

- Diagonal and off-diagonal elements of $[J_5]$:

$$\frac{\partial |V_i|^2}{\partial e_i} = 2e_i$$

$$\frac{\partial |V_i|^2}{\partial e_m} = 0$$

- Diagonal and off-diagonal elements of $[J_6]$:

$$\frac{\partial |V_i|^2}{\partial f_i} = 2f_i$$

$$\frac{\partial |V_i|^2}{\partial f_m} = 0$$

- Fast decouple PFE: $[\Delta|V|] = [B'']^{-1}\left[\dfrac{\Delta Q}{|V_i|}\right]$

- DC load flow: $[\Delta\delta] = [B']^{-1}[\Delta P]$

- Fixed tap-setting transformers: $Y_{ii} = Y_{i1} + Y_{i2} + \dots\dots + \dfrac{Y_{im}}{a^2} + \dots + Y_{in}\, a$

$$Y_{mm} = Y_{m1} + Y_{m2} + \dots\dots + Y_{mi} + \dots + Y_{in}$$

$$Y_{im} = Y_{mi} = -\frac{Y_{im}}{a}$$

- Off-nominal turns ratio transformers:

$$I_i = \frac{(P_i + jQ_i)^*}{V_i^*} - Y_iV_i - \frac{Y_{im}}{a}\left(\frac{1}{a} - 1\right)V_i$$

$$I_m = \frac{(P_m + jQ_m)^*}{V_m^*} - Y_mV_m - Y_{im}\left(\frac{a-1}{a}\right)V_m$$

- TUCL transformers: $bZ_{im} = a\left(1 + \dfrac{a}{\Delta a}\right)$

- Phase shifting transformers: $Y_{ii} = Y_{i1} + Y_{i2} + ... + Y_{im} + ... + Y_{in};\ Y_{im} = \dfrac{y_{im}}{\left(a^2 + b^2\right)}$

- $Y_{mm} = Y_{m1} + Y_{m2} + ... + Y_{mi} + ... + Y_{mn};\ Y_{mi} = -\dfrac{y_{im}}{(a + jb)}$

$$a = \frac{abs\left(V_i\right)}{abs\left(V_j\right)}$$

EXERCISES

Review Questions

7.1 Write a short essay on the power flow problem in electrical power systems.

7.2 Starting with the network performance equations, formulate the PFE in the Y_{bus} frame of reference when the bus voltages and admittance are expressed in rectangular and polar forms respectively.

7.3 Formulate the PFE for a power system network in the Z_{bus} frame of reference.

7.4 Discuss the computational aspects of the power flow problem.

7.5 Describe the iterative technique for solving PFE and explain (a) bus mismatches, and (ii) convergence criteria.

7.6 (a) Explain the methodology for computing power at (i) a slack bus, (ii) load buses, and (iii) voltage controlled buses.

(b) Derive expressions for computing power flows in a transmission line.

7.7 Develop expressions for the N-R method for single dimensional case.

7.8 Develop expressions for the N-R method for n dimensional case.

7.9 Explain how the Gauss–Seidel iterative technique is applied to (i) load buses, and (ii) voltage controlled buses. What is acceleration of convergence?

7.10 Explain triangular factorization as applied in the Gauss elimination method to compute PFE.

7.11 Derive expressions for the elements of a Jacobian matrix in (i) polar form, and (ii) rectangular form.

7.12 Explain clearly the difference between N–R load flow, de-coupled load flow, and fast de-coupled load flow. How does the Jacobian matrix in each case differ?

7.13 (a) State why it is necessary to control flow of power in a system and enumerate the different types of control components and their functions.

(b) Derive expressions, in Y_{bus} and Z_{bus} frames of reference, to simulate transformers (i) with fixed tap settings, (ii) with off-nominal turns ratio, and (iii) with TCUL.

7.14 Derive expressions to simulate phase shifting transformers in power flow studies.

Numerical Problems

7.1 A two-bus three-phase, 50-Hz, 450-MVA, 220-kV power system is connected by a transmission line as shown in Fig. 7.10.

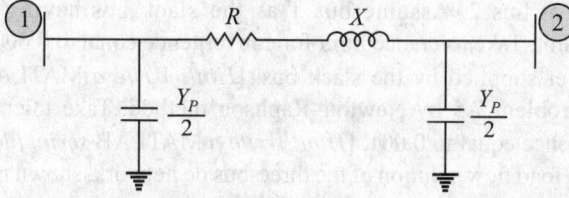

Fig. 7.10

The line is 250 km long and its parameters are as follows: $R = 0.1 \ \Omega/\text{km}$, inductance $L = 1.5 \ \text{mH/km}$, and capacitance $= 0.01 \ \mu\text{F/km}$. Assume a short transmission line and calculate the admittance matrix in (i) actual quantities and (ii) per unit. Assume the system power and voltage rating as the base values.

7.2   In Problem 7.1, calculate the real power transmitted when the voltage at bus 2 is held constant at 1.0 and the power factor of the load is 0.8 lagging. What is the power transmitted if the power factor of the load changes to 0.8 leading? Draw the phasor diagram in both cases. Assume the voltage of bus 1 constant at 1.05 per unit in both cases and $Y_P = 0$.

7.3   In Problem 7.2, determine the power at buses 1 and 2 for the lagging and leading power factors. For the leading power factor of 0.8, discuss the variation in voltage at bus 2 when the load increases. Determine the maximum magnitude of the voltage at bus 2. What is the real power load when the magnitude of the voltage at bus 2 reduces to 0.9 per unit?

7.4   A two-bus ac power circuit is connected by a line whose reactance is $j0.04$ per unit as shown in Fig. 7.11. The complex power inputs and outputs, in per unit, at the two buses are shown in the figure. If the voltage magnitudes at the two buses are maintained at 1.02, determine (i) the power angle, (ii) the reactive power flow in the line, (iii) load and power factor at buses 1 and 2. Assume real power input at the buses equal to 15 pu.

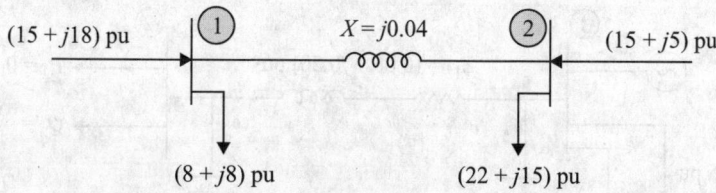

Fig. 7.11

7.5   Figure 7.12 shows a two-bus dc power system connected by a transmission line whose resistance $R = 0.05$ per unit. The load P on bus 2 is 1.0 per unit. Write the load flow equation and solve it by the Gauss–Seidel method to obtain the

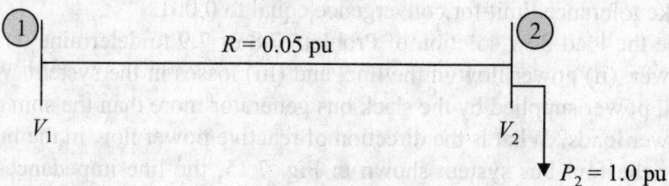

Fig. 7.12

voltage at bus 2. Assume bus 1 as the slack bus having a voltage of 1.0 per unit. Take tolerance limit for convergence equal to 0.0001. Determine the power supplied by the slack bus. [*Hint*: *Write a* MATLAB *script file*]

7.6 Solve Problem 7.5 by Newton–Raphson method. Take tolerance limit for convergence equal to 0.001. [*Hint*: *Write a* MATLAB *script file*]

7.7 Obtain a load flow solution of the three-bus dc network, shown in Fig. 7.13, by the Newton–Raphson method. All data is in per unit. Take tolerance limit for convergence equal to 0.001. From the solution determine (i) slack bus power, (ii) line flows, and (iii) system losses.

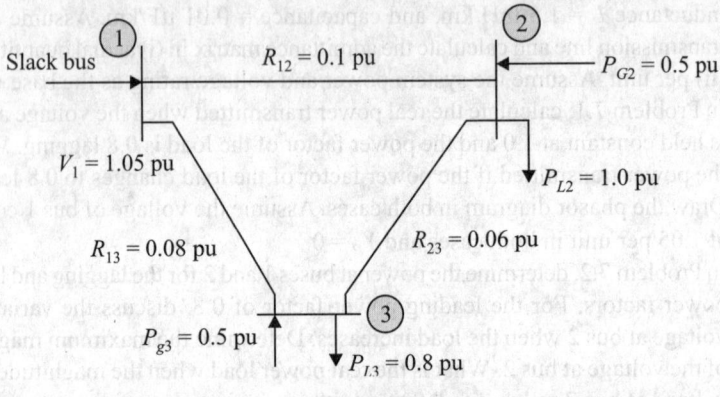

Fig. 7.13

7.8 The per unit impedance and line charging admittance of a line connecting buses 1 and 2 is $Z_{12} = 0.06 + j0.20$ and $Y_{sh} = 0 + j0.025$. Compute the Y_{bus} matrix. The per unit load on the buses is shown in Fig. 7.14. If the slack bus 1 voltage is fixed at $1.05 + j0$, determine the voltage of bus 2 by the Gauss–Seidel method. Take tolerance limit for convergence equal to 0.0001.

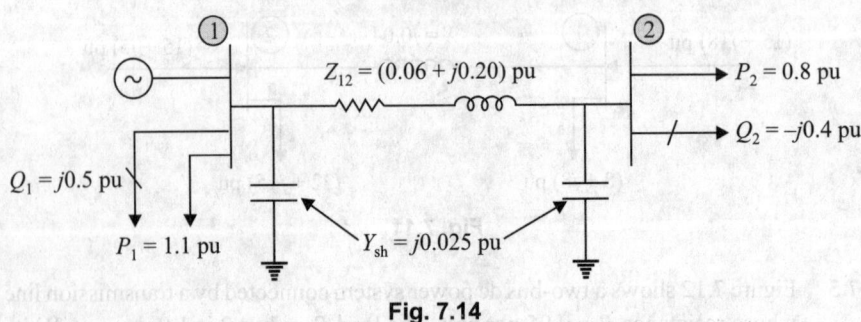

Fig. 7.14

7.9 Use the Newton–Raphson method to determine voltage of bus 2 in Problem 7.8. Take tolerance limit for convergence equal to 0.001.

7.10 Use the load flow solution of Problem 7.8 or 7.9 to determine (i) slack bus power, (ii) power flow in the line, and (iii) losses in the system. Why is the real power supplied by the slack bus generator more than the sum of the real power loads? What is the direction of reactive power flow in the line?

7.11 For the four-bus system shown in Fig. 7.15, the line impedances and line charging admittances in per unit on a 100 MVA base are given in the following table:

Bus nos		Line impedance	Line charging admittance
From	**To**	**Z**	**Y/2**
1	2	$0.04 + j0.06$	$0.0 + j0.030$
1	3	$0.03 + j0.25$	$0.0 + j0.015$
2	3	$0.05 + j0.15$	$0.0 + j0.020$
3	4	$0.06 + j0.10$	$0.0 + j0.010$
4	1	$0.02 + j0.06$	$0.0 + j0.025$

Formulate the Y_{bus} matrix.

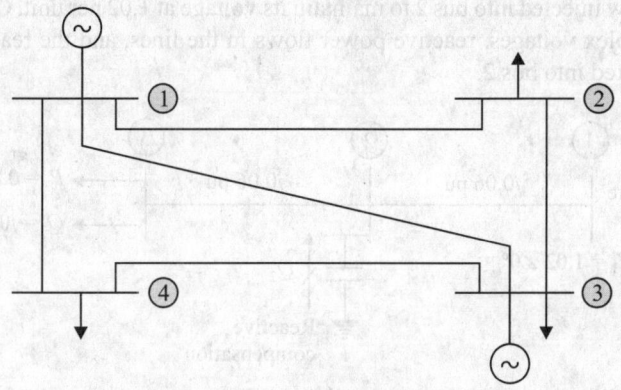

Fig. 7.15

7.12 If the scheduled generations and loads on the buses along with the assumed voltages for the network in Problem 7.11 are as given in the table, write the load flow equations of the system. Use the Gauss–Seidel method to obtain a load flow solution of the system.

Bus no.	Assumed bus voltage	Generation		Load	
		MW	*MVAR*	*MW*	*MVAR*
1	$1.04 + j0.0$ (slack bus)	0	0	0	0
2	$1.04 + j0.0$	0	0	40	30
3	$1.04 + j0.0$	50	40	20	15
4	$1.04 + j0.0$	0	0	60	20

7.13 Find the load flow solution of Problem 7.11 by the Newton–Raphson method. Assume that bus 4 is a voltage-controlled bus with a specified voltage of 1.04 and a real power input of 0.5 pu.

7.14 From the load flow solution of the four-bus network in Problem 7.11, compute (i) the slack bus power, (ii) the line flows and (iii) the total loss in the system.

7.15 Modify the N–R load flow equations to obtain a power flow solution of the power system in Problem 7.11 by using the fast decoupled N–R method. State and explain clearly the assumptions made. Compare the results obtained with the conventional N–R iterative technique.

7.16 The line impedances, in per unit, of a three-bus network are as follows:

Line 1-2 $0.02 + j0.04$
Line 1-3 $0.01 + j0.02$
Line 2-3 $0.02 + j0.05$

The per unit power injections at buses 1 and 2 are respectively given by $S_1 = (0.5 - j0.4)$ per unit and $S_2 = (0.8 - j0.6)$ per unit. Assemble the Z_{bus} matrix by assuming bus 3 to be the slack bus.

7.17 Use the Gauss–Seidel iterative technique in the Z_{bus} matrix frame of reference to obtain a load flow solution of the system. Calculate (i) the slack bus power, (ii) line flows in both directions, and (iii) system losses.

7.18 Figure 7.16 shows a three-bus system connected by short lines. All the data is provided in the figure. (i) Assuming that no reactive compensation is provided at bus 2, compute the complex voltages at buses 2 and 3. (ii) Reactive power is now injected into bus 2 to maintain its voltage at 1.02 per unit. Calculate the complex voltages, reactive power flows in the lines, and the reactive power injected into bus 2.

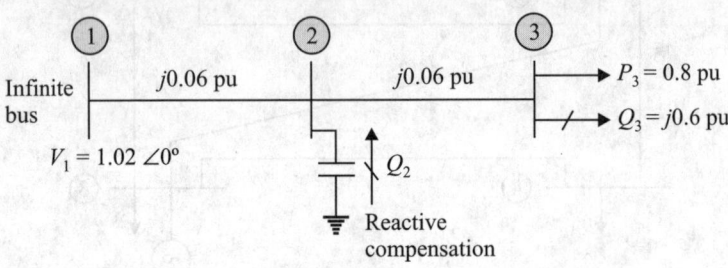

Fig. 7.16

7.19 For the four-bus power system in Problem 7.11, the series line impedances in per unit on 100 MVA and 220 kV base are given in the following table:

Bus nos		Line impedance
From	To	Z
1	2	$0.04 + j0.06$
1	3	$0.03 + j0.25$
2	3	$0.05 + j0.15$
3	4	$0.06 + j0.10$
4	1	?

The generation on bus 3 is $(1.0 + j0.8)$ per unit and the load on bus 4 is $(1.25 + j1.0)$ per unit. The voltages at buses 2 and 3 after a load flow are $1.04 \angle 0.5°$ per unit and $1.0 \angle -0.5°$ respectively. If bus 1 is the designated slack bus and its voltage is $1.04 \angle 0°$, determine (i) slack bus power, (ii) load at bus 2, (iii) voltage at bus 4, (iv) impedance of line 4-1. Show the real and reactive power flows in the lines and calculate the power system efficiency.

Multiple Choice Objective Questions

7.1 For which of the following, power flow studies are not undertaken?
  (a) Steady state
  (b) Transient state
  (c) Normal operating condition
  (d) Balanced three-phase systems
7.2 For which of the following, power flow studies are not performed?
  (a) Ensuring bus voltages are within specified
  (b) Ensuring rated frequency

(c) Ensuring specified power exchange between networks

(d) All of these

7.3 The PFE can be performed in

(a) Y_{bus} frame of reference only

(b) Z_{bus} frame of reference only

(c) either Y_{bus} or Z_{bus} frame of reference

(d) none of these

7.4 Which of the following is specified at a slack bus?

(a) $|V| \delta$ (b) $|V|, S_G$

(c) P_L (d) Q_L

7.5 Based on the specified variables which of the following would classify as voltage controlled bus?

(a) Net power injected is specified.

(b) Voltage magnitude and angle are specified.

(c) Voltage magnitude and real power injected are specified.

(d) None of these.

7.6 Which of the following would be sufficient to result in a solution of the power flow problem?

(a) Formulation of a mathematical model co-relating the bus voltages and powers

(b) Specifying voltage and power constraints at all the buses

(c) Computation of voltage magnitudes and phase angles at each bus

(d) All of these

7.7 The number of unknown variables in PFE required to be solved in an n bus system are

(a) $4n$ (b) $3n$

(c) $2n$ (d) n

7.8 Which of the following are associated with a bus i in a power system network?

(a) $|V_i|$ (b) δ_i

(c) Net injected P_i and Q_i (d) All of these

7.9 Which of the following describes the iterative process?

(a) It is a repetitive process.

(b) It is a direct process.

(c) The number of arithmetic operations are exact.

(d) It is accurate.

7.10 The solution of PFE by Gauss elimination is obtained by

(a) iteration   (b) computing the Jacobian of the matrix

(c) triangular factorization   (d) none of these

7.11 In an n node power system, number of $P-Q$ buses is m_1 and $P-V$ buses is m_2. Which of the following represents the number of unknown bus voltage angles δ required to be solved, by the N – R method, for $P-Q$ and $P-V$ buses?

(a) m_1 (b) $m_1 + m_2$

(c) $m_1 - m_2$ (d) m_2

7.12 Which of the following represents the diagonal elements of $[J_3]$?

(a) $\dfrac{\partial P_i}{\partial \delta_i} = \sum_{m \neq i} |V_i| \, |V_m| \, |Y_{im}| \sin(\theta_{im} - \delta_i + \delta_m)$

(b) $\dfrac{\partial P_i}{\partial |V_i|} = 2|V_i| \, |Y_{im}| \cos \theta_{ii} + \sum_{m \neq i} |V_m| \, |Y_{im}| \cos(\theta_{im} - \delta_i + \delta_m)$

(c) $\dfrac{\partial Q_i}{\partial \delta_i} = \sum_{m \neq i} |V_i|\ |V_m|\ |Y_{im}|\cos(\theta_{im} - \delta_i + \delta_m)$

(d) $\dfrac{\partial Q_i}{\partial |V_i|} = -2|V_i|\ |Y_{im}|\sin\theta_{ii} - \sum_{m \neq i} |V_m|\ |Y_{im}|\sin(\theta_{im} - \delta_i + \delta_m)$

7.13 Which of the following is a reason which makes decoupling of the N–R power flow solution feasible?

(a) X/R ratio is high.

(b) Voltage angle between two adjacent buses is small.

(c) Real power flow is largely governed by the bus voltage angle.

(d) All of these.

7.14 Which of the following is not a characteristic of the N–R method?

(a) Convergence is slow.

(b) Convergence is quadratic.

(c) Initial estimates must be close to the actual solution.

(d) All of these.

7.15 DC load flow method is used to compute, which of the following line flows?

(a) MVA   (b) MW

(c) MVAR   (d) All of these

7.16 Which of the following is applicable to the Gauss–Seidel method?

(a) Convergence is slow.

(b) Arithmetic operations required to complete an iteration are few.

(c) Efficient utilization of computer memory.

(d) All of these

7.17 Which of the following qualifies as an accurate and versatile method for solving the PFE?

(a) Gauss elimination   (b) Gauss–Seidel

(c) Newton–Raphson   (d) All of these

7.18 Which of the following is not used to control the flow of power in a system?

(a) Shunt capacitors and reactors   (b) Series capacitors and reactors

(c) Phase shifting transformers   (d) None of these

7.19 Which of the following represents the mutual admittance term when an off-nominal turns ratio is connected between bus i and m?

(a) Y_{im} (b) $-Y_{im}$

(c) Y_{im}/a (d) $-Y_{im}/a$

7.20 Which of the following gives the mutual impedance between buses i and m when the turns ratio of a TCUL changes to $(a + \Delta a)$?

(a) $a(1 + a/\Delta a)$ (b) $a(1 + \Delta a/a)$

(c) $a/\Delta a$ (d) $\Delta a/a$

Answers

7.1 (b)	7.2 (d)	7.3 (c)	7.4 (a)	7.5 (c)	7.6 (d)
7.7 (c)	7.8 (d)	7.9 (a)	7.10 (c)	7.11 (b)	7.12 (c)
7.13 (d)	7.14 (a)	7.15 (b)	7.16 (d)	7.17 (c)	7.18 (d)
7.19 (d)	7.20 (a)				

CHAPTER 8

Power System Controls

Learning Outcomes

A thorough study of this chapter will enable the reader to
- Understand the requirements to control a power system when it is operating in the steady state with both the real and reactive power loads continually varying
- Become aware of the two basic control loops in synchronous generators and analyse their performance with the help of block diagrams to regulate the voltage and control the frequency when the generators are operating in interconnected systems
- Perceive and simulate the functioning of a speed governing system and a turbo-generator
- Visualize a control area to analyse its performance from the perspective of load frequency control and determine its static and dynamic characteristics in addition to studying its response to addition of integral control
- Understand load frequency control in a multi-area system and analyse a two-area system

OVERVIEW

An electrical power system normally operates under steady-state condition. In this state, it is essential to maintain a power balance in the system. As is clear from the chapter on power flow studies, power balance means that the total active and reactive powers generated at power generating stations must match the sum of active and reactive powers demanded by the system loads and the active and reactive power transmission line losses. Under steady-state operation, both the system frequency and bus voltages are maintained at prescribed constant values. It is essential to maintain a power balance in the system, for the satisfactory performance of machines, industrial processes, etc.

In reality, the system is never under steady state as the active and reactive power load demands of the system are never steady but are continually changing with increase or fall in load demand. Thus the power output of generators must be adjusted at all times so that the power balance is maintained.

The active or real power delivered by a generator is changed by controlling the mechanical power output of a prime mover such as a steam turbine, hydro-turbine, gas turbine, or diesel engine depending on the type of generation. In case of a steam or hydro-turbine, the mechanical power is controlled by opening or closing of valves regulating input of steam or water flow into the turbine.

Thus, if the active power load increases, the active power generation must increase by opening wider steam and/or water valves. While if the active power load decreases, active power generation must decrease and this requires the valve opening to be smaller. In both these cases, the active power imbalance is sensed through the change in generator speeds and/or frequency. When there is excess active power generation, the generator will tend to speed up and the frequency will rise, while when there is deficiency in active power generation, the generator speed and frequency will fall. These deviations in speed and/or frequency from normal speed and/or frequency are used as control signals for causing appropriate valve opening or closing action automatically. The control action in this case is provided by governor-controlled mechanism. It may be noted that frequency constitutes a sensitive indicator of the energy balance in the power system that is normally used as the sensor portion of the load frequency control (LFC), the job of which is to provide such power balance automatically.

Apart from maintaining an active power balance, the maintenance of system frequency close to nominal value is also important. It is important to maintain the variations in frequency within acceptable limits around the nominal value in order to ensure that the synchronous clocks show correct time and the outputs and speeds of frequency-dependent equipment, such as pumps and fans, are maintained at the desired levels. In addition sustained operation at frequencies off the nominal value may lead to damage of the steam turbine blades and failure of the unit due to sub-synchronous resonance. The overall operation of a power system can be much better controlled if the frequency deviations are kept within strict limits.

The control of the voltage level in a power system is an important aspect. Practically all the equipment in operation in a power system are designed for rated, nameplate voltage rating. The performance of the device suffers if the voltage deviates from that value and its life is reduced. For example, the light intensity from a lamp varies strongly with the voltage; the torque of an induction motor varies as the square of the voltage, etc. Further, the active power loss or I^2R loss depends on the active and reactive line flows. The line losses can be minimized by selecting an optimum power flow in terms of active and reactive powers. Reactive power flow depends mainly on the line-end voltages, which can thus become a means of controlling active power loss.

There are different methods of voltage control in a power system. The control actions are based on a common principle that the voltage level of a bus is strongly related to the reactive power injections at that bus. Excitation control of generators can effectively maintain the voltage at the generator buses. Static Var Compensator (SVC) has the capability of continuously injecting or draining reactive power from a bus. Synchronous condensers when connected to a bus can control reactive power fed to the bus. Tap-changing transformers are also installed for control of voltage.

Based on the foregoing, the real and reactive powers are controlled separately. The load frequency control (LFC) loop is employed for controlling the real power and frequency. Automatic voltage regulator (AVR) loop is used for regulating the reactive power and voltage magnitude. The overall approach is to control the frequency and voltage in an interconnected system such that the system generates and supplies energy, as reliably and economically as possible, within the limits of economic feasibility.

8.1 BASIC CONTROL LOOPS IN A GENERATOR

Most large synchronous generators are equipped with two major control loops, namely automatic voltage regulator (AVR) loop and load frequency control (LFC) loop. A schematic block diagram of a synchronous generator equipped with these two control loops is shown in Fig. 8.1. A generator in an interconnected system is also equipped with the AVR and LFC control loops.

8.1.1 Automatic Voltage Regulator Loop

The automatic voltage regulator (AVR) loop (see Fig. 8.2) controls the magnitude of the terminal voltage V. The voltage magnitude $|V|$ is monitored continuously by a potential transformer and rectifier. This dc voltage signal, which is proportional to $|V|$, is compared with a dc reference voltage $|V|_{ref}$ and the error signal voltage Δe is amplified and fed to the exciter as input signal, and this in turn changes the voltage applied to the field winding of the synchronous generator.

8.1.2 Load Frequency Control Loop

The load frequency control (LFC) loop regulates the active power output and the frequency of the generator. It consists of a fast primary loop and a slower secondary loop. The function of the primary loop is to respond to relatively fast changes in load fluctuations that is sensed through changes in speed (or frequency) by the speed governor. A signal ΔP_C is generated which is amplified and transformed into real power signal ΔP_V. This signal ΔP_V regulates the control valves of steam or hydro-turbine unit, thereby changing the generator real power output that matches the fast load changes. The operation of this loop tends to maintain the active power balance. The action time of the primary loop ranges from one to several seconds and performs a coarse type of speed or frequency control.

A secondary loop is a slow acting loop and does fine adjustment of the frequency. In addition, it regulates the tie-line power interchange schedules with adjacent areas with which it is interconnected. The tie-line power deviation ΔP_{tie} and the frequency deviation Δf are detected, mixed, integrated, and added to the primary load frequency control loop for necessary corrective effect on the signal ΔP_C. The secondary loop is not sensitive to fast changes in load and frequency.

8.1.3 Cross-coupling between AVR and LFC Loops

A deficit of active power tends to decrease the system frequency and vice versa. The change in frequency will be uniformly felt throughout the system. However, in this process no resulting changes occur in the voltage levels of the system buses.

A deficit of reactive power tends to decrease the voltage level of a system. This change, however, is not uniform but will be greatest at the buses, where the deficit of reactive power is the greatest. In this case, the active powers of the loads are marginally affected with the change in voltage at the buses. Thus it can be inferred that LFC and AVR loops are decoupled from each other.

In addition, the AVR loop is much faster than the LFC loop. As a result of this, the dynamics of AVR settle down before these effects can make themselves felt in the slower LFC loop. However, in this book the AVR and LFC control loops are discussed independently and cross-coupling between the loops is neglected.

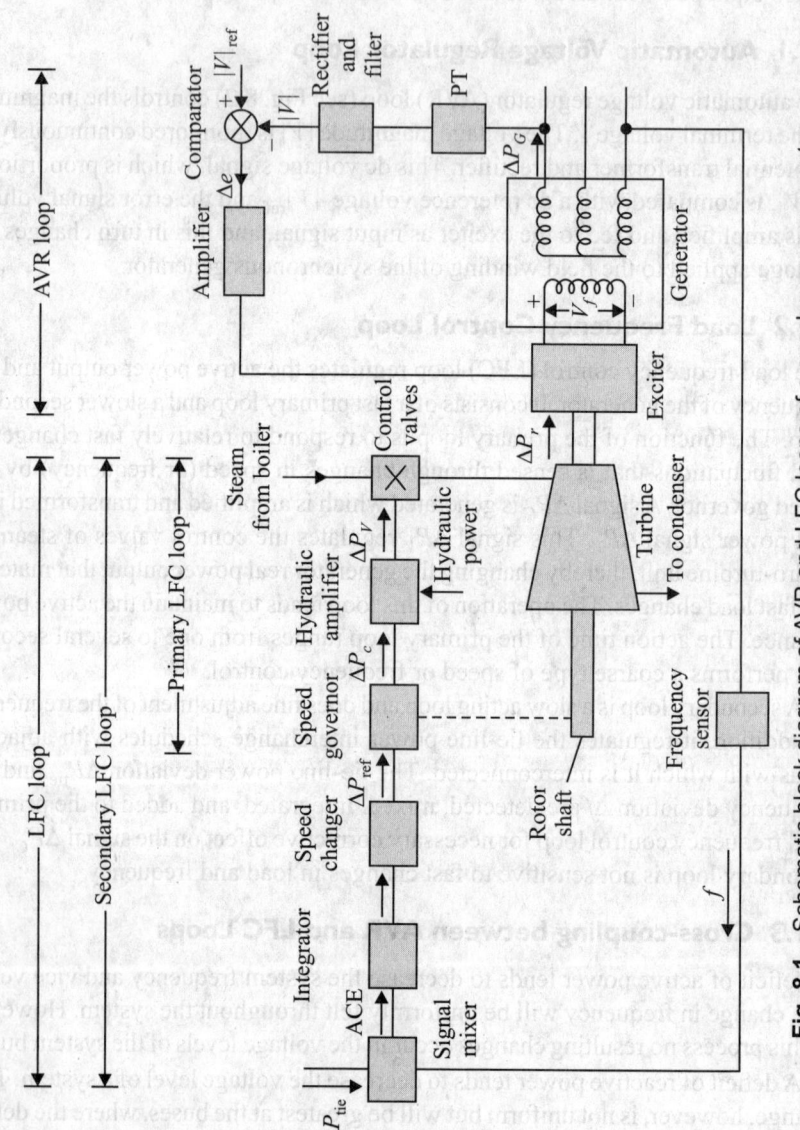

Fig. 8.1 Schematic block diagram of AVR and LFC loops of a synchronous generator

8.2 ANALYSIS OF STEADY-STATE AND TRANSIENT CONDITION

The electrical power system can be either in a steady-state condition or in a transient condition. In the steady-state condition, operating variables will not be constant; they will vary continuously, although these variations occurring about a certain mean value may be so small (usually of the order of a few per cent of normal operating values) that the operating conditions may be practically considered to be in the steady state. These variations are caused basically by continuous load variations and the response of the control devices of the system. Thus the analysis of performance of a power system in the steady state is associated with small deviations of system variables. Under these conditions, the analysis of power system dynamics, called small signal analysis, leads to differential equations of linear type and Laplace transform method can be used. In this analysis, the symbols $\Delta P, \Delta f, \Delta \delta$, and $\Delta |V|$ represent deviations from normal operating values of the variables — power, frequency, angle, and voltage magnitude respectively.

Transient condition in a power system is caused due to large disturbances arising out of a short-circuit and its subsequent clearing, switching off of a heavily loaded transmission line, etc. These result in considerable deviations of the operating variables from their initial values. In these types of large disturbances, the analysis of power system dynamics, called large signal analysis, leads to differential equations of the non-linear type. This type of analysis will be encountered later in the chapter on transient stability studies. In large signal analysis the symbols $P, f, \delta, |V|$ represent actual values of the variables—power, frequency, angle and voltage magnitude respectively.

8.3 AUTOMATIC VOLTAGE REGULATOR

The basic function of the AVR is to maintain a constant voltage magnitude at the generator terminals during normal steady-state operation and when small and slow changes take place in the load. However, the AVR loop is normally designed to function during emergency conditions.

The main constituent of the AVR loop is an *exciter*. In the earlier days, a dc generator mounted on the same shaft as the main synchronous generator was employed as an exciter. In this arrangement, dc power was transferred to the generator field winding through slip rings and brushes. Modern-day exciters are mainly either brushless or static with a good speed of response (rise time less than 0.1 s) and low power rating. Brushless excitation uses an ac generator with rotating rectifiers. In static exciters, the excitation power is obtained from the station service bus, which is rectified by a thyristor bridge and fed to the generator field via slip-rings.

8.3.1 Simulation of an Exciter

Figure 8.2 shows the schematic diagram of an AVR loop. When the terminal voltage magnitude $|V|$ increases, this immediately results in decrease of error voltage e, which in turn causes decrease in values of v_R, i_R, v_f, and i_f. As the generator field current i_f decreases, the magnitude of generator internal emf E and terminal voltage V reduces.

The mathematical model of the exciter will now be developed with the action of the stability compensator initially overlooked. For the comparator and the amplifier, the linearized equations can be written as

$$\Delta|V|_{ref} - \Delta|V| = \Delta e \tag{8.1a}$$

$$\Delta v_R = K_A \Delta e \tag{8.1b}$$

where K_A is the amplifier gain.

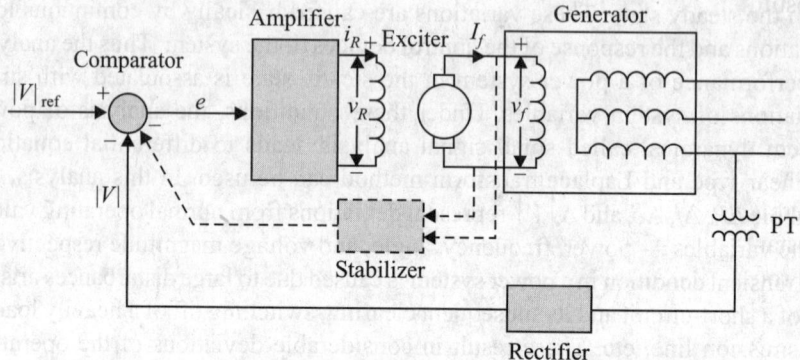

Fig. 8.2 Block diagram representation of the AVR loop

Laplace transformation of Eqs (8.1a) and (8.1b) gives

$$\Delta|V|_{ref}(s) - \Delta|V|(s) = \Delta e(s) \tag{8.2a}$$

$$\Delta v_R(s) = K_A e(s) \tag{8.2b}$$

Equation (8.2b) yields the transfer function gain $G_A(s)$ of the amplifier as

$$G_A(s) = \frac{\Delta v_R(s)}{\Delta e(s)} = K_A \tag{8.3}$$

Equation (8.3) shows that the response of the amplifier is instantaneous. However, in practice there is a delay in response of the amplifier which can be represented by a time constant T_A, and Eq. (8.3) can now be modified as follows

$$G_A(s) \cong \frac{\Delta v_R(s)}{\Delta e(s)} = \frac{K_A}{1 + sT_A} \tag{8.4}$$

The output from the amplifier is fed to the field winding of an exciter whose resistance and inductance are R_e and L_e respectively. The voltage in the exciter field circuit can be written as

$$R_e \times \Delta i_e + L_e \frac{d(\Delta i_e)}{dt} = \Delta v_R \tag{8.5}$$

where i_e is the exciter current.

Equation (8.5) is transformed by using Laplace transformation as follows

$$R_e \times \Delta i_e(s) + L_e s \Delta i_e(s) = \Delta v_R(s)$$

or $$\Delta i_e(s) = \frac{\Delta v_R(s)}{R_e + L_e s} \tag{8.6}$$

If it is assumed that the generator main field voltage is proportional to the exciter current, the expression for the field voltage is written as

$$\Delta v_f = K_1 \, \Delta i_e \tag{8.7}$$

where K_1 is the proportionality constant.

Laplace transformation of Eq. (8.7) yields

$$\Delta v_f(s) = K_1 \, \Delta i_e(s) \tag{8.8}$$

Substituting for $\Delta i_e(s)$ from Eq. (8.6) in Eq. (8.8) gives

$$\Delta v_f(s) = K_1 \frac{\Delta v_R(s)}{R_e + L_e s} = \frac{\dfrac{K_1}{R_e} \Delta v_R(s)}{1 + \dfrac{L_e}{R_e} s} = \frac{K_E \Delta v_R(s)}{1 + sT_E} \tag{8.9}$$

where $K_E = K_1/R_e$ is the exciter gain constant and $T_E = L_e/R_e$ is the exciter time constant.

From Eq. (8.9), the transfer function $G_E(s)$ of the exciter is obtained as follows

$$G_E(s) \cong \frac{\Delta v_f(s)}{\Delta v_R(s)} = \frac{K_E}{1 + sT_E} \tag{8.10}$$

The transfer function blocks of the comparator, amplifier, and exciter portions of the AVR loop are shown in Fig. 8.3. The magnitude of the amplifier time constant T_A is of the order of 0.02–0.10 s while the magnitude of the exciter time constant T_e lies between 0.5–1.0 s.

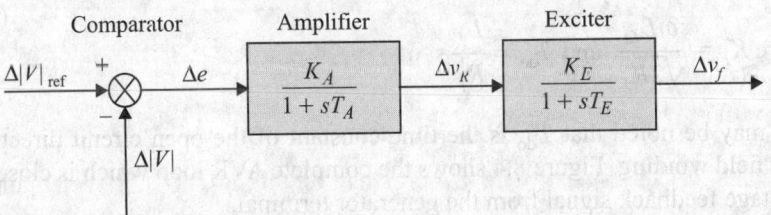

Fig. 8.3 Model of the comparator, amplifier, and exciter portions of an AVR loop

8.3.2 Simulation of a Generator

The terminal voltage $|V|$ of the generator is equal to the internal emf E minus the voltage drop due to internal impedance of the generator. The voltage, therefore, is a function of the load current. When the generator carries no load or is lightly loaded, the terminal voltage $|V|$ is approximately equal to the internal emf $|E|$. If R_f and L_{ff} are the resistance and self inductance respectively of the field winding, the field winding voltage Δv_f by the application of Kirchhoff's voltage law is written as

$$\Delta v_f = R_f \times \Delta i_f + L_{ff} \frac{d(\Delta i_f)}{dt} \tag{8.11}$$

By Faraday's law of electromagnetic induction, the internal generator voltage $|E|$ is expressed as

$$|E| = \frac{\omega\phi_{fa}}{\sqrt{2}}$$

where ϕ_{fa} is the flux linking the rotor field and the armature.

Assume that L_{fa} is the coefficient of mutual inductance between the rotor field and stator armature when their magnetic axes coincide. Thus, ϕ_{fa} can be expressed as

$$\phi_{fa} = L_{fa}i_f$$

Therefore, the internal generator voltage $|E|$ is expressed as

$$|E| = \frac{\omega L_{fa}i_f}{\sqrt{2}} \tag{8.12}$$

Substituting Eq. (8.12) in Eq. (8.11) leads to

$$\Delta v_f = \frac{\sqrt{2}}{\omega L_{fa}}\left[R_f\Delta E + L_f\frac{d(\Delta E)}{dt}\right] \tag{8.13}$$

Laplace transformation of Eq. (8.13) yields the transfer function $G_F(s)$ of the generator field, as follows

$$G_F(s) = \frac{\Delta E(s)}{\Delta v_f(s)} \cong \frac{\Delta|V|(s)}{\Delta v_f(s)} = \frac{\dfrac{\omega L_{fa}}{\sqrt{2}R_f}}{1 + \dfrac{L_f}{R_f}s} = \frac{K_F}{1 + sT'_{do}} \tag{8.14}$$

where $K_F = \dfrac{\omega L_{fa}}{\sqrt{2}R_f}$ and $T'_{do} = \dfrac{L_f}{R_f}$ s

It may be noted that T'_{do} is the time constant of the open circuit direct axis rotor field winding. Figure 8.4 shows the complete AVR loop which is closed by a voltage feedback signal from the generator terminal.

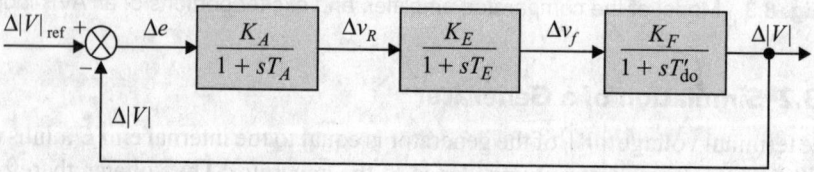

Fig. 8.4 Block diagram representation of a closed AVR loop

By substituting for $\Delta v_f(s)$ from Eq. (8.10) in Eq. (8.14) and then substituting for $\Delta v_R(s)$ from Eq. (8.4) in the resultant equation, the composite transfer function $G_C(s)$ of the AVR loop is obtained as follows

$$G_C(s) = \frac{K_F}{1 + sT'_{do}} \times \frac{K_F}{1 + sT_E} \times \frac{K_A}{1 + sT_A} = \frac{K}{(1 + sT_A)(1 + sT_E)(1 + sT'_{do})} \tag{8.15}$$

where $K = K_A K_E K_F$.

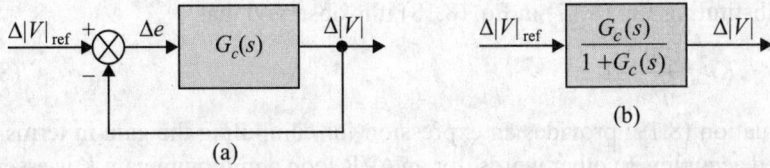

Fig. 8.5 (a) Combined model of the AVR loop (b) Closed loop transfer function

Figure 8.5(a) depicts the block diagram representation of the composite transfer function $G_C(s)$ of the AVR loop. Figure 8.5(b) represents the closed loop transfer function, the derivation of which is obtained in the following manner.

$$\frac{\Delta|V|(s)}{\Delta e(s)} = G_C(s)$$

Substituting for $\Delta e(s)$ from Eq. (8.2) yields

$$\Delta|V|(s) = G_C(s)[\Delta|V|_{\text{ref}}(s) - \Delta|V|(s)]$$

Simplification of the above identity provides the following closed loop transfer function

$$\frac{\Delta|V|(s)}{\Delta|V|_{\text{ref}}(s)} = \frac{G_C(s)}{1 + G_C(s)} \tag{8.16}$$

8.3.3 Analysis of the AVR Loop

An AVR loop is designed to control the magnitude of the generator terminal voltage $|V|$ within a specified static accuracy limit; fast speed of response and stable performance. Therefore, the performance of an AVR loop is analysed on the basis of its (a) static response and (b) dynamic response.

Static response The static accuracy of an AVR loop is defined as, for a constant reference input $\Delta|V|_{\text{ref0}}$, a constant error Δe_0 which must be less than a certain specified percentage p of the reference. Mathematically, the static accuracy is stated as

$$\Delta e_0 = \Delta|V|_{\text{ref0}} - \Delta|V|_0 < \frac{p}{100} \times \Delta|V|_{\text{ref0}} \tag{8.17}$$

With reference to Fig. 8.5(b) and taking $s = 0$ for the constant input, the transfer function is obtained as

$$\Delta|V|_0 = \frac{G_C(0)}{1 + G_C(0)} \times \Delta|V|_{\text{ref0}}$$

Substituting for $\Delta|V|_0$ in Eq. (8.17) leads to

$$\Delta e_0 = \left[1 - \frac{G_C(0)}{1 + G_C(0)}\right] \Delta|V|_{\text{ref0}} = \frac{1}{1 + G_C(0)} \times \Delta|V|_{\text{ref0}}$$

In Eq. (8.15), by substituting $s = 0$, $G_C(0)$ becomes equal to K. Hence,

$$\Delta e_0 = \frac{1}{1 + K} \times \Delta|V|_{\text{ref0}} \tag{8.18}$$

By substituting Eq. (8.17) in Eq. (8.18) it is observed that

$$K > \frac{100}{p} - 1 \tag{8.19}$$

Equation (8.19) provides an expression for computing the gain in terms of a desired accuracy. In other words, for an AVR loop a minimum gain K is essential for it to have a specified static accuracy p.

Dynamic performance   The response of the AVC loop in the time domain t can be obtained by taking inverse Laplace transform of Eq. (8.16) as follows

$$\Delta|V|(t) = L^{-1}\left\{\Delta|V|_{\text{ref}}(s) \times \frac{G_C(s)}{1 + G_C(s)}\right\} \tag{8.20}$$

Any control system, from an engineer's point of view, is usable if it is stable. A linear time invariant system is said to be stable when it provides a bounded output to every bounded input (BIBO).

In terms of linear systems, the stability requirement may be defined in terms of the location of the poles of the closed loop transfer function. The closed loop system transfer function of the AVR loop is given by Eq. (8.16). The denominator $[1 + G_C(s)]$ of Eq. (8.16) when equated to zero is called the characteristic equation. Therefore, the characteristic equation for the AVR loop is written as

$$1 + G_C(s) = 0 \tag{8.21}$$

The roots of the characteristic equation are called poles or eigen values. The bounded system response is achieved if the closed loop poles are located in the left half of the s-plane. In other words, the necessary and sufficient condition for a system to be stable is that all the roots of the characteristic equation must have negative real parts. The stability of a time invariant linear system can be checked by applying the Routh–Hurwitz criterion or by the root-locus method.

The open loop transfer function $G_C(s)$ for the AVR loop is of third order, and therefore, it has three poles s_1, s_2, and s_3. If the three poles are real and distinct, the transient response in the time domain would be of the form

$$A_1 e^{s_1 t}, A_2 e^{s_2 t}, A_3 e^{s_3 t} \tag{8.22}$$

On the other hand, if two of the poles, say s_2 and s_3, are complex conjugate, $\sigma \pm j\omega$, an oscillatory term of the type given below, is contained in the transient response

$$A_1 e^{\sigma t} \sin(\omega t + \beta) \tag{8.23}$$

For the AVR loop to be stable, the transient components must disappear with time. Then, the three poles must lie in the left-hand side of the s-plane. It is also desirable that the time decay of the transient components should be rapid for which it is necessary that poles are located deep into the left-hand side of the s-plane.

The terms A_1, A_2, and A_3 in Eqs (8.22) and (8.23) express the relative size of the transient process. If one or more terms are relatively large then the corresponding poles are called dominant poles. Usually the closer the pole is located to the $j\omega$ axis the more dominant it becomes.

Root locus plot of an AVR loop   The transient performance of a closed loop control system is directly related to the location of closed loop roots of the characteristic equation in the s-plane. It is frequently necessary to adjust one or more of the system parameters in order to obtain suitable root locations. The root loci is the path of the roots of the characteristic equation traced out in the s-plane as a system parameter is changed.

For AVR loop, the location of roots in the s-plane depends upon the parameters K, T_A, T_E, and T'_{do}. Of these parameters only the loop gain K is adjustable. Therefore, it would be appropriate to study the effect of variation of the magnitude of gain K on the location of the poles in the s-plane and hence transient responses of the AVR loop.

The open loop transfer function $G_C(s)$ of Eq. (8.15) has three poles located at $S_1 = (-1/T_A)$, $S_2 = (-1/T_{do})$, $S_3 = (-1/T_E)$. In Fig. 8.6, the open loop poles are marked with a cross sign ($\times$). Since there are three open loop poles, there will be three loci; each emanating from a pole ($\times$). When the gain K has a low value, the poles (marked with a square sign $\square$) are located close to the open-loop poles and are shown at their positions marked a in Fig. 8.6. It is to be noted that since T'_{do} has a relatively high time constant, the pole s_2 is located close to the origin. Thus, s_2 is a dominant pole giving rise to a slowly decaying exponential transient and making the overall response of the loop inadequately slow. Further a low K will lead to an imprecise static response since the inequality in Eq. (8.19) will not be fulfilled.

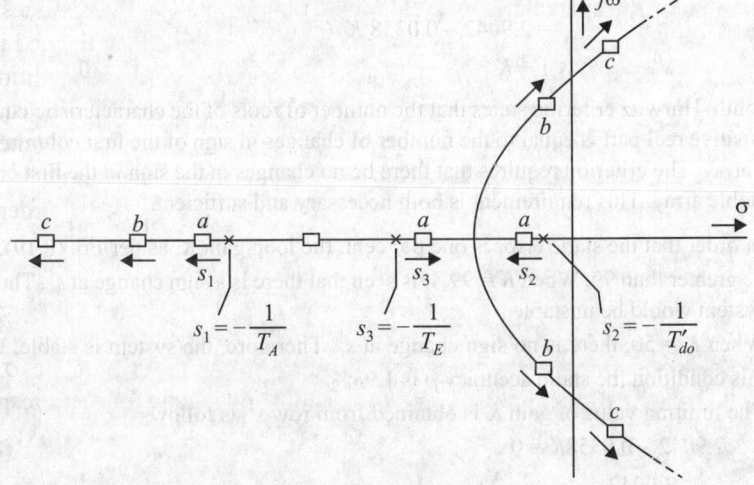

Fig. 8.6 Effect of variation in loop gain K on root loci of an AVR loop

When K is increased, the pole s_2 shifts to the left and the loop response is faster. If the increase in magnitude of K is continued, at some value of K, the poles s_2 and s_3 merge. Further increase in K results in a change in the nature of the poles, that is, they change from negative real poles to conjugate complex poles. Since the pole pair s_2 and s_3, shown at their positions marked b, are the dominant poles, the response of the loop is oscillatory. Any further increase in K causes the pole pair s_2 and s_3 to cross into the right half of the s-plane, shown at their positions marked c, where the poles are complex conjugate with positive real values. The latter causes the loop to become unstable.

Example 8.1   The time constants of the AVR loop in Fig. 8.3 are as follows:

$$T_A = 0.04 \text{ s}, T_E = 0.4 \text{ s, and } T'_{do} = 2.5 \text{ s.}$$

(a) It is desired to have a static accuracy $p = 1\%$. Use the Routh–Hurwitz criterion to determine whether the loop is stable. If the loop gain K is set equal to 50, is the system stable? What is the static accuracy?

(b) At what value of K is the loop marginally stable? Determine the static accuracy for this condition.

Solution   The open loop gain $G_C(s) = \dfrac{K}{(1+0.04s)(1+0.4s)(1+2.5s)}$

The characteristic equation is

$$1 + G_C(s) = 0$$

or

$$1 + \frac{K}{(1+0.04s)(1+0.4s)(1+2.5s)} = 0$$

or

$$0.04s^3 + 1.116s^2 + 2.94s + 1 + K = 0$$

The characteristic equation is a third-order polynomial. The Routh–Hurwitz array for the characteristic equation is as follows:

s^3	0.04	2.94
s^2	1.116	$1+K$
s^1	$\dfrac{1.116 \times 2.94 - 0.04 \times (1+K)}{1.116}$	0
	$= 2.9042 - 0.0358\,K$	
s^0	$1+K$	0

The Routh–Hurwitz criterion states that the number of roots of the characteristic equation with positive real part is equal to the number of changes in sign of the first column of the Routh array. The criterion requires that there be no changes in the sign in the first column for a stable array. This requirement is both necessary and sufficient.

(a) In order that the static error is one per cent, the loop gain K, as per Eq. (8.19), must be greater than 99. When $K = 99$, it is seen that there is a sign change at s^1. Thus, the system would be unstable.

When $K = 50$, there is no sign change at s^1. Therefore, the system is stable. Under this condition the static accuracy p is 1.98%.

(b) The limiting value of gain K is obtained from row s^1 as follows

$$2.9042 - 0.0358K = 0$$

or

$$K = \frac{2.9042}{0.0358} = 81.12$$

Therefore the system is stable when $K < 81.12$. From the s^2 row, for $K = 81.12$, the roots are obtained as

$$1.116s^2 = (81.12 + 1)$$

or

$$s = \pm j\sqrt{\frac{82.12}{1.116}} = \pm j8.58$$

From the foregoing it is observed that when $K = 81.12$, a pair of conjugate poles lie on the imaginary axis of the s-plane and the system is marginally stable. For $K = 81.12$, from Eq. (8.19) the static accuracy p is computed as approximately equal to $[(100/81)-1] = 0.23$.

8.3.4 Stability Compensation

From the foregoing example it is clear that for a low static error p, a high gain K is essential. The latter in turn causes an unsatisfactory dynamic response, leading to an unstable performance of the AVR loop. Thus, the requirement of a low static error and high gain are in conflict with each other. This divergent condition is overcome by including a series and/or feedback stability compensation device in the AVR loop as shown in Fig. 8.2.

The causes of a poor dynamic response of an AVR loop are the high values of cascaded time constants, T_A, T_E, and T'_{do}, which individually have a high magnitude. A compensation circuit, having some form of phase advancement characteristic, would help in resolving the problem of high-valued cascaded time constants. It is assumed that a phase lead compensator is connected in series in Fig. 8.4. If the transfer function of the phase lead compensator is

$$G_{se}(s) = 1 + sT_C$$

the open loop transfer function of the control loop, with the phase compensator in series, in Fig. 8.3 becomes

$$G_C(s) = \frac{K(1 + sT_C)}{(1 + sT_A)(1 + sT_E)(1 + sT'_{do})} \qquad (8.24)$$

Comparing Eq. (8.24) with Eq. (8.15), it is observed that the open loop transfer of the series compensated loop now contains a zero at $s = (-1/T_C)$. By the addition of a series compensator, the static loop gain K remains unchanged, and therefore, the static error is not affected. However, the dynamic response of the loop is improved.

Let the series compensator be considered to be tuned such that its time constant T_C is equal to the time constant T_E of the exciter. The open loop transfer function represented by Eq. (8.24) then becomes

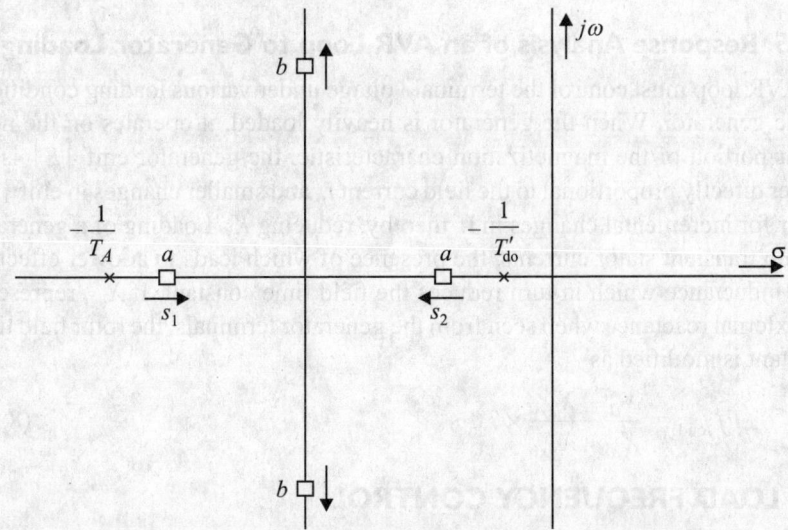

Fig. 8.7 Root loci of a series compensated AVR loop

$$G_C(s) = \frac{K}{(1 + sT_A)(1 + sT'_{do})} \qquad (8.25)$$

It is noted from Eq. (8.25) that there are only two poles compared to the three poles of Eq. (8.15). The poles of the compensated loop are shown in Fig. 8.7.

At low values of K, the pole s_1 still has a negative real part. The dominant pole s_2 produces a sluggish response term. As the gain K reaches high values, both poles s_1 and s_2 merge and produce an oscillatory response represented by the root locus b. Though the poles are complex conjugate, the damping of the oscillatory response does not reduce, as it happened in the uncompensated loop.

Example 8.2   The AVR loop of Example 8.1 is compensated by a series network compensator whose time constant is set equal to 0.4 s. Determine the loop gain K of the system for a stable dynamic response.

Solution   Since the time constant of the series compensator network is equal to the exciter time constant, the open loop gain is given by

$$G_C(s) = \frac{K}{(1 + 0.04s)(1 + 2.5s)}$$

The characteristic equation of the compensated loop is written as

$$1 + G_C(s) = 1 + \frac{K}{(1 + 0.04s)(1 + 2.5s)} = 0$$

or $0.1s^2 + 2.59s + 1 + K = 0$

The Routh–Hurwitz array is computed as follows

s^2	0.1	$1 + K$
s^1	2.59	0
s^0	$1 + K$	

From the s^0 array it is observed that for a stable dynamic response the loop gain K must be greater than -1.

8.3.5 Response Analysis of an AVR Loop to Generator Loading

An AVR loop must control the terminal voltage under various loading conditions of the generator. When the generator is heavily loaded, it operates on the non-linear portion of the magnetization characteristic, the generator emf $|E|$ is no longer directly proportional to the field current i_f, and smaller changes in emf $|E|$ occur for incremental changes in i_f, thereby, reducing K_f. Loading of a generator sets up transient stator currents, the presence of which leads to a lower effective field inductance which in turn reduces the field time constant. If X_{ext} represents the external reactance when seen from the generator terminals, the rotor field time constant is modified as

$$T_{d,\,load} = \frac{X'_d + X_{ext}}{X_d + X_{ext}} \times T'_{do} \text{ s} \qquad (8.26)$$

8.4 LOAD FREQUENCY CONTROL

A load frequency control system (LFC) performs the function of controlling small and slow changes in real power load and frequency when a system is operating in the steady state. A LFC system is also sometimes referred to as automatic generation control (AGC) system. Additionally an LFC system maintains the net interchange of power between pool members.

An LFC system is designed to effectively function during small and slow changes in real power load and frequency. For large imbalances in real power associated with rapid frequency changes that occur during a fault condition, an LFC system is unable to control frequency and is, therefore, ineffective.

In this section, the operation of the LFC system of a generator connected to local load is discussed. Later in the chapter, the case of several generators of a control area supplying load to all parts of the area will be discussed.

8.4.1 Simulation of the Speed Governing System

Figure 8.8 shows a schematic representation of the operating features of a speed governing system. The system has the following major components.

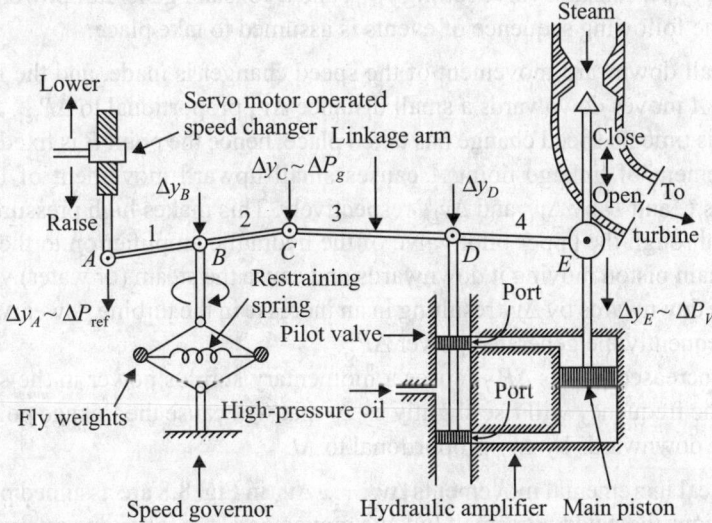

Fig. 8.8   Schematic representation of a speed governing system

Speed governor   It consists of centrifugal fly balls driven directly or through gears by the turbine shaft and provide up and down vertical movement of linkage point B proportional to the change in shaft speed.

Linkage mechanism   The linkage arms 1 and 2 are stiffly coupled, and so are arms 3 and 4. All five linkage points A, B, C, D, and E are free. This mechanism provides movement to the turbine valve through a hydraulic amplifier in proportion to change in shaft speed. Link arm 4 provides a feedback from the turbine valve movement.

Hydraulic amplifier   Very large mechanical forces are needed to position the main valve against the high steam (or hydro) pressure, and these forces are obtained via several stages of hydraulic amplifiers. In the simplified version, only one stage is shown in Fig. 8.8.

Speed changer   The speed changer consists of the servo-motor driven rotating vertical screw arrangement to move the linkage point A up and down. By adjusting the linkage point A, scheduled load at nominal frequency can be adjusted. The servo motor can be operated manually or automatically.

The qualitative analysis of the speed governing system requires an investigation of the incremental movements of the linkage points A to E in the linkage mechanism. The movements of the linkage points are measured in millimetres. However, these movements are calibrated against the resultant power increments. Therefore, the movement in positions of linkage points are assumed to correspond to power increments in megawatt or per unit megawatt as the case maybe. Thus, the change in position Δy_A of point A corresponds to a change in reference power setting ΔP_{ref}. Similarly, Δy_C corresponds to a change in governor output command ΔP_g and Δy_E corresponds to a change in valve position command ΔP_V, which changes the turbine power by ΔP_T.

The simulation of speed governing system is developed for small signal analysis around a nominal steady-state operating condition, characterized by a constant frequency f_0, a constant valve setting $y_{E(0)}$, and a constant generator power output $P_{G(0)}$. The following sequence of events is assumed to take place.

1. A small downward movement of the speed changer is made, and the linkage point A moves downwards a small distance Δy_A proportional to ΔP_{ref}.
2. At this time no speed change has taken place, hence the point B is fixed. Thus, movement of linkage point A causes small upward movement of linkage points C and D by Δy_C and Δy_D, respectively. This makes high pressure oil to flow through the upper pilot valve of the hydraulic amplifier on to the top of the main piston moving it downwards and cause the steam (or water) valve to move downwards by Δy_E resulting in an increase in the turbine power ΔP_T and consequently the generated power ΔP_G.
3. The increased power ΔP_G causes a momentary surplus power in the system, and the frequency will rise slightly by Δf that will cause the linkage point B to move downwards by Δy_B proportional to Δf.

All vertical incremental movements $\Delta y_A, \ldots, \Delta y_E$ in Fig. 8.8 are assumed positive in directions indicated. For small linkage movements, the following relationships can be written

$$\Delta y_C = k_1 \Delta f - k_2 \Delta P_{\text{ref}} \tag{8.27}$$

$$\Delta y_D = k_3 \Delta y_C + k_4 \Delta y_E \tag{8.28}$$

where the constants k_1 and k_2 depend on the lengths of linkage arms 1 and 2 and upon the constants of the speed changer and the speed governor. The constants k_3 and k_4 depend on the lengths of linkage arms 3 and 4.

If it is assumed that the flow of oil into the hydraulic amplifier is proportional to incremental movement Δy_D of the pilot valve, then the volume of oil entering the cylinder is proportional to the time integral of Δy_D. The movement Δy_E can be obtained by dividing the oil volume by the cross-sectional area of the piston as follows:

$$\Delta y_E = k_5 \int_0^t (-\Delta y_D) \, dt \tag{8.29}$$

where constant k_5 depends upon the orifice, cylinder geometries, and oil pressure.

Taking the Laplace transform of Eqs (8.27), (8.28), and (8.29), and eliminating Δy_C and Δy_D, we get

$$\Delta Y_E(s) = \frac{k_2 k_3 \Delta P_{ref(s)} - k_1 k_3 \Delta F(s)}{k_4 + s/k_5} \tag{8.30}$$

where $\Delta Y_E(s)$, $\Delta P_{ref}(s)$, and $\Delta F(s)$ are Laplace transforms of Δy_E, ΔP_{ref}, and Δf, respectively.

Equation (8.30) may be rewritten in the following form

$$\Delta Y_E(s) = \frac{K_{SG}}{1 + sT_{SG}}\left[\Delta P_{ref}(s) - \frac{1}{R}\Delta F(s)\right] = G_{SG}(s)\left[\Delta P_{ref}(s) - \frac{1}{R}\Delta F(s)\right] \tag{8.31}$$

where R is the speed regulation due to governor action $= (k_2/k_1)$, K_{SG} is the static gain of the speed governing mechanism $= (k_2 k_3/k_4)$, T_{SG} is the time constant of the speed governing mechanism $= (1/K_4 K_5)$, and $G_{SG}(s)$ is the transfer function of the speed governing mechanism.

Equation (8.31) is represented in the form of a block diagram in Fig. 8.9.

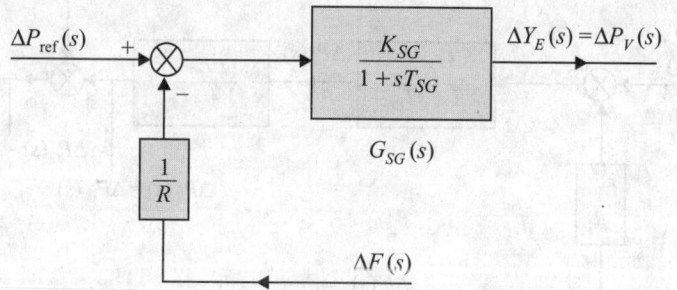

Fig. 8.9 Block diagram representation of a speed governing system

8.4.2 Simulation of the Turbo-generator

When an electrical power system is operating in the steady state, there is an exact balance in the turbine power P_T and the generator electromechanical air gap power P_G. As a result there is no acceleration and power is delivered at a constant speed or frequency. However, when a disturbance occurs, the balance in power between P_T and P_G is upset and the machine accelerates if $\Delta P_T > \Delta P_G$ and the machine will decelerate if $\Delta P_T < \Delta P_G$. When the machine is accelerating, $\Delta P_T - \Delta P_G$ is positive.

The change in turbine power ΔP_T is wholly dependent upon the incremental valve power ΔP_v and the response characteristics of the turbine. In the crudest model, a non-reheat steam turbine may be represented by the transfer function of $G_T(s)$ as

$$G_T(s) = \frac{\Delta P_T(s)}{\Delta X_E(s)} = \frac{\Delta P_T(s)}{\Delta P_V(s)} = \frac{K_T}{1 + sT_T} \tag{8.32}$$

where T_T is the time constant of the turbine and has a value in the range 0.2 to 2 s, and K_T is the gain constant of the turbine.

In the case of hydro turbines, although their design varies due to their dependence on water head yet their response characteristics are similar. The transfer function of a hydro-turbine is expressed as

$$G_T(s) = \frac{1 - 2sT_W}{1 + sT_W} \tag{8.33}$$

where T_W is the time required in seconds for the water to pass through the penstock to reach the turbine. The response time of hydro-turbines is of the order of several seconds.

Any increase in the load demand ΔP_L on the generator is immediately met by a corresponding increase in the generated power ΔP_G. In comparison to the slow variation in turbine power, the adjustment of the generator output ΔP_G to load changes ΔP_L is considered to be instantaneous. Therefore, it is possible to write

$$\Delta P_G = \Delta P_L \tag{8.34}$$

Figure 8.10 shows the transfer function representation of the power control mechanism of a generator with non-reheat turbine.

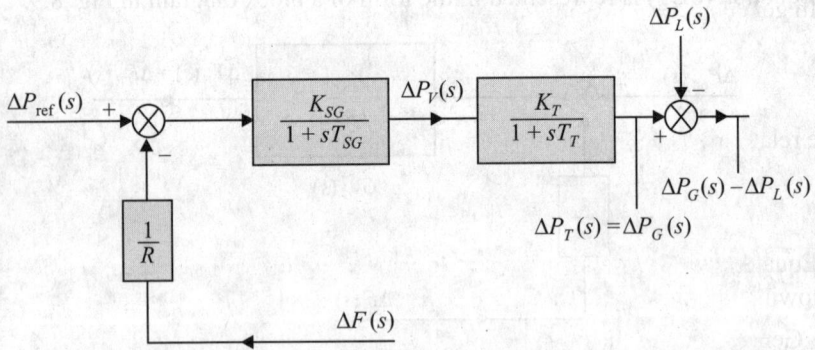

Fig. 8.10 Transfer function representation of power control mechanism of generator

8.4.3 Static Performance of a Speed Governor

The static performance of a speed governor of a generator, delivering power to a single load, can be obtained from an analysis of the LFC loop shown in Fig. 8.10. Let it be assumed that the load suddenly increases by an amount ΔP_L. The speed changer position will not be changed immediately and it can be assumed that $\Delta P_{ref} = 0$. Since there is no immediate change in turbine power, the generator will momentarily be delivering more power than it receives. It can do so by borrowing energy from the stored kinetic energy of the rotating mass of the turbine-generator. So the kinetic energy of the rotating mass is reduced and the speed and frequency will drop. The decrease in speed is sensed by the governor control mechanism shown in Fig. 8.8. The linkage point B moves upwards, and so also the pilot valve, causing the steam valve to open and thus increase generation.

Since more power is now generated, less energy will need to be borrowed from the kinetic energy, and the speed will drop at a decreasing rate. Finally a new steady state is reached characterized by lower speed and a new generation that has increased by the exact amount to offset the load increase.

From the transfer function representation shown in Fig. 8.10, the static equilibrium can be determined. In this case $\Delta P_{ref} = 0$. From Fig. 8.10,

$$\Delta P_G(s) = \frac{K_{SG}K_T}{(1+sT_{SG})(1+sT_T)}\left[-\frac{\Delta F(s)}{R}\right] \tag{8.35}$$

It is very practical to arrange $K_{SG}K_T = 1$. By using final-value theorem, the correlation between the static signals is obtained as

$$\Delta P_{G,0} = -\frac{1}{R}\Delta f_0 \tag{8.36}$$

where subscript 0 refers to static response. This result explains the physical significance of the feedback parameter R, called the speed regulation. It may be observed that if the unit of $\Delta P_{G,0}$ is megawatt, the unit of droop or regulation R is hertz per megawatt. However, in power system analysis, power is expressed in per unit. Hence, if change in frequency is expressed in per unit (that is as a ratio of the base frequency; 50 Hz), the droop R is also expressed in per unit (that is per unit hertz per unit power).

In general, the following relation can be written from Fig. 8.10,

$$\Delta P_G(s) = \frac{K_{SG}K_T}{(1+sT_{SG})(1+sT_T)}\left[\Delta P_{ref}(s) - \frac{\Delta F(s)}{R}\right] \tag{8.37}$$

The relationship between static signals is obtained by letting $s \to 0$, which gives

$$\Delta P_{G,0} = \Delta P_{ref,0} - \frac{1}{R}\Delta f_0 \tag{8.38}$$

Equation (8.38) provides the basis for analysing the static performance for the following operating conditions.

(a) Generator connected to an infinite network: In this case frequency is independent of the power output of the generator. Hence, Eq. (8.38) takes the following form

$$\Delta P_{G,0} = \Delta P_{ref,0} \tag{8.39}$$

Thus, the turbine power output is directly proportional to the reference power and any variation in turbine power output can be easily met by giving an appropriate command to the servo-motor to change the reference power setting.

(b) Generator connected to a finite network with the servo-motor setting held constant: This condition is discussed at the start of this section and the response is given by Eq. (8.36).

(c) Both generator reference power and frequency experience changes simultaneously: It is observed from Eq. (8.38) that for a given setting of the servo-motor of the speed governor ($\Delta P_{ref,0}$ held constant), the variation of frequency against turbine power generation, $\Delta P_{G,0}$ is a drooping straight line, which explains why R is called droop.

Equation (8.38) represents a family of sloping lines in a frequency versus generation graph as shown in Fig. 8.11, each line corresponding to a specific reference power setting.

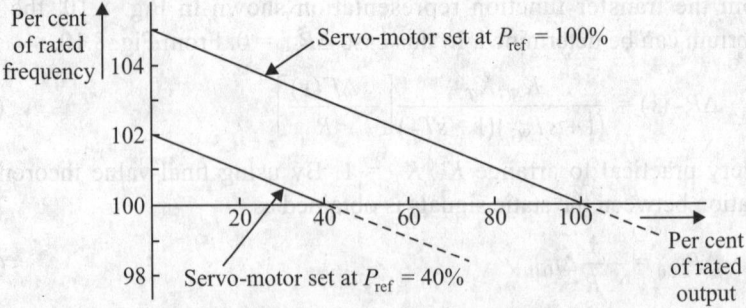

Fig. 8.11 Frequency versus power generation characteristics of a speed governor

Example 8.3 A 110-MW generator is supplying power to a network when the frequency drops by 0.1 Hz. (a) If the network is infinite, (i) what effect will the drop in frequency have on the servo-motor setting? (ii) How can the turbine generation be decreased by 10 MW? (b) If the network is finite, what is the increase in the turbine power if the servo-motor setting remains unchanged? Assume that the speed governor has a droop of 0.05 per unit and the system frequency is 50 Hz.

Solution
(a) Generator is supplying power to an infinite network
   (i) Since the generator is supplying power to an infinite network, frequency changes are independent of generator output.
   (ii) From Eq. (8.39) it is seen that $\Delta P_{G,\,0}$ is proportional to $\Delta P_{ref,\,0}$. Therefore, the turbine generation can be reduced by giving a command to the servo-motor of the speed changer to lower the turbine generation by 10 MW.
(b) Generator is supplying power to a finite network
In the example, droop is expressed in per unit. The droop in hertz per megawatt may be computed as

$$R = \frac{0.05 \times 50}{110} = 0.023 \text{ Hz/MW}$$

Since the servo-motor setting remains unchanged, Eq. (8.36) is used to compute the increase in turbine power. Substituting $\Delta f_0 = -0.1$ in the equation leads to

$$\Delta P_{G,0} = -\frac{1}{0.023}(-0.1) = 4.348 \text{ MW}$$

Example 8.4 Two generators, each rated at 200 MW and 300 MW, are supplying power to a network. Both the generators are loaded at 50% of their individual full rated capacity and the system frequency is 50 Hz. The load on the system decreases by 150 MW and the frequency rises by 0.5 Hz. Compute (i) the droop of each generator and (ii) the droop in per unit. Assume that the load is decreased on each generator in proportion to their individual rating.

Solution   Load shed by the 200-MW generator $= \dfrac{200}{200 + 300} \times 150 = 60 \text{ MW}$

Load shed by the 300-MW generator $= \dfrac{300}{200 + 300} \times 150 = 90 \text{ MW}$

(i) Using Eq. (8.36) yields the droop for the generators

$$R \text{ for the 200-MW generator} = -\frac{0.5}{60} = -0.0083 \text{ Hz/MW}$$

$$R \text{ for the 300-MW generator} = -\frac{0.5}{90} = -0.0056 \text{ Hz/MW}$$

(ii) R in per unit $= \dfrac{0.5/50}{60/200} = \dfrac{0.5/50}{90/300} = 0.0333$ pu

It may be noted that both the generators have equal droop when it is expressed in per unit and the generators share the load changes in proportion to their individual ratings.

8.4.4 Perception of Control Area

A single generator supplying power to a local area has been discussed so far, but in practice such a situation is far from reality. The analyses are not applicable to the more prevalent case of several generators operating in parallel to feed system load.

The LFC loop shown in Fig. 8.12 can be considered to represent the whole system, if individual control loops have the same regulation parameter and individual turbine generators have the same response characteristics. Then, the control of the generators in the system is said to be in unison and the entire system is referred to as a control area.

8.4.5 Incremental Power Balance of Control Area

The operation of the LFC system of a generator connected to local load is discussed till now. However, under the domain of a generator, a service area is to be supplied with power to feed an assortment of loads. It is intended to develop the incremental power-frequency dynamics of a single control area.

It is assumed that the area experiences a real load change of ΔP_L. Due to the action of governor controlled mechanism, the generator power increases by ΔP_G. The surplus power $\Delta P_G - \Delta P_L$ will be accounted for in two ways:

(a) By increasing the rate of rise of area kinetic energy
(b) By an increased load consumption

At nominal frequency f_0 (50 Hz), the stored area kinetic energy $W_{KE,0}$ is given by

$$W_{KE,0} = H \times P_{\text{rated}} \text{ MWs or MJ}$$

where P_{rated} is the MW rating of the generator and H is the inertia constant in seconds.

Parameter H is preferred over $W_{KE,0}$ as it is essentially independent of system size. Typical values of H lie in the range 2–8 s.

Due to the power surplus, the frequency throughout the system increases to a new value $f = f_0 + \Delta f$. Since the kinetic energy is proportional to the square of the frequency, the area kinetic energy of the rotating equipment W_{KE} at frequency f maybe expressed as

$$W_{KE} = W_{KE,0} \left(\frac{f}{f_0} \right)^2 = W_{KE,0} \left(\frac{f_0 + \Delta f}{f_0} \right)^2 \text{ MWs}$$

$$= W_{KE,0} \left[1 + \frac{2\Delta f}{f_0} + \left(\frac{\Delta f}{f_0} \right)^2 \right] \approx W_{KE,0} \left[1 + \frac{2\Delta f}{f_0} \right]$$

$$= H P_{\text{rated}} \left[1 + \frac{2\Delta f}{f_0} \right] \tag{8.40}$$

Then, the rate of rise of kinetic energy is

$$\frac{d}{dt}W_{KE} = \frac{2HP_{\text{rated}}}{f_0}\frac{d}{dt}\Delta f \tag{8.41}$$

All typical loads, being predominantly motor load, experience an increase in frequency by a parameter

$$D = \frac{\partial P_L}{\partial f} \text{ MW/Hz} \tag{8.42}$$

Thus the increase in load can be expressed as

$$\Delta P_L = D\Delta f \tag{8.43}$$

In order to maintain the power balance in the area, the change in generated power output ΔP_G must cater to the initial load, new load changes, and the rate of change in the kinetic energy of the rotating mass. Thus,

$$\Delta P_G - \Delta P_L = \frac{d}{dt}(W_{KE}) + D(\Delta f) \text{ MW} \tag{8.44}$$

Substituting Eq. (8.41) into Eq. (8.44) yields

$$\Delta P_G - \Delta P_L = \frac{2HP_{\text{rated}}}{f_0}\frac{d}{dt}(\Delta f) + D(\Delta f) \text{ MW} \tag{8.45}$$

Equation (8.45) when divided by P_{rated} MW to convert it into per unit becomes

$$\Delta P_G - \Delta P_L = \frac{2H}{f_0}\frac{d}{dt}(\Delta f) + D(\Delta f) \text{ pu MW} \tag{8.46}$$

It may also be noted that in Eq. (8.46) both ΔP_G and ΔP_L are now in per unit MW and D has the unit of per unit MW per Hz.

Laplace transformation of Eq. (8.46) yields

$$\Delta P_G(s) - \Delta P_L(s) = \frac{2H}{f_0}s\Delta F(s) + D\Delta F(s) = \left[\frac{2H}{f_0}s + D\right]\Delta F(s) \tag{8.47}$$

Equation (8.47) can be written in the form

$$\Delta F(s) = \frac{1}{\left[\dfrac{2H}{f_0}s + D\right]}[\Delta P_G(s) - \Delta P_L(s)]$$

$$= G_{PS}(s)[\Delta P_G(s) - \Delta P_L(s)] \tag{8.48}$$

where $\quad G_{PS}(s) = \dfrac{K_{PS}}{1 + sT_{PS}} \tag{8.49}$

$$K_{PS} = \frac{1}{D} \text{ Hz/pu MW} \tag{8.50}$$

$$T_{PS} = \frac{2H}{Df_0} \text{ s} \tag{8.51}$$

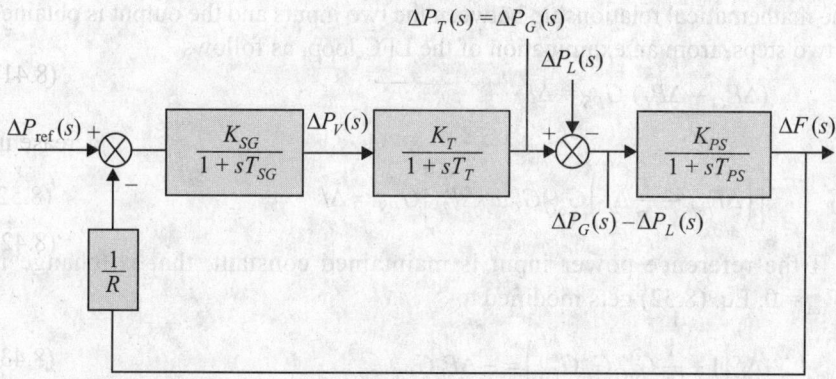

Fig. 8.12 Primary LFC loop for isolated power system

Equation (8.48) provides the transfer function block to be added to close the frequency loop and gives the mathematical relationship to corelate changes in frequency Δf and changes in turbine power ΔP_G. This is the primary closed loop control. Figure 8.12 shows the primary LFC loop of an isolated power system.

Example 8.5 The rated capacity of a power system supplying power to a certain area is 2500 MW and is operating at a normal initial load of 1500 MW. If the inertia constant is 4.0 s and all the generators supplying power to the area have a regulation of 2.0 Hz/ pu MW, determine (i) the initial area load frequency dependence parameter and its value in per unit of rated area capacity, (ii) gain, and (iii) time constant to represent the power system of the area. Assume that the load frequency dependence is linear.

Solution From the assumption in the problem it is clear that for a 1% change in frequency, the initial load will also change by 1% Therefore,

(i) The initial area load frequency dependence parameter is given by

$$D = \frac{\partial P_D}{\partial f} = \frac{1500}{50} = 30 \text{ MW/Hz}$$

Rated area capacity = 2500 MW
Hence, per unit rated area capacity

$$D = \frac{30}{2500} = 0.012 \text{ pu MW/Hz}$$

(ii) Using Eq. (8.50), $K_{PS} = \dfrac{1}{D} = \dfrac{1}{0.012} = 83.33$ Hz/pu MW

(iii) Using Eq. (8.51), $T_{PS} = \dfrac{2H}{Df_0} = \dfrac{2 \times 4.0}{0.012 \times 50} = 13.33$ s

8.5 PERFORMANCE OF PRIMARY LFC LOOP

In this section the primary LFC loop is analysed for its static and dynamic characteristics under the conditions of varying loads.

8.5.1 Static Characteristics

One of the pre-requisites of the LFC loop is that it should maintain constant frequency under varying load conditions. From the LFC loop, shown in Fig. 8.12, it is observed that there are two inputs, namely ΔP_{ref} and ΔP_L while the output is Δf.

The mathematical relationship between the two inputs and the output is obtained in two steps, from an examination of the LFC loop, as follows

$$(\Delta P_G - \Delta P_L)\, G_{PS} = \Delta f$$

or

$$\left[\left(\Delta P_{ref} - \frac{1}{R}\Delta f\right)G_{SG}G_T - \Delta P_L\right]G_{PS} = \Delta f \tag{8.52}$$

If the reference power input is maintained constant, that is, change in $\Delta P_{ref} = 0$, Eq. (8.52) gets modified to

$$\Delta f\left(1 + \frac{1}{R}G_{SG}G_T G_{PS}\right) = -\Delta P_L G_{PS}$$

or

$$\Delta f = -\frac{\Delta P_L G_{PS}}{\left(1 + \dfrac{1}{R}G_{SG}G_T G_{PS}\right)} \tag{8.53}$$

Taking the Laplace transform of Eq. (8.53) yields

$$\Delta F(s) = -\frac{G_{PS}(s)}{\left(1 + \dfrac{1}{R}G_{SG}(s)G_T(s)G_{PS}(s)\right)}\Delta P_L(s) \tag{8.54}$$

Let it be assumed that a constant step load change ΔP_L takes place. Then,

$$\Delta P_L(s) = \frac{\Delta P_L}{s} \tag{8.55}$$

Substituting for $\Delta P_L(s)$ from Eq. (8.55) into Eq. (8.54), and applying final value theorem to Eq. (8.54) yields the static frequency drop

$$\Delta f_0 = \lim_{s\to 0}\left[s\Delta F(s)\right] = -\frac{K_{PS}}{\left(1 + \dfrac{K_{PS}}{R}\right)}\times \Delta P_L = -\frac{\Delta P_L}{D + \dfrac{1}{R}} = -\frac{\Delta P_L}{\beta}\ \text{Hz}$$

$$\tag{8.56}$$

where $\beta = D + \dfrac{1}{R}$ is called area frequency response characteristic (AFRC).

Example 8.6   The system in Example 8.5 sheds its load by 1.2%. Determine (i) the constant step change in load in per unit MW and (ii) the static frequency change in Hz. Is there a decrease or increase in system frequency? (iii) What will be the reduction in frequency if the system load increases from zero to full load? (iv) If the speed generator loop was non-existent or open, what would be the percentage change in normal frequency? Discuss the results.

Solution   Drop in load $= 0.012 \times 2500 = 30$ MW

(i)   Constant step change in load $\Delta P_L = -\dfrac{30}{2500} = -0.012$ pu MW

(ii) $D + \dfrac{1}{R} = 0.012 + \dfrac{1}{2.0} = 0.512$ pu MW/Hz

$$\Delta f_0 = -\frac{(-0.012)}{0.512} = 0.0234 \text{ Hz}$$

There is an increase in frequency equal to 0.0234 Hz.

(iii) An increase in load from zero to full load implies that $\Delta P_L = 1.0$ pu MW. Drop in frequency, therefore is given by

$$\Delta f_0 = -\frac{1.0}{0.512} = -1.95 \text{ Hz}$$

(iv) Opening of the speed governor loop is equivalent to equating R to ∞. Thus,

$$D = 0.012 \text{ pu MW}$$

$$\Delta f_0 = -\frac{(-0.012)}{0.012} = 1.0 \text{ Hz}$$

Percentage change in frequency $= \dfrac{1.0}{50} \times 100 = 2.0\%$

8.5.2 Dynamic Characteristics

The inverse Laplace transform of Eq. (8.54) will give an expression of $\Delta f(t)$. The denominator of Eq. (8.54) is of third order since G_{SG}, G_T, and G_{PS} contain the time constants of T_{SG}, T_T, and T_{PS}. It is seen from Example 8.6 that the system time constant $T_{PS} = 13.33$ s, which is high compared to T_{SG} and T_T. Hence, it would be fair to assume that $T_{SG} = 0$ and $T_T = 0$.

With the above assumptions, Eq. (8.54) can be simplified to

$$\Delta F(s) \approx -\frac{G_{PS}(s)}{\left(1 + \dfrac{1}{R}G_{PS}(s)\right)} \times \frac{\Delta P_L}{s} \tag{8.57}$$

Substituting for G_{PS} from Eq. (8.49) leads to

$$\Delta F(s) \approx -\frac{\dfrac{K_{PS}}{1 + sT_{PS}}}{\left(1 + \dfrac{1}{R} \times \dfrac{K_{PS}}{1 + sT_{PS}}\right)} \frac{\Delta P_L}{s} \tag{8.58}$$

Simplification of Eq. (8.58) and resolving into partial fractions yields

$$\Delta F(s) = -\Delta P_L \frac{RK_{PS}}{(R + K_{PS})}\left[\frac{1}{s} - \frac{1}{s + \dfrac{R + K_{PS}}{RT_{PS}}}\right] \tag{8.59}$$

From Eq. (8.56), $\Delta f_0 = -\Delta P_L \dfrac{RK_{PS}}{R + K_{PS}}$ Hz. Hence Eq. (8.59) becomes

$$\Delta F(s) = \Delta f_0 \left[\frac{1}{s} - \frac{1}{\left(s + \dfrac{R + K_{PS}}{RT_{PS}}\right)}\right] \tag{8.60}$$

Taking inverse Laplace transformation of Eq. (8.60) gives

$$\Delta f(t) = \Delta f_0 \left(1 - e^{-t/\tau}\right) \tag{8.61}$$

where $\tau = \dfrac{RT_{PS}}{R + T_{PS}}$ = time constant of the closed loop system $\qquad$ (8.62)

If data from Example 8.6 is used, then values of $\Delta P_L = -0.012$ pu MW, $R = 2.0$ Hz/pu MW, $K_{PS} = 83.33$ Hz/pu MW, $T_{PS} = 13.33$ s, $\Delta f_0 = 0.02344$ Hz, and $\tau = 1/3.20$ s. Then,

$$\Delta f(t) = 0.02344\left(1 - e^{-3.20t}\right) \text{Hz} \qquad (8.63)$$

It may be noted that since there is a drop of load in the system an increase in frequency will result. On the other hand if the load on the system increases by the same amount ($\Delta P_L = 0.012$ pu MW), a decrease in frequency will be produced. The change in frequency with time is exponential with a static frequency increase of 0.02344 Hz. Figure 8.13 shows the sketch of frequency deviation versus time response of the LFC loop. If the time constants T_{SG} and T_T are taken into account, the frequency-time response of the LFC is oscillatory as shown in Fig. 8.13.

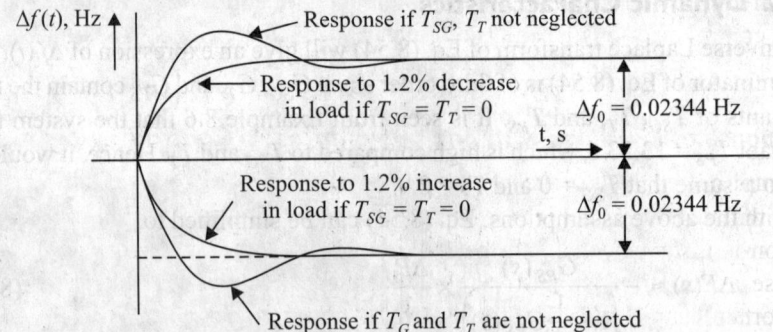

Fig. 8.13 Dyanamic response of the primary LFC loop

8.5.3 Addition of a Secondary Loop

From the discussions in the previous section it is observed that with a primary LFC loop, a change in system load results in a steady-state frequency deviation depending on the governor regulation. The steady-state drop in frequency from no-load to full-load in the illustrative example (Example 8.6) is 1.95 Hz. This much change in frequency cannot be tolerated as the system frequency specification is quite stringent. In order to reduce the frequency deviation to zero, a reset action must be provided. The reset action can be achieved by the addition of an integral controller as the secondary control loop. In the secondary control loop, the frequency error signal, after amplification, is integrated with respect to time and is fed to the speed changer to change the speed set point as shown in Fig. 8.14. The system now modifies to proportional plus integral control which forces the final steady-state frequency deviation to zero.

Mathematically, the change in speed changer position due to integral control action may be expressed as

$$\Delta P_{ref} = -K_I \int \Delta f \, dt \qquad (8.64)$$

where K_I is the amplification factor or gain constant. The unit of K_I is pu MW/s.

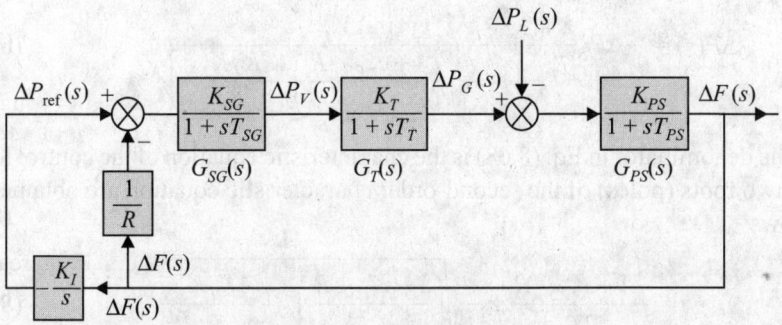

Fig. 8.14 Addition of integral (secondary) control loop

It may be noted that the polarity of the integral controller is negative. The polarity is selected so as to ensure an appropriate signal to the speed changer, that is, a positive frequency error should generate a negative or decrease command. The frequency error signal Δf fed into the integral controller is designated as the area control error (ACE) and is written as

$$ACE = \Delta f \qquad (8.65)$$

Response of LFC Loop with Addition of Integral Control

A simplified analysis of the LFC loop, shown in Fig. 8.13, to a step change in load ΔP_L will now be made. It is assumed that $T_{SG} = 0$ and $T_T = 0$; the speed changer action is instantaneous; and the ACE signal from the integrator is continuous. These assumptions simplify the mathematical analysis without distorting the important characteristics of the output. The errors introduced, as a consequence of these assumptions, affect the transient response and have no influence on the static response.

The Laplace transformation of Eq. (8.64) yields

$$\Delta P_{\text{ref}}(s) = -\frac{K_I}{s}\Delta F(s) \qquad (8.66)$$

For a step load change ΔP_L, the block diagram given in Fig. 8.14 leads to the expression

$$\Delta F(s) = \left(\Delta P_{\text{ref}}(s) - \frac{1}{R}\Delta F(s) - \frac{\Delta P_L}{s}\right)\frac{K_{PS}}{1 + sT_{PS}} \qquad (8.67)$$

Substituting Eq. (8.66) in Eq. (8.67) and simplifying gives

$$\Delta F(s)\left[1 + \left(\frac{K_I}{s} + \frac{1}{R}\right)G_{PS}(s)\right] = -\frac{\Delta P_L}{s}G_{PS}(s)$$

or $$\Delta F(s) = -\frac{\Delta P_L}{s}\frac{G_{PS}(s)}{\left[1 + \left(\frac{K_I}{s} + \frac{1}{R}\right)G_{PS}(s)\right]} = -\frac{\Delta P_L}{s}\frac{1}{\left[\frac{1}{G_{PS}(s)} + \left(\frac{K_I}{s} + \frac{1}{R}\right)\right]}$$

Substituting $G_{PS}(s) = [K_{PS}/(1 + sT_{PS})]$ and simplifying the resultant expression yields the following relation

$$\Delta F(s) = -\Delta P_L \frac{K_{PS}}{T_{PS}} \frac{1}{\left[s^2 + s\left(\frac{1 + K_{PS}/R}{T_{PS}}\right) + \frac{K_I K_{PS}}{T_{PS}}\right]} \tag{8.68}$$

The denominator in Eq. (8.68) is the characteristic equation of the control loop. The two roots (poles) of the second-order characteristic equation are obtained as follows

$$s_1, s_2 = -\frac{1 + K_{PS}/R}{2T_{PS}} \pm \frac{1}{2}\sqrt{\left(\frac{1 + K_{PS}/R}{T_{PS}}\right)^2 - \frac{4K_I K_{PS}}{T_{PS}}} \tag{8.69}$$

It is obvious that the character of the poles depends on the magnitudes of the integral gain K_I. If

$$K_I > \frac{T_{PS}}{K_{PS}}\left(\frac{1 + K_{PS}/R}{2T_{PS}}\right)^2$$

or

$$K_I > \frac{1}{4T_{PS}K_{PS}}\left(1 + \frac{K_{PS}}{R}\right)^2 \tag{8.70}$$

then the denominator can be written in the form $(s + \alpha)^2 + \omega^2$, where α and ω^2 are both positive and real. Then the poles are complex conjugate pair with negative real part. The time response of $\Delta f(t)$ is damped oscillatory.

In the limiting condition when the roots are identical, the magnitude of the critical gain setting $K_{I, \text{crit}}$ is given as

$$K_{I, \text{crit}} = \frac{1}{4T_{PS}K_{PS}}\left(1 + \frac{K_{PS}}{R}\right)^2 \tag{8.71}$$

For magnitudes of gain setting in the sub-critical zone, that is, $K_I < K_{I, \text{crit}}$, it may be observed that the two roots have negative real values. Therefore, the time response of $\Delta f(t)$ of the LFC loop is non-oscillatory and decay exponentially. Figure 8.15 shows the sketch of typical time response of $\Delta f(t)$ for various integral gain settings K_I.

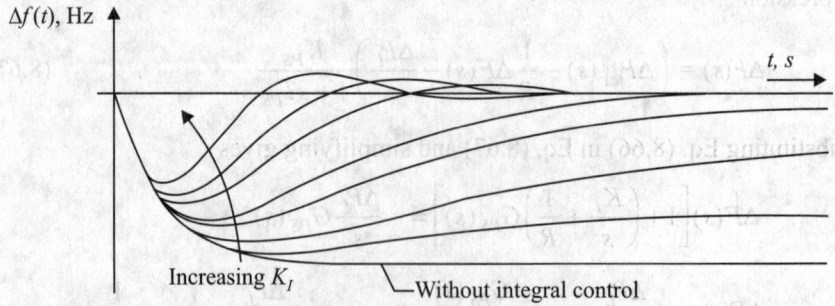

Fig. 8.15 Plot of D$f(t)$ versus time of the LFC loop with the secondary loop included

A careful study of the family of curves in Fig. 8.15 shows that for magnitudes of gain settings of K_I in the sub-critical zone, the frequency time response is non-oscillatory and slow. The effect of this lethargic response is that both the integral of $\Delta f(t)$ and the time error will be comparatively large. In practice such an LFC loop, with a sluggish response, is preferred to a fast response, since it prevents the generator from tracking swift and transitory load variations thus preventing unnecessary wear and tear of the equipment.

The physical action of the ALFC loop shown in Fig. 8.15 is interpreted in the following manner.

When there is a sudden demand for additional load on the system, the primary LFC loop immediately gets activated due to its faster response time. Thus, the response of the system is as depicted by Fig. 8.13. After a short time, depending upon the gain setting K_I, the integral controller (secondary LFC loop) comes into action and restores the frequency to its original value. It may be noted that the higher the gain setting K_I the shorter the response time of the secondary control loop.

8.6 LFC IN MULTI-AREA SYSTEM

In Section 8.4.4, the concept of control area was introduced. The generators closely coupled internally and swinging in unison are said to form a coherent group. Such a system is referred to as a control area. It is possible to let the LFC loop represent the control area. Modern-day electrical power systems can be considered as a number of LFC areas interconnected by means of tie-lines. The load frequency control of a two-area system connected by a tie-line will be studied now.

8.6.1 LFC of a Two-area System

A power system comprising two control areas interconnected by a weak lossless tie-line is considered. Each control area is represented by an equivalent generating unit interconnected by a tie-line with reactance X_{12}. Each area is represented by a voltage source behind an equivalent source reactance.

Under steady-state operation, the transfer of power over the tie-line P_{1-2} can be written as

$$P_{1-2} = \frac{|E_1||E_2|}{X_{12}} \sin(\delta_1 - \delta_2) \tag{8.72}$$

where $|E_1|$ and $|E_2|$ are the magnitudes of the end voltages of control areas 1 and 2 respectively, and δ_1 and δ_2 are the voltage angles of $|E_1|$ and $|E_2|$, respectively.

For small changes $\Delta\delta_1$ and $\Delta\delta_2$ in voltage angles, the change in tie-line power, ΔP_{1-2}, can be shown as

$$\Delta P_{1-2} = \frac{|E_1||E_2|}{X_{12}} \cos(\delta_1 - \delta_2)(\Delta\delta_1 - \Delta\delta_2)$$

or $\qquad \Delta P_{1-2} = T_{12}(\Delta\delta_1 - \Delta\delta_2) \tag{8.73}$

where T_{12} is defined as the synchronizing coefficient and is given by

$$T_{12} = \frac{|E_1||E_2|}{X_{12}} \cos(\delta_1 - \delta_2) \tag{8.74}$$

The frequency deviation Δf is related to the rate of change of internal voltage angle as

$$\Delta f = \frac{1}{2\pi}\frac{d}{dt}(\delta + \Delta\delta) = \frac{1}{2\pi}\frac{d}{dt}(\Delta\delta) \text{ Hz} \tag{8.75}$$

or $\quad \Delta\delta = 2\pi \int_0^t \Delta f \, dt \tag{8.76}$

Expressing Eq. (8.73) in terms of Δf rather than $\Delta\delta$ using Eq. (8.76), gives

$$\Delta P_{1\text{-}2} = 2\pi T_{12}\left[\int_0^t \Delta f_1 dt - \int_0^t \Delta f_2 dt\right] \tag{8.77}$$

Taking Laplace transform of Eq. (8.77)

$$\Delta P_{1\text{-}2}(s) = 2\pi T_{12} \times \frac{1}{s}[\Delta F_1(s) - \Delta F_2(s)] \tag{8.78}$$

The block diagram representation of Eq. (8.78) is shown in Fig. 8.16.

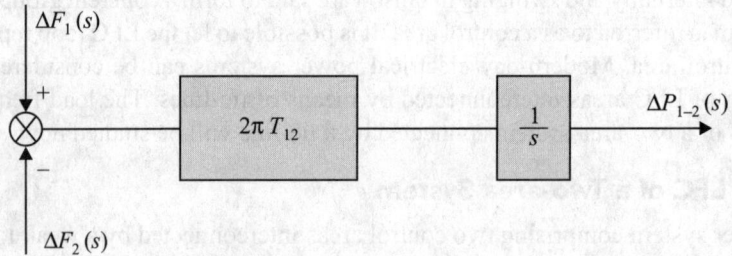

Fig. 8.16 Block diagram of a tie-line connecting two control areas 1 and 2

The flow of power over the tie-line from area 1 to area 2 is conventionally considered positive. As the losses in the tie-line are neglected, the transfer of power over the tie-line is expressed as

$$\Delta P_{1\text{-}2} = -\Delta P_{2\text{-}1} \tag{8.79}$$

8.6.2 Block Diagram Model of a Two-area System

Equation (8.47) provides an expression for maintaining a balance of power in a single control area. With the connection of the control area to another control area via a tie-line, the expression for transfer of power, Eq. (8.79), is required to be incorporated into the power balance equation.

An inspection of Fig. 8.12 shows that Eq. (8.47) is applicable at the second comparator. If two primary LFC loops, similar to those shown in Fig. 8.12, are used to model the two control areas then these can be interconnected by the tie-line as shown in Fig. 8.17.

Readers may note that in a single control area, per unit powers and other parameters H, R, and D are computed with the control area capacity as the base power. However, while computing per unit powers and parameters of two or multiple areas, constituted of varying capacities, a common base power is selected.

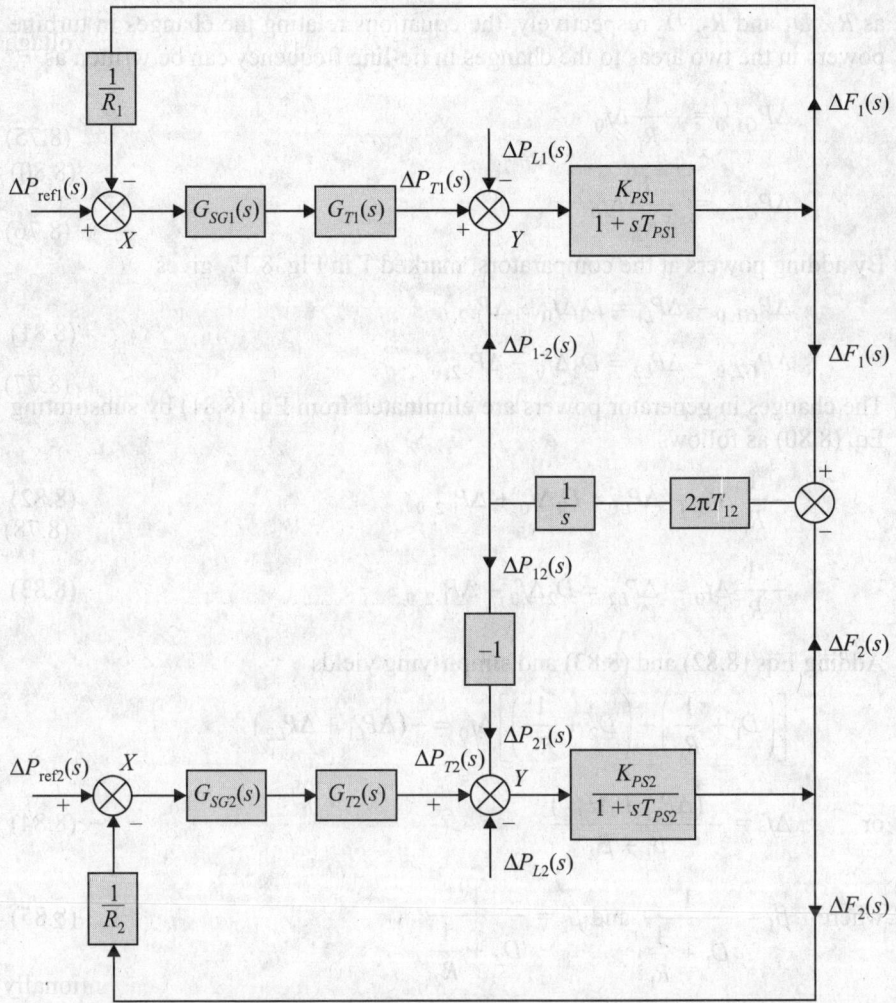

Fig. 8.17 Block diagram simulation of LFC of a two-area control system

8.7 ANALYSIS OF A TWO-AREA SYSTEM

Like the single area control system, the two-area system is also analysed for static response and dynamic response.

8.7.1 Static Response

As in the single area control system, the static response is undertaken by assuming that the positions of the speed changers in the two areas are held constant, that is,

$$\Delta P_{\text{ref 1}} = \Delta P_{\text{ref 2}} = 0$$

Now suppose that as a result of constant step changes in loads ΔP_{L1} and ΔP_{L2} in areas 1 and 2 respectively, the changes in tie-line frequency and power are, respectively, represented by Δf_0 and $\Delta P_{1\text{-}2,\,0}$. The static response is computed by letting $s \to 0$. If the static parameters for the control areas 1 and 2 are designated as R_1, D_1 and R_2, D_2 respectively, the equations relating the changes in turbine powers in the two areas to the changes in tie-line frequency can be written as

$$\Delta P_{G1,0} = -\frac{1}{R_1}\Delta f_0$$

$$\Delta P_{G2,0} = -\frac{1}{R_2}\Delta f_0 \qquad (8.80)$$

By adding powers at the comparators, marked Y in Fig. 8.17, gives

$$\Delta P_{G1,0} - \Delta P_{L1} = D_1 \Delta f_0 + \Delta P_{1\text{-}2,0}$$

$$\Delta P_{G2,0} - \Delta P_{L2} = D_2 \Delta f_0 - \Delta P_{1\text{-}2,0} \qquad (8.81)$$

The changes in generator powers are eliminated from Eq. (8.81) by substituting Eq. (8.80) as follows

$$-\frac{1}{R_1}\Delta f_0 - \Delta P_{L1} = D_1 \Delta f_0 + \Delta P_{1\text{-}2,0} \qquad (8.82)$$

$$-\frac{1}{R_2}\Delta f_0 - \Delta P_{L2} = D_2 \Delta f_0 - \Delta P_{1\text{-}2,0} \qquad (8.83)$$

Adding Eqs (8.82) and (8.83) and simplifying yields

$$\left[\left(D_1 + \frac{1}{R_1}\right) + \left(D_2 + \frac{1}{R_2}\right)\right]\Delta f_0 = -(\Delta P_{L1} + \Delta P_{L2})$$

or $\qquad \Delta f_0 = -\dfrac{(\Delta P_{L1} + \Delta P_{L2})}{\beta_1 + \beta_2} \qquad (8.84)$

where $\quad \beta_1 = \dfrac{1}{D_1 + \dfrac{1}{R_1}}$ and $\beta_2 = \dfrac{1}{D_2 + \dfrac{1}{R_2}} \qquad (8.85)$

Substituting Eq. (8.84) in Eq. (8.82) and simplifying leads to

$$\Delta P_{1\text{-}2,0} = -\Delta P_{2\text{-}1,0} = \frac{\beta_1 \Delta P_{L2} - \beta_2 \Delta P_{L1}}{\beta_1 + \beta_2} \qquad (8.86)$$

8.7.2 Dynamic Response

A mathematical simulation of the dynamic response of the two-area system, shown in Fig. 8.17, will result in a characteristic equation of the seventh order, rigorous analysis of which is time-consuming and cumbersome.

Dynamic analysis is simplified by considering two equal areas, that is, $R_1 = R_2 = R$; neglecting system damping, that is, $D_1 = D_2 = 0$. Also, assuming the response of the speed governing system of the turbine generator to be fast relative to rest of the system, that is, $G_{SG} = G_T = 1.0$.

The following power balance equations can be written directly by referring to Fig. 8.17,

$$\Delta F_1(s) = G_{PS1}(s)\left[-\frac{1}{R_1}\Delta F_1(s) - \Delta P_{L1}(s) - \Delta P_{1\text{-}2}(s)\right] \qquad (8.87)$$

$$\Delta F_2(s) = G_{PS2}(s)\left[-\frac{1}{R_2}\Delta F_2(s) - \Delta P_{L2}(s) - \Delta P_{1-2}(s)\right] \tag{8.88}$$

Knowing that $K_{PS} = 1/D$ and $T_{PS} = \dfrac{2H}{Df_0}s$, and using Eq. (8.49) gives

$$G_{PS1}(s) = G_{PS2}(s) = G_{PS}(s) = \frac{1}{s}\times\frac{f_0}{2H}$$

Also $R_1 = R_2 = R$. Then Eqs (8.87) and (8.88) now take the form

$$\Delta F_1(s) = \frac{1}{s}\times\frac{f_0}{2H}\left[-\frac{1}{R}\Delta F_1(s) - \Delta P_{L1}(s) - \Delta P_{1-2}(s)\right] \tag{8.89}$$

$$\Delta F_2(s) = \frac{1}{s}\times\frac{f_0}{2H}\left[-\frac{1}{R}\Delta F_2(s) - \Delta P_{L2}(s) + \Delta P_{1-2}(s)\right] \tag{8.90}$$

From Eqs (8.78), (8.89), and (8.90), the final relation for change in tie-line power is derived and is given as

$$\Delta P_{1-2}(s) = \frac{\pi f_0 T_{12}}{H}\times\frac{[\Delta P_{L1}(s) - \Delta P_{L2}(s)]}{\left[s^2 + \left(\dfrac{f_0}{2RH}\right)s + \dfrac{2\pi f_0 T_{12}}{H}\right]} \tag{8.91}$$

The denominator of Eq. (8.91) may be converted into the form

$$(s + \alpha)^2 + \omega^2 - \alpha^2 \tag{8.92}$$

where $\alpha = \dfrac{f_0}{4RH}$ and $\omega^2 = \dfrac{2\pi f_0 T_{12}}{H}$ (8.93)

Since both α and ω are positive and real, the time responses of the deviations in frequency and tie-line power are of damped, oscillatory, and stable nature. The frequency of oscillations will be

$$\omega_0 = \sqrt{\omega^2 - \alpha^2} = \sqrt{\frac{2\pi f_0 T_{12}}{H} - \left(\frac{f_0}{4RH}\right)^2} \tag{8.94}$$

The graphs showing the nature of variations of Δf_1, Δf_2, and ΔP_{1-2} are shown in Fig. 8.18.

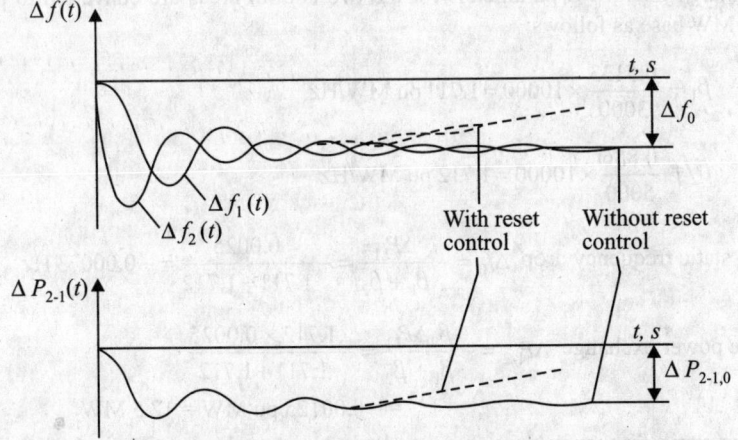

Fig. 8.18 Power and frequency oscillations on a tie-line following load change in area 2

Example 8.7   (a) A power system of capacity 3000 MW is supplying a normal load of 2000 MW in the stand-alone mode. Determine (i) linear frequency dependency parameter D, (ii) area frequency response characteristic (ARFC) parameter β, and (iii) static frequency drop for a load demand of 25 MW. Assume regulation $R = 2.0$ Hz/per unit MW.

(b) The area supplied by the power system in (a) is designated as control area 1 which is now interconnected, via a tie-line, to another control area 2 of capacity 5000 MW. If the per unit value of parameter β is the same as that in (a), (i) determine its per unit values on a 5000-MW base, (ii) determine the static frequency drop and tie-line power change, on a common base of 10000 MW, for a load demand of 25 MW in area 1, and (iii) compare the static frequency drop in control area 1, in pool operation, with the frequency drop in area 1 in stand-alone mode and explain. What is the percentage of load supplied by control area 2 to control area 1 and why?

Solution   (a) Since there is a linear relation between frequency and load,

(i)   The load frequency parameter $D = 2000/50 = 40$ MW/Hz.

If the system capacity is taken as base MW, $D = \dfrac{40}{3000} = 0.0133$ per unit MW/Hz

(ii)   ARFC parameter $\beta = D + \dfrac{1}{R} = 0.0133 + \dfrac{1}{2.0} = 0.5133$ per unit MW/Hz

(iii)   Load demand $\Delta P_L = \dfrac{25}{3000} = 0.0083$ per unit MW

Static frequency drop $\Delta f_0 = -\dfrac{\Delta P_L}{\beta} = -\dfrac{0.0083}{0.5133} = -0.0162$ Hz

(b) (i) ARFC parameter β on a base of 5000 MW is $\dfrac{0.5133}{3000} \times 5000 = 0.856$ pu MW/Hz

(ii)   Load demand of 25 MW in area 1 implies that on a base of 10000 MW,

$$\Delta P_{L1} = \dfrac{25}{10000} = 0.0025 \text{ pu MW}$$

Since there is no load demand in control area 2, $\Delta P_{L2} = 0$.

Similarly, the ARFC parameters for the two control areas are converted to per unit 10000-MW base as follows:

$$\beta_1 = \dfrac{0.5133}{3000} \times 10000 = 1.711 \text{ pu MW/Hz}$$

and $\beta_2 = \dfrac{0.856}{5000} \times 10000 = 1.712$ pu MW/Hz

Hence, static frequency drop $\Delta f_0 = -\dfrac{\Delta P_{L1}}{\beta_1 + \beta_2} = -\dfrac{0.0025}{1.711 + 1.712} = -0.00073$ Hz

Tie-line power exchange $\Delta P_{1\text{-}2} = -\dfrac{\beta_2 \Delta P_{L1}}{\beta_1 + \beta_2} = -\dfrac{1.712 \times 0.0025}{1.711 + 1.712}$

$$= -0.00125 \text{ pu MW} = 12.5 \text{ MW}$$

(iii) The static frequency drop in control area 1 in pool operation is 4.5%, which is (0.00073/0.0162) of the static frequency drop when the system is operating in the stand-alone mode. This is due to 'frequency support' received from control area 2 in the form of receipt of additional power of 12.5 MW via the tie-line.

Control area 2 supplies 50% of the load increase. Since $\beta_1 = \beta_2$,

$$\text{Tie-line power exchange } \Delta P_{1\text{-}2} = \frac{\Delta P_{L1}}{2}$$

Example 8.8 Two power systems in pool operation have the following parameters

$$R = 4.0 \text{ Hz/per unit MW, } H = 4.0 \text{ s, and } f_0 = 50 \text{ Hz.}$$

Assume that the tie-line capacity is 0.2 per unit and is operating at a power angle of 50°.

Solution Synchronizing coefficient $T_{12} = 0.2 \times \cos 50° = 0.1286$ pu

Oscillating frequency is given by

$$f_0 = \frac{1}{2\pi} \sqrt{\frac{2\pi f_0 T_{12}}{H} - \left(\frac{f_0}{4RH}\right)^2}$$

$$= \frac{1}{2\pi} \sqrt{\frac{2\pi \times 50 \times 0.1286}{4.0} - \left(\frac{50}{4 \times 4.0 \times 4}\right)^2}$$

$$= 0.49 \text{ Hz}$$

8.8 TIE-LINE BIAS CONTROL

The response curves in Fig. 8.18 clearly show that some sort of reset control must be applied to the two-area system. A control standard that has been adopted by most operating systems is termed as tie-line bias control. This control is based on the principle that all operating pool members must contribute their share to frequency control in addition to taking care of their own net interchange.

8.8.1 Tie-line Bias Control for Two-area System

In order to apply the control strategy stated earlier to the two-area system, the block diagram shown in Fig. 8.17 for the two-area system is modified, and the modified diagram is shown in Fig. 8.19.

The control error for each area is a linear combination of tie-line power and frequency errors and can be expressed as shown

$$ACE_1 = \Delta P_{1\text{-}2} + B_1 \Delta f_1$$

$$ACE_2 = \Delta P_{2\text{-}1} + B_2 \Delta f_2 = -\Delta P_{1\text{-}2} + B_2 \Delta f_2 \tag{8.95}$$

The signals to the respective speed changers will, therefore, be of the following type

$$\Delta P_{\text{ref1}} = -K_{I1} \int (\Delta P_{1\text{-}2} + B_1 \Delta f_1) \, dt$$

$$\Delta P_{\text{ref2}} = -K_{I2} \int (\Delta P_{2\text{-}1} + B_2 \Delta f_2) \, dt \tag{8.96}$$

where B_1 and B_2 represent tie-line bias parameters; and, K_{I1} and K_{I2} are integrator gains. It may be noted that the minus sign ensures a signal to the speed changer to increase generator output whenever either the change in tie-line power or the frequency error is negative.

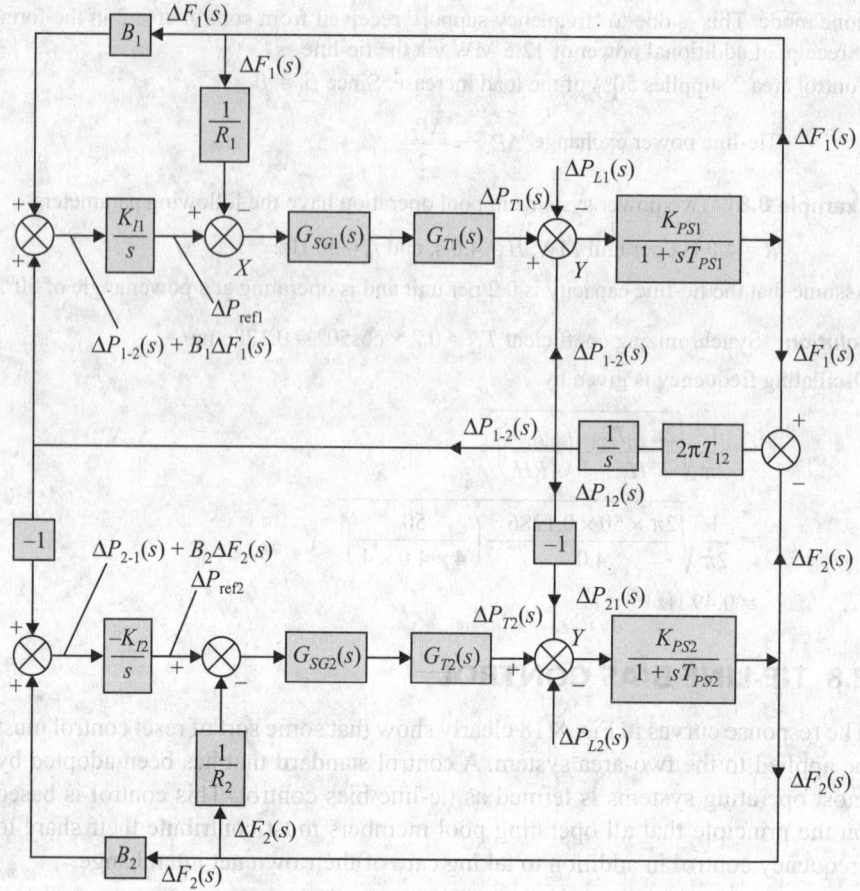

Fig. 8.19 Block diagram simulation of the two-area system

8.8.2 Tie-line Bias Control of Multi-area Systems

In discussing the two-area power system control it was assumed that the two areas are connected by a single tie-line. In practice, however, such a situation is non-existent; a control area is connected through several tie-lines to adjacent control areas, which are all a part of a general pool. In order to write a mathematical expression for the tie-line bias control, consider a control area designated as k and the net interchange of power is equal to the sum of power over all the j interconnecting tie-lines. The area control error (ACE) is indicative of the total interchange of power. Therefore, it is expressed in the following form:

$$\text{ACE}_k = \sum_{i=1}^{i=j} \Delta P_{ij} + B_k \Delta f_k \qquad (8.97)$$

Sampled data techniques are employed to translate Eq. (8.97), into implementation, for tie-line bias control of multi-area systems. At predetermined intervals of time (say, one second), the power data from all the tie-lines is sent to the central energy control centre. At the control centre, the power data is totalled, compared with the scheduled power interchange, and the overall error is computed. ACE is then obtained by summing the overall power error and the biased frequency error. All the area generators which constitute the secondary LFC receive the ACE and initiate corrective action for restoring the tie-line power to its agreed power interchange and frequency to its normal value.

SUMMARY

- When a power system network operates in the steady state, it is required to meet, at all times, the continually varying demands of the real and reactive powers at the specified voltage magnitude and frequency. For these purposes, synchronous generators are equipped with two independent automatic controls, namely (i) AVR loop and (ii) LFC loop. The former controls the generated voltage by varying the field current in the exciter circuit and the latter controls the frequency by changing the speed of the prime mover through a variation in input to it.
- A comparator, an amplifier, and an exciter, which constitute the components of an AVR loop, along with the synchronous generator are modelled based on the well-known principles of control theory and analysed for static and dynamic performance.
- Mathematical models of the speed governing system and the turbo-generator are developed to study the static and dynamic behaviour of the LFC loop.
- LFC in a control area is simulated on the theory of incremental power balance and (i) static and (ii) dynamic characteristics studied.
- Response of the LFC loop to addition of (i) a secondary loop and (ii) integral control is also analysed. Static and dynamic response of an LFC loop in a two-area system connected by a weak lossless tie line is also studied.
- Tie-line bias control based on the principle that each pool member contributes its share to frequency control in addition to maintaining its own net interchange is explained for (i) two-area and (ii) multi-area systems.

SIGNIFICANT FORMULAE

- Closed AVR loop: $G_F(s) = \dfrac{K_F}{1 + sT'_{do}}$

$$K_F = \frac{\omega L_{fa}}{\sqrt{2}R_f}, T'_{do} = \frac{L_{ff}}{R_f}$$

- Closed AVR loop transfer function: $\dfrac{\Delta|V|(s)}{\Delta|V|_{ref}(s)} = \dfrac{G_C(s)}{1 + G_c(s)}$

- Characteristic equation of the AVR loop: $1 + G_C(s) = 0$

- Stability compensation in AVR loop: $G_C(s) = \dfrac{K}{(1 + sT_A)(1 + sT'_{d0})}$

- Transfer function of speed governing system:

$$\Delta Y_E(s) = G_{SG}(s)\left[\Delta P_{tie}(s) - \frac{1}{R}\Delta F(s)\right]$$

- Transfer function of a turbine: $G_T(s) = \dfrac{K_T}{1 + sT_T}$

- Transfer function of a hydro-turbine: $G_T = \dfrac{1-2sT_W}{1+sT_W}$

- Performance of a speed governor:

$$\Delta P_G(s) = \frac{K_{SG}K_T}{(1+sT_{SG})(1+sT_T)}\left[\Delta P_{ref}(s) - \frac{\Delta F(s)}{R}\right]$$

- Transfer function for primary LFC loop for isolated power system:

$$\Delta F(s) = G_{ps}(s)\,[\Delta P_G(s) - \Delta P_L(s)]$$

$$G_{PS}(s) = \frac{K_{PS}}{1+sT_{PS}}, \quad K_{PS} = \frac{1}{D}, \quad T_{PS} = \frac{2H}{Df_0}$$

- Static frequency drop: $\Delta f_0 = \dfrac{\Delta P_L}{\beta}$

- Dynamic response of primary LFC loop: $\Delta f(t) = \Delta f_0\left(1 - e^{-t/\tau}\right)$

- LFC loop response with integral control:

$$\Delta F(s) = -\Delta P_L \frac{K_{PS}}{T_{PS}} \frac{1}{\left[s^2 + s\left(\dfrac{1+K_{PS}/R}{T_{PS}}\right) + \dfrac{K_I K_{PS}}{T_{PS}}\right]}$$

- Roots of characteristic equation:

$$s_1, s_2 = -\frac{1+K_{PS}/R}{2T_{PS}} \pm \frac{1}{2}\sqrt{\left(\frac{1+K_{PS}/R}{T_{PS}}\right)^2 - \frac{4K_I K_{PS}}{T_{PS}}}$$

- Static response of a LFC loop of a two-area system:

$$\Delta P_{1-2,0} = -\Delta P_{2-1,0} = \frac{\beta_1 \Delta P_{L2} - \beta_2 \Delta P_{L1}}{\beta_1 + \beta_2}$$

- Dynamic response of a LFC loop of a two-area system:

$$\Delta P_{1-2}(s) = \frac{\pi f_0 T_{12}}{H} \times \frac{[\Delta P_{L1}(s) - \Delta P_{L2}(s)]}{\left[s^2 + \left(\dfrac{f_0}{2RH}\right)s + \dfrac{2\pi f_0 T_{12}}{H}\right]}$$

- Tie-line bias control for two-area system: $ACE_1 = \Delta P_{1-2} + B_1 \Delta f_1$

$$ACE_2 = \Delta P_{2-1} + B_2 \Delta f_2$$

- Tie-line bias control for multi-area systems: $ACE_k = \sum\limits_{i=1}^{i=j} \Delta P_{ij} + B_k \Delta f_k$

EXERCISES

Review Questions

8.1   (a) Describe the operation of a power system in steady state.

(b) Explain in general terms how voltage and frequency are maintained in such a system.

8.2 Construct the root locus plot for the control loop of Example 8.1. Discuss the stability requirement of the loop and prove that the requirement of high static accuracy is in conflict with a fast dynamic response. How can these conflicting requirements be resolved?

8.3 Derive the transfer function of an AVR loop including a synchronous generator. Draw its block diagram.

8.4 Explain the static and dynamic response of an AVR loop. How can its dynamic response be improved?

8.5 Develop transfer functions for a (i) speed governing system and (ii) turbo-generator.

8.6 Explain the static performance of a speed governing system when a generator is connected to (i) an infinite network, (ii) a finite network, and (iii) when generator reference power and frequency change at the same time.

8.7 Two generators, each having a regulation of R_1 and R_2 operate in parallel to feed a single area power system. If the load sensitivity to frequency is represented by D, prove that the steady-state frequency deviation, for a load demand ΔP_L, is given by

$$\Delta f_0 = -\frac{\Delta P_L}{D + \dfrac{1}{R_1} + \dfrac{1}{R_2}}$$

8.8 Prove that the changes in tie-line power between two control areas, for small changes in voltage angles can be represented by

$$\Delta P_{\text{tie 1-2}} = \frac{|V_1||V_2|}{X_{12}} \cos(\delta_1^0 - \delta_2^0)(\Delta\delta_1 - \Delta\delta_2) \text{ MW}$$

Assume $|V_1|$, δ_1^0 and $|V_2|$, δ_2^0 are the magnitudes of the end voltages and the voltage angles of control areas 1 and 2 respectively.

8.9 Derive an expression for the change of tie-line power and frequency when the two control areas have equal parameters. In such a system, what is the percentage load taken up by control area 2 when the step change load occurs only in area 1? Discuss the results and show that the multi-area system operation is beneficial.

Numerical Problems

8.1 For the control loop shown in Fig. 8.20, determine (i) open loop composite transfer function, (ii) closed loop transfer function, and (iii) characteristic equation. The transfer function of each component is given in the figure.

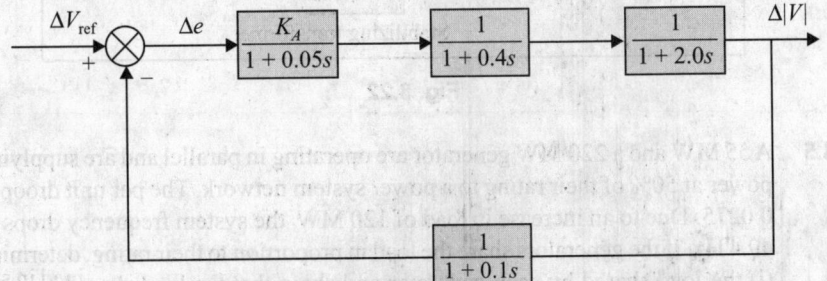

Fig. 8.20

8.2 Develop the Routh–Hurwitz array for the characteristic equation computed in Problem 8.1 and therefrom calculate the range of the amplifier gain for the control loop to remain stable.

8.3 The secondary loop of a stabilizing transformer is connected in the amplifier circuit as shown in Fig. 8.21. If the resistance, inductance, and mutual inductance of the primary loop of the transformer are represented by R, L, and M respectively, derive an expression for the transfer function of the transformer. If it is assumed that $R \gg L$, how does the transfer function of the transformer change? Assume that the amplifier has high input impedance.

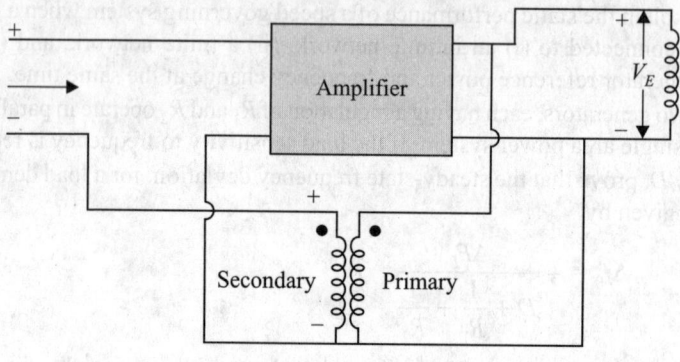

Fig. 8.21

8.4 An AVR loop contains both a phase lead compensator and a stabilizing transformer as shown in Fig. 8.22. If the loop gain is K, (i) compute the open loop composite transfer function of the loop. (ii) Write down the characteristic equation from the open loop composite transfer function. (iii) Use MATLAB for plotting (a) the root locus of the open loop transfer function, and (b) the root locus of the characteristic equation for varying values of K. Comment on the stability of the loop.

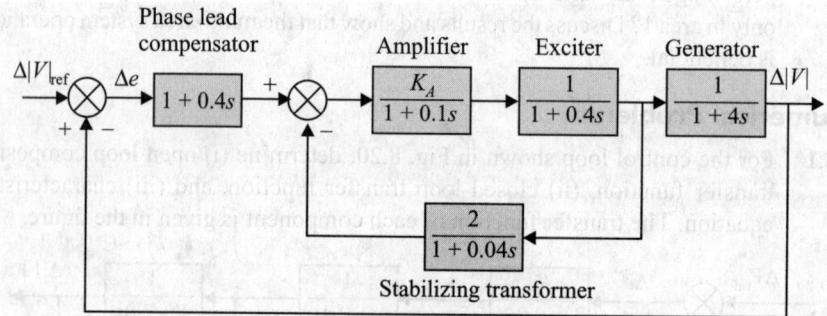

Fig. 8.22

8.5 A 55 MW and a 220 MW generator are operating in parallel and are supplying power at 50% of their rating to a power system network. The per unit droop is 0.0275. Due to an increase in load of 120 MW, the system frequency drops to 49.4 Hz. If the generators share the load in proportion to their rating, determine (i) the load shared by each generator and show that the load shared by each generator is proportional to its rating, and (ii) the droop in each generator.

8.6 A 300-MW turbine generator has a speed regulation of 0.045 per unit on its own rated capacity as base. Determine the increase in power output when the frequency drops from normal 50 Hz to a steady state value of 49.95 Hz.

8.7 (a) Assume that the transfer function of the LFC loop shown in Fig. 8.10 is operating under open loop condition. Determine $[\Delta P_T(s)/\Delta P_{ref}(s)]$. Assume $T_{SG}=0.1$ s and $T_T=0.4$ s.

(b) At $t = 0$ s, the servo-motor setting of the speed governor is changed and an independent increment $\Delta P_{ref} = 1.0$ is applied. Express the turbine power changes as a function of time and plot the turbine power response versus time.

8.8 Two generating stations are supplying a power to a single area power system. The stations are operating in parallel and each has a rating of 1000 MW and 800 MW. The regulation of each station is 0.06 per unit and 0.04 per unit, respectively. At nominal frequency, the 1000-MW and 800-MW generating stations are supplying a load of 500 MW and 400 MW each. The load is increased by 100 MW. (i) Determine the steady-state frequency deviation, new frequency, and the new load supplied by each generating station. Assume the load is independent of frequency. (ii) Repeat (i) by assuming that for a frequency change of 1%, the load varies by 1.2%. What is the total generation? Explain why the increase in generation does not match the load demand.

8.9 Two control area systems are connected by a tie-line whose synchronizing power coefficient is 1.5 per unit. The parameters, on a system base power of 1000 MW, of the two areas are shown in a tabular form here.

Parameter	Control area 1	Control area 2
Speed regulation R	0.04	0.055
Load frequency coefficient D	0.75	0.90
Inertia constant H	4.0 s	5.0 s

Compute the tie-line power flow and the new steady-state frequency in the two areas following a load demand of 150 MW in area 2. What is the MW power supplied by area 1. Assume that the normal system frequency is 50 Hz.

8.10 Two equal control areas have the following parameters: $R = 3.5$ Hz/pu MW, $H = 4.5$s, normal operating frequency $f_0 = 50$ Hz. If the synchronizing coefficient T_0 is equal to 0.2, determine the damping coefficient α and angular frequency ω. Use MATLAB to plot the tie-line power deviation response.

Multiple Choice Objective Questions

8.1 Which of the following applies when a power system operates in the steady state?

(a) Real power load is varying.

(b) Reactive power load is varying.

(c) Both real and reactive powers are varying.

(d) None of these.

8.2 Which of the following is a reason to maintain power frequency in a network within specified limits?

(a) Synchronous clocks show correct time.

(b) Power balance is maintained.

(c) Bus voltage magnitudes are maintained.

(d) All of these.

8.3 Which of the following is undertaken when the reactive power demand in a network increases?

(a) Exciter voltage is increased.

(b) Input to the generator is increased.

(c) Input to the prime mover is increased.

(d) None of these.

8.4 Which of the following is used for voltage control in a power system?

(a) Static var compensation

(b) Synchronous condensers

(c) Excitation control of generators

(d) All of these

8.5 Which of the following is not a function of the AVR loop?

(a) Reactive power control   (b) Frequency control

(c) Voltage magnitude control   (d) All of these

8.6 Which of the following will apply if the power frequency of a network is to be maintained within specified limits?

(a) Generated real power must balance real power demand.

(b) Generated reactive power must balance reactive power demand.

(c) Generated real power must balance real power demand plus line losses.

(d) All of these.

8.7 Which of the following is a reason to decouple the AVR and LFC loops when there is a shortage of reactive power?

(a) Change in bus voltages marginally affects the active power loads.

(b) AVR loop response is faster than the LFC loop.

(c) LFC loop regulates the active power output.

(d) All of these.

8.8 Which of the following is a simulation of the closed loop AVR?

(a) $\dfrac{K_E}{1+sT_E}$ (b) $G_C(s)$

(c) $\dfrac{G_C(s)}{1+G_c(s)}$ (d) $\dfrac{G_C(s)}{1-G_c(s)}$

8.9 For the transient components in the response of an AVR loop to decay rapidly, which of the following will apply to the roots of its characteristic equation?

(a) Lie deep in the positive half of the s-plane

(b) Lie deep in the negative half of the s-plane

(c) Must have complex conjugate roots

(d) None of these

8.10 Which of the following happens when a series compensator loop is added to an AVR loop?

(a) K remains unchanged.   (b) K increases.

(c) K decreases.   (d) K becomes 0.

8.11 Which of the following states the Routh–Hurwitz criterion for stability, when applied to the first column of the Routh array?

(a) There is no change in sign.

(b) Number of changes equals the number of poles.

(c) Number of changes is one more than the number of poles.

(d) Number of changes is one less than the number of poles.

8.12 Which is the full form of BIBO?
    (a) Basic input basic output   (b) Basic input bounded output
    (c) Bounded input basic output   (d) Bounded input bounded output

8.13 For which of the following an LFC system will function effectively?
    (a) Small and slow changes in real power load and frequency
    (b) Large and rapid changes in real power loads and frequency
    (c) Small and slow changes in voltage magnitudes and reactive power load demands
    (d) Large and rapid changes in voltage magnitudes and reactive power load demands

8.14 Which of the following is the transfer function of a hydro-turbine?

    (a) $\dfrac{K_T}{1+sT_T}$ (b) $\dfrac{1-2sT_W}{1+sT_W}$

    (c) $\dfrac{K_T}{1-sT_T}$ (d) $\dfrac{1+2sT_W}{1-sT_W}$

8.15 The static performance analysis of a speed governor system can be applied to which of the following?
    (a) Generator is connected to an infinite power system.
    (b) Generator is connected to a finite power system.
    (c) Reference generator power and frequency vary at the same time.
    (d) All of these.

8.16 Which of the following is the transfer function of an LFC loop for an isolated power system?
    (a) $G_{PS}(s)\,\Delta P_G(s)$ (b) $G_{PS}(s)\,\Delta P_L(s)$
    (c) $G_{PS}(s)[\Delta P_G(s) - \Delta P_L(s)]$ (d) $G_{PS}(s)[\Delta P_G(s) + \Delta P_L(s)]$

8.17 The integral controller is imparted a negative polarity to ensure an appropriate signal to the speed changer such that positive frequency error generates a
    (a) positive command   (b) negative command
    (c) zero command   (d) None of these

8.18 Which of the following represents the nature of the deviations in frequency and tie-line power?
    (a) Damped   (b) Oscillatory
    (c) Stable   (d) All of these

8.19 Which of the following represents the control error for tie-line bias control in a two-area system?
    (a) $ACE_1 = \Delta P_{1\text{-}2}$ (b) $ACE_1 = B_1\,\Delta f_1$
    (c) $ACE_1 = \Delta P_{1\text{-}2} + B_1\,\Delta f_1$ (d) None of these

Answers

8.1 (c)	8.2 (a)	8.3 (a)	8.4 (d)	8.5 (b)	8.6 (c)
8.7 (d)	8.8 (c)	8.9 (b)	8.10 (a)	8.11 (a)	8.12 (d)
8.13 (a)	8.14 (b)	8.15 (d)	8.16 (c)	8.17 (b)	8.18 (d)
8.19 (c)					

CHAPTER 9

Symmetrical Fault Analyses

Learning Outcomes

A study of this chapter will enable the reader to
- Know the types of symmetrical and unsymmetrical faults which can occur to disturb the normal balanced steady state operation of a three-phase power system
- Understand in broad terms the impact of a fault on the power system equipment and how it is protected from disruption and damage
- Recognize the significance of power system fault analyses
- Mathematically simulate a power network based on an understanding of the simplifying assumptions
- Apply Thevenin's and superposition theorems to determine bus voltages and current flows during fault and post-fault conditions

OVERVIEW

The normal operation of a power system is a balanced steady-state three-phase operation. This condition can be temporarily disturbed by a number of undesirable and unavoidable incidents. A short-circuit or fault is said to occur if the insulation of the system fails at any location due to system overvoltages caused by lightning or switching surges, insulation contamination (salt spray or pollution), wind damage, trees falling across lines, birds shorting the lines, small animals entering switchgear, breakage of conductor caused by wind and ice loading, etc.

When short-circuit occurs, high currents, that is, currents several times that of normal operating currents, flow in the system depending on the nature and location of the fault. The fault currents, if allowed to persist, may cause thermal damage to the equipment. Windings and busbars may also suffer mechanical damage due to high electromagnetic forces during faults. It is, therefore, necessary to remove faulty sections of a power system from service as soon as possible.

The fault currents are sensed through relays, which take about half a cycle, or so. With proper coordination among relays, faulty sections of the power system are isolated quickly by actuation of the circuit breakers in another three to four cycles. Standard extra high voltage (EHV) protective equipment is designed to clear faults within three cycles. Low voltage (LV) protective equipment operates more slowly and may take 5–20 cycles.

Fault studies form an important part of power system analysis. The problem consists of determining bus voltages and line currents during various types of faults. Based on these studies the short-circuit MVA at various points in the network can be calculated. These short-circuit levels provide the basis for specifying interrupting capacities of circuit breakers. Changes in the configuration of the transmission network alter both the short-circuit levels and the short-circuit currents in the system. Hence when any major modifications to the power system are made, these computations must be repeated to determine the adequacy of the protective equipment.

Faults on power systems are divided into three-phase balanced faults and unbalanced faults. The different types of unbalanced faults are single line-to-ground fault, line-to-line fault, and double line-to-ground fault, which are dealt with in Chapter 10. The magnitude of the fault currents depends on the internal impedance of the generators plus the impedance of the intervening circuit. The reactance of a generator under short-circuit condition is not constant, as mentioned in Chapter 5. For the purpose of fault studies, the generator behaviour can be divided into three periods: the subtransient period, lasting only for the first few cycles; the transient period, covering a relatively longer time; and the steady-state period. In this chapter, three-phase balanced faults are discussed. The bus impedance matrix by the building algorithm is employed for the systematic computation of bus voltages and line currents during a three-phase fault. MATLAB programs to compute the node voltages and line currents, under a three-phase fault condition have been developed and included.

9.1 BALANCED THREE-PHASE FAULTS

A three-phase symmetrical short-circuit is caused by the application of three equal fault impedances Z_f between each phase and ground. The short-circuit is said to be solid if $Z_f = 0$. This is the most severe fault to which a system can be subjected and such a fault occurs infrequently. Because the network is balanced, it is solved on a per-phase basis. The other two phases carry identical currents except for the phase shift.

In Section 5.5.1, short-circuit of an unloaded synchronous machine was discussed and the equivalent circuit of a synchronous machine for transient studies was developed. It was also established that the reactance of the synchronous generator under short-circuit conditions is a time-varying quantity and the representation of the synchronous machine for short-circuit analysis was dependent on time elapsing from the incidence of fault. For the first few cycles of the short-circuit current, the subtransient reactance X''_d, for the next about 25 Hz, the transient reactance X'_d, and after that the synchronous reactance X_d should be used. Generally, the subtransient reactance is used for determining the interrupting capacity of the circuit breakers. In fault studies, required for relay setting and coordination, transient reactance is used. The modification of the equivalent circuit, to account for the initial load current supplied by the generator was also discussed in Section 5.5.2.

The following assumptions are made in the calculation of fault currents.

- Transformers are represented by their leakage reactances. Winding resistances, shunt admittances, and D-Y phase shifts are neglected.
- Transmission lines are represented by their equivalent series reactances. Series resistances and shunt admittances are neglected.

- Synchronous machines are represented by constant-voltage sources behind subtransient reactances. Armature resistance, saliency, and saturation are neglected.
- All non-rotating impedance loads are neglected.
- Induction motors are either neglected (especially for small motors rated less than 50 hp) or represented in the same manner as synchronous machines.

In practice, these assumptions should not be made for all cases. For example, in distribution systems, resistances of primary and secondary distribution lines may in some cases significantly reduce fault current magnitudes.

A three-phase fault represents a structural network change equivalent to that caused by the addition of three equal impedances Z_f at the place of fault. The changes of voltages and currents that will result from this structural change can be conveniently analysed by Thevenin's theorem.

Thevenin's theorem states that the changes that take place in the voltages and currents in the network due to addition of impedance between two nodes of the network are identical with those voltages and currents that would be caused by Thevenin's voltage V_{TH} in series with Thevenin's impedance Z_{TH}, where V_{TH} is equal to pre-fault voltage that existed between the nodes and Z_{TH} is the equivalent impedance looking into the network at the two nodes with all other active sources set to zero, that is, with all the independent voltage sources short-circuited, and all the independent current sources open-circuited, of course, leaving behind their internal impedances.

The post-fault currents and voltages in the network can be obtained by superimposing these changes on the pre-fault current and voltages. The application of Thevenin's theorem in fault computations is demonstrated in the following example.

Example 9.1 A single-line diagram consisting of a synchronous generator feeding a synchronous motor through a transmission line is shown in Fig. 9.1. All impedances indicated in the diagram are expressed in per unit on a common 100-MVA base. The motor is drawing 50 MW at 0.95 power factor lagging at rated voltage when a symmetrical three-phase fault occurs at the motor terminals. Determine the per unit values of (i) subtransient fault current, (ii) subtransient fault current in the generator and the motor neglecting pre-fault load current, and (iii) subtransient fault current in the generator and the motor including pre-fault load current.

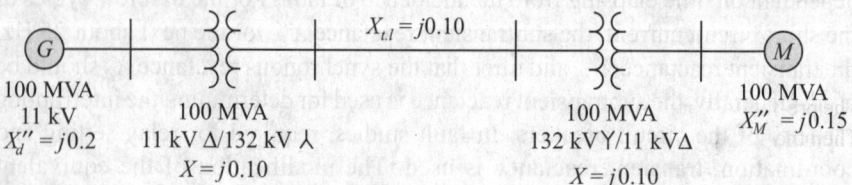

100 MVA   100 MVA   100 MVA   100 MVA

11 kV

$X_d'' = j0.2$ 11 kV Δ/132 kV λ   132 kVΥ/11 kVΔ $X_M'' = j0.15$

      $X = j0.10$ $X = j0.10$

Fig. 9.1 Single-line diagram for Example 9.1

Solution (i) Figure 9.2(a) shows the equivalent circuit of the system shown in Fig. 9.1, where the voltages E_G'' and E_M'' are the pre-fault internal voltages behind the subtransient reactances of the machines. The pre-fault voltage V_F at the point of fault is shown as a generator having the same terminal voltage V_F. The generator voltage V_F has no effect on the current flowing from the generator to the motor before the fault occurs. In calculating the subtransient fault current, E''_G and E''_M are assumed to be constant-voltage sources. The closing of switch S represents the fault.

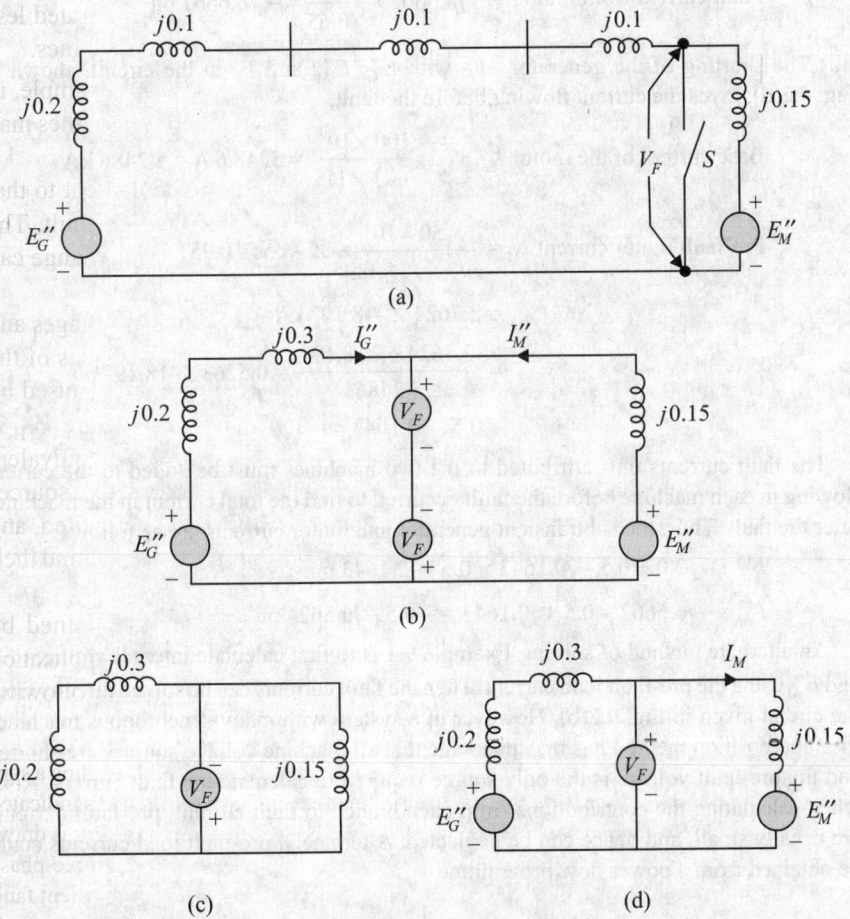

Fig. 9.2   Application of superposition to three-phase short-circuit

Thevenin's impedance as viewed from the fault is

$$Z_{\mathrm{TH}} = jX_{\mathrm{TH}} = j\frac{(0.15)(0.50)}{(0.15+0.50)} = j0.1154 \text{ pu}$$

The pre-fault voltage at the motor terminal is the rated voltage of the motor, that is, 1.0 pu. Then the Thevenin's voltage is

$$V_{\mathrm{TH}} = V_f = 1.0 \angle 0° \text{ pu}$$

The subtransient fault current is then

$$I_F = \frac{1.0\angle 0°}{j0.1154} = -j\,8.6667 \text{ pu}$$

(ii) Adding in series with V_F another generator having an emf of equal magnitude but 180° out of phase with V_F gives the circuit shown in Fig. 9.2(b). The principle of superposition applied by first shorting E''_G, E''_M, and $V_{F'}$, shown in Fig. 9.2(c), gives the currents found by distributing the fault current between the two generators inversely as the impedances of their circuits.

Fault current from generator $= -j8.6667 \times \dfrac{j0.15}{j0.65} = -j2.0$ pu

Fault current from motor $= -j8.6667 \times \dfrac{j0.50}{j0.65} = -j6.6667$ pu

(iii) The shorting of the generator $-V_F$ with E''_G, E''_M, and V_F in the circuit, shown in Fig. 9.2(d), gives the current flowing before the fault.

Base current of the motor $I_{\text{base, }M} = \dfrac{100 \times 10^3}{\sqrt{3} \times 11} = 5248.6$ A $= 5.2486$ kA

Pre-fault motor current $I_M = \dfrac{50 \times 10^3}{\sqrt{3} \times 11 \times 0.95} \angle -\cos^{-1} 0.95$

$$= 2.7624 \angle -18.19° \text{ kA}$$

$$= \frac{2.7624 \angle -18.19°}{5.2486} = 0.5263 \angle -18.19° \text{ pu}$$

$$= 0.5 - j0.1643 \text{ pu}$$

The fault currents thus attributed to the two machines must be added to the current flowing in each machine before the fault occurred to find the total current in the machines after the fault. Thus, the subtransient generator and motor currents are as follows:

$I''_G = -j2.0 + 0.5 - j0.1643 = 0.5 - j2.1643$ pu

$I''_M = -j6.6667 - 0.5 + j0.1643 = -0.5 - j6.5024$ pu

An alternate method of solving Example 9.1 is to first calculate internal voltages E''_G and E''_M using the pre-fault load current. Then the fault currents can be solved directly from the circuit given in Fig. 9.2(b). However in a system with many synchronous machines the superposition method has the advantage that all machine voltage sources are shorted and the pre-fault voltage is the only source required to calculate the fault current. Also, when calculating the contributions from each branch to fault current, pre-fault currents are usually small, and hence can be neglected. Alternately, pre-fault load currents could be obtained from a power flow programme.

9.2 SYSTEMATIC THREE-PHASE FAULT COMPUTATION USING BUS IMPEDANCE MATRIX

To carry out short-circuit computations, using the digital computer, a systematic general computational procedure is developed. The network reduction technique used in the preceding example can be used only in textbook examples of low dimensionality. A technique applicable to an *n*-bus system will now be developed.

Consider the part of the *n*-bus system shown in Fig. 9.3. The system is assumed to be operating under balanced steady-state condition. Synchronous machines are represented by constant-voltage sources behind subtransient reactances. Transmission lines are represented by their equivalent series reactances. Series resistances and shunt admittances are neglected. Transformers are represented by their leakage reactances, neglecting the winding resistances, shunt admittances, and Δ-Y phase shifts. All non-rotating impedance loads are neglected. Induction motors are represented in the same manner as synchronous machines. However, small induction motors are neglected.

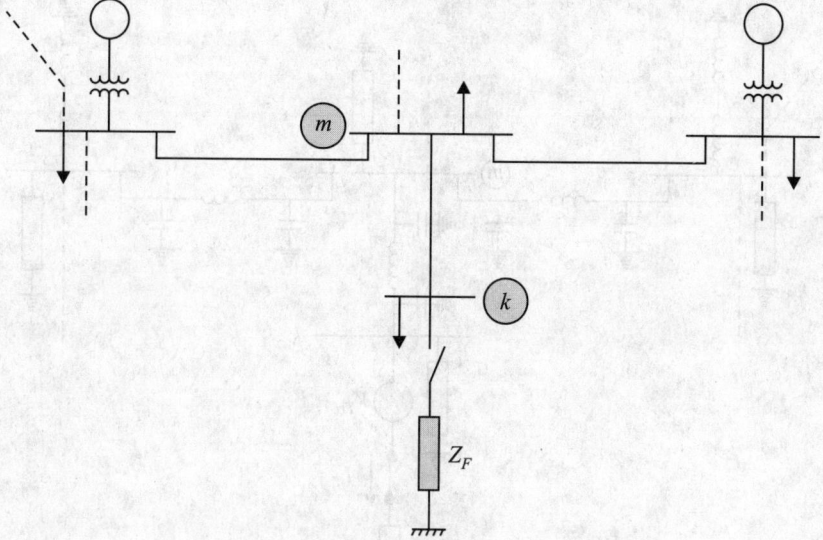

Fig. 9.3 Part of a general *n*-bus system

A symmetrical, three-phase short-circuit is applied at bus k through fault impedance Z_f. The pre-fault bus voltages can be obtained from the power flow solution described in Chapter 7, and constitutes an n-dimensional vector

$$V_{\text{bus, 0}} = \begin{bmatrix} V_{1,0} \\ \vdots \\ V_{k,0} \\ \vdots \\ V_{n,0} \end{bmatrix} \tag{9.1}$$

The changes in the network bus voltages caused due to a fault with fault impedance Z_f at bus k can be obtained by application of Thevenin's theorem. On zeroing all active sources and representing all components by their appropriate impedances, the Thevenin's equivalent network is obtained as shown in Fig. 9.4. The effect of the fault is as usual represented by the Thevenin voltage $V_{\text{TH}} = V_{k,0}$. In this network, the bus voltages represent the changes caused by the fault. These changes are represented by the n-dimensional Thevenin voltage vector.

$$V_{\text{TH}} = \begin{bmatrix} \Delta V_1 \\ \vdots \\ \Delta V_k \\ \vdots \\ \Delta V_n \end{bmatrix} \tag{9.2}$$

Thevenin's theorem now states that the post-fault bus voltages which constitute the components of the n vector $V_{\text{bus}, F}$ are obtained by superposition, and are given by

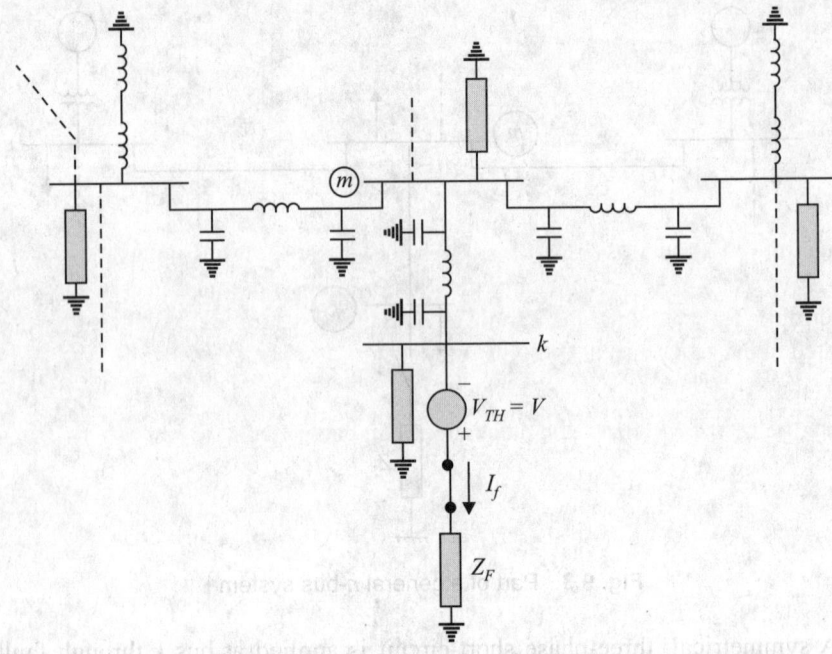

Fig. 9.4 Thevenin's equivalent of the network given in Fig. 9.3

$$V_{bus, F} = V_{bus, 0} + V_{TH} \tag{9.3}$$

The *n-bus* system shown in Fig. 9.4 can be represented by a Z_{bus} matrix of the order $n \times n$. The relationship between bus voltages and injected bus currents is given as

$$V_{bus} = Z_{bus}I_{bus} \tag{9.4}$$

In the network shown in Fig. 9.4, bus currents are injected at bus k only, and the injected current at bus k is $-I_f$. Thus, the bus current vector during fault becomes

$$I_{bus} = I_{bus, F} = \begin{bmatrix} 0 \\ \vdots \\ -I_f \\ \vdots \\ 0 \end{bmatrix} \tag{9.5}$$

and the bus voltage vector during fault is

$$V_{bus} = V_{TH} \tag{9.6}$$

From Eq. (9.4), the following can be written

$$V_{TH} = Z_{bus}I_{bus, F} \tag{9.7}$$

Substituting Eq. (9.7) in Eq. (9.3) yields

$$V_{bus, F} = V_{bus, 0} + Z_{bus}I_{bus, F} \tag{9.8}$$

The vector Eq. (9.8) written in terms of its elements becomes

$$
\begin{bmatrix} V_{1,F} \\ \vdots \\ V_{k,F} \\ \vdots \\ V_{n,F} \end{bmatrix} = \begin{bmatrix} V_{1,0} \\ \vdots \\ V_{k,0} \\ \vdots \\ V_{n,0} \end{bmatrix} + \begin{bmatrix} Z_{11} & \cdots & Z_{1k} & \cdots & Z_{1n} \\ \vdots & \vdots & \vdots & \vdots & \vdots \\ Z_{k1} & \cdots & Z_{kk} & \cdots & Z_{kn} \\ \vdots & \vdots & \vdots & \vdots & \vdots \\ Z_{n1} & \cdots & Z_{nk} & \cdots & Z_{nn} \end{bmatrix} \begin{bmatrix} 0 \\ \vdots \\ -I_f \\ \vdots \\ 0 \end{bmatrix} \tag{9.9}
$$

Equation (9.9) contains $n + 1$ unknowns: the n post-fault bus voltages $V_{1,F}$, ..., $V_{k,F}$, ..., $V_{n,F}$ and the fault current I_f. One additional equation is needed for determining $n + 1$ unknowns. From Fig. 9.4, the post-fault bus voltage at bus k is related to the fault current through

$$ V_{k,F} = Z_f I_f \tag{9.10} $$

From the vector Eq. (9.9), the n equations can be written as

$$ V_{1,F} = V_{1,0} - Z_{1k} I_f $$
$$ \vdots $$
$$ V_{k,F} = V_{k,0} - Z_{kk} I_f \tag{9.11} $$
$$ \vdots $$
$$ V_{n,F} = V_{n,0} - Z_{nk} I_f $$

Substituting Eq. (9.10) into the kth equation of Eq. (9.11) and simplifying yields the following expression of the fault current

$$ I_f = \frac{V_{k,0}}{Z_f + Z_{kk}} \tag{9.12} $$

Since $V_{k,0}$ is assumed as known, the fault current can be determined. The post-fault bus voltages at all other buses can be determined by substituting the value of I_f into Eq. (9.11). Thus,

$$ V_{i,f} = V_{i,0} - \frac{Z_{ik}}{Z_f + Z_{kk}} V_{k,0} \qquad \text{for } i \neq k \tag{9.13} $$

$$ V_{k,F} = \frac{Z_f}{Z_f + Z_{kk}} V_{k,0} \tag{9.14} $$

In case $Z_f = 0$, that is, the three-phase fault is solid or bolted, the fault current and post-fault bus voltages become

$$ I_f = \frac{V_{k,0}}{Z_{kk}} \tag{9.15} $$

$$ V_{i,F} = V_{i,0} - V_{i,0} - \frac{Z_{ik}}{Z_{kk}} V_{k,0} \qquad \text{for } i \neq k \tag{9.16} $$

$$ V_{k,F} = 0 \tag{9.17} $$

Once the values for the fault current I_f and the n post-fault bus voltages $V_{i,F}$ are obtained, the post-fault currents in all branches of the network can be computed. For the line connecting buses k and m with impedance z_{km}, the short-circuit current in the line flowing from bus k to bus m is given by

$$ I_{km} = \frac{V_{k,F} - V_{m,F}}{z_{km}} \tag{9.18} $$

Example 9.2 Figure 9.5 shows a synchronous generator feeding a motor through transformers and a transmission line. Compute Z_{bus} matrix by using the MATLAB program function [Zbus]=zeebus(nelemnt,ind) and determine the fault current for a three-phase fault at bus 1. Also compute the current flowing in the transmission line under the fault condition. All reactances, in pu, are shown in the figure. Assume equal constant voltages of $1.02 \angle 0°$ pu behind subtransient generator and motor reactances.

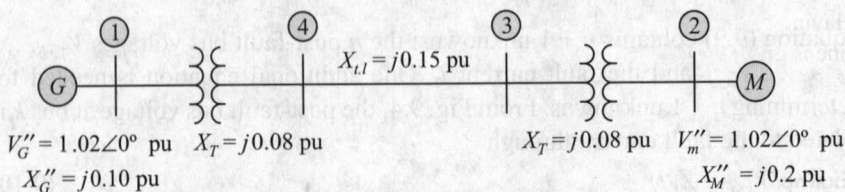

$V_G'' = 1.02\angle 0°$ pu $\quad X_T = j0.08$ pu $\qquad\qquad\qquad X_T = j0.08$ pu $\quad V_m'' = 1.02\angle 0°$ pu
$X_G'' = j0.10$ pu $\qquad\qquad\qquad\qquad\qquad\qquad\qquad\qquad\qquad\qquad X_M'' = j0.2$ pu

Fig. 9.5 Single-line diagram for Example 9.2

Solution Since buses 3 and 4 are not to be faulted, the two transformer and transmission line reactances can be combined and the circuit redrawn as a two-bus system. The combined circuit is shown in Fig. 9.6.

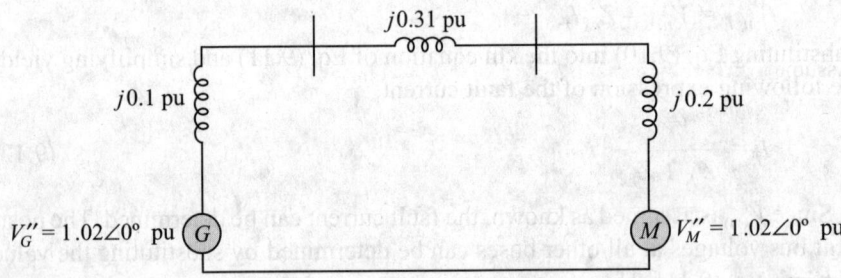

$V_G'' = 1.02\angle 0°$ pu $\quad$ (G) $\qquad\qquad\qquad\qquad\qquad\qquad$ (M) $V_M'' = 1.02\angle 0°$ pu

Fig. 9.6 Equivalent circuit for Example 9.2

```
>> nelemnt=3;ind=0;
>> [Zbus]=zeebus(nelemnt,ind)
bus number p0
bus number q1
impedance0.10i
type of bus1
bus number p0
bus number q2
impedance0.2i
type of bus1
bus number p1
bus number q2
impedance0.31i
type of bus3
Zbus =

      0 + 0.0836i      0 + 0.0328i
      0 + 0.0328i      0 + 0.1344i
```

Subtransient fault current at bus 1 is obtained by using Eq. (9.15) as follows:

$$I_f = \frac{V_F}{Z_{11}} = \frac{1.02\angle 0°}{j0.0836} = -j12.201 \text{ pu}$$

Employing Eq. (9.16) the voltages at bus 1 and bus 2 are obtained as follows:

$$V_1 = \left(1 - \frac{Z_{11}}{Z_{11}}\right)1.02\angle0° = 0$$

$$V_2 = \left(1 - \frac{Z_{21}}{Z_{11}}\right)1.02\angle0° = \left(1 - \frac{j0.0328}{j0.0836}\right)\angle0° = 0.6077 \text{ pu}$$

Having computed the voltages at bus 1 and bus 2, the current flowing in the transmission line is given by

$$I_{21} = \frac{V_2 - V_1}{X_{21}} = \frac{0.6077}{j0.31} = -j1.9603 \text{ pu}$$

Example 9.3   For the power system whose equivalent circuit is shown in Fig. 9.7, compute the bus voltages and the branch currents for a three-phase fault on bus 1. Assume the fault impedance $Z_f = j0.2$. Repeat the problem for a three-phase short-circuit on bus 2. The Z_{bus} matrix in pu for the network is

$$[Z_{\text{bus}}] = j\begin{bmatrix} 0.0776 & 0.0448 & 0.0597 \\ 0.0448 & 0.1104 & 0.0806 \\ 0.0597 & 0.0806 & 0.2075 \end{bmatrix}$$

Assume a pre-fault constant voltage of $1.0\angle0°$ pu.

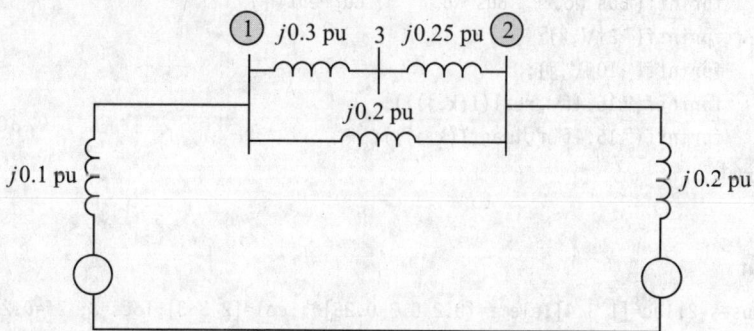

Fig. 9.7   Equivalent circuit for Example 9.3

Solution   A MATLAB program `function fault3faze(Zbus,nfbuses,loc,elemz, col,locs,Zf)` is used to compute the bus voltages and currents for three phase faults on buses 1 and 2. Along with the system Z_{bus} matrix, the program inputs required are as follows:

nfbuses	The number of buses to be faulted
loc	A row vector which gives the start of each row
elemz	A row vector which stores the line impedances
col	A row vector which provides the column number
nbus	Number of buses in the system
locs	Gives the size of the loc vector

It may be noted that the last five inputs are the same as used in the program `function [Ybus]=admitmat1(loc,elemz,col,nbus,locs)` discussed in Section 6.4.3.

```
function fault3faze (Zbus,nfbuses,loc,elemz,col,locs,Zf);
for n=1:nfbuses
  p=input('number of bus to be faulted');
```

```
Vf=input('fault bus voltage');
If=Vf/(Zbus(p,p)+Zf);
fprintf('Bus No.        Fault Current\n');
fprintf('%2i\',p);
fprintf(             '%15.4f\',real(If));
fprintf(             '%15.4f\n',imag(If));
for k=1:size(Zbus)
   V(k)=Vf-Zbus(k,p)*If;
   fprintf('Bus No.    Bus Voltage\n');
   fprintf('%2i\',k);
   fprintf(             '%15.4f\',real (V(k)));
   fprintf(             '%15.4f\n',imag (V(k)));
end
kk=1;
for k=1:locs-1
   add=loc(k+1)-loc(k);
   for m=1:add
      j=col(kk);
      I(k,j)=(V(k)-V(j))/elemz(kk);
      kk=kk+1;
      fprintf('Bus No.    Bus No.      Current\n')
      fprintf('%2i\',k);
      fprintf('%10i\',j);
      fprintf('%15.4f\',real(I(k,j)));
      fprintf('%15.4f\n',imag(I(k,j)));
   end
end
end
```

Output

```
>> nfbuses=2;loc=[1 3 4];elemz=[0.2 0.3 0.25]*i;col=[2 3 3];locs=3; Zf=0.2i;
>> fault3faze (Zbus,nfbuses,loc,elemz,col,locs,Zf)
number of bus to be faulted1
fault bus voltage1.0
Bus No.     Fault Current
1           0.0000              -3.6022
Bus No.     Bus Voltage
1           0.7204               0.0000
Bus No.     Bus Voltage
2           0.8387               0.0000
Bus No.     Bus Voltage
3           0.7849               0.0000
Bus No.     Bus No.          Current
1           2                0.0000              0.5914
Bus No.     Bus No.          Current
1           3                0.0000              0.2151
Bus No.     Bus No.          Current
2           3                0.0000             -0.2151
```

```
number of bus to be faulted2
fault bus voltage1.0
Bus No.      Fault Current
2            0.0000              -3.2212
Bus No.      Bus Voltage
1            0.8558              0.0000
Bus No.      Bus Voltage
2            0.6442              0.0000
Bus No.      Bus Voltage
3            0.7404              0.0000
Bus No.          Bus No.        Current
1                2              0.0000      -1.0577
Bus No.          Bus No.        Current
1                3              0.0000      -0.3846
Bus No.          Bus No.        Current
2                3              0.0000   0.3846
```

SUMMARY

- Power system networks are analysed for three-phase symmetrical short-circuits by the application of three equal impedances Z_f between each phase and ground.
- Network being balanced is simulated on a per phase basis by making assumptions that (i) transformers are represented by their leakage reactances and D-Y phase shifts are neglected, (ii) transmission lines are represented by their equivalent series reactances, (iii) synchronous machines are represented by constant voltages behind sub-transient reactances, (iv) small induction motors are represented in the same way as synchronous machines, and (v) all non-rotating loads are neglected.
- Thevenin's equivalent for the network is used to compute bus voltages and current flows. Post-fault currents and voltages are determined by superposing these on the pre-fault currents and voltages.

SIGNIFICANT FORMULAE

- Fault current: $I_f = \dfrac{V_{k,0}}{Z_f + Z_{kk}}$

- Post-bus voltages: $\begin{bmatrix} V_{1,F} \\ \vdots \\ V_{k,F} \\ \vdots \\ V_{n,F} \end{bmatrix} = \begin{bmatrix} V_{1,0} \\ \vdots \\ V_{k,0} \\ \vdots \\ V_{n,0} \end{bmatrix} + \begin{bmatrix} Z_{11} & \cdots & Z_{1k} & \cdots & Z_{1n} \\ \vdots & \vdots & \vdots & \vdots & \vdots \\ Z_{k1} & \cdots & Z_{kk} & \cdots & Z_{kn} \\ \vdots & \vdots & \vdots & \vdots & \vdots \\ Z_{n1} & \cdots & Z_{nk} & \cdots & Z_{nn} \end{bmatrix} \begin{bmatrix} 0 \\ \vdots \\ -I_f \\ \vdots \\ 0 \end{bmatrix}$

- Post-fault current flows: $\dfrac{V_{k,F} - V_{m,F}}{z_{km}}$

EXERCISES

Review Question

9.1 Derive mathematical expressions to simulate a three-phase symmetrical fault on a balanced three-phase system operating in the steady state. Assume an impedance of Z_f Ω between each phase and ground. Clearly state the assumptions made in the simulation.

Numerical Problems

9.1 A three-phase dead short-circuit occurs on bus 2 of the power system network shown in Fig. 9.8. Assemble the Z_{bus} matrix with the help of MATLAB program zeebus. Determine the fault current and the bus voltages from the first principle.

9.2 Run the MATLAB program fault3faze and determine the bus voltages and network currents for a dead three-phase fault on bus 1 of the power system network shown in Fig. 9.8.

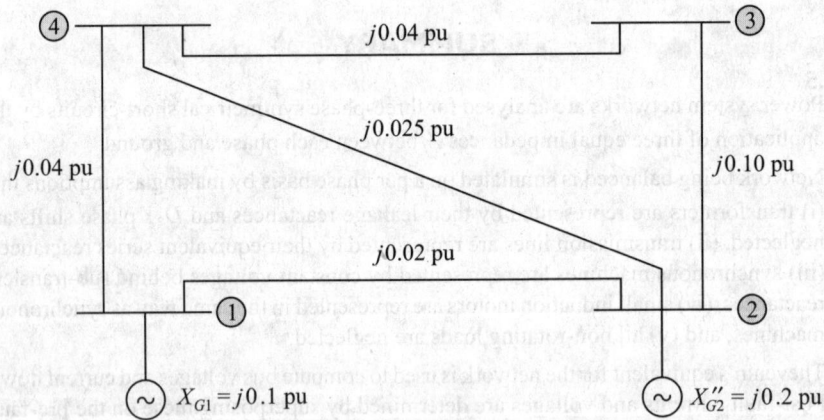

Fig. 9.8 Single line diagram for Problem 9.1

9.3 Repeat Problem 9.2, assuming that the fault impedance $Z_f = j0.10$ pu.

9.4 The power system network shown in Fig. 9.9 has the following equipment ratings:

Generator G_1:	500 MVA, 11 kV, $X'' = 0.15$ pu
Transformer T_1:	500 MVA, 11/400 kV, star–delta, $X = 0.08$ pu
Generator G_2:	400 MVA, 11kV, $X'' = 0.12$ pu
Transformer T_2:	300 MVA, 11/400 kV, star–delta, $X = 0.10$ pu
Generator G_3:	300 MVA, 11 kV, $X'' = 0.10$ pu
Transformer T_3:	300 MVA, 22/400 kV, $X = 0.10$ pu

Transmission lines:

1-4	$X = j40$ Ω
2-4	$X = j50$ Ω
3-4	$X = j30$ Ω

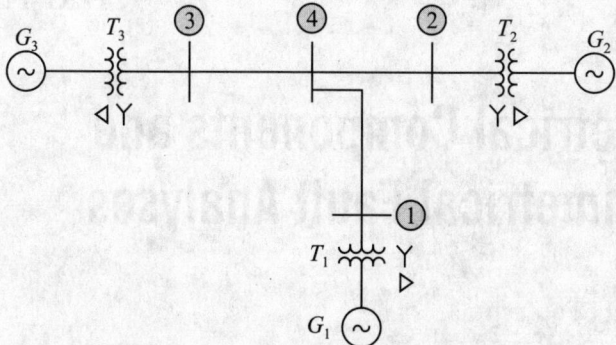

Fig. 9.9 Single line diagram for Problem 9.4

By assuming a system power base of 500 MVA and a voltage base of 400 kV for the transmission lines, convert all the system parameters into per unit values and draw the single line diagram for the network. A dead three-phase fault occurs on bus 1 of the network. Compute (i) the subtransient fault current in kA, (ii) the bus voltages, and (iii) currents in lines 1-4, 2-4, and 3-4. Neglect pre-fault load current and assume a pre-fault voltage of 404 kV.

9.5 Repeat Problem 9.4 for a three-phase fault on bus 4.

CHAPTER 10

Symmetrical Components and Unsymmetrical Fault Analyses

Learning Outcomes

A thorough study of this chapter will enable the reader to
- Recognize that unbalance in a power system network is brought about by unequal loads and unsymmetrical faults
- Categorize unsymmetrical faults into shunt and series faults and enumerate them according to their frequency of occurrence
- Apply symmetrical components to resolve unbalanced networks into three uncoupled balanced symmetric networks
- Develop sequence parameters and sequence networks of generators, transformers, and transmission lines
- Identify the boundary conditions and proceed systematically to perform steady-state analyses of the faulted network
- Become adept at applying the superposition principle to determine the distribution of currents and voltages in the network by working backwards and reconstructing its topology

OVERVIEW

As already stated in previous chapters, a balanced three-phase power system can be replaced by its single-phase equivalent network for the purposes of analyses. The results obtained from such an equivalent network can be applied to the three-phase power system, since it is assumed that complete phase symmetry, or balance, is maintained and the quantities for the remaining phases differ from the first phase by ±120°. However, three-phase power systems in practice are seldom balanced. The fault analysis that was presented in Chapter 9 considered balanced three-phase symmetrical short-circuits where per phase approach is applicable.

The unbalance is introduced due to unequal loads and unsymmetrical short-circuits. Different types of unsymmetrical faults in decreasing order of frequency of occurrence are single line-to-ground fault, line-to-line fault, and double line-to-ground fault. The other types of faults include one-conductor open and two-conductor open faults, which can occur when conductors break or when one or two phases of a circuit breaker remain inadvertently open.

The method of symmetrical components is a powerful technique for analysing unbalanced three-phase systems. By the method of symmetrical components, the unbalanced three-phase network is resolved into three uncoupled balanced sequence networks. Furthermore, for the unbalanced three-phase systems, the sequence networks are connected only at the points of unbalance. As a result, the sequence networks for many cases of unbalanced faults in three-phase systems are relatively easy to analyse. The results of sequence networks can be superposed to obtain three-phase network results.

In this chapter, the concept of symmetrical components is introduced. The application of symmetrical components for the computation of fault currents and voltages of faulted power systems is discussed in detail. A MATLAB program for computing currents and voltages to analyse a power system for the category of shunt faults is included. A general and systematic computational approach for unsymmetrical short-circuits is also presented.

10.1 SYMMETRICAL COMPONENTS

In 1918, C.L. Fortescue introduced the concept of symmetrical components. He stated that a system of n unbalanced set of phasors can be resolved into sets of $n - 1$ balanced phasor systems of different phase sequences plus one zero-phase sequence system. He defined a zero-phase sequence system as one in which all phasors are of equal magnitude and angle, or they are all identical.

The application of the above for a more practical three-phase system is considered here. According to Fortescue's theorem, three unbalanced phasors of a three-phase system can be resolved into three balanced systems of phasors as follows:

1. Positive-sequence components consisting of three phasors equal in magnitude, displaced in phase from each other by 120°, and having the same phase sequence as the original phasors.
2. Negative-sequence components consisting of three phasors equal in magnitude, displaced in phase from each other by 120°, and having phase sequence opposite to the original phasors.
3. Zero-sequence components consisting of three phasors equal in magnitude, and with zero phase displacement from each other.

Mathematically, if V_a, V_b, and V_c are three unbalanced voltage phasors, then these can be written as

$$V_a = V_{a0} + V_{a1} + V_{a2}$$
$$V_b = V_{b0} + V_{b1} + V_{b2}$$
$$V_c = V_{c0} + V_{c1} + V_{c2}$$

(10.1)

In Eq. (10.1), the suffixes 1, 2, and 0 indicate positive (+ve), negative (–ve), and zero sequence components respectively. Figure 10.1 shows three sequence components of unbalanced voltage phasors V_a, V_b, and V_c, namely positive sequence components V_{a1}, V_{b1}, and V_{c1}; negative sequence components V_{a2}, V_{b2}, and V_{c2}; and zero sequence components V_{a0}, V_{b0}, and V_{c0}.

Using the phase operator a, the three sequence phasors can be mathematically represented as follows:

$$V_{a1} = V_{a1}\angle 0°, \ V_{a2} = V_{a2}\angle 0°, \ V_{a0} = V_{a0}\angle 0°$$
$$V_{b1} = a^2 V_{a1}, \ V_{b2} = a V_{a2}, \ V_{b0} = V_{a0}$$

(10.2)

$$V_{c1} = aV_{a1}, \ V_{c2} = a^2V_{a2}, \ V_{c0} = V_{a0}$$

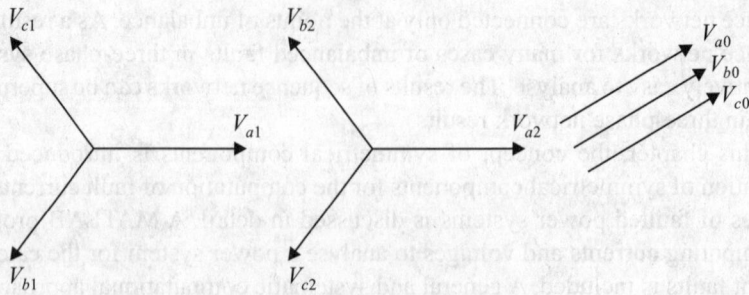

Fig. 10.1 Sequence components

Substitution of Eq. (10.2) in Eq. (10.1) gives

$$\begin{aligned}
V_a &= V_{a0} + V_{a1} + V_{a2} \\
V_b &= V_{a0} + a^2 V_{a1} + a V_{a2} \\
V_c &= V_{a0} + a V_{a1} + a^2 V_{a2}
\end{aligned}$$
(10.3)

In matrix form, Eq. (10.3) is written as

$$\begin{bmatrix} V_a \\ V_b \\ V_c \end{bmatrix} = \begin{bmatrix} 1 & 1 & 1 \\ 1 & a^2 & a \\ 1 & a & a^2 \end{bmatrix} \begin{bmatrix} V_{a0} \\ V_{a1} \\ V_{a2} \end{bmatrix} = A \begin{bmatrix} V_{a0} \\ V_{a1} \\ V_{a2} \end{bmatrix}$$
(10.4)

where A is defined as the symmetrical components transformation matrix and is given by

$$A = \begin{bmatrix} 1 & 1 & 1 \\ 1 & a^2 & a \\ 1 & a & a^2 \end{bmatrix}$$
(10.5)

If $V_{abc} = \begin{bmatrix} V_a \\ V_b \\ V_c \end{bmatrix}$ and $V_{012} = \begin{bmatrix} V_{a0} \\ V_{a1} \\ V_{a2} \end{bmatrix}$
(10.6a)

then $V_{abc} = AV_{012}$
(10.6b)

In order to obtain the sequence components, Eq. (10.6b) is written as

$$V_{012} = A^{-1} V_{abc}$$
(10.7a)

where

$$A^{-1} = \frac{1}{3} \begin{bmatrix} 1 & 1 & 1 \\ 1 & a & a^2 \\ 1 & a^2 & a \end{bmatrix}$$

Equation (10.7a) in expanded form is given by

$$\begin{bmatrix} V_{a0} \\ V_{a1} \\ V_{a2} \end{bmatrix} = \frac{1}{3} \begin{bmatrix} 1 & 1 & 1 \\ 1 & a & a^2 \\ 1 & a^2 & a \end{bmatrix} \begin{bmatrix} V_a \\ V_b \\ V_c \end{bmatrix}$$
(10.7b)

Corresponding expressions for phase currents can be obtained from Eqs (10.6b) and (10.7a) respectively as follows

$$I_{abc} = AI_{012} \tag{10.8}$$

$$I_{012} = A^{-1}I_{abc} \tag{10.9a}$$

Equation (10.9a) in expanded form is given by

$$\begin{bmatrix} I_{a0} \\ I_{a1} \\ I_{a2} \end{bmatrix} = \frac{1}{3} \begin{bmatrix} 1 & 1 & 1 \\ 1 & a & a^2 \\ 1 & a^2 & a \end{bmatrix} \begin{bmatrix} I_a \\ I_b \\ I_c \end{bmatrix} \tag{10.9b}$$

Example 10.1 Calculate the symmetrical components of phase voltages, when phase voltages are given as $V_a = 100\angle 0°$ V, $V_b = 150\angle 120°$ V, and $V_c = 210\angle 240°$ V.

Solution In computing the voltage sequence components, Eq. (10.7a) and MATLAB have been employed.

$V_a = 100\angle 0° = (100 + j0)$ V, $V_b = 150\angle 120° = (-75 + j129.90)$ V,

$V_c = 210\angle 240° = (-105 - j181.865)$ V

≫ V= [100; -75+129.90i; -105 - 181.865i]

V =

  1.0e+002 *

  1.0000

  -0.7500 + 1.2990i

  -1.0500 - 1.8187i

≫ a = (-0.5+0.866i);

≫ A= [1 1 1; 1 a^2 a;1 a a^2]

A =

  1.0000   1.0000   1.0000

  1.0000 -0.5000 - 0.8660i -0.5000 + 0.8660i

  1.0000 -0.5000 + 0.8660i -0.5000 - 0.8660i

≫ VS=inv (A)*V

VS =

  1.0e+002 *

  -0.2667 - 0.1732i

  -0.2667 + 0.1732i

  1.5334 - 0.0000i

From the last output, the sequence voltages can be obtained as follows:

$$V_{a0} = 1.0e + 002 *(-0.2667 - 0.1732i) = (-26.67 - j17.32) = 31.80\angle 33° \text{ V}$$

$$V_{a1} = 1.0e + 002 *(-0.2667 + 0.1732i) = (-26.67 + j17.32) = 31.80\angle -33° \text{ V}$$

$$V_{a2} = 1.0e + 002 *(1.5334 - 0.0000i) = (153.34 - j0.0) = 153.34\angle 0° \text{ V}$$

Example 10.2 In a three-phase network, the sequence components of current are $I_{a0} = 0$ A, $I_{a1} = 5.78\angle -30°$ A, and $I_{a2} = 5.78\angle 45°$ A. Calculate the line currents.

Solution With the help of Eq. (10.8) and MATLAB functions, the line currents are calculated as follows:

$I_{a0} = 0$ A, $I_{a1} = 5.78\angle{-30°} = (5.006 - j2.89)$ A, $I_{a2} = 5.78\angle{45°} = (4.087 + j4.087)$ A

```
>> IS= [0; 5.006-2.89i; 4.087+4.087i];
>> a= (-0.5+0.866i);
>> A= [1 1 1; 1 a^2 a; 1 a a^2];
>> IP=A*IS
IP =
9.0930 + 1.1970i
-10.5884 - 1.3945i
1.4958 + 0.1975i
```

From the last output, the line currents are as follows:

$$I_a = (9.0930 + 1.1970i) = (9.0930 + j1.1970) = 9.17\angle{7.5°}\text{ A}$$
$$I_b = (-10.5884 - 1.3945i) = (-10.5884 - j1.3945) = 10.68\angle{7.5°}\text{ A}$$
$$I_c = (1.4958 + 0.1975i) = (1.4958 + j0.1975) = 1.509\angle{7.52°}\text{ A}$$

10.1.1 Coupling between Sequence Currents

The analyses of three-phase balanced power systems are greatly simplified by the use of per-phase analysis. This is possible due to zero coupling between phase currents. However, such a decoupling does not exist when the three-phase systems are unbalanced or faulted. The advantage of introducing symmetrical components, which induct nine variables, would be lost if some coupling existed between the sequence currents. Therefore, it is imperative to establish the coupling characteristics between sequence currents. The non-existence of coupling (or zero coupling) between the sequence currents will make the analyses of generators, transformers, and transposed transmission lines feasible on per sequence component basis.

If V_a, V_b, V_c, and I_a, I_b, I_c, are actual phase voltages and currents respectively in a three-phase power system, then the total power S is given by

$$S = P + jQ = V_a I_a^* + V_b I_b^* + V_c I_c^* = V_{abc}^t I_{abc}^*$$

where I_{abc}^* is the conjugate of phase current vector.

Substituting from Eqs (10.6a) and (10.8), the power becomes

$$S = P + jQ = V_{abc}^t I_{abc}^* = V_{012}^t A^t A I_{012}^*$$

Since A is symmetric, that is, $A^t = A$, it can easily be shown that $AA^* = 3U$, where U is a unit matrix and is given by

$$U = \begin{bmatrix} 1 & 0 & 0 \\ 0 & 1 & 0 \\ 0 & 0 & 1 \end{bmatrix}$$

Hence,

$$S = P + jQ = 3V_{012}^t I_{012}^* = 3(V_{a0} I_{a0}^* + V_{a1} I_{a1}^* + V_{a2} I_{a2}^*) \qquad (10.10)$$

From Eq. (10.10), it is observed that there are no cross products of sequence components of voltages and currents, for example, $V_{a0} I_{a1}^*$, $V_{a1} I_{a2}^*$, etc. Therefore, it can be concluded that there is zero coupling between the sequence currents. The total power, thus, in a three-phase, unbalanced power system is three times the sum of the symmetrical sequence component powers.

10.2 COMPUTATION OF SEQUENCE COMPONENTS OF UNEQUAL SERIES IMPEDANCES

As already stated, one of the causes of balanced power systems becoming unbalanced is due to the system experiencing an unsymmetrical fault. In this section, effect of unequal series impedances in a three-phase network is discussed.

Figure 10.2 shows a part of a three-phase system having three unequal impedances Z_a, Z_b, and Z_c in series in phases a, b, and c respectively. The mutual impedances between the phases are also unequal, that is, $Z_{ab} \neq Z_{ba}$, $Z_{bc} \neq Z_{cb}$, and $Z_{ca} \neq Z_{ac}$.

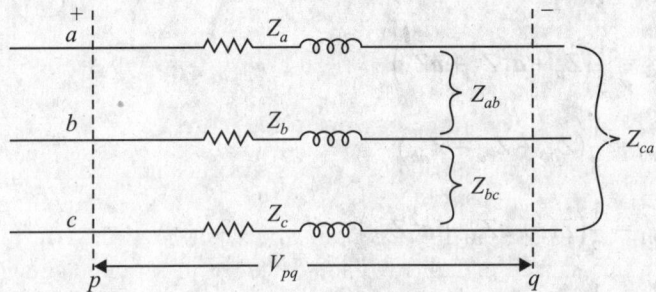

Fig. 10.2 Three-phase system with unequal impedances

In matrix form, the voltage drops in different phases between p and q can be written as

$$V_{pq} = \begin{bmatrix} V_{pqa} \\ V_{pqb} \\ V_{pqc} \end{bmatrix} = \begin{bmatrix} Z_a & Z_{ab} & Z_{ac} \\ Z_{ba} & Z_b & Z_{bc} \\ Z_{ca} & Z_{cb} & Z_c \end{bmatrix} \begin{bmatrix} I_a \\ I_b \\ I_c \end{bmatrix} \tag{10.11}$$

Transforming phase quantities in Eq. (10.11) into symmetrical components using Eqs (10.6) and (10.8) gives

$$AV_{pq012} = \begin{bmatrix} Z_a & Z_{ab} & Z_{ac} \\ Z_{ba} & Z_b & Z_{bc} \\ Z_{ca} & Z_{cb} & Z_c \end{bmatrix} AI_{012} \tag{10.12}$$

or

$$V_{pq012} = A^{-1} \begin{bmatrix} Z_a & Z_{ab} & Z_{ac} \\ Z_{ba} & Z_b & Z_{bc} \\ Z_{ca} & Z_{cb} & Z_c \end{bmatrix} AI_{012}$$

$$V_{pq012} = \boldsymbol{Zpq}_{012} \boldsymbol{I}_{012} \tag{10.13}$$

where

$$\boldsymbol{Z}_{pq012} = A^{-1} \begin{bmatrix} Z_a & Z_{ab} & Z_{ac} \\ Z_{ba} & Z_b & Z_{bc} \\ Z_{ca} & Z_{cb} & Z_c \end{bmatrix} A \tag{10.14}$$

On performing the indicated transformations in Eq. (10.14), Z_{pq012} can be written as

$$Z_{pq012} = \begin{bmatrix} (Z_{s0} + 2Z_{m0}) & (Z_{s2} - Z_{m2}) & (Z_{s1} - Z_{m1}) \\ (Z_{s1} - Z_{m1}) & (Z_{s0} - Z_{m0}) & (Z_{s2} + 2Z_{m2}) \\ (Z_{s2} - Z_{m2}) & (Z_{s1} + 2Z_{m1}) & (Z_{s0} - Z_{m0}) \end{bmatrix} \tag{10.15}$$

where

$$Z_{s0} = \frac{1}{3}(Z_a + Z_b + Z_c)$$

$$Z_{s1} = \frac{1}{3}(Z_a + aZ_b + a^2 Z_c) \tag{10.16a}$$

$$Z_{s2} = \frac{1}{3}(Z_a + a^2 Z_b + aZ_c)$$

$$Z_{m0} = \frac{1}{3}(Z_{bc} + Z_{ca} + Z_{ab})$$

$$Z_{m1} = \frac{1}{3}(Z_{bc} + aZ_{ca} + a^2 Z_{ab}) \tag{10.16b}$$

$$Z_{m2} = \frac{1}{3}(Z_{bc} + a^2 Z_{ca} + aZ_{ab})$$

Equations (10.15) and (10.16) provide a general transformation of three-phase impedances to sequence impedances.

In a special case, if $Z_a = Z_b = Z_c$ and $Z_{ab} = Z_{bc} = Z_{ca}$, then Eqs (10.16a) and (10.16b) takes the following form:

$$Z_{s0} = Z_a$$
$$Z_{s1} = Z_{s2} = 0$$
$$Z_{s2} = Z_{m1} = Z_{m2} = 0$$

Thus Eq. (10.13) can be written as

$$\begin{bmatrix} V_{pq0} \\ V_{pq1} \\ V_{pq2} \end{bmatrix} = \begin{bmatrix} Z_a & 0 & 0 \\ 0 & Z_a & 0 \\ 0 & 0 & Z_a \end{bmatrix} \begin{bmatrix} I_{a0} \\ I_{a1} \\ I_{a2} \end{bmatrix} \tag{10.17a}$$

or

$$V_{pq0} = Z_a I_{a0}$$
$$V_{pq1} = Z_a I_{a1} \tag{10.17b}$$
$$V_{pq2} = Z_a I_{a2}$$

From Eq. (10.17) it is observed that in balanced series impedances or balanced star-connected loads, with zero mutual coupling, unbalanced currents produce symmetrical sequence voltage drops which are due to the symmetrical sequence currents only. However, it is also observed from Eqs (10.15) and (10.16) that if the impedances are unequal the voltage drop for any one sequence, computed by using Eq. (10.13), is due to all the three sequence component currents. Such a situation arises for an unbalanced star-connected load, since the q ends (Fig. 10.2) of phases a, b, and c can be connected to form a star point.

In transmission lines, when current flows in a conductor in one phase, it induces voltages in the conductors of the other two phases, thereby giving rise to mutual coupling. However, because of the manner in which the reactance is calculated, the effect of the mutual coupling is eliminated. In other words, the effect of mutual reactance in a transmission line, computed on the basis of full transposition, is accounted for in the self-inductance of the line. Thus, in a transposed transmission line, series impedances of the three phases are equal and the mutual impedance is zero. Therefore, it can be concluded that a particular sequence component current causes voltage drop in a transmission line of that sequence component only.

In other words, the positive sequence voltage drop will be due to positive sequence current only and the impedance causing the voltage drop is called the positive sequence impedance Z_1. Similarly, negative and zero sequence drops will be respectively due to negative and zero sequence currents alone and the corresponding impedances causing these drops are called negative sequence impedance Z_2 and zero sequence impedance Z_0.

In order to employ symmetrical components, it is necessary to compute the sequence impedances of power system elements such as alternators, transformers, and transmission lines, which are discussed in the following sections.

Example 10.3 Determine the sequence impedance matrix and develop the sequence network for the circuit shown in Fig. 10.3. Draw its sequence network. The phase and neutral impedances are shown in the figure.

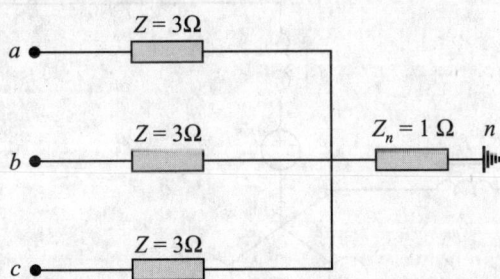

Fig. 10.3 Circuit diagram for Example 10.3

Solution The phase impedance matrix can be written as

$$Z_{abc} = \begin{bmatrix} (Z+Z_n) & Z_n & Z_n \\ Z_n & (Z+Z_n) & Z_n \\ Z_n & Z_n & (Z+Z_n) \end{bmatrix} = \begin{bmatrix} 4 & 1 & 1 \\ 1 & 4 & 1 \\ 1 & 1 & 4 \end{bmatrix}$$

Using the transformation of Eq. (10.14) and MATLAB, the sequence matrix Z_{012} can be computed as follows:

$$Z_{012} = A^{-1}Z_{abc}A = A^{-1} \begin{bmatrix} (Z+Z_n) & Z_n & Z_n \\ Z_n & (Z+Z_n) & Z_n \\ Z_n & Z_n & (Z+Z_n) \end{bmatrix} A$$

```
>> inv(A)*Z*(A)
ans =
      6.0000      0.0000    + 0.0000i      0.0000    + 0.0000i
      0           3.0000    + 0.0000i     -0.0000    - 0.0000i
    - 0.0000      0         - 0.0000i      3.0000    - 0.0000i
```

From the above,

$$\mathbf{Z}_{012} = \begin{bmatrix} 6 & 0 & 0 \\ 0 & 3 & 0 \\ 0 & 0 & 3 \end{bmatrix} = \begin{bmatrix} (Z+Z_n) & 0 & 0 \\ 0 & Z & 0 \\ 0 & 0 & Z \end{bmatrix}$$

The sequence network is drawn in Fig. 10.4.

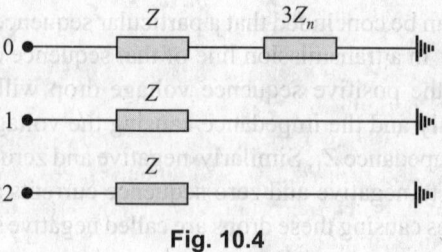

Fig. 10.4

10.2.1 Sequence Parameters and Sequence Networks of Generators

Figure 10.5 shows a three-phase unloaded star-connected generator. The star point is earthed through an impedance of Z_n Ω. The induced voltages in each phase are E_a, E_b, and E_c.

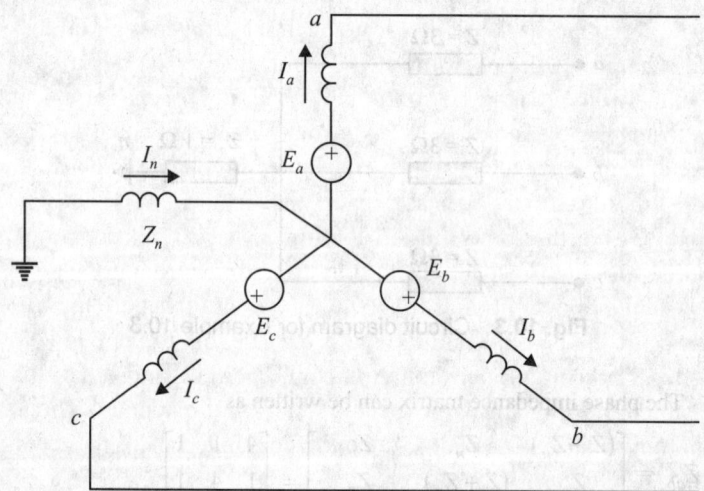

Fig. 10.5 Three-phase, star-connected generator

When a fault occurs at the generator terminals, currents I_a, I_b, and I_c flow in all the three phases. If the fault involves the ground, then the current I_n flows into the neutral of the generator. One or two of the line currents may be zero, thereby producing unbalanced conditions. The unbalanced currents can be transformed into their symmetrical components. The voltage drops due to sequence current components can be written as

$$V_{a0} = -I_{a0}Z_0$$

$$V_{a1} = E_a - I_{a1}Z_1 \tag{10.18a}$$

$$V_{a2} = -I_{a2}Z_2$$

where Z_0, Z_1, and Z_2 are zero, positive, and negative sequence impedances respectively, and E_a is the induced phase voltage of the generator. If the resistance is neglected then Eq. (10.18a) may be written as

$$V_{a0} = -jI_{a0}X_0$$
$$V_{a1} = E_a - jI_{a1}X_1 \qquad\qquad (10.18b)$$
$$V_{a2} = jI_{a2}X_2$$

where X_0, X_1, and X_2 are zero, positive, and negative sequence reactances respectively, of the generator.

It may be noted that generators are designed to supply positive sequence three-phase voltages and no negative sequence or zero sequence voltages are supplied by a generator. Since the generator is on no-load, the positive sequence voltage is the no-load terminal voltage to neutral.

The reactance X_1 is due to the mmf produced by the positive sequence current. Since this mmf is stationary with respect to the rotor, $X_1 = X_d = X_q$ for round rotor machines. Even in salient pole machines where $X_d \neq X_q$, the approximation can be adopted since I_{a1} lags E_a by almost 90°, that is, I_{a1} is almost equal to the positive sequence direct axis current in phase a and hence the effect of the positive sequence quadrature current in phase a can be ignored. The mmf produced by the negative sequence current in the armature rotates in the opposite direction of the dc field. As such it is traversing rapidly over the rotor face. Therefore, the position of the mmf, due to the negative sequence current, vis-à-vis the machine's direct and quadrature axes varies constantly. Based on the above explanation, X_2 is normally taken as equal to $(X_d + X_q)/2$. In the case of zero sequence currents, since the currents are in phase, the mmfs produced in the three windings are space displaced by 120° and attain their maximum values at the same time. If the mmfs of each phase had an exactly sinusoidal distribution in space, their sum around the armature would be zero at every instant and every point. Since no flux would exist around the air gap, the reactance of the phase winding would only be due to the end connections and leakage. Since in an operational machine, perfect sinusoidal distribution is not feasible, X_0 is higher than expected in an ideal machine having an exact sinusoidal mmf distribution. Thus, it can be concluded that in a generator, X_1, X_2, and X_0 have different values.

Using Eqs (10.18) positive, negative, and zero sequence networks can now be drawn and are shown in Figs 10.6(a), (b), and (c) respectively. In Fig. 10.6(c), observe that the three zero sequence currents flow through Z_n, hence the zero sequence impedance between the neutral and ground is made $3Z_n$. Thus, the total zero sequence impedance Z_{0T} of the network is

$$Z_{0T} = 3Z_n - Z_0 \qquad\qquad (10.19)$$

Note that in the positive and negative sequence networks, since the neutral is at ground potential, the reference bus in both cases is the generator neutral. In case of zero sequence network, however, the ground is taken as the reference bus because the generator neutral is no longer at ground potential.

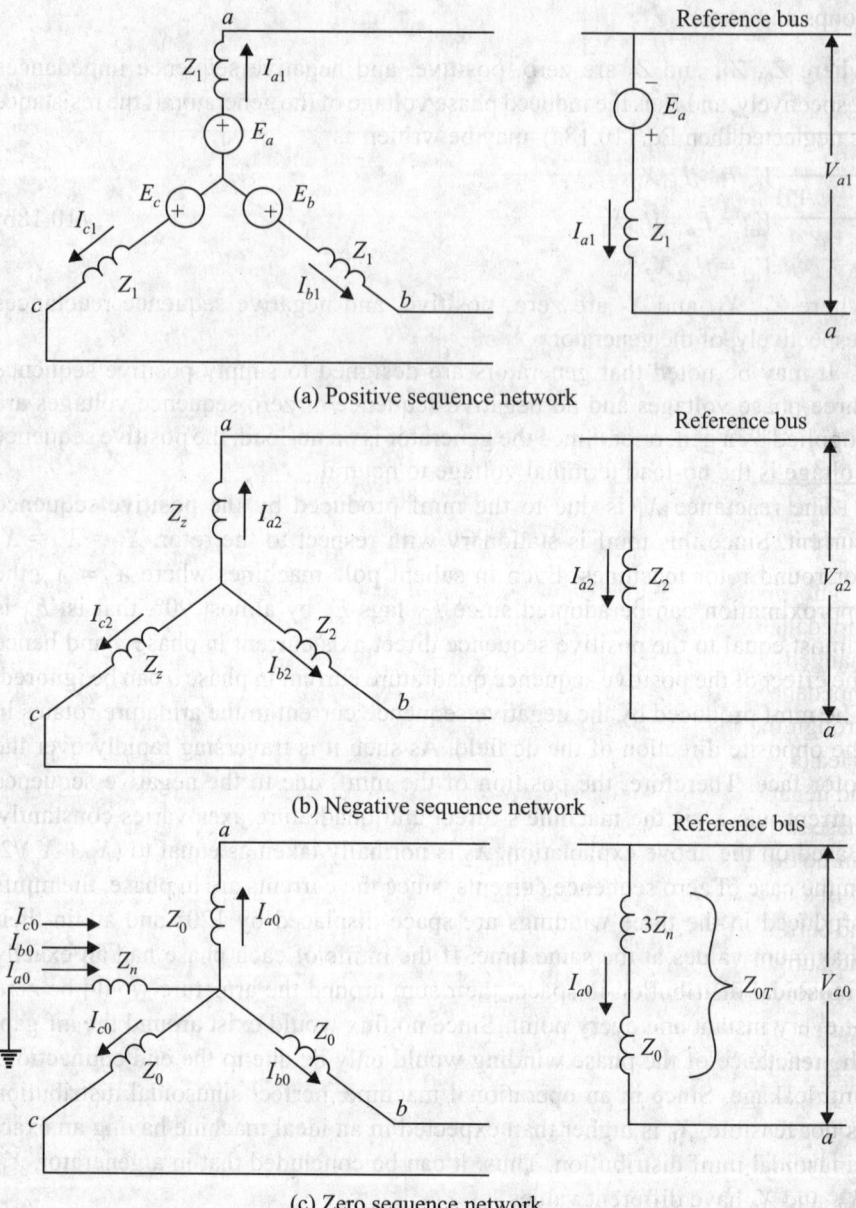

(a) Positive sequence network

(b) Negative sequence network

(c) Zero sequence network

Fig. 10.6 Positive, negative, and zero sequence networks for a generator

10.2.2 Sequence Parameters and Sequence Networks of Transformers

Three-phase transformers comprise an integral part of the present-day AC power system. The primary and secondary voltages bear a definite phase relationship, which depends on how the windings are connected, that is, in delta or star. In order to facilitate their operation and simulate their simulation for fault analyses, transformers are grouped into vector (phasor) groups* which show the phase shift between the primary and secondary voltages. Table 10.1 summarizes the vector groups and phase shifts for various types of transformer connections.

Table 10.1 Summary of vector groups and phase shifts for different types of transformer connections

Winding connections		Vector (phasor) group	Phase shift
Primary	Secondary		
Star	Delta	Yd1	− 30°
	Alternate delta	Yd11	+ 30°
Star	Star	Yy0	0°
Delta	Delta	Dd0	0°
	Alternate delta	Dd6	180°
Delta	Star	Dy0	0°
	Alternate star	Dy5	−150°

* For additional details, please refer to the authors' other title *Basic Electrical Engineering* (2nd edition), pp. 384–390, published by OUP.

The positive sequence impedance of a transformer equals the leakage impedance. As the transformer is a static device, the leakage impedance is not changed if the phase sequence is changed. Thus, positive and negative sequence impedances of a transformer are identical. Also, if zero sequence current flows through the transformer, the phase impedance to zero sequence current is equal to leakage impedance. Thus $Z_0 = Z_1 = Z_2$, where Z_0, Z_1, and Z_2 are zero, positive, and negative sequence impedance, respectively, of the transformer. However, the phase shift is different for positive and negative sequence currents. The phase shift can be determined by the following rule.

In a transformer, if the phase shift is α degrees for positive sequence voltages and currents, then the phase shift will be $-\alpha$ degrees for negative sequence voltages and currents.

In developing the sequence networks of transformers, the shunt admittance branch of the transformer has been neglected, since the magnitude of the magnetizing current is very low and its exclusion does not introduce significant errors. The positive and negative sequence networks are simple circuits with the sequence impedance connected between the reference bus and the neutral.

The zero sequence networks of three-phase transformers is dependent on the type of transformer connections and whether a path is available to the flow of zero sequence currents. When the core reluctance is neglected in a transformer, the primary mmf exactly balances the secondary mmf. This means that the current can flow in one winding, provided the current flows in the other winding as well. Based on this, the zero sequence networks have been derived for the following five types of connections.

Type 1 (star–star with neutral grounded on one side) If neutral on one side is grounded, zero sequence current cannot flow in either winding. The absence of path through one winding prevents current flow in the other. An open circuit condition prevails for the two parts of the system connected by the transformer. Figure 10.7 shows the transformer symbol, type of connection, and equivalent zero sequence network. The letters A and B identify corresponding points on the connection diagram and equivalent circuit.

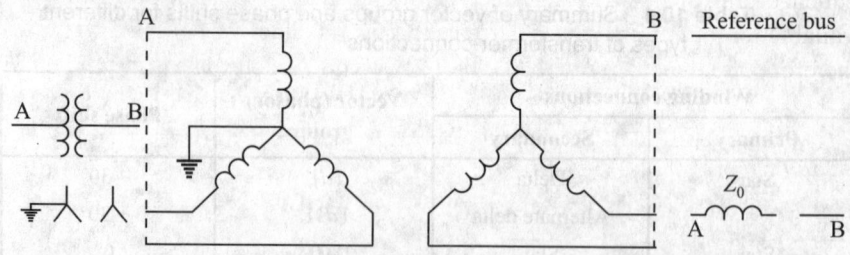

Fig. 10.7   Star–star with neutral grounded on one side

Type 2 (star–star with both neutrals grounded)   With both neutrals grounded, a path is available for the zero sequence currents to flow. In this case, the zero sequence reactance is connected to the points A and B in two sides of the transformer as shown in Fig. 10.8.

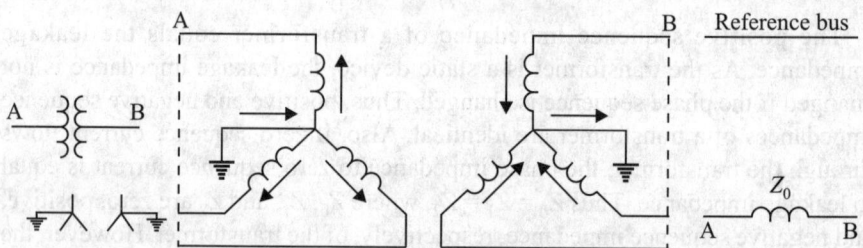

Fig. 10.8   Star–star with neutrals grounded on both sides

Type 3 (star–delta with neutral on star side grounded)   Since the neutral on star side is grounded, a path is available for the zero sequence currents to flow through the ground. The induced currents on the delta side circulate within the delta only. Hence, an open circuit condition prevails between the line and reference bus on the delta side. On the star side, the reference bus is connected to the line to show that a path exists between the equivalent leakage reactance of the transformer and the reference bus. In case the neutral is grounded through an impedance Z_n, its effect is indicated by including impedance equal to $3Z_n$ in series with the equivalent leakage reactance of the transformer. Figure 10.9 shows the circuit diagram for such a connection.

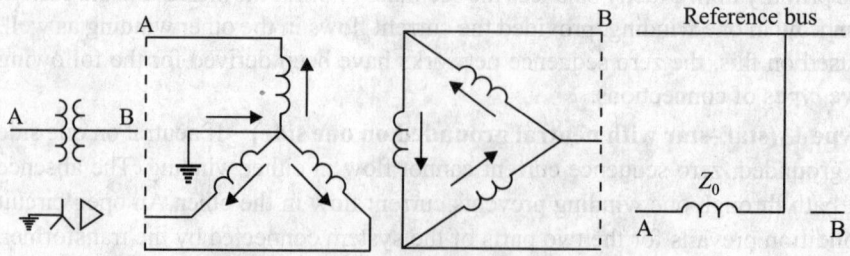

Fig. 10.9   Star–delta with neutral grounded

Type 4 (star–delta with star side ungrounded) Figure 10.10 shows the zero sequence equivalent circuit for this type of connections. It is seen that no path is available for the zero sequence currents to flow on either side of the transformer. Therefore, the impedance to the flow of zero sequence currents is infinite.

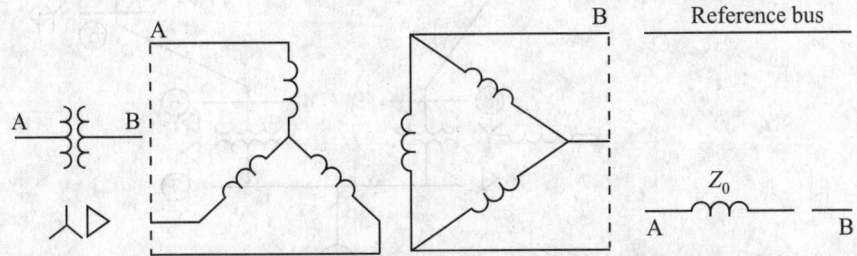

Fig. 10.10 Star–delta with neutral ungrounded

Type 5 (delta–delta connection) In this type of a connection, the zero sequence currents can circulate within the transformer windings. No zero sequence current can leave the delta terminals. Figure 10.11 shows the equivalent zero sequence network.

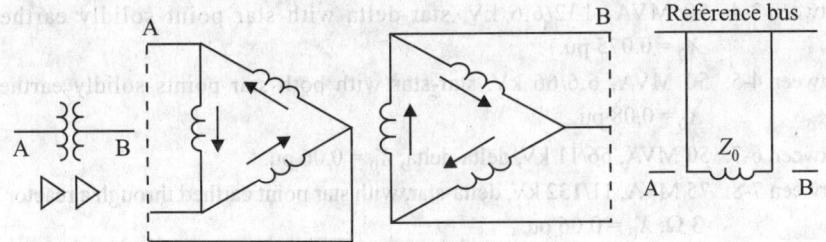

Fig. 10.11 Delta–delta connection

10.2.3 Sequence Parameters and Sequence Networks for Transmission Lines

Like a three-phase transformer, a transmission line is also a linear, static, and symmetrical network. Hence, the positive and negative sequence impedances are equal. In case of zero sequence currents flowing in a transmission line, each phase carries the same current. As such the magnetic field produced by zero sequence currents is different from that set up by the positive or negative sequence currents. Due to the difference in the magnetic fields in the two cases, the zero sequence impedance of a transmission line is two to three times the positive sequence impedance. The return path for the zero sequence currents is through the neutral conductor or the ground or both.

Example 10.4 The zero sequence reactances for the various components of the power system shown in Fig. 10.12 are as follows:

Generator 1 Star-connected 100 MVA, 11 kV, $X_0 = 0.08$ pu with star point earthed through a reactor of 3.0 Ω.

Generator 2 Star-connected 50 MVA, 11 kV, $X_0 = 0.05$ pu with star point isolated.

Motor Star-connected 25 MVA, 6.6 kV, $X_0 = 0.05$ pu with star point solidly earthed.

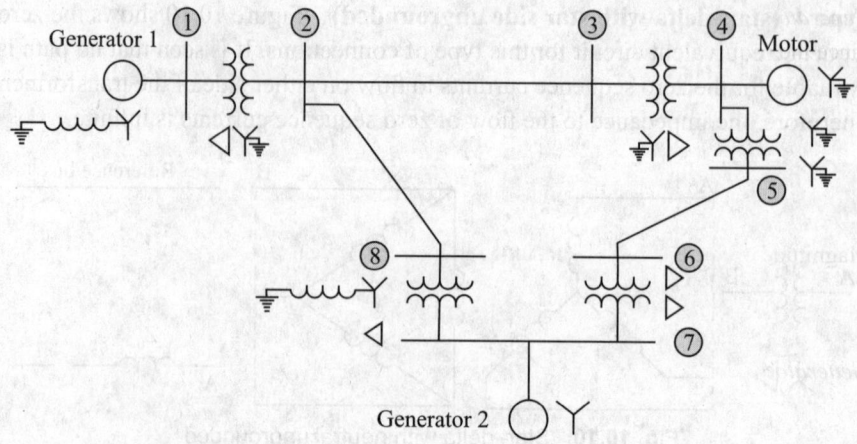

Generator 1

Motor

Generator 2

Fig. 10.12 Single-line diagram of power system for Example 10.4

Transformers

Between 1-2: 100 MVA, 11/132 kV, delta-star with star point solidly earthed, $X_0 = 0.1$ pu.

Between 3-4: 50 MVA, 132/6.6 kV, star-delta with star point solidly earthed, $X_0 = 0.075$ pu.

Between 4-5: 50 MVA, 6.6/66 kV, star-star with both star points solidly earthed. $X_0 = 0.08$ pu.

Between 6-7: 50 MVA, 66/11 kV, delta-delta, $X_0 = 0.06$ pu.

Between 7-8: 75 MVA, 11/132 kV, delta-star, with star point earthed through a reactor of 3 Ω; $X_0 = 0.06$ pu.

Line 2-3: zero sequence reactance $X_0 = 300$ Ω

Line 2-8: zero sequence reactance $X_0 = 250$ Ω

Line 5-6: zero sequence reactance $X_0 = 200$ Ω

Draw the zero sequence networks for the system.

Solution Based on the foregoing discussion, the zero sequence network of the power system is drawn in Fig. 10.13.

Select 100 MVA as the power base for the system. The base voltage for each section of the system is the rated voltage. Per unit base values of zero sequence reactance of the components of the system are computed as follows:

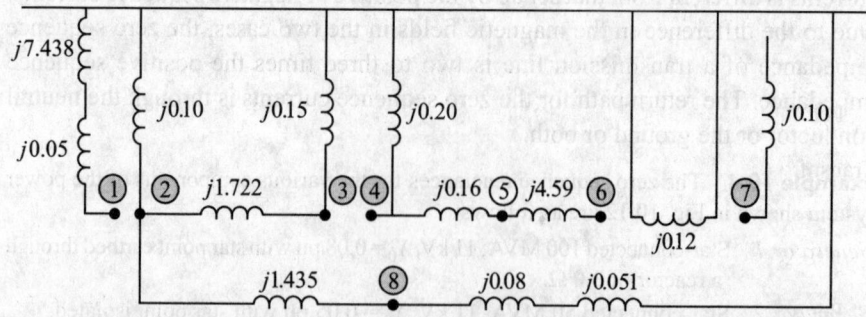

Fig. 10.13 Zero sequence equivalent circuit for Example 10.4

Generator 1

Since the machine rating is the same as the system power and voltage base $X_0 = j0.08$ pu.

$$Z_{base} \text{ in generator 1 circuit} = \frac{(11)^2}{100} = 1.21 \ \Omega$$

$$\text{Per unit value of the grounding reactor} = \frac{3}{1.21} = j2.479$$

Magnitude of the reactor to be included in the zero sequence network

$$= 3 \times j2.479$$
$$= j7.438$$

Generator 2

$$\text{Per unit reactance on system base } X_0 = j0.05 \times \frac{100}{50} = j0.10$$

Motor Per unit reactance on system base $X_0 = j0.05 \times \frac{100}{25} = j0.20$

Transformers

Between 1-2: $X_0 = 0.1$ pu

Between 3-4: $X_0 = j0.075 \times \frac{100}{50} = j0.15$

Between 4-5: $X_0 = j0.08 \times \frac{100}{50} = j0.16$

Between 6-7: $X_0 = j0.076 \times \frac{100}{50} = j0.12$

Between 7-8: $X_0 = j0.06 \times \frac{100}{75} = j0.08$

$$\text{For grounding reactor } Z_{base} = \frac{(132)^2}{100} = 174.24 \ \Omega$$

$$\text{Per unit reactance} = \frac{3.0}{174.24} = j0.017$$

Value of pu reactance to be included in the zero sequence network

$$= 3 \times j0.017 = j0.051$$

Transmission line 2-3:

$$Z_{base} \text{ for transmission line} = \frac{(132)^2}{100} = 174.24 \ \Omega$$

$$\text{Per unit value of line reactance } X_0 = j\frac{300}{174.24} = j1.722$$

Transmission line 2-8:

$$Z_{base} = \frac{(132)^2}{100} = 174.24 \ \Omega$$

$$\text{Per unit value of line reactance} = j\frac{250}{174.24} = j1.435$$

Transmission line 5-6:

$$Z_{base} = \frac{(66)^2}{100} = 43.56 \ \Omega$$

Per unit value of line reactance $= j\frac{200}{43.56} = j4.591$

10.3 SYNTHESIS OF SEQUENCE NETWORKS

As noted in the previous section, sequence networks replicate the unbalanced three-phase system by the zero, positive, and negative sequence impedances. In passive three-phase networks, positive and negative sequence networks are identical, except that in the former a positive sequence voltage source is included to represent the internal voltage of the generator. Zero sequence networks are usually different from positive and negative sequence networks. It may sometimes be an open-circuit zero sequence network. Sequence networks for rotating machines are all different.

Since linearity is assumed in drawing the sequence networks, each of the networks can be represented by its Thevenin's equivalent between the two terminals, one terminal is the fault point F and the second terminal is the reference or zero-potential bus N. Figure 10.14 shows the representation of Thevenin's equivalent of each sequence network. In the case of the positive sequence network, the pre-fault voltage to neutral at the point of application of the fault is the Thevenin's equivalent voltage V_f. The Thevenin's equivalent impedance Z_1 is the impedance between the points F and N of the positive sequence network with all internal emfs short-circuited. Since during pre-fault condition no negative and zero sequence currents are flowing, the pre-fault voltage between points F and N is zero in the negative and zero sequence networks. The impedances Z_2 and Z_0 are the impedances between the points F and N of the negative and zero sequence networks respectively.

It is observed from Fig. 10.14 that the sequence currents flowing into the fault point F, which is the point of unbalance, are taken as positive. The sequence voltages have also been taken to be positive from N to F, that is, there is a voltage rise in sequence voltages from N to F. The fault impedance or unbalanced connection is always shown outside the boxes at point F.

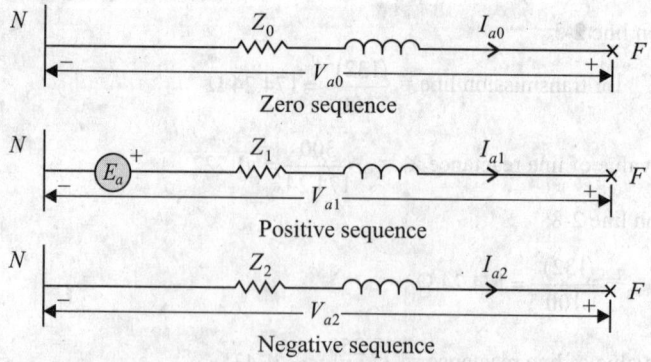

Fig. 10.14   Thevenin's equivalent sequence networks

Each of the sequence networks can be analysed by using Thevenin's theorem to obtain the sequence currents and voltages. The sequence currents and voltages so obtained are synthesized to obtain the phase currents and voltages.

10.4 CLASSIFICATION OF UNSYMMETRICAL FAULTS

Unsymmetrical faults in power systems are of two types, namely shunt faults and series faults. A shunt fault is an unbalance between phases or between phase and neutral such as single line-to-ground (SLG), line-to-line (LL), double line-to-ground (2LG), and three-phase (3ϕ) fault. Experience shows that 70 per cent of the faults occurring on transmission lines are of SLG type. The 3ϕ fault, though a symmetrical fault, has been included here because it produces a shunt unbalance. The 3ϕ fault is next to the SLG fault from the point of view of general interest.

A series fault is an unbalance in the line impedances and does not involve the neutral or ground, such as unequal series impedances, one line open (1LO), two lines open (2LO).

10.5 ANALYSES OF UNSYMMETRICAL FAULTS

The analyses of a faulted network require the development of zero, positive, and negative sequence networks and determining their interconnections based on the type of fault. The sequence network is then solved to obtain the sequence currents and voltages, which determine the phase quantities.

10.5.1 Five Steps to Steady-state Fault Analyses

The following five steps form the basis for analyses of faults:

Step 1: Draw a fully labelled single circuit diagram, indicating therein, the phase connections up to the fault point, marking the assumed polarities and positive directions of current flows respectively, and the impedances. It may be noted that the direction of flow of current is assumed positive from the fault point to ground and phase voltages are assumed as voltage drops from the phase to ground. On both sides of the fault point, the system consists of balanced impedances and the Thevenin's equivalent, looking into the fault point, can be computed. Figure 10.15 shows a circuit diagram at the fault point.

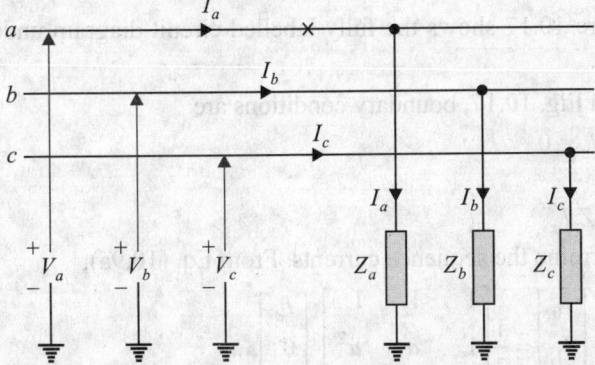

Fig. 10.15 Representation of a power system under a fault condition

Step 2: Depending on the type of fault, write the boundary conditions relating known currents and voltages.

Step 3: By using the transformation A or A^{-1} convert the known voltages and current quantities from phase frame (*a-b-c*) system to sequence frame (0-1-2).

Step 4: Determine the proper connection of the F and N terminals of sequence networks satisfying the relationship between the sequence currents and voltages in step 3.

Step 5: Interconnect and draw the sequence networks satisfying the current and voltage conditions of the sequence components.

The application of the above five steps is demonstrated in the analyses of power systems for various types of faults in the following section.

10.6 SHUNT FAULTS

Shunt faults are an important class of unsymmetrical faults. Analysis of these faults is now discussed.

10.6.1 Single Line-to-ground Fault

An SLG fault occurs on phase *a* of a three-phase power system and is shown in Fig. 10.16. The fault impedance between the fault point F and ground is Z_f. The following steps are used to analyse the fault.

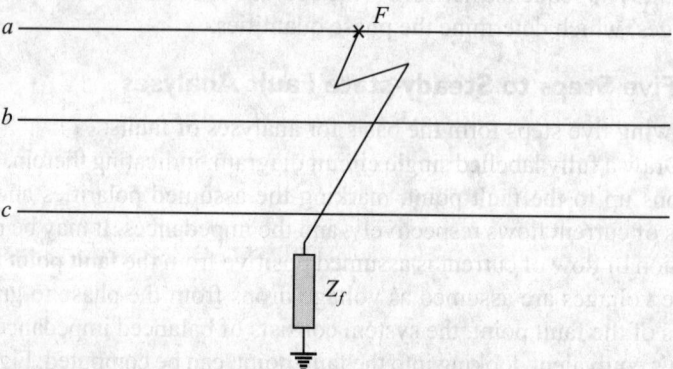

Fig. 10.16 SLG fault on phase *a* of a three-phase system

Step 1: Figure 10.17 shows the fully labelled circuit diagram under SLG fault condition.

Step 2: From Fig. 10.17, boundary conditions are

$$I_b = I_c = 0 \qquad (10.20a)$$

and

$$V_a = Z_f I_a \qquad (10.20b)$$

Step 3: Determine the sequence currents. From Eq. (10.9a),

$$\begin{bmatrix} I_0 \\ I_1 \\ I_2 \end{bmatrix} = \frac{1}{3} \begin{bmatrix} 1 & 1 & 1 \\ 1 & a & a^2 \\ 1 & a^2 & a \end{bmatrix} \begin{bmatrix} I_a \\ 0 \\ 0 \end{bmatrix}$$

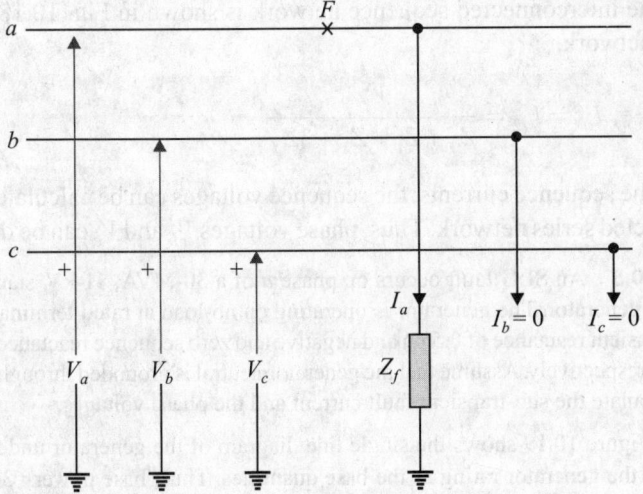

Fig. 10.17 Labelled circuit diagram for an SLG fault at F

or
$$I_0 = I_1 = I_2 = \frac{1}{3} I_a \tag{10.21a}$$

Hence, $V_a = 3Z_f I_{a1}$

or
$$V_{a0} + V_{a1} + V_{a2} = 3Z_f I_{a1} \tag{10.21b}$$

Step 4: It is observed from Eq. (10.21a) that all the sequence currents are equal. Therefore, the positive, negative, and zero sequence Thevenin's equivalent circuits have to be connected in series. Equation (10.21b) shows that the sum of the sequence voltage drops is equal to $3Z_f I_{a1}$. Hence, an external impedance of three times the fault impedance is to be added to the sequence network.

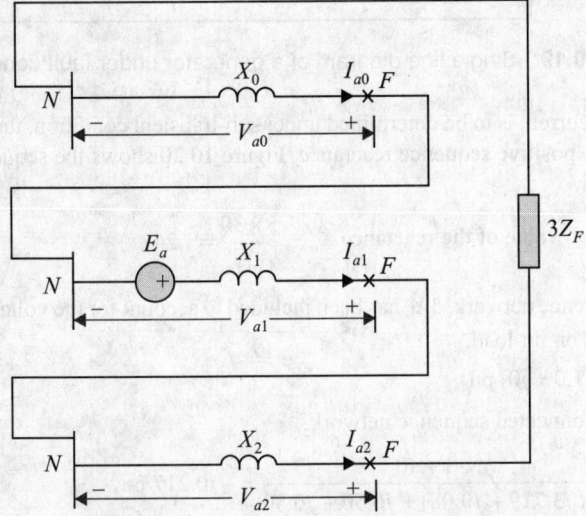

Fig. 10.18 Interconnected sequence network

Step 5: The interconnected sequence network is shown in Fig. 10.18. From the sequence network,

$$I_{a0} = I_{a1} = I_{a2} = \frac{V_f}{Z_0 + Z_1 + Z_2 + 3Z_f} \qquad (10.21c)$$

Knowing the sequence currents, the sequence voltages can be calculated from the interconnected series network. Thus, phase voltages V_b and V_c can be determined.

Example 10.5 An SLG fault occurs on phase *a* of a 30-MVA, 11-kV, star-connected, three-phase generator. The generator is operating on no-load at rated terminal voltage. It has a subtransient reactance of 0.5 pu and negative and zero sequence reactances of 0.30 pu and 0.08 pu respectively. Assume that the generator neutral is grounded through a reactance of 5 Ω. Calculate the sub-transient fault current and the phase voltages.

Solution Figure 10.19 shows the single line diagram of the generator under fault condition. Take the generator rating as the base quantities. Thus, base power = 30 MVA and base voltage = 11 kV.

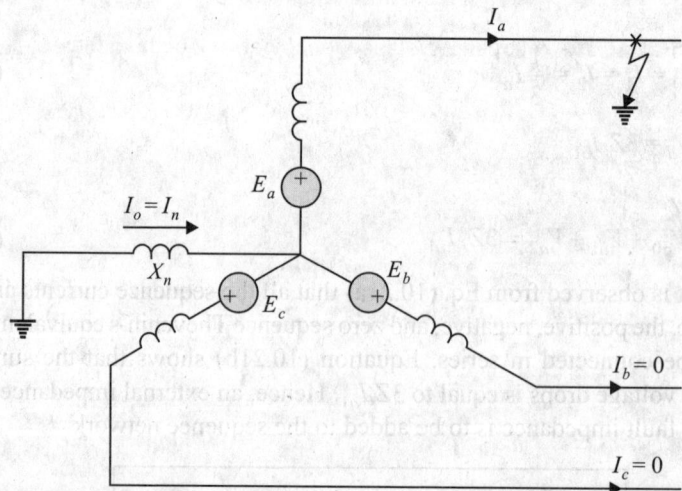

Fig. 10.19 Single line diagram of a generator under fault condition

Since the fault current is to be determined under sub-transient condition, the sub-transient reactance is the positive sequence reactance. Figure 10.20 shows the sequence networks of the generator.

$$\text{Per unit value of the reactance } X_n = \frac{5 \times 30}{(11)^2} = 1.24$$

In the zero sequence network, $3X_n$ has been included to account for the voltage drop. Since the generator is on no-load,

$$E_a = (1.0 + j0) \text{ pu}$$

From the interconnected sequence network,

$$I_{a1} = \frac{1 + j0}{(j3.719 + j0.08) + j0.50 + j0.30} = -j0.217 \text{ pu}$$

Sub-transient phase current

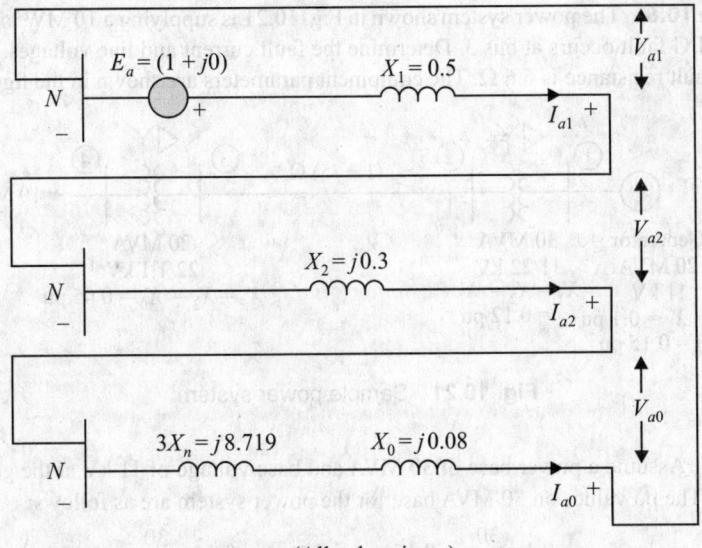

(All values in pu)

Fig. 10.20 Interconnected sequence network of the generator

$I_a = 3 \times (-j0.217) = -j0.652$ pu

$V_{a0} = -I_{a0}(X_0 + 3X_n) = -(-j0.217) \times j(0.08 + 3.719) = -0.824$ pu

$V_{a1} = E_a - I_{a1}X_1 = (1.0 + j0) - (-j0.217)(j0.5) = 0.892$ pu

$V_{a2} = -I_{a2}X_2 = -(-j0.217)(j0.3) = -0.065$ pu

Line-to-ground voltages are as follows

$$\begin{bmatrix} V_a \\ V_b \\ V_c \end{bmatrix} = \begin{bmatrix} 1 & 1 & 1 \\ 1 & a^2 & a \\ 1 & a & a^2 \end{bmatrix} \begin{bmatrix} -0.824 \\ 0.892 \\ -0.065 \end{bmatrix} = \begin{bmatrix} 0.0030 \\ -1.2375 - 0.8288i \\ -1.2375 + 0.8288i \end{bmatrix}$$

Line-to-line voltages are given by

$V_{ab} = V_a - V_b = 1.2375 + j0.8288 = 1.489\angle 33.81°$ pu

$V_{bc} = V_b - V_c = 0 - j1.6576 = 1.6576 \angle -270°$ pu

$V_{ca} = V_c - V_a = -1.2375 + j0.8288 = 1.489\angle 146.19°$ pu

$I_{\text{base}} = \dfrac{30 \times 1000}{\sqrt{3} \times 11} = 1574.638$ A

Actual sub-transient fault current in line $a = -j0.217 \times 1574.638 = -j341.696$ A

Actual line voltages are given by

$V_{ab} = 1.489\angle 33.81° \times \dfrac{11}{\sqrt{3}} = 9.457\angle 33.81°$ kV

$V_{bc} = 1.6576 \angle -270° \times \dfrac{11}{\sqrt{3}} = 10.527\angle -270°$ kV

$V_{ca} = 1.489\angle 146.19° \times \dfrac{11}{\sqrt{3}} = 9.457\angle 146.19°$ kV

Example 10.6   The power system shown in Fig. 10.21 is supplying a 10 MW load at 1.1 kV. An SLG fault occurs at bus 3. Determine the fault current and line voltages. Assume that the fault resistance is 6.6 Ω. The equipment parameters are shown in the figure.

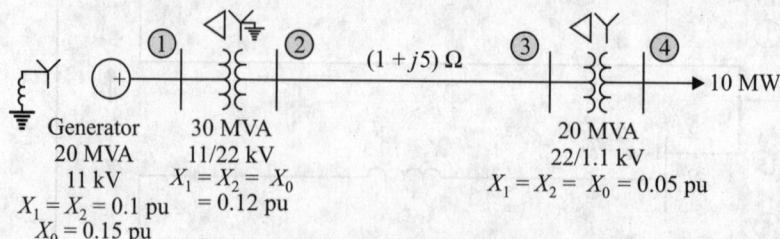

Fig. 10.21   Sample power system

Solution   Assume a power base of 30 MVA and base voltage of 11 kV at the generator terminal. The pu values on 30 MVA base for the power system are as follows:

Generator $X_1 = X_2 = j0.1 \times \dfrac{30}{20} = j0.15$ pu and $X_0 = j0.15 \times \dfrac{30}{20} = j0.225$ pu

Transformer 1-2 $X_1 = X_2 = X_0 = j0.12$ pu

Transmission line 2-3 $Z_{base} = \dfrac{(22)^2}{30} = 16.133$ Ω

Hence, pu impedance $= \dfrac{1 + j5.0}{16.1333} = 0.062 + j0.31$

Transformer 3-4 $X_1 = X_2 = X_0 = j0.05 \times \dfrac{30}{20} = j0.075$ pu

If load is to be represented as a series resistance, assume that P MW load is supplied at a voltage of V kV. Then, equivalent resistance R of the load is given by

$$R = \frac{V^2}{P} \ \Omega$$

We know that Z_{base}, in this case $R_{base} = \dfrac{(\text{Base kV})^2}{\text{Base MVA}}$

Thus,

$$R_{pu} = \frac{R\ \Omega}{R_{base}} = \frac{V^2}{P} \times \frac{\text{Base MVA}}{(\text{Base kV})^2} = \frac{\dfrac{V^2}{(\text{Base kV})^2} \times \text{Base MVA}}{P}$$

$$= \frac{(\text{pu kV})^2 \times \text{Base MVA}}{P}$$

$$= \frac{(1.0)^2 (30)}{10} = 3.0 \text{ pu}$$

Per unit value of the fault resistance $= \dfrac{6.6 \times 30}{(22)^2} = 0.409$

Load current I_L in pu $= \dfrac{10}{30} = 0.333$

The positive, negative, and zero sequence networks are shown in Fig. 10.22.

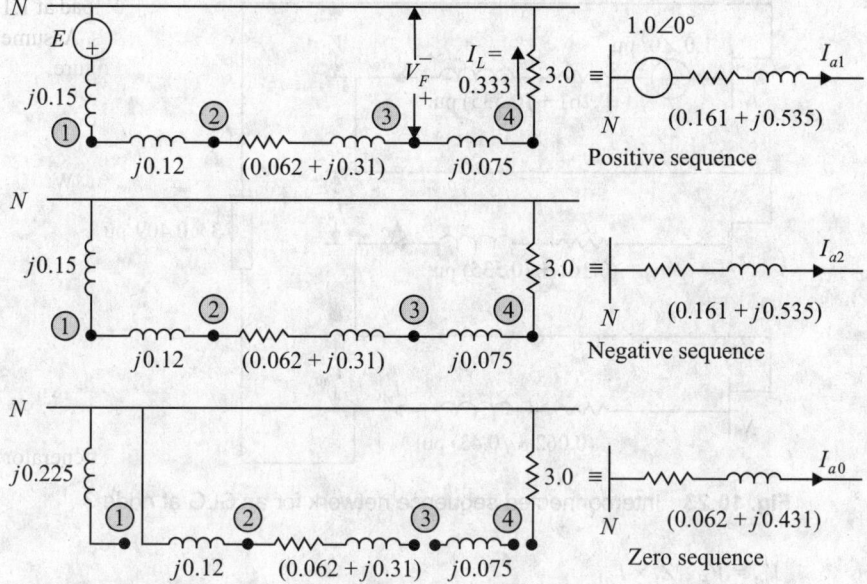

Fig. 10.22 Positive, negative, and zero sequence networks

Assuming a pre-fault voltage of 1.0 pu at node 4, the Thevenin's equivalent voltage at node 3

$$V_F = 1.0 + j0 + 0.333 \times (j0.075) = 1.0 + j0.025 = 1.0\angle 1.432°$$

Assuming that V_F is the reference phasor at the fault point, i.e., node 3,

$$V_F = 1.0\angle 0° \text{ pu}$$

The equivalent Thevenin's positive and negative sequence impedances looking into the fault point arc

$$\frac{(j0.15 + j0.12 + 0.062 + j0.31) \times (3.0 + j0.075)}{(3.062 + j0.655)}$$

$$= \frac{(0.062 + j0.58) \times (3.0 + j0.075)}{(3.062 + j0.655)}$$

$$= \frac{0.583\angle 83.898° \times 3\angle 1.432°}{3.131\angle 12.074°} = 0.559\angle 73.256° = (0.161 + j0.535) \text{ pu}$$

The equivalent Thevenin's zero sequence impedance is given by

$$j0.12 + 0.062 + j0.31 = (0.062 + j0.43) = 0.434 \angle 81.795° \text{ pu}$$

The interconnected network for an SLG at node 3 is shown in Fig. 10.23.
From the interconnected network I_{a1} is

$$\frac{1.0\angle 0°}{1.611 + j1.5} = \frac{1}{2.201\angle 42.957} = 0.454\angle -42.957$$

$$= (0.333 - j0.310)$$

$$I_a = 3 \times I_{a1} = 1.362\angle -42.957° = (0.999 - j0.930) \text{ pu}$$

$$V_{a0} = -Z_0 \times I_{a0} = -0.434\angle 81.795° \times 0.454\angle -42.957°$$

$$= -0.197\angle 38.838°$$

$$= (-0.153 - j0.124) \text{ pu}$$

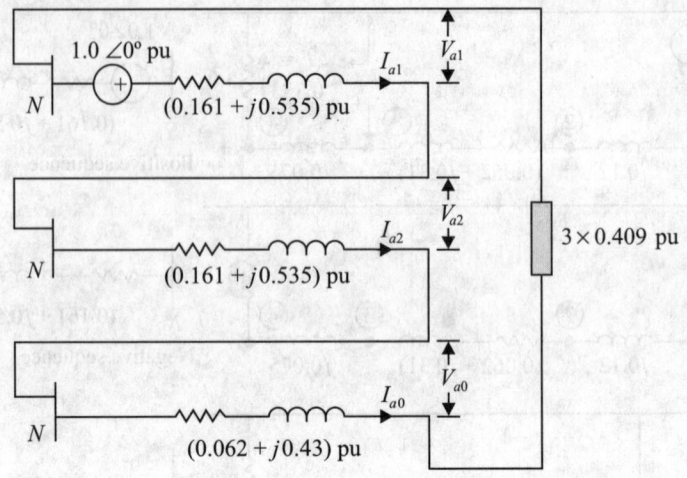

Fig. 10.23 Interconnected sequence network for an SLG at node 3

$V_{a1} = V_F - Z_1 \times I_{a1}$

$= 1.0 + j0 - 0.559\angle 73.256° \times 0.454\angle -42.957°$

$= 1.0 + j0 - 0.254\angle 30.299° = 1.0 + j0 - 0.219 - j0.128$

$= (0.781 - j0.128)$ pu

$V_{a2} = -0.559\angle 73.256° \times 0.454\angle -42.957° = -0.254\angle 30.299°$

$= (-0.219 - j0.128)$ pu

Base current $I_{\text{base}} = \dfrac{30 \times 1000}{\sqrt{3} \times 22} = 787.319$ A

Fault current $I_a = 787.319 \times 1.362 \angle -42.957° = 1072.328\angle -42.957°$ A

$$\begin{bmatrix} V_a \\ V_b \\ V_c \end{bmatrix} = \begin{bmatrix} 1 & 1 & 1 \\ 1 & a^2 & a \\ 1 & a & a \end{bmatrix} \begin{bmatrix} -0.153 - j0.124 \\ 0.781 - j0.128 \\ -0.219 - j0.128 \end{bmatrix} = \begin{bmatrix} 0.4090 - 0.3800i \\ -0.4340 - 0.8620i \\ -0.4340 + 0.8700i \end{bmatrix}$$

The line-to-line voltages at node 3 are computed as

$V_{ab} = V_a - V_b = 0.4090 - j0.3800 - (-0.4340 - j0.8620) = 0.843 + j0.482$

$= 0.971\angle 29.76°$ pu

$V_{bc} = V_b - V_c = -0.4340 - j0.8620 - (-0.434 + j0.870) = 0 - j1.732$

$= 1.732\angle -90°$ pu

$V_{ca} = V_c - V_a = -0.434 + j0.870 - (0.4090 - j0.3800) = -0.843 + j1.25$

$= 1.508\angle -56°$ pu

10.6.2 Line-to-line Fault

A line-to-line (LL) fault occurs on phase b and phase c of a three-phase power system and Z_f is the fault impedance. Analysis of this fault can be done by the following steps.

Step 1: Figure 10.24 shows a circuit diagram wherein a line-to-line fault has occurred between phase b and phase c. The fault impedance is Z_f.

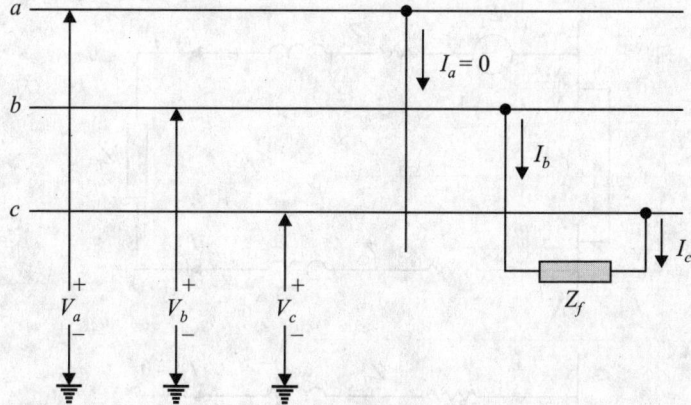

Fig. 10.24 Line-to-line fault between phase *b* and phase *c*

Step 2: From Fig. 10.24, boundary conditions are as follows:

$$I_b = -I_c \tag{10.22}$$

and

$$I_a = 0 \tag{10.23}$$

$$V_b - V_c = I_b Z_f \tag{10.24}$$

Step 3: Applying the boundary conditions, the phase quantities may be transformed to sequence components. Substituting the boundary conditions from Eqs (10.22) and (10.23), Eq. (10.9b) is written as

$$\begin{bmatrix} I_{a0} \\ I_{a1} \\ I_{a2} \end{bmatrix} = \frac{1}{3} \begin{bmatrix} 1 & 1 & 1 \\ 1 & a & a^2 \\ 1 & a^2 & a \end{bmatrix} \begin{bmatrix} 0 \\ I_b \\ -I_b \end{bmatrix} = j\frac{1}{\sqrt{3}} \times \begin{bmatrix} 0 \\ I_b \\ -I_b \end{bmatrix} \tag{10.25}$$

The two sides of Eq. (10.24) are rewritten, in terms of sequence components with the help of Eqs (10.3) and (10.8) as

$$V_{a0} + a^2 V_{a1} + aV_{a2} - (V_{a0} + aV_{a1} + a^2 V_{a2}) = Z_f(I_{a0} + a^2 I_{a1} + aI_{a2}) \tag{10.26}$$

Substituting from Eq. (10.25), Eq. (10.26) is modified as

$$(a^2 - a) V_{a1} + (a - a^2) V_{a2} = Z_f(a^2 - a)I_{a1}$$

$$V_{a1} - V_{a2} = Z_f I_{a1} \tag{10.27}$$

Step 4: Equations (10.25) and (10.27) provide the basis for interconnecting the sequence component networks. Since $I_{a0} = 0$, the zero sequence component network will be on open circuit. The positive and negative sequence components are to be connected so as to satisfy the voltage condition in Eq. (10.27) and sequence current condition of $I_{a1} = -I_{a2}$.

Step 5: The interconnected sequence networks are shown in Fig. 10.25.

From the interconnected sequence network, the sequence current can be computed as follows

$$I_{a1} = \frac{E_a}{Z_1 + Z_2 + Z_f} \text{ A} \tag{10.28}$$

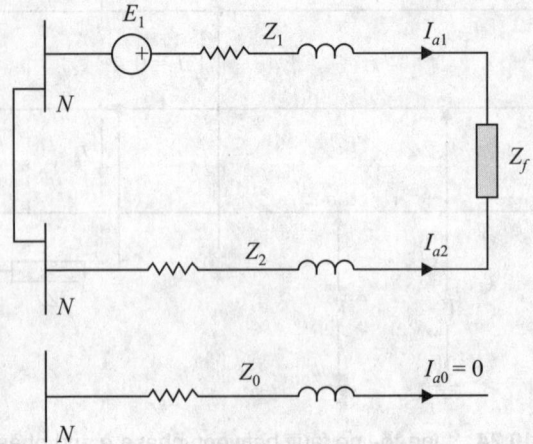

Fig. 10.25   Interconnected sequence networks for LL fault

Example 10.7   For the power system shown in Fig. 10.21, calculate the phase voltages and currents, if phases b and c, at node 3, experience a short-circuit through a resistance of 6.6 Ω.

Solution   The sequence networks shown in Fig. 10.22 remain unchanged. Figure 10.26 shows the interconnected positive and negative sequence networks to satisfy the LL fault conditions.
From Fig. 10.26, the positive and negative sequence currents are given by

$$I_{a1} = -I_{a2} = \frac{1.0\angle0°}{(0.161+j0.535)+(0.161+j0.535)+0.409}$$

$$= \frac{1.0\angle0°}{0.731+j1.07} = \frac{1.0\angle0°}{1.296\angle55.66°} = 0.772\angle-55.66° = (0.435-j0.637) \text{ pu}$$

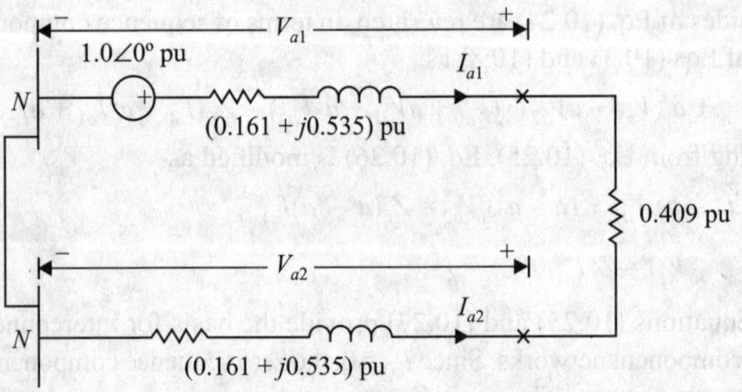

Fig. 10.26   Interconnected sequence network satisfying LL fault conditions

From the interconnected sequence network,

$$V_{a0} = 0$$
$$V_{a1} = E_a - Z_1 I_{a1} = 1.0 - 0.559\angle73.256° \times 0.772\angle55.66°$$
$$= 1.0 - 0.431\angle17.596° = 1.0 - (0.411 + j0.130)$$
$$= (0.589 - j0.130) \text{ pu}$$

$$V_{a2} = -Z_2 I_{a2} = -0.559\angle73.256° \times (-0.772\angle55.66°)$$

$$= 0.431\angle17.596° = (0.411 + j0.130)\ \text{pu}$$

The phase voltages are computed as follows:

$$\begin{bmatrix} V_a \\ V_b \\ V_c \end{bmatrix} = \begin{bmatrix} 1 & 1 & 1 \\ 1 & a^2 & a \\ 1 & a & a^2 \end{bmatrix} \begin{bmatrix} 0 \\ 0.589 - j0.130 \\ 0.411 + j0.130 \end{bmatrix} = \begin{bmatrix} 1.0000 \\ -0.7251 - 0.1542i \\ -0.2748 + 0.1542i \end{bmatrix}$$

From the foregoing,

$$V_a = 1.0\angle0°$$

$$V_b = -0.7251 - j0.1542 = 0.741\angle12.0°\ \text{pu}$$

$$V_c = -0.2748 + j0.1542 = 0.315\angle29.30°\ \text{pu}$$

The line-to-line voltages are obtained as follows:

$$V_{ab} = V_a - V_b = 1.0 + j0 - (-0.7251 - j0.1542)$$

$$= (1.7251 + j0.1542)\ \text{pu}$$

$$V_{bc} = V_b - V_c = -0.7251 - j0.1542 - (-0.2748 + j0.1542)$$

$$= (0.4503 - j0.308)\ \text{pu}$$

$$V_{ca} = V_c - V_a = -0.2748 + j0.1542 - 1.0 + j0$$

$$= (-1.2748 + j0.1542)\ \text{pu}$$

It is left to the reader to check $V_{bc} = V_b - V_c = Z_f I_b$
From Eq. (10.25),

$$I_b = -I_c = -j\sqrt{3} \times (0.435 - j0.637) = -1.103 - j0.753$$

$$-1.336\angle34.321°\ \text{pu}$$

Actual phase current $I_b = -I_c = 787.319 \times 1.336\angle34.321° = 1051.858\ \text{A}$

10.6.3 Double Line-to-ground Fault

Assume that a double line-to-ground (2LG) fault occurs on phases b and c of a power system. Steps for analysis are as follows:

Step 1: Figure 10.27 shows the network under the fault condition. Z_f and Z_g are the fault and ground impedances, respectively.

Step 2: By inspection of Fig. 10.27, the following boundary conditions can be defined:

$$I_a = 0 \tag{10.29}$$

$$V_b = Z_f I_b + Z_g(I_b + I_c) = (Z_f + Z_g) I_b + Z_g I_c \tag{10.30}$$

$$V_c = Z_f I_c + Z_g(I_b + I_c) = Z_g I_b + (Z_f + Z_g) I_c \tag{10.31}$$

Step 3: Transformation of Eqs (10.29) to (10.31) into sequence quantities is accomplished as follows:

$$I_{a0} + I_{a1} + I_{a2} = 0 \tag{10.32}$$

Subtracting Eq. (10.31) from Eq. (10.30) yields

$$V_b - V_c = Z_f(I_b - I_c) \tag{10.33}$$

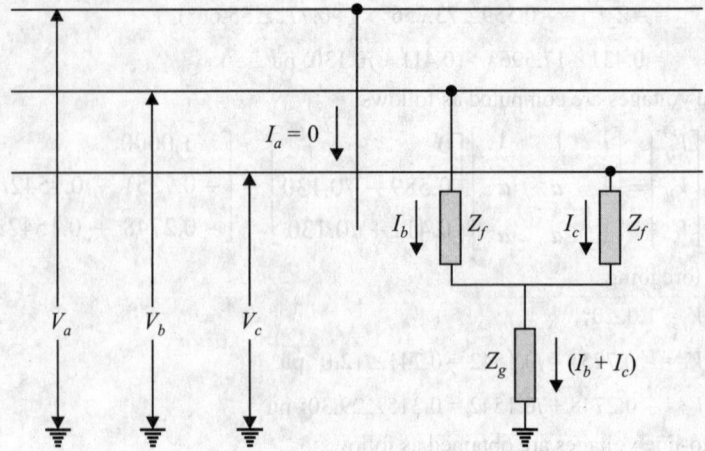

Fig. 10.27 Double line-to-ground fault between phases *b* and *c* of a power system

Using Eqs (10.4) and (10.8) respectively to obtain the phase voltage and current differences in terms of sequence components, Eq. (10.33) is written as

$$V_{a0} + a^2 V_{a1} + aV_{a1} - (V_{a0} + aV_{a1} + a^2 V_{a2})$$
$$= Z_f[I_{a0} + a^2 I_{a1} + aI_{a2} - (I_{a0} + aI_{a1} + a^2 I_{a2})]$$

or

$$-j\sqrt{3}\,(V_{a1} - V_{a2}) = -j\sqrt{3}\,Z_f(I_{a1} - I_{a2})$$

then,

$$V_{a1} - Z_f I_{a1} = V_{a2} - Z_f I_{a2} \tag{10.34}$$

Similarly add Eqs (10.30) and (10.31) and use Eqs (10.4) and (10.8) to transform the phase voltages and currents into sequence components of voltage and current respectively. Therefore,

$$V_b + V_c = (Z_f + 2Z_g)(I_b + I_c)$$

or

$$[2V_{a0} - (V_{a1} + V_{a2})] = (Z_f + 2Z_g)[2I_{a0} - (I_{a1} + I_{a2})] \tag{10.35}$$

Equation (10.35) on simplification yields the following equation

$$2V_{a0} - 2Z_f I_{a0} - 4Z_g I_{a0}$$
$$= V_{a1} + V_{a2} - Z_f(I_{a1} + I_{a2}) - 2Z_g\,(I_{a1} + I_{a2}) \tag{10.36}$$

Substituting for $(I_{a1} + I_{a2})$ from Eq. (10.32) and using Eq. (10.34), Eq. (10.36) simplifies to

$$V_{a0} - Z_f I_{a0} - 3Z_g I_{a0} = V_{a1} - Z_f I_{a1} = V_{a2} - Z_f I_{a2} \tag{10.37}$$

Step 4: Equations (10.32) and (10.37) define the method of connecting the sequence networks. Equation (10.32) shows that the sum of sequence currents at a point must be zero; hence, the sequence buses must form a node point. From Eq. (10.37), it is evident that the sequence voltages become equal if external impedances of $3Z_g$ and Z_f is included in series in the zero sequence network and Z_f is included in series, in both the positive and negative sequence networks.

Step 5: The connection diagram of the sequence networks satisfying Eqs (10.32) and (10.37) are shown in Fig. 10.28.

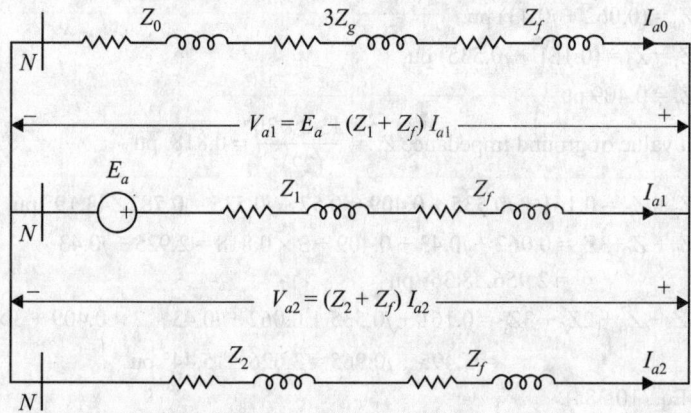

Fig. 10.28 Interconnected sequence networks for 2LG fault

From Fig. 10.28, it is observed that

$$I_{a1} = \frac{E_a}{\left(Z_1 + Z_f\right) + \dfrac{\left(Z_2 + Z_f\right)\left(Z_0 + Z_f + 3Z_g\right)}{\left(Z_2 + Z_0 + 2Z_f + 3Z_g\right)}} \tag{10.38a}$$

Once I_{a1} is computed, I_{a2} can be determined as follows:

$$I_{a2} = \frac{\left(Z_0 + Z_f + 3Z_g\right)}{\left(Z_2 + Z_0 + 2Z_f + 3Z_g\right)} \times I_{a1} \tag{10.38b}$$

The zero sequence current component is determined by using Eq. (10.32) or alternatively from the interconnected sequence network in Fig. 10.28.

$$I_{a0} = \frac{\left(Z_2 + Z_f\right)}{\left(Z_2 + Z_0 + 2Z_f + 3Z_g\right)} \times I_{a1} \tag{10.38c}$$

Example 10.8 For the power system shown in Fig. 10.21, determine the phase currents and voltages if the system experiences a double line-to-ground fault at node 3. Assume a fault impedance of 6.6 Ω and ground impedance of 13.2 Ω.

Solution The sequence networks remain the same. The interconnected sequence networks for the fault condition are shown in Fig. 10.29.

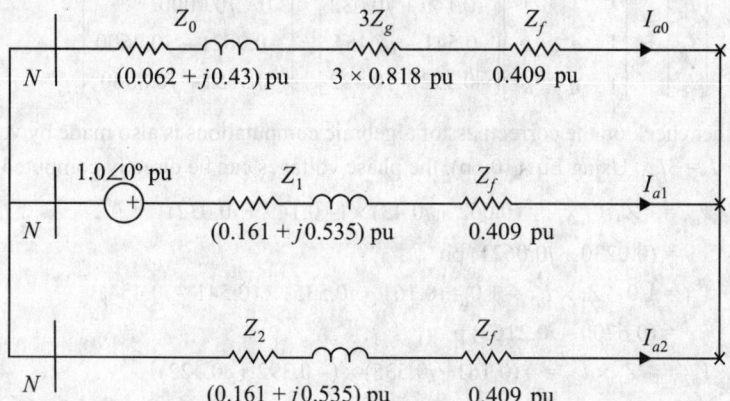

Fig. 10.29 Interconnected sequence network for a 2LG fault

$Z_0 = (0.062 + j0.43)$ pu,

$Z_1 = Z_2 = (0.161 + j0.535)$ pu,

$Z_f = 0.409$ pu

Per unit value of ground impedance $Z_g = \dfrac{13.2 \times 30}{(22)^2} = 0.818$ pu

$Z_2 + Z_f = 0.161 + j0.535 + 0.409 = 0.57 + j0.535 = 0.782\angle43.19°$ pu

$Z_0 + Z_f + 3Z_g = 0.062 + j0.43 + 0.409 + 3 \times 0.818 = 2.925 + j0.43$

$\qquad\qquad\qquad = 2.956\angle8.36°$ pu

$Z_2 + Z_0 + 2Z_f + 3Z_g = 0.161 + j0.535 + 0.062 + j0.43 + 2 \times 0.409 + 3 \times 0.818$

$\qquad\qquad\qquad\qquad = 3.495 + j0.965 = 3.626\angle15.44°$ pu

Applying Eq. (10.38a),

$$I_{a1} = \cfrac{1.0\angle0°}{0.161 + j0.535 + 0.409 + \cfrac{(0.782\angle43.19°)\ (2.956\angle8.36°)}{3.626\angle15.44°}}$$

$$= \frac{1.0\angle0°}{0.57 + j0.535 + 0.638\angle36.11°} = \frac{1.0\angle0°}{0.57 + j0.535 + 0.515 + j0.376}$$

$$= \frac{1.0\angle0°}{1.085 + j0.911} = \frac{1.0\angle0°}{1.417\angle40.02°}$$

$$= 0.706\angle-40.02° = 0.541 - j0.454$$

$Z_0 + Z_f + 3Z_g = 0.062 + j0.43 + 0.409 + 3 \times 0.818 = 2.925 + j0.43$

$\qquad\qquad\qquad = 2.956\angle8.36°$

$$I_{a2} = -\frac{2.956\angle8.36°}{3.626\angle15.44°} \times 0.706\angle-40.02° = -0.576\angle-47.1°$$

$$= -0.392 + j0.422 \text{ pu}$$

$$I_{a0} = -\frac{0.782\angle43.19°}{3.626\angle15.44°} \times 0.706\angle-40.02 = -0.152\angle-12.27°$$

$$= -0.149 + j0.032 \text{ pu}$$

Note that $I_{a0} + I_{a1} + I_{a2} = 0$, thereby verifying the algebraic computations.

The phase currents are obtained by transformation given by Eq. (10.8) as follows:

$$\begin{bmatrix} I_a \\ I_b \\ I_c \end{bmatrix} = \begin{bmatrix} 1 & 1 & 1 \\ 1 & a^2 & a \\ 1 & a & a^2 \end{bmatrix} \begin{bmatrix} -0.149 + j0.032 \\ 0.541 - j0.454 \\ -0.392 + j0.422 \end{bmatrix} = \begin{bmatrix} 0 \quad j0.0000 \\ -0.9821 - j0.7600 \\ 0.5351 + j0.8560 \end{bmatrix}$$

Another check on the correctness of algebraic computations is also made by verifying that $I_b + I_c = 3I_{a0}$. Using Eq. (10.6b), the phase voltages can be directly computed.

$V_{a0} = -Z_0 \times I_{a0} = -(0.062 + j0.43) \times (-0.149 + j0.032)$

$\qquad = (0.0230 + j0.0621)$ pu

$V_{a1} = 1.0 - Z_1 \times I_{a1} = 1.0 - (0.161 + j0.535) \times (0.541 - j0.454)$

$\qquad = (0.6700 - j0.2163)$ p

$V_{a2} = -Z_1 \times I_{a2} = -(0.161 + j0.535) \times (-0.392 + j0.422)$

$\qquad = (0.2889 + j0.1418)$ pu

$$\begin{bmatrix} V_a \\ V_b \\ V_c \end{bmatrix} = \begin{bmatrix} 1 & 1 & 1 \\ 1 & a^2 & a \\ 1 & a & a^2 \end{bmatrix} \begin{bmatrix} 0.0230 + j0.0621 \\ 0.6700 - j0.2163 \\ 0.2889 + j0.1418 \end{bmatrix} = \begin{bmatrix} 0.9819 - j0.0125 \\ -0.7665 - j0.2307 \\ -0.1463 + j0.4294 \end{bmatrix}$$

$V_{ab} = V_a - V_b = 1.7484 + j0.2182 = 1.7620\angle 7.11°$ pu

$V_{bc} = V_b - V_c = -0.6202 - j0.6601 = 0.9058\angle -133.22°$ pu

$V_{ca} = V_c - V_a = -1.1282 + j0.4419 = 1.2117\angle 158.61°$ pu

10.6.4 Three-phase Fault

Although the frequency of a three-phase (3ϕ) fault occurring in a power system is considered to be of the order of 5% only, yet power systems have to be analysed for 3ϕ faults for the following reasons.

1. A 3ϕ fault is the severest of all faults in power systems except in the case of a SLG fault, where generator neutrals are solidly earthed or earthed through low values of impedances on the star side of a delta–star transformer with the star point grounded.

2. For determining the interrupting capacity of circuit breakers.

3. Owing to the likelihood of any other fault in a power system degenerating into a 3ϕ fault if not cleared immediately.

4. It is the simplest type of fault to compute and provides sufficient information in the absence of complete data about a power system.

The steps for analyses are as follows:

Step 1: Figure 10.30 shows a 3ϕ fault to ground in a power system. Z_f and Z_g are the impedances, respectively, between the individual phases and neutral-to-ground.

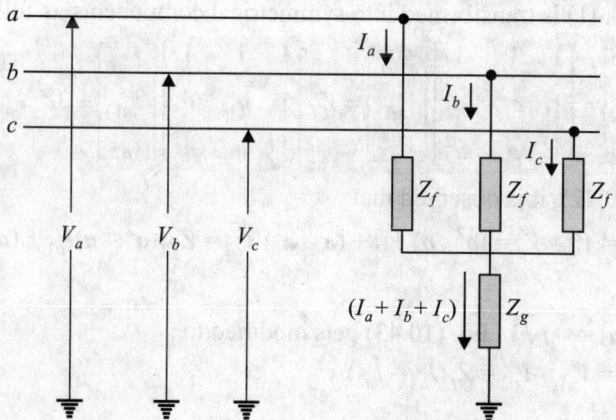

Fig. 10.30 A three-phase to ground fault on a power system

It is observed from the circuit diagram in Fig. 10.30 that this type of a fault represents a balanced load condition. Since the voltages V_a, V_b, and V_c are balanced, the currents also are balanced. Thus,

$$3I_{a0} = I_a + I_b + I_c = 0$$

and

$$\begin{bmatrix} I_{a0} \\ I_{a1} \\ I_{a2} \end{bmatrix} = \frac{1}{3} \begin{bmatrix} 1 & 1 & 1 \\ 1 & a & a^2 \\ 1 & a^2 & a \end{bmatrix} \begin{bmatrix} I_a \\ I_b \\ I_c \end{bmatrix} = \begin{bmatrix} 0 \\ I_{a1} \\ 0 \end{bmatrix}$$

or

$$I_{a1} = I_a, I_{a2} = I_{a0} = 0$$

From the foregoing, it becomes evident that since this type of a fault is a typical case of a balanced load, the connection diagram for the sequence networks can be derived directly and is shown in Fig. 10.31. However, for the sake of completeness a full analysis is presented in the following section.

Step 2: From Fig. 10.30, the following boundary conditions can be obtained

$$I_a + I_b + I_c = 0 \tag{10.39}$$

$$V_a = Z_f \times I_a + Z_g (I_a + I_b + I_c) \tag{10.40a}$$

$$V_b = Z_f \times I_b + Z_g (I_a + I_b + I_c) \tag{10.40b}$$

$$V_c = Z_f \times I_c + Z_g (I_a + I_b + I_c) \tag{10.40c}$$

Step 3: From Eq. (10.39), it can be concluded that

$$3I_{a0} = I_a + I_b + I_c = 0$$

After substituting the value of $I_a + I_b + I_c$ from Eq. (10.39), Eqs (10.40a), (10.40b), and (10.40c), in matrix form, can be written as

$$\begin{bmatrix} V_a \\ V_b \\ V_c \end{bmatrix} = Z_f \begin{bmatrix} I_a \\ I_b \\ I_c \end{bmatrix} + 3Z_g I_{a0} \tag{10.41}$$

Equation (10.41) is transformed into symmetrical components as follows

$$\begin{bmatrix} V_a \\ V_b \\ V_c \end{bmatrix} = \begin{bmatrix} 1 & 1 & 1 \\ 1 & a^2 & a \\ 1 & a & a^2 \end{bmatrix} \begin{bmatrix} V_{a0} \\ V_{a1} \\ V_{a2} \end{bmatrix} = Z_f \begin{bmatrix} 1 & 1 & 1 \\ 1 & a^2 & a \\ 1 & a & a^2 \end{bmatrix} \begin{bmatrix} I_{a0} \\ I_{a1} \\ I_{a2} \end{bmatrix} + 3Z_g I_{a0} \tag{10.42}$$

From Eq. (10.42), it is observed that

$$V_{bc} = V_b - V_c = (a^2 - a) V_{a1} + (a - a^2) V_{a2} = Z_f [(a^2 - a)I_{a1} + (a - a^2)I_{a2}] \tag{10.43}$$

Since $(a^2 - a) = -j\sqrt{3}$, Eq. (10.43) gets modified to

$$V_{bc} = V_{a1} - V_{a2} = Z_f(I_{a1} - I_{a2})$$

or

$$V_{a1} - Z_f I_{a1} = V_{a2} - Z_f I_{a2} \tag{10.44}$$

Again from Eq. (10.42),

$$V_a + V_b = (2V_{a0} - aV_{a1} - a^2 V_{a2}) = Z_f(2I_{a0} - aI_{a1} - a^2 I_{a2}) + 6Z_g I_{a0} \tag{10.45}$$

Simplifying Eq. (10.45) by using the relation $1 + a + a^2 = 0$, we get

$$2(V_{a0} - Z_f I_{a0} - 3Z_g I_{a0}) = -(V_{a1} - Z_f I_{a1}) \tag{10.46}$$

Again from Eq. (10.42), the following may be written

$$V_b + V_c = (2V_{a0} - V_{a1} - V_{a2}) = Z_f(2I_{a0} - I_{a1} - I_{a2}) + 6Z_g I_{a0}$$

or $2(V_{a0} - Z_f I_{a0} - 3Z_g I_{a0}) = (V_{a1} - Z_f I_{a1}) + (V_{a2} - Z_f I_{a2})$

If the relationship of Eq. (10.44) is used, the above equation takes the form

$$V_{a0} - Z_f I_{a0} - 3Z_g I_{a0} = V_{a1} - Z_f I_{a1} \qquad (10.47)$$

Step 4: For this type of a fault the sequence currents are non-existent.

Step 5: In order to draw the sequence networks, Eqs (10.44), (10.46), and (10.47) are required to be examined. Equations (10.44) and (10.47) suggest that the interconnected sequence network will be the same as shown in Fig. 10.31 with $(Z_f + 3Z_g)$ connected in series with the zero sequence network, and Z_f connected in series with both the positive and negative sequence networks. On the other hand, from Eqs (10.46) and (10.47), it is observed that

$$V_{a1} - Z_f I_{a1} = -2(V_{a0} - Z_f I_{a0} - 3Z_g I_{a0}) \qquad (10.48a)$$
$$V_{a1} - Z_f I_{a1} = V_{a0} - Z_f I_{a0} - 3Z_g I_{a0} \qquad (10.48b)$$

Equations (10.48a) and (10.48b) are satisfied if $I_{a0} = 0$ and $V_{a0} = 0$, which in turn requires that the right hand side of Eq. (10.47) be equal to zero, that is,

$$V_{a1} - Z_f I_{a1} = V_{a2} - Z_f I_{a2} = 0$$

Based on the above, the connection of the sequence networks will be as shown in Fig. 10.31.

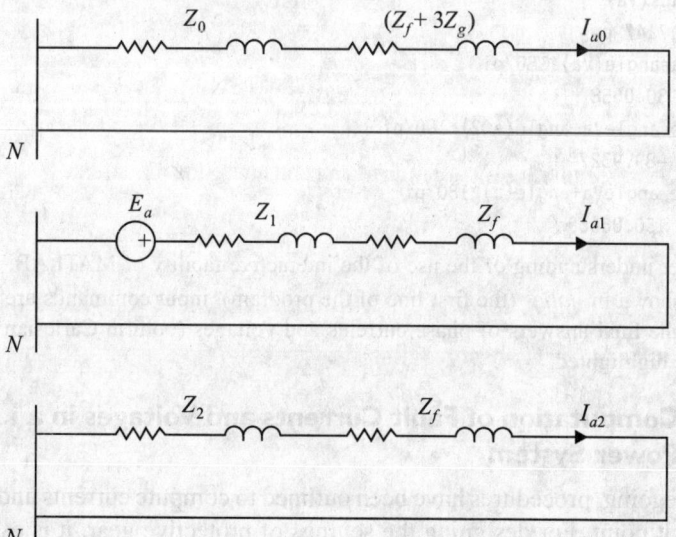

Fig. 10.31 Sequence network for a three-phase fault on a power system

It is to be noted that since the positive sequence network contains a voltage source, the zero and negative sequence networks are passive networks. Thus, $I_{a1} = I_a$ and is given by

$$I_{a1} = I_a = \frac{E_a}{Z_1 + Z_f} \qquad (10.49)$$

Example 10.9   The power system shown in Fig. 10.21 experiences a 3ϕ fault at node 3. Determine the phase currents and voltages under the fault condition. Assume a fault impedance of $Z_f = 6.6 \, \Omega$.

Solution   The per unit fault impedance $Z_f = 0.409$
   From Eq. (10.49)

$$I_a = \frac{1.0}{(0.161 + j0.535) + 0.409}$$

Employing the interactive facility of MATLAB, the phase currents and voltages can be directly computed as follows:

```
>> Z1=0.161+0.535i;Zf=0.409;
>> Ia=1.0/(Z1+Zf)
Ia = 0.9327 - j0.8754 pu
>> absIa=abs(Ia)
absIa = 1.2792 pu
>> angleIa=angle(Ia)*180/pi
angleIa = -43.1858°
>> angleIb=angleIa+(angle(a^2)*180/pi)
angleIb = -163.1844°
>> angleIc=angleIa+(angle(a)*180/pi)
angleIc = 76.8149°
>> Va=1.0-Zf*Ia
Va = 0.6185 + j0.3581 pu
>> absVa=abs(Va)
absVa = 0.7147 pu
>> angleVa=angle(Va)*180/pi
angleVa = 30.0658°
>> angleVb=angleVa+angle(a^2)*180/pi
angleVb = -89.9327°
>> angleVc=angleVa+angle(a)*180/pi
angleVc = 150.0666°
```

For a better understanding of the use of the interactive facility of MATLAB, input data has been shown in *italics* (the first line of the program), input commands are in normal font, and the final answers of phase currents and voltages (both in Cartesian and polar forms) are highlighted.

10.6.5 Computation of Fault Currents and Voltages in a Power System

In the foregoing, procedures have been outlined to compute currents and voltages at the fault point. For designing the settings of protective gear, it is essential to know the magnitudes of currents and voltages in the entire power system being influenced by the fault.

   In deriving the mathematical model to represent the fault condition, Thevenin's positive, negative, and zero sequence networks are developed which are interconnected depending on the boundary conditions. The mathematical model is developed by ignoring the pre-fault loading conditions, in the system, which may not be unreasonable since the fault currents have a much higher magnitude than the load current. The pre-fault loading conditions can be incorporated in fault conditions; this will be outlined later in the chapter.

The drawback of developing Thevenin's positive, negative, and zero sequence networks is that the original topology of the system is lost. Hence, in order to determine the fault currents and voltages in the whole system, it is essential to work backwards and reconstruct the topology and employ the principle of superposition to determine the distribution of currents and voltages. The method is demonstrated by computing currents and voltages in Example 10.10.

Example 10.10 For the system shown in Fig. 10.21, determine (a) phase current in the transmission line between nodes 2-3, (b) phase current into the transformer between nodes 3-4, (c) phase voltages at node 2; when an SLG occurs at node 3. Assume $Z_f = 6.6 \ \Omega$.

Solution From Example 10.6,

$$I_{a1} = I_{a2} = I_{a0} = (0.333 - j0.310) \text{ pu}$$

$$I_a = 3 \times I_{a1} = (0.999 - j0.930) \text{ pu}$$

$$V_F = 1.0$$

The positive, negative, and zero sequence networks of the system are shown in Fig. 10.22. From the positive sequence network it is seen that positive sequence component of current between nodes 2-3 is given by

$$\frac{(3.0 + j0.075) \times (0.333 - j0.310)}{(j0.15 + j0.12 + 0.062 + j0.31 + 3.0 + j0.075)} = (0.2588 - j0.3509) \text{ pu}$$

Positive sequence component of current between nodes 3-4 is

$$I_{a1} - (0.2588 - j0.3509) = (0.0742 + j0.0409) \text{ pu}$$

Since the negative sequence impedance network is similar to the positive sequence network, negative sequence component of current between nodes 2-3

$$= (0.2588 - j0.3509) \text{ pu}$$

Negative sequence component of current between nodes 3-4

$$= (0.0742 + j0.0409) \text{ pu}$$

Zero sequence current between nodes 2-3 = $(0.333 - j0.31)$ pu
Zero sequence current between nodes 3-4 = 0 pu
Using Eq. (10.8) the phace currents between nodes 2-3 are

$$\begin{bmatrix} I_a \\ I_b \\ I_c \end{bmatrix} = \begin{bmatrix} 1 & 1 & 1 \\ 1 & a^2 & a \\ 1 & a & a^2 \end{bmatrix} \begin{bmatrix} 0.333 - j0.31 \\ 0.2588 - j0.3509 \\ 0.2588 - j0.3509 \end{bmatrix} = \begin{bmatrix} 0.8446 - j1.0118 \\ 0.0772 + j0.0409 \\ 0.0772 + j0.0409 \end{bmatrix}$$

Similarly, the phase currents between nodes 3-4 will be

$$\begin{bmatrix} I_a \\ I_b \\ I_c \end{bmatrix} = \begin{bmatrix} 1 & 1 & 1 \\ 1 & a^2 & a \\ 1 & a & a^2 \end{bmatrix} \begin{bmatrix} 0.0 \\ 0.0742 + j0.0409 \\ 0.0742 + j0.0409 \end{bmatrix} = \begin{bmatrix} 0.1484 + j0.0818 \\ -0.0742 \quad j0.0409 \\ -0.0742 \quad j0.0409 \end{bmatrix}$$

Positive sequence voltage at node 2

$$= 1.0 + (0.062 + j0.31) \times (0.2588 - j0.3509) = 1.1248 + j0.0585 \text{ pu}$$

Negative sequence voltage at node 2

$$= (0.062 + j0.31) \times (0.2588 - j0.3509) = 0.1248 + j0.0585 \text{ pu}$$

Zero sequence voltage at node 2

$$= (0.062 + j0.31) \times (0.333 - j0.31) = 0.1167 + j0.0840$$

Hence, phase voltages at node 2 are given by

$$\begin{bmatrix} V_a \\ V_b \\ V_c \end{bmatrix} = \begin{bmatrix} 1 & 1 & 1 \\ 1 & a^2 & a \\ 1 & a & a^2 \end{bmatrix} \begin{bmatrix} 0.1167 + j0.0840 \\ 1.1248 + j0.0585 \\ 0.1248 + j0.0585 \end{bmatrix} = \begin{bmatrix} 1.3663 + j0.2010 \\ -0.5081 - j0.8405 \\ -0.5081 + j0.8915 \end{bmatrix}$$

10.7 SERIES FAULTS

A group of unbalanced conditions which do not involve any connection between lines or between lines and neutral may occur. These are categorized as series faults and there is an unbalance in series impedance. The various types of conditions which bring about a series unbalance are as follows:

(i)   Unequal series impedances
(ii) One line conductor open (1LO)
(iii) Two line conductors open (2LO)

In case of unequal series impedances, there is no fault point as such. The unequal impedances between two points F and F' cause unbalance as shown in Fig. 10.32(a). The current direction is assumed to be from F to F' and a voltage drop takes place in the direction of current flow. The sequence networks contain the symmetrical portions of the system, looking back to the left of F and to the right of F'. The symmetrical portions may or may not be interconnected. The unbalanced portion is isolated outside the sequence network. The sequence networks for series faults between F and F' are shown in Fig. 10.32(b).

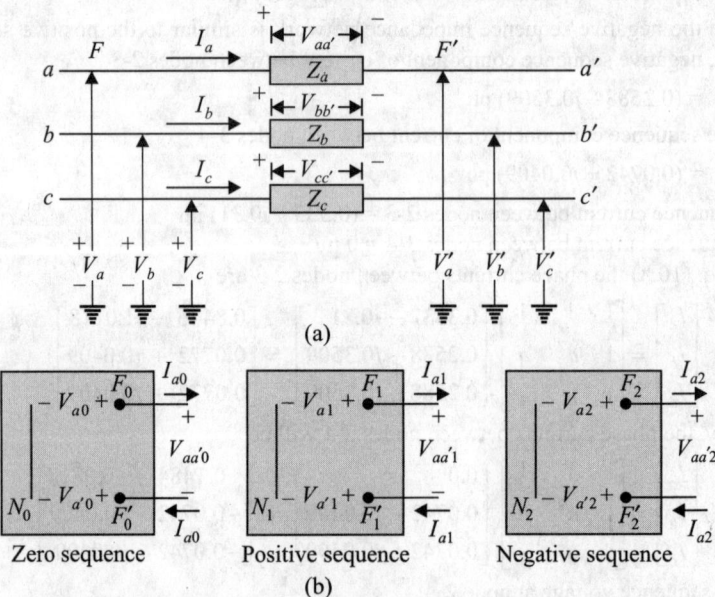

Fig. 10.32   (a) Power system network with unequal series impedances
(b) Sequence networks with unequal series impedances

10.7.1 Unequal Series Impedances

Step 1: Figure 10.33 shows a power system network with series unbalance. The unbalance is because series impedance of phase a is Z_a, whereas series impedances of phases b and c are the same and are equal to Z_b.

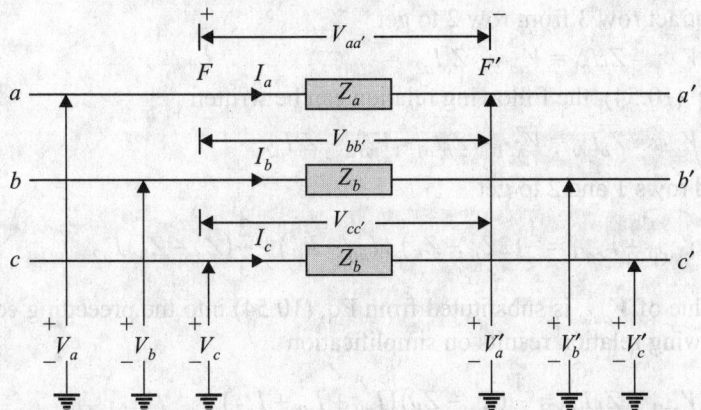

Fig. 10.33 Power system network with unequal series impedances

Step 2: The boundary conditions can be derived by an examination of Fig. 10.33 and are enunciated as follows:

$$
\begin{bmatrix} V_{aa'} \\ V_{bb'} \\ V_{cc'} \end{bmatrix} = \begin{bmatrix} V_a \\ V_b \\ V_c \end{bmatrix} - \begin{bmatrix} V_{a'} \\ V_{b'} \\ V_{c'} \end{bmatrix} = \begin{bmatrix} Z_a & 0 & 0 \\ 0 & Z_b & 0 \\ 0 & 0 & Z_b \end{bmatrix} \begin{bmatrix} I_a \\ I_b \\ I_c \end{bmatrix}
\tag{10.50}
$$

In concise form, Eq. (10.50) is written as

$$
V_{abc} - V_{a'b'c'} = Z_{abc} I_{abc}
\tag{10.51}
$$

Step 3: The phase voltages, currents, and impedances in Eq. (10.51) are transformed into sequence components as follows:

$$
V_{aa'-012} = V_{012} - V'_{02} = Z_{012} I_{012}
\tag{10.52a}
$$

The sequence impedance Z_{012} is given by

$$
Z_{012} = A^{-1} Z_{abc} A
\tag{10.52b}
$$

That is,

$$
Z_{012} = \begin{bmatrix} 1 & 1 & 1 \\ 1 & a^2 & a \\ 1 & a & a^2 \end{bmatrix}^{-1} \begin{bmatrix} Z_a & 0 & 0 \\ 0 & Z_b & 0 \\ 0 & 0 & Z_b \end{bmatrix} \begin{bmatrix} 1 & 1 & 1 \\ 1 & a^2 & a \\ 1 & a & a^2 \end{bmatrix}
\tag{10.53a}
$$

$$
= \begin{bmatrix} (Z_a + 2Z_b) & (Z_a - Z_b) & (Z_a - Z_b) \\ (Z_a - Z_b) & (Z_a + 2Z_b) & (Z_a - Z_b) \\ (Z_a - Z_b) & (Z_a - Z_b) & (Z_a + 2Z_b) \end{bmatrix}
\tag{10.53b}
$$

Step 4: The sequence currents for this condition do not exist. However, Eq. (10.52) can be used to evolve the interconnection of the sequence networks. By subtracting row 2 from row 1, the following relation is obtained

$$
V_{aa'0} - V_{aa'1} = Z_b(I_{a0} - I_{a1})
$$

or $\quad V_{aa'0} - Z_b I_{a0} = V_{aa'1} - Z_b I_{a1}$
\tag{10.54}

Now subtract row 3 from row 2 to get

$$V_{aa'1} - Z_b I_{a1} = V_{aa'2} - Z_b I_{a2}$$

From Eq. (10.54), the following relation can be written

$$V_{aa'0} - Z_b I_{a0} = V_{aa'1} - Z_b I_{a1} = V_{aa'2} - Z_b I_{a2} \qquad (10.55)$$

Now add rows 1 and 2 to get

$$V_{aa'0} + V_{aa'1} = \frac{1}{3}(2Z_a + Z_b)(I_{a0} + I_{a1}) + \frac{2}{3}(Z_a - Z_b)I_{a2}$$

If the value of $V_{aa'0}$ is substituted from Eq. (10.54) into the preceding equation, the following relation results on simplification

$$V_{aa'1} - Z_b I_{a1} = \frac{1}{3}(Z_a - Z_b)(I_{a0} + I_{a1} + I_{a2}) \qquad (10.56)$$

Equations (10.55) and (10.56) suffice to draw the interconnected two port sequence networks.

Step 5: It can be observed from Eqs (10.55) and (10.56) that an impedance Z_b is to be connected in series with each of the positive, negative, and zero sequence networks and the combination is connected in parallel across an impedance equal to $(Z_a - Z_b)/3$. The interconnected two-port sequence network satisfying the conditions of Eqs (10.55) and (10.56) is shown in Fig. 10.34.

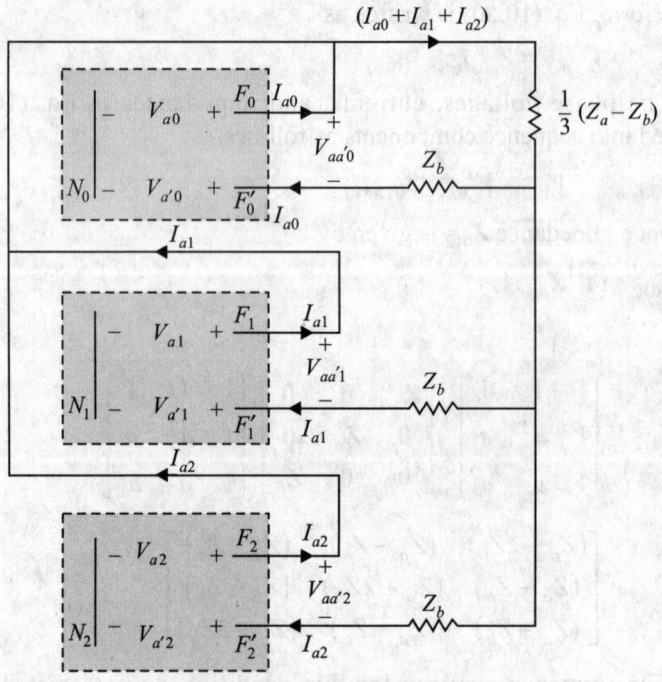

Fig. 10.34 Sequence network connection for series unbalance condition

10.7.2 One Line Conductor Open (1LO)

An open conductor fault is in series with the line. The analysis of such a fault is given below.

Step 1: The power system network for the phase a open line conductor condition is similar to that shown in Fig. 10.34.

Step 2: The boundary conditions for this type of a series unbalance are $Z_a = \infty$ and Z_b has a finite value.

Steps 3 and 4: The one line open conductor condition is a special case of the series unbalance condition discussed in Section 10.7.1. From an examination of Fig. 10.33, it can be concluded that the impedance $(Z_a - Z_b)/3$, which is connected in parallel across the sequence networks, is infinite.

Step 5: The interconnected two-port sequence network for the one line open condition is shown in Fig. 10.35 without the shunt branch.

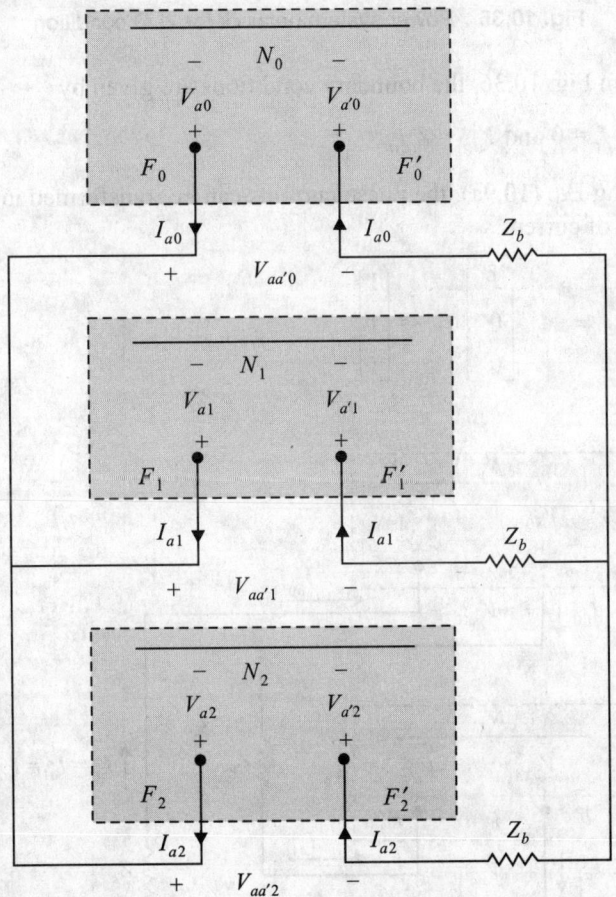

Fig. 10.35 Interconnected sequence networks for 1LO condition

10.7.3 Two Line Conductors Open (2LO)

Two conductors open is in series with the line. The circuit diagram for this condition is shown in Fig. 10.36.

Step 1: This condition is simulated by making $Z_b = Z_c = \infty$.

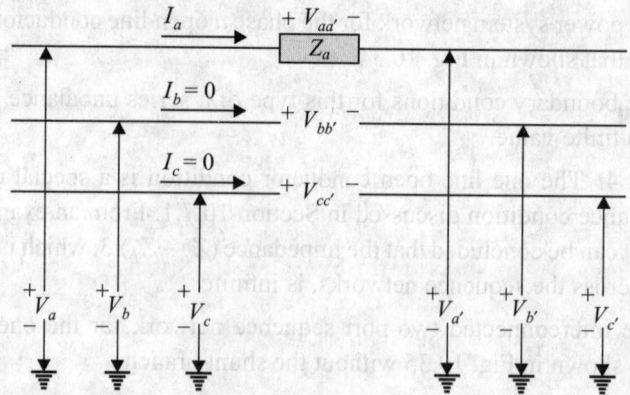

Fig. 10.36 Power system network for 2LO condition

Step 2: From Fig. 10.36, the boundary conditions are given by

$$I_b = I_c = 0 \text{ and } V_{aa'} = Z_a I_a \qquad (10.57)$$

Step 3: Using Eq. (10.9a), the phase currents can be transformed into sequence components of current.

$$\begin{bmatrix} I_0 \\ I_1 \\ I_2 \end{bmatrix} = A^{-1} \begin{bmatrix} I_a \\ 0 \\ 0 \end{bmatrix} = \frac{I_a}{3} \begin{bmatrix} 1 \\ 1 \\ 1 \end{bmatrix}$$

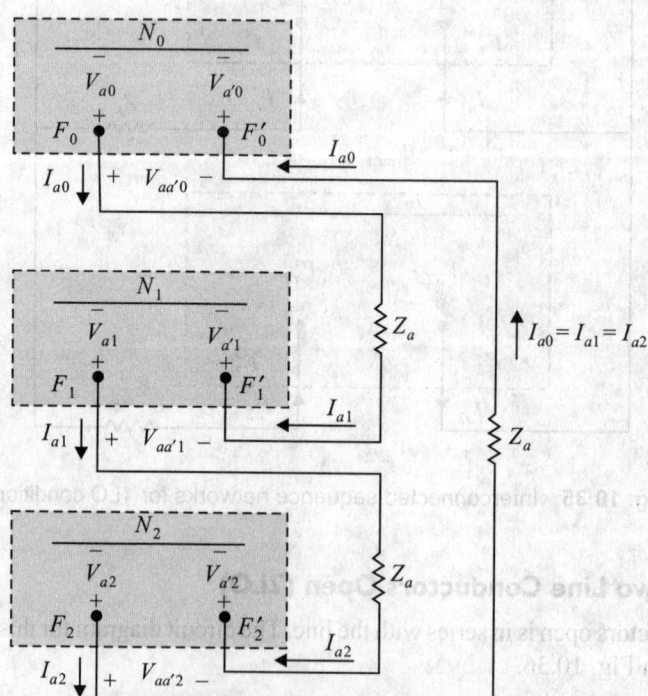

Fig. 10.37 Interconnected sequence networks for 2LO condition

Thus, $\qquad I_0 = I_1 = I_2 = \dfrac{I_a}{3}$ $\hfill$ (10.58a)

Rewriting the voltage relation of Eq. (10.57) in terms of the sequence components, the following equation results

$$\left(V_{aa'0} + V_{aa'1} + V_{aa'2}\right) = Z\left(I_{a0} + I_{a1} + I_{a2}\right)$$

or $\qquad \left(V_{aa'0} - Z_a I_{a0}\right) + \left(V_{aa'1} - Z_a I_{a1}\right) + \left(V_{aa'2} - Z_a I_{a2}\right) = 0$ $\hfill$ (10.58b)

Step 4: Equation (10.58) provide the basis for connecting the two-port sequence networks. From Eq. (10.58a) it is observed that since the sequence currents are equal, the sequence networks have to be connected in series. On the other hand, Eq. (10.58b) shows that an external impedance Z_a is to be connected in series with each of the sequence networks.

Step 5: Based on Eq. (10.58), the interconnected two-port sequence networks, reflecting the 2LO condition are drawn in Fig. 10.37.

10.8 MATLAB PROGRAM

The MATLAB program given here is an executable function file called faultcal.m. The first line in the program is the function definition line. The variables within the [ ] define the output variables and the variables within ( ) define the input variables. In the present case, the output variables are the phase and line voltages, phase currents, and sequence components of voltages and currents.

The input variables consist of sequence, fault, and ground impedances and voltage at the fault point. The input variable type specifies the kind of fault to be analysed, that is, for SLG, type = 1; for LL, type = 2; for 2LG, type = 3; and for 3ϕ, type = 4.

```
function [Ia,Ib,Ic,Va,Vb,Vc,Vab,Vbc,Vca,Iaz,Ia1,Ia2,Vaz,Va1,Va2]
=faultcal(type,Vf,Zz,Z1,Z2,Zf,...
  Zg)
%Program for analysing shunt fault conditions in power systems
a=(-0.5+0.866i);
A=[1 1 1;1 a^2 a;1 a a^2];
if type == 1;
   Iaz=Vf/(Zz+Z1+Z2+3*Zf);
   Ia1=Vf/(Zz+Z1+Z2+3*Zf);
   Ia2=Vf/(Zz+Z1+Z2+3*Zf);
else
   end
   if type == 2;
   Iaz=0;
   Ia1=Vf/(Z1+Z2+Zf);
   Ia2=-Ia1;
else
   end
if type == 3;
   Ia1=Vf/((Z1+Zf)+((Z2+Zf)*(Zz+Zf+3*Zg)/(Z2+Zz+2*Zf+3*Zg)));
   Ia2=-(Zz+Zf+3*Zg)*Ia1/(Z2+Zz+2*Zf+3*Zg);
   Iaz=-(Z2+Zf)*Ia1/(Z2+Zz+2*Zf+3*Zg);
else
   end
```

```
if type == 4;
  Ia1=Vf/(Z1+Zf);
  Ia2=0;
  Iaz=0;
else
  end
  Vaz=-Zz*Iaz;
  Va1=Vf-Z1*Ia1;
  Va2=-Z2*Ia2;
  I=[Iaz;Ia1;Ia2];
  V=[Vaz;Va1;Va2];
  II=A*I;
  Ia=II(1);
  Ib=II(2);
  Ic=II(3);
  VV=A*V;
  Va=VV(1);
  Vb=VV(2);
  Vc=VV(3);
  Vab=Va-Vb;
  Vbc=Vb-Vc;
  Vca=Vc-Va;
```

The reader will do well to input the data of the sample power system shown in Fig. 10.21 and execute the function file faultcal.m for a few of the examples solved earlier.

10.9 UNSYMMETRICAL SHORT-CIRCUIT COMPUTATION USING BUS IMPEDANCE MATRIX

In this section, a computational algorithm is developed for the computation of unsymmetrical faults using bus impedance matrix. Figure 10.38 shows an n-bus system subjected to an unbalanced fault at bus k. Prior to the incidence of fault, it is operating in steady-state balanced operation. All pre-fault bus voltages are assumed as known and so are the power flows.

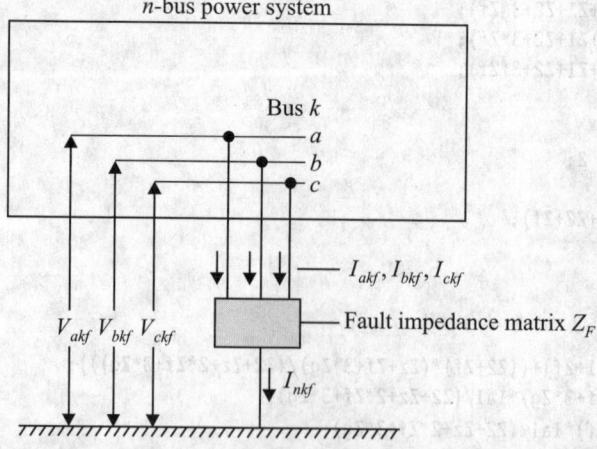

Fig. 10.38 Unsymmetrical fault situation at bus k of an n-bus system

10.9.1 Fault Shunt Symmetrical Component Impedance Matrix $Z_{S,(F)}$

The nature of the fault at bus k is assumed to be fully known. The general unbalance case is shown in Fig. 10.39. By assigning proper values to the impedances Z_a, Z_b, Z_c, and Z_g, practically any fault case can be simulated. For example, by setting

$$Z_a = Z_g = 0 \text{ and } Z_b = Z_c = \infty$$

a solid single line-to-ground short-circuit on phase a is created. By setting

$$Z_a = \infty \text{ and } Z_b = Z_c = Z_g = 0$$

a solid double line-to-ground short-circuit is simulated.

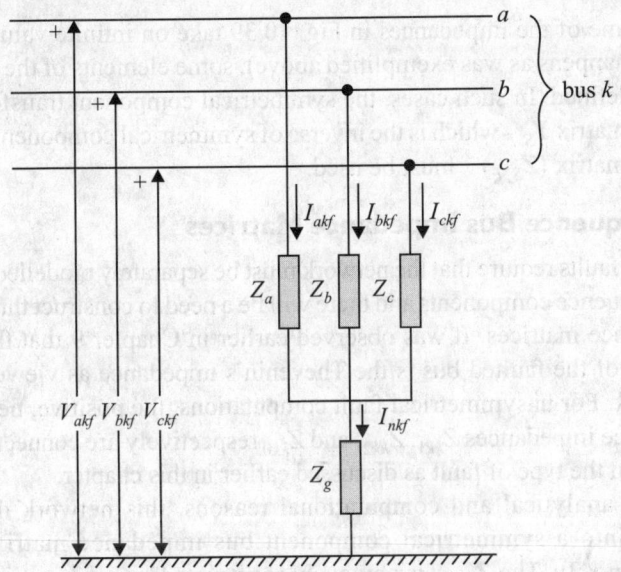

Fig. 10.39 A general unbalanced fault case

In symmetrical fault analysis, explained in Chapter 9, the fault current was determined from a knowledge of the fault impedance Z_f. In the case of unsymmetrical faults, the post-fault currents and bus voltages at the faulted bus are fully determined from the knowledge of a fault impedance matrix $Z_{S,(F)}$. These currents and voltages are indicated in Fig. 10.39. For the circuit in this figure, the relationships between voltages and currents in vector form can be written as

$$
\begin{bmatrix} V_{ak\,f} \\ V_{bk\,f} \\ V_{ck\,f} \end{bmatrix} =
\begin{bmatrix} Z_a + Z_g & Z_g & Z_g \\ Z_g & Z_b + Z_g & Z_g \\ Z_g & Z_g & Z_c + Z_g \end{bmatrix}
\begin{bmatrix} I_{ak\,f} \\ I_{bk\,f} \\ I_{ck\,f} \end{bmatrix}
\tag{10.59}
$$

or $\quad V_{abc,\,k(F)} = Z_F I_{abc,\,k(F)}$ $\hfill$ (10.60)

where $\quad V_{abc,k(F)} = \begin{bmatrix} V_{ak\,f} \\ V_{bk\,f} \\ V_{ck\,f} \end{bmatrix} ; I_{abc,k(F)} = \begin{bmatrix} I_{ak\,f} \\ I_{bk\,f} \\ I_{ck\,f} \end{bmatrix} ; Z_F = \begin{bmatrix} Z_a + Z_g & Z_g & Z_g \\ Z_g & Z_b + Z_g & Z_g \\ Z_g & Z_g & Z_c + Z_g \end{bmatrix}$

$\hfill$ (10.61)

Upon symmetrical component transformation, Eq. (10.60) yields

$$AV_{S, k(F)} = Z_F AI_{S, k(F)} \tag{10.62}$$

and on pre-multiplying by A^{-1}, Eq. (10.62) becomes

$$V_{S, k(F)} = A^{-1} Z_F AI_{S, k(F)} = Z_{S, (F)} I_{S, k(F)} \tag{10.63}$$

where $Z_{S,(F)} = A^{-1} Z_F A$

$$= \begin{bmatrix} Z_a + Z_b + Z_c & Z_a + a^2 Z_b + a Z_c & Z_a + a Z_b + a^2 Z_c \\ Z_a + a Z_b + a^2 Z_c & Z_a + Z_b + Z_c & Z_a + a^2 Z_b + a Z_c \\ Z_a + a^2 Z_b + a Z_c & Z_a + a Z_b + a^2 Z_c & Z_a + Z_b + Z_c + 9 Z_g \end{bmatrix} \tag{10.64}$$

When some of the impedances in Fig. 10.39 take on infinite values (and this commonly happens as was exemplified above), some elements of the $Z_{S, F}$ matrix become undefined. In such cases, the symmetrical component transformed fault admittance matrix $Y_{S, F}$ which is the inverse of symmetrical component shunt fault impedance matrix $(Z_{S, F})^{-1}$ must be used.

10.9.2 Sequence Bus Impedance Matrices

Unbalanced faults require that the network must be separately modelled for each of the three sequence components and there will be a need to construct three different bus impedance matrices. It was observed earlier in Chapter 9 that the diagonal element Z_{kk} of the faulted bus is the Thevenin's impedance as viewed from the faulted bus k. For unsymmetrical fault computations, the positive, negative, and zero sequence impedances Z_{kk1}, Z_{kk2}, and Z_{kk0} respectively are connected together depending on the type of fault as discussed earlier in this chapter.

For both analytical and computational reasons, this network data is best assembled into a symmetrical component bus impedance matrix, $Z_{S,bus}$ of dimension $3n \times 3n$. The $Z_{S, bus}$ is obviously constructed by n submatrices, each of dimension 3×3. These matrices are as follows:

$$Z_{S, bus} = \begin{bmatrix} Z_{S11} & \cdots & Z_{S1k} & \cdots & Z_{S1n} \\ \vdots & & \vdots & & \vdots \\ Z_{Sk1} & \cdots & Z_{Skk} & \cdots & Z_{Skn} \\ \vdots & & \vdots & & \vdots \\ Z_{Sn1} & \cdots & Z_{Snk} & \cdots & Z_{Snn} \end{bmatrix} \tag{10.65}$$

where $Z_{Skn} = \begin{bmatrix} Z_{kn1} & 0 & 0 \\ 0 & Z_{kn2} & 0 \\ 0 & 0 & Z_{kn0} \end{bmatrix} \tag{10.66}$

10.9.3 Sequence Bus Voltage and Current Vectors

For an n-bus system operated in a balanced mode, the n bus voltages and currents can be summarized in n-dimensional vectors V_{bus} and I_{bus}, respectively. In case of unsymmetrical faults, each bus voltage and current can be decomposed into three symmetrical components. The n-bus system will thus be described by $3n$ voltage and current phasors, which can be arranged into $3n$-dimensional symmetrical component bus voltage and bus current vectors. These matrices are

$$V_{S,\,bus} = Z_{S,\,bus}\,I_{S,\,bus} \tag{10.67}$$

where

$$V_{S,\,bus} = \begin{bmatrix} V_{11} \\ V_{12} \\ V_{10} \\ \cdots \\ \vdots \\ \cdots \\ V_{k1} \\ V_{k2} \\ V_{k0} \\ \cdots \\ \vdots \\ \cdots \\ V_{n1} \\ V_{n2} \\ V_{n0} \end{bmatrix} = \begin{bmatrix} V_{S,\,1} \\ \vdots \\ V_{S,\,k} \\ \vdots \\ V_{S,\,n} \end{bmatrix} \quad \text{and} \quad I_{S,\,bus} = \begin{bmatrix} I_{11} \\ I_{12} \\ I_{10} \\ \cdots \\ \vdots \\ \cdots \\ I_{k1} \\ I_{k2} \\ I_{k0} \\ \cdots \\ \vdots \\ \cdots \\ I_{n1} \\ I_{n2} \\ I_{n0} \end{bmatrix} = \begin{bmatrix} I_{S,\,1} \\ \vdots \\ I_{S,\,k} \\ \vdots \\ I_{S,\,n} \end{bmatrix} \tag{10.68}$$

10.9.4 Short-circuit Computations

The unbalanced fault was assumed to occur on a system which is in a pre-fault balanced steady state. Thus, all pre-fault bus voltages contain only positive sequence components, that is, the pre-fault symmetrical component bus voltages can be written in vector form as

$$V_{S,\,bus,\,0} = \begin{bmatrix} V_{1(0)} \\ 0 \\ 0 \\ \cdots \\ \vdots \\ \cdots \\ V_{k(0)} \\ 0 \\ 0 \\ \cdots \\ \vdots \\ \cdots \\ V_{n(0)} \\ 0 \\ 0 \end{bmatrix} \tag{10.69}$$

Using the Thevenin's theorem, the post-fault bus voltages can be obtained from the following matrix equation

$$V_{S,\,\text{bus}(F)} = V_{S,\,\text{bus}(0)} + Z_{S,\,\text{bus}} I_{S,\,\text{bus}(F)} \tag{10.70}$$

In this equation, the fault current vector $I_{S,\,\text{bus}(F)}$ has as its components the fault current injected $-I_{S,\,k(F)}$, at bus k where the fault occurs. Thus,

$$I_{S,\,\text{bus}(F)} = \begin{bmatrix} 0 \\ \vdots \\ -I_{S,\,k(F)} \\ \vdots \\ 0 \end{bmatrix} \tag{10.71}$$

The above vector equations have $3n$ components.

The n vector components of Eq. (10.70) can be written as

$$V_{S,\,1(F)} = V_{S,\,1(0)} - Z_{S,\,1k}\,I_{S,\,k(F)}$$

$$\cdots\cdots\cdots\cdots\cdots\cdots\cdots\cdots\cdots\cdots\cdots$$

$$V_{S,\,k(F)} = V_{S,\,k(0)} - Z_{S,\,kk}\,I_{S,\,k(F)} \tag{10.72}$$

$$\cdots\cdots\cdots\cdots\cdots\cdots\cdots\cdots\cdots\cdots\cdots$$

$$V_{S,\,n(F)} = V_{S,\,n(0)} - Z_{S,\,nk}\,I_{S,\,k(F)}$$

Equating the expression of $V_{S,\,k\,(F)}$ given by the kth equation of Eq. (10.72) and Eq. (10.63),

$$V_{S,\,k(0)} - Z_{S,\,kk}\,I_{S,\,k(F)} = Z_{S,\,(F)}\,I_{S,\,k(F)} \tag{10.73}$$

Then the fault current $I_{S,\,k(F)}$ at bus k may be determined from

$$I_{S,\,k(F)} = (Z_{S,\,(F)} + Z_{S,\,kk})^{-1} V_{S,\,k(\,0)} \tag{10.74}$$

It may be noted that Eq. (10.74) is a three-dimensional vector equation. Substituting the value of fault current in Eq. (10.72), the post-fault bus voltages are obtained as

$$V_{S,\,m(F)} = V_{S,\,m(0)} - Z_{S,\,mk}(Z_{S,\,(F)} + Z_{S,\,kk})^{-1} V_{S,\,k(0)} \tag{10.75}$$

And for the faulted bus k, the bus voltage is

$$V_{S,\,k(F)} = Z_{S,\,(F)}(Z_{S,\,(F)} + Z_{S,\,kk})^{-1} V_{S,\,k(0)} \tag{10.76}$$

With the fault current and all bus voltages known the short-circuit currents can now be computed.

SUMMARY

- The advantage of symmetrical components is that a three-phase unbalanced power system can be resolved into three balanced uncoupled sequence circuits and analysed independently.
- Generators under normal operation produce only positive sequence three-phase voltages. However, under a fault condition, currents flow in all the three phases leading to positive, negative, and zero sequence reactances; all have different magnitudes.
- The positive and negative sequence networks of transformers are simple circuits with sequence impedances connected between the neutral and reference bus. However, the zero sequence networks depend on whether a path is available for the flow of zero sequence current.

- Since transmission lines, like the transformers, are linear, symmetric, and static devices posses equal positive and negative reactances. Zero sequence reactance, however, is two to three times the positive sequence reactance.
- Power networks can be analysed for unsymmetrical shunt and series faults systematically by (i) drawing a labelled single line diagram, (ii) identifying the boundary conditions, (iii) transforming phase quantities into sequence components, (iv) establishing the proper connections between the fault point and neutral, and (v) appropriately connecting the sequence networks.
- The fault currents and voltages in the system are then obtained by working backwards and reconstructing its topology and applying the superposition principle.
- The bus impedance approach is also employed for fault analyses by constructing a symmetrical component bus impedance matrix of dimension $3n \times 3n$ consisting of n sub-matrices, each of dimension 3×3. The sequence bus voltage and current vectors are similarly arranged into $3n$ dimensional symmetrical voltage bus and current vectors.

SIGNIFICANT FORMULAE

- Transformation matrix: $A = \begin{bmatrix} 1 & 1 & 1 \\ 1 & a^2 & a \\ 1 & a & a^2 \end{bmatrix}$

- Sequence components transformation: $V_{012} = A^{-1} V_{abc}$ and $I_{012} = A^{-1} I_{abc}$
- Complex sequence power: $S = P + jQ = 3 V_{012}^t I_{012}^*$
$$= 3 (V_{a0} I_{a0}^* + V_{a1} I_{a1}^* + V_{a2} I_{a2}^*)$$
- Sequence voltage drop in a three-phase system: $V_{pq\,012} = Z_{pq\,012} I_{012}$

where $Z_{pq\,012} = A^{-1} \begin{bmatrix} Z_a & Z_{ab} & Z_{ac} \\ Z_{ba} & Z_b & Z_{bc} \\ Z_{ca} & Z_{cb} & Z_c \end{bmatrix} A$

- Sequence voltage drop in a generator: $V_{a0} = -jI_{a0}X_0$; $V_{a1} = E_a - jI_{a1}X_1$;
$$V_{a2} = -jI_{a2}X_2$$

- SLG fault: $I_b = I_c = 0$; Sequence currents $I_{a0} = I_{a1} = I_{a2} = \dfrac{V_F}{Z_0 + Z_1 + Z_2 + 3Z_F}$

- LL fault: $I_b = -I_c$; $I_a = 0$; $V_b - V_c = I_b Z_f$; $I_{a1} = \dfrac{E_a}{Z_1 + Z_2 + Z_f}$

- 2LG fault: $I_a = 0$; $V_b = Z_f I_b + Z_f (I_b + I_c) = (Z_f + Z_g) I_b + Z_g I_c$;
$$V_c = Z_f I_c + Z_g (I_b + I_c) = Z_g I_b + (Z_f + Z_g) I_c;$$

$$I_{a1} = \dfrac{E_a}{(Z_1 + Z_f) + \dfrac{(Z_2 + Z_f)(Z_0 + Z_f + 3Z_g)}{(Z_2 + Z_0 + 2Z_f + 3Z_g)}};$$

$$I_{a2} = \dfrac{(Z_0 + Z_f + 3Z_g)}{(Z_2 + Z_0 + 2Z_f + 3Z_g)} \times I_{a1};$$

$$I_{a0} = \frac{(Z_2 + Z_f)}{(Z_2 + Z_0 + 2Z_f + 3Z_g)} \times I_{a1}$$

- 3φ fault: $I_a + I_b + I_c = 0$; $V_a = Z_f \times I_a + Z_g(I_a + I_b + I_c)$;

$$V_b = Z_f \times I_b + Z_g(I_a + I_b + I_c);$$
$$V_c = Z_f \times I_c + Z_g(I_a + I_b + I_c);$$

$$I_{a1} = I_a = \frac{E_a}{Z_1 + Z_f}$$

- Unequal series impedance fault: $V_{abc} - V_{a'b'c'} = Z_{abc}I_{abc}$; $V_{aa'-012}$

$$= V_{012} - V'_{012} = Z_{012} I_{012}; Z_{012} = A^{-1} Z_{abc} A$$

- 1LO fault: $Z_a = \infty$; Z_b has a finite value. Special case of series unbalance condition.

- 2LO fault: $I_b = I_c = 0$; $V_{aa'} = Z_a I_a$; $I_0 = I_1 = I_2 = \dfrac{I_a}{3}$;

$$(V_{aa'0} - Z_a I_{a0}) + (V_{aa'1} - Z_a I_{a1}) + (V_{aa'2} - Z_a I_{a2}) = 0$$

EXERCISES

Review Questions

10.1   State the concept of symmetrical components and explain how they can be applied to practical three-phase systems.

10.2   Prove that there is no coupling between sequence currents.

10.3   A three-phase system has series impedances $Z_a \neq Z_b \neq Z_c$ and mutual impedances given by $Z_{ab} \neq Z_{ba}$, $Z_{bc} \neq Z_{cb}$ and $Z_{ca} \neq Z_{ac}$. Derive expressions to determine the sequence voltage drops. Also show that when the series impedances are equal and there is no mutual coupling, the sequence voltage drops are $V_{pq0} = Z_a I_{a0}$, $V_{pq1} = Z_a I_{a1}$ and $V_{pq2} = Z_a I_{a2}$.

10.4   Derive expressions for the sequence parameters of a synchronous generator. Also draw the corresponding sequence networks.

10.5   Draw, with justification, the zero sequence networks for the following types of three-phase transformer connections: (i) Δ–Δ connection, (ii) Y–Δ connection with neutral not grounded ; (iii) Y–Δ with neutral grounded, (iv) Y–Y with the neutrals grounded on both sides, and (v) Y–Y with neutral grounded on one side.

10.6   (a) Categorize the various types of unsymmetrical faults and state the order of frequency of occurrence of shunt faults.

(b) Write a note on the synthesis of sequence networks.

10.7   How will the interconnections of the sequence networks change, if the neutral point *n* is connected to ground through Z_g? Derive an expression for the neutral voltage.

10.8   Determine the relation between neutral current I_n and I_{a0} for an unbalanced three-phase, star-connected load with the neutral point grounded.

10.9   Enumerate the different types of series faults. Derive expressions for computing sequence voltages, for unequal series impedance faults. Draw a diagram to show the sequence network circuit connections.

10.10 Explain the bus impedance matrix approach for computing unsymmetrical short-circuit fault on a power system network.

10.11 Repeat exercise 10.10 for a one line conductor open fault.

10.12 Repeat exercise 10.10 for a two lines conductors open fault.

Numerical Problems

10.1 Compute the symmetrical components, if the phase currents are $I_a = 15\angle 0°$A, $I_b = 20 \angle 120°$A, and $I_c = 30 \angle -120°$A.

10.2 The sequence components of voltages in a circuit are $V_{an1} = 100\angle 90°$ V, $V_{an2} = 50\angle 0°$ V, and $V_{an0} = 50\angle 180°$ V. Determine the phase to neutral components of voltages.

10.3 A star-connected, balanced load of 10 Ω each has the following voltages across its terminals; $V_{ab} = 200$ V, $V_{bc} = 220$ V, and $V_{ca} = 180$ V. Calculate the symmetrical components of line and phase voltages. From the symmetrical components of line voltages, determine the line currents.

[**Note:** Assume $V_{ca} = V_{ca}\angle 120°$, use the cosine law for the triangle to calculate the phase angles of the other line voltages]

10.4 In Problem 10.3, compute the total power consumed from the symmetrical components of voltages and currents. Verify the answer.

10.5 For the network shown in Fig. 10.40, draw the per unit single line diagram on a system base of 100 MVA. For the transmission lines, select a voltage base of 220 kV.

   Generator G_1: 50 MVA, 10.6 kV, $X = 0.12$ pu
   Generator G_2: 30 MVA, 6.6 kV, $X = 0.10$ pu
   Motor M: 25 MVA, 10 kV, $X = 0.08$ pu
   T_1: 75 MVA, 11/220 kV, $X = 0.15$ pu
   T_2: 50 MVA, 11/220 kV, $X = 0.15$ pu
   T_3: 30 MVA, 6.6/220 kV, $X = 0.15$ pu
   Line 1: $(50 + j100)\ \Omega$
   Line 2: $(25 + j50)\ \Omega$
   Line 3: $(45 + j60)\ \Omega$
   Load: 30 MVA, 11 kV at 0.8 power factor lagging

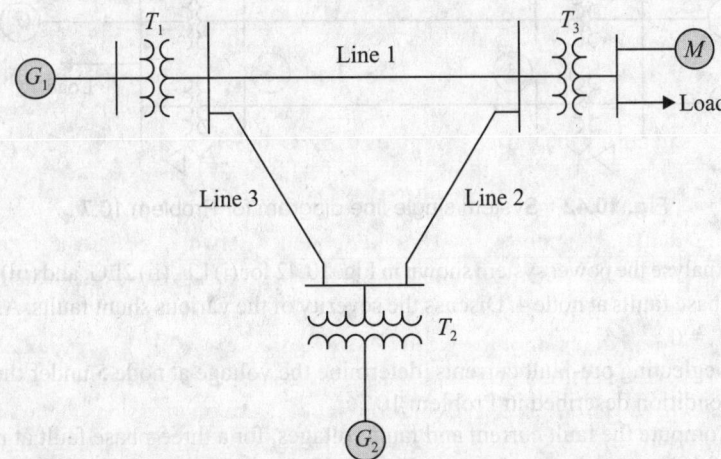

Fig. 10.40 System single line diagram for Problem 10.6

10.6   For the circuit shown in Fig. 10.41, determine the interconnections of the sequence networks.

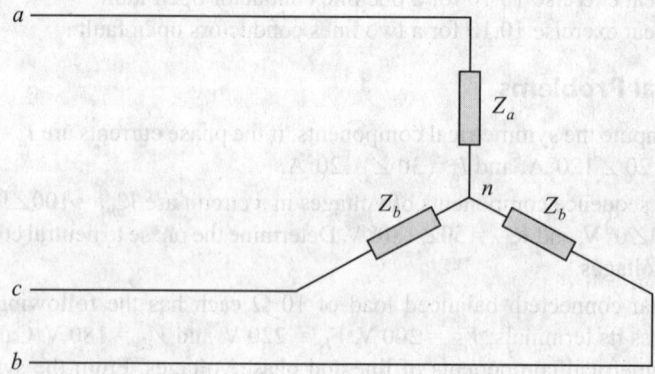

Fig. 10.41   Circuit diagram for Problem 10.6

10.7   An SLG fault occurs at node 4 of the power system shown in Fig. 10.42. Determine the fault current, in amperes, sequence voltages, phase and line voltages at the point of fault. Assume a fault resistance $Z_f = 0.05$ pu.

Generator 1: 40 MVA, 6.6 kV, $X_1 = X_2 = 0.10$ pu, $X_0 = 0.20$ pu

Generator 2: 60 MVA, 6.6 kV, $X_1 = X_2 = 0.15$ pu, $X_0 = 0.20$ pu

Generator 3: 50 MVA, 11 kV, $X_1 = X_2 = 0.12$ pu, $X_0 = 0.15$ pu

Transformer 1-2: 50 MVA, 6.6/132 kV, $X_1 = X_2 = X_0 = 0.10$ pu

Transformer 1-5: 70 MVA, 6.6/220 kV, $X_1 = X_2 = X_0 = 0.08$ pu

Transformer 3-4: 30 MVA, 11/132 kV, $X_1 = X_2 = X_0 = 0.06$ pu

Transformer 6-4: 25 MVA, 11/220 kV, $X_1 = X_2 = X_0 = 0.075$ pu

Transmission line 2-3: $(30 + j90)\ \Omega$

Transmission line 5-6: $(50 + j80)\ \Omega$

Load: 50 MVA at 0.8 power factor lagging at 11 kV.

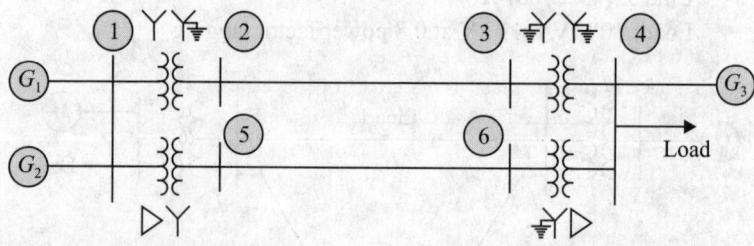

Fig. 10.42   System single line diagram for Problem 10.7

10.8   Analyse the power system shown in Fig. 10.42 for (i) LL, (ii) 2LG, and (iii) three-phase faults at node 4. Discuss the severity of the various shunt faults. Assume $Z_g = 0$.

10.9   Neglecting pre-fault currents, determine the voltage at node 5 under the fault condition described in Problem 10.7.

10.10   Compute the fault current and fault voltages, for a three-phase fault at node 1 of the power system shown in Fig. 10.42.

10.11 Determine the distribution of the fault currents and fault voltages at all the nodes of the power system shown in Fig. 10.42 for the condition in Problem 10.10.

10.12 Calculate the fault currents and voltages for a 2LG fault in the middle of the transmission line 5-6 of the power system shown in Fig. 10.42. Assume $Z_f = (5 + j0)\ \Omega$ and $Z_g = (10 + j0)\ \Omega$. How will the fault currents and voltages change for $Z_g = \infty$?

10.13 An LL fault occurs at 25% of the distance from node 2 in transmission line 2-3 of the power system shown in Fig. 10.42. Compute the fault currents and voltages at the fault point $Z_f = (5 + j0)\ \Omega$. What happens if Z_f is made equal to zero?

Multiple Choice Objective Questions

10.1 Which of the following is the purpose of conducting fault studies?
 (a) To determine fault voltages (b) To determine fault currents
 (c) Short circuit MVA (d) All of these

10.2 Which of the following is not a correct assumption for computing fault currents when a balanced three-phase short-circuit occurs?
 (a) Synchronous machines are represented by constant voltage source behind X'_d.
 (b) Synchronous machines are represented by constant voltage source behind X''_d.
 (c) Small induction motors are represented in the same manner as synchronous machines.
 (d) Phase shift in *D-Y* connected transformers is ignored.

10.3 Which of the following is not true when a three-phase fault occurs?
 (a) Structural network change is equivalent to the addition of three equal impedances Z_f at the place of fault.
 (b) Thevenin's theorem is applicable for network analyses under fault condition.
 (c) Superposition principle is applicable for post-fault network analyses.
 (d) None of these.

10.4 Which of the following is not correct?
 An unbalanced three-phase power system can be resolved into
 (a) Three balanced phasor systems of different phase sequences
 (b) Two balanced phasor systems of different phase sequences plus one zero sequence phase system
 (c) One balanced phasor system plus two zero sequence phase systems
 (d) All of these

10.5 Which of the following represents a zero sequence phase system?
 (a) Three phasors equal in magnitude, displaced in phase from each other by 120°, and having the same phase sequence as the original phasors
 (b) Three phasors equal in magnitude, displaced in phase from each other by 120°, and having phase sequence opposite to the original phasors
 (c) Three phasors equal in magnitude, and with zero phase displacement from each other
 (d) None of these

10.6 Which of the following is indicative of the condition that there is no coupling between the sequence components?

(a) $3\left(V_{a0}I_{a0}^* + V_{a1}I_{a1}^* + V_{a2}I_{a2}^*\right)$ (b) $3\left(V_{a0}I_{a0}^* + V_{a1}I_{b1}^* + V_{a2}I_{b2}^*\right)$

(c) $3\left(V_{a0}I_{a0}^* + V_{a1}I_{c1}^* + V_{a2}I_{c2}^*\right)$ (d) $3\left(V_{a0}I_{b0}^* + V_{b1}I_{c1}^* + V_{b2}I_{a1}^*\right)$

10.7 Which of the following represents the value of a^3?

(a) $-0.5 + j0.8660$ (b) $-0.5 - j0.866$

(c) 1.0 (d) None of these

10.8 Which of the following represents the symmetrical transformation matrix?

(a) $\begin{bmatrix} 1 & 1 & 1 \\ 1 & a & a^2 \\ 1 & a^2 & a \end{bmatrix}$ (b) $\begin{bmatrix} 1 & 1 & 1 \\ 1 & a^2 & a \\ 1 & a & a^2 \end{bmatrix}$

(c) $\begin{bmatrix} 1 & 1 & 1 \\ a^2 & a & 1 \\ a & a^2 & 1 \end{bmatrix}$ (d) $\begin{bmatrix} 1 & a^2 & a \\ 1 & a & a^2 \\ 1 & 1 & 1 \end{bmatrix}$

10.9 For which of the following unbalanced currents will produce only symmetrical sequence voltage drops?

(a) $Z_a \neq Z_b \neq Z_c$ and $Z_{ab} \neq Z_{bc} \neq Z_{ca}$

(b) $Z_a = Z_b = Z_c$ and $Z_{ab} \neq Z_{bc} \neq Z_{ca}$

(c) $Z_a = Z_b = Z_c$ and $Z_{ab} = Z_{bc} = Z_{ca}$

(d) $Z_a = Z_b = Z_c$ and $Z_{ab} = Z_{bc} = Z_{ca} = 0$

10.10 Which of the following represents the zero sequence network of a generator?

(a) $3Z_n - Z_0$ (b) $3Z_n + Z_0$

(c) $Z_n - 3Z_0$ (d) $Z_n + 3Z_0$

10.11 In which of the following the positive, negative, and zero sequence impedances have different magnitudes?

(a) transformers   (b) transmission lines

(c) generators   (d) all of these

10.12 Which of the following is not a shunt fault?

(a) SLG   (b) LL

(c) 3φ (d) None of these

10.13 Which of the following has the highest percentage of occurrence in a power network?

(a) 2LG   (b) SLG

(c) LL   (d) 3φ

10.14 Which of the following is the severest and has the lowest percentage of occurrence?

(a) 2LG   (b) SLG

(c) LL   (d) 3φ

10.15 Which of the following represents the condition for a 3φ fault?

(a) $I_a + I_b + I_c = 0$ (b) $3I_{a0} = 0$

(c) $I_{a1} = I_a = \dfrac{E_a}{Z_1 + Z_f}$ (d) All of these

10.16 Which of the following is not a boundary condition for an SLG fault on phase 'a' of a power system?

(a) $I_a = 0$ (b) $I_b = 0$

(c) $I_c = 0$ (d) None of these

10.17 What is the type of fault when the boundary conditions for a fault on phase
'a' of a power system are $I_b = -I_c$, $I_a = 0$, and $V_b - V_c = 0$?
(a) 3φ
(b) SLG
(c) LL
(d) 2LG

10.18 Which of the following characterizes series faults?
(a) No connection between lines
(b) No connection between lines and neutral
(c) Imbalance in series impedance
(d) All of these

10.19 The boundary conditions in a power system are $I_b = I_c = 0$ and $V_{aa'} = Z_a I_a$.
Which type of fault is indicated?
(a) Unequal series impedance
(b) 1LO
(c) 2LO
(d) None of these

Answers

10.1 (d)	10.2 (a)	10.3 (d)	10.4 (b)	10.5 (c)	10.6 (a)
10.7 (c)	10.8 (b)	10.9 (d)	10.10 (a)	10.11 (c)	10.12 (d)
10.13 (b)	10.14 (d)	10.15 (d)	10.16 (a)	10.17 (c)	10.18 (d)
10.19 (c)					

CHAPTER 11

Power System Stability

Learning Outcomes

A thorough study of this chapter will aid the reader to
- Generally familiarize with the changes that occur when the steady-state operation of a complex interconnected power system is disturbed
- Visualize the consequences of large and small disturbances in a power system and distinguish between transient and steady-state stability
- Develop the swing equation for a synchronous generator based on rotational mechanics, and understand the significance of inertia constant
- Enumerate the assumptions, develop the classical model for stability analysis, and apply the equal area criterion, for different types of disturbances, to determine the stability of a synchronous generator connected to an infinite bus
- Skilfully solve the swing equation by applying different numerical techniques
- Simulate the multi-machine transient problem and familiarize with the methods of solution
- Simulate the steady-state stability problem and apply Lyapunov theorems to determine stability
- Understand, generally, the various techniques for improving transient and steady-state stability

OVERVIEW

An electric power system is constituted of several power elements such as generators, transformers, power transmission lines, distribution networks, and loads, as well as control elements such as automatic voltage regulators of synchronous machines, automatic load frequency control mechanism, protective relays, and circuit breakers. All these units combine to form a system and exhibit properties, as a system, which are quite different from the properties of these individual units.

Under normal conditions (termed as the steady-state), the average electrical speeds of all the generators remain the same throughout the system. In the steady-state condition, there is equilibrium between the input mechanical torque and output electrical torque of each machine. This is termed as the synchronous operation of the system. If the system is perturbed by a disturbance, small or large, it affects the synchronous operation of the system. The disturbance upsets the equilibrium between mechanical input torque and electrical output torque of each machine, resulting in acceleration or deceleration of the machines according to the laws of rotating bodies. If a generator temporarily runs faster than another machine, the rotor angular position of the machine relative to the slower machine will advance. The resulting angular difference transfers part of the load from the slow machine to the fast machine depending on the power–angle relationship, and this tends to reduce the speed difference and hence the angular separation. The power–angle characteristic being sinusoidal is non-linear. Beyond a certain limit, an increase in angular separation is accompanied by a decrease in power transfer; this further increases the angular separation and leads to instability.

The tendency of a power system to develop restoring forces equal to or greater than the disturbing forces to maintain synchronous running of generators is known as *stability*. For any given situation, the stability of the system depends on whether or not the deviations in angular positions of the rotors result in sufficient restoring forces.

When a machine loses synchronism with other machines of the system, its rotor angle will undergo wide variations, voltage and frequency may deviate widely from normal values, and the protection system isolates the unstable machine from the system. Loss of synchronism may occur between two or more groups of machines while maintaining synchronism within each group, after its separation from others.

The stability problem is broadly divided into two classes—rotor angle stability and voltage stability. The rotor angle stability problem is further subdivided into two parts: steady-state stability and transient stability. In this chapter, all aspects of rotor angle stability are presented, and voltage stability is discussed in Chapter 12.

Steady-state stability is the ability of a system to restore the initial condition after a small disturbance or to reach a condition very close to the initial condition when the disturbance is still present. Small random changes in load or generation can be termed as small disturbances. Transient stability is a condition that characterizes the dynamics of a power system subjected to a fault, the initial state preceding the fault being a balanced one. Power system faults, which result in a sudden dip in bus voltages and require immediate remedial action in the form of clearing of the fault, can be termed as large disturbances. A system is said to possess transient stability if after the fault it is capable of maintaining synchronous operation and returning to its initial state or a state close to it. Transient stability of a system depends on various factors such as time of fault clearance, type and location of fault, time of reclosing after fault clearance, and also on the stability improvement measures available. The maximum amount of steady-state power that the system can transmit for specified operating conditions without losing synchronism after being subjected to a fault is known as the transient stability limit of the system. By the nature of power system operation such transient stability limits are many depending on the type of situation one has on hand. The planning, design, and operation of an electric power system should ensure that the system is stable for all contingencies that are expected to occur.

It is to be noted that steady-state stability is a function of operating conditions and the system dynamics can be analysed using linearized equations around the initial operating point. The transient stability however is a function of both the operating condition and the disturbance(s). Transient stability analysis is thus considerably complicated as non-linearities in the system cannot be ignored.

Transient stability studies are carried out when new generating and transmitting facilities are planned. These studies provide the changes in bus voltages and power flows during and immediately following disturbances. These studies are helpful in determining the relaying and protective system needed, critical fault clearing times of circuit breakers, and power transfer capability between systems. The steady-state stability analysis of a power system involves determining its critical operating condition, the nature of the transient process and its quality for the given automatic voltage regulator, and choosing the type of automatic voltage regulator and its parameters for the stable operation of the system.

In this chapter, the swing equation of a generator rotor is derived. The simulation of the transient stability problem for a single machine connected to an infinite bus is developed. The application of equal area criterion used for the quick prediction of stability of a single machine connected to an infinite bus for different contingency conditions is described. Numerical methods for solving the swing equation(s) are discussed. The formulation of the multi-machine transient stability problem and the computational algorithm for solving the same is described. Finally, the simulation of a steady-state stability problem of a simple power system is discussed.

11.1 TRANSIENT STABILITY

The transient stability problem of a present-day power system usually concerns the transmission of power from one group of synchronous machines to another. During disturbances, the machines of each group swing together and are said to form a coherent group. For the purpose of analysis, each group can be replaced by one equivalent synchronous machine and the transient stability investigations may be carried out by assuming the system to be represented by a two-machine system. If synchronism is lost, the machines of each group stay together, although they go out of step with another group. In case of a two-machine system representation, when the capacity of one equivalent machine is greater than the other equivalent machine by at least ten times, then it may be represented by an infinite bus (constant voltage source with zero internal impedance and infinite inertia).

Transient stability problems can be subdivided into first swing and multi-swing stability problems. During the first swing, which lasts for about one second, the prime mover input to the generator and its voltage behind transient reactance is assumed constant. Multi-swing stability that extends over longer periods must consider effects of turbine-governor and excitation systems and more detailed representation of synchronous machines.

11.2 ASSUMPTIONS COMMONLY MADE IN STABILITY STUDIES

In transient stability investigations, the following simplifying assumptions are made.

1. The synchronous machine is represented by a constant voltage E' behind direct axis transient reactance X'_d (determined from pre-fault power flow study) as discussed in Section 5.5.3.

2. Power input P_i from governor and turbine remains essentially constant equal to the pre-fault value during the entire transient period.
3. The mechanical angle δ_i of each rotor coincides with the electrical phase angle of voltage behind the transient reactance E_i'.
4. Damping or asynchronous power is negligible.
5. The network parameters are modelled as lumped circuits and remain fixed during stability studies as the variation in frequency during transient period is negligible.
6. Loads are converted into equivalent admittances based on the voltages at the buses calculated from the pre-fault power flow study.
7. Synchronous power may be calculated from the steady-state solution of network to which the machines are connected.

The above assumptions simplify the transient stability analysis although more accurate studies involve the use of more detailed modelling, relaxing one or more of the assumptions.

11.3 SWING EQUATION

The equation of motion of the rotor of a synchronous machine is determined by the laws of rotation and can be written as

$$ J_m \alpha_m = J_m \frac{d^2\theta_m}{dt^2} = T_a = T_i - T_u \tag{11.1} $$

where J_m is the total moment of inertia of the rotor masses in kgm^2, α_m is the rotor angular acceleration in mechanical rad/s^2, θ_m is the angular displacement of rotor with respect to stationary reference axis in rad/s, T_a is the net accelerating torque in Nm, T_i is the mechanical or shaft torque supplied by the prime mover minus the retarding torque due to rotational losses in Nm, and T_u is the net electromagnetic torque that accounts for the total power output of the generator plus electrical losses in Nm, and t is the time in seconds.

It may be noted that T_i and T_u are both positive for generator action, and for motor action both T_i and T_u are negative. In the steady state, T_i equals T_u, the accelerating torque T_a is zero, and the machine rotor runs at constant speed (synchronous speed). When T_i is greater than T_u, the accelerating torque T_a is positive and the rotor speed increases. Similarly, when T_i is less than T_u, then T_a is negative and the rotor speed decreases.

The angular position θ_m is measured with respect to the stationary reference axis on the stator, and it is an absolute measure of the rotor angle. Thus at constant speed, it continuously increases with time at synchronous speed. Since the rotor speed relative to the synchronous speed is of interest, it is more convenient to measure the rotor angular position with respect to a synchronously rotating reference axis. Accordingly, the rotor angular position is defined as

$$ \theta_m = \omega_{sm} t + \delta_m \tag{11.2} $$

where ω_{sm} is the synchronous angular speed of the rotor in mechanical rad/s and δ_m is the rotor angular position with respect to synchronously rotating reference in mechanical radians

Differentiating Eq. (11.2) with respect to time t yields rotor angular velocity as

$$\omega_m = \frac{d\theta_m}{dt} = \omega_{sm} + \frac{d\delta_m}{dt} \tag{11.3}$$

and the rotor acceleration is

$$\alpha_m = \frac{d^2\theta_m}{dt^2} = \frac{d^2\delta_m}{dt^2} \tag{11.4}$$

Substituting Eq. (11.4) in Eq. (11.1) gives

$$J_m \frac{d^2\delta_m}{dt^2} = T_a = T_i - T_u \tag{11.5}$$

It is convenient to work with power rather than torque. Hence, multiplying both sides of Eq. (11.5) by the angular velocity ω_m leads to

$$J_m \omega_m \frac{d^2\delta_m}{dt^2} = \omega_m T_a = \omega_m T_i - \omega_m T_u \tag{11.6}$$

Since angular velocity multiplied by torque is equal to the power, Eq. (11.5) may be written in terms of power as

$$M \frac{d^2\delta_m}{dt^2} = P_a = P_i - P_u \tag{11.7}$$

where $M = J_m \omega_m$ is the angular momentum of the rotor; and is called the inertia constant, P_i is the mechanical power input from the prime mover minus the mechanical losses in watts, P_u is the electrical power output plus the electrical losses in watts, and P_a is the accelerating power which accounts for any unbalance between P_i and P_u in watts.

Equation (11.7) is called the swing equation. An equation of this form may be written for each machine of the system.

11.3.1 The Inertia Constant

Sometimes the available information regarding the angular momentum of a machine takes the form of the value of its stored kinetic energy at rated speed. The value of inertia constant M from the data has to be calculated. The inertia constant M is related to the kinetic energy of the rotating mass, W_{KE}. The kinetic energy is

$$W_{KE} = \frac{1}{2} J_m \omega_m^2 = \frac{1}{2} M \omega_m \tag{11.8}$$

or

$$M = \frac{2W_{KE}}{\omega_m} \quad \text{Js/mechanical radian} \tag{11.9}$$

The unit of M is joule-seconds per mechanical radian.

It may be noted that the inertia constant M is strictly not a constant, as ω_m does not equal synchronous speed under all conditions of operation. However, since ω_m does not change significantly from synchronous speed when the machine is stable, M is evaluated at synchronous speed and is equal to $J_m \omega_{sm}$. This value of M is known as the inertia constant of the machine, and Eq. (11.9) is modified as

$$M = \frac{2W_{KE}}{\omega_{sm}} \text{ Js/mechanical radian} \tag{11.10}$$

In data supplied by machine manufacturers for stability studies, another constant related to inertia, H constant, is often encountered. H constant is defined as

$$H = \frac{\text{Stored kinetic energy in megajoules at synchronous speed}}{\text{Machine rating in MVA}} \tag{11.11a}$$

$$= \frac{\frac{1}{2}J_m\omega_{sm}^2}{S_{rated}} = \frac{\frac{1}{2}M\omega_{sm}}{S_{rated}} \text{ MJ/MVA} \tag{11.11b}$$

where S_{rated} is the three-phase rating of the machine in MVA.

The inertia constant H has the dimension of time expressed in seconds. The value of H varies within a narrow range. For steam turbo-generators, the value of H lies in the range 3 to 10 MJ per MVA, while for hydro generators it lies in the range 2 to 4 MJ per MVA. Solving Eq. (11.11b) for M gives

$$M = \frac{2H}{\omega_{sm}} S_{rated} \text{ MJ/mechanical radian} \tag{11.12}$$

and substituting Eq. (11.11) in Eq. (11.7), the swing equation is obtained as

$$\frac{2H}{\omega_{sm}} \frac{d^2\delta_m}{dt^2} = \frac{P_a}{S_{rated}} = \frac{P_i - P_u}{S_{rated}} \tag{11.13}$$

in which P_a, P_i, and P_u are expressed in MW, the angle δ_m in the numerator is expressed in mechanical radians, and ω_{sm} in the denominator is expressed in mechanical radians per second.

For a synchronous generator with P poles, the power angle δ, the synchronous electrical radian frequency ω_s, and the electrical angular acceleration α are given by

$$\delta = \frac{P}{2}\delta_m \tag{11.14}$$

$$\omega_s = \frac{P}{2}\omega_{sm} \tag{11.15}$$

$$\alpha = \frac{P}{2}\alpha_m \tag{11.16}$$

Substituting Eqs (11.14) to (11.16) in Eq. (11.13), yields

$$\frac{2H}{\omega_s} \frac{d^2\delta}{dt^2} = P_{a,\text{pu}} = P_{i,\text{pu}} - P_{u,\text{pu}} \tag{11.17}$$

where $P_{a,\text{pu}}$, $P_{i,\text{pu}}$, and $P_{u,\text{pu}}$ are the per unit accelerating, mechanical, and electrical powers respectively. The subscript pu may be omitted as the power is understood to be in per unit. For a system with electrical frequency f, Eq. (11.17) with the rotor angle δ in electrical radians becomes

$$\frac{H}{\pi f} \frac{d^2\delta}{dt^2} = P_a = P_i - P_u \text{ pu} \tag{11.18}$$

and when the rotor angle δ is in electrical degrees, Eq. (11.17) becomes

$$\frac{H}{180f} \frac{d^2\delta}{dt^2} = P_a = P_i - P_u \text{ pu} \tag{11.19}$$

Example 11.1 A 50-Hz, 100-MVA, four-pole, synchronous generator has an inertia constant of 3.5 s and is supplying 0.16 pu power on a system base of 500 MVA. The input to the generator is increased to 0.18 pu. Determine (i) the kinetic energy stored in the moving parts of the generator and (ii) the acceleration of the generator. If the acceleration continues for 7.5 cycles, calculate (iii) the change in rotor angle and (iv) the speed in rpm at the end of the acceleration.

Solution (i) Kinetic energy stored $= H \times$ machine rating $= 3.5 \times 100 = 350$ MJ
(ii) System base $= 500$ MVA
Then the accelerating power $P_a = (P_i - P_u) = (0.18 - 0.16) \times 500 = 10$ MW
Using Eq. (11.12),

$$M = \frac{2H}{\omega_s} \times S_{\text{rated}} = \frac{2 \times 3.5}{360 \times 50} \times 100 = \frac{3.5}{90}$$

Using Eq. (11.13) yields,

$$\frac{3.5}{90} \times \frac{d^2\delta}{dt^2} = P_a = 10$$

$$\therefore \quad \frac{d^2\delta}{dt^2} = \frac{10 \times 90}{3.5} = 257.143° \text{ electrical/s}^2 = 4.488 \text{ radians/s}^2 \tag{11.1.1}$$

(iii) Acceleration period in seconds $= \frac{7.5}{50} = 0.15$ s

$$N_s = \frac{120f}{P} = \frac{120 \times 50}{4} = 1500 \text{ rpm}$$

Equation (11.1.1) is rewritten as

$$2 \frac{d\delta}{dt} \times \frac{d^2\delta}{dt^2} \times dt = 2 \frac{d\delta}{dt} \times dt \times 4.488 \tag{11.1.2}$$

Integrating Eq. (11.1.2) with respect to δ gives

$$\left(\frac{d\delta}{dt}\right)^2 = 8.976\delta + K$$

where K is the integration constant.

At time $t = 0$, $\dfrac{d\delta}{dt} = 0$, hence $K = 0$.

Then, $\quad \dfrac{d\delta}{dt} = \sqrt{8.976\delta} = 2.996\sqrt{\delta} \tag{11.1.3}$

$$\frac{d\delta}{\sqrt{\delta}} = 2.996 \, dt$$

Integration of the preceding equation leads to

$$\delta = (2.996t)^2 + K$$

If at $t = 0$, the rotor angle is represented by $\delta^{(0)}$, then

$$K = \delta^{(0)}$$

and $\qquad \delta = (2.996t)^2 + \delta^{(0)}$

Change in rotor angle $= \delta - \delta^{(0)} = (2.996t)^2 = (2.996 \times 0.15)^2 = 0.202$ rad

$$= 11.57° \text{ electrical}$$

Using Eq. 11.1.3, rotor velocity is calculated as

$$\frac{d\delta}{dt} = 2.996\sqrt{\delta} = 2.996\sqrt{0.202} = 1.347 \text{ rads/s}$$

$$= \frac{1.347}{\pi\left(\dfrac{P}{2}\right)} \times 60 = 12.858 \text{ rpm}$$

Speed of the rotor at the end of the acceleration period

$$= N_s + 12.858 = 1500 + 12.858 = 1512.858 \text{ rpm}$$

Example 11.2 A two-pole, 50-MVA, 11-kV generator is supplying full load at 0.8 power factor lagging. If the inertia constant of the moving parts of the generator is 6.0 MJ/MVA, calculate the energy stored when the generator is running at the synchronous speed of 3000 rpm. If the net input to the generator is suddenly increased to 62000 metric HP, calculate the acceleration produced.

Solution   Kinetic energy stored at the synchronous speed $= H \times S_{\text{rated}} = 50 \times 6 = 300$ MJ

Power output $= 50 \times 0.8 = 40$ MW

$$\text{Power input} - \frac{62000 \times 735.5}{10^6} = 45.601 \text{ MW}$$

Accelerating power $= 45.601 - 40.00 = 5.601$ MW

$$f = \frac{2 \times 3000}{120} = 50 \text{ Hz}$$

$$M = \frac{2H}{\omega_s} \times S_{\text{rated}} = \frac{2 \times 6}{360 \times 50} \times 50 = 0.0333 \text{ MJ s/rad}$$

Using Eq. (11.13) gives

$$0.0333 \frac{d^2\delta}{dt^2} = 5.601$$

$$\frac{d^2\delta}{dt^2} = \frac{5.601}{0.0333} = 168.20° \text{elec.}/s^2 = 168.20 \times \frac{\pi}{180} = 2.94 \text{ elec. rads/s}^2$$

11.4 CLASSICAL SYSTEM MODEL FOR STABILITY ANALYSIS

The simplest model for stability analysis is the classical model of a single machine connected through a transmission line to an infinite bus as shown in Fig. 11.1(a). Here the single generator represents a single machine equivalent of a power plant consisting of several generators. The infinite bus by definition represents a bus with a fixed voltage source. The magnitude, frequency, and phase of the voltage are unaltered by the changes in the output of the generator. The equivalent circuit of the system shown in Fig. 11.1(a) is given in Fig. 11.1(b). The synchronous machine is represented by a constant voltage E' behind direct axis transient reactance X'_d as discussed in Section 11.1, and X_E is the total external reactance viewed from the generator terminals. The voltage V_B of the infinite bus has constant magnitude and constant phase angle $0°$. To simplify analysis, losses are neglected here.

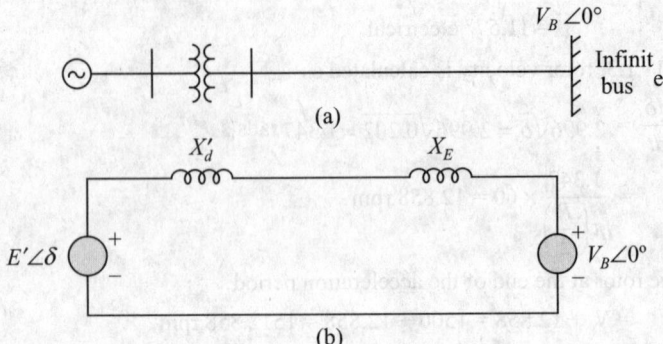

Fig. 11.1   (a) Simplified synchronous machine model for transient stability
(b) Equivalent circuit of the system

The equivalent reactance between the machine internal voltage and the infinite bus is $X_{eq} = X'_d + X_E$. From Eq. (5.22a), the electrical power output P_u of the synchronous generator to the infinite bus is

$$P_u = \frac{(|E'||V_B|)}{X_{eq}} \sin\delta = P_{max}\sin\delta \qquad (11.20)$$

where $P_{max} = (|E'||V_B|)/X_{eq}$ is the peak value of the electrical power output of the generator and δ is the phase angle (power angle) of the machine with respect to the infinite bus. Then the swing equation of a synchronous machine connected to an infinite bus, under transient condition becomes

$$M\frac{d^2\delta}{dt^2} = P_i - P_u = P_i - P_{max}\sin\delta = P_a \qquad (11.21)$$

11.5 EQUAL AREA CRITERION

Stability of a synchronous machine connected to an infinite bus after a large disturbance can be determined from the solution of swing equation of the machine, and from the results so obtained the variation of power angle δ of the machine is plotted with respect to time, called the *swing curve*. If the swing curve shows that the power angle δ of the machine increases without limit then the system is unstable. On the other hand, after the disturbance including switching has occurred, the angle δ reaches a maximum value and thereafter decreases, the system is stable. Equal area criterion is a simple graphical method to quickly determine qualitatively whether stability will be maintained or not. Although the equal area criterion is applicable only for a single machine connected to an infinite bus, yet it is a useful technique since it provides an insight into the rotational dynamics of the machine.

The swing of the machine after a disturbance can be written from Eq. (11.21) as

$$M \frac{d^2\delta}{dt^2} = P_a = P_i - P_u \qquad (11.22)$$

Multiplying both sides of Eq. (11.22) by $\frac{2}{M}\frac{d\delta}{dt}$ gives

$$2\frac{d\delta}{dt}\frac{d^2\delta}{dt^2} = 2\frac{P_a}{M}\frac{d\delta}{dt} \qquad (11.23)$$

or $$\frac{d}{dt}\left[\left(\frac{d\delta}{dt}\right)^2\right] = 2\frac{P_a}{M}\frac{d\delta}{dt}$$

then $$d\left[\left(\frac{d\delta}{dt}\right)^2\right] = 2\frac{P_a}{M}d\delta \qquad (11.24)$$

and integrating Eq. (11.24), yields

$$\left(\frac{d\delta}{dt}\right)^2 = \frac{2}{M}\int_{\delta_0}^{\delta} P_a d\delta \qquad (11.25)$$

or $$\Delta\omega = \left(\frac{d\delta}{dt}\right) = \sqrt{\frac{2}{M}\int_{\delta_0}^{\delta} P_a d\delta} \qquad (11.26)$$

When the machine comes to rest with respect to the infinite bus, a condition which may be taken to indicate stability is

$$\Delta\omega = \frac{d\delta}{dt} = 0$$

when $$\int_{\delta_0}^{\delta_m} P_a d\delta = 0 \qquad (11.27)$$

The integral of Eq. (11.27) may be graphically interpreted as the area under a curve of P_a plotted against δ between limits δ_0, the initial angle, and δ_m, the final angle as shown in Fig. 11.2. Since P_a equals $P_i - P_u$, the integral of Eq. (11.27) may be interpreted also as the area between the curve of P_i versus δ and the curve of P_u versus δ. Since P_i is assumed constant, the curve of P_i versus δ is a horizontal line. The curve of P_u versus δ is a sinusoid. The area to be equal to zero must consist of a positive area A_1, for $P_i > P_u$, and an equal and opposite negative area A_2, for $P_i < P_u$. Hence, the name 'equal area criterion for stability'.

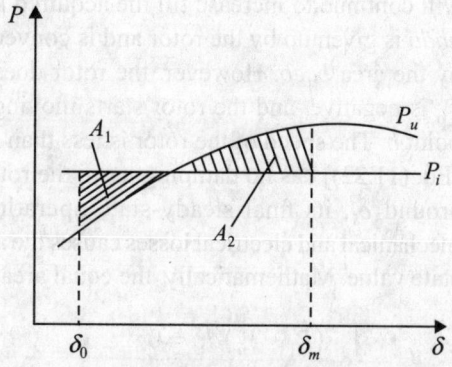

Fig. 11.2 The equal area criterion for stability

The areas A_1 and A_2 may be interpreted in terms of kinetic energy. The work done on a rotating body by a torque T acting through an angle $\delta_0 - \delta_m$ is

$$W = \int_{\delta_0}^{\delta_m} T d\delta \qquad (11.28)$$

and this work increases the kinetic energy of the body. The accelerating power P_a is proportional to the torque, as the speed is assumed to be nearly constant. Hence, the work done on the machine to accelerate it, which appears as kinetic energy, is proportional to area A_1. When the accelerating power becomes negative and the machine is retarded, this kinetic energy is given up; and, when it is all given up, the machine returns to its original speed. This occurs when $A_2 = A_1$. The kinetic energies involved in this explanation are fictitious, being calculated in terms of the relative speed rather than the actual speed.

In the next section, the application of the equal area criterion will be illustrated for a synchronous machine connected to an infinite bus, subjected to disturbances of different types. The machine is assumed to operate initially at the equilibrium point δ_0, with mechanical power input P_{i0}, which equals electrical power output P_{u0}.

11.5.1 Sudden Increase in Power Input

Suppose a step change in power input from P_{i0} to P_{i1} occurs at $t = 0$ as shown on the power angle characteristic given in Fig. 11.3. Due to rotor inertia, the rotor position cannot change instantaneously at $t = 0$, and the rotor relative angular velocity $\Delta\omega = 0$. Thus, $\delta(0^+) = \delta(0^-) = \delta_0$, and $P_u(0^+) = P_u(0^-) = P_{u0}$; where 0^+ and 0^- indicate the instant of time just after and just before the instant $t = 0$. Since, $P_i(0^+) = P_{i1}$, then $P_{i1} > P_u(0^+)$, and the acceleration power $P_a(0^+)$ is positive. The rotor begins to accelerate, the rotor relative angular velocity $\Delta\omega$ increases, and the rotor angle δ increases. When δ reaches δ_1, $P_u = P_{i1}$, $P_a = 0$, but $\Delta\omega$ is positive. The total kinetic energy gained by the rotor at $\delta = \delta_1$ is proportional to the area *abda* (Fig. 11.3), and δ continues to increase. When $\delta > \delta_1$, $P_{i1} < P_u$, P_a is negative, and the rotor decelerates. The kinetic energy gained by the rotor during the accelerating period is transformed to potential energy as the rotor moves from point *b* to point *c*. On reaching point *c*, the rotor loses all its kinetic energy gained during the acceleration, and hence relative speed $\Delta\omega$ becomes zero and the rotor speed is equal to the synchronous speed N_S. In other words, the rotor angle will continue to increase till the acquired kinetic energy represented by the area *abda* is given up by the rotor and is converted to potential energy as represented by the area *bceb*. However, the rotor does not stop at point *c*, since at $\delta = \delta_m$, P_a is negative, and the rotor starts moving backward and again approaches the point *b*. The speed of the rotor is less than N_S and $\Delta\omega < 0$. As the swing equation [Eq. (11.22)] has no damping term, the rotor angle δ would continue to oscillate around δ_1, its final steady-state operating point. However, damping due to mechanical and electrical losses causes the angle δ to stabilize at δ_1, its final steady-state value. Mathematically, the equal area criterion can be expressed as follows:

$$\int_{\delta_1}^{\delta_m} (P_{i1} - P_u)\,d\delta = 0$$

and, for the present case

$$\int_{\delta_0}^{\delta_1} (P_{i1} - P_u)\,d\delta = \text{Area } abda = \text{Area } A_1 \tag{11.29a}$$

$$\int_{\delta_1}^{\delta_m} (P_{i1} - P_u)\,d\delta = \text{Area } bceb = \text{Area } A_2 \tag{11.29b}$$

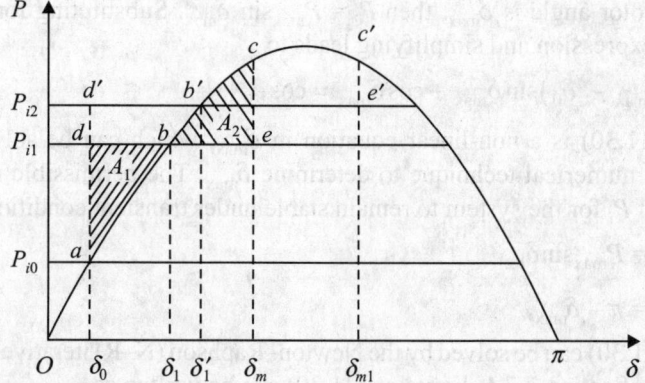

Fig. 11.3   Equal area criterion for sudden increase in power input

The equal area criterion can be employed to determine stability limit P_i for the maximum additional step change in power input to a generator initially having power input P_{i0}. If the prime mover input is increased by shifting the constant input line horizontally upwards from P_{i1} to P_{i2}, then the accelerating area A_1 will be increased, represented by area $ab'd'a$, which equals the increased decelerating area A_2, represented by area $b'e'c'b'$, and the maximum value of δ increases to δ_{m1}. The limit of stability will be reached by increasing the input line upwards to P_{im} such that the maximum value of rotor angle reaches δ_{max}, the intersection of the line P_{im} and the power angle curve for $90° < \delta < 180°$, as shown in Fig. 11.4. It can be observed from Fig. 11.4 that $\delta_{max} = \pi - \delta_1$. If P_i is increased beyond this point, area A_2 will be less than area A_1, and the system becomes unstable.

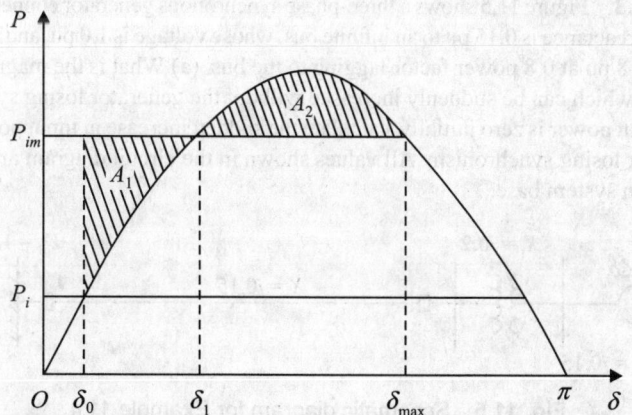

Fig. 11.4   Determination of stability limit for step change in power input

The condition $A_2 = A_1$ in Fig. 11.3 is satisfied mathematically as follows:

$$P_{im} \times (\delta_1 - \delta_0) - \int_{\delta_0}^{\delta_1} P_{max} \sin \delta \, d\delta = \int_{\delta_1}^{\delta_{max}} P_{max} \sin \delta \, d\delta - P_i (\delta_{max} - \delta_1)$$

Integrating the preceding equation and simplifying the result gives

$$(\delta_{max} - \delta_0) P_i = P_{max} (\cos \delta_0 - \cos \delta_{max})$$

When the rotor angle is δ_{max}, then $P_i = P_{max}\sin\delta_{max}$. Substituting for P_i in the preceding expression and simplifying leads to

$$(\delta_{max} - \delta_0)\sin\delta_{max} + \cos\delta_{max} = \cos\delta_0 \qquad (11.30)$$

Equation (11.30) is a non-linear equation in δ_{max}, which can be solved by an appropriate numerical technique to determine δ_{max}. The permissible maximum power input P_i for the system to remain stable under transient condition is

$$P_i = P_{max}\sin\delta_1 \qquad (11.31)$$

where $\delta_1 = \pi - \delta_{max}$ $\qquad (11.32)$

Equation (11.30) can be solved by the Newton–Raphson (N–R) iterative technique described in Section 7.7.1. Equation (11.30) can be written as

$$f(\delta) = \cos\delta_0 - (\delta_{max} - \delta_0)\sin\delta_{max} - \cos\delta_{max} \qquad (11.33)$$

The iterative algorithm suitable for writing a MATLAB function, for obtaining the value of δ_m by the N–R method is expressed as

$$\Delta\delta_m^k = \frac{f(\delta_{max}^k)}{\dfrac{df(\delta_{max}^k)}{d\delta_{max}^k}} \qquad (11.34)$$

$$\delta_{max}^{k+1} = \delta_{max}^k + \Delta\delta_{max}^k \qquad (11.35)$$

where k is the iteration count. The solution of $f(\delta_m)$ is obtained when

$$\left|\delta_{max}^{k+1} - \delta_{max}^k\right| \leq \text{tolerance} \qquad (11.36)$$

Example 11.3 Figure 11.5 shows a three-phase synchronous generator connected through a line whose reactance is 0.15 pu to an infinite bus, whose voltage is 1.0 pu, and is delivering real power 0.8 pu at 0.8 power factor lagging to the bus. (a) What is the magnitude of the power input which can be suddenly increased without the generator losing synchronism? (b) If the input power is zero initially, calculate the sudden increase in input power without the generator losing synchronism. All values shown in the circuit diagram are in per unit on a common system base.

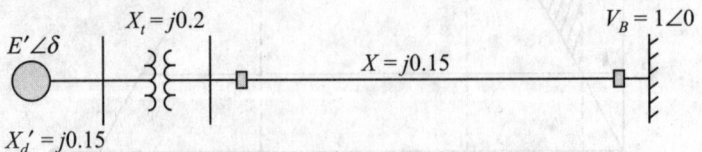

Fig. 11.5 Schematic diagram for Example 11.4

Solution The MATLAB function `pinstab` for determining the transient stability when input power is suddenly increased is described here. The input data required is: voltage of the infinite bus, reactances of the line, transformer, and generator, power factor, and tolerance. The inputs for generator output P_u and the initial estimate of $\ddot{a}_m$ are made during the execution of the function.

```
function [S,pin,deltam,itr,delta1] = pinstab(V,Xl,Xt,Xd,pf,type,tolr);
% Program to compute sudden increase in input to a generator
```

```
% Developed by the authors
% Computation of voltage behind transient reactance
Xtot=Xl+Xt+Xd;
Pu=input ('Generator output power')
phi=acos(pf)*180/pi;
Qu=Pu*tan(phi*pi/180);
if type == 0;
   S=Pu+i*Qu
else
   S=Pu-i*Qu
end
I=conj(S)/conj(V);
Edash=V+I*(i*Xtot);
Edash=abs(Edash);
% Computation of initial rotor angle delta0
delta0=asin(Pu*Xtot/(Edash*V));
% Computation of deltam by N-R method
itr=0;
deltam=input ('initial estimate of deltam')
ndeltam=0;
diff=abs(ndeltam-deltam);
while diff > tolr;
   itr=itr+1;
   fdeltam = cos(delta0)-(deltam-delta0)*sin(deltam)- cos(deltam);
   dfdeltam = (deltam-delta0)*cos(deltam);
   ndeltam = deltam + fdeltam/dfdeltam;
   diff = abs(ndeltam-deltam);
   deltam = ndeltam;
end
delta1 = pi-deltam;
Pin = (Edash*V/Xtot)*sin(delta1)
deltam = deltam*180/pi
delta1 = delta1*180/pi
itr
```

The output from the function pinstab is as follows:

```
≫ V=1.0;Xl=0.15;Xt=0.2;Xd=0.15;pf=0.8;type=0;tolr=0.001;
≫ [S,pin,deltam,itr,delta1] = pinstab(V,Xl,Xt,Xd,pf,type,tolr);
Generator output power0.8
Pu = 0.8000
S = 0.8000 + 0.6000i
initial estimate of deltam120*pi/180 [δ_m is in radians]
deltam = 2.0944
Pin = 2.2093 [Magnitude of the power input without the generator
                loosing synchronism]
deltam = 125.6941 [δ_m is in degrees]
delta1 = 54.3059 [δ_1 is in degrees]
itr = 3
≫ [S,pin,deltam,itr,delta1] = pinstab(V,Xl,Xt,Xd,pf,type,tolr);
```

```
(b) Generator output power0
Pu = 0
S = 0
initial estimate of deltam120*pi/180 [δₘ is in radians]
deltam = 2.0944
Pin = 1.4492 [Magnitude of the power input without the generator
             loosing synchronism]
deltam = 133.5635 [δₘ is in degrees]
delta1 = 46.4365 [δ₁ is in degrees]
itr = 4
```

For further understanding of the application of the equal area criterion, the readers are advised to verify the results graphically.

11.5.2 Sudden Decrease in Power Output due to Three-phase Fault

Consider a synchronous generator operating in steady state with a constant mechanical power input P_i and delivering power to the infinite bus. A three-phase fault of transient nature occurs at the generator bus for a time duration t_c seconds and after the fault clearance, the electrical power output is restored with no disconnection of line.

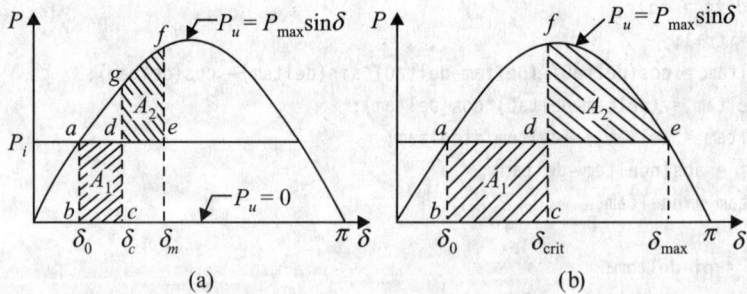

Fig. 11.6 Sudden decrease in power output due to three-phase fault

The power angle characteristic of the generator is shown in Fig. 11.6(a). As losses are neglected, the output power P_u delivered to the infinite bus equals constant input P_i, and the initial power angle is δ_0 is shown at point a. When the fault occurs at time $t = 0$, the power output $P_u = 0$, that is, no power is transmitted to the infinite bus, and the power angle characteristic for the faulted condition corresponds to the horizontal axis. But the input mechanical power is unaltered. Then the accelerating power $P_a = P_i$ = constant, and the machine accelerates. The rotor relative angular velocity $\Delta\omega > 0$, and the rotor angle advances.

When the fault is cleared at the rotor angle δ_c, the power output is $P_u = P_{max} \sin\delta_c$, as indicated by point g in Fig. 11.6(a). At this point, there is a decelerating torque on the rotor, yet the rotor angle δ continues to increase due to the excess kinetic energy in the rotating parts, while the rotor relative angular velocity $\Delta\omega$ continues to be positive. At point f, when the rotor angle δ_m is reached, the accelerating area $abcda$ shown as A_1 equals the decelerating area $defgd$ shown as A_2 in Fig. 11.6(a). The magnitude of $\Delta\omega$ becomes zero at point f, and the rotor angle starts to decrease from point f, as the power output $P_u > P_i$. Finally the machine will oscillate around its steady operating angle δ_0.

The equal area criterion can be employed in this case to determine the critical fault clearing angle δ_{crit}. It can be noted from Fig. 11.6(a) that as fault clearing angle δ_c is increased, the value of δ_m also increases and the stability is maintained as long as the equal area criterion is satisfied. The angle $\delta_c = \delta_{crit}$ when the maximum value of rotor angle δ_m reaches a value δ_{max}, or point e is at the intersection of constant input power line P_i and the electrical power output curve P_u as shown in Fig. 11.6(b). At this point the area A_1 equals area A_2 and the value of $\delta_{max} = \pi - \delta_0$.

From Fig. 11.6(b), applying equal area criterion at critical clearing angle gives

$$A_1 = abcda = P_i\left(\delta_{crit} - \delta_0\right) = A_2 = defd$$

$$= \int_{\delta_{crit}}^{\pi - \delta_0} P_{max} \sin \delta d\delta - P_i\left(\delta_{max} - \delta_{crit}\right)$$

or $\quad P_i\left(\delta_{crit} - \delta_0\right) = P_{max}\left[\cos\delta_{crit} - \cos(\pi - \delta_0)\right] - P_i\left(\pi - \delta_0 - \delta_{crit}\right)$ (11.37)

When the machine is operating synchronously, constant power input $P_i = P_{max} \sin \delta_0$. Substituting the value of P_i in Eq. (11.37) and simplifying gives

$$\cos\delta_{crit} = (\pi - 2\delta_0)\sin\delta_0 - \cos\delta_0 \tag{11.38}$$

It may be noted that to determine the critical fault clearing time t_{crit}, the swing equation has to be solved. For this particular case of three-phase short-circuit when output power $P_u = 0$, the analytical solution for critical clearing time can be obtained. Using Eq. (11.22), the swing equation during fault with $P_u = 0$ may be written as

$$M\frac{d^2\delta}{dt^2} = P_i \tag{11.39}$$

Integrating Eq. (11.39) yields

$$\frac{d\delta}{dt} = \omega = \frac{P_i}{M}t + K \tag{11.40}$$

where K is an integration constant.

At $t = 0$, $\dfrac{d\delta}{dt} = 0$. Hence, $K = 0$. $\tag{11.41}$

Then $\quad \dfrac{d\delta}{dt} = \omega = \dfrac{P_i}{M}t$ $\tag{11.42}$

Again integration of Eq. (11.42) yields

$$\delta = \frac{P_i}{2M}t^2 + K_1 \tag{11.43}$$

At $t = 0$, $\delta = \delta_0$, hence $K_1 = \delta_0$. Therefore, the rotor angle is written as

$$\delta = \frac{P_i}{2M}t^2 + \delta_0 \tag{11.44}$$

The critical fault clearing time t_{crit} may be obtained from Eq. (11.44), taking $\delta = \delta_{crit}$. Thus,

$$\delta_{crit} = \frac{P_i}{2M}t_{crit}^2 + \delta_0 \tag{11.45}$$

and the critical fault clearing time t_{crit} is written as

$$t_{\text{crit}} = \sqrt{\frac{2M(\delta_{\text{crit}} - \delta_0)}{P_i}} \qquad (11.46)$$

Example 11.4 In Example 11.3, a temporary three-phase fault occurs at the sending end of the line which is cleared after five cycles and the line remains intact. Determine whether the generator will lose synchronism or not. (i) If the generator remains stable, calculate the maximum swing of the rotor. (ii) Compute the critical angle and critical clearing time of the fault.

Solution $\phi = -\cos^{-1}(0.8) = -36.87°$

$S = (0.8 + j0.6)$ pu

Current supplied to the infinite bus

$$I = \frac{S^*}{V^*} = \frac{(0.8 - j0.6)}{(1.0 - j0)} = 0.8 - j0.6 = 1.0\angle -36.87° \text{pu}$$

Voltage behind transient reactance = $E' = V + jIX_{\text{tot}}$

$$= (1.0 + j0) + j1.0\angle -36.87 \times (0.5)$$
$$= 1 + 0.5\angle 90° - 36.87° = 1 + 0.5\angle 53.13°$$
$$= 1.30 + j0.4 \text{ pu}$$

$$|E'| = \sqrt{(1.3)^2 + (0.4)^2} = 1.3601 \text{ pu}$$

$$M = \frac{H}{\pi f} = \frac{6.0}{\pi \times 50} = 0.0382 \text{ pu}$$

Assume the rotor angle position to be δ_0 before the disturbance occurs.

$$P_u = \frac{E'V}{X_{\text{tot}}} \sin \delta_0$$

$$\delta_0 = \sin^{-1}\left(\frac{0.8 \times 0.5}{1.3601 \times 1.0}\right) = 0.2985 \text{ rads} = 17.1033°$$

Fault duration = 5 cycles = 0.1 s

(i) Assume that the rotor angle is equal to d when the fault is cleared. Thus,

$$\delta = \frac{P_{i0}}{2M}t^2 + \delta_0 = \frac{0.8}{2 \times 0.0382} \times (0.1)^2 + 0.2985 = 0.1047 + 0.2985 = 0.4032 \text{ rads}$$

Accelerating area $A_1 = P_{i0}(\delta - \delta_0) = 0.8(0.4032 - 0.2985) = 0.0838$

Assume that the maximum rotor position is δ_m after the fault is cleared.

Decelerating area $A_2 = \int_{\delta}^{\delta_m} \frac{1.3601 \times 1.0}{0.5} \sin \delta d\delta - P_{i0}(\delta_m - \delta)$

$$= 2.7202[\cos(0.4032) - \cos \delta_m] - 0.8(\delta_m - 0.4032)$$
$$= 2.5021 - 2.7202 \cos \delta_m - 0.8\delta_m + 0.3226$$

For the machine to remain stable, A_1 should be equal to A_2, that is,

$$2.7202 \cos \delta_m + 0.8\delta_m = 2.8247$$

In order to obtain the value of δ_m, the preceding equation is required to be solved iteratively. Thus, $f(\delta_m) = 2.8247 - 2.7202 \cos \delta_m - 0.8\delta_m$

$$\frac{df(\delta_m)}{d\delta_m} = 2.7202 \sin \delta_m$$

```
function [deltam] = stabnr(tolr);
% Program for solving non-linear eqs for application of equal area criterion
% Program developed by the authors
```

```
itr = 0;
ndeltam = 0;
deltam = input ('initial estimate of deltam')
diff = abs(ndeltam–deltam);
while diff > tolr;
   itr = itr+1;
   fdeltam = 2.7202*cos(deltam)+0.8*deltam–2.8247;
   dfdeltam = 2.7202*sin(deltam);
   ndeltam = deltam + fdeltam/dfdeltam;
   diff = abs(ndeltam–deltam);
   deltam = ndeltam;
end
     deltam=deltam*180/pi
```

The program function [deltam] = stabnr (tolr) is executed as follows:

```
      >>> tolr = 0.001,
      >>> [deltam] = stabnr (tola);
      initial estimate of deltam60*pi/180
      deltam = 1.0472         % input angle in radians
      deltam = 23.2363°       % input angle in degrees
```

Maximum swing of the rotor angle is $\delta_m = 23.2363°$. Since δ_m is less than $(\pi - \delta_0)$, the system will remain stable.

(ii) Assume that the critical clearing angle is δ_c.

Accelerating area $A_1 = P_{i0} (\delta_c - \delta_0) = 0.8(\delta_c - 0.2985)$

Decelerating area $A_2 = \int_{\delta_c}^{\delta_m} \dfrac{1.3601 \times 1.0}{0.5} \sin\delta \, d\delta - P_{i0}(\delta_m - \delta_c)$

Upon integrating and substituting $\delta_m = \pi - \delta_0$ in the preceding equation, the following expression is obtained

$$A_2 = 2.7202[\cos\delta_c + \cos\delta_0] - 0.8(\pi - \delta_0 - \delta_c)$$

$$= 2.7202\cos\delta_c + 2.5999 - 2.2745 + 0.8\delta_c$$

$$= 2.7202\cos\delta_c + 0.8\delta_c - 0.3254$$

For the system to remain stable

$$2.7202\cos\delta_c + 0.8\delta_c - 0.3254 = 0.8 \,(\delta_c - 0.2985)$$

$$2.7202\cos\delta_c = -0.8 \times 0.2985 + 0.32514 \ = 0.0866$$

$$\delta_c = \cos^{-1}\left(-\dfrac{0.0866}{2.7202}\right) = 1.5390 \text{ rad} = 88.1782°$$

The critical clearing time is obtained by using Eq. (11.46) as follows

$$t_{\text{crit}} = \sqrt{\dfrac{2M(\delta_c - \delta_0)}{P_{i0}}} = \sqrt{\dfrac{2 \times 0.0382(1.5390 - 0.2985)}{0.8}}$$

$$= 0.3442 \text{ s} = 17.21 \text{ cycles}$$

11.5.3 Sudden Loss of Partial Transmission Capability

A synchronous generator supplying power to an infinite bus through a double-circuit transmission line is shown in Fig. 11.7. Input power P_{i0} to the generator is assumed constant. Let it be assumed that the infinite bus voltage is $V_B\angle 0°$, the voltage behind transient reactance of the machine is $E' \angle \delta_0 °$. If the reactance of each line is X and the transient reactance of the machine is X'_d, the power output, when the machine is under steady-state operation is given by

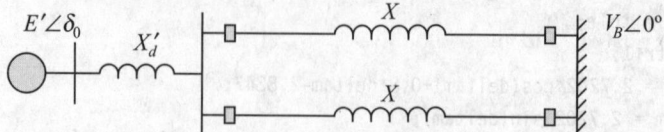

Fig. 11.7 A synchronous machine connected to an infinite bus

$$P_{uA} = P_{\text{max A}} \sin \delta_0 = \frac{|E'| \, |V_B|}{\left(X'_d + \dfrac{X}{2} \right)} \sin \delta_0 \qquad (11.47)$$

When one transmission line is suddenly switched off, the power output of the machine is given by

$$P_{uB} = \frac{|E'| \, |V_B|}{\left(X'_d + X \right)} \sin \delta_0 = P_{\text{max A}} \sin \delta_0 \qquad (11.48)$$

Figure 11.8 shows a plot of the power angle characteristics for different operating conditions. Curves A and B represent the power output for pre-fault condition and one line switched out condition respectively. During pre-fault condition, the power input is P_{i0} and the rotor angle is δ_0 as indicated by point a in Fig. 11.8. When one line is switched off, the operating point shifts to b on curve B. At point b, the rotor relative angular velocity $\Delta\omega = 0$, the constant mechanical input exceeds the electrical power output and the rotor accelerates which increases the rotor angle; and the operating point shifts along the curve B. At point c, $P_{i0} = P_{uB}$, thus $P_a = 0$, $\Delta\omega > 0$, and due to the angular momentum of the moving parts, the rotor angle continues to advance. Beyond b, $P_{i0} < P_{uB}$, and the machine begins to retard and $\Delta\omega$ continues to be greater than zero but its magnitude decreases. At point e, $\Delta\omega = 0$, $P_{i0} < P_{uB}$, and the condition of equal area criterion is met, that is, area $A_1 = abca = $ area $A_2 = cdec$. At point e, the rotor angle starts falling back and travels along the curve B towards point c, and $\Delta\omega$ becomes negative under the influence of the decelerating torque. The rotor will oscillate around the new operating point c the steady-state operating point and finally settle at this angle.

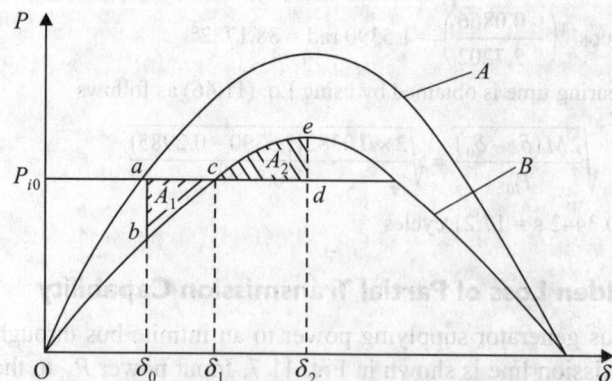

Fig. 11.8 Equal area criterion for partial loss of transmission capability

11.5.4 Three-phase Fault at the Middle of One Line of the Double-circuit Line

For the system shown in Fig. 11.7, assume that the fault occurs at the middle of one line of the double-circuit transmission line. Assume input power P_{i0} to the generator constant and the generator is operating under steady-state condition supplying power to the infinite bus with a power angle δ_0. During the fault, reduced power is transferred from the generator to the infinite bus due to an increase in equivalent transfer reactance. Curves A, B, and C in Fig. 11.9, represent power output P_u versus δ of the generator for pre-fault, post-fault, and during fault conditions, respectively.

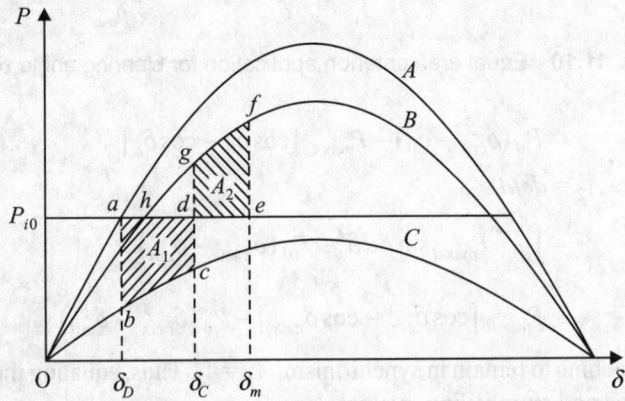

Fig. 11.9 Three-phase fault at the middle of one line of double-circuit line

On the occurrence of the fault, the operating point shifts from point a on the pre-fault curve A to point b on the during fault curve C. At this point $\Delta\omega = 0$, the constant mechanical input exceeds the electrical power output and the rotor accelerates which increases the rotor angle; and the operating point moves towards c along the curve C. If at rotor position δ_c the fault is cleared by opening the faulted line, the operating point c will shift to point g on the post-fault output curve B. At point g, the net power is decelerating and the previously excess kinetic energy stored during accelerating period will be brought down to zero at point f on curve B and $\Delta\omega = 0$. The shaded area A_1 (*abeda*) equals area A_2 (*defgd*). At point f the rotor will fall back due to decelerating power and will travel backward along curve B. The generator rotor will oscillate to the new steady-state operating point h, the point of intersection of P_{i0} and the post-fault power angle curve B.

If fault clearance angle δ_c is delayed, the rotor reaches critical clearing angle δ_{crit} and any further increase in δ_c will make the accelerating area A_1 greater than the decelerating area A_2. Figure 11.10 shows the application of the equal area criterion for determining the critical clearing angle.

It is observed from Fig. 11.10 that δ_{crit} should be so positioned between δ_0 and δ_{max}, so that area A_2 is equal to area A_1, where $\delta_{max} = \pi - \delta_0$.

Area $A_1 = abcda$

$$= P_{i0}\left(\delta_{crit} - \delta_0\right) - \int_{\delta_0}^{\delta_c} P_{\max C}\sin\delta\,d\delta$$

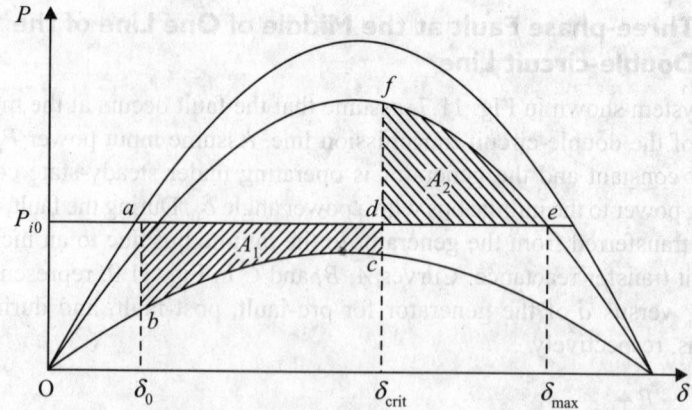

Fig. 11.10 Equal area criterion application for clearing angle δ_{crit}

$$= P_{i0}\left(\delta_{\text{crit}} - \delta_0\right) - P_{\max C}\left[\cos\delta_0 - \cos\delta_c\right]$$

Area $A_2 = defd$

$$= \int_{\delta_{\text{crit}}}^{\delta_{\max}} P_{\max B}\sin\delta d\delta - P_{i0}\left(\delta_{\max} - \delta_{\text{crit}}\right)$$

$$= P_{\max B}\left[\cos\delta_{\text{crit}} - \cos\delta_{\max}\right] - P_{i0}\left(\delta_{\max} - \delta_{\text{crit}}\right)$$

For the machine to remain in synchronism, $A_1 = A_2$. Thus, equating the preceding two equations and simplifying, we get

$$\cos\delta_{\text{crit}} = \frac{P_{i0}\left(\delta_{\max} - \delta_0\right) + P_{\max B}\cos\delta_{\max} - P_{\max C}\cos\delta_0}{P_{\max B} - P_{\max C}} \tag{11.49}$$

Example 11.5 A three-phase, 50-Hz, synchronous generator is delivering 0.9 pu real power to an infinite bus via the transmission circuit shown in Fig. 11.11(a). All values shown in the circuit diagram are in per unit on a common system base. A temporary three-phase fault occurs in the middle of line 2. Determine the rotor angle position before the fault occurs. Also compute the critical clearing angle if the fault is cleared by opening the faulted line. Assume $H = 4.5$ MJ/MVA.

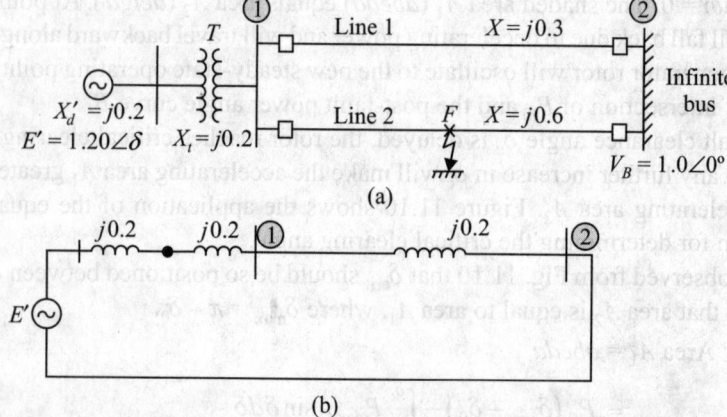

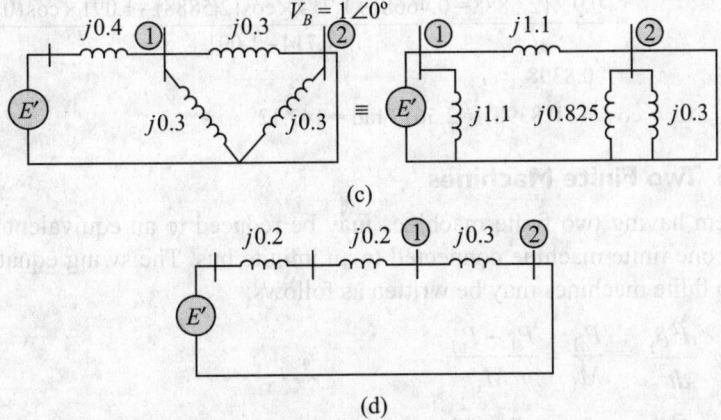

(c)

(d)

Fig. 11.11 (a) Circuit diagram; (b) Pre-fault equivalent circuit;
(c) Equivalent circuit during fault; (d) Post-fault equivalent circuit

Solution *Pre-fault condition* The equivalent circuit for the pre-fault condition is shown in Fig. 11.11(b).

Transfer reactance $X = 0.2 + 0.2 + \dfrac{0.3 \times 0.6}{0.9} = 0.6$ pu

Input power P_{i0} = output power P_u = 0.9 pu

Rotor angle $\delta_0 = \sin^{-1}\left(\dfrac{0.9 \times 0.6}{1.2 \times 1.0}\right) = 26.74° = 0.4668$ rads

Maximum power $P_{maxA} = \dfrac{1.2 \times 1.0}{0.6} = 2.0$ pu

Fault condition The equivalent circuit for the faulted condition is shown in Fig. 11.11(c).

The transfer reactance is obtained by the star–delta transformation in the following manner:

$$X = \frac{0.4 \times 0.3 + 0.3 \times 0.3 + 0.3 \times 0.4}{0.3} = 1.1 \text{ pu}$$

Maximum power $P_{maxC} = \dfrac{1.2 \times 1.0}{1.1} = 1.091$ pu

Post-fault condition The equivalent circuit for the post-fault condition is shown in Fig. 11.11(d).

Transfer reactance $X = 0.2 + 0.2 + 0.3 = 0.7$ pu

Maximum power $P_{maxB} = \dfrac{1.2 \times 1.0}{0.7} = 1.714$ pu

$$\delta_{max} = \left[180° - \sin^{-1}\left(\frac{P_{i0}}{P_{maxB}}\right)\right] = \left[180° - \sin^{-1}\left(\frac{0.9}{1.714}\right)\right]$$

$$= (180° - 31.67°) = 148.33° = 2.5888 \text{ rads}$$

As per Eq. (11.49),

$$\cos\delta_{crit} = \frac{P_{i0}(\delta_{max} - \delta_0) + P_{maxB}\cos\delta_{max} - P_{maxC}\cos\delta_0}{P_{maxB} - P_{maxC}}$$

$$= \frac{0.9 \times (2.5888 - 0.4668) + 1.714 \times \cos(2.5888) - 1.091 \times \cos(0.4668)}{1.714 - 1.091}$$

$$= -0.8398$$

$$\delta_{crit} = \cos^{-1}(-0.8398) = 2.5677 \text{ rad} = 147.12°$$

11.5.5 Two Finite Machines

A system having two finite machines may be reduced to an equivalent system having one finite machine connected to an infinite bus. The swing equations of the two finite machines may be written as follows:

$$\frac{d^2\delta_1}{dt^2} = \frac{P_{a1}}{M_1} = \frac{P_{i1} - P_{u1}}{M_1} \tag{11.50}$$

$$\frac{d^2\delta_2}{dt^2} = \frac{P_{a2}}{M_2} = \frac{P_{i2} - P_{u2}}{M_2} \tag{11.51}$$

Subtracting Eq. (11.51) from Eq. (11.50) gives

$$\frac{d^2(\delta_1 - \delta_2)}{dt^2} = \frac{d^2\delta_{12}}{dt^2} = \frac{P_{a1}}{M_1} - \frac{P_{a2}}{M_2} \tag{11.52}$$

where $\delta_{12} = \delta_1 - \delta_2$.

Multiplying both the sides of Eq. (11.52) by $\dfrac{M_1 M_2}{M_1 + M_2}$ gives

$$\frac{M_1 M_2}{M_1 + M_2} \frac{d^2\delta_{12}}{dt^2} = \frac{M_2 P_{a1} - M_1 P_{a2}}{M_1 + M_2}$$

$$= \frac{M_2 P_{i1} - M_1 P_{i2}}{M_1 + M_2} - \frac{M_2 P_{u1} - M_1 P_{u2}}{M_1 + M_2} \tag{11.53}$$

which in compact form may be written as

$$M \frac{d^2\delta_{12}}{dt} = P_a = P_i - P_u \tag{11.54}$$

Equation (11.54) is identical to the swing equation of a single machine connected to the infinite bus where the equivalent input P_i, the equivalent output P_u, and the equivalent inertia constant are given by

$$P_i = \frac{M_2 P_{i1} - M_1 P_{i2}}{M_1 + M_2} \tag{11.55}$$

$$P_u = \frac{M_2 P_{u1} - M_1 P_{u2}}{M_1 + M_2} \tag{11.56}$$

$$M = \frac{M_1 M_2}{M_1 + M_2} \tag{11.57}$$

11.6 SOLUTION OF SWING EQUATION

The solution of a swing equation determines δ versus t for a synchronous generator. A graph of the solution is known as a swing curve. Inspection of swing curves of all the machines of a system will show whether the machines will remain in synchronism after a disturbance. The computations of swing curves are carried over a period long enough to determine whether δ will increase without limit or reach a maximum and then start to decrease.

A number of different methods are available for the numerical evaluation of second-order differential equations in step-by-step computations for small increments of the independent variable. The more elaborate methods are practical only when the computations are performed on a digital computer. The step-by-step method used for hand calculation is necessarily simpler than some of the methods recommended for digital computers. For multi-machine systems, numerical methods suitable for digital computation are used.

11.6.1 Step-by-step Method

In the step-by-step method for hand calculation, the change in the angular position of the rotor during a short interval of time is computed by making the following assumptions:

1. The accelerating power P_a computed at the beginning of an interval is constant from the middle of the preceding interval to the middle of the interval considered.
2. The angular velocity is constant throughout any interval at the value computed for the middle of the interval. However neither of the assumptions are true, since δ is changing continuously and both P_a and angular velocity ω are functions of δ. As the time interval is decreased, the computed swing curve approaches the true curve. Figure 11.12 will help in visualizing the assumptions.

In this method, the solution of the swing equation is obtained at distinct uniform intervals of time Δt which is also referred to as the step length.

Consider the nth interval which begins at $t = (n-1)\Delta t$. The angular position at this instant is δ_{n-1}. The acceleration α_{n-1} at this instant is assumed to be constant from $t = (n-3/2)\Delta t$ to $t = (n-1/2)\Delta t$ as shown in Fig. 11.12(a). During this time period, the change in speed that occurs is computed as

$$\Delta\omega_{n-1/2} = \omega_{n-1/2} - \omega_{n-3/2} = \frac{d^2\delta}{dt^2}\Delta t = \frac{\Delta t}{M}P_{a\,(n-1)} \qquad (11.58)$$

Then the speed at the end of the time period is

$$\omega_{n-1/2} = \omega_{n-3/2} + \Delta\omega_{n-1/2} \qquad (11.59)$$

Substituting Eq. (11.58) in Eq. (11.59) yields

$$\omega_{n-1/2} = \omega_{n-3/2} + \frac{\Delta t}{M}P_{a\,(n-1)} \qquad (11.60)$$

Thus the change in speed is assumed to occur as a step at the middle of the time period, that is, at $t = (n-1)\Delta t$, which is the same instant when the acceleration was computed as shown in Fig. 11.12(b). From $t = (n-1)\Delta t$ to $t = n\Delta t$, that is, the nth interval, the speed will be constant at the value $\omega_{n-1/2}$. The change in δ during the nth interval is

$$\Delta\delta_n = \Delta t \times \omega_{n-1/2} = \Delta t \left[\omega_{n-3/2} + \frac{\Delta t}{M}P_{a\,(n-1)}\right] \qquad (11.61)$$

By analogy to Eq. (11.61), the change in δ during the $(n-1)$th interval is

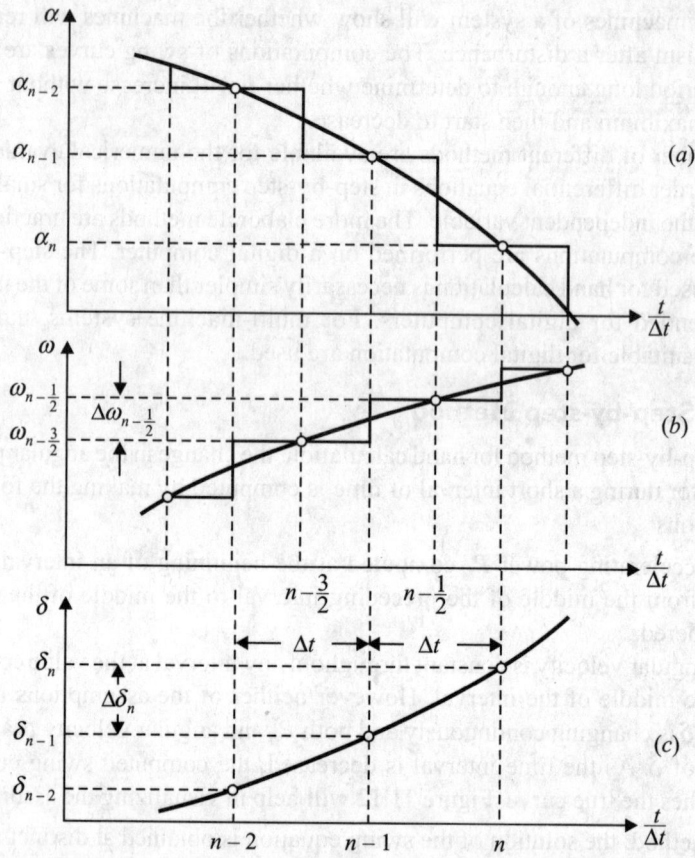

Fig. 11.12 Plot of acceleration, speed, and angular position versus time; step-by-step method of solution of the swing equation

$$\Delta\delta_{n-1} = \Delta t \times \omega_{n-3/2} \tag{11.62}$$

Substituting Eq. (11.62) in Eq. (11.61) gives

$$\Delta\delta_n = \Delta\delta_{n-1} + \frac{(\Delta t)^2}{M} P_{a(n-1)} = \Delta\delta_{n-1} + k P_{a(n-1)} \tag{11.63}$$

and the angular position at the end of the *n*th interval, as shown in Fig. 11.12(c), is

$$\delta_n = \delta_{n-1} + \Delta\delta_n \tag{11.64}$$

where $\quad k = \dfrac{(\Delta t)^2}{M} \tag{11.65}$

Equations (11.63) and (11.64) are repeatedly used to obtain the rotor angle against time by selecting a suitable time step Δt. An interval of 0.05 s is usually satisfactory. At the instant when a fault occurs, it is assumed that time $t = 0$. Just before the fault occurs, that is, at $t = 0^-$ s, the accelerating power $P_{a0^-} = 0$, since the machine is operating in synchronism and the rotor angle is the initial steady-state rotor position δ_0. At $t = 0^+$ immediately after the fault has occurred, if the accelerating power is denoted by P_{a0^+}, the average accelerating power at the beginning of the interval ($t = 0$) is $\dfrac{P_{a0^+}}{2}$.

Similarly, for representing any change in switching condition during the stability analysis, the accelerating power P_a to be used is the average of the accelerating power just before and just after the change in switching condition.

Example 11.6 In Example 11.5, the three-phase fault at the middle of the line is cleared in 0.2 s. Compute the swing curve using the step-by-step method.

Solution Inertia constant $M = \dfrac{H}{180 f} = \dfrac{4.5}{180 \times 50} = 0.0005$ pu

Initial conditions

Power input $P_u = 0.9$ pu

Power output $= P_{maxA} \sin\delta_0 = 2.0 \sin\delta_0 = P_u$

Rotor angle $\delta_0 = \sin^{-1}\left(\dfrac{0.9}{2.0}\right) = 26.74°$

During fault condition

$$P_u = 1.091 \sin \delta; \text{ then } P_{maxC} = 1.091$$

Fault cleared condition

$$P_u = 1.714 \sin \delta; \text{ then } P_{maxB} = 1.714$$

Take the time interval as $\Delta t = 0.05$ s. The steps of computation for each point are as follows:

$$P_{a(n-1)} = P_i - P_{u(n-1)} = -P_{u(n-1)} \text{ pu power}$$

$$\frac{(\Delta t)^2}{M} P_{a(n-1)} = \frac{(0.05)^2}{0.0005} P_{a(n-1)} = 5 P_{a(n-1)} \text{ elec. degrees}$$

$$\Delta\delta_n = \Delta\delta_{n-1} + 5 P_{a(n-1)} \text{ elec. degrees}$$

$$\delta_n = \delta_{n-1} - \Delta\delta_n \text{ elec. degrees}$$

The step-by-step computation of the swing curve is shown in Table 11.1.

Tabe 11.1 Step-by-step computation of the swing curve

t (sec)	P_{max} (pu)	$\sin\delta$	P_u (pu)	P_a (pu)	$5 P_{a(n-1)}$ (° elect)	$\Delta\delta$ (° elect)	δ (° elect)
0^-	2.00	0.450	0.900	0.000			26.74
0^+	1.091	0.450	0.491	0.409			26.74
0 avg	1.5455			0.2045	1.022		
						1.0225	
0.05	1.091	0.466	0.508	0.392	1.96		27.76
						2.98	
0.10	1.091	0.511	0.557	0.343	1.715		30.74
						4.695	
0.15	1.091	0.580	0.6325	0.2675	1.337		35.43
						6.0325	
0.20^-	1.091	0.662	0.722	0.178			41.46
0.02^+	1.714	0.662	1.135	−0.235			41.46
0.20 avg				0.815	−0.0285	−0.1425	
						5.89	
0.25	1.714	0.736	1.261	−0.361	−1.805		47.35
						4.085	

(Contd)

Table 11.1 (*Contd*)

t (sec)	P_{max} (pu)	$\sin\delta$	P_u (pu)	P_a (pu)	$5P_{a(n-1)}$ (° elect)	$\Delta\delta$ (° elect)	δ (° elect)
0.35	1.714	0.782	1.340	−0.440	−2.20		51.43
						1.885	
0.40	1.714	0.802	1.375	−0.475	−2.375		53.31
						−0.49	
0.45	1.714	0.797	1.366	−0.466	−2.33		52.82
						−2.82	
0.50	1.714	0.766	1.313	−0.413	−2.065		50.00
						−4.88	
0.55	1.714	0.709	1.215	−0.315	−1.575		45.12
						−6.455	
0.60	1.714	0.625	1.071	−0.171	−0.855		38.67
						−5.6	
0.65	1.714	0.546	0.936	−0.036	−0.180		33.07
						−5.78	
0.70	1.714	0.458	0.785	0.115	0.575		27.29
						−5.21	
0.75	1.714	0.376	0.644	0.256	1.280		22.08
						−3.93	
0.80	1.714	0.311	0.533	0.367	1.835		18.15
						−2.10	
0.85	1.714	0.276	0.473	0.427	2.135		16.05
						0.035	
0.90	1.714	0.276	0.473	0.427	2.135		16.02
						2.17	
0.95	1.714	0.312	0.535	0.465	2.325		18.19
						4.49	
1.00	1.714	0.386	0.661	0.339	1.70		22.68
						6.19	
1.05	1.714	0.483	0.828	0.072	0.36		28.87
						6.55	
1.10	1.714	0.58	0.994	−0.094	0.47		35.42
						7.02	
1.05							42.44

11.6.2 Modified Euler Method

There are a large number of integration algorithms out of which the modified Euler method possesses inherent simplicity and requires less computation per step time interval. The second-order swing equation of a machine is written as two first-order differential equations in the following form:

$$\frac{d\delta}{dt} = \omega - \omega_s = \Delta\omega \tag{11.66a}$$

$$\frac{d\Delta\omega}{dt} = \frac{1}{M}(P_i - P_u) = \frac{1}{M}P_a \tag{11.66b}$$

where $\omega_s = 2\pi f$ = synchronous speed

In the application of the modified Euler method, the first estimates of rotor angle δ and machine relative speed $\Delta\omega$ at time $(t + \Delta t)$ are obtained by using the derivatives at the beginning of the step as

$$\delta_{i+1}^{f} = \delta_i + \left. \frac{d\delta}{dt} \right|_{\Delta\omega_i} \Delta t$$

$$\Delta\omega_{i+1}^{f} = \Delta\omega_i + \left. \frac{d\Delta\omega}{dt} \right|_{\delta_i} \Delta t \qquad (11.67)$$

The first estimates of δ_{i+1}^{f} and $\Delta\omega_{i+1}^{f}$ calculated from Eq. (11.67) are now used to determine the derivatives at the end of the step as shown here

$$\left. \frac{d\delta}{dt} \right|_{\Delta\omega_{i+1}^{f}} = \Delta\omega_{i+1}^{f}$$

$$\left. \frac{d(\Delta\omega)dt}{dt} \right|_{\delta_{i+1}^{f}} = \left. \frac{1}{M} P_a \right|_{\delta_{i+1}^{f}} \qquad (11.68)$$

The average values of the derivatives are used to compute the new corrective values of the rotor angle and the rotor velocity as follows:

$$\delta_{i+1}^{n} = \delta_i + \frac{\left. \dfrac{d\delta}{dt} \right|_{\Delta\omega_i} + \left. \dfrac{d\delta}{dt} \right|_{\Delta\omega_{i+1}^{f}}}{2} \Delta t$$

$$\Delta\omega_{i+1}^{n} = \Delta\omega_i + \frac{\left. \dfrac{d(\Delta\omega)}{dt} \right|_{\delta_i} + \left. \dfrac{d(\Delta\omega)}{dt} \right|_{\delta_{i+1}^{f}}}{2} \Delta t \qquad (11.69)$$

where superscripts f and n refer to, respectively, the first estimate and new corrected values of the variables.

11.6.3 MATLAB Functions for Solution of Non-linear Differential Equations

MATLAB provides a very useful collection of ordinary differential equation (ODE) solvers. The functions, called ode23 and ode45, respectively employ the second/third and fourth/fifth order Runge–Kutta numerical techniques for the solution of non-linear differential equations.

The following four steps demonstrate the procedure for the use of ode23 or ode45 for obtaining the solution of a differential equation.

Step 1: Introduce new variables to formulate all ODEs of order two and above as first-order ODEs. The original ODEs are recast in terms of the new variables as first-order ODEs. In vector form, the equations are written as

$$\begin{bmatrix} \dot{x}_1 \\ \dot{x}_2 \\ \vdots \\ \dot{x}_n \end{bmatrix} = \begin{bmatrix} f_1(x_1, x_2, \ldots, x_n, t) \\ f_2(x_1, x_2, \ldots, x_n, t) \\ \vdots \\ f_n(x_1, x_2, \ldots, x_n, t) \end{bmatrix} \qquad (11.70)$$

or $\qquad \dot{x} = f(x,t)$ $\hfill$ (11.71)

where $x = [x_1, x_2, \ldots, x_n]^t$

Step 2: From the given input (x, t) write a MATLAB function to compute $f_1, f_2, \ldots, f_n$. Since x is a column vector, that is, $x = [x_1, x_2, \ldots, x_n]^t$ care must be taken to ensure that the MATLAB function also returns the derivative of x as a column vector.

Step 3: Use ode23 or ode45 to solve the MATLAB function developed in *Step 2*. The developed MATLAB function is used as an input for ode23 or ode45 as the case maybe. The syntax for using ode23 or ode45 is the same and is explained in the following examples.

Step 4: The solution output matrix contains n columns which correspond to the n number of ordinary differential equations whose solution is being sought. It is essential to know which variable corresponds to which column for an analysis of the problem. Alternately, the output variables can be plotted against the independent time variable t.

Compared to the power of the ODE solvers (functions ode 23 and ode 45) to provide quick solutions of the swing equations and their utility in power system studies, other numerical techniques pale into insignificance. The utility of ode23 and ode45 is demonstrated with the help of the following examples.

Example 11.7 Determine the current in the RL circuit shown in Fig. 11.13 when the switch S is closed at $t = 0$ s. The applied voltage $e(t)$ varies as follows:

$$e(t) = 8t \qquad 0 \le t \le 0.25$$

$$e(t) = 2.0 \qquad t > 0.25$$

Assume that the current in the circuit is zero when the switch S is closed at $t = 0$ s. All other parameters are shown in the circuit diagram given in Fig. 11.13.

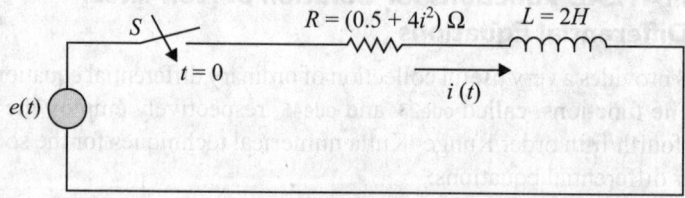

Fig. 11.13 Circuit diagram for Example 11.7

Solution Step 1: From the circuit diagram, the first-order differential equation for current is written by applying the laws of circuit analysis as follows:

At any time t, the current in the circuit is given by the following differential equation

$$L\frac{di}{dt} + Ri = e(t) \qquad (11.7.1)$$

Substituting the values for R and L in Eq. (11.7.1) yields

$$2\frac{di}{dt} + \left(0.5 + 4i^2\right) i = e\ (t)$$

or
$$\frac{di}{dt} = 0.5e(t) - \left(0.25 + 2i^2\right)i \qquad (11.7.2)$$

Equation (11.7.2) is already a first-order equation.

Step 2: Equation (11.7.2) is the given function in (i, t) and the MATLAB function rlckt given below computes i. The function rlckt is saved as an M-file named **rlckt.m**.

```
function idot = rlckt(t,i);
if t <= 0.25;
  e=4*t;
else
  e=1.0;
end
idot=e-(0.25+2*i^2)*i;
```

Step 3: The following commands in the command window help to solve the differential equation for the given *RL* circuit.

```
>> tspan=[0 1];    % specifies the time period; t =0 to t = 1.0 sec.
>> i0=0;           % specifies the initial value of current; i = 0 at t = 0
>> [t,i]=ode23('rlckt',tspan,i0)
            % rlckt is the function in (t, i) which contains the ODEs to
            % be computed. Note (' ') within which the function name
            % is enclosed. On execution, the ode23 returns the
            % following output.
```

t =	i =
0	0
0.0792	0.0125
0.1584	0.0495
0.2110	0.0875
0.2416	0.1143
0.2722	0.1435
0.3029	0.1728
0.4029	0.2651
0.5029	0.3514
0.6029	0.4295
0.7029	0.4978
0.8029	0.5552
0.9029	0.6018
1.0000	0.6374

```
>> plot(t,i,'k-');                          % plots t vs. i
>> xlabel('Time t in secs.'),ylabel('Current i in amperes');   % Labels the axes
>> hold;                                    % Invokes hold-on for overlay plots.
Current plot held
>> [t,i]=ode45('rlckt',tspan,i0);
>> plot(t,i,'k*');
>> plotedit;          % Invokes the edit command for modifying and annotating
                      % the plotted figures.
```

Step 4: The variation of current obtained as an output in *step 3* is plotted in Fig. 11.14.

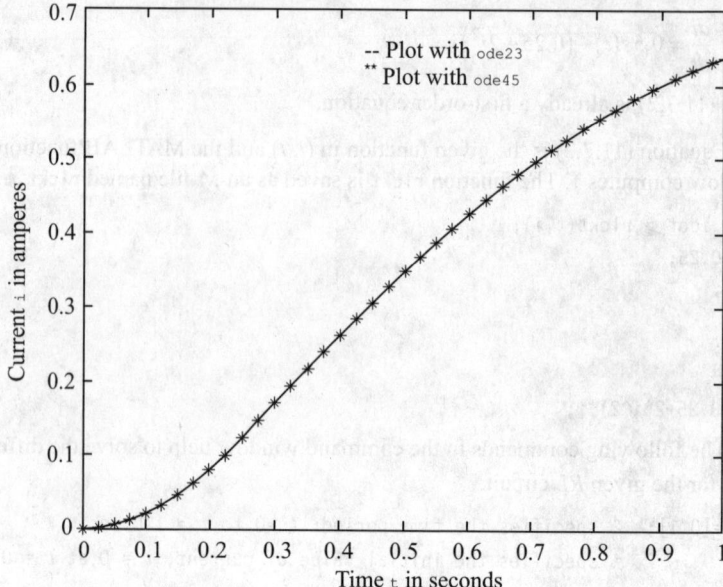

Fig. 11.14 Plot of current versus time for Example 11.8

Example 11.8 Plot the variation in rotor angle and speed versus time of the machine in Example 11.6, if the fault is cleared in 0.2 s by opening the line.

Solution From Example 11.6, the following data may be utilized to plot variation in rotor angle and speed of the machine:

Initial rotor angle $\delta_0 = 26.74°$

Inertia constant $M = 0.02865$ pu

Maximum power output during fault condition $P_m = 1.091$ pu

Maximum power output after fault is cleared $P_m = 1.714$ pu

Step 1: Equation (11.13) which represents the swing equation of a synchronous machine is a second-order differential equation. The MATLAB **ode** solver requires that the second-order equation be reduced to first-order equations. Thus Eq. (11.13) is written as two first-order differential equations in the following manner:

$$\omega = \frac{d\delta}{dt} = \dot{\delta}$$

$$\omega = \frac{d\omega}{dt} = \frac{d^2\delta}{dt^2} = \frac{P_m \sin\delta}{M}$$

In matrix form, the equations are written as

$$\begin{bmatrix} \dot{\delta} \\ \ddot{\delta} \end{bmatrix} = \begin{bmatrix} \omega \\ \dfrac{P_m \sin\delta}{M} \end{bmatrix}$$

Step 2: It is now required to develop a MATLAB function which would compute δ from the given vector input against time. The MATLAB function deltadot computes δ. It is written and saved as rtrangl.m file.

```
function deltadot = rtrangl(t,delta);          % Input data is time and
                                               % [δ; δ̇ (ω)]
M=0.02865;      % Inputs the value of inertia constant
```

```
if t <= 0.2;
    Pm=1.091;
else
    Pm=1.714;
End
deltadot=[delta(2);(0.9-Pm*sin(delta(1)))/M];
```

Step 3: In order to obtain a solution with the help of ode23 or ode45 and plot the output, it is necessary to write a set of commands. These commands can be written in a script file (as in Example 11.7) or as a MATLAB function. The MATLAB function swingcurve is written and has been saved as an M-file named pwrang1.m.

```
function swingcurve = pwrang1(tspan,delta0);   % Input time period and
                                               % initial rotor angle δ₀
[t,delta]=ode23('rtrang1',tspan,delta0);       % Execute ode23 solver
x=delta(:,1);y=delta(:,2);        % x = first column of delta, y = second
                                  % column of delta
plot(t,x,'k.');                   % Plots t versus x
xlabel('Time t in sec.');
ylabel('Rotor angle in degrees');
title ('Plot of rotor angle versus time');
figure(2)                         % Opens a new window for the figure
plot(t,y,'k-');                   % Plots t versus y
xlabel('Time t in sec.');
ylabel('Rotor speed in radians/sec.');
title ('Plot of rotor speed versus time');
```

The program is executed by calling the function swingcurve which is saved in the file pwrang1.m as follows:

```
>> tspan=[0 2.0];                 % Define the time span over which rotor
                                  % angle and speed is to be plotted.
>> delta0=[0.4668;0];             % Inputs pre-fault conditions
>> swingcurve = pwrang1(tspan,delta0);
```

Step 4: Figure 11.5 shows the variation of rotor angle and speed of the synchronous generator with respect to time.

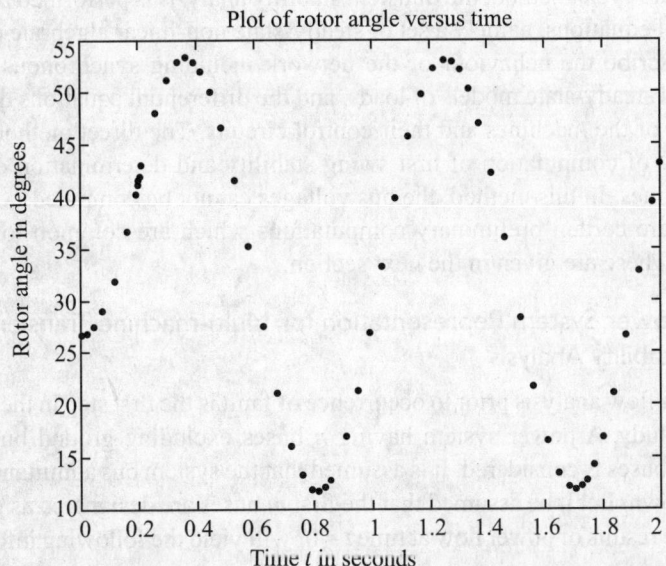

Plot of rotor angle versus time

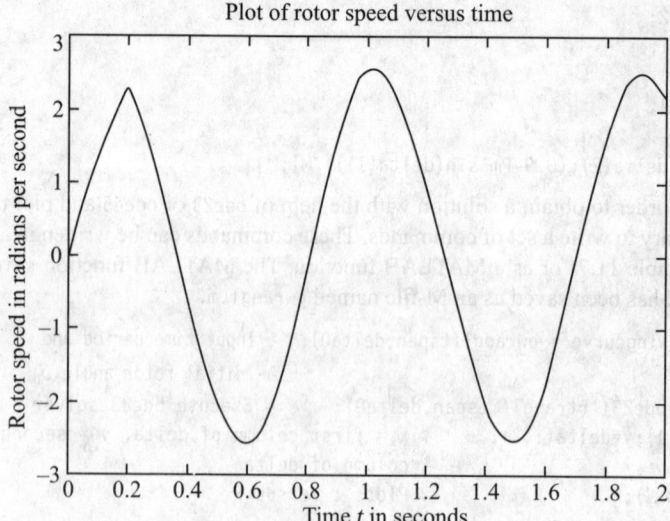

Fig. 11.15   Plot of δ and $\Delta\omega$ versus time

11.7 TRANSIENT STABILITY ANALYSES OF MULTI-MACHINE POWER SYSTEMS

The transient stability analysis of a multi-machine system requires the solution of the swing equations for all the generators of the system. The numerical integration methods discussed in Section 11.6 can be used to solve the swing equations. However, power outputs computations of the generators are to be carried out from the solution of algebraic network equations.

There are two methods for solving the transient stability problem of multi-machine systems: (i) alternate cycle solution of the network equations and the swing equations; (ii) the direct method involving the solution of swing equations. In the alternate cycle method, the transient stability analysis is performed by solving two sets of equations, namely a set of steady-state non-linear algebraic equations which describe the behaviour of the network including synchronous machine model and steady-state models of loads, and the differential equations describing the swing of the machines and their control circuits. The direct method offers a quick way of computation of first swing stability and determination of critical clearing times. In this method, the bus voltages cannot be computed.

There are certain preliminary computations which are common to both the methods. These are given in the next section.

11.7.1 Power System Representation for Multi-machine Transient Stability Analysis

The power flow analysis prior to occurrence of fault is the first step in the transient stability study. A power system having n buses excluding ground bus with m generator buses is considered. It is assumed that the system bus admittance matrix Y_{bus} is known. Let it be assumed that the first m buses are designated as generator buses. The results of power flow at time $t = 0^-$ will yield the following information:

$$P_{Li}, Q_{Li} \qquad\qquad \text{for } i = 1, 2, ..., n$$
$$P_{Gi}, Q_{Gi} \qquad\qquad \text{for } i = 1, 2, ..., m$$
$$V_i = |V_i| \angle \psi_i \qquad\qquad \text{for } i = 1, 2, ..., n$$

where P, Q are active and reactive powers, respectively, at the buses; subscripts G and L refer respectively to generation and load; V is the voltage phasor; and ψ the phase angle of the bus voltage.

The loads are converted into equivalent admittance y_{Li} between ith bus and ground bus as

$$y_{Li} = \frac{P_{Li} - jQ_{Li}}{|V_i|^2} \qquad \text{for } i = 1, 2, ..., n \qquad\qquad (11.72)$$

With the addition of elements y_{Li} the Y_{bus} gets modified as

$$Y_{\text{bus, mod}} = Y_{\text{bus}} + \text{diag}\,(y_{Li}) \qquad \text{for } i = 1, 2, ..., n \qquad\qquad (11.73)$$

Thus only the diagonal elements of Y_{bus} get modified.

All the generator voltages behind the direct axis transient reactances can then be obtained as

$$E'_i = V_i + jX'_{di} \frac{P_{Gi} - jQ_{Gi}}{V_i^*} = |E'_i| \angle \delta_i \quad \text{for } i = 1, 2, ..., m \qquad (11.74)$$

where $|E'_i|$ is the voltage behind transient reactance of ith machine, and δ_i is its phase angle, and X'_{di} is the sub-transient reactance of the machine.

A power system for transient stability study with all loads converted into admittances and the voltage behind transient reactances is represented in Fig. 11.16.

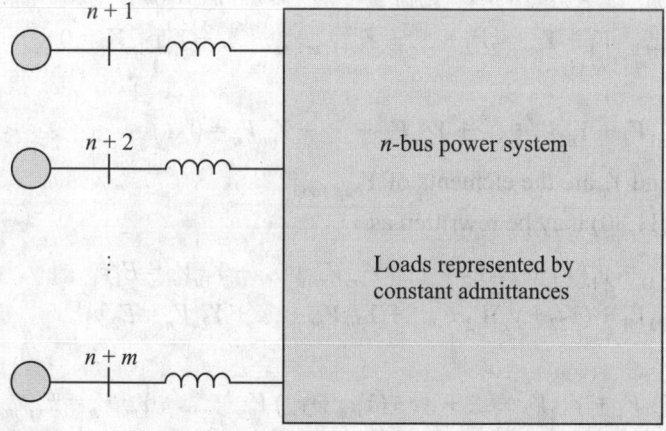

Fig. 11.16 Representation of a power system for transient stability study

11.7.2 Alternate Cycle Solution Method

In this method, the network equations are solved to compute bus voltages during various operating conditions such as faulted conditions and post-fault conditions.

$$I_{\text{bus}} = Y_{\text{bus, mod}}\, V_{\text{bus}} \qquad\qquad (11.75)$$

that is,

$$
\begin{bmatrix} I_1 \\ I_2 \\ \vdots \\ I_m \\ I_{m+1} \\ \vdots \\ I_n \end{bmatrix} = \boldsymbol{Y}_{\text{bus, mod}} \begin{bmatrix} V_1 \\ V_2 \\ \vdots \\ V_m \\ V_{m+1} \\ \vdots \\ V_n \end{bmatrix} \tag{11.76}
$$

The injected currents at the generator buses are given by

$$
I_i = \frac{\left(E_i' - V_i\right)}{jX_{di}'} = \left(E_i' - V_i\right)y_i \qquad \text{for } i = 1, 2, \ldots, m \tag{11.77}
$$

$$
I_i = 0 \qquad\qquad \text{for } i = m+1, m+2, \ldots, n \tag{11.78}
$$

where

$$
y_i = \frac{1}{jX_{di}'} \qquad\qquad \text{for } i = 1, 2, \ldots, m \tag{11.79}
$$

In expanded form Eq. (11.75) can be written as

$$
\begin{aligned}
Y_{11}V_1 + Y_{12}V_2 + \ldots + Y_{1m}V_m + \ldots + Y_{1n}V_n &= (E_1' - V_1)\,y_1 \\
Y_{21}V_1 + Y_{22}V_2 + \ldots + Y_{2m}V_m + \ldots + Y_{2n}V_n &= (E_2' - V_2)\,y_2 \\
\vdots \\
Y_{m1}V_1 + Y_{m2}V_2 + \ldots + Y_{mm}V_m + \ldots + Y_{mn}V_n &= (E_m' - V_m)y_m \\
Y_{m+1,1}V_1 + Y_{m+1,2}V_2 + \ldots + Y_{m+1,m}V_m + \ldots + Y_{m+1n}V_n &= 0 \\
\vdots \\
Y_{n1}V_1 + Y_{n2}V_2 + \ldots + Y_{nm}V_m + \ldots + Y_{nn}V_n &= 0
\end{aligned} \tag{11.80}
$$

where Y_{ii} and Y_{ij} are the elements of $\boldsymbol{Y}_{\text{bus, mod}}$.

Equation (11.80) may be rewritten as

$$
\begin{aligned}
(Y_{11} + y_1)\,V_1 + Y_{12}V_2 + \ldots + Y_{1m}V_m + \ldots + Y_{1n}V_n &= E_1' y_1 \\
Y_{21}V_1 + (Y_{22} + y_2)V_2 + \ldots + Y_{2m}V_m + \ldots + Y_{2n}V_n &= E_2' y_2 \\
\vdots \\
Y_{m1}V_1 + Y_{m2}V_2 + \ldots + \ldots + (Y_{mm} + y_m)\,V_m + \ldots + Y_{mn}V_n &= E_m' y_m \\
Y_{m+1,1}V_1 + \ldots + Y_{m+1,m}V_m + \ldots + Y_{m+1,n}V_n &= 0 \\
\vdots \\
Y_{n1}V_1 + \ldots + \ldots + Y_{nn}V_n &= 0
\end{aligned} \tag{11.81}
$$

Knowing the values of E_m', where $m = 1, 2, \ldots, m$, the bus voltages V_i, where $i = 1, 2, \ldots, n$, can be computed by Gauss–Seidel iterative technique. Then, the electrical power output of each machine may be computed from the following equation

$$P_{ui} = \mathrm{Re}\left[E_i'\left(\frac{E_i' - V_i}{jX_{di}'}\right)^* \right] \qquad \text{for } i = 1, 2, \ldots, m \qquad (11.82)$$

Knowing the electrical power output, the solution of swing equations for the machines can be solved using either the step-by-step method or modified Euler method as discussed in Section 11.6.

11.7.3 Direct Method

In this method all the buses except the internal machine buses (the buses behind the transient reactances) are eliminated, and a multi-port representation of the internal nodes of the generators is obtained. Using the self and transfer admittance parameters of the reduced network, electrical power output of the generators can be obtained involving rotor angles δ_i, where $i = 1, 2, \ldots, m$. The procedure is outlined here.

The admittance matrix $Y_{\text{bus, mod}}$ of Eq. (11.73) is augmented by including the transient reactances of the generators. Let $Y_{\text{bus, mod}}$ after inclusion of load impedances be partitioned as

$$Y_{\text{bus, mod}} = \begin{array}{c} \\ m \\ n-m \end{array} \begin{array}{cc} m & n-m \\ \left[\begin{array}{cc} Y_1 & Y_2 \\ Y_3 & Y_4 \end{array}\right] \end{array} \qquad (11.83)$$

where submatrix Y_1 is of the order $m \times m$ and corresponds to the buses where generators are connected, and Y_2, Y_3, Y_4 are the other submatrices.

The machine admittance matrix y may be written as

$$y = \begin{bmatrix} 1/jX_{d1}' & & & \\ & 1/jX_{d2}' & & \\ & & \ddots & \\ & & & 1/jX_{dm}' \end{bmatrix} = \begin{bmatrix} y_1 & & & \\ & y_2 & & \\ & & \ddots & \\ & & & y_m \end{bmatrix} \qquad (11.84)$$

Then, the augmented bus admittance matrix, $Y_{\text{bus, aug}}$, with ground as reference, would be represented as

$$Y_{\text{bus, aug}} = \begin{array}{c} \\ m \\ m \\ n-m \end{array} \begin{array}{ccc} m & m & n-m \\ \left[\begin{array}{ccc} y & -y & 0 \\ -y & Y_1 + y & Y_2 \\ 0 & Y_3 & Y_4 \end{array}\right] \end{array} \qquad (11.85)$$

The matrix $Y_{\text{bus, aug}}$ is reduced by applying Kron's reduction formula eliminating all buses except the internal machine buses (the buses behind the transient reactances). Before applying the network reduction process, the fault condition must be incorporated in the augmented matrix $Y_{\text{bus, aug}}$. For a symmetrical three-phase fault to ground at bus k, the row and column elements corresponding to bus k are set to zero before applying network reduction. Similarly, the switching operation of opening of a line, following a fault clearance, is simulated by setting to zero the off-diagonal element between the two buses connected by the line to zero and modifying the diagonal elements of the corresponding buses. The reduced bus admittance matrix $Y_{\text{bus, red}}$ will be of the order $m \times m$.

In a stability analysis, three reduced matrices are required to be computed to represent pre-fault, during fault, and post-fault switching conditions in a power system.

The generator electrical power output for each machine is computed by employing Eq. (7.19) as discussed in Section 7.2. Thus, Eq. (7.19) after appropriate modifications takes the following form and represents the electrical power output of the kth generator in a power system made up of m generators.

$$P_{uk} = \left| E'_k \right| \sum_{j=1}^{m} \left| Y_{ij} \right| \left| E'_j \right| \cos \left(\delta_i - \delta_j - \theta_{ij} \right) \quad \text{for } i = 1, 2, \ldots, m \quad (11.86)$$

The swing equation as given by Eq. (11.7) is required to be solved for all the generators in the system. Hence, the swing equation for a generator k is written as

$$M_k \frac{d^2 \delta_k}{dt^2} = P_{ik} - P_{uk} \quad \text{for } i = 1, 2, \ldots, m \quad (11.87)$$

Since in a transient stability analysis, action of the governor is neglected, the input power P_{ik} is taken equal to the pre-fault electrical power of the generator and is assumed to remain constant throughout the study. The electrical power output P_{uk} is computed from Eq. (11.86) by employing constant voltage behind direct axis transient reactance and using the appropriate reduced $Y_{\text{bus, red}}$ matrices for the during fault or post-fault condition, as the case may be. The mechanical rotor angle of a generator coincides with the angle of its voltage behind direct axis transient reactance. The initial rotor angle of a generator, when the system is operating in synchronism, is calculated from the load flow data. Equation (11.87) is a non-linear differential equation and is solved with the help of numerical integration techniques which are described in the earlier section.

11.8 STEADY-STATE STABILITY STUDIES

Steady-state or small-signal stability is the ability of the power system to maintain synchronism when subjected to small disturbances. A disturbance is considered to be small, if the equations that describe the dynamic characteristics of the system can be linearized for the purpose of analysis. Instability that may result can be of two forms: (i) steady increase in generator rotor angle due to lack of synchronizing torque or (ii) rotor oscillations of increasing amplitude due to lack of sufficient damping torque. In today's practical power systems, the small-signal stability problem is usually one of insufficient damping of system oscillations. Small signal analysis assists in the design of a power system.

Lets us now study the steady-state performance of a single machine connected to a large system through transmission lines. A general system configuration is shown in Fig. 11.17(a). Analysis of such a system is extremely useful in understanding basic effects and concepts. The system shown in Fig. 11.17(a) may be reduced to the form shown in Fig. 11.17(b) by using Thevenin's equivalent of the transmission network external to the machine and the adjacent transmission. Because of the large size of the system to which the machine is supplying power, dynamics associated with the machine will cause virtually no change in the voltage and frequency of Thevenin's voltage V_B. Such a voltage source of constant voltage and constant frequency is referred to as an infinite bus. Figure 11.17(c) shows the system representation with the generator represented by the classical model and all resistances neglected. Here E' is the voltage behind X'_d. Its magnitude is assumed to remain constant at the pre-disturbance value. Let δ be the angle by which E' leads the infinite bus voltage V_B.

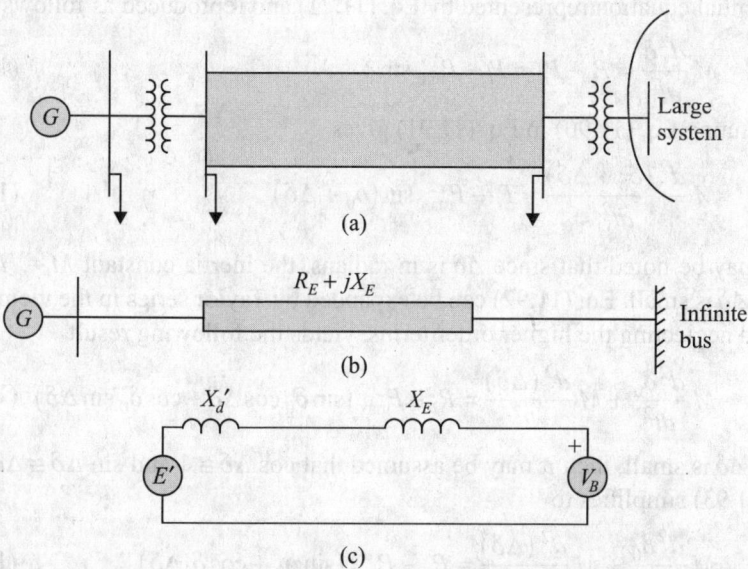

(a)

(b)

(c)

Fig. 11.17 Single machine connected to a large system through transmission lines: (a) general layout; (b) equivalent system; (c) classical model

11.8.1 Simulation of the Steady-state Problem

Using Eq. (11.20), the power transfer from the generator to infinite bus may be written as

$$P_u = \frac{|E'||V_B|}{X_{eq}} \sin \delta = P_{max} \sin \delta \tag{11.88}$$

where X_{eq} is the transfer reactance between E' and V_B, and δ is the angle by which E' leads V_B.

Figure 11.18 is a plot of power transferred versus load angle δ. It is also observed that maximum power transfer occurs at $\delta = \pi/2$ and is given by

$$P_{max} = \frac{[E']|V_B|}{X_{eq}} \text{ pu} \tag{11.89}$$

The maximum power transfer limit is called the steady-state power limit and is inherent to a system. When resistance is included then the magnitude of the steady-state power limit is less than the steady-state power limit obtained by neglecting the resistance and it occurs at a load angle $\delta < \pi/2$.

11.8.2 Linearization for Steady-state Stability Studies

Assume that the system shown in Fig. 11.17(c) is operating in steady-state with rotor angle δ_0, when it experiences a small disturbance ΔP at time $t = 0$ s, and the system is set into oscillations, and the power angle deviates from its steady-state position δ_0 by a small value $\Delta\delta$ radians such that

$$\delta = \delta_0 + \Delta\delta \tag{11.90}$$

The oscillations occurring in the system are expressed by the second-order differential equation represented by Eq. (11.21) and reproduced as follows:

$$M\frac{d^2\delta}{dt^2} = P_i - P_u = P_i - P_{max}\sin\delta \tag{11.91}$$

Substituting Eq. (11.90) in Eq. (11.91) gives

$$M\frac{d^2(\delta_0 + \Delta\delta)}{dt^2} = P_i - P_{max}\sin(\delta_0 + \Delta\delta) \tag{11.92}$$

It may be noted that since $\Delta\delta$ is in radians, the inertia constant $M = H/\pi f$ s. Since $\Delta\delta$ is small, Eq. (11.92) can be expanded by Taylor series in the vicinity of δ_0, and neglecting the higher order terms, yields the following result

$$M\frac{d^2\delta_0}{dt^2} + M\frac{d^2(\Delta\delta)}{dt^2} = P_i - P_{max}(\sin\delta_0\cos\Delta\delta + \cos\delta_0\sin\Delta\delta) \tag{11.93}$$

Since $\Delta\delta$ is small, then it may be assumed that $\cos\Delta\delta \cong 1$ and $\sin\Delta\delta \cong \Delta\delta$, and Eq. (11.93) simplifies to

$$M\frac{d^2\delta_0}{dt^2} + M\frac{d^2(\Delta\delta)}{dt^2} = P_i - P_{max}(\sin\delta_0 + \cos\delta_0\Delta\delta) \tag{11.94}$$

But at time $t = 0^-$, when the system is in the initial operating state

$$M\frac{d^2\delta_0}{dt^2} = P_i - P_{max}\sin\delta_0 \tag{11.95}$$

Hence Eq. (11.94) can be written as

$$M\frac{d^2(\Delta\delta)}{dt^2} + P_{max}\cos\delta_0\Delta\delta = 0 \tag{11.96}$$

or $$Mp^2(\Delta\delta) + P_S(\Delta\delta) = 0$$

or $$\left[Mp^2 + P_S\right]\Delta\delta = 0 \tag{11.97}$$

where the operator $p = d/dt$ and $P_S = P_{max}\cos\delta_0$, called the synchronizing coefficient.

Figure 11.18 shows that the synchronizing coefficient is the slope of the power angle curve at δ_0. The solution of Eq. (11.97) provides the unknown quantity $\Delta\delta$. For $\Delta\delta$ to have a non-zero value, $[Mp^2 + P_S]$ must be equal to zero. Hence,

$$Mp^2 + P_S = 0 \tag{11.98}$$

Equation (11.98) is the linear differential equation and is called the characteristic equation of the system. It has two roots and its solution in terms of $\Delta\delta$ is given by

$$\Delta\delta = K_1 e^{p_1 t} + K_2 e^{p_2 t} \tag{11.99}$$

where $$p_{1,2} = \pm j\sqrt{\frac{P_S}{M}} \tag{11.100}$$

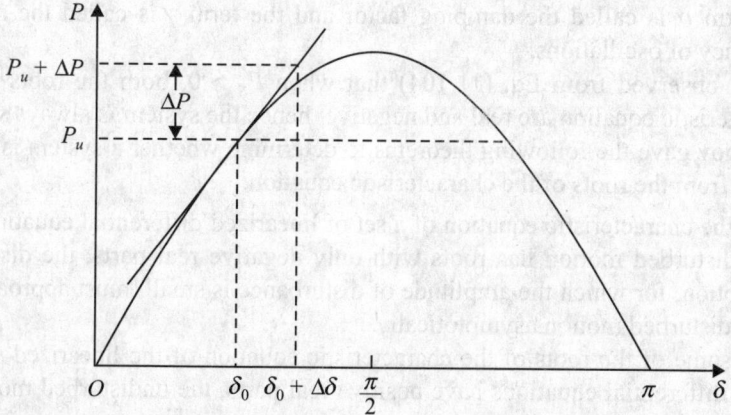

Fig. 11.18 Linearization of the power transfer curve for small disturbances

From the foregoing, it is clear that the nature of oscillations is governed by the roots of Eq. (11.100). Alternately stated, the type of oscillations is determined by the sign of P_S. For example, if P_S is positive ($P_S > 0$), the two roots are imaginary. Since the roots lie on the imaginary axis of the s-plane (complex plane), the system is set into continuous oscillations with a natural frequency given by

$$\gamma = \sqrt{\frac{P_S}{M}} \quad \text{rad/s} \tag{11.101}$$

In the preceding discussion, damping torque has been ignored. Damping can be accounted for by assuming damping torque to be proportional to the derivative of the rotor angle and introducing a term in Eq. (11.97) which represents the linearized equation of motion. Therefore, the linearized equation of motion with damping is written as

$$Mp^2(\Delta\delta) + P_d p(\Delta\delta) + P_S(\Delta\delta) = 0$$

$$[Mp^2 + P_d p + P_S](\Delta\delta) = 0 \tag{11.102}$$

where P_d is the coefficient of damping power. The characteristic equation is given by

$$Mp^2 + P_d p + P_S = 0 \tag{11.103}$$

Equation (11.103) has two roots that are given by

$$P_{1,2} = \frac{-P_d \pm \sqrt{P_d^2 - 4MP_S}}{2M} = -\frac{P_d}{2M} \pm \frac{\sqrt{P_d^2 - 4MP_S}}{2M}$$

$$= \alpha \pm j\gamma \tag{11.104}$$

where

$$\alpha = -\frac{P_d}{2M} \tag{11.105}$$

and

$$\gamma = \frac{\sqrt{P_d^2 - 4MP_S}}{2M} = \sqrt{\frac{P_d^2}{4M^2} - \frac{4MP_S}{4M^2}} = \sqrt{\alpha^2 - \frac{P_S}{M}} \tag{11.106}$$

The term α is called the damping factor and the term γ is called the natural frequency of oscillations.

It is observed from Eq. (11.104) that when $P_S > 0$, both the roots of the characteristic equation are real and negative; hence the system is always stable. Lyapunov gave the following theorems to determine whether a system is stable or not, from the roots of the characteristic equation.

(a) If the characteristic equation of a set of linearized differential equations for a disturbed motion has roots with only negative real parts, the disturbed motion, for which the amplitude of disturbance is small, must approach the undisturbed motion asymptotically.

(b) If some of the roots of the characteristic equation of the linearized system of differential equations have positive real parts, the undisturbed motion is unstable.

These theorems cannot be applied when the roots have zero real parts. However, with some special methods, which are beyond the scope of this book, it is possible to obtain their solution. Lyapunov's theorems, nevertheless, are sufficient for solving most of the practical problems.

Example 11.9 A 50-Hz synchronous generator is supplying 0.8 pu real power at 0.8 lagging power factor to an infinite bus via a transmission line whose reactance is 0.4 pu. If the direct axis transient reactance of the generator is 0.2 pu and the inertia constant $H = 10$ MJ/MVA, determine (i) the steady-state power limit, (ii) synchronizing power coefficient, (iii) the frequency of free oscillations, and (iv) the time period of free oscillations. Assume the infinite bus voltage equal to $1.0 \angle 0°$

Solution Transfer reactance between the infinite bus and the voltage behind transient reactance is

$$X_{eq} = 0.4 + 0.2 = 0.6 \text{ pu}$$

Magnitude of the generator current $I = \dfrac{0.8}{1.0 \times 0.8} = 1.0$ pu

Since the power factor angle $\varphi = \cos^{-1}(0.8) = 36.87°$, the current lags the infinite bus voltage by $36.87°$. Figure 11.19 shows the phasor relationship between the infinite bus voltage V_B and the voltage behind generator transient reactance E'.

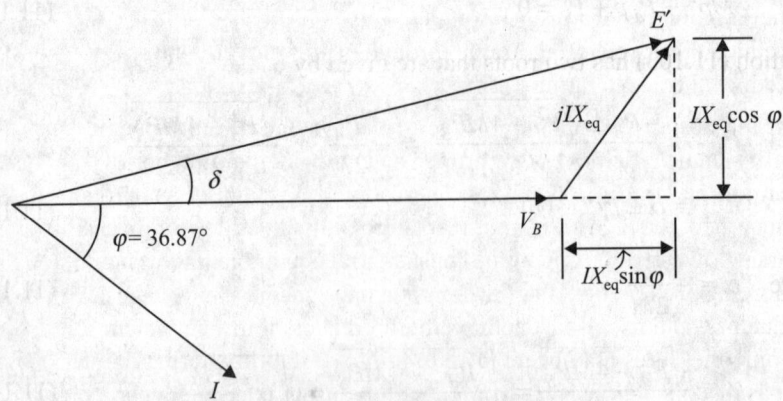

Fig. 11.19 Phasor diagram

From the phasor diagram

$$E' = \sqrt{\left[(1.0+1.0\times0.2\times0.6)^2 + (1.0\times0.2\times0.8)^2\right]} = 1.1314 \text{ pu}$$

$$\tan\delta = \frac{1.0\times0.2\times0.8}{1.0+1.0\times0.2\times0.6} = 0.1429$$

$$\delta = 0.1419 \text{ rads} = 8.13°$$

Thus, the initial operating angle $\delta_0 = 8.13°$

(i) Steady-state power limit $P_{\max} = \dfrac{|E'||V_B|}{X_{eq}} = \dfrac{1.0\times1.13}{0.6} = 1.8833$ pu

(ii) Synchronizing power coefficient

$$P_{\max}\cos\delta_0 = 1.8833 \times \cos(8.13°) = 1.8644 \text{ pu}$$

(iii) $M = \dfrac{H}{\pi f} = \dfrac{10}{\pi \times 50}$

$$\gamma = \sqrt{\frac{P_S}{M}} = \sqrt{\frac{1.8644 \times \pi \times 50}{10}} = 5.4116 \text{ rad/s}$$

$$f = \frac{\gamma}{2\pi} = \frac{5.4116}{2\times\pi} = 0.8613 \text{ Hz}$$

(iv) Time period of free oscillations $T = \dfrac{1}{f} = \dfrac{1}{0.8613} = 1.1610$ s

11.9 SOME TECHNIQUES FOR IMPROVING TRANSIENT STABILITY

The various techniques outlined here are discussed as technical problems for improving transient stability. The ultimate analysis to select a particular technique is an issue of the gains achieved from improving the transient stability against the costs involved in implementing the technique, but the gains should also justify the cost of the technique.

Methods of improving transient stability try to achieve one or more of the following effects:

- Minimize the fault severity and duration
- Increase the restoring synchronizing forces
- Reduce the accelerating torque through prime-mover control
- Reduce the accelerating torque by applying artificial load

At the design and planning stage of the system, the improvement in system stability can be achieved by proper selection of parameters of synchronous generators and transmission lines. These include, as far as a synchronous machine is concerned, reduced values for reactances, higher values for mechanical inertia constant, lowering of exciter time constants, raising the ceiling values of the exciter, and forcing of excitation. For transmission lines, increase in voltage, use of bundled conductors, and use of fast acting circuit breakers improve system stability.

Some measures may be adopted to improve stability during regular operation such as disconnection of part of generation during post-fault operation, resorting to three-phase and single-phase automatic reclosure, and automatic load rejection. Additional measures may be required in some cases to improve transient stability, which involves additional expenditure. Some of these measures are as follows:

(a) Earthing of transformer neutral through resistance or reactance
(b) Series compensation of lines
(c) Methods to alter the transfer admittance during post-fault operation such as bang-bang line reactance control and dynamic braking
(d) Construction of intermediate switching stations
(e) Construction of additional lines

Once a power system is commissioned, the option regarding the generator and the transmission line parameters is already exercised. Thus, in case the system stability has to be improved due to system expansion, then the additional measures listed above have to be made use of if the routine or operational measures available are not sufficient to achieve the desired degree of stability.

A task force appointed by IEEE in 1978 has classified the measures available for transient stability improvement into two broad categories, namely the primary controls and the discrete supplementary controls. According to this classification, the primary controls are the speed governors and the excitation systems which are continuous in operation and exert a control effort proportional to the output error. The protective relays are termed discrete controls as these operate only on the incidence of predetermined class of disturbances. By a combination of the continuous and the discrete controls, most of the generating units are controlled under all probable operating conditions. The IEEE task force introduced the term *discrete supplementary control* to refer to this special class of controls. Some of these controls are as follows:

(a) Dynamic braking
(b) High-speed circuit breaker reclosing
(c) Independent control of excitation
(d) Discrete control of excitation
(e) Controlled system separation and load shedding
(f) Series capacitor insertion
(g) Power modulation by direct-current lines
(h) Turbine bypass valving
(i) Momentary and sustained fast valving
(j) Generator tripping

11.9.1 Dynamic Braking

Dynamic braking, which is one of the most effective discrete supplementary controls available for improvement of transient stability, is recommended in situations such as close-in line faults on generating stations coupled with a line breaker failure. Dynamic braking is the concept of applying an artificial electric load to a portion of the generation transmission load complex to correct a temporary imbalance between power generated and power delivered. One form of dynamic braking involves the switching in of shunt resistors for about 0.5 s following a fault to reduce the accelerating power of nearby generators and remove the kinetic energy gained during the fault.

To date, braking resistors have been applied only to hydraulic generating stations remote from load centres. Hydraulic units in comparison to thermal units are quite rugged; therefore, they can withstand the sudden shock from the switching in of resistors without any adverse effect on the units.

The dynamic brake may be applied either as a series resistance connected between the generator and transformer terminals, or as a resistor connected between ground and neutral of the wye-connected high-voltage winding of the delta–wye generator step-up transformer. The series braking resistor may be applied at the neutral end of each phase of a wye-connected transformer. The braking effect is produced by the braking resistance connected between the neutral of the wye-connected transformer and ground only during the grounded fault conditions. The series braking resistance is active primarily during the fault period and acts to reduce the fault accelerating power. Conversely, the shunt brake becomes active after the fault is cleared and stays on for a longer period of time. Series brakes are of preventive type, whereas shunt brakes are of restorative type. A single shunt brake can be used in some cases for protection against instability of an entire area whereas at least one series brake must be included in each transmission line for which a fault would be destabilized. Hence the dynamic brake has been generally studied and applied as a shunt resistive load connected at a generation site, with the intent of slowing down the turbine-generator shaft after clearing a severe electric fault.

With the switched form of braking resistors, the switching times should be based on detailed simulations. If the resistors remain connected for too long, there is a possibility of instability on the back swing.

11.9.2 High-speed Circuit Breaker Reclosing

High-speed circuit breaker reclosing is used for rapid line reconnection to improve stability. Automatic reclosing is used because 80 per cent of transmission line faults are transient in nature, and the faults are deionized in a short time. Successful reclosure is beneficial because it keeps generating units online for transient transmission line faults, thus minimizing outage of generating units. However, high-speed reclosing may produce high transient shaft torques, cause voltage dips, and increase circuit breaker duty.

Independent pole operation of circuit breakers is used where the design criterion is to guard a three-phase fault compounded with a breaker failure. The independent pole operation of circuit breakers reduces a three-phase fault to a single line-to-ground fault (if one pole of the breaker is struck), or to a double line-to-ground fault (if two poles of the breaker are struck). These breakers are among the least expensive stability aids and independent breaker operation is most efficient at higher voltages, where equipment costs are high.

11.9.3 Independent Control of Excitation

Considerable improvements in transient stability are possible through rapid temporary increase in generator excitation, which increases the internal voltage of the generator and this in turn increases the synchronizing power. The effectiveness of this type of control depends on the ability of the excitation system to quickly increase the field voltage to the highest possible value. High initial response excitation systems with high ceiling voltages are most suitable. Ceiling voltages are limited from the point of view of rotor insulation. Ceiling voltage for a thermal unit is limited to 2.5 to 3.0 times the rated load field voltage.

For transient stability improvement, fast exciter response sometimes leads to lesser damping of oscillations. To overcome this, power system stabilizer (PSS) enables the high response excitation system to damp out the system oscillations.

11.9.4 Discrete Control of Excitation

The concept of discrete control arises because of the need to satisfy the conflict between the requirement for transient stability and those for steady-state stability under conditions of small oscillations. Discrete control improves transient stability by controlling the generator excitation, so that the terminal voltage is maintained near the maximum permissible value of about 1.15 pu over the entire positive swing of the rotor angle. The excitation system uses a signal proportional to the change in angle during the transient period of about 2 s. The rotor relative angular velocity signal, $\Delta\omega$, provides continuous control to maintain small signal stability under normal operation.

11.9.5 Controlled System Separation and Load Shedding

Controlled system separation and load shedding generally represents a last measure towards saving loss of generation in one generating area within a large inter-connected system. The main objective is to achieve near-equilibrium condition between generation and load following loss of generation.

11.9.6 Series Capacitor Insertion

Series capacitor insertion reduces the net transmission line inductive reactance between the sending and receiving ends with the aim of increasing power transfer capability of the lines. During transient disturbances, additional series capacitor insertion improves first swing transient stability. One problem with series capacitor compensation is the possibility of sub-synchronous resonance with the nearby turbo alternators. This aspect must be analysed carefully and appropriate preventative measures must be taken.

Traditionally, series capacitors have been used to compensate for very long overhead lines. Recently, there has been an increasing recognition of the advantages of compensating shorter, but heavily loaded, lines by using series capacitors.

For transient stability applications, the use of switched series capacitors offers some advantages. On detection of a fault or power swing, a series capacitor bank can be switched in and then removed about 0.5 s later. Such a switched bank can be located in a substation where it can serve several lines.

11.9.7 Power Modulation by Direct-current Lines

Power flow on dc transmission lines can be modulated by controlling the converters at the ends of the line. The converters can be controlled to achieve discrete power level changes thereby improving transient stability. A problem associated with such a control is that a fault disturbance in one system will appear as a sudden load change on the other system.

11.9.8 Turbine Bypass Valving

Turbine bypass valving allows turbine-generator to perform full load rejection while the boiler and reheater have remained at full flow operation, thus reloading of the turbine is possible after a time lag of as much as 15 minutes.

11.9.9 Momentary and Sustained Fast Valving

Momentary fast valving is rapid closure of intercept valves only and immediate full reopening at a slower rate. For strong tie-lines, tie stability is improved whereas for weaker ties the stability may be lost in the second swing. Sustained fast valving is the rapid closure of intercept valves with simultaneous repositioning of control valves and immediate partial opening followed by full reopening of intercept valves at a predetermined rate. Thus the plant output can be reduced to a desired level without a unit trip out, and full output may be restored within a few minutes.

11.9.10 Generator Tripping

Generation tripping is used for certain disturbances and transmission line outages. Generator tripping is also used in interconnected system operation in order to reduce the loading of critical system elements following a disturbance, thus improving system dynamic performance. The impact of generator trip out on prime mover and energy supply system must be considered carefully.

11.10 IMPROVING STEADY-STATE STABILITY LIMIT

The problem of steady-state stability is usually one of insufficient damping of system oscillations. The use of power system stabilizers to control generator excitation is the most cost-effective method for steady-state stability improvement of power systems. In addition, supplementary stabilizing signals, HVDC converter controls, and static var compensators may be used to improve damping of system oscillations.

Steady-state power limit is increased by selecting higher transmission voltages for new systems and upgrading the voltages on the existing systems. Steady-state power limit is also increased by reducing the system transfer reactances by additional parallel transmission lines. The use of bundled phase or oversized conductors in transmission lines reduces the series reactance of transmission lines. Series line inductances are compensated by series capacitors.

SUMMARY

- Stability is the attribute of a power system to return to its initial steady state operating point or close to it following a disturbance.
- Stability studies are categorized as (i) transient and (ii) steady state. The former is a condition which is brought about due to a fault in the system and the latter condition is due to small random changes in generation or load.
- Transient stability analyses are performed by developing the swing equation based on simplifying assumptions and applying the equal area criterion for different operating conditions such as sudden increase in power, three-phase fault leading to sudden loss in power output, and so on.

- To solve the swing equation and plot the swing curve (δ versus t), step-by step or modified Euler method is used.
- Transient stability analysis of a multi-machine power system requires (i) power flow analysis of the system prior to the disturbance, (ii) simulation of generators as voltage sources behind direct axis transient reactances, and (iii) modifying the bus admittance matrix to include generator transient reactances and loads at buses as shunt admittances.
- The steady-state stability problem is simulated by linearizing the swing equation at the point of operation and developing the characteristic equation.
- Lyapunov theorems are applied to determine stability of the system.
- Several techniques are applied to improve both transient and steady-state stability of a system.

SIGNIFICANT FORMULAE

- Swing equation: $M \dfrac{d^2\delta}{dt^2} = P_a = P_i - P_u$

- Inertia constant: $M = \dfrac{H}{180 f}$

- Equal area criterion for stability: $\displaystyle\int_{\delta_0}^{\delta_m} P_a \, d\delta = 0$

- Critical fault clearing time: $t_{\text{crit}} = \sqrt{\dfrac{2M(\delta_{\text{crit}} - \delta_0)}{P_i}}$

- Equivalent power input of two finite machines: $P_i = \dfrac{M_2 P_{i1} - M_1 P_{i2}}{M_1 + M_2}$

- Equivalent power output of two finite machines: $P_u \dfrac{M_2 P_{u1} - M_1 P_{u2}}{M_1 + M_2}$

- Equivalent inertia constant of two finite machines: $M = \dfrac{M_1 M_2}{M_1 + M_2}$

- Solution of the swing equation by the step-by-step method:

$$\Delta\delta_n = \Delta\delta_{n-1} + \dfrac{(\Delta t)^2}{M} P_{a(n-1)} \qquad \delta_n = \delta_{n-1} + \Delta\delta_n$$

- Solution of the swing equation by modified Euler method:

$$\delta_{i+1}^n = \delta_i + \dfrac{\left.\dfrac{d\delta}{dt}\right|_{\Delta\omega_i} + \left.\dfrac{d\delta}{dt}\right|_{\Delta\omega_{i+1}^f}}{2} \Delta t$$

$$\Delta\omega_{i+1}^n = \Delta\omega_i + \dfrac{\left.\dfrac{d(\Delta\omega)}{dt}\right|_{\delta_i} + \left.\dfrac{d(\Delta\omega)}{dt}\right|_{\delta_{i+1}^f}}{2} \Delta t$$

- Representation of load as equivalent admittance: $y_{Li} = \dfrac{P_{Li} - jQ_{Li}}{|V_i|^2}$

- Voltage behind transient reactance: $E'_i = V_i + j X'_{di} \dfrac{P_{Gi} - jQ_{Gi}}{V_i^*}$

- Power transfer from the generator to infinite bus:

$$P_u = \frac{|E'|\,|V_B|}{X_{eq}} \sin\delta = P_{max} \sin\delta$$

- Characteristic equation for steady state stability: $Mp^2 + P_d p + P_S = 0$
  Roots of the characteristic equation:

$$p_{1,\,2} = -\frac{P_d}{2M} \pm \frac{\sqrt{P_d^2 - 4MP_S}}{2M} = \alpha \pm j\gamma$$

EXERCISES

Review Questions

11.1 Distinguish between transient and steady-state stability and discuss the need for performing stability analyses of power systems.

11.2 (a) Based on the laws of rotation, develop an equation to simulate the motion of the rotor of a synchronous machine.
(b) Derive an expression to show the correlation between M and H.

11.3 Explain the application of equal area criterion to determine stability of a synchronous machine connected to an infinite bus through a transmission line.

11.4 How is the equal area criterion applied when there is a sudden (a) increase in power input, and (b) decrease in power output due to a three-phase fault?

11.5 Derive an expression for power transfer between two synchronous generators having a transfer admittance of $Y_{12} \angle \theta_{12}$ pu between the internal voltages of the machines. Show that the steady-state power limit is less than the steady-state power limit when resistance is neglected. What is the power angle at which maximum power transfer occurs?

11.6 (a) Discuss the possible solutions of the disturbed motion of a synchronous generator, connected to an infinite bus, when subjected to a small disturbance represented by $\Delta\delta$. Assume damping to be proportional to $d\delta/dt$.
(b) Briefly explain how steady-state stability limit can be improved.

11.7 Write an essay on the methods of improving transient stability.

11.8 Write short notes on: (a) Dynamic braking, (b) High-speed circuit breaker reclosing, and (c) Series capacitor insertion.

Numerical Problems

11.1 Prove that a group of coherently swinging machines can be represented by a single machine whose inertia constant is the sum of the inertia constants of the individual machines. State any assumption(s) made.
Three synchronous generators have the following MVA ratings and inertia constants:

Generator 1	$G = 200$ MVA	$H = 4.0$ MJ/MVA
Generator 2	$G = 500$ MVA	$H = 3.0$ MJ/MVA
Generator 3	$G = 750$ MVA	$H = 5.0$ MJ/MVA

If the generators are a part of a synchronously operating power system whose base MVA is 1000, determine (i) equivalent H and (ii) equivalent M.

11.2 The moment of inertia of a 250-MVA generator is 75000 kgm^2. If the operating frequency is 50 Hz and the generator has two poles, compute (i) kinetic energy of the rotating parts, (ii) H, and (iii) M.

11.3 A two-pole, three-phase, 50-MVA, 11-kV generator is supplying rated power at 0.8 lagging power factor to an 11-kV bus. Due to a fault the generator output is reduced to 40%. Compute (i) accelerating power and (ii) acceleration at the time of fault. Assume that the kinetic energy stored in the moving parts of the generator is 175 MJ.

11.4 In Problem 11.3, if the generated voltage is 11 kV and the synchronous reactance of the generator is 2 Ω/phase, calculate the rotor angle just before loss of generation. If the acceleration due to loss of generation continues for 5 cycles, determine (i) rotor angle and (ii) rotor speed in rpm at the end of the acceleration period.

11.5 Two synchronous machines are connected as shown in Fig. 11.20. All values in the diagram are in per unit on a common system base. If the inputs and outputs to the two machines are represented by P_{1i}, P_{1u}, P_{2i}, and P_{2u} respectively, derive expressions for input and output power for the equivalent machine. If $H_1 = 4.5$ MJ/MVA and $H_2 = 6.0$ MJ/MVA, calculate the inertia constant, input power, and the maximum power of the equivalent machine. Assume $P_{1i} = 1.25$ pu, $P_{2i} = 1.1$ pu. Write the swing equation of the equivalent machine.

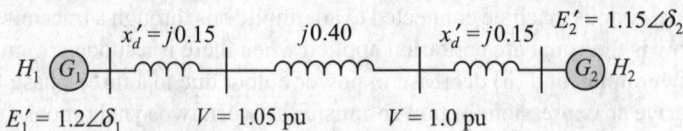

Fig. 11.20 Two machine system for Problem 11.5

11.6 A synchronous generator is supplying power to an infinite bus via two parallel lines as shown in Fig. 11.21. All values in the circuit diagram are in per unit on a common system base. If the power frequency is 50 Hz and the inertia constant of the generator is 4 MJ/MVA, calculate the voltage behind transient reactance and write the swing equation. Assume that the machine is delivering a power of 0.8 at a power factor of 0.85 lagging.

A three-phase fault occurs at the generator end through a reactance of 0.05 per unit. Determine the accelerating power and the acceleration at the time of fault.

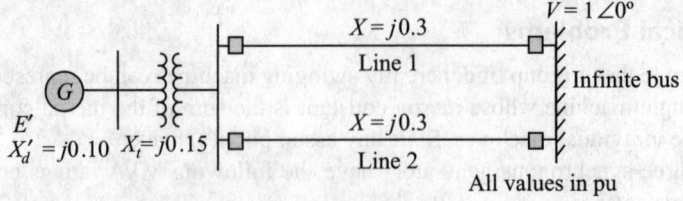

Fig. 11.21 Schematic diagram of a system for Problem 11.6

11.7 In problem 11.6, a three-phase fault occurs at the sending of one of the lines, which is cleared after 10 cycles with the faulted line remaining intact. Establish whether the system is stable or not. If the system is stable, determine the maximum rotor angle following the three-phase fault.

11.8 The generator in Problem 11.7 is delivering 1.2 pu power at unity power factor when a three-phase fault occurs in the middle of one of the parallel lines. The fault is cleared by opening the faulted line. Determine the critical clearing angle. Use MATLAB graphics to plot the power output versus power angle for the pre-fault, during fault, and post-fault conditions.

11.9 A synchronous generator is supplying a real power of 1.0 pu to an infinite bus as shown in Fig. 11.22. A temporary three-phase fault occurs in line 2, at one-tenth the distance from the infinite bus end. (i) What is the rotor angle when the generator is operating synchronously? (ii) What is the generator output, accelerating power, and acceleration when the fault occurs? (iii) If the fault is cleared after 15 cycles, by opening of the faulted line, compute the rotor angle, decelerating power, and deceleration immediately after the faulted line

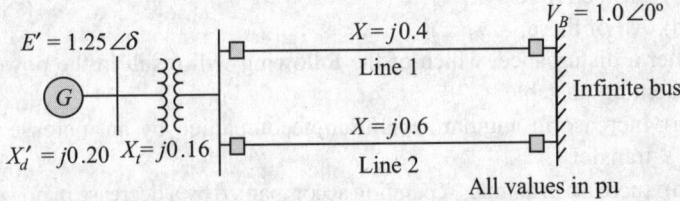

Fig. 11.22 Schematic diagram of a system for Problem 11.9

is opened. Why is the generator decelerating? Assume the power frequency to be 50 Hz and the inertia constant of the generator to be 3.5 MJ/MVA. All values in the circuit diagram are in per unit on a common system base.

11.10 Write a program to use the MATLAB ODE solver to solve and plot the following differential equation:

$$\frac{dy}{dt} = t^3 - y^2$$

for $0 \leq t \leq 1.0$. Assume that at $t_0 = 0$, $y_0 = 0.5$.

11.11 The following equation represents the swinging of a synchronous generator, connected to the infinite bus, during a fault.

$$\frac{d^2\delta}{dt^2} = P_a$$

Write (a) the swing equation as two first-order differential equations, and (b) a MATLAB function to compute the state derivatives, that is, $\dot{\delta}$ and $\dot{\omega}$. Use ode 23 or ode 45 to solve the second-order differential equation. Plot rotor angle versus time. Assume that the power angle prior to the disturbance is 30° and the fault is cleared in 0.25 s. Take $P_a = 0.75 + 0.3\sin\delta$ pu during the fault, and $P_a = 0.75 + 1.5\sin\delta$ during the post-fault period.

11.12 A 50-Hz synchronous generator is connected to an infinite bus of voltage $1.0\angle0°$ via two parallel lines each having a reactance of 0.4 pu. If the generator delivers 0.75 pu real power at 0.85 lagging power factor to the infinite bus, compute (i) the damping factor and (ii) the natural frequency of oscillations when a disturbance of $\Delta\delta = 12.5°$ occurs due to a temporary opening and immediate closing of one of the circuit breakers. Assume the following data for the generator: transient reactance = 0.4 pu, inertia constant H = 8 MJ/MVA, and coefficient of power damping = 0.15. Write a MATLAB function to plot rotor angle and frequency versus time.

Multiple Choice Objective Questions

11.1 For which of the following disturbance, is a power system required to be analysed for steady-state stability?
    (a) Small random change in generation
    (b) Sudden loss in generation
    (c) Sudden decrease in power output
    (d) All of these

11.2 Which of the following is a component of a power system?
    (a) Automatic load frequency control mechanism
    (b) Power transmission lines
    (c) Protective relays
    (d) All of these

11.3 After a disturbance, which of the following will result in the power system becoming unstable?
    (a) Increase in angular separation accompanied by an increase in power transfer
    (b) Increase in angular separation accompanied by a decrease in power transfer
    (c) Transfer of part of the load from the slow machine to the fast machine
    (d) None of these

11.4 Transient stability studies are conducted to determine
    (a) relaying and protective system settings
    (b) power transfer capability between systems
    (c) critical fault clearing times of circuit breakers
    (d) all of these

11.5 Transient stability analyses are performed for which of the faults?
    (a) SLG   (b) 2LG
    (c) 3φ   (d) LL

11.6 Which of the following is not a correct assumption for simulating the transient stability problem?
    (a) P_i remains constant at the pre-fault value during the transient period.
    (b) Synchronous machine is represented by a constant voltage source behind sub-transient reactance.
    (c) Loads are converted into equivalent admittances at the pre-fault voltage values.
    (d) Damping power is negligible.

11.7 Which is the range of the inertia constant H, in MJ/MVA, of a hydrogenerator?
    (a) 2 to 4   (b) 4 to 6
    (c) 6 to 8   (d) 8 to 10

11.8 Which of the following represents the value of M?

    (a) $M = \dfrac{2H}{\omega_{Sm}} S_{rated}$ (b) $\dfrac{H}{\pi f}$

    (c) $\dfrac{H}{180 f}$ (d) All of these

11.9 Which of the following represents the value of the maximum power transferred?

(a) $\dfrac{|E'|\,|V_B|}{X_{eq}}$

(b) $\dfrac{|E'|\,|E'|}{X_{eq}}$

(c) $\dfrac{|V_B|\,|V_B|}{X_{eq}}$

(d) None of these

11.10 Which of the following indicates transient stability of a system?

(a) $\dfrac{d\delta}{dt}=0$

(b) $\Delta\omega=0$

(c) $\int_{\delta_0}^{\delta_m} P_a\, d\delta=0$

(d) All of these

11.11 Which of the following does not describe the equal area criterion?
(a) It is graphical.
(b) It is analytical.
(c) It is qualitative.
(d) It is approximate.

11.12 Which of the following determines the critical fault clearing time?

(a) $\sqrt{\dfrac{2M\delta_{\text{crit}}}{P_i}}$

(b) $\sqrt{\dfrac{2M\delta_0}{P_i}}$

(c) $\sqrt{\dfrac{2M(\delta_{\text{crit}}+\delta_0)}{P_i}}$

(d) $\sqrt{\dfrac{2M(\delta_{\text{crit}}-\delta_0)}{P_i}}$

11.13 Which of the following represents the equivalent input P_i?

(a) $\dfrac{M_2 P_{i1}-M_1 P_{i2}}{M_1+M_2}$

(b) $\dfrac{M_2 P_{i1}+M_1 P_{i2}}{M_1+M_2}$

(c) $\dfrac{M_1 P_{i1}-M_2 P_{i2}}{M_1+M_2}$

(d) $\dfrac{M_1 P_{i1}+M_2 P_{i2}}{M_1+M_2}$

11.14 Which of the following does not apply to the characteristic equation in the steady-state stability analysis of a system?
(a) It is non-linear.
(b) It is a differential equation.
(c) It is of the second order.
(d) All of these.

11.15 Which of the following roots of characteristic equation require special methods to apply Lyapunov theorems?
(a) Roots have only negative real parts.
(b) Some roots have positive real parts.
(c) Roots have zero real parts.
(d) All of these.

11.16 Which of the following improves the steady-state stability limit?
(a) Adding transmission lines in parallel
(b) Using oversized conductors
(c) Higher levels of transmission voltages
(d) All of these

Answers

11.1 (a)	**11.2** (d)	**11.3** (b)	**11.4** (d)	**11.5** (c)	**11.6** (b)
11.7 (a)	**11.8** (d)	**11.9** (a)	**11.10** (d)	**11.11** (b)	**11.12** (d)
11.13 (a)	**11.14** (a)	**11.15** (c)	**11.16** (d)		

CHAPTER 12

Voltage Stability

Learning Outcomes

A thorough study of this chapter will aid the reader to
- Understand various aspects of power system stability and distinguish between them
- Comprehend voltage stability and visualize the impact of large and small disturbances
- Classify time frame based voltage stability and comprehend the processes
- Based on the fundamentals of power flow in a radial feeder develop P–V and V–Q characteristics and make use of them for analysing transient voltage stability,

OVERVIEW

Traditionally, the power system stability problem has been one of maintaining synchronous operations and is limited to rotor angle stability. Since power systems rely on synchronous machines for generation of electrical power, a necessary condition of satisfactory system operation is that all the synchronous machines must run in step, or in synchronism. This aspect of stability is influenced by the dynamics of generator rotor angles and power-angle relationships. This has been discussed in Chapter 11.

Instability may also occur without loss of synchronism. For example, a system consisting of a synchronous generator feeding an induction motor load through a transmission line can become unstable because of the collapse of load voltage. In this case, synchronism is not an issue; instead, the stability and control of load voltage is of concern. This form of instability can also occur in loads covering a large area supplied by a power system. The voltage instability problem once associated with weak systems and long lines are now a source of concern in developed networks as a result of heavier loadings. This phenomenon is generally characterized as an inadequate VAR support at critical network buses. Thus, voltage instability phenomena are the ones in which the receiving end voltage decreases well below its normal value and does not come back even after setting restoring mechanisms such as VAR compensators, or continues to oscillate for lack of damping against the disturbances. Voltage collapse is the process by which the voltage falls to a low, unacceptable value as a result of an avalanche of events accompanying voltage instability.

In recent years, voltage instability has been responsible for several collapses of major networks worldwide. In some cases, rotor angles and frequency remained constant, while voltage continued to decay to a critical value, causing protection equipment to isolate the network. In other cases, rotor angles and the frequency swings accompanied the voltage decay. The increasing incidence of voltage collapse experienced by some utilities during the last decade has drawn attention of the practising engineers to the study of voltage stability problem.

12.1 VOLTAGE STABILITY AND VOLTAGE COLLAPSE

Voltage stability is the ability of a power system to maintain steady acceptable voltages at all buses in the system under normal operating conditions and after being subjected to a disturbance. A system enters a state of voltage instability when a disturbance such as increase in load demand or change in system condition causes a progressive and uncontrollable drop in voltage. The main cause of instability is the inability of a power system to meet the demand for reactive power.

A criterion for voltage stability is that at a given operating condition for every bus in the system, the bus voltage magnitude increases as the reactive power injection at the same bus is increased. A system is voltage unstable if, for at least one bus in the system, the bus voltage magnitude decreases as the reactive power injection at the same bus is increased. In other words, a system is voltage stable if $V\text{-}Q$ sensitivity is positive for every bus and voltage unstable if $V\text{-}Q$ sensitivity is negative for at least one bus.

Progressive drop in voltage can also be associated with rotor angles going out of step. For example, the gradual loss of synchronism of machines, as rotor angles between two groups of machines approach or exceed 180°, would result in very low voltages at intermediate points in the network. However, in contrast, sustained fall of voltage, which is related to voltage instability, occurs where rotor angle stability is not an issue.

Voltage instability is essentially a local phenomenon. However, its consequences may have a widespread impact. Voltage instability at a bus, if not corrected, affects the adjacent buses. Such a cascading effect of voltage instability results in unusually low voltages or a blackout in a part of the system. Thus, voltage collapse is the result of widespread impact of voltage instability.

12.1.1 Definitions

Voltage stability is a subset of overall power system stability. The definitions adopted by CIGRE Task Force and IEEE Committee Report are as follows:

A power system in a given operating state is small disturbance voltage stable if, following any small disturbance, voltages near loads are identical or close to the pre-disturbance values.

A power system at a given operating state and subject to a given disturbance is voltage stable if voltages near loads approach post-disturbance equilibrium values. The disturbed state is within the region of attraction of the stable post-disturbance equilibrium.

A power system at a given operating state and subject to a given disturbance undergoes voltage collapse if post-disturbance equilibrium voltages are below acceptable limits. Voltage collapse may be total (blackout) or partial.

Voltage instability is the absence of voltage stability and results in progressive voltage decrease (or increase).

12.1.2 Classification

Voltage stability may be classified into two categories: *large disturbance voltage stability* and *small disturbance voltage stability*.

Large disturbance voltage stability is concerned with a system's ability to control voltages following large disturbances such as system faults, loss of generation, or circuit contingencies. Determination of large-disturbance stability requires the examination of the non-linear dynamic performance of a system over a period of time sufficient to capture the interactions of the devices such as under-load tap-changers and generator field current limiters. The study period of interest may extend from a few seconds to tens of minutes. Therefore, long-term dynamic simulations are required for analysis.

A criterion for large disturbance voltage stability is that following a given disturbance and following system-control actions, voltages at all buses reach acceptable steady-state levels.

Small disturbance voltage stability is concerned with a system's ability to control voltages following small perturbations like incremental changes in system load. This form of stability is determined by the characteristics of load, continuous controls, and discrete controls at a given instant of time. The basic processes contributing to small-disturbance voltage instability are essentially of a steady-state nature. Therefore, static analysis can be effectively used for its study.

A criterion for small-disturbance voltage stability is that at a given operating condition for every bus in the system, the bus voltage magnitude increases as the reactive power injection at the same bus is increased. A system is voltage unstable if for at least one bus in the system the bus voltage magnitude $|V|$ decreases as the reactive power injection (Q) at the same bus is increased.

12.2 TYPES OF VOLTAGE STABILITY ANALYSES

Some of the reasons due to which voltage stability studies of power systems, in modern times, have gained in importance are as follows:

- In larger power systems, generation is centralized, large electrical distances between generation and load, and lesser number of voltage controlled buses.
- Shunt compensation is extensively employed.
- Demands of economics have forced operation of systems nearer to their limits.
- Frequent line and generator outages lead to voltage instability.
- Several happenings in developed countries like Belgium, France, Japan, Sweden, USA, and so on.

The type of a disturbance leading to possible voltage instability in a system determines the form of voltage stability analyses. Based on the type of disturbance, an appropriate technique is employed to simulate the voltage stability problem. Voltage stability is categorized as

(i) Large disturbance voltage stability

(ii) Small disturbance voltage stability

12.2.1 Large Disturbance Voltage Stability

As the name implies, this type of voltage stability is concerned with large disturbances such as system faults, loss of generation, circuit contingencies like loss of line, and so on. An analysis of large disturbances voltage stability requires an assessment of the interactions of equipment such as generator, field current limiters, and onload tap-changers (OLTCs). In other words, large disturbance voltage stability studies require an analysis of the dynamic performance of a system whose simulation is not only non-linear, but also has to be investigated from a few seconds to tens of minutes.

A system can be said to be voltage stable following a large disturbance, if after all system control actions the voltages at all buses have attained satisfactory steady-state voltage levels.

12.2.2 Small Disturbance Voltage Stability

Small disturbance voltage stability is conducted to analyse the ability of a power system to maintain steady voltages following small perturbations due to gradual changes in system load. At a given moment of time, steady-state voltage stability is governed by system load characteristics, and continuous and discrete control devices.

A condition for small disturbance voltage stability may be stated that for a given operating condition for every bus in the system, the voltage magnitude of the bus must increase when the reactive power injection at that bus is increased. Differently stated, a system will be voltage unstable if for at least one bus in a system the voltage magnitude, $|V|$, of the bus decreases when the injected reactive power Q is increased. Thus for a voltage stable system, the $|V|$–Q characteristic must have a positive sensitivity for all system buses. A system will be voltage unstable if $|V|$–Q characteristic is negative even for one bus.

Since small disturbance voltage instability is a result of small changes, the fundamental process, therefore, is steady in nature. Consequently, static analysis techniques suffice for establishing stability margins, determining parameters which impact on stability and studying various possible system conditions for different post-contingency settings.

12.3 TIME FRAME BASED VOLTAGE STABILITY ANALYSIS

Voltage instability in a system is usually accompanied by power angle instability. In fact, one type of instability may lead to the other type and it may not be possible to establish precedence of one type over the other. However, in order to develop suitable design and operating methods, it is important to distinguish between angle and voltage stabilities so as to understand the basic reasons of the problems.

In an attempt to analyse extreme variations in power flows, voltages, and frequency, produced by severe system disturbances, it is necessary to understand the actions of slow processes, control and protection devices, which cannot be simulated as in the customary transient stability analyses. For the purpose, the following types of time frame based voltage stability studies are categorized.

Transient or short-term voltage stability phenomenon has a short duration of up to 10 seconds and is quite similar to the transient rotor angle stability. During this period dynamics of automatic voltage regulators (AVRs), excitation systems, and turbine governors come into play along with induction motors, electronically operated loads, and HVDC interconnections. If the system is stable it will withstand the disturbance, allow it to die out and enter into a slow long-term phase where onload tap-changers (OLTCs), limiters, boilers, etc. are operating.

During transient voltage instability, there is little that an operations engineer in the control room can do to take corrective action to arrest the current–voltage instability. The corrective actions are undertaken through protective devices which function by isolating the unstable part of the system and safeguarding the largest part.

Longer-term voltage stability   is further subdivided into (a) long-term or post-transient and (b) longer-term voltage instability problems. The time frame of the disturbance in the former type is longer compared with transient voltage stability, ranging typically between 10 seconds to a few minutes, while in the latter category its time range is from a few minutes to tens of minutes.

Long-term voltage stability problems can be noticed in heavily loaded systems having large electrical distance between a generator and a load. A large power transfer from far off generating stations, or a massive load build up, or a sudden large disturbance may prompt long-term voltage instability. Given a long enough timescale, appropriate controller intervention, like reactive power compensation or load shedding, may be feasible to forestall such a type of voltage instability.

Rapid build up of load, of the order of megawatts per minute, requiring fast transfer of massive power increase may lead to the longer-term voltage instability problem. Voltage instability problems can be avoided by operator intervention through timely load shedding and initiation of appropriate reactive power equipment.

In the following section, techniques to study the voltage stability problem due to a small disturbance are described; a discussion of the methods to study large disturbance voltage instability problem being out of the scope of the present text has, therefore, not been included.

12.4 TRANSIENT VOLTAGE STABILITY

Power flow through a radial line best explains the fundamentals of transient voltage stability. Expressions for complex power flow in a radial line have been developed in Section 4.9.2 and are reproduced below for convenience.

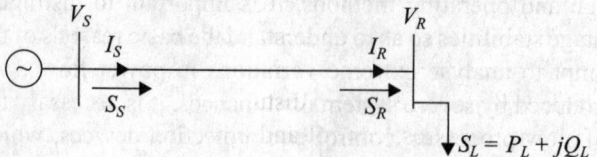

Fig. 12.1 Radial short-length line supplying a complex power load

The expressions for active power and reactive power at the sending end and the receiving end of the radial line shown in Fig. 12.1 are as follows:

$$P_R = P_L = \frac{|V_S|\,|V_R|}{X}\sin\delta \tag{12.1}$$

$$Q_S = \frac{|V_S|^2}{X} - \frac{|V_S|\,|V_R|}{X}\cos\delta \tag{12.2}$$

$$Q_R = -\frac{|V_R|^2}{X} + \frac{|V_R|\,|V_S|}{X}\cos\delta \tag{12.3}$$

As can be seen from Eq. (12.1), active power flow is governed by the power angle δ and it flows from the leading voltage bus to the lagging voltage bus. If it is assumed that δ is small, then $\cos\delta \approx 1$ and it can be easily seen from Eq. (12.3) that Q_R is equal to $[|V_R|(|V_S| - |V_R|)/X]$. This shows that the reactive power flow depends on the difference in magnitudes of the bus voltages and flows from the higher voltage bus to the lower voltage bus.

Before proceeding to discuss the transient voltage stability, it would be helpful to understand the intricacies of the above relationships during different operating conditions through the example given below.

Example 12.1 Voltages at the sending and receiving ends of a radial line are 1.05 and 1.0 pu, respectively. Determine the sending end and receiving end reactive powers for (i) $\delta = 10°$, (ii) $\delta = 20°$, (iii) $\delta = 30°$, and (iv) $\delta = 40°$. Discuss the results.

Solution Here $V_S = 1.05$ pu and $V_R = 1.0$. Assume the line to be loss less and has an inductive reactance of X pu. The sending and receiving end reactive powers are computed by using Eqs (12.2) and (12.3) below.

S. No.	δ	10°	20°	30°	40°
1	Q_S	0.0685/X	0.1158/X	0.1932/X	0.2982/X
2	Q_R	0.0340/X	− 0.0133/X	− 0.0907/X	− 0.1957/X

Discussion: The receiving end voltage V_R is within the specified ± 5% requirement. When δ is small (= 10°), the reactive power flows from the sending end to receiving end, and the line loss is $(0.0685 - 0.0340)/X$ as is to be expected. When δ increases to 20°, despite reactive power flowing into the line from the sending end, there is a demand of $- 0.0133/X$ pu reactive power at the receiving end and the line loss is now $(0.1158 + 0.0133)/X$. Thus, instead of reactive power flowing from the sending end to the receiving end, it is being consumed in the line. In other words, rather than transmitting reactive power, the line is draining power from the system. Also, increase in delta is an indication that the real power demand is being met, however, the belief that the reactive power will flow from higher voltage bus to the lower voltage bus does not hold good.

For obtaining generalized plots applicable to all values of X, Eqs (12.1) and (12.3) are normalized by making the following substitutions $v = [|V_R|/|V_S|]$, $p = (P_R \times X)/|V_S|^2$, and $q = (Q_R \times X)/|V_S|^2$. Substitution of the normalizing terms yields

$$p = v\sin\delta \tag{12.4}$$

$$q = -v^2 + v\cos\delta \tag{12.5}$$

In order to eliminate δ, squaring and adding Eqs (12.4) and (12.5), and rearranging yields

$$v^4 + (2q - 1)v + \left(p^2 + q^2\right) = 0$$

or, $$v^2 = \frac{-(2q-1) \pm \sqrt{(2q-1)^2 - 4(p^2 + q^2)}}{2} = \frac{-(2q-1) \pm \sqrt{1 - 4q - 4p^2}}{2} \quad (12.6)$$

12.5 TECHNIQUES FOR TRANSIENT VOLTAGE STABILITY STUDIES

Several methods are available for studying the voltage stability problem. Two of the commonly used methods are described here.

12.5.1 P–V Curve Method

Reference to Eq. (12.6) shows that v^2 will have a real solution if

$$(1 - 4q - 4p^2) \geq 0 \quad (12.7)$$

The inequality in Eq. (12.7) gets modified by assuming constant power factor load, that is, $q/p = k$, as follows

$$p \leq \frac{1}{2}\left(\sqrt{1 + k^2} - k\right) \quad (12.8)$$

Equation (12.6) is also modified in terms of the constant power factor k and is expressed as follows

$$v^2 = \frac{-(2pk - 1) \pm \sqrt{1 - 4pk - 4p^2}}{2} = 0.5 - pk \pm \sqrt{0.25 - pk - p^2} \quad (12.9)$$

Hence, Eq. (12.8) will yield a value of p corresponding to every value of k, which will lead to two values of v from Eq. (12.9). The two voltage values are obtained as follows

$$v_1 = \sqrt{0.5 - pk + \sqrt{0.25 - pk - p^2}} \quad (12.10a)$$

$$v_2 = \sqrt{0.5 - pk - \sqrt{0.25 - pk - p^2}} \quad (12.10b)$$

For the voltages in Eqs (12.10) to be positive, the value of the terms under the square root should always be positive. It is easily seen that v_1 will always be positive. For v_2 to have a positive value, the condition required to be met is

$$\left(0.5 - pk - (0.25 - pk - p^2)^{0.5}\right)^{0.5} \geq 0$$

or, $$p^2(1 + k^2) \geq 0 \quad (12.11)$$

From Eq. (12.11) it is seen that voltage v_2 will always have a positive value.

By representing load as a constant power factor type, Eqs (12.10) are employed to plot active power p in pu versus voltage v in pu for fixed values of k and are shown in Fig. 12.2. The shape of P–V curve is similar to that of a parabola. Equation (12.8) establishes the maximum value of p and is called the critical operating or knee point and is shown in Fig. 12.2. The locus of critical points is shown in Fig. 12.2 by a dashed line, and normally the operating points above the critical points denote acceptable operating conditions. The knee point of P–V curve indicates critical loading point. At the knee of the P–V curve, the voltage drops rapidly with an increase in load demand at the receiving end. Power flow solution fails to converge beyond this limit, which is indicative of instability. When a system is operating close to the critical point, any sudden decrease in load power factor which corresponds to an increase in reactive load power may shift the operating point from a stable operating condition to an unstable one falling below the critical locus line leading to voltage instability. The distance between the operating point and the knee point of the curve gives the stability margin.

The P–V method provides an estimate of the real power margin prior to the critical voltage instability point. In radial systems, changes in active power consumption are monitored against voltage at a critical bus. In case of large inter-connected networks, voltage of a representative bus or a critical bus is examined against the total active load in the load area.

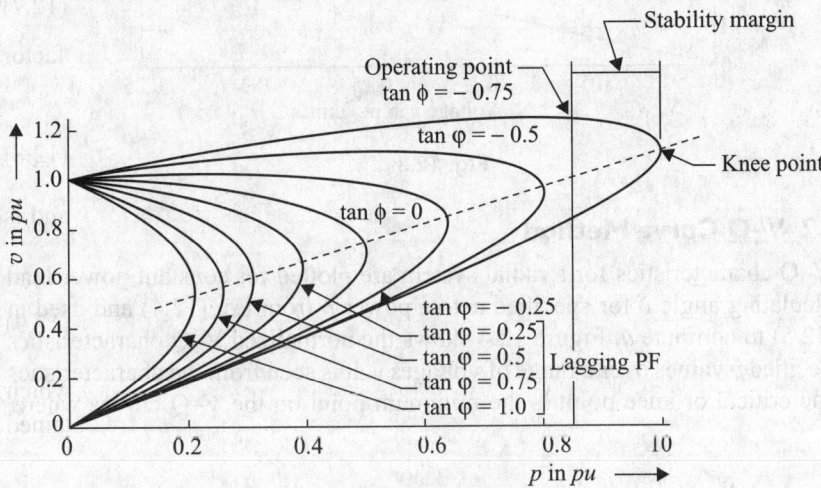

Fig. 12.2 P–V plots for normalized values of p and v

Example 12.2 The 220 kV radial feeder, shown in Fig. 12.1, is supplying a load at a power factor of 0.78. Plot the P–V curve and determine the critical MW load and the critical voltage (kV) of the line. Assume a line reactance of 60 Ω.

Solution The power factor angle of the load $\varphi = \cos^{-1}(0.78) = 38.74°$

Constant power factor load factor $k = \tan(\varphi) = 0.8$

Employing the inequality of Eqs (12.8) and (12.10) for the two voltage values, the P–V curve is plotted using MATLAB and is shown in Fig. 12.3.

From the plot,

Critical power $p = 0.24$ pu

Critical voltage $v = 0.55$ pu

Using the normalizing factor $p = P_R X / |V_S|^2$ the critical power

$$P_R = \frac{0.24 \times (220)^2}{60} = 193.6 \text{ MW}$$

From the normalizing factor $v = |V_R|/|V_S|$ the critical voltage $|V_R| = 0.55 \times 220 = 121$ kV

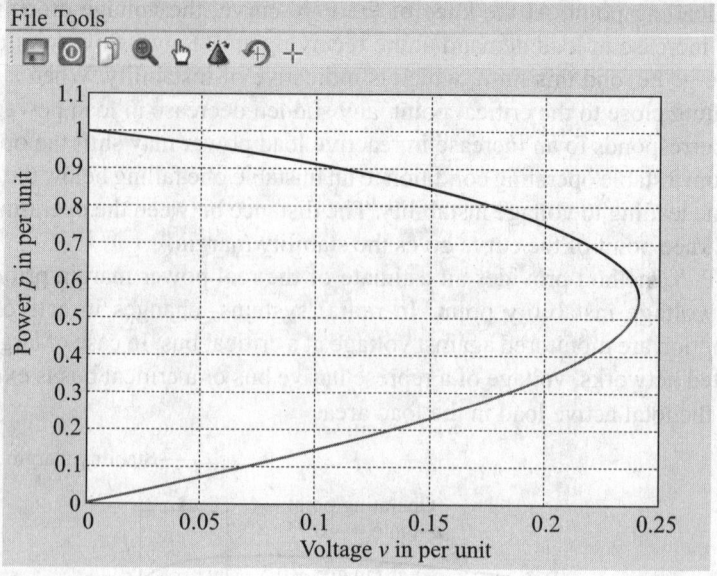

Fig. 12.3

12.5.2 V–Q Curve Method

The V–Q characteristics for a radial system are plotted for constant power load by calculating angle δ for specified active power p from Eq. (12.4) and used in Eq. (12.5) to compute q. Figure 12.4 shows the normalized V–Q characteristics for specified p values over a range of voltages v. It is seen from the characteristics that the critical or knee point is the minimum point on the V–Q curves where,

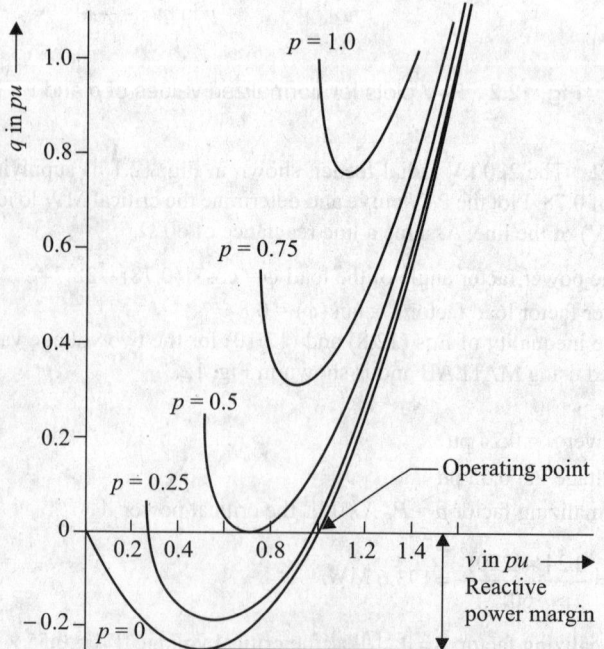

Fig. 12.4 Normalized V–Q characteristics for specified p values

$dq/dv = 0$. The knee point of the V–Q curve represents the voltage stability limit. To the right of the critical (minimum) point, the slope dq/dv is positive and therefore represents a stable operation region. The section to the left of the critical point represents an unstable operation region since dq/dv has a negative slope. In the region where dq/dv is negative, stable operation is feasible with a reactive power compensation whose polarity is opposite to that of the normal direction and which has enough control range and a high Q/V range.

The description of the voltage stability problem, though simplistic, helps to visualize various features of power system stability. In interconnected complex networks, the problem of voltage instability leading to voltage collapse is complex and dependent on several factors such as limits of generator reactive power capability, characteristics of reactive compensating devices, strength of transmission system, power transfer levels, and load characteristics. Further, if the action of different control systems and protective devices are not properly coordinated, the problem gets compounded.

Example 12.3 Plot the V–Q characteristic for 0.8 per unit active power flow in the radial feeder of Example 12.2 and therefrom determine the critical MVAR and the critical voltage.

Solution For the specified active power and voltage, the power angle δ is computed from Eq. (12.4). The power angle δ is then used in Eq. (12.5) to compute reactive power q for a range of voltages v. The required V–Q characteristic is plotted by making use of MATLAB and is shown in Fig. 12.5.

From the V–Q characteristic in Fig. 12.5, critical reactive power q is estimated at 0.38 pu and critical voltage at 0.95 pu.

As before, from the normalizing factors, critical reactive power

$$Q = \frac{q|V_S|^2}{X} = \frac{0.38 \times 220^2}{60} = 306.53 \text{ MVAR}$$

and critical voltage $|V_R| = 0.95 \times 220 = 209$ kV

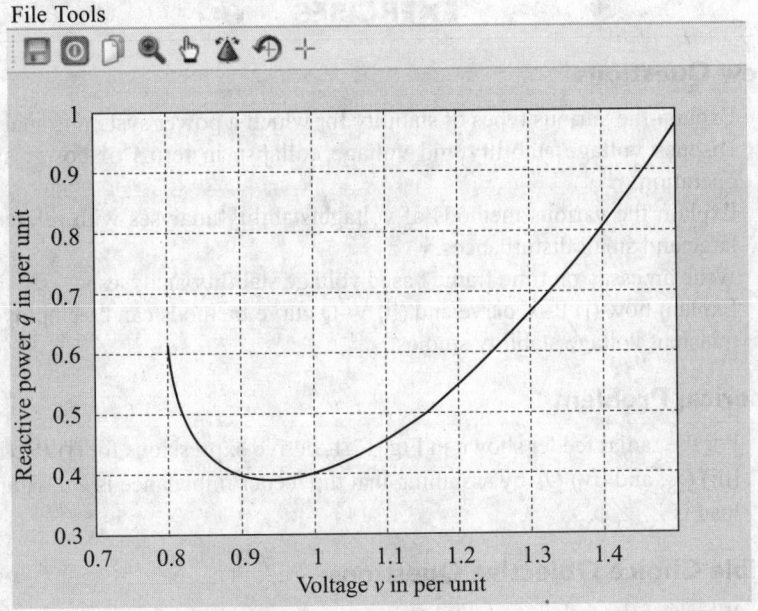

File Tools

Fig. 12.5

SUMMARY

- Voltage stability is load driven and the need to conduct such studies has arisen mainly due to large complex interconnected networks operating close to their limits.
- Voltage instability in power systems is brought about due to (i) a large disturbance and (ii) a small disturbance.
- Small disturbance or transient voltage stability is performed to check whether a system will maintain steady voltages following gradual variation in system load.
- A system is voltage stable if $dQ/d|V|$ maintains a positive slope at all its buses.
- Based on normalized expressions for power flow in a radial feeder, P–V and V–Q characteristics are developed which are employed to determine critical active and reactive powers along with critical voltages.

SIGNIFICANT FORMULAE

- Active power: $P_R = P_L = \dfrac{|V_S|\,|V_R|}{X}\sin\delta$

- Reactive powers: $Q_S = \dfrac{|V_S|^2}{X} - \dfrac{|V_S|\,|V_R|}{X}\cos\delta$

$$Q_R = -\dfrac{|V_R|^2}{X} + \dfrac{|V_R|\,|V_S|}{X}\cos\delta$$

- Normalized – Active power $p = v\sin\delta$
  - Reactive power $q = -v^2 + v\cos\delta$
  - Voltage $v^2 = \dfrac{-(2q-1)\pm\sqrt{1-4q-4p^2}}{2}$

EXERCISES

Review Questions

12.1 Explain the various types of stability for which a power system is analysed.

12.2 Discuss voltage stability and voltage collapse in terms of power system operation.

12.3 Explain the various methods of voltage stability analyses with reference to large and small disturbances.

12.4 Write an essay on time frame based voltage stability analyses.

12.5 Explain how (i) P–V curve and (ii) V–Q curve methods can be employed for transient voltage stability studies.

Numerical Problem

12.1 For the radial feeder shown in Fig. 12.1, derive expressions for (i) P_S, (ii) P_R, (iii) Q_S, and (iv) Q_R by assuming that the feeder impedance is $Z \angle \theta$, and the load is $Z_L \angle \varphi$.

Multiple Choice Objective Questions

12.1 Which of the following stability is not generator driven?
   (a) Rotor angle   (b) Frequency

(c) Voltage   (d) None of these

12.2 Which of the following is a reason for power systems not being able to operate with redundancies?
   (a) Environment conservation   (b) Different loading patterns
   (c) Greater usage of electricity   (d) All of these

12.3 A power system is voltage stable if it is able to maintain voltages within
   (a) ± 5%   (b) ± 6.5%
   (c) ± 8%   (d) ± 10%

12.4 Which of the following is a reason for attention being focussed on voltage stability studies of power systems?
   (a) Lesser number of voltage buses
   (b) Economic compulsions
   (c) Frequent line and generator outages
   (d) All of these

12.5 In which of the following conditions, a system will be transient voltage stable?
   (a) dq/dv is negative on all the buses.
   (b) dq/dv is positive on all the buses.
   (c) dq/dv is negative on at least one bus.
   (d) dq/dv is positive on at least one bus.

12.6 The dynamics of which of the following devices do not come into play during transient period of voltage stability studies?
   (a) AVRs   (b) Turbine governors
   (c) OLTCs   (d) Excitation systems

12.7 Which of the following is the correct expression for reactive power at the receiving end?

   (a) $-\dfrac{|V_R|^2}{X}+\dfrac{|V_R|\,|V_S|}{X}\cos\delta$ (b) $\dfrac{|V_S|^2}{X}-\dfrac{|V_S|\,|V_R|}{X}\cos\delta$

   (c) $-\dfrac{|V_S|^2}{X}-\dfrac{|V_S|\,|V_R|}{X}\cos\delta$ (d) $-\dfrac{|V_R|^2}{X}-\dfrac{|V_R|\,|V_S|}{X}\cos\delta$

12.8 Which of the following will lead to an imaginary solution of v^2?
   (a) $(1-4q-4p^2)>0$ (b) $(1-4q-4p^2)=0$
   (c) $(1-4q-4p^2)<0$ (d) None of these

12.9 Which of the following conditions is met at the nose point in the V–Q characteristics?

   (a) $\dfrac{dq}{dv}=0$ (b) positive $\dfrac{dq}{dv}$

   (c) negative $\dfrac{dq}{dv}$ (d) None of these

12.10 In which of the following, voltage instability will lead to voltage collapse?
   (a) Weak transmission system   (b) Uncoordinated control systems
   (c) Unmatched characteristics of reactive compensating devices
   (d) All of these

Answers

12.1 (c)	12.2 (d)	12.3 (a)	12.4 (d)	12.5 (b)	12.6 (c)
12.7 (a)	12.8 (c)	12.9 (a)	12.10 (d)		

CHAPTER 13

Contingency Analysis Techniques

Learning Outcomes

A careful study of this chapter will assist the reader to
- Understand the consequences of power supply failure and the need for examining the problem of security in power systems
- Know the different categories of security levels and familiarize with the preventive actions to be taken when a security level is breached
- Simulate mathematical models of power systems (including interconnected networks), based on simplifying approximations, understand loop impedance matrix formation and network reduction techniques for contingency analyses
- Understand how to develop current injection and current shift distribution factors
- Perform analyses for single and compound outages
- Comprehend the approximate power flow method and apply it to simulate and analyse contingencies in power systems

OVERVIEW

The desirable attributes in an electrical power system are economic operation, minimum damage to the environment, and security of energy supply. A power system supply is considered to be secure when the probability of power supply failure is low. A power supply failure occurs due to the tripping of a transmission line or an equipment malfunction and if this trend is not arrested, it may lead to the blackout of a region due to the cascading effect.

Clearly, economy of operation and security of the energy management system (EMS) are in conflict with each other; higher the security of power supply more costly will be its operation. Hence, power systems are operated at minimum cost with assured levels of security of supply. The latter ensures mitigation of emergency conditions.

EMS security is constituted of (i) system monitoring by employing state estimation techniques and (ii) contingency analysis along with corrective action measures.

In system monitoring, online data, as loads and generation vary, is gathered on real-time basis at the remote terminal units (RTUs) and transmitted over microwave links to a central digital computer, which analyses the same by employing statistical techniques and estimates the state variables such as the system bus voltage magnitudes and voltage phase angles of the system. Additionally, information in respect of frequency, generation outputs, transformer taps, etc. is also acquired and transmitted to the central computer. Based on an analysis of the large database, the central computer directs the EMS managers or load dispatchers to take corrective action for overloads or out of limit voltages.

13.1 SECURITY IN A POWER SYSTEM

A power system is classified into various levels of security. When an electrical power system is operating in the normal steady state, the frequency and bus voltage magnitudes are maintained within the maximum and minimum specified values by maintaining a balance between the real and reactive powers generated and the real and reactive load demands. Along with upholding a balance between generation and load demand, another precondition for normal operation in the steady state is to ensure that the loads on the generators and transformers are maintained within their rated capacities, in addition to guaranteeing that the transmission lines also do not exceed both their steady-state and thermal state limits.

In the normal steady-state operation, a certain level of security is provided by maintaining some amount of excess generation in the shape of spinning reserve. When the margin of generation falls, due to an increase in demand, below a chosen threshold level or a disturbance is expected to happen, the security level is lowered to the alert level (level 1). Even in the alert level of security, the system continues to operate in synchronism with all equalities and inequalities still being followed. However, in order to restore the original operating normal condition, automatic load control action is initiated to re-establish the generation margin and/or to eliminate the disturbance.

If the pre-emptive control action falls short or if a severe disturbance happens, the system goes into the emergency state (level 2) in which one or some of the constituents are overloaded. At this point, it is essential that emergency control action be commenced and the system returns to the normal steady state or the alert state. Emergency control actions involve rescheduling generation, disconnection of faulted portion of the system, or even load shedding.

Should the emergency action fail, the system enters an extreme state (level 3) when the system splits into sections, leading eventually to generator tripping and consequential power failure or blackout. The duration of such a cascading of events, which transform the system from normal operating condition to a state of emergency, could characteristically be from a few seconds to several minutes. The restorative process, wherein the system is reinstated to its normal operating state, is much slower and could take a few hours or even a few days.

Power system security, therefore, is envisaged as a measure of the robustness of the system to withstand disturbances. System security is the process of detection of emergency operating conditions, which manifest themselves in the form of limit violations, in its pre-disturbance or post-disturbance operating condition. System security appraisal is a procedure to identify such emergency operating conditions which consist of (i) system monitoring and (ii) contingency analysis.

System monitoring relies on real-time online data acquisition which enables the EMS managers to control circuit breakers, disconnect switches, and change transformer taps by remote control whenever violations are detected in the operating conditions. In an emergency condition, the changes in a network are fast and system monitoring cannot be relied upon to ensure a secure system.

Contingency analysis consists of a digital program which simulates an existing power system and predicts the bus voltage magnitudes and line currents for a possible generator or a transmission line tripping. This information enables the EMS manager to establish whether the system can be operated safely, following a particular generator or a transmission line tripping. Thus, contingency analysis ensures that no single contingency will result in overloads or limit violations of voltage magnitudes, thereby guaranteeing a secure power system.

13.2 APPROXIMATIONS IN CONTINGENCY ANALYSIS

In a power system, whenever a circuit breaker switches-in or switches-out a transmission line or a transformer, the bus voltage magnitudes and line currents are reallocated. As already stated a contingency analysis program computes the new steady-state values of bus voltage magnitudes and line currents following the switching operation. Unlike fault computations in a system, the interest of the EMS manager, in contingency analysis, centres on knowing whether voltage limits have been exceeded or whether current overload levels have occurred. Hence, the computations are not required to be accurate, which permits approximations in simulations for contingency analysis of a power system. Therefore, in view of the foregoing, the following approximations are assumed.

(a) Purely reactive network models are employed by neglecting resistances of the system components.
(b) Linearity is presumed, thereby making the principle of superposition feasible.
(c) Shunt capacitances in transmission lines and off-nominal taps in transformers are excluded.
(d) Loads are simulated as constant current injections into various buses, in both Y_{bus} and Z_{bus} systems. The use of bus current injections simplifies the computations.
(e) Switching out a line from a system model is simulated by representing the line, connected between the two buses, by the negative of its series reactance.

Prior to proceeding with a description of the contingency analysis, it is necessary to develop the concept of compensating currents to simulate the effect of addition or removal of multiple lines in a power system without modifying its Z_{bus} matrix and introduce tools for modelling interconnected networks.

13.3 SIMULATION OF ADDITION AND REMOVAL OF MULTIPLE LINES IN A POWER SYSTEM

Consider that the Z_{bus} matrix for the network of n buses, shown in Fig. 13.1, has been assembled by the method described in Section 6.8. In Fig. 13.1, I_1, I_2, I_3, ..., I_n, represent the per unit current injections in buses 1, 2, 3, ..., n, respectively and V_1, V_2, V_3, ..., V_n, are the corresponding per unit bus voltages with respect to the reference, as a consequence of the current injections.

Assume that the bus voltages V_1, V_2, V_3, ..., V_n set up in the system as a result of the constant current injections I_1, I_2, I_3, ..., I_n are known and can be expressed by Eq. (6.5) as follows:

$$V_{\text{bus}} = Z_{\text{bus}} I_{\text{bus}} \qquad (13.1)$$

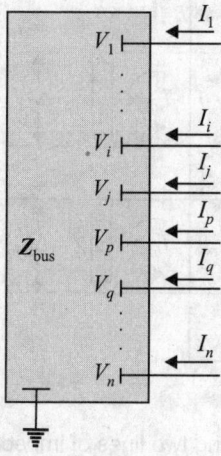

Fig. 13.1 Bus voltages in a network of *n*-buses due to current injections at the buses

In the expanded matrix form, this relation may be written as

$$
\begin{bmatrix} V_1 \\ .. \\ .. \\ V_i \\ V_j \\ V_p \\ V_q \\ .. \\ .. \\ V_n \end{bmatrix}
=
\begin{bmatrix}
Z_{11} & ..... & ..... & ..... & Z_{1i} & Z_{1j} & Z_{1p} & Z_{1q} & ..... & Z_{1n} \\
..... & ..... & ..... & ..... & ..... & ..... & & & ..... & ..... & ..... \\
..... & ..... & ..... & ..... & ..... & ..... & i & ..... & ..... & i \\
Z_{i1} & ..... & ..... & ..... & Z_{ii} & Z_{ij} & Z_{ip} & Z_{iq} & ..... & Z_{in} \\
Z_{j1} & ..... & ..... & ..... & Z_{ji} & Z_{jj} & Z_{jp} & Z_{jq} & ..... & Z_{jn} \\
Z_{p1} & ..... & ..... & ..... & Z_{pi} & Z_{pj} & Z_{pp} & Z_{pq} & ..... & Z_{pn} \\
Z_{q1} & ..... & ..... & ..... & Z_{qi} & Z_{qj} & Z_{qp} & Z_{qq} & ..... & Z_{qn} \\
..... & & & & & & & & & \\
..... & & & & & & & & & \\
Z_{n1} & ..... & ..... & ..... & Z_{ni} & Z_{nj} & Z_{np} & Z_{nq} & ..... & Z_{nn}
\end{bmatrix}
\begin{bmatrix} I_1 \\ .. \\ .. \\ I_i \\ I_j \\ I_p \\ I_q \\ .. \\ .. \\ I_n \end{bmatrix}
$$

$$(13.2)$$

It is now desired to connect two lines of impedances Z_a and Z_b between buses i and j and buses p and q, respectively, as shown in Fig. 13.2.

Clearly, as a result of connecting the impedances between the two pairs of buses, voltages at all the buses will change. Let the new values of the bus voltages be represented by a column vector V'.

$$V' = \left[V'_1,\ V'_2,\ ...,\ V'_i,\ V'_j,\ V'_p,\ V'_q,\ ...,\ V'_n \right]^t$$

If it is assumed that current flowing from bus i to bus j is I_a and from bus p to bus q is I_b respectively, the relation between the new bus voltages and currents, in terms of the line impedances can be written as

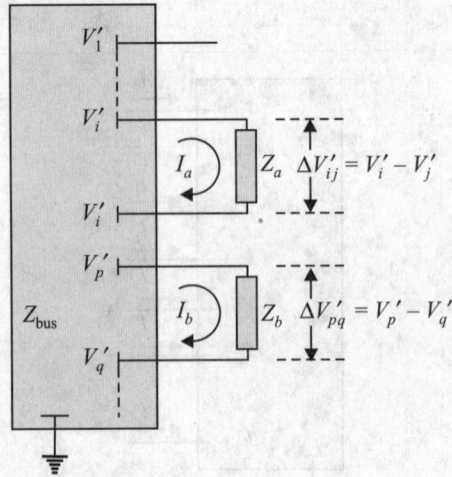

Fig. 13.2 Connecting two lines of impedances Z_a and Z_b between two pairs of buses

$$V'_i - V'_j = Z_a I_a \tag{13.3a}$$

$$V'_p - V'_q = Z_b I_b \tag{13.3b}$$

Equations (13.3) can be written in matrix form as

$$
\begin{array}{c}
\quad 1.. \quad i \quad j \quad p \quad q.. \quad\quad n \\[4pt]
\begin{array}{c} a \\ b \end{array}
\begin{bmatrix}
0.. & 1 & -1 & 0 & 0 & \cdots & 0 \\
0.. & 0 & 0 & 1 & -1 & \cdots & 0
\end{bmatrix}
\end{array}
\begin{bmatrix} V'_1 \\ \vdots \\ V'_i \\ V'_j \\ V'_p \\ V'_q \\ \vdots \\ V'_n \end{bmatrix}
= AV' =
\begin{bmatrix} Z_a & 0 \\ 0 & Z_b \end{bmatrix}
\begin{bmatrix} I_a \\ I_b \end{bmatrix}
\tag{13.4}
$$

In Eq. (13.4), A is the element (branch) to node (bus) matrix (see Section 6.2.2) which shows the addition of the two new branches between buses i-j and p-q respectively. Figure 13.3 shows the equivalent circuit of the network in which the new injected currents I_a and I_b combine with the currents of the original network to set up new bus voltages represented by

$$V' = \begin{bmatrix} V'_1, & V'_2, & ..., & V'_i, & V'_j, & V'_p, & V'_q, & ..., & V'_n \end{bmatrix}^t$$

Therefore, it can be noted that the new equivalent constant current injections produce an effect which is equivalent to that of connecting the branch impedances

Z_a and Z_b. In this manner, it is observed that the modification of the Z_{bus} matrix is avoided by injecting equivalent currents I_a and I_b. For this reason, the equivalent injected currents are designated as compensating currents.

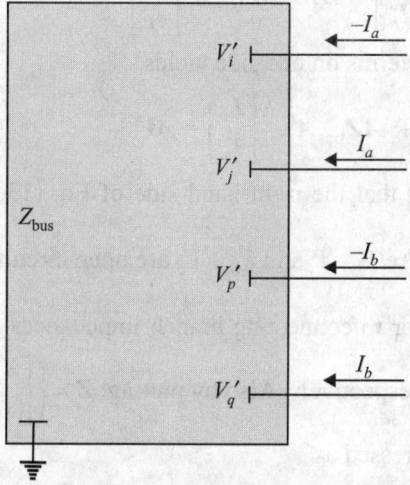

Fig. 13.3 Equivalent circuit with line impedances replaced by constant injected currents

The changes in bus voltages due to the compensating currents can be written in matrix form as follows

$$\Delta V = V' - V = Z_{bus} I_C \tag{13.5}$$

where V is the voltage vector of existing bus voltages, and I_C is the compensating current vector which can be expressed as

$$I_C = \begin{matrix} 1 \\ \cdots \\ i \\ j \\ p \\ q \\ \cdots \\ n \end{matrix} \begin{bmatrix} 0 \\ -I_a \\ I_a \\ -I_b \\ I_b \\ 0 \end{bmatrix} = \begin{matrix} 1 \\ \cdots \\ i \\ j \\ p \\ q \\ \\ n \end{matrix} \begin{array}{cc} a & b \\ \begin{bmatrix} 0 & 0 \\ -1 & 0 \\ 1 & 0 \\ 0 & -1 \\ 0 & 1 \\ 0 & 0 \\ 0 & 0 \end{bmatrix} \end{array} \begin{bmatrix} I_a \\ I_b \end{bmatrix} = -A^t \begin{bmatrix} I_a \\ I_b \end{bmatrix} \tag{13.6}$$

Substitution of Eq. (13.6) in Eq. (13.5) eliminates I_C and yields the voltage changes in terms of the compensating currents I_a and I_b as follows:

$$\Delta V = V' - V = -Z_{bus} A^t \begin{bmatrix} I_a \\ I_b \end{bmatrix} \tag{13.7}$$

Rearranging Eq. (13.7) gives

$$V' = V + \Delta V = V - Z_{bus} A^t \begin{bmatrix} I_a \\ I_b \end{bmatrix} \tag{13.8}$$

Pre-multiplying Eq. (13.8) by A and substituting for the product AV' from Eq. (13.4) yields

$$\begin{bmatrix} Z_a & 0 \\ 0 & Z_b \end{bmatrix} \begin{bmatrix} I_a \\ I_b \end{bmatrix} = AV - AZ_{bus} A^t \begin{bmatrix} I_a \\ I_b \end{bmatrix}$$

Gathering the current terms on one side yields

$$\left(\begin{bmatrix} Z_a & 0 \\ 0 & Z_b \end{bmatrix} + AZ_{bus} A^t \right) \begin{bmatrix} I_a \\ I_b \end{bmatrix} = AV \tag{13.9}$$

It can be observed that the right hand side of Eq. (13.9) can be written as

$$AV = \begin{bmatrix} V_i - V_j \\ V_p - V_q \end{bmatrix}$$ where $V_i - V_j$ and $V_p - V_q$ are open circuit voltage drops in the

original network (prior to connecting branch impedances Z_a and Z_b) between

bus pairs i-j and p-q respectively. Also by putting $Z = \left(\begin{bmatrix} Z_a & 0 \\ 0 & Z_b \end{bmatrix} + AZ_{bus} A^t \right)$,

Eq. (13.9) can be expressed as

$$Z \begin{bmatrix} I_a \\ I_b \end{bmatrix} = \begin{bmatrix} V_i - V_j \\ V_p - V_q \end{bmatrix}$$

or

$$\begin{bmatrix} I_a \\ I_b \end{bmatrix} = Z^{-1} \begin{bmatrix} V_i - V_j \\ V_p - V_q \end{bmatrix} \tag{13.10}$$

In Eq. (13.10), the open circuit voltages at buses i, j, p, and q can be computed by employing Eq. (13.2) if not already known. The significance of putting

$Z = \left(\begin{bmatrix} Z_a & 0 \\ 0 & Z_b \end{bmatrix} + AZ_{bus} A^t \right)$ is seen when $A Z_{bus} A^t$ is computed as follows

$$AZ_{bus}A^t = \begin{array}{c} a \\ b \end{array} \begin{bmatrix} 1 & -1 & 0 & 0 \\ 0 & 0 & 1 & -1 \end{bmatrix} \begin{array}{c} i \\ j \\ p \\ q \end{array} \begin{bmatrix} Z_{ii} & Z_{ij} & Z_{ip} & Z_{iq} \\ Z_{ji} & Z_{jj} & Z_{jp} & Z_{jq} \\ Z_{pi} & Z_{pj} & Z_{pp} & Z_{pq} \\ Z_{qi} & Z_{qj} & Z_{qp} & Z_{qq} \end{bmatrix} \begin{array}{c} i \\ j \\ p \\ q \end{array} \begin{bmatrix} 1 & 0 \\ -1 & 0 \\ 0 & 1 \\ 0 & -1 \end{bmatrix}$$

$$= \begin{bmatrix} \{(Z_{ii} - Z_{ij}) - (Z_{ji} - Z_{jj})\} & \{(Z_{ip} - Z_{iq}) - (Z_{jp} - Z_{jq})\} \\ \{(Z_{pi} - Z_{pj}) - (Z_{qi} - Z_{qj})\} & \{(Z_{pp} - Z_{pq}) - (Z_{qp} - Z_{qq})\} \end{bmatrix} \tag{13.11}$$

In computing the product $AZ_{bus} A^t$ in Eq. (13.11), for the sake of clarity and understanding, all the elements of both the A and Z_{bus} matrices have not been depicted. Figures 13.4 and 13.5 show the Thevenin's equivalent circuits looking into the buses i-j and p-q in Fig. 13.2.

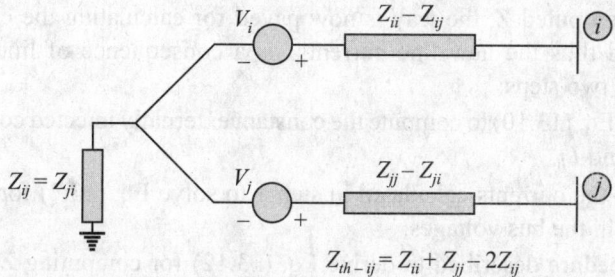

Fig. 13.4 Thevenin's equivalent circuit looking into buses *i-j*

A scrutiny of the 2×2 matrix in Eq. (13.11) shows that its diagonal terms are the Thevenin's equivalent impedances of the buses *i-j* and *p-q* as shown in Figs 13.4 and 13.5 respectively. Further, it may also be observed that the elements of the 2×2 matrix follow a pattern and can be directly computed from the $\mathbf{Z}_{bus}$ matrix of the original network.

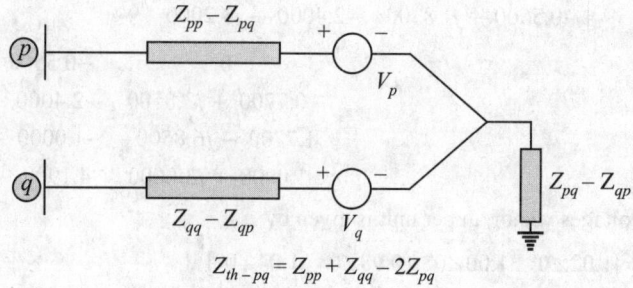

Fig. 13.5 Thevenin's equivalent circuit looking into buses *p-q*

It can be easily observed that $\mathbf{Z} = \begin{bmatrix} Z_a & 0 \\ 0 & Z_b \end{bmatrix} + \mathbf{A}\mathbf{Z}_{bus}\mathbf{Z}$ can now be written as

$$\mathbf{Z} = \begin{bmatrix} Z_a & 0 \\ 0 & Z_b \end{bmatrix} + \mathbf{A}\mathbf{Z}_{bus}\mathbf{A}^t$$

$$= \begin{array}{c} a \\ b \end{array} \begin{bmatrix} \{(Z_{ii}-Z_{ij})-(Z_{ji}-Z_{jj})\}+Z_a & \{(Z_{ip}-Z_{iq})-(Z_{jp}-Z_{jq})\} \\ \{(Z_{pi}-Z_{pj})-(Z_{qi}-Z_{qj})\} & \{(Z_{pp}-Z_{pq})-(Z_{qp}-Z_{qq})\}+Z_b \end{bmatrix}$$

$$\overset{a}{} \qquad\qquad\qquad\qquad \overset{b}{}$$

(13.12)

Equation (13.12) helps to draw the following important conclusions to study the effect of addition or removal of lines:

(i) The loop impedance $\mathbf{Z}$ can be computed directly from the columns of i, j, p, and q of the original $\mathbf{Z}_{bus}$ impedance of the network.

(ii) The effect of adding lines to the network can be simulated by adding their impedances to the respective diagonal elements of $\mathbf{Z}$.

(iii) Similarly, the effect of removing lines can be simulated by adding the negative of the line impedances to the respective diagonal elements of $\mathbf{Z}$.

Having computed Z, the way is now paved for calculating the changed bus voltages and thus the new line currents, as a consequence of line additions/removals, in two steps:

Step 1: Use Eq. (13.10) to compute the constant externally injected compensating currents I_a and I_b.

Step 2: Use the currents calculated in step 1 to solve Eq. (13.7) for computing the changes in the bus voltages.

The procedure described to derive Eq. (13.12) for computing Z can be extended to addition/removal of N lines. In such a case, the order of the Z matrix will be $N \times N$.

Example 13.1 The Y_{bus} admittance matrix, in per unit on a base impedance of 100 ohms shown here represents a four-bus hypothetical network of Example 6.3.

$$Y_{bus} = \begin{bmatrix} 1.1400 - j4.1900 & -0.5900 + j2.3500 \\ -0.5900 + j2.3500 & 3.7600 - j9.4000 \\ 0 & -0.7700 + j3.8500 \\ -0.5500 + j1.8300 & -2.4000 + j3.2000 \end{bmatrix}$$

$$\begin{bmatrix} 0 & -0.5500 + j1.8300 \\ -0.7700 + j3.8500 & -2.4000 + j3.2000 \\ 1.7700 - j6.8500 & -1.0000 + j3.0000 \\ -1.0000 + j3.0000 & 4.1900 - j8.0300 \end{bmatrix}$$

The bus voltages vector, in per unit is given by

$$V = [1.02\angle0° \quad 1.00\angle0° \quad 0.98\angle0° \quad 1.04\angle0°]^t.$$

Determine (i) the Thevenin's impedances looking into buses 1-2 and 3-4 and (ii) the new values of the bus voltages when one line having a reactance of $j0.4$ per unit is connected between buses 1-2 and a second line of reactance $j0.3$ per unit is connected between buses 3-4. Neglect resistances of the lines.

Solution Using MATLAB command the Y_{bus} matrix is inverted as follows:
```
>> Zb=inv(Y)        % MATLAB command for matrix inversion
Zbus =
```

$$\begin{bmatrix} 4.2211 + j0.4412 & 4.1857 + j0.2501 & 4.1713 + j0.2158 & 4.1483 + j0.1759 \\ 4.1857 + j0.2501 & 4.2231 + j0.3076 & 4.1974 + j0.2461 & 4.1562 + j0.1744 \\ 4.1713 + j0.2158 & 4.1974 + j0.2461 & 4.2173 + j0.3495 & 4.1577 + j0.1738 \\ 4.1483 + j0.1759 & 4.1562 + j0.1744 & 4.1577 + j0.1738 & 4.1593 + j0.1728 \end{bmatrix}$$

Since resistance is to be neglected, the imaginary portion of $[Z_{bus}]$ is extracted.
```
>> Zbus=imag(Zb)    %MATLAB command for extracting only the imaginary
                    components
```

$$Z_{bus} = j \times \begin{bmatrix} 0.4412 & 0.2501 & 0.2158 & 0.1759 \\ 0.2501 & 0.3076 & 0.2461 & 0.1744 \\ 0.2158 & 0.2461 & 0.3495 & 0.1738 \\ 0.1759 & 0.1744 & 0.1738 & 0.1728 \end{bmatrix}$$

It is seen as per the statement $i = 1$, $j = 2$, $p = 3$, and $q = 4$.
The Thevenin's impedances are computed as follows:

$$Z_{th\,1-2} = j\{Zbus(1,\,1) - Zbus(1,\,2) - (Zbus(1,\,2) - Zbus(2,\,2))\}$$

$$= j\{0.4412 - 0.2501 - (0.2502 - 0.3076)\} = j0.2486$$

$$Z_{th\,3-4} = j\{Zbus(3,\,3) - Zbus(3,\,4) - (Zbus(4,\,3) - Zbus(4,\,4))\}$$

$$= j\{0.3495 - 0.1738 - (0.1738 - 0.1728)\} = j0.1747$$

The elements of the $2 \times 2\ \mathbf{Z}$ matrix are computed by using Eq. (13.12) as follows:

```
>> z(1,1)=Zbus(1,1)-Zbus(1,2)-(Zbus(1,2)-Zbus(2,2))+0.4;
>> z(1,2)=Zbus(1,3)-Zbus(1,4)-(Zbus(2,3)-Zbus(2,4));
>> z(2,1)=Zbus(3,1)-Zbus(3,2)-(Zbus(4,1)-Zbus(4,2));
>> z(2,2)=Zbus(3,3)-Zbus(3,4)-(Zbus(4,3)-Zbus(4,4))+0.3;
>> z=[z(1,1) z(1,2);z(2,1) z(2,2)]
```

$$\mathbf{z} = j \begin{bmatrix} 0.6846 & -0.0318 \\ -0.0318 & 0.4747 \end{bmatrix}$$

From the bus voltages, the difference vector is given by

$$V = \begin{bmatrix} V_1 - V_2 \\ V_3 - V_4 \end{bmatrix} = \begin{bmatrix} 1.02\angle 0° - 1.00\angle 0° \\ 0.98\angle 0° - 1.04\angle 0° \end{bmatrix} = \begin{bmatrix} 0.0200 \\ -0.0600 \end{bmatrix}$$

The compensating current vector I_C is calculated as shown

```
>> IC = inv(z)*V
```

$$I_C = \begin{bmatrix} I_1 \\ I_2 \end{bmatrix} = \begin{bmatrix} 0 - j0.0247 \\ 0 + j0.1247 \end{bmatrix}$$

The branch to bus matrix $A = \begin{bmatrix} 1 & -1 & 0 & 0 \\ 0 & 0 & 1 & -1 \end{bmatrix}$

The changes in bus voltages as per Eq. (13.7) are computed as follows:

```
>> delV=-Zbus*A'*IC
```

$$\Delta V = \begin{bmatrix} 0.0005 \\ 0.0103 \\ 0.0226 \\ 0.0001 \end{bmatrix}$$

The new bus voltages as a consequence of addition of two line impedances are

$$\begin{bmatrix} V_1' \\ V_2' \\ V_3' \\ V_4' \end{bmatrix} = \begin{bmatrix} 1.02\angle 0° \\ 1.00\angle 0° \\ 0.98\angle 0° \\ 1.04\angle 0° \end{bmatrix} + \begin{bmatrix} 0.0005 \\ 0.0103 \\ 0.0226 \\ 0.0001 \end{bmatrix} = \begin{bmatrix} 1.0205 \\ 1.0103 \\ 1.0026 \\ 1.0401 \end{bmatrix}$$

Example 13.2 Compute the bus voltages in Example 13.1 when (i) a line of reactance $j\ 0.4$ per unit between buses 1 and 2 along with a line of reactance $j\ 0.3$ per unit between buses 3 and 4 are removed, (ii) a line of reactance $j\ 0.4$ per unit between buses 1 and 2 is removed and a line of reactance $j\ 0.3$ per unit between buses 3 and 4 is added, and (iii) a line of reactance $j\ 0.4$ per unit between buses 1 and 2 is added and a line of reactance $j\ 0.3$ per unit between buses 3 and 4 is removed.

Solution It may be noted that when lines are added or removed, only the diagonal elements of Z change.

(i) When both the lines are removed

$Z(1, 1) = j\,0.2486 - j\,0.4 = -j\,0.1514$ and $Z(2, 2) = j\,0.1747 - j\,0.3 = -j\,0.1253$

$$Z = j\begin{bmatrix} -0.1514 & -0.0318 \\ -0.0318 & -0.1253 \end{bmatrix}$$

From Example 13.1

$$V = \begin{bmatrix} 0.0200 \\ -0.0600 \end{bmatrix}$$

The compensating current vector I_C is given by

```
>> IC=inv(z)*V
```

$$I_C = \begin{bmatrix} I_1 \\ I_2 \end{bmatrix} = \begin{bmatrix} +j0.2458 \\ -j0.5412 \end{bmatrix}$$

The changes in bus voltages as per Eq. (13.7) are

```
>> delv=- Zbus*A'*IC
```

$$\Delta V = \begin{bmatrix} 0.0254 \\ -0.0529 \\ -0.1025 \\ -0.0002 \end{bmatrix}$$

The new bus voltages as a consequence of the removal of lines between buses 1-2 and 3-4 are

$$\begin{bmatrix} V_1' \\ V_2' \\ V_3' \\ V_4' \end{bmatrix} = \begin{bmatrix} 1.02\angle0° \\ 1.00\angle0° \\ 0.98\angle0° \\ 1.04\angle0° \end{bmatrix} + \begin{bmatrix} 0.0254 \\ -0.0529 \\ -0.1025 \\ -0.0002 \end{bmatrix} = \begin{bmatrix} 1.0454 \\ 0.9471 \\ 0.8775 \\ 1.0402 \end{bmatrix}$$

(ii) When a line between buses 1-2 is removed and a line between buses 3-4 is added

$$Z(1, 1) = j\,0.2486 - j\,0.4 = -j\,0.1514 \text{ and } Z(2, 2) = j\,0.1747 + j\,0.3 = j\,0.4747$$

$$Z = j\begin{bmatrix} -0.1514 & -0.0318 \\ -0.0318 & 0.4747 \end{bmatrix}$$

```
>> IC=inv(z)*V
```

$$I_C = \begin{bmatrix} I_1 \\ I_2 \end{bmatrix} = \begin{bmatrix} j0.1041 \\ j0.1334 \end{bmatrix}$$

```
>> delv=-Zbus*A'*IC
```

$$\Delta V = \begin{bmatrix} 0.0252 \\ 0.0036 \\ 0.0203 \\ 0.0003 \end{bmatrix}$$

The new bus voltages as a consequence of the removal of the lines between buses 1-2 and addition of a line between buses 3-4 are

$$\begin{bmatrix} V_1' \\ V_2' \\ V_3' \\ V_4' \end{bmatrix} = \begin{bmatrix} 1.02\angle0° \\ 1.00\angle0° \\ 0.98\angle0° \\ 1.04\angle0° \end{bmatrix} + \begin{bmatrix} 0.0252 \\ 0.0036 \\ 0.0203 \\ 0.0003 \end{bmatrix} = \begin{bmatrix} 1.0452 \\ 1.0036 \\ 1.0003 \\ 1.0403 \end{bmatrix}$$

(iii) When a line between buses 1-2 is added and a line between buses 3-4 is removed

$Z(1, 1) = j\,0.2486 + j\,0.4 = j\,0.6486$ and $Z(2, 2) = j\,0.1747 - j\,0.3 = -j\,0.1253$

$$Z = j \begin{bmatrix} 0.6486 & -0.0318 \\ -0.0318 & -0.1253 \end{bmatrix}$$

≫ IC=inv(z)*V

$$I_C = \begin{bmatrix} I_1 \\ I_2 \end{bmatrix} = \begin{bmatrix} -j0.0536 \\ -j0.4652 \end{bmatrix}$$

≫ delv=-Z$_{bus}$*A'*IC

$$[\Delta V] = \begin{bmatrix} -0.0288 \\ -0.0303 \\ -0.0801 \\ -0.0005 \end{bmatrix}$$

The new bus voltages as a consequence of the removal of the lines between buses 1-2 and addition of a line between buses 3-4 are

$$\begin{bmatrix} V_1' \\ V_2' \\ V_3' \\ V_4' \end{bmatrix} = \begin{bmatrix} 1.02\angle 0° \\ 1.00\angle 0° \\ 0.98\angle 0° \\ 1.04\angle 0° \end{bmatrix} + \begin{bmatrix} -0.0288 \\ -0.0505 \\ -0.0801 \\ -0.0005 \end{bmatrix} = \begin{bmatrix} 0.9912 \\ 0.9495 \\ 0.8999 \\ 1.0395 \end{bmatrix}$$

13.4 SIMULATION OF TIE-LINES IN INTERCONNECTED POWER SYSTEMS

Modern-day tradition is to interconnect systems operated by different power companies through tie-lines and this increases the reliability of electric energy supply. It is, therefore, important to introduce the concepts of modelling interconnected networks for addition or removal of tie-lines for contingency analysis.

The modelling of power networks, interconnected through tie-lines, adopts the piecewise methods for their solutions. The piecewise methods are based on the assumption that each network is simulated by a linear model (that is, the relationship between currents and corresponding voltages is linear) for its analysis by the individual operating authority.

Figure 13.6 shows a four-bus power system A interconnected through two tie-lines of impedances Z_a and Z_b with another three-bus power network B. Both the power systems are independent except for the interconnections and are earthed, thereby providing a common reference point.

Initially it is assumed that both the systems operate in the stand-alone mode, and use the Z_{bus} of their own networks for analysing their systems. These results can be later modified, by each system, to take advantage of the interconnections between the two networks.

From Fig. 13.6, it may be noted that buses 3 and 4 of system A are connected to buses 5 and 6 of system B. The current injections and the voltages at each bus, shown in the figure, are assumed known and represent the networks in the stand-alone mode. Mathematically, the relationship between the bus voltages and injected currents with the tie-lines open may be written as

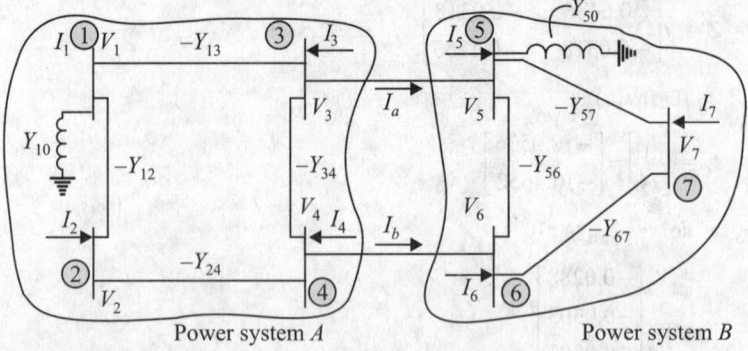

Fig. 13.6 Two power system networks connected through tie-lines

$$
V = \begin{bmatrix} V_1 \\ \vdots \\ V_4 \\ \hline V_5 \\ \vdots \\ V_7 \end{bmatrix} = \begin{bmatrix} \mathbf{Z}_{\text{bus}}^A & 0 \\ \hline 0 & \mathbf{Z}_{\text{bus}}^B \end{bmatrix} \begin{bmatrix} I_1 \\ \vdots \\ I_5 \\ I_6 \\ \vdots \\ I_7 \end{bmatrix}
\tag{13.13}
$$

where $\mathbf{Z}_{\text{bus}}^A$ is the 4×4 bus impedance matrix of system A. Similarly, $\mathbf{Z}_{\text{bus}}^B$ is the 3×3 bus impedance matrix of system B.

The task is to determine the new injected bus currents and the resulting changed bus voltages, in the two power systems, when they are interconnected via two tie-lines. Assume that the impedances of the tie-lines connected between buses 3-5 and 4-6 are Z_a and Z_b respectively, while the currents flowing through them are respectively, I_a and I_b. Also assume that the changed voltages are represented by V_1', V_2', V_3', and V_4' in system A and by V_5', V_6', and V_7' in power system B due to the injected currents I_a and I_b. The effect of the tie-line currents can be simulated by tearing the interconnected network shown in Fig. 13.6 into two pieces at the tie-lines and applying Eq. (13.12). Figure 13.7 shows the two equivalent power systems along with the injected currents.

The impedance matrix Z can be computed by substituting $i = 3, j = 5, p = 4$, and $q = 6$ in Eq. (13.12). Thus,

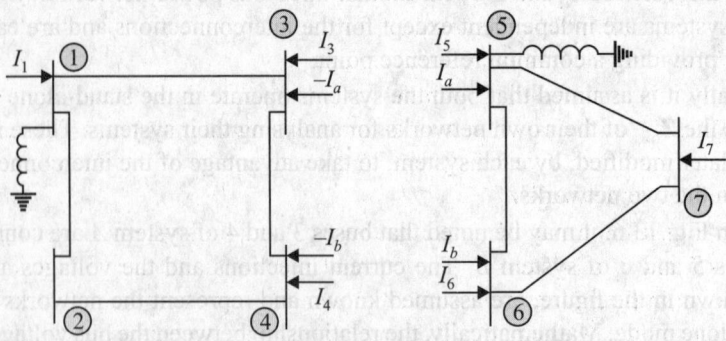

Fig. 13.7 Equivalent system networks along with the injected currents I_a and I_b

$$Z = \begin{array}{c} a \\ b \end{array} \left[\begin{array}{c|c} Z_{33}^A + Z_{55}^B + Z_a & Z_{34}^A + Z_{56}^B \\ \hline Z_{43}^A + Z_{65}^B & Z_{44}^A + Z_{66}^B + Z_b \end{array} \right] \tag{13.14}$$

For the individual system A, the bus voltages V_3 and V_4 are known. Similarly for the system B, the bus voltages V_5 and V_6 are also known. By using Eq. (13.10), the tie-line currents are computed as follows

$$\begin{bmatrix} I_a \\ I_b \end{bmatrix} = \mathbf{Z}^{-1} \begin{bmatrix} V_3 - V_5 \\ V_4 - V_6 \end{bmatrix} = \begin{array}{c} a \\ b \end{array} \left[\begin{array}{cc} Z_{33}^A + Z_{55}^B + Z_a & Z_{34}^A + Z_{56}^B \\ Z_{43}^A + Z_{65}^B & Z_{44}^A + Z_{66}^B + Z_b \end{array} \right]^{-1} \begin{bmatrix} V_3 - V_5 \\ V_4 - V_6 \end{bmatrix}$$
$$\tag{13.15}$$

From Eq. (13.15), the Thevenin's equivalent circuits for the power systems A and B can be perceived directly. Figure 13.8 shows the Thevenin's equivalent circuits for the two power systems. The paths traced by the tie-line currents I_a and I_b through the reference bus, are also shown. The tie-line to bus impedance matrix A is given by

$$A = \begin{array}{c} \\ a \\ b \end{array} \begin{array}{cccccccc} & 1 & 2 & 3 & 4 & 5 & 6 & 7 \\ & \left[\begin{array}{cccc|ccc} 0 & 0 & 1 & 0 & -1 & 0 & 0 \\ 0 & 0 & 0 & 1 & 0 & -1 & 0 \end{array} \right] \end{array}$$

It may be observed that matrix A shows the incidence of tie-lines on the buses of systems A and B.

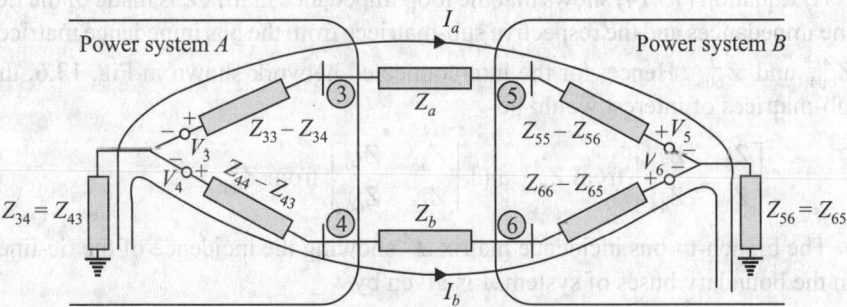

Fig. 13.8 Thevenin's equivalent circuits of the interconnected systems showing the paths of the tie-line currents I_a and I_b through the reference bus

Assume that due to the injected tie-line currents I_a and I_b, the changed bus voltages in system A are designated by V_1', V_2', V_3', V_4' and similarly the new bus voltages in system B are represented by V_5', V_6', and V_7'. By using Eq. (13.7), the new bus voltages are computed as follows

$$\begin{bmatrix} V_1' \\ .. \\ .. \\ V_4' \\ \hline V_5' \\ .. \\ V_7' \end{bmatrix} = \begin{bmatrix} V_1 \\ .. \\ .. \\ V_4 \\ \hline V_5 \\ .. \\ V_7 \end{bmatrix} - \begin{bmatrix} \mathbf{Z}_{\text{bus}}^A & 0 \\ 0 & \mathbf{Z}_{\text{bus}}^B \end{bmatrix} \begin{array}{c} a \\ b \end{array} \left[\begin{array}{cccc|ccc} 0 & 0 & 1 & 0 & -1 & 0 & 0 \\ 0 & 0 & 0 & 1 & 0 & -1 & 0 \end{array} \right]^t \begin{bmatrix} I_a \\ I_b \end{bmatrix}$$

$$\tag{13.16a}$$

Alternatively, the piece-wise method may also be applied to obtain the new system voltages as follows:

$$
\begin{bmatrix} V_1' \\ V_2' \\ V_3' \\ V_4' \end{bmatrix} = \begin{bmatrix} V_1 \\ V_2 \\ V_3 \\ V_4 \end{bmatrix} - \mathbf{Z}_{\text{bus}}^A \begin{matrix} a \\ b \end{matrix}\begin{bmatrix} 0 & 0 & 1 & 0 \\ 0 & 0 & 0 & 1 \end{bmatrix}^t \begin{bmatrix} -I_a \\ -I_b \end{bmatrix}
\tag{13.16b}
$$

and

$$
\begin{bmatrix} V_5' \\ V_6' \\ V_7' \end{bmatrix} = \begin{bmatrix} V_5 \\ V_{56} \\ V_7 \end{bmatrix} - \mathbf{Z}_{\text{bus}}^B \begin{matrix} a \\ b \end{matrix}\begin{bmatrix} -1 & 0 & 0 \\ 0 & -1 & 0 \end{bmatrix}^t \begin{bmatrix} I_a \\ I_b \end{bmatrix}
\tag{13.16c}
$$

In computing the new bus voltages, when the two networks are interconnected through tie-lines, the major task is the formation of the loop impedance matrix $\mathbf{Z}$. The computation of the inverse of the loop impedance matrix $\mathbf{Z}$, however, is simple since its order is dependent on the number of tie-lines and is small.

13.4.1 Formation of the Loop Impedance Matrix Z

Mathematically, the loop impedance matrix $\mathbf{Z}$ may be formed by using the piece-wise method and applying Eq. (13.11) in the following manner:

(i) Equation (13.14) shows that the loop impedance matrix $\mathbf{Z}$ is made of the tie-line impedances and the respective sub-matrices from the bus impedance matrices $\mathbf{Z}_{\text{bus}}^A$ and $\mathbf{Z}_{\text{bus}}^B$. Hence, for the interconnected network shown in Fig. 13.6, the sub-matrices of interest would be

$$
\begin{bmatrix} Z_{33} & Z_{34} \\ Z_{43} & Z_{44} \end{bmatrix} \text{ from } \mathbf{Z}_{\text{bus}}^A \text{ and } \begin{bmatrix} Z_{55} & Z_{56} \\ Z_{65} & Z_{66} \end{bmatrix} \text{ from } \mathbf{Z}_{\text{bus}}^B
$$

The branch-to-bus incidence matrix A^A showing the incidence of the tie-lines on the boundary buses of system A is given by

$$
A^A = \begin{matrix} & 3 & 4 \\ a & \\ b & \end{matrix}\begin{bmatrix} 1 & 0 \\ 0 & 1 \end{bmatrix}
$$

$$
\mathbf{Z}^A = A^A \begin{bmatrix} Z_{33} & Z_{34} \\ Z_{43} & Z_{44} \end{bmatrix} A^{A^t}
\tag{13.17a}
$$

Similarly, for system B

$$
A^B = \begin{matrix} & 5 & 6 \\ a & \\ b & \end{matrix}\begin{bmatrix} -1 & 0 \\ 0 & -1 \end{bmatrix}
$$

$$
\mathbf{Z}^B = A^B \begin{bmatrix} Z_{55} & Z_{56} \\ Z_{65} & Z_{66} \end{bmatrix} A^{A^t}
\tag{13.17b}
$$

The loop incidence matrix $\mathbf{Z}$, therefore, is computed as

$$Z = Z^A + \begin{bmatrix} Z_a & 0 \\ 0 & Z_b \end{bmatrix} + Z^B \qquad (13.17c)$$

In a generalized form, the sub-matrices removed from the original Z_{bus}^A and Z_{bus}^B system impedance matrices may be expressed as

$$\begin{array}{c} \quad i \quad\;\; j \quad\;\; k \\ \begin{matrix} i \\ j \\ k \end{matrix} \begin{bmatrix} Z_{ii} & Z_{ij} & Z_{ik} \\ Z_{ji} & Z_{jj} & Z_{jk} \\ Z_{ki} & Z_{kj} & Z_{kk} \end{bmatrix} \end{array} \text{ taken out from } Z_{bus}^A \text{ matrix of system } A \text{ and}$$

$$\begin{array}{c} p \quad\;\; q \quad\;\; r \quad\;\; s \\ \begin{bmatrix} Z_{pp} & Z_{pq} & Z_{pr} & Z_{ps} \\ Z_{qp} & Z_{qq} & Z_{qr} & Z_{qs} \\ Z_{rp} & Z_{rq} & Z_{rr} & Z_{rs} \\ Z_{sp} & Z_{sq} & Z_{sr} & Z_{ss} \end{bmatrix} \end{array} \text{ taken out from } Z_{bus}^B \text{ matrix of power system } B.$$

In the two sub-matrices, i, j, and k are the boundary buses of system A and p, q, r, and s are the boundary buses of system B. The two systems are interconnected via tie-lines at the boundary buses.

Inverse of the above sub-matrices leads to the bus admittance matrices. These bus admittance matrices symbolize the systems A and B connected via tie-lines at the boundary buses. These equivalent admittance sub-matrices are called *Ward equivalents*. The mathematical computation of the loop impedance matrix Z may physically be interpreted in the following manner.

From the Thevenin's equivalent circuit of the interconnected power systems in Fig. 13.8 and a perusal of Eq. (13.14), the following conclusions can be drawn:

(i) The diagonal elements of the loop impedance matrix Z are the sum of the impedances in the loop or the path traversed by the tie-line currents starting from the reference bus of system A to the reference bus of the system B. For example, in Fig. 13.8, the sum of the impedances of the path traversed by tie-line current I_b (starting from the reference bus of system A) is $\{Z_{43} + (Z_{44} - Z_{43}) + Z_b + (Z_{66} - Z_{65}) + Z_{65}\} = Z_{44} + Z_{66} + Z_b$, which as per Eq. (13.14) is the diagonal element of Z.

(ii) The off-diagonal elements of Z are represented by those impedances through which tie-line currents are flowing and are the cause for producing new system voltages when systems are interconnected. Referring to Fig. 13.8, it is observed that tie-line currents I_a and I_b are flowing through $Z_{34} = Z_{43}$ and $Z_{56} = Z_{65}$. Hence, the off-diagonal elements in Z are equal to $Z_{34} + Z_{56}$ or $Z_{43} + Z_{65}$.

Example 13.3 In Fig. 13.9, the four-bus power system A is interconnected via three tie-lines to another four-bus power system B. The bus voltages, in per unit, of the individual power systems in the stand alone mode, in the steady state, are shown in the figure. Use the piece-wise method to determine the bus voltages of the entire interconnected network when the three tie-lines are simultaneously closed. The line and tie-line reactances shown in the figure are in per unit.

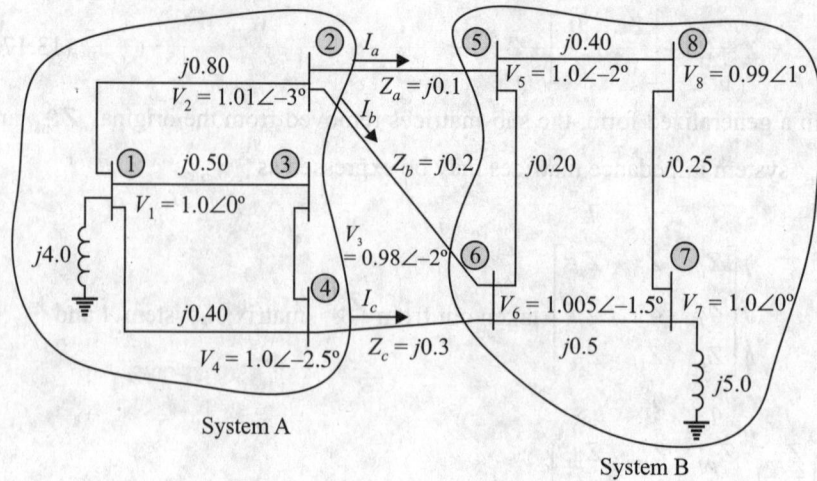

Fig. 13.9 Four-bus system of Example 13.3

Solution In solving the example, the versatility of the MATLAB facility has been made use of. Each MATLAB command has been explained alongside.

```
≫ Ybusa=zeros(4,4);Ybusb=zeros(4,4);  % Generates two 4 × 4 null matrices
≫ Ybusa(1,1)=1/(i*4.0)+1/(i*0.8)+1/(i*0.5)+1/(i*0.4);Ybusa(1,2)=-1/(i*0.8);
≫ Ybusa(2,1)=Ybusa(1,2);Ybusa(1,3)=-1/(i*0.5);Ybusa(3,1)=Ybusa(1,3);
≫ Ybusa(1,4)=-1/(i*0.4);Ybusa(4,1)=Ybusa(1,4);Ybusa(2,2)=1/(i*0.8);
≫ Ybusa(3,3)=1/(i*0.5)+1/(i*0.1);Ybusa(3,4)=-1/(i*0.1);Ybusa(4,3)
  =Ybusa(3,4);
≫ Ybusa(4,4)=1/(i*0.1)+1/(i*0.4);
  % Computation of bus admittance matrix for power system 'A'.
≫ Ybusa
```

```
Ybusa =    0 - 6.0000i     0 + 1.2500i     0 + 2.0000i     0 + 2.5000i
           0 + 1.2500i     0 - 1.2500i     0               0
           0 + 2.0000i     0               0 - 12.0000i    0 + 10.0000i
           0 + 2.5000i     0               0 + 10.0000i    0 - 12.5000i
```

```
≫ Ybusb(1,1)=1/(i*0.2)+1/(i*0.4);Ybusb(1,2)=-1/(i*0.2);Ybusb(2,1)
  =Ybusb(1,2);
≫ Ybusb(1,4)=-1/(i*0.4);Ybusb(4,1)=Ybusb(1,4);Ybusb(2,2)=1/(i*0.2)+1/
  (i*0.5);
≫ Ybusb(2,3)=-1/(i*0.5);Ybusb(3,2)=Ybusb(2,3);
≫ Ybusb(3,3)=1/(i*5.0)+1/(i*0.5)+1/(i*0.25);Ybusb(3,4)=-1/(i*0.25);
≫ Ybusb(4,3)=Ybusb(3,4);Ybusb(4,4)=1/(i*0.25)+1/(i*0.4);
  % Computation of bus admittance matrix for power system 'B'.
≫ Ybusb
```

```
Ybusb =    0 - 7.5000i     0 + 5.0000i     0               0 + 2.5000i
           0 + 5.0000i     0 - 7.0000i     0 + 2.0000i     0
           0               0 + 2.0000i     0 - 6.2000i     0 + 4.0000i
           0 + 2.5000i     0               0 + 4.0000i     0 - 6.5000i
```

```
≫ Zbusa=inv(Ybusa)        % Computation of bus impedance matrix of
                          power system 'A'.
```

```
Zbusa =     0 + 4.0000i     0 + 4.0000i     0 + 4.0000i     0 + 4.0000i
            0 + 4.0000i     0 + 4.8000i     0 + 4.0000i     0 + 4.0000i
            0 + 4.0000i     0 + 4.0000i     0 + 4.2500i     0 + 4.2000i
            0 + 4.0000i     0 + 4.0000i     0 + 4.2000i     0 + 4.2400i
>> Zbusb=inv(Ybusb)                  % Computation of bus impedance matrix of
                                     power system 'B'.
Zbusb =     0 + 5.3370i     0 + 5.2407i     0 + 5.0000i     0 + 5.1296i
            0 + 5.2407i     0 + 5.3148i     0 + 5.0000i     0 + 5.0926i
            0 + 5.0000i     0 + 5.0000i     0 + 5.0000i     0 + 5.0000i
            0 + 5.1296i     0 + 5.0926i     0 + 5.0000i     0 + 5.2037i
>> Zbusa(1,:)=[];Zbusa(:,1)=[];      % Deleting row and column 1 from the
                                     impedance matrix of power system 'A'
                                     for formulating the impedance matrix
                                     at the boundary buses
>> Zbusa
Zbusa =     0 + 4.8000i     0 + 4.0000i     0 + 4.0000i
            0 + 4.0000i     0 + 4.2500i     0 + 4.2000i
            0 + 4.0000i     0 + 4.2000i     0 + 4.2400i
>> Zbusb(:,3:4)=[]; Zbusb([3 4],:)=[];       % Deleting rows and columns
                                             3 & 4 from the impedance matrix
                                             of power system 'B' for
                                             formulating the impedance
                                             matrix at the boundary buses
>> Zbusb
Zbusb =     0 + 5.3370i     0 + 5.2407i
            0 + 5.2407i     0 + 5.3148i
>> AA=[1 0 0;1 0 0;0 0 1];       % Forming tie-line to bus incidence matrix at
                                 the boundary buses of power system 'A'.
>> Zbounda=AA*Zbusa*AA'           % Computing impedance sub-matrix of power system
                                  'A' as per Eq. (13.17a)
Zbounda =   0 + 4.8000i     0 + 4.8000i     0 + 4.0000i
            0 + 4.8000i     0 + 4.8000i     0 + 4.0000i
            0 + 4.0000i     0 + 4.0000i     0 + 4.2400i
>> AB=[-1 0;0 -1;0 -1];          % Forming tie-line to bus incidence matrix
                                 at the boundary buses of power system 'B'.
>> Zboundb=AB*Zbusb*AB'           % Computing impedance sub-matrix of
                                  power system 'B' as per Eq. (13.17b)
Zboundb =   0 + 5.3370i     0 + 5.2407i     0 + 5.2407i
            0 + 5.2407i     0 + 5.3148i     0 + 5.3148i
            0 + 5.2407i     0 + 5.3148i     0 + 5.3148i
>> Ztie=[i*0.1 0 0;0 i*0.2 0;0 0 i*0.3];     % Writing the tie-line
                                             diagonal impedance matrix.
>> Ztie
Ztie =      0 + 0.1000i     0               0
            0               0 + 0.2000i     0
            0               0               0 + 0.3000i
>> Z=Zbounda+Ztie+Zboundb        % Computation of loop impedance matrix
                                 [Z] as per Eq. (13.17c)
```

```
Z =           0 + 10.2370i      0 + 10.0407i      0 + 9.2407i
              0 + 10.0407i      0 + 10.3148i      0 + 9.3148i
              0 + 9.2407i       0 + 9.3148i       0 + 9.8548i
>> VA=[1.0*exp(i*pi*0/180);1.01*exp(i*pi*(-3.0)/180);...
0.98*exp(i*pi*(-2.0)/180);1.0*exp(i*pi*(-2.5)/180)]
```

> % Writing the bus voltages vector of power system 'A', in complex form, prior to interconnecting power systems 'A' and 'B' via tie-lines

```
VA =  1.0000
      1.0086 - 0.0529i
      0.9794 - 0.0342i
      0.9990 - 0.0436i
>> VB=[1.0*exp(i*pi*(-2.0)/180);1.005*exp(i*pi*(-1.5)/180);...
1.0*exp(i*pi*0/180);0.99*exp(i*pi*1.0/180)]
```

> % Writing the bus voltages vector of power system 'B', in complex form, prior to inter-connecting power systems 'A' and 'B' via tie- lines

```
VB = 0.9994 - 0.0349i
     1.0047 - 0.0263i
     1.0000
     0.9898 + 0.0173i
>> Vdiff=[VA(2)-VB(1);VA(2)-VB(2);VA(4)-VB(2)]
```

> % Computing the voltages difference vector between the boundary buses.

```
Vdiff =    0.0092 - 0.0180i
           0.0040 - 0.0266i
          -0.0056 - 0.0173i
>> Itie=inv(Z)*Vdiff          % Computing the tie-line currents
Itie =     0.0160 - 0.0146i
          -0.0206 + 0.0065i
           0.0027 + 0.0081i
>> Abusa=[0 1 0 0;0 1 0 0;0 0 0 1];
```

> % Writing the tie-line to bus incidence matrix for power system 'A'.

```
>> VAdash=VA-inv(Ybusa)*Abusa'*Itie
```

> % Computing the new bus voltages in power system 'A' after connecting the tie-lines as per Eq.(13.16b)

```
VAdash =  1.0001 + 0.0075i = 1.0001∠0.4325°
          1.0022 - 0.0416i = 1.0031∠-2.3795°
          0.9811 - 0.0272i = 0.9815∠-1.5876°
          1.0011 - 0.0367i = 1.0017∠-2.1006
>> Abusb=[-1 0 0 0;0 -1 0 0;0 -1 0 0];
```

> % Writing the tie-line to bus incidence matrix for power system 'B'.

```
>> VBdash=VB-inv(Ybusb)*Abusb'*(-Itie)
```

> % Computing the new bus voltages in power system 'B' after connecting the tie-lines as per Eq.(13.16c)

```
VBdash =  0.9981 - 0.0266i = 0.9984∠-1.5239°
          1.0058 - 0.0151i = 1.0059∠-0.8596°
          1.0001 + 0.0094i = 1.0001∠0.5406°
          0.9894 + 0.0263i = 0.9897∠1.5224°
```

Example 13.4 For the network of Example 13.3, determine the Ward equivalents, at the boundary buses for (i) power system A and (ii) power system B. Draw the Ward equivalent admittance circuits. Also draw the Thevenin's equivalent circuit for the interconnected network of Example 13.3.

Solution (i) The impedance matrix at the boundary buses is obtained from the bus impedance matrix of power system A as follows:

$$\begin{array}{cc} 0 + 4.8000i & 0 + 4.0000i \\ 0 + 4.0000i & 0 + 4.2400i \end{array}$$

The inverse of the boundary bus matrix yields the Ward equivalent for power system A.

$$\begin{array}{cc} 0 - 0.9743i & 0 + 0.9191i \\ 0 + 0.9191i & 0 - 1.1029i \end{array}$$

(ii) Similarly the impedance matrix at the boundary buses, is obtained from the bus impedance matrix of power system B as given here

$$\begin{array}{cc} 0 + 5.3370i & 0 + 5.2407i \\ 0 + 5.2407i & 0 + 5.3148i \end{array}$$

Again the inverse of the boundary bus matrix leads to the Ward equivalent for power system B as shown here

$$\begin{array}{cc} 0 - 5.9043i & 0 + 5.8220i \\ 0 + 5.8220i & 0 - 5.9290i \end{array}$$

The Ward equivalents for the two-system networks are given in Fig. 13.10.

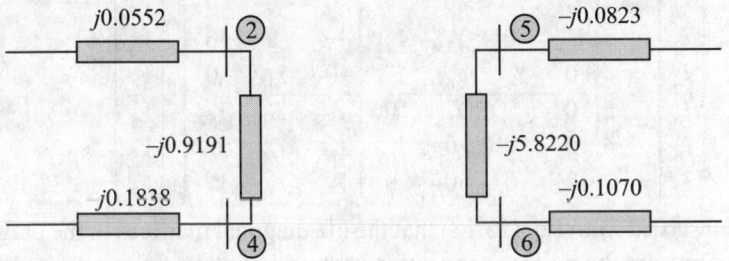

Fig. 13.10 Ward equivalents for power systems A and B

The Thevenin's equivalent circuit, for the interconnected systems, looking into the buses 2-4 and 5-6, is drawn (Fig. 13.11) from the bus impedance matrices of the power systems A and B respectively.

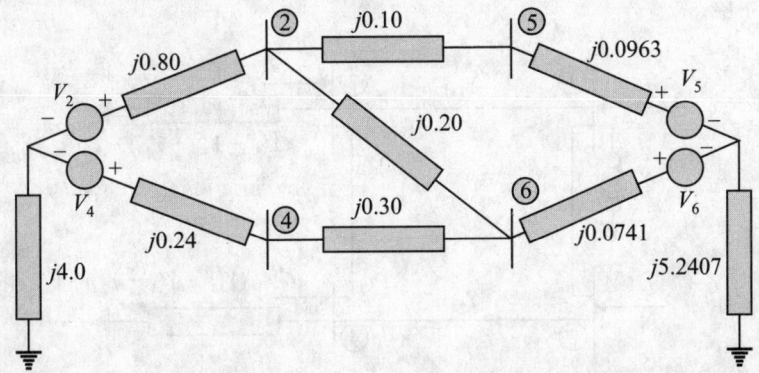

Fig. 13.11 Thevenin's equivalent circuit for the interconnected power systems

13.5 NETWORK REDUCTION FOR CONTINGENCY ANALYSIS

In interconnected power system operation, while undertaking contingency analysis, it is often expedient to represent the portion of the network required to be analysed in detail and the adjoining system by an equivalent network. In this section, application of network reduction for contingency analysis is discussed.

Consider the interconnected network shown in Fig. 13.6. It is desired to perform contingency analysis of the interconnected network with (i) system A retained in detail and system B and the tie-lines represented by an equivalent at the boundary buses 3-4 and (ii) system B represented by an equivalent at the boundary buses 5-6.

It is assumed that an ac power flow check of the interconnected power system has been performed prior to undertaking the contingency analysis. Therefore, it is also understood that from the load flow data, generation inputs and load outputs have been converted into equivalent current injections at all the buses and the bus voltages V_1, V_2, V_3, V_4, V_5, V_6, and V_7 are the numerical values obtained from the ac power flow solution of the interconnected network. The node admittance equations for the interconnected system are written as

$$
\begin{bmatrix} I_1 \\ I_2 \\ I_3 \\ I_4 \\ \hline I_5 \\ I_6 \\ I_7 \end{bmatrix} =
\begin{array}{c} 1 \\ 2 \\ 3 \\ 4 \\ 5 \\ 6 \\ 7 \end{array}
\begin{bmatrix}
Y_{11} & Y_{12} & Y_{13} & 0 & 0 & 0 & 0 \\
Y_{21} & Y_{22} & 0 & Y_{24} & 0 & 0 & 0 \\
Y_{31} & 0 & Y_{33} & Y_{34} & Y_{35} & 0 & 0 \\
0 & Y_{42} & Y_{43} & Y_{44} & 0 & Y_{46} & 0 \\
\hline
0 & 0 & Y_{53} & 0 & Y_{55} & Y_{56} & Y_{57} \\
0 & 0 & 0 & Y_{64} & Y_{65} & Y_{66} & Y_{67} \\
0 & 0 & 0 & 0 & Y_{75} & Y_{76} & Y_{77}
\end{bmatrix}
\begin{bmatrix} V_1 \\ V_2 \\ V_3 \\ V_4 \\ \hline V_5 \\ V_6 \\ V_7 \end{bmatrix}
\qquad (13.18)
$$

It is observed from Eq. (13.18) that the off-diagonal matrices in the partitioned matrix represent the tie-line connections between the two systems A and B.

Case I: System A is retained in detail and the remaining network is represented by an equivalent circuit at the boundary buses 3-4.

By respectively using Eqs (6.60) and (6.61), nodes 5, 6, and 7 are eliminated and the current injections at the eliminated nodes redistributed. The Ward equivalent circuit of power system B, including the tie-lines, at buses 3 and 4 is shown in Fig. 13.12.

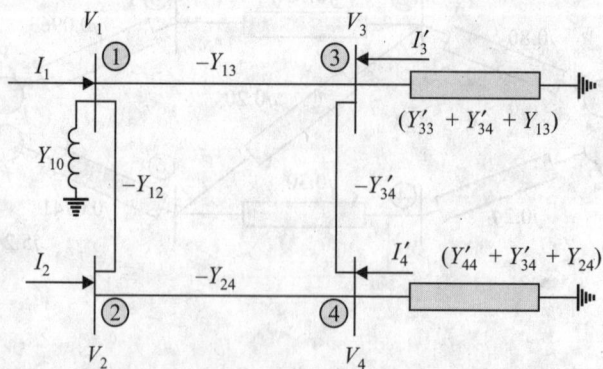

Fig. 13.12 Ward equivalent circuit of system B, including tie-lines, at buses 3 and 4

The node admittance equations, for the reduced network are, therefore, given by

$$
\begin{bmatrix} I_1 \\ I_2 \\ I_3' \\ I_4' \end{bmatrix} = \begin{matrix} & \begin{matrix} 1 & 2 & 3 & 4 \end{matrix} \\ \begin{matrix} 1 \\ 2 \\ 3 \\ 4 \end{matrix} & \begin{bmatrix} Y_{11} & Y_{12} & Y_{13} & 0 \\ Y_{21} & Y_{22} & 0 & Y_{24} \\ Y_{31}' & 0 & Y_{33}' & Y_{34}' \\ Y_{41}' & Y_{42}' & Y_{43}' & Y_{44}' \end{bmatrix} \end{matrix} \begin{bmatrix} V_1 \\ V_2 \\ V_3 \\ V_4 \end{bmatrix}
$$ (13.19)

In Eq. (13.19), the injected currents at buses 3 and 4 are the sums of currents at these buses and actual currents distributed as a consequence of the elimination of buses 5, 6, and 7. By taking the inverse of the admittance matrix in Eq. (13.19), the bus equations in impedance form are obtained as follows:

$$
\begin{bmatrix} V_1 \\ V_2 \\ V_3 \\ V_4 \end{bmatrix} = \begin{matrix} & \begin{matrix} 1 & 2 & 3 & 4 \end{matrix} \\ \begin{matrix} 1 \\ 2 \\ 3 \\ 4 \end{matrix} & \begin{bmatrix} Z_{11} & Z_{12} & Z_{13} & Z_{14} \\ Z_{21} & Z_{22} & Z_{23} & Z_{24} \\ Z_{31} & Z_{32} & Z_{33} & Z_{34} \\ Z_{41} & Z_{42} & Z_{43} & Z_{44} \end{bmatrix} \end{matrix} \begin{bmatrix} I_1 \\ I_2 \\ I_3' \\ I_4' \end{bmatrix}
$$ (13.20)

where

$$
\mathbf{Z}_{\text{bus}} = \begin{bmatrix} Z_{11} & Z_{12} & Z_{13} & Z_{14} \\ Z_{21} & Z_{22} & Z_{23} & Z_{24} \\ Z_{31} & Z_{32} & Z_{33} & Z_{34} \\ Z_{41} & Z_{42} & Z_{43} & Z_{44} \end{bmatrix} = \begin{bmatrix} Y_{11} & Y_{12} & Y_{13} & 0 \\ Y_{21} & Y_{22} & 0 & Y_{24} \\ Y_{31}' & 0 & Y_{33}' & Y_{34}' \\ Y_{41}' & Y_{42}' & Y_{43}' & Y_{44}' \end{bmatrix}^{-1}
$$

In Eq. (13.20), $\mathbf{Z}_{\text{bus}}$ is the impedance matrix of the reduced network and is representative of the entire interconnected power system consisting of power systems A and B and the tie-lines. The application of $\mathbf{Z}_{\text{bus}}$ to compute distribution factors to perform contingency analysis of system A is described in subsequent sections.

Case II: System A is retained in detail and system B is represented by an equivalent circuit at the boundary buses 5-6.

Again, with the help of Eq. (6.60) node 7 is eliminated and the injected current I_7 at node 7 is redistributed to the retained nodes 5 and 6 by employing Eq. (6.61). The Ward equivalent circuit at buses 5 and 6 is drawn in Fig. 13.13.

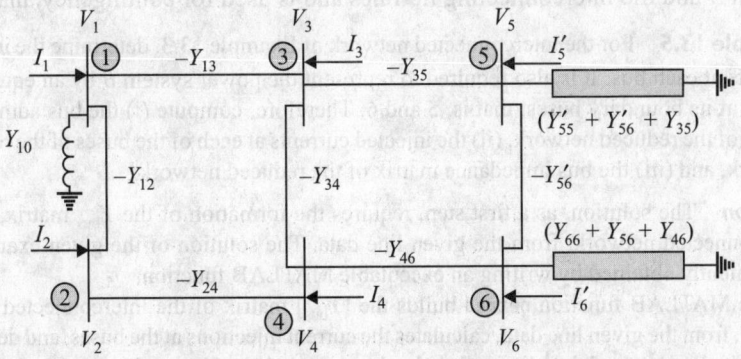

Fig. 13.13 Ward equivalent circuit of system B at the boundary buses 5 and 6

The node admittance equations for the reduced network shown in Fig. 13.10 are written as follows:

$$
\begin{bmatrix} I_1 \\ I_2 \\ I_3 \\ I_4 \\ I_5' \\ I_6' \end{bmatrix}
=
\begin{matrix} 1 \\ 2 \\ 3 \\ 4 \\ 5 \\ 6 \end{matrix}
\begin{bmatrix}
Y_{11} & Y_{12} & Y_{13} & 0 & 0 & 0 \\
Y_{21} & Y_{22} & 0 & Y_{24} & 0 & 0 \\
Y_{31} & 0 & Y_{33} & Y_{34} & Y_{35} & 0 \\
0 & Y_{42} & Y_{43} & Y_{44} & 0 & Y_{46} \\
0 & 0 & Y_{53} & 0 & Y_{55}' & Y_{56}' \\
0 & 0 & 0 & Y_{64} & Y_{65}' & Y_{66}'
\end{bmatrix}
\begin{bmatrix} V_1 \\ V_2 \\ V_3 \\ V_4 \\ V_5 \\ V_6 \end{bmatrix}
\tag{13.21}
$$

Here too, node currents I_5' and I_6' in Eq. (13.21) respectively represent the actual currents at these nodes plus the portion of node current I_7, which gets distributed to nodes 5 and 6 as a result of its elimination. Similar to Eq. (13.20), the bus equations in impedance form are obtained from Eq. (13.21) and are as follows:

$$
\begin{bmatrix} V_1 \\ V_2 \\ V_3 \\ V_4 \\ V_5 \\ V_6 \end{bmatrix}
=
\begin{matrix} 1 \\ 2 \\ 3 \\ 4 \\ 5 \\ 6 \end{matrix}
\begin{bmatrix}
Z_{11} & Z_{12} & Z_{13} & Z_{14} & Z_{15} & Z_{16} \\
Z_{21} & Z_{22} & Z_{23} & Z_{24} & Z_{25} & Z_{26} \\
Z_{31} & Z_{32} & Z_{33} & Z_{34} & Z_{35} & Z_{36} \\
Z_{41} & Z_{42} & Z_{43} & Z_{44} & Z_{45} & Z_{46} \\
Z_{51} & Z_{52} & Z_{53} & Z_{54} & Z_{55} & Z_{56} \\
Z_{61} & Z_{62} & Z_{63} & Z_{64} & Z_{65} & Z_{66}
\end{bmatrix}
\begin{bmatrix} I_1 \\ I_2 \\ I_3 \\ I_4 \\ I_5' \\ I_6' \end{bmatrix}
\tag{13.22}
$$

where

$$
\boldsymbol{Z}_{\text{bus}} =
\begin{bmatrix}
Z_{11} & Z_{12} & Z_{13} & Z_{14} & Z_{15} & Z_{16} \\
Z_{21} & Z_{22} & Z_{23} & Z_{24} & Z_{25} & Z_{26} \\
Z_{31} & Z_{32} & Z_{33} & Z_{34} & Z_{35} & Z_{36} \\
Z_{41} & Z_{42} & Z_{43} & Z_{44} & Z_{45} & Z_{46} \\
Z_{51} & Z_{52} & Z_{53} & Z_{54} & Z_{55} & Z_{56} \\
Z_{61} & Z_{62} & Z_{63} & Z_{64} & Z_{65} & Z_{66}
\end{bmatrix}
=
\begin{bmatrix}
Y_{11} & Y_{12} & Y_{13} & 0 & 0 & 0 \\
Y_{21} & Y_{22} & 0 & Y_{24} & 0 & 0 \\
Y_{31} & 0 & Y_{33} & Y_{34} & Y_{35} & 0 \\
0 & Y_{42} & Y_{43} & Y_{44} & 0 & Y_{46} \\
0 & 0 & Y_{53} & 0 & Y_{55}' & Y_{56}' \\
0 & 0 & 0 & Y_{64} & Y_{65}' & Y_{66}'
\end{bmatrix}^{-1}
$$

Likewise $\boldsymbol{Z}_{\text{bus}}$ in Eq. (13.22) represents the reduced network constituted of system A and the interconnecting tie-lines and is used for contingency analysis.

Example 13.5 For the interconnected network of Example 13.3, determine the injected currents at each bus. It is also required to represent the power system B by an equivalent circuit at its boundary buses, that is, 5 and 6. Therefore, compute (i) the bus admittance matrix of the reduced network, (ii) the injected currents at each of the buses of the reduced network, and (iii) the bus impedance matrix of the reduced network.

Solution The solution, as a first step, requires the formation of the $\boldsymbol{Y}_{\text{bus}}$ matrix, of the interconnected network, from the given line data. The solution of the given example is conveniently obtained by writing an executable MATLAB function.

The MATLAB function netred builds the $[\boldsymbol{Y}_{\text{bus}}]$ matrix of the interconnected power system, from the given line data, calculates the current injections at the buses, and develops the equivalent network by eliminating the desired number of buses. Network reduction is performed by Kron's reduction as described in Section 6.6.

```
function[Yred,Idis,Zbus]=netred(node,lines,lndata,V);
% Program for network reduction for contingency analysis
Y=zeros(node,node);
for k = 1 : lines;
    I=lndata(k,1);J=lndata(k,2);
    Y(I,I)=Y(I,I)+1/lndata(k,3);
    if J ~= 0;
        Y(J,J)=Y(J,J)+1/lndata(k,3);
        Y(I,J)=-1/lndata(k,3);
        Y(J,I)=Y(I,J);
    else
    end
end
Y
Iinj=Y*V
n=input('No. of nodes to be eliminated');
Y11d=zeros(node-n,node-n);
Y22d=zeros(n,n);
Y12d=zeros(node-n,n);
I1d=zeros(node-n,1);
I2d=zeros(n,1);
for I = 1:node-n;
    for J = 1:node-n;
        Y11d(I,J)=Y(I,J);
    end
    I1d(I)=Iinj(I);
end
Y11d
I1d
for I = 1:node-n;
    for J = 1:n;
        Y12d(I,J)=Y(I,J+node-n);
    end
end
Y12d
for I = 1:n;
    for J = 1:n;
        Y22d(I,J)=Y(I+node-n,J+node-n);
    end
    I2d(I)=Iinj(I+node-n);
end
Y22d
I2d
Yred=Y11d-(Y12d*inv(Y22d)*(Y12d)')
Idis=I1d-(Y12d*inv(Y22d)*I2d)
Zbus=inv(Yred)
```

This MATLAB function is executed through a function file called netred and the program is saved as a file named netred.m. The input parameters are as follows:

node	The total number of nodes in the interconnected network.
lines	The total number of lines, including shunt connections, in the interconnected network.
lndata	The line data, that is, the adjacent bus numbers to which the line is connected and its reactance is fed as a matrix whose dimension is (lines $\times 3$). The reference is designated by 0 for inputting any line data between a bus and the reference.
V	Bus voltages, as a column vector.

The execution of the MATLAB function netred yielded the following output:

```
≫ node=8;lines=13; % Input: No. of nodes and no. of lines
≫ lndata=[1 2 i*0.8;1 3 i*0.5;1 4 i*0.4;1 0 i*4.0;...  % Input: line data
2 5 i*0.1;2 6 i*0.2;3 4 i*0.1;4 6 i*0.3;5 6 i*0.2;...
5 8 i*0.4;6 7 i*0.5;7 8 i*0.25;7 0 i*5.0];
≫ V=[1.0;1.01*exp(i*(-3.0)*pi/180);0.98*exp(i*(-2.0)*pi/180);...
1.0*exp(i*(-2.5)*pi/180);1.0*exp(i*(-2.0)*pi/180);...
1.005*exp(i*(-1.5)*pi/180);1.0*exp(i*(-0.0)*pi/180);...
0.99*exp(i*(1.0)*pi/180)];              %Input: bus voltages
≫ [Yred,Idis,Zbus]=netred(node,lines,lndata,V);% Calls function 'netred'
Y =                       % Output [Y_bus] matrix of interconnected network
Columns 1 through 4
```

0 − 6.0000i	0 + 1.2500i	0 + 2.0000i	0 + 2.5000i
0 + 1.2500i	0 −16.2500i	0	0
0 + 2.0000i	0	0 −12.0000i	0 +10.0000i
0 + 2.5000i	0	0 +10.0000i	0 −15.8333i
0	0 +10.0000i	0	0
0	0 + 5.0000i	0	0 + 3.3333i
0	0	0	0
0	0	0	0

```
Columns 5 through 8
```

0	0	0	0
0 +10.0000i	0 + 5.0000i	0	0
0	0	0	0
0	0 + 3.3333i	0	0
0 −17.5000i	0 + 5.0000i	0	0 + 2.5000i
0 + 5.0000i	0 −15.3333i	0 + 2.0000i	0
0	0 + 2.0000i	0 − 6.2000i	0 + 4.0000i
0 + 2.5000i	0	0 + 4.0000i	0 − 6.5000i

```
Iinj =          % Computation of current injections in pu at the buses
                0.2435 − 0.2828i
               -0.3784 − 0.1228i
                0.0258 + 0.2376i
               -0.2609 − 0.1754i
                0.0062 + 0.0947i
                0.1808 − 0.0345i
               -0.0165 − 0.2313i
                0.1996 + 0.0645i
No. of nodes to be eliminated2          % Queries the number of nodes to
                                          be eliminated
Y11d =    % Extraction of ⌊Y'₁₁⌋ sub-matrix. (See section 6.6)
```

Columns 1 through 4

```
        0 - 6.0000i      0 + 1.2500i      0 + 2.0000i      0 + 2.5000i
        0 + 1.2500i      0 -16.2500i      0                0
        0 + 2.0000i      0                0 -12.0000i       0 +10.0000i
        0 + 2.5000i      0                0 +10.0000i       0 -15.8333i
        0                0 +10.0000i      0                0
        0                0 + 5.0000i      0                0 + 3.3333i
```

Columns 5 through 6

```
0                 0
0 +10.0000i       0 + 5.0000i
0                 0
0                 0 + 3.3333i
0 -17.5000i       0 + 5.0000i
0 + 5.0000i       0 -15.3333i
```

I1d =   % Extraction of $\lfloor I'_1 \rfloor$ current vector. (See section 6.6)

```
0.2435 - 0.2828i
-0.3784 - 0.1228i
0.0258 + 0.2376i
-0.2609 - 0.1754i
0.0062 + 0.0947i
0.1808 - 0.0345i
```

Y12d =   % Extraction of $\lfloor Y'_{12} \rfloor$ sub-matrix. (See section 6.6)

```
        0                0
        0                0
        0                0
        0                0
        0                0 + 2.5000i
        0 + 2.0000i      0
```

Y22d =   % Extraction of $\lfloor Y'_{22} \rfloor$ sub-matrix. (See section 6.6)

```
        0 - 6.2000i      0 + 4.0000i
        0 + 4.0000i      0 - 6.5000i
```

I2d =   % Extraction of $\lfloor I'_2 \rfloor$ sub-matrix. (See section 6.6)

```
   -0.0165 - 0.2313i
    0.1996 + 0.0645i
```

Yred =   % Computation of $[Y_{bus}]$ matrix of reduced network. [See Eq.(6.60)]

Columns 1 through 4

```
        0 - 6.0000i      0 + 1.2500i      0 + 2.0000i      0 + 2.5000i
        0 + 1.2500i      0 -16.2500i      0                0
        0 + 2.0000i      0                0 -12.0000i       0 +10.0000i
        0 + 2.5000i      0                0 +10.0000i       0 -15.8333i
        0                0 +10.0000i      0                0
        0                0 + 5.0000i      0                0 + 3.3333i
```

Columns 5 through 6

```
        0                0
        0 +10.0000i      0 + 5.0000i
        0                0
        0                0 + 3.3333i
        0 -19.0947i      0 + 4.1770i
        0 + 4.1770i      0 -16.4033i
```

```
Idis =      % Computation of injected bus current in pu in reduced network.
            [See Eq.(6.61)]
        0.2435 - 0.2828i
       -0.3784 - 0.1228i
        0.0258 + 0.2376i
       -0.2609 - 0.1754i
        0.1267 + 0.0407i
        0.2377 - 0.1370i
```

Zbus =   % Computation of $[Z_{bus}]$ matrix of reduced network

Columns 1 through 4

```
        0 + 0.4088i      0 + 0.1228i      0 + 0.3007i      0 + 0.2791i
        0 + 0.1228i      0 + 0.1734i      0 + 0.1163i      0 + 0.1150i
        0 + 0.3007i      0 + 0.1163i      0 + 0.4119i      0 + 0.3341i
        0 + 0.2791i      0 + 0.1150i      0 + 0.3341i      0 + 0.3451i
        0 + 0.0899i      0 + 0.1138i      0 + 0.0884i      0 + 0.0881i
        0 + 0.1170i      0 + 0.1052i      0 + 0.1259i      0 + 0.1276i
```

Columns 5 through 6

```
        0 + 0.0899i      0 + 0.1170i
        0 + 0.1138i      0 + 0.1052i
        0 + 0.0884i      0 + 0.1259i
        0 + 0.0881i      0 + 0.1276i
        0 + 0.1308i      0 + 0.0859i
        0 + 0.0859i      0 + 0.1408i
```

The reader will do well to execute the function netred to develop the equivalent circuit and compute the Z_{bus} matrix by eliminating power system *B* up to the boundary buses of power system *A*, that is, buses 2 and 4.

13.6 CONTINGENCY ANALYSIS

Contingency analysis is a digital program which investigates the changes in bus voltages and line flows due to probable outages or tripping of a transformer and/ or line or lines. Exclusion of a line or a transformer from service is termed as an outage or tripping. Outages can be unforeseen (due to a fault, weather conditions, etc.) or scheduled. Scheduled outages are performed for repair and maintenance purposes. In such a case, the equipment is de-energized and removed from service by opening the relevant circuit breakers.

Since contingency analysis is limited to identifying severe conditions resulting in insecure or vulnerable conditions, after the system has attained steady-state conditions, following outage(s), immense precision is not warranted. Contingency analysis, therefore, may be approached by adopting the linear Z_{bus} technique or approximate ac power flow method. The linear Z_{bus} technique is akin to the fast decoupled power flow described in Section 7.8.5.

The basic input data is obtained from an initial load flow study or state estimate of the power system. Simulation of contingencies assumes constant steady-state bus voltages and models the network by equivalent constant current injections, which are obtained by converting the loads and generator inputs into currents at steady-state values of the bus voltages. The network branches are represented by series impedances, for the purpose of explaining the concepts of contingency analysis. In case shunt elements are neglected, one of the system buses is selected as a reference bus to avoid isolation of the neutral.

13.6.1 Current-injection Distribution Factors

The current flows in all the lines and all the bus voltages in a power system network undergo a change when an extra current is injected into any one bus. Consider that a current ΔI_k is injected into bus k of a power system of n-buses. The changes in bus voltages as a result of the additional current injection may be written as

$$
\begin{bmatrix} \Delta V_1 \\ \cdots \\ \cdots \\ \Delta V_i \\ \Delta V_j \\ \cdots \\ \Delta V_n \end{bmatrix}
=
\begin{array}{c} 1 \\ \cdots \\ k \\ i \\ j \\ \cdots \\ n \end{array}
\begin{bmatrix}
Z_{11} & \cdots & \cdots & Z_{1k} & \cdots & \cdots & Z_{1n} \\
\cdots & \cdots & \cdots & Z_{2k} & \cdots & \cdots & \cdots \\
\cdots & \cdots & \cdots & Z_{kk} & \cdots & \cdots & \cdots \\
\cdots & \cdots & \cdots & Z_{ik} & \cdots & \cdots & \cdots \\
\cdots & \cdots & \cdots & Z_{jk} & \cdots & \cdots & \cdots \\
\cdots & \cdots & \cdots & \cdots & \cdots & \cdots & \cdots \\
Z_{n1} & \cdots & \cdots & Z_{nk} & \cdots & \cdots & Z_{nn}
\end{bmatrix}
\begin{bmatrix} 0 \\ 0 \\ \Delta I_k \\ 0 \\ 0 \\ 0 \\ 0 \end{bmatrix}
=
\begin{bmatrix} Z_{1k} \\ Z_{2k} \\ Z_{kk} \\ Z_{ik} \\ Z_{jk} \\ \vdots \\ Z_{nk} \end{bmatrix} \Delta I_k
$$

(13.23)

Equation (13.23) shows that only the kth column of the Z_{bus} impedance matrix is needed to compute the changes in bus voltages due to the extra injected current at bus k. Hence,

$$\Delta V_i = Z_{ik}\Delta I_k \text{ and } \Delta V_j = Z_{jk}\Delta I_k$$

Assume that the impedance of the line connecting buses i-j is represented by Z_l. The change in the current flow, in the line connecting buses i-j, is given by

$$\Delta I_{ij} = \frac{\Delta V_i - \Delta V_j}{Z_l} = \frac{Z_{ik} - Z_{jk}}{Z_l}\Delta I_k$$

or
$$\frac{\Delta I_{ij}}{\Delta I_k} = \frac{Z_{ik} - Z_{jk}}{Z_l} \qquad (13.24)$$

Equation (13.24) defines the *current-injection distribution factor*. The current-injection distribution factor, which is designated as K_{ij-k}, may be defined as the ratio of the change in current produced in a line flow to the additional current injected in a bus in a power system. Therefore,

$$K_{ij-k} = \frac{\Delta I_{ij}}{\Delta I_k} = \frac{Z_{ik} - Z_{jk}}{Z_l} \qquad (13.25)$$

13.6.2 Current-shift Distribution Factors

In the normal steady-state operation of a power system, the overload on a line is relieved by reducing the current on that line and increasing the current, by a corresponding amount, on another line. Assume that the changes in current injections at two buses p and q are represented by ΔI_p and ΔI_q respectively. The changes in line currents connecting buses i-j, due to currents ΔI_p and ΔI_q, from Eq. (13.25) can be expressed as

$$\Delta I_{ij-p} = K_{ij-p}\Delta I_p \text{ and } \Delta I_{ij-q} = K_{ij-q}\Delta I_q$$

The total change in line current, by the superposition principle, can be written as

$$\Delta I_{ij} = \Delta I_{ij-p} + \Delta I_{ij-q} = K_{ij-p}\Delta I_p + K_{ij-q}\Delta I_q \tag{13.26}$$

Equation (13.26) shows how shifting the current injection from one bus to another, affects the line flows in the system. The current-injection distribution factors are also referred to as *current-shift distribution factors*.

13.6.3 Single Outage Contingency Analysis

Single outage contingency analysis involves the simulation of a transformer or a transmission line outage and appraising the system line flows and bus voltages to avoid line overloads and extreme voltages which may lead to blackouts due to cascading effects. Figure 13.14 shows the Thevenin's equivalent circuit, at two buses *p-q* of a power system, prior to an outage.

With the system operating in a steady-state condition, the pre-outage current in the line connected between buses *p-q* is given by

$$I_{pq} = \frac{V_p - V_q}{Z_a} \tag{13.27}$$

where V_p and V_q are the respective voltages of buses p and q prior to the outage.

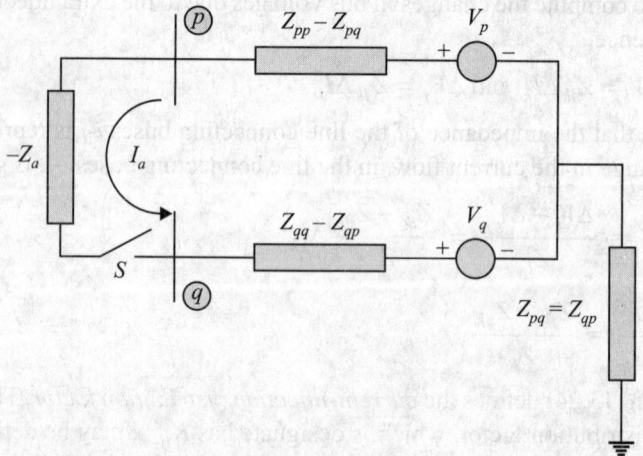

Fig. 13.14 Thevenin's equivalent circuit looking into buses
p and *q* prior to an outage

The outage between the buses can be simulated by adding impedance $-Z_a$ (which is the negative of the impedance of the line connected between *p-q* and is to be removed from service) by closing the switch *S*. When the switch *S* is closed, the circulating loop current I_a is given by

$$I_a = \frac{V_p - V_q}{Z_{th-pq} - Z_a} \tag{13.28a}$$

where V_p and V_q are the respective voltages at buses p and q after the outage, and Z_{th-pq} is the Thevenin's equivalent impedance looking into the buses at *p-q*.

By referring to Fig. 13.8 and recalling that $Z_{pq} = Z_{qp}$, the Thevenin's equivalent impedance, looking into the buses p-q is given by

$$Z_{th-pq} = Z_{pp} + Z_{qq} - 2Z_{pq}$$

Substituting for the Thevenin's equivalent impedance in Eq. (13.28a) leads to

$$I_a = \frac{V_p - V_q}{\left(Z_{pp} + Z_{qq} - 2Z_{pq}\right) - Z_a} \tag{13.28b}$$

The effect of the circulating loop current I_a on the system bus voltages may also be simulated by viewing I_a as two compensating currents, that is, $\Delta I_p = -I_a$ and $\Delta I_q = I_a$. Substituting these values in Eq. (13.26) provides an expression for the current changes in the line connected between buses i-j.

$$\Delta I_{ij} = -K_{ij-p}I_a + K_{ij-q}I_a$$

Substituting for the distribution factors from Eq. (13.25) in the preceding expression results in

$$\Delta I_{ij} = -\frac{Z_{ip} - Z_{jp}}{Z_l}I_a + \frac{Z_{iq} - Z_{jq}}{Z_l}I_a = -\frac{\left[(Z_{ip} - Z_{iq}) - (Z_{jp} - Z_{jq})\right]}{Z_l}I_a \tag{13.29}$$

Substitution of Eq. (13.28a) in Eq. (13.29) leads to

$$\Delta I_{ij} = -\frac{\left[(Z_{ip} - Z_{iq}) - (Z_{jp} - Z_{jq})\right]}{Z_l}\left\{\frac{(V_p - V_q)}{Z_{th-pq} - Z_a}\right\}$$

and substituting for $(V_p - V_q)$ in the preceding expression from Eq. (13.27) yields the following equation

$$\Delta I_{ij} = -\frac{Z_a}{Z_l}\left\{\frac{(Z_{ip} - Z_{iq}) - (Z_{jp} - Z_{jq})}{Z_{th-pq} - Z_a}\right\}I_{pq} \tag{13.30}$$

or

$$\frac{\Delta I_{ij}}{I_{pq}} = L_{ij-pq} = -\frac{Z_a}{Z_l}\left\{\frac{(Z_{ip} - Z_{iq}) - (Z_{jp} - Z_{jq})}{Z_{th-pq} - Z_a}\right\} \tag{13.31}$$

In Eq. (13.31), L_{ij-pq} is called the *line-outage distribution factor* and is defined as a fraction of the line current, connected between buses p-q, prior to the outage. Equation (13.31) also shows that the line-outage distribution factor can be determined from the elements of the $\mathbf{Z}_{bus}$ matrix and the line data.

Similar line-outage distribution factors can be defined for various other lines in the system. For example, the equation for the line-outage distribution factor for a line connected between two buses r-s can be written as

$$L_{rs-pq} = -\frac{Z_a}{Z_{l-rs}}\left\{\frac{(Z_{rp} - Z_{rq}) - (Z_{sp} - Z_{sq})}{Z_{th-pq} - Z_a}\right\} \tag{13.32}$$

In Eq. (13.32), Z_{l-rs} is the impedance of the line connecting buses r-s. Therefore, from the foregoing it can be noted that tables of line-outage factors can be computed beforehand to model the impact of single-line outages. Equations (13.31) and

(13.32) also show that if I_{ij} and I_{rs} are the pre-outage currents, the new values of the currents, following an outage of the line connected between buses p-q, in the lines connected between buses i-j and r-s respectively can be expressed as

$$I'_{ij} = I_{ij} + \Delta I_{ij} = I_{ij} + L_{ij-pq} I_{pq}$$

$$I'_{rs} = I_{rs} + \Delta I_{rs} = I_{rs} + L_{rs-pq} I_{pq}$$

$$(13.33)$$

Equation (13.33) provides the basis to establish whether a line is overloaded, by comparing the new values of the line currents, following the tripping of a particular line, with the pre-outage values.

Example 13.6 For the power system network shown in Fig. 13.15, the line between buses 3-4 is taken out of service for maintenance. Determine for all the lines in the system (i) line-outage distribution factors and (ii) new values of line currents. The lines between buses 1-2, 1-3, and 1-4 each carry a load of 25% and each of the lines between buses 2-3 and 2-4 are loaded to 75% of their rated capacities, prior to tripping the line between buses 3-4 out of service. Calculate the loading on these lines after tripping of the above line. The line data and bus voltages, in per unit, prior to line tripping is shown in the figure.

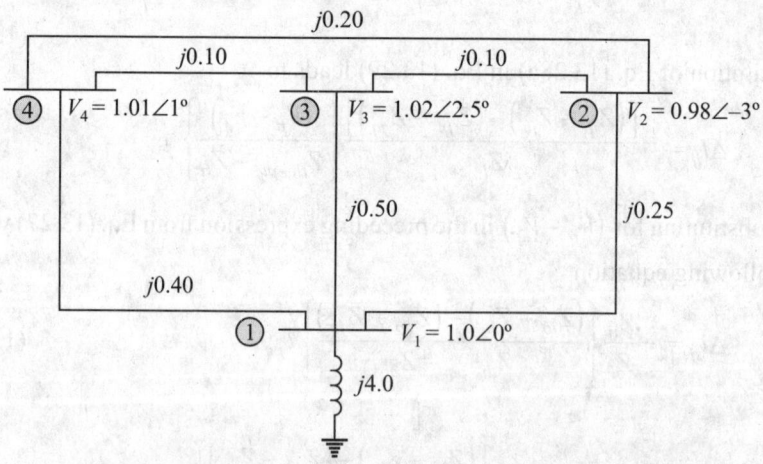

Fig. 13.15 Power system network of Example 13.6

Solution From the line data, $Z_a = j0.1$ per unit. In order to calculate the Thevenin's equivalent circuit impedance, looking into the buses at 3-4, the bus impedance matrix Z_{bus} is required to be formulated. From the given line data Z_{bus} is computed as follows:

```
≫Ybus=zeros(4,4); % Creation of a 4 × 4 null matrix
≫Ybus(1,1)=1/(i*4.0)+1/(i*0.25)+1/(i*0.5)+1/(i*0.4);Ybus(1,2)=-1/
(i*0.25);
≫Ybus(2,1)=Ybus(1,2); Ybus(1,3)=-1/(i*0.5);Ybus(3,1)=Ybus(1,3);
≫Ybus(1,4)=-1/(i*0.4); Ybus(4,1)=Ybus(1,4);
≫Ybus(2,2)=1/(i*0.25)+1/(i*0.1)+1/(i*0.2);Ybus(2,3)=-1/(i*0.10);
≫Ybus(3,2)=Ybus(2,3);Ybus(2,4)=-1/(i*0.20);Ybus(4,2)=Ybus(2,4);
≫Ybus(3,3)=1/(i*0.5)+1/(i*0.1)+1/(i*0.1);Ybus(3,4)=-1/(i*0.10);
≫Ybus(4,3)=Ybus(3,4);Ybus(4,4)=1/(i*0.4)+1/(i*0.2)+1/(i*0.1);
≫Ybus                    %[Ybus] matrix from the given line data
```

```
Ybus =
        0 - 8.7500i      0 + 4.0000i      0 + 2.0000i      0 + 2.5000i
        0 + 4.0000i      0 - 19.0000i     0 + 10.0000i     0 + 5.0000i
        0 + 2.0000i      0 + 10.0000i     0 - 22.0000i     0 + 10.0000i
        0 + 2.5000i      0 + 5.0000i      0 + 10.0000i     0 - 17.5000i
  » Z=inv(Ybus)          % Computation of bus impedance matrix Z by taking
                         inverse of [Ybus]matrix
Z =
        0 + 4.0000i      0 + 4.0000i      0 + 4.0000i      0 + 4.0000i
        0 + 4.0000i      0 + 4.1348i      0 + 4.1064i      0 + 4.0993i
        0 + 4.0000i      0 + 4.1064i      0 + 4.1454i      0 + 4.1135i
        0 + 4.0000i      0 + 4.0993i      0 + 4.1135i      0 + 4.1504i
```

The Thevenin's equivalent circuit impedance looking into the buses 3-4 is given by

$$Z_{th-34} = Z_{33} + Z_{44} - 2Z_{34} = j4.1454 + j4.1504 - 2 \times j4.1135 = j0.0688 \text{ per unit}$$

```
  » V=[1.0*exp(i*0*pi/180);0.98*exp(i*(-3.0)*pi/180);1.02*exp(i*(2.5)*
pi/180);...
  1.01*exp(i*(-1.0)*pi/180)];      % Complex bus voltages input prior to
                                   line tripping

  » V
V =
      1.0000
      0.9787 - 0.0513i
      1.0190 + 0.0445i
      1.0098 - 0.0176i
```

Line currents prior to the tripping of the line, by using Eq. (13.27) are calculated as follows:

Line current (bus 1 to bus 2) $= \dfrac{V_1 - V_2}{Z_{l1-2}} = [1.0 - (0.9787 - 0.0513i)]/(i * 0.25)$

$$= 0.2052 - 0.0854i = 0.2222\angle - 22.5937°$$

Line current (bus 1 to bus 3) $= \dfrac{V_1 - V_3}{Z_{l1-3}} = [1.0 - (1.0190 + 0.0445i)]/(i * 0.5)$

$$= -0.0890 + 0.0381i = 0.0968\angle - 23.1753°$$

Line current (bus 1 to bus 4) $= \dfrac{V_1 - V_4}{Z_{l1-4}} = (1.0-(1.0098 - 0.0176i))/(i * 0.4)$

$$= 0.0441 + 0.0246i = 0.0505\angle 29.1872°$$

Line current (bus 2 to bus 3) $= \dfrac{V_2 - V_3}{Z_{l2-3}}$

$$= [0.9787 - 0.0513i - (1.0190 + 0.0445i)]/(i * 0.1)$$

$$= -0.9578 + 0.4037i = 1.0394\angle -22.8548°$$

Line current (bus 2 to bus 4) $= \dfrac{V_2 - V_4}{Z_{l2-4}}$

$$= [0.9787 - 0.0513i - (1.0098 - 0.0176i)]/(i * 0.2)$$

$$= -0.1683 + 0.1559i = 0.2295\angle - 42.8161°$$

Line current (bus 3 to bus 4) $= \dfrac{V_3 - V_4}{Z_{l3-4}}$

$$= [(1.0190 + 0.0445i) - (1.0098 - 0.0176i)]/(i*0.1)$$

$$= 0.6212 - 0.0918i = 0.6279\angle -8.4091°$$

(i)   The line-outage distribution factors for the five in-service lines are computed from Eq. (13.31) as follows:

≫ Za=-i*0.1;   % Sets Za equal to the negative of the impedance of the line being tripped

≫ Zden=Z$_{th-34}$ -Za   % Computes the denominator within brackets in Eq. (13.31)

Zden =
0 + 0.1688i

$$L_{12-34} = -\frac{Z_{1-34}}{Z_{1-12}}\left\{\frac{Z_{13} - Z_{14} - (Z_{23} - Z_{24})}{Z_{th-34} - Z_{1-34}}\right\}$$

= Za/(i*0.25)*((Zbus(1,3)-Zbus(1,4))-(Zbus(2,3)-Zbus(2,4)))/Zden

= 0.0168

$$L_{13-34} = -\frac{Z_{1-34}}{Z_{1-13}}\left\{\frac{(Z_{13} - Z_{14}) - (Z_{33} - Z_{34})}{Z_{th-34} - Z_{1-34}}\right\}$$

= Za/(i*0.5)*((Zbus(1,3)-Zbus(1,4))-(Zbus(3,3)-Zbus(3,4)))/Zden

= 0.0378

$$L_{14-34} = -\frac{Z_{1-34}}{Z_{1-14}}\left\{\frac{(Z_{13} - Z_{14}) - (Z_{43} - Z_{44})}{Z_{th-34} - Z_{1-34}}\right\}$$

= Za/(i*0.4)*((Zbus(1,3)-Zbus(1,4))-(Zbus(4,3)-Zbus(4,4)))/Zden

= -0.0547

$$L_{23-34} = -\frac{Z_{1-34}}{Z_{1-23}}\left\{\frac{(Z_{23} - Z_{24}) - (Z_{33} - Z_{34})}{Z_{th-34} - Z_{1-34}}\right\}$$

= Za/(i*0.1)*((Zbus(2,3)-Zbus(2,4))-(Zbus(3,3)-Zbus(3,4)))/Zden

= 0.1469

$$L_{24-34} = -\frac{Z_{1-34}}{Z_{1-24}}\left\{\frac{(Z_{23} - Z_{24}) - (Z_{43} - Z_{44})}{Z_{th-34} - Z_{1-34}}\right\}$$

= Za/(i*0.2)*((Zbus(2,3)-Zbus(2,4))-(Zbus(4,3)-Zbus(4,4)))/Zden

= -0.1303

(ii)   By using Eq. (13.33), the new values of the currents in the in-service lines are computed as follows:

$I'_{12} = I_{12} + L_{12-34}\, I_{34}$ = I12+L12*I34 = 0.2156 - 0.0869i = 0.2325∠-21.9556°

$I'_{13} = I_{13} + L_{13-34}I_{34}$ = I13+L13*I34 = -0.0655 + 0.0346i = 0.0741∠-27.8581°

$I'_{14} = I_{14} + L_{14-34}I_{34}$ = I14+L14*I34 = 0.0101 + 0.0296i = 0.0313∠71.1464°

$I'_{23} = I_{23} + L_{23-34}I_{34}$ = I23+L23*I34 = -0.8665 + 0.3902i = 0.9504∠-24.2434°

$I'_{24} = I_{24} + L_{24-34}I_{34}$ = I24+L24*I34 = -0.2493 + 0.1679i = 0.3006∠-33.965°

The current flows in the various lines, prior to and after the tripping of the line between buses 3-4 are tabulated as follows:

Line connection	Prior to tripping	Post tripping
Between buses 1-2	$0.2222\angle-22.5937°$	$0.2325\angle-21.9556°$
Between buses 1-3	$0.0968\angle-23.1753°$	$0.0741\angle-27.8581°$
Between buses 1-4	$0.0505\angle29.1872°$	$0.0313\angle71.1464°$
Between buses 2-3	$1.0394\angle-22.8548°$	$0.9504\angle-24.2434°$
Between buses 2-4	$0.2295\angle-42.8161°$	$0.3006\angle-33.965°$

The loading on each line after tripping of the line between buses 3-4 is calculated as follows:

```
Line between buses 1-2 = » 0.2325/(4*0.2222)*100 = 26.1540%
Line between buses 1-3 = » 0.0741/(4*0.0968)*100 = 19.1349%
Line between buses 1-4 = » 0.0313/(4*0.0505)*100 = 15.5093%
Line between buses 2-3 = » 0.9504/(1.33333*1.0394)*100 = 68.5755%
Line between buses 2-4 = » 0.3006/(1.33333*0.2295)*100 = 98.2430%
```

From the foregoing load computations, it is observed that none of the lines are overloaded post outage of line between buses 3-4.

Example 13.7 For the power system in Example 13.6, calculate the power at each bus. If 50% of the real power at bus 3 is shifted to bus 1, compute the change in current flow in the line connecting buses 3-2. Verify the result by performing an ac power flow study on the system.

Solution In order to compute the power at each bus, first injected currents are calculated by using the relation $I_{bus} = Y_{bus}V_{bus}$. Thus,

```
» Iinj=Ybus*V               % Computation of injected bus current vector[I]
Iinj =
        0.1602 - 0.2727i
       -1.3313 + 0.6450i
        1.6680 - 0.5336i
       -0.4969 - 0.0887i
```

Complex bus powers are calculated from the relation $S = VI*$ as shown below

```
» S1=V(1)*conj(Iinj(1))       % Complex power at bus 1
S1 =  0.1602 + 0.2727i
» S2=V(2)*conj(Iinj(2))       % Complex power at bus 2
S2 = -1.3359 - 0.5630i
» S3=V(3)*conj(Iinj(3))       % Complex power at bus 3
S3 = 1.6760 + 0.6180i
» S4=V(4)*conj(Iinj(4))       % Complex power at bus 4
S4 = -0.5003 + 0.0984i
```

By employing Eq. (13.25), the current distribution factors are computed when 50% of real power is shifted from bus 3 to bus 1. Hence;

```
» K321=(Zbus(3,1)-Zbus(2,1))/(i*0.1)     % Computation of K₃₂₋₁
K321 = 0
» K323=(Zbus(3,3)-Zbus(2,3))/(i*0.1)     % Computation of K₃₂₋₃
K323 = 0.3901
```

Shifting of real power in per unit is equivalent to the shifting of real component of the injected current from bus 3 to bus 1.

```
>> delIinj1=0.8340;
```
% Additional current $\Delta I_1 = \dfrac{1.6680}{2} = 0.8340$ injected

% into bus 1 due to shifting of 50% of real power
% from bus 3 to bus 1

```
delIinj3 = -0.8340
```
% Portion of current $\Delta I_3 = -\dfrac{1.6680}{2} = -0.8340$

% reduced from bus 1 due to shifting of 50% of real
% power from bus 3 to bus 1

The change in line current ΔI_{32} connecting buses 3-2, due to shifting of real power from bus 3 to 1 is determined by using Eq. (13.26) as follows:

```
>> delI32=K321*delIinj1+K323*delIinj3
delI32 = -0.3253                       % Determination of ΔI₃₂.
>> I32d=I32+delI32                      % Computation of new line current I'₃₂ = I₃₂
                                        % + ΔI₃₂ connecting buses 3-2 due to shifting
                                        % of real power from bus 3 to bus 1
I32d = 0.6325 - 0.4037i = 0.7504∠-32.5503°                                    (13.7.1)
```

Due to the shifting of power from bus 3 to bus 1, the injected currents at these two buses have changed. In order to perform an ac power flow, it is essential to determine the new bus voltages with the new injected bus currents. Hence,

```
Iinj =                      % Computation of new injected bus current vector.
    0.9942 - 0.2727i
   -1.3313 + 0.6450i
    0.8340 - 0.5336i
   -0.4969 - 0.0887i
>> Vnew=inv(Ybus)*Iinj       % Computation of new bus voltages.
Vnew =
    1.0000 - 0.0000i
    0.9787 - 0.1400i
    1.0190 - 0.0768i
    1.0098 - 0.1123i
>> I32new=(Vnew(3)-Vnew(2))/(i*0.1)    % Computation of new current flow
                                       % in the line connected between buses
                                       % 3-2, from the ac power flow data.
I32new = 0.6325 - 0.4037i = 0.7504 ∠-32.5503°                                (13.7.2)
```

By comparing (13.7.1) and (13.7.2) it is observed that the current-shift distribution factors lead to similar results as obtained from an ac power flow study.

13.6.4 Compound Outages Contingencies Analysis

Compound or multiple contingencies in a power system network are brought about by (i) sequential tripping of lines and (ii) tripping of a line while exploring the possibility of relieving an overloaded line, by a transfer in generation. The techniques required for the treatment of these two conditions, for multiple contingencies analysis in a system, are different and are described in the following subsections.

Sequential Tripping of Lines

Sequential tripping condition in a system occurs when a second line trips and goes out of service, when one line has already suffered an outage. Consider that a line connecting buses p-q in a power system trips when a line connecting buses i-j is already out of service. If the impedance of the lines between buses i-j and p-q is Z_a and Z_b respectively, the effect due to the tripping of the second line can be simulated by substituting the value of $\boldsymbol{Z}$ from Eq. (13.12) in Eq. (13.10) and replacing the line impedances by their negative values as follows:

$$
\begin{bmatrix} V_i - V_j \\ V_p - V_q \end{bmatrix} =
$$

$$
\begin{matrix} & a & b & \\ a \\ b \end{matrix}
\begin{bmatrix} \{(Z_{ii}-Z_{ij})-(Z_{ji}-Z_{jj})\}-Z_a & \{(Z_{ip}-Z_{iq})-(Z_{jp}-Z_{jq})\} \\ \{(Z_{pi}-Z_{pj})-(Z_{qi}-Z_{qj})\} & \{(Z_{pp}-Z_{pq})-(Z_{qp}-Z_{qq})\}-Z_b \end{bmatrix}
\begin{bmatrix} I_a \\ I_b \end{bmatrix}
$$

$$(13.34)$$

The terms within brackets in the diagonal elements of $\boldsymbol{Z}$ in Eq. (13.34) are the Thevenin's equivalents impedances [see Eq. (13.15)] at buses i-j and p-q, respectively. Thus, substituting

$$ Z_{th-ij} = \{(Z_{ii}-Z_{ij})-(Z_{ji}-Z_{jj})\} \text{ and } Z_{th-pq} = \{(Z_{pp}-Z_{pq})-(Z_{qp}-Z_{qq})\} $$

in Eq. (13.34) yields

$$
\begin{matrix} & a & b & \\ \end{matrix}
\begin{bmatrix} V_i - V_j \\ V_p - V_q \end{bmatrix} =
\begin{matrix} a \\ b \end{matrix}
\begin{bmatrix} Z_{th-ij}-Z_a & \{(Z_{ip}-Z_{iq})-(Z_{jp}-Z_{jq})\} \\ \{(Z_{pi}-Z_{pj})-(Z_{qi}-Z_{qj})\} & Z_{th-pq}-Z_b \end{bmatrix}
\begin{bmatrix} I_a \\ I_b \end{bmatrix}
$$

$$(13.35)$$

It may also be remembered that V_i, V_j, V_p, and V_q are the respective voltages of buses i, j, p, and q prior to the line voltages. Therefore, the current flows I_{ij} and I_{pq} in the lines connecting buses i-j and p-q, prior to the line outages, are respectively given by

$$ I_{ij} = \frac{V_i - V_j}{Z_a} $$

and

$$ I_{pq} = \frac{V_p - V_q}{Z_b} \tag{13.36} $$

The sequential use of Eqs (13.35) and (13.36) help to define the procedure for developing distribution factors to simulate the effects of outages of multiple lines in a power network. The step-wise procedure is elucidated here.

Step 1: Derivation of an expression to compute the injected currents I_a and I_b in terms of the pre-outage line currents I_{ij} and I_{pq}:

Divide rows 1 and 2, in Eq. (13.35), by their respective diagonal elements. The resultant expression is

$$
\begin{bmatrix} \dfrac{V_i - V_j}{Z_{th-ij} - Z_a} \\ \dfrac{V_p - V_q}{Z_{th-pq} - Z_b} \end{bmatrix} =
$$

$$
\begin{array}{c}
 \\
a \\
b
\end{array}
\left[
\begin{array}{cc}
\overset{a}{1} & \overset{b}{\dfrac{\{(Z_{ip}-Z_{iq})-(Z_{jp}-Z_{jq})\}}{Z_{th-ij}-Z_a}} \\[2em]
\dfrac{\{(Z_{pi}-Z_{pj})-(Z_{qi}-Z_{qj})\}}{Z_{th-pq}-Z_b} & 1
\end{array}
\right]
\left[
\begin{array}{c}
I_a \\[1em]
I_b
\end{array}
\right]
$$

$$(13.37)$$

Comparing the off-diagonal terms in Eq. (13.37) with Eq. (13.31), it is observed

that
$$
\frac{\{(Z_{ip}-Z_{iq})-(Z_{jp}-Z_{jq})\}}{Z_{th-ij}-Z_a} = -\frac{Z_b}{Z_a}L_{ij-pq}
$$

and
$$
\frac{\{(Z_{pi}-Z_{pj})-(Z_{qi}-Z_{qj})\}}{Z_{th-pq}-Z_b} = \frac{Z_a}{Z_b}L_{pq-ij}
$$

Substituting the preceding equations for the off-diagonal terms in Eq. (13.37) yields the following expression

$$
\left[
\begin{array}{c}
\dfrac{V_i-V_j}{Z_{th-ij}-Z_a} \\[2em]
\dfrac{V_p-V_q}{Z_{th-pq}-Z_b}
\end{array}
\right]
=
\begin{array}{c}
 \\
a \\
b
\end{array}
\left[
\begin{array}{cc}
\overset{a}{1} & \overset{b}{-\dfrac{Z_b}{Z_a}L_{ij-pq}} \\[2em]
-\dfrac{Z_a}{Z_b}L_{pq-ij} & 1
\end{array}
\right]
\left[
\begin{array}{c}
I_a \\[1em]
I_b
\end{array}
\right]
$$

$$(13.38)$$

Writing Eq. (13.38) as two simultaneous equations gives

$$
\frac{V_i-V_j}{Z_{th-ij}-Z_a} = I_a - \frac{Z_b}{Z_a}L_{ij-pq}I_b
$$

$$(13.39a)$$

$$
\frac{V_p-V_q}{Z_{th-pq}-Z_b} = -\frac{Z_a}{Z_b}L_{pq-ij}I_a + I_b
$$

$$(13.39b)$$

On multiplying Eq. (13.39b) by $(Z_b/Z_a)/I_{ij-pq}$ and adding to Eq. (13.39a), eliminates I_b. The expression for I_a is then given by

$$
I_a = \frac{1}{1-(L_{ij-pq})(L_{pq-ij})}\left\{\frac{V_i-V_j}{Z_{th-ij}-Z_a}+\frac{Z_b}{Z_a}L_{ij-pq}\left(\frac{V_p-V_q}{Z_{th-pq}-Z_b}\right)\right\}
$$

$$(13.40a)$$

Similarly I_b is given by

$$
I_b = \frac{1}{1-(L_{ij-pq})(L_{pq-ij})}\left\{\frac{Z_a}{Z_b}L_{pq-ij}\frac{V_i-V_j}{Z_{th-ij}-Z_a}+\left(\frac{V_p-V_q}{Z_{th-pq}-Z_b}\right)\right\}
$$

$$(13.40b)$$

From Eq. (13.36), $V_i-V_j = Z_aI_{ij}$ and $V_p-V_q = Z_bI_{pq}$; and substituting these in Eqs (13.40a) and (13.40b) leads to

$$I_a = \frac{1}{1-\left(L_{ij-pq}\right)\left(L_{pq-ij}\right)} \left\{ \frac{Z_a}{Z_{th-ij}-Z_a} I_{ij} + \frac{Z_b^2}{Z_a} \frac{L_{ij-pq}}{Z_{th-pq}-Z_b} I_{pq} \right\}$$

(13.41a)

$$I_b = \frac{1}{1-\left(L_{ij-pq}\right)\left(L_{pq-ij}\right)} \left\{ \frac{Z_a^2}{Z_b} \frac{L_{pq-ij}}{Z_{th-ij}-Z_a} I_{ij} + \frac{Z_b}{Z_{th-pq}-Z_b} I_{pq} \right\}$$

(13.41b)

In matrix form, the preceding results are written as

$$\begin{bmatrix} I_a \\ I_b \end{bmatrix} = \frac{1}{1-\left(L_{ij-pq}\right)\left(L_{pq-ij}\right)} \begin{bmatrix} \dfrac{Z_a}{Z_{th-ij}-Z_a} & \dfrac{Z_b^2}{Z_a}\dfrac{L_{ij-pq}}{Z_{th-pq}-Z_b} \\ \dfrac{Z_a^2}{Z_b}\dfrac{L_{pq-ij}}{Z_{th-ij}-Z_a} & \dfrac{Z_b}{Z_{th-pq}-Z_b} \end{bmatrix} \begin{bmatrix} I_{ij} \\ I_{pq} \end{bmatrix}$$

(13.42)

It may be recalled that in Eq. (13.42), currents I_a and I_b represent the compensating current injections, into a regular power system network, to simulate the effect of simultaneous outages. On the other hand, I_{ij} and I_{pq} are the current flows in the lines, prior to line outages, connecting buses i-j and p-q respectively.

Step 2: Computation of changes in current flows in the lines due to I_a and I_b:

The effects of line tripping are simulated by imagining that current of magnitude $-I_a$ is injected into bus i and I_a is injected into bus j. Similarly, it is assumed that current of magnitude $-I_b$ is injected into bus p and I_b is injected into bus q. The expressions for voltage drops, at the other buses in the system, due to these injected currents are written from Eq. (13.23) as follows

$$\Delta V_r = \left(Z_{ri}-Z_{rj}\right)I_a + \left(Z_{rq}-Z_{rp}\right)I_b$$

(13.43a)

$$\Delta V_s = \left(Z_{si}-Z_{sj}\right)I_a + \left(Z_{sq}-Z_{sp}\right)I_b$$

(13.43b)

If the impedance of the line connected between buses i-j is represented by Z_{l-rs}, the current flow in the line is given by

$$\Delta I_{rs} = \frac{\Delta V_r - \Delta V_s}{Z_{l-rs}}$$

Substituting Eq. (13.43) and gathering the like terms, yields

$$\Delta I_{rs} = \frac{\left(Z_{ri}-Z_{rj}\right)-\left(Z_{si}-Z_{sj}\right)}{Z_{l-rs}} I_a + \frac{\left(Z_{rq}-Z_{rp}\right)-\left(Z_{sq}-Z_{sp}\right)}{Z_{l-rs}} I_b \quad (13.44)$$

The line outage distribution factors are written from Eq. (13.32) as follows

$$L_{rs-ij} = -\frac{Z_a}{Z_{l-rs}} \left\{ \frac{\left(Z_{ri}-Z_{rj}\right)-\left(Z_{si}-Z_{sj}\right)}{Z_{th-ij}-Z_a} \right\}$$

(13.45a)

$$L_{rs-pq} = -\frac{Z_b}{Z_{l-rs}} \left\{ \frac{\left(Z_{rp}-Z_{rq}\right)-\left(Z_{sp}-Z_{sq}\right)}{Z_{th-pq}-Z_b} \right\}$$

(13.45b)

Substitution of Eq. (13.45) in Eq. (13.44) gives the change in line current in terms of the distribution factors. The result of the substitution, in matrix form is as follows

$$\Delta I_{rs} = \begin{bmatrix} \dfrac{Z_{th-ij} - Z_a}{Z_a} L_{rs-ij} & \dfrac{Z_{th-pq} - Z_b}{Z_b} L_{rs-pq} \end{bmatrix} \begin{bmatrix} I_a \\ I_b \end{bmatrix} \tag{13.46}$$

Step 3: Derivation of an expression for computation of changes in line currents in terms of pre-outage line flows:

Equation (13.46) determines the changes in line currents in terms of current injections at the buses i-j and p-q. Substitution for I_a and I_b, from Eq. (13.42) in Eq. (13.46) yields

$$\Delta I_{rs} = \begin{bmatrix} \dfrac{Z_{th-ij} - Z_a}{Z_a} L_{rs-ij} & \dfrac{Z_{th-pq} - Z_b}{Z_b} L_{rs-pq} \end{bmatrix} \times$$

$$\dfrac{1}{1-\left(L_{ij-pq}\right)\left(L_{pq-ij}\right)} \begin{bmatrix} \dfrac{Z_a}{Z_{th-ij} - Z_a} & \dfrac{Z_b^2}{Z_a}\dfrac{L_{ij-pq}}{Z_{th-pq} - Z_b} \\ \dfrac{Z_a^2}{Z_b}\dfrac{L_{pq-ij}}{Z_{th-ij} - Z_a} & \dfrac{Z_b}{Z_{th-pq} - Z_b} \end{bmatrix} \begin{bmatrix} I_{ij} \\ I_{pq} \end{bmatrix}$$

On multiplying the two matrices and simplifying the product, the following equation results.

$$\Delta I_{rs} = \dfrac{1}{1-\left(L_{ij-pq}\right)\left(L_{pq-ij}\right)}$$

$$\begin{bmatrix} \left(L_{rs-ij} + L_{rs-pq}L_{pq-ij}\right) & \left(L_{rs-pq} + L_{rs-ij}L_{ij-pq}\right) \end{bmatrix} \begin{bmatrix} I_{ij} \\ I_{pq} \end{bmatrix}$$

$$= \dfrac{1}{1-\left(L_{ij-pq}\right)\left(L_{pq-ij}\right)} \begin{bmatrix} L'_{rs-ij} & L'_{rs-pq} \end{bmatrix} \begin{bmatrix} I_{ij} \\ I_{pq} \end{bmatrix} \tag{13.47}$$

where L'_{rs-ij} and L'_{rs-pq} are called the effective distribution factors and are given by

$$L'_{rs-ij} = L_{rs-ij} + L_{rs-pq}L_{pq-ij}$$

and

$$L'_{rs-pq} = L_{rs-pq} + L_{rs-ij}L_{ij-pq}$$

Equation (13.47) thus provides an expression for computing the changes in line flows, in terms of the distribution factors and pre-outage line flows, due to the tripping of a second line when one line is already tripped. Further, it is also observed that distribution factors, due to the outage of a second line, are determined from the already calculated distribution factors for a single line outage.

Proposed Shift in Current and a Line Outage

When power is shifted from one generation bus, say p, to another generation bus, say q, current shifting occurs in lines. On the other hand, if a line connected between buses i-j carrying a current I_{ij} goes out of service, it is possible that another line connecting buses r-s may get overloaded. The overload on the line connecting buses r-s can be relieved by reducing the current injected into a bus p in the system and correspondingly increasing the injected current into another bus q. The effect on the overloaded line is independent of the sequence in which the line tripping and the projected current shift occur, since the Z_{bus} matrix is linear. It is more expedient, for the purpose of analysis, to assume that the proposed current shift is followed by a tripping of the line connecting buses i-j.

Equation (13.26) forms the basis of analysis. Assume that a current shift is proposed to be made from bus p to bus q and that the current distribution factors, prior to the proposed current shift for the lines connected between buses r-s and i-j are given by K_{rs-p}, K_{rs-q} and K_{ij-p}, K_{ij-q}, respectively. Using Eq. (13.26), the changes in the current flows in the two lines can be written as

$$\Delta I_{rs} = K_{rs-p}\Delta I_p + K_{rs-q}\Delta I_q \tag{13.48a}$$

$$\Delta I_{ij} = K_{ij-p}\Delta I_p + K_{ij-q}\Delta I_q \tag{13.48b}$$

In Eq. (13.48), ΔI_p and ΔI_q are the proposed shift in current injections at buses p and q, respectively. If the current flows in the lines r-s and i-j, prior to the proposed current shift, are I_{rs} and I_{ij} respectively, the changed line currents, as a consequence of the proposed current shifts are represented by

$$I'_{rs} = I_{rs} + \Delta I_{rs} \tag{13.49a}$$

$$I'_{ij} = I_{ij} + \Delta I_{ij} \tag{13.49b}$$

At this juncture, if the line connecting buses i-j is tripped, the supplementary change in current flow in the line connecting buses r-s is given by

$$\Delta I'_{rs} = L_{rs-ij}I'_{ij} = L_{rs-ij}(I_{ij} + \Delta I_{ij}) \tag{13.50}$$

The term $\Delta I'_{rs}$ in Eq. (13.50) expresses the change in current flow in the line connecting buses r-s, due to the combined effect of proposed current shift in buses p and q and the tripping of the line connected between buses i-j. The total current in the line is, therefore, given by

$$I''_{rs} = I'_{rs} + \Delta I'_{rs}$$

Substituting for I'_{rs} from Eq. (13.49a) and for $\Delta I'_{rs}$ from Eq. (13.50) yields the following result

$$I''_{rs} = \left(I_{rs} + \Delta I_{rs}\right) + L_{rs-ij}\left(I_{ij} + \Delta I_{ij}\right) = \left\{I_{rs} + L_{rs-ij}I_{ij}\right\} + \Delta I_{rs} + L_{rs-ij}\Delta I_{ij}$$

$$\tag{13.51}$$

Comparing the bracketed part of Eq. (13.51) with Eq. (13.33) it can be concluded that it represents the portion of the current flowing in the line connected between buses r-s when the line connected between buses i-j is taken out of service and there is no current shift taking place.

By rewriting Eq. (13.51) differently a very important result emerges, which is as follows

$$\Delta I_{rs}^{sh} = I_{rs}'' - \left\{ I_{rs} + L_{rs-ij}I_{ij} \right\} = \Delta I_{rs} + L_{rs-ij}\Delta I_{ij} \tag{13.52}$$

The right hand side of Eq. (13.52) symbolizes the resultant change in the overloaded line due to the current (that is generation) shift in the system with the line connecting buses *i-j* disconnected from the network. On substituting from Eq. (13.48) in Eq. (13.52) and reorganizing, the resultant expression leads to

$$\Delta I_{rs}^{sh} = \Delta I_{rs} + L_{rs-ij}\Delta I_{ij}$$

$$= \left(K_{rs-p}\Delta I_p + K_{rs-q}\Delta I_q \right) + L_{rs-ij}\left(K_{ij-p}\Delta I_p + K_{ij-q}\Delta I_q \right)$$

$$= \left(K_{rs-p} + L_{rs-ij}K_{ij-p} \right)\Delta I_p + \left(K_{rs-q} + L_{rs-ij}K_{ij-q} \right)\Delta I_q$$

or $\qquad \Delta I_{rs}^{sh} = K_{rs-p}' \Delta I_p + K_{rs-q}' \Delta I_q \tag{13.53}$

where $K_{rs-p}' = K_{rs-p} + L_{rs-ij}K_{ij-p}$ and $K_{rs-q}' = K_{rs-q} + L_{rs-ij}K_{ij-q}$

The terms K_{rs-p}' and K_{rs-q}' are the modified generation shift distribution factors and can be conveniently determined from the factors computed for the customary power system configuration. The modified generation shift distribution factors take into account the outage of the line connecting buses *i-j* which already exists. Equation (13.53) enables the system operating engineer to establish the quantum of release in overload possible by the proposed shift in generation (current) from one bus to another.

Example 13.8 For the system of Example 13.6 compute the change in current, in the line connecting buses 1-4, when lines connected between buses 3-4 and 2-3 are tripped simultaneously.

Solution The change in line current connecting buses 1-4 is obtained from Eq. (13.47) by making the following substitutions

$$r = 1, s = 4, p = 3, q = 4, i = 2, j = 3$$

Hence,

$$\Delta I_{14} = \frac{1}{1 - (L_{23-34})(L_{34-23})} \left[L_{14-23}' \quad L_{14-34}' \right] \begin{bmatrix} I_{23} \\ I_{34} \end{bmatrix} \tag{13.8.1}$$

where $\quad L_{14-23}' = L_{14-23} + L_{14-34}L_{34-23}$ and $L_{14-34}' = L_{14-34} + L_{14-23}L_{23-34}$

Following the procedure in Example 13.6, the line outage distribution factors, L_{14-23} and L_{34-23} are computed as follows:

```
≫ Zb=-i*0.1;
  Zik-23 = Z22+ Z33 - 2Z23 = 4.1348i + 4.145i - 2×4.1064i = 0.0674i
≫ Zden1=Zth23-Zb
Zden1 = 0 + 0.1674i per unit
≫ L1423=Zb/(i*0.4)*((Z(1,2)-Z(1,3))-(Z(4,2)-Z(4,3)))/Zden1
L1423 = -0.0212
≫ L3423=Zb/(i*0.1)*((Z(3,2)-Z(3,3))-(Z(4,2)-Z(4,3)))/Zden1
L3423 = 0.1481
```

Next the elements L_{11} and L_{12} of the row matrix in Eq. (13.8.1) are computed in the following manner:

>> L$_{11}$=(L1423+L1434*L3423)/(1−L2334*L3423)
L$_{11}$ = −0.0300
>> L$_{12}$=(L1434+L1423*L2334)/(1−L2334*L3423)
L$_{12}$ = −0.0591

Using the values of current flows in lines connected between buses 2-3 and 3-4 from Example 13.6, the new current flow in the line connecting buses 1-4 due to simultaneous tripping is obtained by substituting for L_{11} and L_{12} in the equation (13.8.1) as follows

>> I14d = I14+(L$_{11}$*I23+L$_{12}$*I34)
I14d = 0.0361 + 0.0179i = 0.0403∠26.37°

For a complete understanding of the procedure, it is advisable that readers obtain the current flow in a line connecting another pair of buses, say 3-4, by assuming simultaneous tripping of lines, say connected between buses 2-3 and 2-4.

13.7 APPROXIMATE POWER FLOW METHOD FOR SIMU-LATING CONTINGENCIES

The fast decoupled power flow solution method, which forms the basis of contingency analysis, is described in Section 7.8.5. The simulation of contingencies, also known as the decoupled model, is based on the following approximations.

(i) The line resistance is neglected. In other words, the system is loss free and the lines are modelled by their series reactance.

(ii) The voltage magnitudes of all the buses in the system are held constant at 1.0 per unit.

(iii) The voltage angles between adjacent buses are small. In order to explain the implication of the assumption, suppose that the voltage angles, in radians, between two adjacent buses p-q are δ_p and δ_q respectively. Since δ_p and δ_q are assumed to be small, $\cos\delta_p = \cos\delta_q$ and $\sin\delta_p = \delta_p$ and $\sin\delta_q = \delta_q$.

If the reactance of the line connected between buses p-q is jX_a per unit, and the bus voltages are $V_p\angle\delta_p$ and $V_q\angle\delta_q$ per unit, the current I_{pq} per unit flowing in the line is given by

$$I_{pq} = \frac{V_p\angle\delta_p - V_q\angle\delta_q}{jX_a} = \frac{V_p\left(\cos\delta_p + j\sin\delta_p\right) - V_q\left(\cos\delta_q + j\sin\delta_q\right)}{jX_a}$$

Rewriting this equation, by applying the preceding assumptions, simplifies it to the following form

$$I_{pq} \cong \frac{\delta_p - \delta_q}{X_a} \text{ per unit} \tag{13.54}$$

The real power flow in the line connecting buses i-j is written as

$$P_{pq} = \frac{V_p V_q}{X_a}\sin\left(\delta_p - \delta_q\right) \cong \frac{\delta_p - \delta_q}{X_a} \text{ per unit} \tag{13.55}$$

Equations (13.54) and (13.55) show that the approximate value of current flow and power flow in the line connecting buses p-q, in per unit, in the dc power flow simulation have equal values, which is due to the application of the three assumptions. Accepting the current and power flow values as accurate, within the limits of the three assumptions, the current flow in another line connected between buses *i-j* can be written as

$$I_{ij} = \frac{\delta_i - \delta_j}{jX_l} \tag{13.56}$$

In Eq. (13.56), jX_l is the per unit reactance of the line connected between buses *i-j*. If the bus voltage angles change, the change in current flow in the line connecting buses *i-j* can be written as

$$\Delta I_{ij} = \frac{\Delta(\delta_i - \delta_j)}{jX_l} = \frac{\Delta\delta_i - \Delta\delta_j}{jX_l} \tag{13.57}$$

Based on the similarities between Eqs (13.54) and (13.55), the following expression for change in power is obtained.

$$\Delta P_{ij} = \frac{\Delta(\delta_i - \delta_j)}{jX_l} = \frac{\Delta\delta_i - \Delta\delta_j}{jX_l}$$

Since the system is assumed loss free [see assumption (i)], impedances in Eq. (13.31) can be replaced by their respective reactances. Thus Eq. (13.31) takes the following form

$$\frac{\Delta I_{ij}}{I_{pq}} = L_{ij-pq} = -\frac{X_a}{X_l}\left\{\frac{\left(X_{ip} - X_{iq}\right) - \left(X_{jp} - X_{jq}\right)}{X_{th-pq} - X_a}\right\} \tag{13.58}$$

Based on the association between per unit current and power flows in the lines, the power distribution factor, when a line trips, for a line connecting buses *i-j* in Eq. (13.58) can be expressed as

$$\frac{\Delta P_{ij}}{P_{pq}} = L_{ij-pq} = -\frac{X_a}{X_l}\left\{\frac{\left(X_{ip} - X_{iq}\right) - \left(X_{jp} - X_{jq}\right)}{X_{th-pq} - X_a}\right\} \tag{13.59}$$

From Eq. (13.59), the following conclusions, within the applicability of the assumptions made, can be drawn.

(i) The distribution factors, due to an outage, by the dc power flow method are fundamentally the same as those derived by the Z_{bus} matrix method.

(ii) Thus, power distribution factors as computed from Eq. (13.59) and the current shift distribution factors as calculated from Eq. (13.31) are numerically equal.

(iii) The decoupled power flow method enables the pre-computation of tables of power/current distribution factors from the actual measured power flows in the lines of an operational power system network.

Example 13.9 For the power system in Example 13.6, compute the change in power flow in the line connected between buses 3-2 when the line connected between buses 3-4 trips while considering a shift of 50% real power from bus 3 to 1. Use the approximate dc power flow simulation.

Solution From Example 13.7, the real power at bus 3 = 1.6760 per unit. Hence,

$$\Delta P_3 = -\frac{1.6760}{2} = -0.838 \text{ per unit and } \Delta P_1 = \frac{1.6760}{2} = 0.838 \text{ per unit}$$

By substituting $r = 3$, $s = 2$, $i = 3$, $j = 4$, $p = 3$, and $q = 1$ in Eq. (13.53), the modified generation shift distribution factors are determined as follows:

$$K'_{32-3} = K_{32-3} + L_{32-34}K_{34-3} \text{ and } K'_{32-1} = K_{32-1} + L_{32-34}K_{34-1}$$

The data from previous examples is used to calculate the various distribution factors as follows:

```
>> L3234=Za/(i*0.1)*((Zbus(3,3)-Zbus(3,4))-(Zbus(2,3)-Zbus(2,4)))/Zden
L3234 = -0.1471          % Computation of line outage distribution factor
```
L_{32-34}
```
>> K323=(Zbus(3,3)-Zbus(2,3))/(i*0.1)
K323 = 0.3901            % Computation of generation shift distribution
                           factor K_{32-3}
>> K321=(Zbus(3,1)-Zbus(2,1))/(i*0.1)
K321 = 0                 % Computation of generation shift distribution
                           factor K_{32-1}
>> K341=(Zbus(3,1)-Zbus(4,1))/(i*0.1)
K341 = 8.8818e-015       % Computation of generation shift distribution
                           factor K_{34-1}
>> K343=(Zbus(3,3)-Zbus(4,3))/(i*0.1)
K343 = 0.3191            % Computation of generation shift distribution
                           factor K_{34-3}
>> K323d=K323+L3234*K343
K323d = 0.3431           % Computation of modified generation shift dis-
                           tribution factor K'_{32-3}
>> K321d=K321+L3234*K341
K321d = -1.3061e-015     % Computation of modified generation shift dis-
                           tribution factor K'_{32-1}
>> delP32=K323d*(-1.6760/2)+K321d*(1.6760/2)
delP32 = -0.2875         % Computation of change in power flow in the line
                           connecting buses 3-2 due to a shift in power from
                           bus 3 to 1 and tripping of line between buses 3-4.
```

It is, thus, noted that the power flow in the line connecting buses 3-4 reduces by 0.2875 per unit as a consequence of generation shift and tripping of line.

SUMMARY

- Contingency analyses of power system networks are undertaken to ensure security of power supply at economic cost.
- Based on simplifying approximations, simulation of contingency analysis assumes that Z_{bus} matrix of the network and ac load flow of the system are known.
- Thevenin's equivalents, along with the loop impedance matrix Z, are utilized to simulate bus voltage changes due to addition and removal of multiple lines in power networks, including interconnected power systems.
- The network reduction technique simplifies contingency analysis.
- Effect of contingencies like single and multiple outages, sequential line tripping along with the outcome of current shifting in lines during the shifting of power from one generation bus to another is studied by computing pre- and post- contingency current flows in lines.
- Approximate ac power flow method is a handy tool to simulate contingencies.

SIGNIFICANT FORMULAE

- Compensating current vector: $I_C = -A^t \begin{bmatrix} I_a \\ I_b \end{bmatrix}$

- Bus voltage change: $\Delta V = V' - V = -Z_{\text{bus}} A^t \begin{bmatrix} I_a \\ I_b \end{bmatrix}$

- Thevenin equivalent looking into at buses i-j: $Z_{th-ij} = Z_{ii} + Z_{jj} - 2Z_{ij}$

- Loop impedance matrix:

$$Z = AZ_{\text{bus}} A^t \begin{array}{c} a \qquad\qquad\qquad\qquad b \\ \left[\begin{array}{c|c} \{(Z_{ii} - Z_{ij}) - (Z_{ji} - Z_{jj})\} & \{(Z_{ip} - Z_{iq}) - (Z_{jp} - Z_{jq})\} \\ \hline \{(Z_{pi} - Z_{pj}) - (Z_{qi} - Z_{qj})\} & \{(Z_{pp} - Z_{pq}) - (Z_{qp} - Z_{qq})\} \end{array} \right] \end{array}$$

- Tie-line currents: $\begin{bmatrix} I_a \\ I_b \end{bmatrix} = \begin{array}{c} a \\ b \end{array} \begin{bmatrix} Z_{33}^A + Z_{55}^B + Z_a & Z_{34}^A + Z_{56}^B \\ Z_{43}^A + Z_{65}^B & Z_{44}^A + Z_{66}^B + Z_b \end{bmatrix}^{-1} \begin{bmatrix} V_3 - V_5 \\ V_4 - V_6 \end{bmatrix}$

- New system voltages in an interconnected power system:

$$\begin{bmatrix} V_1' \\ .. \\ .. \\ V_4' \\ V_5' \\ .. \\ V_7' \end{bmatrix} = \begin{bmatrix} V_1 \\ .. \\ .. \\ V_4 \\ V_5 \\ .. \\ V_7 \end{bmatrix} - \begin{bmatrix} Z_{\text{bus}}^A & 0 \\ \hline 0 & Z_{\text{bus}}^B \end{bmatrix} \times \begin{array}{c} a \\ b \end{array} \begin{bmatrix} 0 & 0 & 1 & 0 & -1 & 0 & 0 \\ 0 & 0 & 0 & 1 & 0 & -1 & 0 \end{bmatrix}^t \begin{bmatrix} I_a \\ I_b \end{bmatrix}$$

- Current shift distribution factor: $K_{ij-k} = \dfrac{\Delta I_{ij}}{\Delta I_k} = \dfrac{Z_{ik} - Z_{jk}}{Z_l}$

- Change in line current: $\Delta I_{ij} = \Delta I_{ij-p} + \Delta I_{ij-q} = K_{ij-p} \Delta I_p + K_{ij-q} \Delta I_q$

- Single-line outage distribution factor:

$$\frac{\Delta I_{ij}}{I_{pq}} = L_{ij-pq} = -\frac{Z_a}{Z_l} \left\{ \frac{(Z_{ip} - Z_{iq}) - (Z_{jp} - Z_{jq})}{Z_{th-pq} - Z_a} \right\}$$

- Single outage new line currents: $I_{ij}' = I_{ij} + L_{ij-pq} I_{pq}$;

$$I_{rs}' = I_{rs} + \Delta I_{rs} = I_{rs} + L_{rs-pq} I_{pq}$$

- Compound outages distribution factors:

$$L_{rs-ij} = -\frac{Z_a}{Z_{l-rs}}\left\{\frac{(Z_{ri}-Z_{rj})-(Z_{si}-Z_{sj})}{Z_{th-ij}-Z_a}\right\}$$

$$L_{rs-pq} = -\frac{Z_b}{Z_{l-rs}}\left\{\frac{(Z_{rp}-Z_{rq})-(Z_{sp}-Z_{sq})}{Z_{th-pq}-Z_b}\right\}$$

- Changes in line currents due to compound outages:

$$\Delta I_{rs} = \frac{1}{1-(L_{ij-pq})(L_{pq-ij})}\begin{bmatrix}L'_{rs-ij} & L'_{rs-pq}\end{bmatrix}\begin{bmatrix}I_{ij}\\I_{pq}\end{bmatrix}$$

- Modified generation shift factors: $K'_{rs-p} = K_{rs-p} + L_{rs-ij}K_{ij-p}$;

$$K'_{rs-q} = K_{rs-q} + L_{rs-ij}K_{ij-q}$$

- Amount of possible release for possible overload: $\Delta I_{rs}^{sh} = K'_{rs-p}\Delta I_p + K'_{rs-q}\Delta I_q$

- Power distribution factor: $\dfrac{\Delta P_{ij}}{P_{pq}} = L_{ij-pq} = -\dfrac{X_a}{X_l}\left\{\dfrac{(X_{ip}-X_{iq})-(X_{jp}-X_{jq})}{X_{th-pq}-X_a}\right\}$

EXERCISES

Review Questions

13.1 Write an essay on the importance of maintaining security in a power system.

13.2 (a) Explain why it is not necessary to make accurate computations for contingency analyses.
(b) Explain the approximations made in simulating contingencies in power systems.

13.3 (a) Two lines of impedances Z_a and Z_b are added to buses *i-j* and *p-q* respectively, and one line of impedance Z_c is removed from buses *k-l* from a power system network of *n* buses. Derive, from first principle, the expression for computing **Z**.
(b) Explain how the loop impedance matrix **Z** can be used to study the effect of addition or removal of lines.

13.4 Derive an expression for the compensating currents I_a, I_b, and I_c, in terms of the pre-outage bus voltages, when three simultaneous line outages take place in a power system network.

Numerical Problems

13.1 For the network shown in Fig. 13.16, determine the equivalent injected currents and the new bus voltages when (i) the line between buses 1-3 is removed and a line of reactance *j*0.4 per unit is added between buses 1-4 and (ii) lines of reactance *j*0.5, *j*1.25, and *j*0.5 per unit are simultaneously added to the original network between buses 2-3, 3-4, and 4-5 respectively. All line reactances and bus voltages, in per unit, are indicated in the figure.

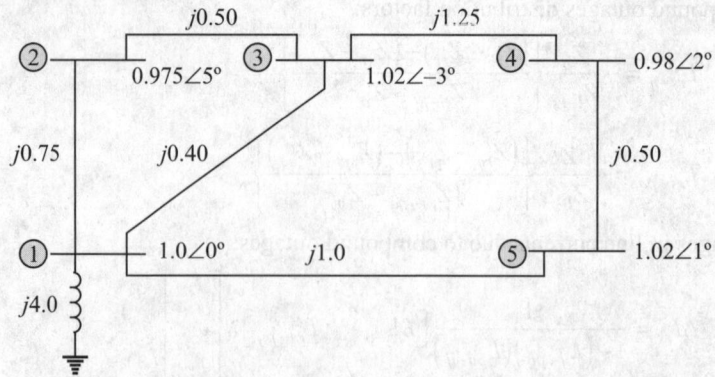

Fig. 13.16 Power system network for Problem 13.2

13.2 For the system of Example 13.6, the line between buses 1-4 is tripped. Compute (i) the Thevenin's equivalent circuit looking into the buses 1-4, (ii) the distribution factors, (iii) the new values of the current flows in the remaining lines, and (iii) bus voltages post outage of the line.

13.3 In the power system network of Example 13.6, the line connected between buses 2-4 is a double-circuit line whose combined reactance is $j0.2$ per unit. One circuit of the double-circuit line is tripped. Calculate the new values of the current flows and bus voltages in the lines after tripping of the circuit.

13.4 Designate the power system in Problem 13.1 as system A and the power system in Example 13.6 as system B. Next assume that the buses 4 and 5 in system A are interconnected via two tie-lines with the buses 3 and 4 in power system B. If the reactance of the tie-line between bus 4 of A and bus 3 of B is $j0.2$ per unit and that of tie-line between bus 5 of A and bus 4 of B is $j0.4$ per unit, determine the Thevenin's equivalent impedance at the boundary buses of (i) system A and (ii) system B. Draw the Thevenin's equivalent circuit for the interconnected network.

13.5 In Problem 13.4, compute the impedance matrices at the boundary buses of power system (i) A and (ii) B. Calculate the loop impedance matrix Z of the interconnected network. From the Z matrix show that its diagonal elements are the sum of the impedances in the loop traced by the tie-line currents starting from one reference bus to the other reference bus and its off-diagonal elements are represented by those impedances through which tie-line currents are flowing and are the cause of voltage drops at the buses.

13.6 Use the piece-wise method to determine the bus voltages after system A is connected to system B via the two tie-lines in Problem 13.5.

13.7 Determine the Ward equivalents, at the boundary buses for system (i) A and (ii) B in Problem 13.6. Draw the Ward equivalent circuits.

13.8 Prove the following identity:

$$\left[\left(L_{rs-ij} + L_{rs-pq}L_{pq-ij}\right)\left(L_{rs-pq} + L_{rs-ij}L_{ij-pq}\right)\right] =$$

$$\left[\frac{Z_{th-ij} - Z_a}{Z_a}L_{rs-ij} \quad \frac{Z_{th-pq} - Z_b}{Z_b}L_{rs-pq}\right] \times \begin{bmatrix} \dfrac{Z_a}{Z_{th-ij} - Z_a} & \dfrac{Z_b^2}{Z_a}\dfrac{L_{ij-pq}}{Z_{th-pq} - Z_b} \\ \dfrac{Z_a^2}{Z_b}\dfrac{L_{pq-ij}}{Z_{th-ij} - Z_a} & \dfrac{Z_b}{Z_{th-pq} - Z_b} \end{bmatrix}$$

13.9 Write a MATLAB script file to compute all possible line outage factors. Test the program by inputting the line data of Problem 13.2.

13.10 The network shown in Fig. 13.17 represents a five buses and six lines power system. Determine (i) the line outage distribution factors for all possible line trippings and (ii) generation shift distribution factors for calculating the changes in current flows in the lines between buses 5-3 and 5-4 when real power is shifted from bus 5 to bus 1. All data in per unit is shown in Fig. 13.17.

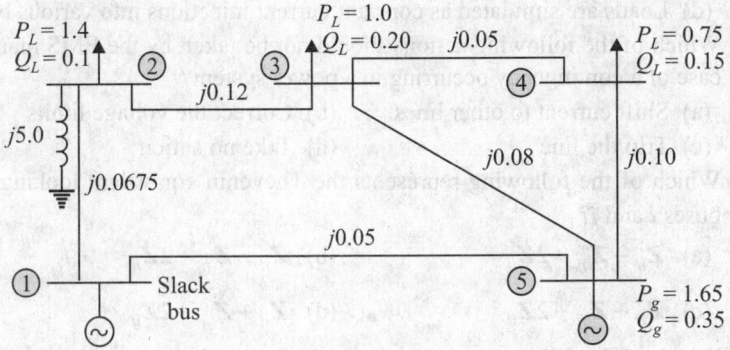

Fig. 13.17 Network for Problem 13.10

13.11 In the power system of Problem 13.10 it is proposed to shift 40% real power from bus 5 to bus 1. Determine the new current in the line connected between buses 5-3. Verify by performing an ac power flow of the system.

13.12 Based on the approximate decoupled power flow model, determine the change in power flow, in Problem 13.11, in the line connected between buses 5-1 when a proposed shift of 40% real power from bus 5 to bus 1 leads to tripping of the line between buses 5-4.

Multiple Choice Objective Questions

13.1 Which of the following is the full form of EMS?
(a) Electricity Management System
(b) Energy Management System
(c) Energy Marketing System
(d) Efficient Management System

13.2 Which of the following are constituents of EMS?
(a) System monitoring   (b) Contingency analyses
(c) Remedial action   (d) All of these

13.3 Which of the following is transmitted as additional information when assessing the state of a power system?
(a) Bus voltage magnitudes   (b) Voltage phase angles
(c) Frequency   (d) None of these

13.4 Which of the following level describes the emergency state in a power system?
(a) Level 1   (b) Level 2
(c) Level 3   (d) None of these

13.5 Which of the following is used to simulate a power system for performing contingency analysis?

(a) Digital program   (b) Analogue program

(c) Real system   (d) None of these

13.6   Which of the following is not an approximation in simulating a power system for contingency analysis?

(a) Switching out of a line is represented by its series impedance/admittance.

(b) Linearity is presumed.

(c) Shunt capacitances of transmission lines are neglected.

(d) Loads are simulated as constant current injections into various buses.

13.7   Which of the following actions should not be taken by the EMS manager in case of a contingency occurring in a power system?

(a) Shift current to other lines   (b) Correct the voltage limits

(c) Trip the line   (d) Take no action

13.8   Which of the following represents the Thevenin equivalent looking into at buses i and j?

(a) $Z_{ii} - Z_{jj} - 2Z_{ij}$ (b) $Z_{ii} - Z_{jj} + 2Z_{ij}$

(c) $Z_{ii} + Z_{jj} + 2Z_{ij}$ (d) $Z_{ii} + Z_{jj} - 2Z_{ij}$

13.9   Which of the following represents the loop impedance matrix Z?

(a) $\begin{bmatrix} Z_a & 0 \\ 0 & Z_b \end{bmatrix} + AZ_{bus}Z$

(b) $\begin{bmatrix} Z_a & 0 \\ 0 & Z_b \end{bmatrix} + AZ_{bus}A^t$

(c) $\begin{array}{c} a \\ b \end{array} \begin{bmatrix} \{(Z_{ii} - Z_{ij}) - (Z_{ji} - Z_{jj})\} + Z_a & \{(Z_{ip} - Z_{iq}) - (Z_{jp} - Z_{jq})\} \\ \{(Z_{pi} - Z_{pj}) - (Z_{qi} - Z_{qj})\} & \{(Z_{pp} - Z_{pq}) - (Z_{qp} - Z_{qq})\} + Z_b \end{bmatrix}$

$\qquad\qquad\qquad\qquad a \qquad\qquad\qquad\qquad\qquad\qquad b$

(d) All of these

13.10   Which of the following gives the loop incidence matrix Z of an interconnected network?

(a) $Z^A + Z^B$ (b) $A^A \begin{bmatrix} Z_{33} & Z_{34} \\ Z_{43} & Z_{44} \end{bmatrix} A^B$

(c) $A^B \begin{bmatrix} Z_{33} & Z_{34} \\ Z_{43} & Z_{44} \end{bmatrix} A^A$ (d) None of these

13.11   The network developed by the network reduction technique of an interconnected power system for contingency analysis is called

(a) Ward equivalent   (b) Thevenin equivalent

(c) Norton equivalent   (d) None of these

13.12   The purpose of performing contingency analyses is to identify

(a) transmission lines to be tripped

(b) transformers to be disconnected

(c) severe conditions

(d) all of these

13.13 Which of the following happens when an extra current is injected into any one bus of a power system?
   (a) Both current and voltage in all the buses are changed.
   (b) Both current and voltage change at the bus into which current is injected.
   (c) Only voltage changes at the bus into which current is injected.
   (d) Both current and voltage change at all the buses except the bus into which current is injected.

13.14 Which of the following power flow methods is appropriate for contingency analyses?
   (a) Gauss–Seidal   (b) Newton–Raphson
   (c) Fast decoupled   (d) Triangular factorization

Answers

13.1 (b)	13.2 (d)	13.3 (c)	13.4 (b)	13.5 (a)	13.6 (a)
13.7 (c)	13.8 (d)	13.9 (d)	13.10 (d)	13.11 (a)	13.12 (c)
13.13 (a)	13.14 (c)				

CHAPTER 14

State Estimation Techniques

Learning Outcomes

A focussed study of the chapter will enable the reader to
- Understand the significance of state estimation of an electrical power system and know the quantities required to be measured for its evaluation
- Visualize generally the functioning of a centralized data acquisition system
- Grasp and apply the statistical technique of weighted least square estimation for estimating power system variables
- Define and distinguish between important statistical terms such as state variables, measurement variables, redundancy and redundancy factor, noise in measurements, and weighting factors and variance
- Recognize the limitations of the line power flow estimator and simulate the state estimator model for computing state variables of a power system based on an understanding of the Gaussian probability density function
- Become skilful at applying mathematical tools to identify and treat bad data to arrive at suitable estimates of the power system

OVERVIEW

An estimation of the state of a power system requires the measurement of electrical quantities, such as real and reactive power flows in transmission lines and real and reactive power injections at the buses. State estimation is a data processing scheme to find the best state vectors, using weighted least square method to fit a scatter of data.

Online monitoring of the generation and transmission data had been used for load frequency control and optimal dispatch of power. With the increase in complexity of power systems due to interconnections, security, and reliability of bulk generation of power and its transmission over extra high-voltage lines has become an added facet of power networks. State estimation is of paramount importance in avoiding system failures and regional blackouts and plays an important role in monitoring and control of power systems.

Centralized automatic control of power system dispatch gathers information from the system through remote terminal units (RTUs), which sample network analogue variables and convert the signals to digital form. The signals are periodically interrogated for the latest values and transmitted by telephone and microwave communication link to the control centre. If enough measurements could be obtained continuously, accurately, and reliably, this should provide all the information to the control centre. However, a perfect data acquisition system is often technically and economically not possible. So the control centre must depend on measurements that are incomplete, inaccurate, delayed, and unreliable. Using state estimation, the available information can be processed and the best possible estimate of the true state of the system can be found.

14.1 DATA ACQUISITION

The remote terminal units (RTUs) collect sample system analogue variables, convert the signals to digital format and transmit the same, over telephone and microwave links to a central computer control system for processing and decision-making. Figure 14.1 shows a schematic block diagram of a centralized data acquisition system. The central computer system performs the following two main functions.

1. It processes the real-time raw data into a more useful form. The processing of raw data involves computation and statistical testing of the state parameters, identifying gross errors and automatically filtering out the same.
2. Providing control decisions for computer control of the power system or human intervention.

If the data transmitted by the RTUs was continuous, accurate, and reliable, it would be enough for controlling a power system. However, the data from the RTUs is in the form of samples and there is a time lag between RTUs and between samples from the same RTU. Additionally, there is redundancy and errors in the data collected. The role of a data acquisition system, also called a *state estimator*, is to process the available data and to provide the finest possible estimates of the state of a system.

14.2 ROLE OF A STATE ESTIMATOR

The state estimator is a computer program which processes real-time system data and computes magnitudes of bus voltages and bus voltage phase angles. As already stated there is redundancy in the data collected. For example, transmission line voltage is measured by using step-down transformers on each phase R, Y, and B. But if the system is balanced it would suffice to measure voltage on any one phase. Further, in a substation, the transmission line voltage may be monitored on the line side of a circuit breaker also, which would bring in more redundancy when all the lines are in service. Similarly, single-phase watt and var meters, along with current transformers in all the phases introduce more redundancies. From the perspective of accuracy, a state estimator is required to process all measurements.

Since the state estimator is required to process all data which is not only redundant but is also huge, it is important that the state estimator detects bad or inaccurate data. This is achieved by employing statistical techniques and designing state estimators which have well-defined error limits and are based on the number, types, and accuracy of measurements.

In the steady-state operation of a system, the sudden opening of one of the phases of a transmission line is reflected in the power flow in the two healthy phases much less than the average power flow indicated by the last state estimate.

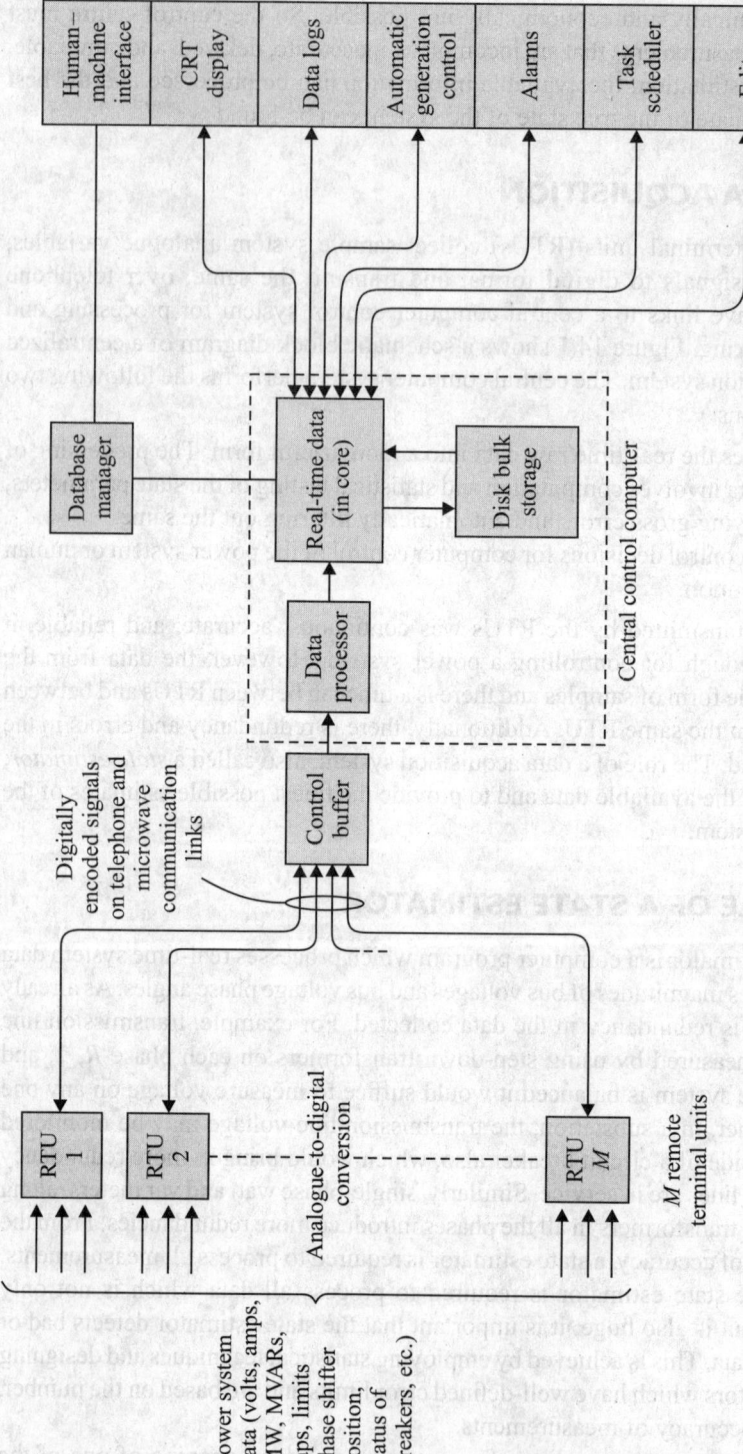

Fig. 14.1 Schematic block diagram of a centralized data acquisition system

The state estimator is required to detect such a change in network configuration and convey a signal indicating the change in circuit configuration and to prepare the operator for corrective action on the first data scan.

The third and final task of the state estimator is to approximate power flows and voltages at a bus whose measurements are not available because of RTU failure or breakdown of a telephone or a communication link. Under such a condition, the state estimator is called upon to make available a set of measurements to replace missing or defective data. In such a case observability, that is, the minimum measurements required for calculating the state and how to improve state estimates by additional estimates is the critical challenge.

14.3 RATIONALE OF STATE ESTIMATION

The objectives and probable applications of state estimation maybe stated as follows:

Online applications

(i) Provides a large real-time database for online security analysis, inconsistency detection, fault diagnosis, and related display functions.

(ii) Facilitates real-time power flow optimization, thereby leading to a minimization of supplementary sensing and telemetry hardware, which results in a reduction in capital expenditure.

(iii) Builds in excellence in a number of control, dispatch, and record keeping functions by improving the quality of data.

Offline applications

(i) Planning and design of the state estimation database to acquire information for central control and dispatch in the most cost-effective manner.

(ii) Other probable applications are in the areas of systems operation, information system planning and design.

14.4 METHOD OF LEAST SQUARES FOR STATE ESTIMATION

The state estimation of a power system based on statistical techniques determine a set of true (but unknown) randomly varying vector x from another randomly varying vector y of known and related numerical values. Before proceeding to describe the method of least squares, it would be appropriate to define the related terms.

State variables State variables are a set of non-superfluous variables which completely illustrate the operation (state) of a system. In a power system, the magnitude of voltage and voltage phase angle at a bus are the state variables and are designated by the vector x of dimension n, where n represents the number of state variables.

Measurement variables Measurement variables are a set of physical quantities obtained from instruments readings. In power systems, these physical quantities could be active and reactive power flows at the buses, active and reactive power flows in the lines, and the voltage magnitude and the phase angles at the buses (the state variables themselves). Measurement variables are denoted by the vector z, regardless of the physical quantities being measured. If the number of measurement variables is m, the dimension of the measurement vector z is m.

Redundancy and redundancy factor Normally, the number of measurement variables is greater than the number of state variables. The difference between the number of measurements m and the number of state variables n, that is, $(n - m)$ is called redundancy or degrees of freedom. The redundancy factor is defined as the ratio of the number of measurement variables to the number of state variables, that is, m/n. In order to achieve the desired accuracy in the estimation of state variables, the magnitude of the redundancy factor should have a range of 1.5 to 2.8.

Noise Since the vector of measured variables is constituted of physical quantities, it is never free of unavoidable inaccuracies. Therefore, physical quantities contain errors which are designated as noise and are represented as components of the vector ζ. Statistically, noise is assumed to have a zero mean over a large number of measurements.

From the foregoing, the relation between the measured variables and state variables can be written as

$$z = Hx + \zeta$$

or

$$
\begin{bmatrix} z_1 \\ z_2 \\ z_3 \\ \vdots \\ \vdots \\ z_m \end{bmatrix}
=
\begin{bmatrix}
h_{11} & h_{12} & h_{13} & \cdots & h_{1n} \\
h_{21} & h_{22} & h_{23} & \cdots & h_{2n} \\
h_{31} & h_{32} & h_{33} & \cdots & h_{3n} \\
\vdots & \vdots & \vdots & & \vdots \\
\vdots & \vdots & \vdots & & \vdots \\
h_{m1} & h_{m2} & h_{m3} & \cdots & h_{mn}
\end{bmatrix}
\begin{bmatrix} x_1 \\ x_2 \\ x_3 \\ \vdots \\ \vdots \\ x_n \end{bmatrix}
+
\begin{bmatrix} \zeta_1 \\ \zeta_2 \\ \zeta_3 \\ \vdots \\ \vdots \\ \zeta_m \end{bmatrix}
\tag{14.1}
$$

In Eq. (14.1), H is a known coefficient matrix of dimension $m \times n$ and ζ represents a noise (random error) vector of dimension m, having a zero mean. In linear measurements, the elements of the H matrix are constant and can be computed from the circuit resistances. If z_{true} represents the true values measurement vector, Eq. (14.1) can be written as

$$\zeta = z - z_{\text{true}} = z - Hx \tag{14.2}$$

Equation (14.2) represents the relation between noise and actual measurements z and true values $z_{\text{true}} \cong Hx$ of the system state variables. True values of the state variables z_{true}, however, are not known. It is not possible to establish the true values of the state and noise vectors but it is possible to compute estimates of the elements of the state vector $\hat{x}$ and noise vector $\hat{\zeta}$ from the approximations of x and ζ respectively. Substituting for these values in Eq. (14.2) yields

$$\hat{\zeta} = z - \hat{z} = z - H\hat{x} = z - H\left(\hat{x} - x\right) \tag{14.3}$$

Before proceeding to define a criterion, it will be useful to note that measurements from instruments are directly related to their accuracy. Therefore, it is prudent to assign greater importance to measurements from instruments which are more accurate. This is easily accomplished by providing higher weighting factors to measurements from instruments of known greater accuracy.

It is now feasible to define a criterion which can be used for computing the approximate values of the elements of the vector $\hat{x}$ from which the elements of the vectors $\hat{\zeta}$ and $\hat{z}$ will be computed. Assume that the chosen criterion is to minimize the algebraic sums of the squares of the errors. Based on the assumed criterion, the performance function f given in the following equation is required to be minimized.

$$f = \sum_{i=1}^{m} w_i \, \zeta_i^2 \tag{14.4}$$

where w_i is the weighting factor assigned to ith reading, depending on the instrument accuracy, and m represents the number of measurements.

The criterion to minimize the algebraic sums of the squares of the errors has been selected, because taking the direct algebraic sums of errors would lead to the cancellation of positive and negative errors, thereby leading to erroneous estimates.

Since instruments which possess greater accuracy will provide more accurate measurements, it is prudent to make certain that such measurements are treated more positively. Therefore, such measurements carry a higher weighting factor. An empirical formula which takes care of the instrument's accuracy makes it feasible to compute the weighting factor as follows:

$$w_i = \frac{50 \times 10^{-6}}{\left[c_1 |S_i| + c_2 \, (\text{FSD}) \right]^2} \tag{14.5}$$

where w_i is the weighting factor for ith measurement, $|S_i|$ the absolute value of the ith measurement, say real and reactive power flow, FSD the full-scale deflection of the instruments, such as watt and var meters, c_1 the accuracy constant, in decimal, of the measurement (for real and reactive power flow measurements typical values of c_1 are 0.01 or 0.02), and c_2 a constant to account for the transducer and analogue-to-digital converter accuracy. Typical value of c_2 is 0.0025 or 0.005.

The numerator in Eq. (14.5) is the normalization constant for succeeding measurements. The weighting factor w_i is computed every time a new measurement is recorded.

For f to be minimum, $(\partial f/\partial x)$ should be zero. Thus,

$$\left.\frac{\partial f}{\partial x_1}\right|_{\hat{x}} = 2\left[w_1\zeta_1 \frac{\partial \zeta_1}{\partial x_1} + w_2\zeta_2 \frac{\partial \zeta_2}{\partial x_1} + \cdots + w_m\zeta_m \frac{\partial \zeta_m}{\partial x_1} \right]_{\hat{x}} = 0$$

$$\left.\frac{\partial f}{\partial x_2}\right|_{\hat{x}} = 2\left[w_1\zeta_1 \frac{\partial \zeta_1}{\partial x_2} + w_2\zeta_2 \frac{\partial \zeta_2}{\partial x_2} + \cdots + w_m\zeta_m \frac{\partial \zeta_m}{\partial x_2} \right]_{\hat{x}} = 0 \tag{14.6}$$

$$\vdots$$

$$\left.\frac{\partial f}{\partial x_m}\right|_{\hat{x}} = 2\left[w_1\zeta_1 \frac{\partial \zeta_1}{\partial x_m} + w_2\zeta_2 \frac{\partial \zeta_2}{\partial x_m} + \cdots + w_m\zeta_m \frac{\partial \zeta_m}{\partial x_m} \right]_{\hat{x}} = 0$$

In matrix form Eq. (14.6) is expressed as

$$\begin{bmatrix} \dfrac{\partial \zeta_1}{\partial x_1} & \dfrac{\partial \zeta_2}{\partial x_1} & \dfrac{\partial \zeta_3}{\partial x_1} & \cdots & \dfrac{\partial \zeta_m}{\partial x_1} \\[2mm] \dfrac{\partial \zeta_1}{\partial x_2} & \dfrac{\partial \zeta_2}{\partial x_2} & \dfrac{\partial \zeta_3}{\partial x_2} & \cdots & \dfrac{\partial \zeta_m}{\partial x_2} \\[2mm] \vdots & \vdots & \vdots & & \vdots \\[2mm] \dfrac{\partial \zeta_1}{\partial x_n} & \dfrac{\partial \zeta_2}{\partial x_n} & \dfrac{\partial \zeta_3}{\partial x_n} & \cdots & \dfrac{\partial \zeta_m}{\partial x_n} \end{bmatrix}_{\hat{x}} \begin{bmatrix} w_1 & & & \\ & w_2 & & \\ & & \ddots & \\ & & & w_m \end{bmatrix} \begin{bmatrix} \zeta_1 \\ \zeta_2 \\ \vdots \\ \zeta_m \end{bmatrix} = \begin{bmatrix} 0 \\ 0 \\ \vdots \\ 0 \end{bmatrix} \tag{14.7}$$

The notation $|_{\hat{x}}$ is used to indicate that the equations are evaluated from the elements of state estimate vector $\hat{x}$ since the real values of the state variables are not known. The values of the estimated errors $\hat{\zeta}_i$ also replace the actual errors ζ_i since the latter are not known. The former are computed from the values of the elements of the state estimate vector $\hat{x}$, which is known. In linear measurements, the elements of the partial derivatives matrix are constants and are given by the elements of the H matrix of Eq. (14.1). Thus, Eq. (14.7) can be written in the following form:

$$H'W\,\hat{\zeta} = 0 \tag{14.8}$$

where $H' = \begin{bmatrix} h_{11} & h_{21} & h_{31} & \cdots & h_{m1} \\ h_{12} & h_{22} & h_{32} & \cdots & h_{m2} \\ h_{13} & h_{23} & h_{33} & \cdots & h_{m3} \\ \vdots & \vdots & \vdots & & \vdots \\ h_{1n} & h_{2n} & h_{3n} & \cdots & h_{mn} \end{bmatrix}$

and W is a diagonal weighting matrix of order m and has the form given in Eq. (14.7). Substituting from Eq. (14.3) in Eq. (14.8) leads to

$$H'W\hat{\zeta} = H'W(z - H\hat{x}) = 0$$

or $\quad H'W(H\hat{x}) = H'Wz \tag{14.9}$

Further simplification of Eq. (14.9) by substituting $G = H'WH$ and solving for $\hat{x}$ leads to the following expression

$$\hat{x} = G^{-1}H'Wz$$

or $\quad \begin{bmatrix} \hat{x}_1 \\ \hat{x}_2 \\ \vdots \\ \hat{x}_n \end{bmatrix} = G^{-1}H'Wz \tag{14.10}$

It may be noted from Eq. (14.10) that the elements of vector $\hat{x}$ are the weighted least squares of the state variables. G is symmetric and is frequently referred to as the *gain matrix*. Further, since $G = H'WH$, its inverse is a composite inverse of the product of the matrices H', W, and H, that is, $G^{-1} = (H'WH)^{-1}$.

Equation (14.10) provides the basis for computing the estimated values of variables $\hat{x}_n$ by applying the weighted least squares technique. It is expected that the values of the variables so obtained will closely approximate the true values x_i of the state variables.

Example 14.1   The measurements recorded on the five meters, in the dc circuit shown in Fig. 14.2 are: $z_1 = 2.00$ A, $z_2 = -0.250$ A, $z_3 = 1.200$ A, $z_4 = 11.00$ V, and $z_5 = -1.00$ V. Compute (i) the coefficient matrix H, (ii) the gain matrix G, and (iii) the estimated elements V_1 and V_2 of the state vector by

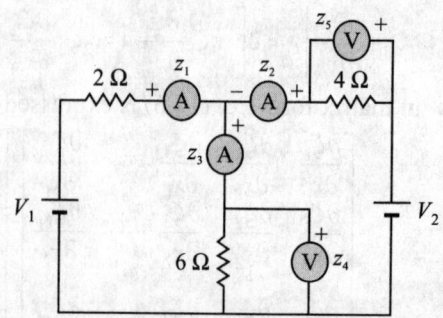

Fig. 14.2   Circuit diagram for Example 14.1

the method of weighted least squares. Assume that ammeters 1 and 2 are 100% accurate, while ammeter 3 and the two voltmeters are 75% accurate.

Solution   (i) The coefficient matrix H relating the measured quantities with the voltage sources is derived by employing the principle of superposition and is as follows:

With the voltage source V_2 short-circuited, it is easily seen that the currents indicated by the ammeters due to voltage source V_1, will depend on the inverse of the circuit resistances. Thus,

$$z_1 = 1/(2 + 6 \times 4/(6+4)) = 0.2273, \ z_2 = \left(\frac{6}{10}\right) \times 0.2273 = 0.1364, \text{ and } z_3 = \left(\frac{4}{10}\right) \times 0.2273$$

$= 0.0909$. Consequently, the voltages indicated by the two voltmeters will be $z_4 = 0.0909 \times 6 = 0.5455$, and $z_5 = 0.1364 \times 4 = 0.5455$. Similarly, when the voltage source V_1 is short-circuited z_1, z_2, z_3, z_4, and z_5 will have the following values: $z_2 = 1/(4 + 6 \times 2/8) = 0.1818$,

$$z_1 = \left(\frac{6}{8}\right) \times 0.1818 = 0.1364, \ z_3 = \left(\frac{2}{8}\right) \times 0.1818 = 0.0455, \ z_4 = 0.0455 \times 6 = 0.2727, \text{ and}$$

$z_5 = 0.1818 \times 4 = 0.7273$.

From the foregoing, the elements of the (5×2) H matrix are written by keeping in mind how the instruments are connected. Hence, $h_{11} = 0.2273$, $h_{21} = -0.1364$, $h_{31} = 0.0909$, $h_{41} = 0.5455$, $h_{51} = -0.5455$, $h_{12} = -0.1364$, $h_{22} = 0.1818$, $h_{32} = 0.0455$, $h_{42} = 0.2727$, and $h_{52} = 0.7273$. In matrix form, H is written as

$$H = \begin{bmatrix} 0.2273 & -0.1364 \\ -0.1364 & 0.1818 \\ 0.0909 & 0.0455 \\ 0.5455 & 0.2727 \\ -0.5455 & 0.7273 \end{bmatrix}$$

(ii) Using the relation $G = H^t W H$ yields the symmetrical gain matrix

$$G = \begin{bmatrix} 0.2273 & -0.1364 & 0.0909 & 0.5455 & -0.5455 \\ -0.1364 & 0.1818 & 0.0455 & 0.2727 & 0.7273 \end{bmatrix} \begin{bmatrix} 100 & & & & \\ & 100 & & & \\ & & 75 & & \\ & & & 75 & \\ & & & & 75 \end{bmatrix} \begin{bmatrix} 0.2273 & -0.1364 \\ -0.1364 & 0.1818 \\ 0.0909 & 0.0455 \\ 0.5455 & 0.2727 \\ -0.5455 & 0.7273 \end{bmatrix}$$

$$G = \begin{bmatrix} 52.2727 & -23.8636 \\ -23.8636 & 50.5682 \end{bmatrix}$$

(iii) Elements of the state vector are obtained by using Eq. (14.10). From the provided data,

$$z = \begin{bmatrix} 2.00A \\ -0.25A \\ 1.20A \\ 11.00V \\ -1.00V \end{bmatrix}$$

$$\hat{x} = G^{-1} H^t W z$$

$$= \begin{bmatrix} 52.2727 & -23.8636 \\ -23.8636 & 50.5682 \end{bmatrix}^{-1}$$

$$\begin{bmatrix} 0.2273 & -0.1364 & 0.0909 & 0.5455 & -0.5455 \\ -0.1364 & 0.1818 & 0.0455 & 0.2727 & 0.7273 \end{bmatrix} \times$$

$$\begin{bmatrix} 100 & & \cdots & & \\ & 100 & . & . & \\ \vdots & . & 75 & . & \vdots \\ & . & . & 75 & \\ & & \cdots & & 75 \end{bmatrix} \times \begin{bmatrix} 2.00A \\ -0.25A \\ 1.20A \\ 11.00V \\ -1.00V \end{bmatrix}$$

$$= \begin{bmatrix} 15.0034V \\ 9.9027V \end{bmatrix} = \begin{bmatrix} V_1 \\ V_2 \end{bmatrix}$$

14.5 ESTIMATION OF POWER SYSTEM STATE VARIABLES BY THE WEIGHTED LEAST SQUARE ESTIMATION (WLSE) TECHNIQUE

In the previous section, application of WLSE technique was discussed to obtain estimates of currents and voltages. The WLSE technique, except for demonstrating its formulation is of little consequence, in estimating the state of an ac power system. In the latter type of network, the measurement equations are non-linear and their solutions require iterative methods. This section explains the modifications in the WLSE technique and its application in estimating the state variables of an electrical power system.

In an ac power system, the RTUs examine (i) magnitudes of bus voltages, (ii) complex power, that is, MW and MVAR flows in transmission lines, and (iii) real and reactive power loadings of generators and transformers. Additionally, the RTUs also monitor the status of the network configuration which changes whenever a switching operation occurs in the system devices, such as circuit breakers and transformer taps. This real-time data and network status report is periodically scrutinized by the RTUs and fed over telephonic and microwave links to the central control computer for processing by the state estimator. It is easy to perceive that the output data from a conventional load flow program and a state estimator are similar. However, the latter differs from the former on the following counts, namely, (a) the input data to the state estimator is (i) scanned for bad data which is filtered out when detected and (ii) drawn from a large data and (b) the computation technique is different.

14.5.1 Simulation of the State Estimator for Power Flow in Transmission Line

The assumptions made in developing the state estimator model for power flow in a transmission line are as follows:

(i) A balanced three-phase steady-state system operation
(ii) The measuring and metering errors are known and defined, that is, accuracy of measuring instruments as a percentage of the true physical values is known, and full scale range of instruments is known
(iii) Discretisation errors due to conversion of analogue quantities to digital signals for telemetering to the central computer centre are known.

Figure 14.3 shows a partial circuit of a transmission line network in a power system. In the diagram, a generator is feeding a bus p, from which a number of transmission lines emanate. Measuring equipment at points 1, 2, and 3 monitor instantaneous currents and voltages, and at points 4, 5, 6, and 7 the real and reactive power flows in the lines are scrutinized. After checking for three-phase balance, the products of instantaneous values of currents and voltages are transformed into serial digital data by the RTU for onward transmission to the central control computer.

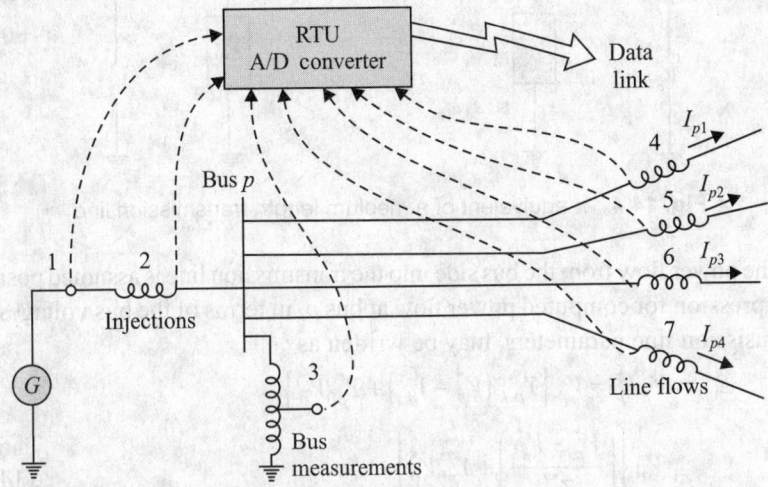

Fig. 14.3 Partial circuit diagram of a transmission line in a power system

Assume that S_{li} represents the complex power (MW + jMVAR) measured on a line designated by l; i in the suffix S_{li} indicates the measurement count. The weighting factor in Eq. (14.5) is modified and is written as follows:

$$w_i = \frac{50 \times 10^{-6}}{\left[c_1 |S_{li}| + c_2 (\text{FSD}) \right]^2}$$

If the number of measurements is m, the performance function f correlating the weighted least squares state estimator is written as

$$f = \sum_{i=1}^{m} w_i |S_{li} - S_{ci}|^2 \tag{14.11}$$

where S_{ci} represents the power flow computed from the estimated values of the state vector $\hat{x}$.

Minimization of the performance function f in Eq. (14.11) will provide the best state vector $\hat{x}$. Note that the state variables in ac power systems are complex quantities. Therefore, the state vector is complex and has the dimensions of $(N-1)$ or, $2(N-1)$ real vectors for an N bus system. Similar to power flow computations, one bus in the system is required to be specified as a slack bus in order to meet the requirement of the imposed power constraints in the system.

In order to compute the power flow S_{ci}, assume that the transmission line is categorized as a medium length line (< 160 km) and therefore can be represented by an equivalent π-section. Figure 14.4 shows the schematic representation of a medium length transmission line connecting the two buses p and k. The power flow measurements S_{li} close to the buses include the power flowing to the shunt elements of the line.

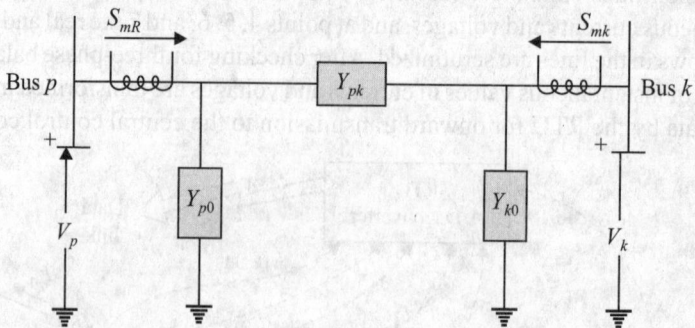

Fig. 14.4 π-equivalent of a medium length transmission line

If the power flow from the bus side into the transmission line is assumed positive, the expression for computed power flow at bus p, in terms of the bus voltages and transmsission line parameters, may be written as

$$S_{cp} = V_p I_p^* = V_p \left\{ Y_{pk}^* \left(V_p^* - V_k^* \right) + Y_{p0}^* V_p^* \right\}$$

$$= V_p \left\{ \left(\frac{V_p^* - V_k^*}{Z_{pk}^*} \right) + Y_{p0}^* V_p^* \right\} \tag{14.12}$$

The power flow from bus p into the transmission line cannot be computed by using Eq. (14.12) since the state vector $\hat{x}$, whose elements are the complex bus voltages, is not known.

If the voltages at buses p and k are the ith measurements, the right hand side of Eq. (14.12) can be equated to the ith measurement of power line flow, that is, S_{li}. Therefore,

$$S_{li} = V_p \left\{ \left(\frac{V_p^* - V_k^*}{Z_{pk}^*} \right) + Y_{p0}^* V_p^* \right\}$$

or $\qquad V_{pl} - V_{kl} = V_{pkl} = \dfrac{Z_{pk}}{E_p^*} S_{li}^* - Z_{pk} Y_{p0} V_p \tag{14.13}$

The power flow in a line is largely governed by the difference in voltages between the buses. If vector V_m is assumed to represent measured voltage differences linked with each power flow measurement, Eq. (14.13) in matrix form is expressed as

$$V_m = \begin{bmatrix} V_{1pl} - V_{1kl} \\ V_{2pl} - V_{2kl} \\ \vdots \\ V_{mpl} - V_{mkl} \end{bmatrix} = H^{-1} S_m^* - K \tag{14.14}$$

where H is a diagonal matrix of order $m \times m$ and is equal to (V_p^* / Z_{pk}) and K is a vector of dimension $(N - 1)$ complex state variables or $2(N - 1)$ real state variables. K is represented by $Z_{pk} Y_{p0} V_p$.

The performance index *f* represented by Eq. (14.11) is the state estimator and may be put together in matrix form as follows

$$f = \sum_{i=1}^{m} w_i |S_{li} - S_{ci}|^2 = [S_m - S_c]^{*t} W [S_m - S_c] \tag{14.15}$$

where S_m is the measurement vector of order *m*, *W* is the diagonal matrix of weighting factors of order $m \times m$, and S_c is a vector of calculated power flows.

From Eq. (14.14), the measurements vector is expressed as $S_m = (HV_m + HK)^*$. In Eq. (14.14), if the measurements voltages difference vector V_m is replaced by the calculated voltages difference vector V_c, the vector for calculated power flows S_c is obtained as

$$S_c = (HV_c + HK)^*$$

Substituting for S_m and S_c in Eq. (14.15) leads to

$$f = [(HV_m + HK) - (HV_c + HK)]^{*t} W [(HV_m + HK) - (HV_c + HK)]$$

or $\qquad f = [V_m - V_c]^{*t} H^* WH [V_m - V_c] \tag{14.16}$

The calculated voltages difference vector V_c can be expressed in terms of the *n* buses in the system as shown here.

Measurements

$\downarrow \quad \rightarrow$ Buses

$$V_c = \begin{matrix} 1 & 2 & 3 & 4 & & & n \end{matrix}$$

$$V_c = \begin{bmatrix} 1 & -1 & 0 & 0 & \cdot & \cdot & 0 \\ 0 & 0 & 0 & 0 & \cdot & -1 & 0 \\ -1 & 0 & 1 & 0 & \cdot & \cdot & 0 \\ \cdot & \cdot & \cdot & \cdot & \cdot & \cdot & 0 \\ & & & & & & 1 \\ & & & & & & -1 \end{bmatrix} \begin{bmatrix} V_1 \\ V_2 \\ - \\ - \\ V_n \end{bmatrix}$$

$$V_c = \widehat{A}V = BV_b + bV_S \tag{14.17}$$

Matrix $\widehat{A}$ in Eq. (14.17) has the dimensions of $m \times n$ where the rows *m* represents measurements and *n* the number of buses in the system. The +1 and –1 elements in each row in the matrix correspond to the line flows measured at each end of the line. The matrix $\widehat{A}$, therefore, is a double branch to the bus incidence matrix. The matrix $\widehat{A}$ when partitioned, as shown in Eq. (14.17), yields a matrix *B* which corresponds to the $(n - 1)$ system bus voltages V_b and a vector *b* which corresponds to the slack bus voltage V_S.

Substituting for V_c from Eq. (14.17) into Eq. (14.16) yields

$$f = [V_m - BV_b - bV_S]^{*t} H^* WH [V_m - BV_b - bV_S] \tag{14.18}$$

In order to minimize the performance index *f*, assume that the product $H^* WH$ is constant and differentiate this product to the bus voltage vector V_b as follows:

$$\frac{\partial f}{\partial V_b} = -B^t H^* WH [V_m - BV_b - bV_S] - [V_m - BV_b - bV_S]^{*t} H^* WHB = 0$$

(14.19)

Using either conjugate form, an expression correlating the bus voltage vector V_b with the measured voltage vector V_m and the slack bus voltage V_S is written by gathering like terms in Eq. (14.19) and is expressed as

$$[B^t H^* WHB]V_b = B^t H^* WH [V_m - bV_S]$$

or

$$[B^t DB]V_b = B^t D[V_m - bV_S]$$

(14.20)

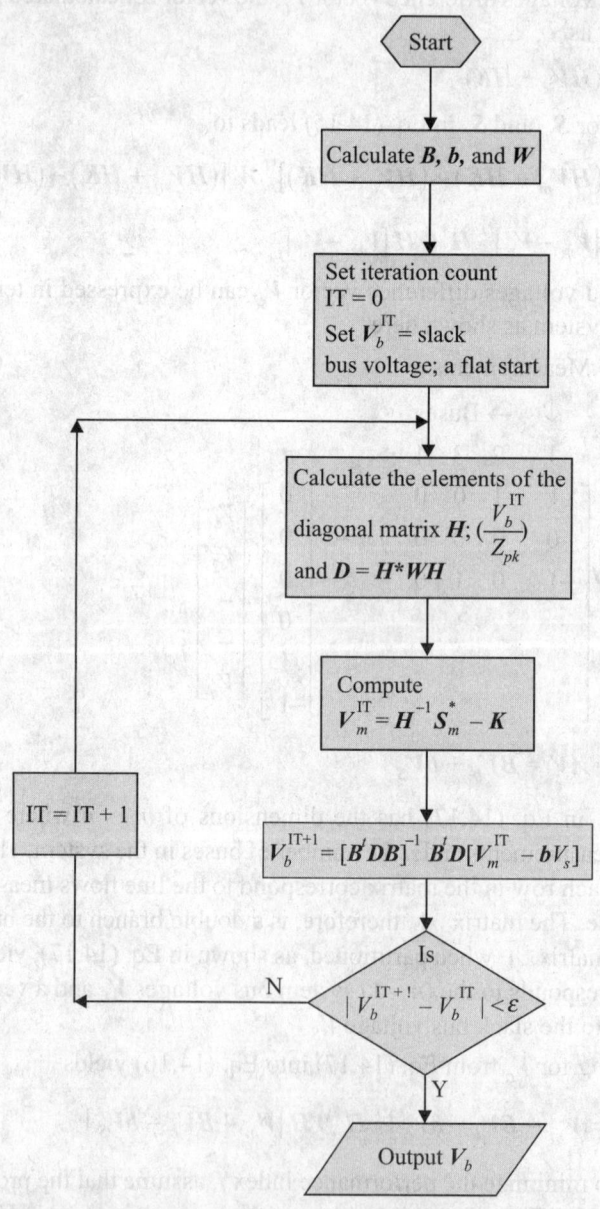

Fig. 14.5 Flow chart for estimating the state variables from the line power flow estimator

where $D = H^*WH$ = diagonal matrix of order m.

Equation (14.20) provides an expression for the power flow line estimator. Since H is a diagonal matrix whose elements (V_p^*/Z_{pk}) are dependent on bus voltage, the solution to Eq. (14.20) can be obtained iteratively. The iterative form of Eq. (14.20) may be expressed as

$$\left[B^t DB \right] V_b^{\text{IT}+1} = B^t D \left[V_b^{\text{IT}} - bV_S \right]$$

or
$$V_b^{\text{IT}+1} = \left[B^t DB \right]^{-1} B^t D \left[V_m^{\text{IT}} - bV_S \right] \qquad (14.21)$$

where IT represents the iteration count.

Observe that in Eq. (14.21), V_m^{IT} is computed from V_b^{IT}. The iterative process is initiated by assuming all bus voltages equal to the slack bus voltage. The flow chart for estimating the state variables (bus voltages) using the line power flow estimator in Eq. (14.21) is shown in Fig. 14.5. From the flow chart in Fig. 14.5, the following can be observed.

(i) At initialization, IT = 0, the bus voltages V_b^0 are set equal to the slack bus voltage.

(ii) The initial $m \times m$ diagonal matrix H is computed from the initial estimates of bus voltages V_b^0. For computing the subsequent diagonal elements, the previous values of the bus voltages V_b^{IT} are used.

(iii) V_m^{IT} measurement data and bus voltages V_b^{IT} are used for computing the voltage difference vector.

(iv) The power flow line estimator is used to obtain new estimates of bus voltages, that is, $V_b^{\text{IT}+1}$ and the iterative process is continued till two consecutive absolute values of the bus voltages are within a prescribed tolerance ε. The usual acceptable tolerance limit is taken as 0.0001 per unit.

Example 14.2 The complex power flows measured in the lines of the power system shown in Fig. 14.6 are given in the accompanying table. The line data, in per unit, is as follows:

Line 1-2: $Z_{12} = 0.05 + j0.20$; Line 2-3: $Z_{23} = 0.075 + j0.25$

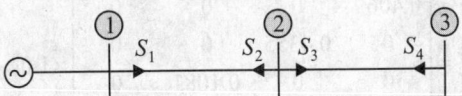

Fig. 14.6 Three-bus power system for Example 14.2

Assume an accuracy of 2.5% for the MW and MVAR meters, each of which has a full-scale deflection of 100. Take the conversion and transducer errors to be 0.5%. Estimate the system variables by using the line power flow estimator represented by Eq. (14.21). Neglect transmission line charging and shunt terms.

Complex power flow measurements

From bus no.	To bus no.	Complex power flow
1	2	$S_1 = 0.50 - j\,0.12$
2	1	$S_2 = -0.48 + j0.10$
2	3	$S_3 = 0.80 - j0.40$
3	2	$S_4 = -0.78 + j0.38$

Solution

First Iteration

Step 1: Computation of diagonal weighting factors matrices W, H, D, B and vector b.

In this case, 2.5 % is the accuracy of the measurements and the conversion and transducer errors are 0.5 %. Hence, $c_1 = 0.025$ and $c_2 = 0.005$. The weighting factors are computed by using Eq. (14.5). Thus, weighting factor for line 1-2 is given by

$$w_1 = \frac{50 \times 10^{-6}}{[0.025 \times 100 \times |0.50 - j0.12| + 0.005 \times 100]^2} = 0.1568 \times 10^{-4}$$

Other weighting factors are similarly determined and are as follows

$$w_2 = 0.1679 \times 10^{-4} \qquad w_3 = 0.0668 \times 10^{-4} \qquad w_4 = 0.0702 \times 10^{-4}$$

Thus, the matrix W for weighting factors is written as

$$W = \begin{bmatrix} 0.1568 & 0 & 0 & 0 \\ 0 & 0.1679 & 0 & 0 \\ 0 & 0 & 0.0668 & 0 \\ 0 & 0 & 0 & 0.0702 \end{bmatrix} \times 1.0e - 0.04$$

$$H = \begin{bmatrix} \dfrac{V_1^*}{Z_{12}} & 0 & 0 & 0 \\ 0 & \dfrac{V_2^*}{Z_{12}} & 0 & 0 \\ 0 & 0 & \dfrac{V_2^*}{Z_{23}} & 0 \\ 0 & 0 & 0 & \dfrac{V_3^*}{Z_{23}} \end{bmatrix}$$

Assume that $V_1 = V_2 = V_3 = 1.05$ per unit. Hence,

$$H = \begin{bmatrix} 1.2353 - j4.9412 & 0 & 0 & 0 \\ 0 & 1.2353 - j4.9412 & 0 & 0 \\ 0 & 0 & 1.1560 - j3.8532 & 0 \\ 0 & 0 & 0 & 1.1560 - j3.8532 \end{bmatrix}$$

$$D = H^*WH = \begin{bmatrix} 0.4069 & 0 & 0 & 0 \\ 0 & 0.4355 & 0 & 0 \\ 0 & 0 & 0.1081 & 0 \\ 0 & 0 & 0 & 0.1136 \end{bmatrix} \times 1.0e - 0.03$$

The double branch to bus incidence matrix $\hat{A}$ can be obtained by an inspection of the topology of the network as follows:

$$\hat{A} = \begin{bmatrix} 1 & -1 & 0 \\ -1 & 1 & 0 \\ 0 & 1 & -1 \\ 0 & -1 & 1 \end{bmatrix}$$

If bus 1 is the slack bus, the bus voltage matrix B and the slack bus matrix b can be obtained by partitioning matrix $\hat{A}$ at column 1 as follows:

$$B = \begin{bmatrix} -1 & 0 \\ 1 & 0 \\ 1 & -1 \\ -1 & 1 \end{bmatrix} \text{ and } b = \begin{bmatrix} 1 \\ -1 \\ 0 \\ 0 \end{bmatrix}$$

Step 2: Computation of the products of matrices $B^t D$ and $B^t DB$

$$B^t D = \begin{bmatrix} -0.4069 & 0.4355 & 0.1081 & -0.1136 \\ 0 & 0 & -0.1081 & 0.1136 \end{bmatrix} \times 1.0e - 0.03$$

$$B^t DB = \begin{bmatrix} 0.0011 & -0.0002 \\ -0.0002 & 0.0002 \end{bmatrix}$$

By comparing the products of $B^t DB$ with $B^t D$, it is observed that the diagonal terms in the $B^t DB$ matrix of order 2×2, are the sums of the elements of the D matrix of the lines terminating on the bus. For example, the first diagonal term of $B^t DB$ matrix, which represents bus 2 at which both the lines terminate is constituted of all the four elements of D, that is, $\{(0.4069 + 0.4355 + 0.1081 + 0.1136) \times 10^{-3} = 0.0011\}$. Similarly, for bus 3 where only one line terminates, the second diagonal term of $B^t DB$ matrix is constituted of the third and fourth diagonal terms of D, that is, $\{(0.1081 + 0.1136) \times 10^{-3} = 0.0002\}$. The off-diagonal terms of $B^t DB$ matrix are made up of the negatives of the diagonal terms on the line connecting buses 2 and 3, that is, $\{-(0.1081 - 0.1136) \times 10^{-3} = -0.0002\}$.

Step 3: Computation of $V_m = H^{-1} S_m^* - K$. It may be noted that K is equal to zero since $Y_{i0} = 0$.

$$V_m = \begin{bmatrix} 1.2353 - j4.9412 & 0 & 0 & 0 \\ 0 & 1.2353 - j4.9412 & 0 & 0 \\ 0 & 0 & 1.1560 - j3.8532 & 0 \\ 0 & 0 & 0 & 1.1560 - j3.8532 \end{bmatrix}^{-1}$$

$$\begin{bmatrix} 0.50 - j0.12 \\ -0.48 + j0.10 \\ 0.80 - j0.40 \\ -0.78 + j0.38 \end{bmatrix}^*$$

$$= \begin{bmatrix} 0.0010 + j0.1010 \\ -0.0038 - j0.0962 \\ -0.0381 + j0.2190 \\ 0.0348 - j0.2129 \end{bmatrix}$$

Step 4: Computation of $V_b = \left[B^t DB \right]^{-1} B^t D [V_m - bV_S]$

$$V_b = \begin{bmatrix} 0.0011 & -0.0002 \\ -0.0002 & 0.0002 \end{bmatrix}^{-1} \begin{bmatrix} -0.4069 & 0.4355 & 0.1081 & -0.1136 \\ 0 & 0 & -0.1081 & 0.1136 \end{bmatrix} \times 1.0e - 003$$

$$\times \left\{ \begin{bmatrix} 0.0010 + j0.1010 \\ -0.0038 - j0.0962 \\ -0.0381 + j0.2190 \\ 0.0348 - j0.2129 \end{bmatrix} - \begin{bmatrix} 1 \\ -1 \\ 0 \\ 0 \end{bmatrix} \times (1.05 + j0.0) \right\}$$

$$V_b = \begin{bmatrix} V_2 \\ V_3 \end{bmatrix} = \begin{bmatrix} 1.0476 - j0.0985 \\ 1.0840 - j0.3144 \end{bmatrix} \text{ per unit}$$

At the end of first iteration, the bus voltages are

$$\begin{bmatrix} V_1 \\ V_2 \\ V_3 \end{bmatrix} = \begin{bmatrix} 1.05 + j0 \\ 1.0476 - j0.0985 \\ 1.0840 - j0.3144 \end{bmatrix} \text{ per unit}$$

Second iteration H is recomputed with the new set of voltages, that is,

$$\begin{bmatrix} V_1 \\ V_2 \\ V_3 \end{bmatrix} = \begin{bmatrix} 1.05 + j0 \\ 1.0476 - j0.0985 \\ 1.0840 - j0.3144 \end{bmatrix} \text{ per unit}$$

With the new H matrix, V_m is calculated as in step 3 of the first iteration. Step 4 is then revisited to obtain the new values of the bus voltages. At the end of second iteration, the values of the bus voltages vector are as follows:

$$\begin{bmatrix} V_1 \\ V_2 \\ V_3 \end{bmatrix} = \begin{bmatrix} 1.05 + j0 \\ 1.0429 - j0.0980 \\ 1.0379 - j0.3073 \end{bmatrix} \text{ per unit}$$

Iterating to convergence The process of iteration is continued till the absolute values of the bus voltages are within the specified tolerance, i.e., ≤ 0.0001. The bus voltages at the end of each iteration are shown in the following table.

Iteration count	Voltage at bus 1	Voltage at bus 2	Voltage at bus 3
0	$1.05 + j0.0$	$1.05 + j0.0$	$1.05 + j0.0$
1	$1.05 + j0.0$	$1.0476 - j0.0985$	$1.0840 - j0.3144$
2	$1.05 + j0.0$	$1.0429 - j0.0980$	$1.0379 - j0.3073$
3	$1.05 + j0.0$	$1.0429 - j0.0982$	$1.0376 - j0.3128$
4	$1.05 + j0.0$	$1.0429 - j0.0982$	$1.0365 - j0.3123$
5	$1.05 + j0.0$	$1.0429 - j0.0982$	$1.0370 - j0.3127$
6	$1.05 + j0.0$	$1.0429 - j0.0982$	$1.0370 - j0.3126$

14.5.2 Limitations of the Line Power Flow Estimator

The line power flow estimator suffers from the following drawbacks.

(i) It is based only on measurements of the power flows in the lines. Therefore, it does not acknowledge other measurements such as real and reactive powers at a bus.

(ii) It does not recognize voltage and current magnitudes measured without phase angles.

(iii) It does not accept measurements of varying accuracies and ranges of real and reactive powers at a point.

(iv) It does not account for measurements corrupted by noise.

Before proceeding to develop a mathematical model of a power system state estimator that takes into account all the earlier mentioned drawbacks, it is prudent to introduce the basic principles associated with statistical measurements and errors.

14.6 STATISTICAL ERRORS AND BAD DATA RECOGNITION

In the online operation of a power system, its true state can only be determined by making repeated measurements of the same quantity over a period of time.

Such repeated measurements under controlled conditions exhibit certain statistical characteristics which help in assessing the true values of the state variables.

14.6.1 Gaussian or Normal Probability Density Function

The continuous curve obtained by plotting a large (theoretically infinite) number of measured values of a function against the relative frequency of their happening is called the *Gaussian* or *normal probability density* function and is mathematically written as

$$p(z) = \frac{1}{\sigma\sqrt{2\pi}}\,\varepsilon^{-\frac{1}{2}\left(\frac{z-\mu}{\sigma}\right)^2} \tag{14.22}$$

where σ is called the standard deviation and μ is called the mean.

Figure 14.7 shows a plot of the function $p(z)$ versus the random variable z along the x-axis. The random variable z represents the values of the measured quantity. As can be seen from the graph, the normal probability function is bell shaped.

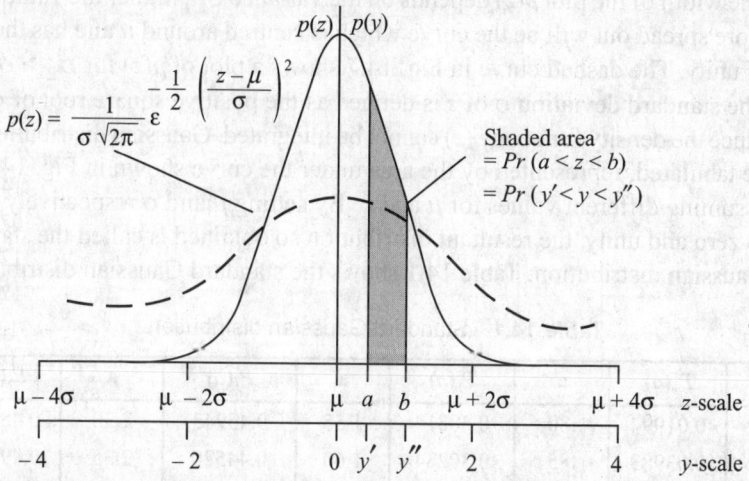

Fig. 14.7 Plot of Gaussian probability density function $p(z)$ versus the random variable z

Properties of the Gaussian probability density function

(i) The plot of the Gaussian probability density function is used to establish the probability of the random variable z acquiring values within a specified range. For example, the probability P_r that the variable z will have values within the range a to b is mathematically written as

$$P_r(a < z < b) = \int_a^b p(z)\,dz = \frac{1}{\sigma\sqrt{2\pi}}\int_a^b \varepsilon^{-\frac{1}{2}\left(\frac{z-\mu}{\sigma}\right)^2}\,dz \tag{14.23}$$

Equation (14.23) is represented by the shaded area in Fig. 14.7 and it provides the probability of z having a value lying between a and b.

(ii) When z varies between $-\infty$ and $+\infty$, the area under the curve is unity, which means that the probability of z lying between randomly large extreme values is 100%.

(iii) Equation (14.23) implies that if the parameters μ and σ are known, the function $p(z)$ is fully determined. Further, it can be observed from the graph in Fig. 14.7 that $p(z)$ has the maximum value when $z = \mu$. Therefore, if $E[z]$ represents the expected value of z, then μ may be written as

$$\mu = E[z] = \int_{-\infty}^{+\infty} zp(z)dz \tag{14.24}$$

It may be observed that z is symmetrically bunched about around the expected value μ and the latter is called the mean value of z.

(iv) Variance of z, designated by σ^2, is defined as the square of the deviations from its mean value. Thus,

$$\sigma^2 = \left[(z-\mu)^2\right] = \int_{-\infty}^{+\infty} (z-\mu)^2 p(z)dz \tag{14.25}$$

The width of the plot $p(z)$ depends on the variance σ^2; higher the value of σ^2 more spread out will be the curve which is centred around μ and has the area of unity. The dashed curve in Fig. 14.7 shows a plot of $p(z)$ for $\sigma_2^2 > \sigma_1^2$.

(v) The standard deviation σ of z is defined as the positive square root of σ_2^2.

(vi) Since the density function $p(z)$ cannot be integrated, Gaussian distribution can be tabulated, represented by the area under the curve shown in Fig. 14.7, by assuming different values for μ and σ. By setting μ and σ respectively equal to zero and unity, the resultant distribution so obtained is called the standard Gaussian distribution. Table 14.1 shows the standard Gaussian distribution.

Table 14.1   Standard Gaussian distribution

a	$P_r(a)$	a	$P_r(a)$	a	$P_r(a)$	a	$P_r(a)$
.05	0.01994	.80	0.28814	1.55	0.43943	2.30	0.48928
.10	0.03983	.85	0.30234	1.60	0.44520	2.35	0.49061
.15	0.05962	.90	0.31594	1.65	0.45053	2.40	0.49180
.20	0.07926	.95	0.32894	1.70	0.45543	2.45	0.49286
.25	0.09871	1.00	0.34134	1.75	0.45994	2.50	0.49379
.30	0.11791	1.05	0.35314	1.80	0.46407	2.55	0.49461
.35	0.13683	1.10	0.36433	1.85	0.46784	2.60	0.49534
.40	0.15542	1.15	0.37493	1.90	0.47128	2.65	0.49597
.45	0.17364	1.20	0.38493	1.95	0.47441	2.70	0.49653
.50	0.19146	1.25	0.39435	2.00	0.47726	2.75	0.49702
.55	0.20884	1.30	0.40320	2.05	0.47982	2.80	0.49744
.60	0.22575	1.35	0.41149	2.10	0.48214	2.85	0.49781
.65	0.24215	1.40	0.41924	2.15	0.48422	2.90	0.49813
.70	0.25804	1.45	0.42647	2.20	0.48610	2.95	0.49841
.75	0.27337	1.50	0.43319	2.25	0.48778	3.00	0.49865

Pr(a) is the value of $\dfrac{1}{\sqrt{2\pi}} \int_{0}^{a} \varepsilon^{-\frac{1}{2}y^2} dy$

By using the transformation expression $y = (z - \mu)/\sigma$, a Gaussian distribution function $p(y)$ is obtained which enables the computation of Gaussian distribution, for other values of μ and σ by simply rescaling the x-axis. The transformation, thus, avoids the need to calculate Gaussian distribution for other different values of μ and σ.

14.6.2 Correlation between Error (Noise) and Measurement

Assume that the errors introduced in two samples of measurement, taken at different times are represented by ζ_i and by ζ_j respectively. Also assume that the errors are independent Gaussian random variables with zero mean (that is, the error in each measurement can acquire a positive or a negative value of the same magnitude) and their variances are σ_i^2 and σ_j^2, respectively. These randomly varying errors are said to be independent when the expectation $E(\zeta_i \zeta_j) = 0$ for $i \neq j$. This implies that noise is not correlated with the measurements. In other words, the noise introduced in one measurement does not compare with the noise introduced in another measurement. This is not entirely true when interpreted from a practical perspective of operation of power systems. Consider the case of an error introduced in voltage measured by a step-down potential transformer. Clearly measurements made for MW and MVAR that use this voltage as one of the inputs for calculation of real and reactive powers will get affected. Therefore, the statement that noise is not correlated with the measurements, though mathematically correct, is not physically valid.

14.6.3 Weighting Factors and Variances

If the number of measurements is represented by m, the noise (or error) vector ζ will be of dimension m. Thus,

$$\zeta \zeta^t = \begin{bmatrix} \zeta_1 \\ \zeta_2 \\ \zeta_3 \\ - \\ - \\ \zeta_m \end{bmatrix} \begin{bmatrix} \zeta_1 & \zeta_2 & \zeta_3 & \cdots & \zeta_m \end{bmatrix} = \begin{bmatrix} \zeta_1^2 & \zeta_1\zeta_2 & \zeta_1\zeta_3 & - & - & \zeta_1\zeta_m \\ \zeta_2\zeta_1 & \zeta_2^2 & \zeta_2\zeta_3 & - & - & \zeta_2\zeta_m \\ \zeta_3\zeta_1 & \zeta_3\zeta_2 & \zeta_3^2 & - & - & \zeta_3\zeta_m \\ - & - & - & - & - & - \\ - & - & - & - & - & - \\ \zeta_m\zeta_1 & \zeta_m\zeta_2 & \zeta_m\zeta_3 & - & - & \zeta_m^2 \end{bmatrix}$$

(14.26)

Since the randomly varying errors are not correlated, that is, $E(\zeta_i \zeta_j) = 0$, for $i \neq j$, the expected value of each off-diagonal term in Eq. (14.26) is zero. The diagonal terms are non-zero. Therefore, the expected value $E(\zeta_i \zeta_j)$ may be written as

$$E\left[\zeta\zeta^t\right] = \begin{bmatrix} E\left[\zeta_1^2\right] & & & & \\ & E\left[\zeta_2^2\right] & & & \\ & & E\left[\zeta_3^2\right] & & \\ & & & \ddots & \\ & & & & E\left[\zeta_m^2\right] \end{bmatrix}$$

(14.27a)

The diagonal elements in the $E\left[\zeta\zeta^t\right]$ matrix match with the variance, that is, $E\left[\zeta^2\right] = \sigma^2$. Hence $E\left[\zeta\zeta^t\right]$ is represented by the symbol R and is written as

$$
R = E\left[\zeta\zeta^t\right] = \begin{bmatrix} \sigma^2 & & & & \\ & \sigma^2 & & & \\ & & \sigma^2 & & \\ & & & \ddots & \\ & & & & \sigma^2 \end{bmatrix} \tag{14.27b}
$$

From Eq. (14.27b) it is seen that R is a square diagonal matrix of order m and is called the *variance matrix*.

Referring back to Eq. (14.1), it can be noted that the measurement vector z is made up of a constant term Hx, which is really the true value designated by z_{true}, and the Gaussian random variable ζ. The effect of adding $z_{\text{true}} = (h_{11}x_1 + h_{12}x_2 + \cdots + h_{1m}x_n)$, for example, to the random variable ζ_1 is to shift the Gaussian random distribution curve to the right (if z_{true} is positive) or to the left (if z_{true} is negative) by an equal amount without distorting the shape of the curve. Therefore, it is fair to state that the measurement z_1 also exhibits a Gaussian probability density function whose mean value μ_1 is equal to z_{true} and whose variance is σ_1^2. Generalizing the above, it can be said that for any measurement i, both the measured value z_1 and the random error (noise) variable σ_i display Gaussian probability density distribution. However, it is very difficult to measure noise statistics and obtain Gaussian probability density distribution.

In writing the performance function f, [Eq. (14.4)], higher weighting factors were assigned to meters of higher accuracy. Based on the same principle, if a weighting factor w_1 is chosen as the reciprocal of the corresponding variance σ_i^2, the $m \times m$ diagonal matrix W of weighting factors is written as

$$
W = \begin{bmatrix} w_1 & & & \\ & w_2 & & \\ & & \ddots & \\ & & & w_m \end{bmatrix}
$$

Replacing the diagonal elements in the W matrix by $w_i = \dfrac{1}{\sigma_i^2}$ yields

$$
W = \begin{bmatrix} \dfrac{1}{\sigma_1^2} & & & \\ & \dfrac{1}{\sigma_2^2} & & \\ & & \ddots & \\ & & & \dfrac{1}{\sigma_m^2} \end{bmatrix} \tag{14.28}
$$

Comparing Eqs (14.28) and (14.27), we get

$$W = R^{-1} = \begin{bmatrix} \dfrac{1}{\sigma_1^2} & & & \\ & \dfrac{1}{\sigma_2^2} & & \\ & & \ddots & \\ & & & \dfrac{1}{\sigma_m^2} \end{bmatrix} \tag{14.29}$$

The gain matrix $G = H^t W H$ may now be represented as follows:

$$G = H^t R^{-1} H \tag{14.30}$$

In conformity with Eq. (14.5), the variance σ_i^2 is correlated with the measurement z_i as follows:

$$\frac{1}{\sigma_i^2} = \frac{3}{[c_1 |Z_i| + c_2 (\text{FSD})]^2} \tag{14.31}$$

Equation (14.31) is formulated in the presence of noise and is similar to Eq. (14.5) except for the numerator.

14.6.4 Attributes of WLSE Technique

In Eq. (14.10) when $z = Hx + \zeta$ is substituted, the following result is obtained

$$\hat{x} = G^{-1} H^t W (Hx + \zeta) = G^{-1} H^t W H x + G^{-1} H^t W \zeta$$

Recalling that $G = H^t W H$, the preceding expression simplifies to

$$\hat{x} = x + G^{-1} H^t W \zeta$$

or

$$\hat{x} - x = \begin{bmatrix} \hat{x}_1 - x_1 \\ \hat{x}_2 - x_2 \\ \hat{x}_3 - x_3 \\ \vdots \\ \hat{x}_n - x_n \end{bmatrix} = G^{-1} H^t W \zeta \tag{14.32}$$

The reader will do well to verify the dimensions of each matrix in Eq. (14.32) and ensure that the dimension of the product matrix is conformable for multiplication. Substituting the value for $(\hat{x} - x)$ from Eq. (14.32) into Eq. (14.3) leads to

$$\hat{z} = z - \hat{z} = z - H\hat{x} = z - HG^{-1} H^t W \zeta$$

$$= \left(I - HG^{-1} H^t W \right) \zeta \tag{14.33}$$

where I is the identity matrix. If it is assumed that a and b are constants, it can be proved with the help of Eq. (14.24) that

$$E[az + b] = aE[z] + b$$

Hence, in terms of the expected values, Eq. (14.33) can be expressed as

$$\begin{bmatrix} E[\hat{x}_1 - x_1] \\ E[\hat{x}_2 - x_2] \\ \vdots \\ E[\hat{x}_n - x_n] \end{bmatrix} = \begin{bmatrix} E[\hat{x}_1] - x_1 \\ E[\hat{x}_2] - x_2 \\ \vdots \\ E[\hat{x}_n] - x_n \end{bmatrix} = G^{-1}H^tR^{-1} \begin{bmatrix} E[\zeta_1] \\ E[\zeta_2] \\ \vdots \\ E[\zeta_m] \end{bmatrix} = \begin{bmatrix} 0 \\ 0 \\ \vdots \\ 0 \end{bmatrix} \qquad (14.34)$$

Equation (14.34) provides a relationship between the Gaussian randomly varying ζ with zero expected values and the unknown but definite true values of the state variables. Hence, from Eq. (14.34), it can be concluded that the technique of weighted least squares estimate determines an expected value of each state variable which is equal to the true value. Mathematically, the last statement is expressed as

$$E[\hat{x}_i] = x_1, \text{ where } i = 1, 2, 3, \ldots, n \qquad (14.35)$$

Equation (14.35) shows that by employing WLSE, it is possible to obtain state variables which represent their true values. Similarly Eq. (14.33) may also be written as

$$E \begin{bmatrix} \hat{\zeta}_1 \\ \hat{\zeta}_2 \\ \vdots \\ \hat{\zeta}_m \end{bmatrix} = E \begin{bmatrix} z_1 - \hat{z}_1 \\ z_2 - \hat{z}_2 \\ \vdots \\ z_m - \hat{z}_m \end{bmatrix} = (I - HG^{-1}H^tW) \begin{bmatrix} \zeta_1 \\ \zeta_2 \\ \vdots \\ \zeta_m \end{bmatrix} = \begin{bmatrix} 0 \\ 0 \\ \vdots \\ 0 \end{bmatrix} \qquad (14.36)$$

Knowing that $z_i = z_{itrue} + \zeta_i$, from Eq. (14.36), it can be concluded that

$$E[\hat{z}_i] = E[z_i] = z_{itrue} \qquad (14.37)$$

Just as in the case of state variables being estimated equal to their true values, by the technique of WLSE, from Eq. (14.37) it can be concluded that WLSE enables the computation of measurements which represent the true measurements.

14.6.5 Identification and Treatment of Bad Data

The attributes of the WLSE technique to be able to approximate state variables and measurements which equal their true values will hold correct, provided the system model and measurements are precise. In other words, there is no bad (or coarsely flawed) data. Hence, the need to detect and treat bad data arises.

In Eq. (14.33), replace W by R^{-1} and post-multiply it by $\hat{\zeta}^t$. The resultant expression is as follows:

$$\hat{\zeta}\hat{\zeta}^t = (z - \hat{x})(z - \hat{x})^t = [I - HG^{-1}H^tR^{-1}]\zeta\zeta^t[I - R^{-1}HG^{-1}H^t] \qquad (14.38)$$

On comparing Eq. (14.38) with the expressions derived earlier, it is observed that the term $\zeta\zeta^t$ on the left side is a square matrix similar to the one in Eq. (14.26), the terms $[I - HG^{-1}H^tR^{-1}]$ and $[I - R^{-1}HG^{-1}H^t]$ on the right hand side are transpose of each other and are made up of constants only, and the term $\zeta\zeta^t$ on the right hand side is again a square matrix similar to the one in Eq. (14.26) and its expected value from Eq. (14.27) is R.

Based on the foregoing, Eq. (14.38) can be written in terms of the expected value as follows:

$$E\left[\widehat{\zeta}\widehat{\zeta}^t\right] = \left[I - HG^{-1}H^tR^{-1}\right]E\left[\widehat{\zeta}\ \widehat{\zeta}^t\right]\left[I - R^{-1}HG^{-1}H^t\right]$$

$$= \left[I - HG^{-1}H^tR^{-1}\right]R\left[I - R^{-1}\ HG^{-1}H^t\right]$$

$$= \left[I - HG^{-1}H^tR^{-1}\right]\left[R - HG^{-1}H^t\right]$$

$$= \left[I - HG^{-1}H^tR^{-1}\right]\left[I - HG^{-1}H^tR^{-1}\right]R \qquad (14.39)$$

Using the property that $\left[I - HG^{-1}H^tR^{-1}\right]\left[I - HG^{-1}H^tR^{-1}\right] = \left[I - HG^{-1}\right.$ $\left. H^tR^{-1}\right]$, Eq. (14.39) is written as

$$E\left[\widehat{\zeta}\widehat{\zeta}^t\right] = \left[I - HG^{-1}H^tR^{-1}\right]R = R - HG^{-1}H^t = R' \qquad (14.40)$$

where $R' = R - HG^{-1} H^t$ is a diagonal square matrix and is called the covariance matrix. It has the same form as Eq. (14.27a). Thus,

$$E\left[\widehat{\zeta}_i^2\right] = R_i' \square = \begin{bmatrix} R_1' & & & & \\ & R_2' & & & \\ & & R_3' & & \\ & & & \ddots & \\ & & & & R_m' \end{bmatrix} \qquad (14.41)$$

If $\widehat{\zeta}_i$ is represented by $(z_i - \widehat{z}_i)$, the diagonal elements of the matrix in Eq. (14.41) are given by

$$R_i' = E\left[\widehat{\zeta}_i^2\right] = E\left[(z_i - \widehat{z}_i)^2\right] \qquad (14.42)$$

Recalling that ζ_i has a zero mean value, Eq. (14.42) provides an expression for the variance of ζ_i. Thus, the standard deviation is given by $\sqrt{R_i'}$. Dividing Eq. (14.42) by R_i' leads to

$$E\left[\left(\frac{\widehat{\zeta}_i}{\sqrt{R_i'}}\right)^2\right] = E\left[\left(\frac{z_i - \widehat{z}_i}{\sqrt{R_i'}}\right)^2\right] = 1 \qquad (14.43)$$

Equation (14.43) expressing the deviation of the estimated error $\widehat{\zeta}$ from its zero mean can be written as

$$\frac{\widehat{\zeta}_i - 0}{\sqrt{R_i'}} = \frac{z_i - \widehat{z}_i}{\sqrt{R_i'}} \qquad (14.44)$$

Equation (14.44) is of the same form $y = (z - \mu)/\sigma$. Thus, Eq. (14.44) yields the expression for the standard Gaussian random distribution with $\mu = 0$ and variance equal to unity.

Remembering that $w_i = 1/\sigma_i^2$ and replacing the measurement error ζ_i by the estimated error $\widehat{\zeta}$ (since the former is unknown) in Eq. (14.4) yields the following estimated performance function

$$\widehat{f} = \sum_{i=1}^{m} \frac{\widehat{\zeta}_i^2}{\sigma_i^2}$$

Substituting for $\widehat{\zeta}_i$ in the preceding equation gives

$$\widehat{f} = \sum_{i=1}^{m} \frac{\widehat{\zeta}_i^2}{\sigma_i^2} = \sum_{i=1}^{m} \frac{(z_i - \widehat{z}_i)^2}{\sigma_i^2} \qquad (14.45)$$

Equation (14.45) is a probability distribution function. It is essential to calculate the mean value of $\widehat{f}$ to obtain the values (areas) by employing the standardized tabulated results. Introduce the calculated variance R_i' into Eq. (14.45) as shown here.

$$\boldsymbol{E}\left[\widehat{f}\right] = \boldsymbol{E}\left[\sum_{i=1}^{m} \frac{\widehat{\zeta}_i^2}{\sigma_i^2}\right] = \boldsymbol{E}\left[\sum_{i=1}^{m} \frac{R_i}{\sigma_i^2} \frac{\widehat{\zeta}_i^2}{R_i}\right] \qquad (14.46)$$

Using Eq. (14.43), Eq. (14.46) reduces to

$$\boldsymbol{E}\left[\widehat{f}\right] = \sum_{i=1}^{m} \frac{R_i'}{\sigma_i^2} \qquad (14.47a)$$

and

$$\boldsymbol{E}\left[\widehat{f}\right] = \boldsymbol{E}\left[\sum_{i=1}^{m} \frac{R_i}{\sigma_i^2} \frac{\widehat{\zeta}_i^2}{R_i}\right] = \boldsymbol{E}\left[\frac{R_i}{\sigma_i^2}\left(\frac{z_i - \widehat{z}_i}{\sqrt{R_i'}}\right)^2\right] = \boldsymbol{E}\left[\left(\frac{z_i - \widehat{z}_i}{\sigma_i^2}\right)^2\right]$$

$$(14.47b)$$

With the help of a numerical example, it will be shown that the mean value of the performance function f is an integer constant and is equal to the redundancy, that is, the difference between the number of measurements m and the number of independent variables n. The difference $(m - n)$ is also referred to as the degrees of freedom.

Chi-square distribution and chi-square test
If the statistical theory is applied to the sum of the weighted squares function represented by Eq. (14.45), it is easily observed that the function $\widehat{f}$ shows a *chi-square* distribution similar to the standard Gaussian distribution of $\widehat{\zeta}_i$. The *chi-square* distribution is mathematically indicated by $\chi_{k\alpha}^2$ where χ is the Greek alphabet, pronounced as 'ki', and $k = (m - n)$, the degrees of freedom, and α represents the area under the curve $\chi_{k\alpha}^2$. The term $(1 - \alpha)$ relates to the confidence level or the accuracy of the calculated value. Figure 14.8 shows a plot of the probability density function $p(\chi_k^2)$, for a specified value of k, of the chi-square distribution $\chi_{k\alpha}^2$.

It is observed from Fig. 14.8 that probability density function is not symmetrically distributed. However, the area under the curve, as is to be expected is unity. For large values of k $(k > 30)$, the χ^2 distribution closely resembles the standard Gaussian distribution.

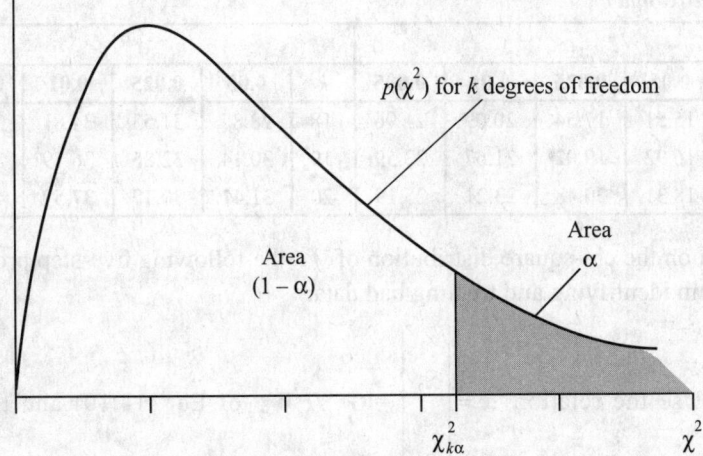

Fig. 14.8 Plot of the probability density function $p(\chi_k^2)$ of the chi-square distribution

A specified area α, for a given value of k, under the curve $\chi_{k\alpha}^2$ indicates the probability that the function $\hat{f}$ exceeds $\chi_{k\alpha}^2$. Conversely, the probability that the function $\hat{f}$ will attain a value less than $\chi_{k\alpha}^2$ is represented by the remaining area, that is, $(1 - \alpha)$ under the curve. The last may be mathematically written as

$$P_r\left(\hat{f} < \chi_{k\alpha}^2\right) = (1 - \alpha)$$

The last paragraph provides the basis for identifying bad data and may be stated as follows:

For a specified value of degrees of freedom $(k - m - n)$ and a desired probability α, if the calculated value of the function $\hat{f}$ is less than the critical value corresponding to α, the data is not a bad data and can be accepted to be accurate. Mathematically, the foregoing translates as follows:

$$\hat{f} < \chi_{k\alpha}^2$$

Table 14.2 provides the chi-square distribution of $\hat{f}$ for specified values of k and α.

Table 14.2 Chi-square distribution of $\hat{f}$

k	α 0.05	0.025	0.01	0.005	k	α 0.05	0.025	0.01	0.005
1	3.84	5.02	6.64	7.88	11	19.68	21.92	24.73	26.76
2	5.99	7.38	9.21	10.60	12	21.03	23.34	26.22	28.30
3	7.82	9.35	11.35	12.84	13	22.36	24.74	27.69	29.82
4	9.49	11.14	13.28	14.86	14	23.69	26.12	29.14	31.32
5	11.07	12.83	15.09	16.75	15	25.00	27.49	30.58	32.80
6	12.59	14.45	16.81	18.55	16	26.30	28.85	32.00	34.27
7	14.07	16.01	18.48	20.28	17	27.59	30.19	33.41	35.72

(Contd)

Table 14.2 (Contd)

	α					α			
k	0.05	0.025	0.01	0.005	k	0.05	0.025	0.01	0.005
8	15.51	17.54	20.09	21.96	18	28.87	31.53	34.81	37.16
9	16.92	19.02	21.67	23.59	19	30.14	32.85	36.19	38.58
10	18.31	20.48	23.21	25.19	20	31.41	30.17	37.57	40.00

Based on the chi-square distribution of $\hat{f}$, the following five-step procedure is useful in identifying and treating bad data:

Step 1: Use the relation $\hat{x} = \begin{bmatrix} \hat{x}_1 \\ \hat{x}_2 \\ \vdots \\ \hat{x}_n \end{bmatrix} = G^{-1}H^tWz$ of Eq. (14.10) and the raw

measurements data z_1 to compute $\hat{x}_i$.

Step 2: Compute the estimated values of $\hat{z}_i$ by using the relation $\hat{z} = H\hat{x}$ where the vector $\hat{x}$ is obtained from step 1. Hence, compute the estimated errors from the relation $\hat{\zeta}_i = z_i - \hat{z}_i$.

Step 3: Compute the estimated performance function $\hat{f} = \sum \left(\hat{\zeta}_i^2 / \sigma_i^2 \right)$.

Step 4: For the specified degrees of freedom k and the corresponding desired probability α, determine the critical value from Table 14.2. Determine whether the computed value of $\hat{f}$ in step 3 is less than the critical value. If less, it may be concluded that there is no measured raw data which is bad and the estimates of the state variables can be accepted as accurate; if not, it may be concluded that there is some bad data which needs to be identified.

Step 5: Use the relation $(\hat{\zeta}_i - 0)/\sqrt{R_i'} = (z_i - \hat{z}_i)/\sqrt{R_i'}$ of Eq. (14.44) to compute the largest standardized error. Exclude the raw measurement corresponding to the largest standardized error and re-compute the performance function $\hat{f}$ and the state estimates. Check for bad data by verifying whether the inequality $\hat{f} < \chi_{k\alpha}^2$ is satisfied. If satisfied, the bad data has been eliminated and the estimates of the state variables can be accepted as accurate.

While the chi-square test makes it convenient to detect bad data, its identification cannot be easily accomplished. Since in real-time estimation of power system variables the number of degrees of freedom is very high, it is feasible to abandon a group of measurements which match with the largest standardized residuals. This criterion may ease the task of identifying gross errors, but it does not ensure that bad measurements are invariably associated with the largest standardized errors.

Example 14.3 Use the data in Example 14.1 to compute the expected value of the performance function $\hat{f}$ of Eq. (14.45).

Solution To perform the various matrix operations MATLAB is used in the computations. First of all R is computed from the weighting factors as follows:

$$R = \begin{bmatrix} 0.0100 & 0 & 0 & 0 & 0 \\ 0 & 0.0100 & 0 & 0 & 0 \\ 0 & 0 & 0.0133 & 0 & 0 \\ 0 & 0 & 0 & 0.0133 & 0 \\ 0 & 0 & 0 & 0 & 0.0133 \end{bmatrix}$$

Using Eq. (14.40), $R' = \left[I - HG^{-1}H'R^{-1} \right] R$ is computed in two steps.

$$HG^{-1}H'R^{-1} = H*\text{inv}(G)*H' *\text{inv}(R)$$

$$= \begin{bmatrix} 0.1015 & -0.0691 & 0.0243 & 0.1457 & -0.2073 \\ -0.0691 & 0.0716 & 0.0019 & 0.0112 & 0.2148 \\ 0.0324 & 0.0025 & 0.0262 & 0.1569 & 0.0075 \\ 0.1943 & 0.0149 & 0.1569 & 0.9415 & 0.0448 \\ -0.2765 & 0.2864 & 0.0075 & 0.0448 & 0.8593 \end{bmatrix}$$

In order to compute the expected value of the performance function $\hat{f}$ of Eq. (14.45), only the diagonal elements of the preceding matrix are required. Hence,

$$Rtemp = HG^{-1}H'R^{-1} = \begin{bmatrix} 0.1015 & 0 & 0 & 0 & 0 \\ & 0.0716 & 0 & 0 & 0 \\ 0 & 0 & 0.0262 & 0 & 0 \\ 0 & 0 & 0 & 0.9415 & 0 \\ 0 & 0 & 0 & 0 & 0.8593 \end{bmatrix}$$

Since, $R' = [I - HG^{-1} H' R^{-1}] R$, it can be observed that
$$R' = [I - Rtemp] R$$

Using this relation, it is clear that Eq. (14.47a) can be written as

$$E\left[\hat{f} \right] = \sum_{i=1}^{m} \frac{R_i'}{\sigma_i^2} = \sum_{i=1}^{m} \frac{\left[I_i - Rtemp_i \right] R_i}{\sigma_i^2} = \sum_{i=1}^{m} \frac{\left[I_i - Rtemp_i \right] \sigma_i^2}{\sigma_i^2}$$

$$= \sum_{i=1}^{m} \left[I_i - Rtemp_i \right] \text{ since } R_i = \sigma_i^2. \text{ Hence,}$$

$$\sum_{i=1}^{m} \frac{R_i'}{\sigma_i^2} = \sum_{i=1}^{m} [I_i - Rtemp_i]$$

$$= (1 - 0.1015) + (1 - 0.0716) + (1 - 0.0262) + (1 - 0.9415) + (1 - 0.8593)$$
$$= 2.999 = 3$$

In the example, the number of measurements $m = 5$ and the number of independent state variables $n = 2$. Hence, the degrees of freedom $= (m - n) = 3$. Therefore, it may be stated that Eq. (14.47b), at all times, is numerically equal to the number of degrees, i.e.,

$$E\left[\hat{f} \right] = E\left[\left(\frac{z_i - \hat{z}_i}{\sigma_i^2} \right)^2 \right] = m - n$$

Example 14.4 Apply the chi-square test to the raw measurements in Example 14.1 and determine bad data, if any. Identify the resultant bad data and compute the state variables from the resultant data. Assume $\alpha = 0.01$.

Solution To solve this example, the five-step procedure outlined earlier will be used.
Step 1: Compute from the raw measurements (already accomplished in Example 14.1).

Step 2:

```
» zh=H*xh          % Computation of ẑᵢ by using the relation ẑ = Hx̂
```

$$zh = \begin{bmatrix} 2.0595 \\ -0.2454 \\ 1.8141 \\ 10.8844 \\ -0.9817 \end{bmatrix}$$ % zh symbol for representing $\hat{z}$

```
» eh=(z-zh)        % Computation of ζ̂ᵢ = zᵢ - ẑᵢ
```

$$eh = \begin{bmatrix} -0.0595 \\ -0.0046 \\ -0.6141 \\ 0.1156 \\ -0.0183 \end{bmatrix}$$ % eh symbol for $\hat{\zeta}_i$

Step 3:

```
» fh=100*eh(1)^2+100*eh(2)^2+75*eh(3)^2+75*eh(4)^2+75*eh(5)^2
```

% Computation of the estimated performance function $\hat{f} = \sum \dfrac{\hat{\zeta}_i^2}{\sigma_i^2}$

```
fh = 29.6642                              % fh symbol for f̂
```

Step 4: The number of degrees of freedom $k = m - n = 5 - 2 = 3$. From Table 14.2, it is observed that for $k = 3$ and $\alpha = 0.01$, the estimated value of the performance function $\hat{f}$ is greater than the critical value of 11.35. Therefore, these measurements contain some bad data.

Step 5:

```
≫ eh(1)/sqrt((1-Rd(1,1))*100)=    % Estimation of standardized error  ζ̂₁/√R₁′
```

$\gg$ eh(1)/sqrt((1-Rd(1,1))*100)=   % Estimation of standardized error $\dfrac{\hat{\zeta}_1}{\sqrt{R_1'}}$

```
-0.0060                           % eh symbol for standardized error
```

$\gg$ eh(2)/sqrt((1-Rd(2,2))*100)   % Estimation of standardized error $\dfrac{\hat{\zeta}_2}{\sqrt{R_2'}}$

```
= -4.5980e-004
```

$\gg$ eh(3)/sqrt((1-Rd(3,3))*75)   % Estimation of standardized error $\dfrac{\hat{\zeta}_3}{\sqrt{R_3'}}$

```
= -0.0714
```

$\gg$ eh(4)/sqrt((1-Rd(4,4))*75)   % Estimation of standardized error $\dfrac{\hat{\zeta}_4}{\sqrt{R_4'}}$

```
= 0.0133
```

$\gg$ eh(5)/sqrt((1-Rd(5,5))*75)   % Estimation of standardized error $\dfrac{\hat{\zeta}_5}{\sqrt{R_5'}}$

```
= -0.0021
```

It is observed that the standardized error $\hat{\zeta}_4 / \sqrt{R_4'}$ has the highest magnitude. Hence, the performance function $\hat{f}$ is to be re-computed by rejecting data related to measurement z_4.

```
≫ H(4,:)=[]                % Elimination of row 4 in matrix H
```

$$H = \begin{bmatrix} 0.2273 & -0.1364 \\ -0.1364 & 0.1818 \\ 0.0909 & 0.0455 \\ -0.5455 & 0.7273 \end{bmatrix}$$ % Resultant 4 × 2 H matrix

```
≫ R(4,:)=[];                    % Elimination of row 4 in matrix R
≫ R(:,4)=[]                     % Elimination of column 4 in matrix R
```

$$R = \begin{bmatrix} 0.0100 & 0 & 0 & 0 \\ 0 & 0.0100 & 0 & 0 \\ 0 & 0 & 0.0133 & 0 \\ 0 & 0 & 0 & 0.0133 \end{bmatrix} \qquad \text{% Resultant } R \text{ matrix}$$

```
≫ W(4,:)=[];                    % Elimination of row 4 in matrix W
≫ W(:,4)=[]                     % Elimination of column 4 in matrix W
```

$$W = \begin{bmatrix} 100 & 0 & 0 & 0 \\ 0 & 100 & 0 & 0 \\ 0 & 0 & 75 & 0 \\ 0 & 0 & 0 & 75 \end{bmatrix} \qquad \text{% Resultant } W \text{ matrix}$$

```
≫ G=H'*W*H                      % Computation of G matrix with measurement bad
                                  data z₄ eliminated
```

$$G = \begin{bmatrix} 29.9587 & -35.0207 \\ -35.0207 & 44.9897 \end{bmatrix}$$

```
≫ z(4)=[]                       % Elimination of bad data z₄
```

$$z = \begin{bmatrix} 2.0000 \\ -0.2500 \\ 1.2000 \\ -1.0000 \end{bmatrix}$$

Step 1:

```
≫ xh=inv(G)*H'*W*z              % Computation of x̂ after elimination of bad
                                  data from the raw measurements
```

$$xh = \begin{bmatrix} 12.5691 \\ 7.9553 \end{bmatrix}$$

Step 2:

```
≫ zh=H*xh                       % Computation of ẑᵢ by using the relation Ẑ = H x̂
```

$$zh = \begin{bmatrix} 1.7718 \\ -0.2676 \\ 1.5043 \\ -1.0702 \end{bmatrix}$$

```
≫ eh=z−zh                       % Computation of ζ̂ᵢ = zᵢ − ẑᵢ
```

$$eh = \begin{bmatrix} 0.2282 \\ 0.0176 \\ -0.3043 \\ 0.0702 \end{bmatrix}$$

Step 3:

```
≫ fh=100*eh(1)^2+100*eh(2)^2+75*eh(3)^2+75*eh(4)^2
```

$$\text{% Computation of the estimated performance function } \hat{f} = \sum \frac{\hat{\zeta}_i^2}{\sigma_i^2}$$

```
fh = 12.5505
```

Step 4: Again from Table 14.2, it is observed that for $k = 2$ and $\alpha = 0.01$, the estimated value of the performance function $\hat{f}$ is greater than the critical value of 9.21. It is obvious that the measurement data is inaccurate and the measuring equipment needs to be more precise.

14.7 POWER SYSTEM STATE ESTIMATOR IN NOISY ENVIRONMENT

As already stated, voltage magnitudes and phase angles are the two state variables of an electrical power system. Unlike the voltage magnitudes, which can be directly measured, the relative voltage angles cannot be measured directly. However, the latter are computed by the state estimator which processes the online system data.

The real-time input to the state estimator is constituted of online data (bus voltages, real and reactive bus powers, and line flows) and status information associated with the system switching devices, such as circuit breakers and transformer taps. The state estimator operates on data which is akin to the conventional power flow data. However, the volume of data used by the state estimator (in terms of the number of actual measurements performed in practical state estimation) exceeds by far the data required for a system power flow study for the purpose of planning and design. Redundancy in data is a necessary and desirable feature in state estimation to take care of malfunctioning in data acquisition equipment and erroneous measurements. Thus, in state estimation the number of equations, whose solution is sought, is invariably in excess of the number of unknown state variables.

Raw measurements data is never used directly in state estimation. The state estimator first processes the raw data to identify gross bad measurements, filters such measurements, and then performs the computations to obtain average estimates of the state variables.

14.7.1 Simulation of the State Estimator

The equations correlating the m measurements to the n state variables can be written as

$$\zeta = z - Hx$$

or

$$
\begin{bmatrix} \zeta_1 \\ \zeta_2 \\ \vdots \\ \zeta_m \end{bmatrix}
=
\begin{bmatrix} z_1 \\ z_2 \\ \vdots \\ z_m \end{bmatrix}
-
\begin{bmatrix}
h_{11} & h_{12} & \cdots & h_{1n} \\
h_{21} & h_{22} & \cdots & h_{2n} \\
\vdots & \vdots & & \vdots \\
h_{m1} & h_{m2} & \cdots & h_{mn}
\end{bmatrix}
\begin{bmatrix} x_1 \\ x_2 \\ \vdots \\ x_n \end{bmatrix}
\qquad (14.48)
$$

Equation (14.48) is similar to Eq. (14.1) except that the vector ζ represents Gaussian random variable noise terms and the coefficient matrix H comprises non-linear functions which correlate the state variables x_i with the measured quantities z_i. The performance function $\hat{f}$ in terms of the sum of squares of errors and variances is expressed as

$$
\hat{f} = \sum_{i=1}^{m} \frac{\hat{\zeta}_i^2}{\sigma_i^2} = \sum_{i=1}^{m} \frac{\left\{ \begin{bmatrix} z_1 \\ z_2 \\ \vdots \\ z_m \end{bmatrix} - \begin{bmatrix} h_{11} & h_{12} & \cdots & h_{1n} \\ h_{21} & h_{22} & \cdots & h_{2n} \\ \vdots & \vdots & & \vdots \\ h_{m1} & h_{m2} & \cdots & h_{mn} \end{bmatrix} \begin{bmatrix} x_1 \\ x_2 \\ \vdots \\ x_n \end{bmatrix} \right\}^2}{\sigma_i^2}
$$

$$(14.49)$$

Just as in the case of Eq. (14.7), the condition for the performance function $\widehat{f}$ to be a minimum can be written from Eq. (14.49) as follows:

$$
\left[\begin{array}{cccc}
\dfrac{\partial h_1}{\partial x_1} & \dfrac{\partial h_2}{\partial x_1} & \cdots & \dfrac{\partial h_m}{\partial x_1} \\[2mm]
\dfrac{\partial h_1}{\partial x_2} & \dfrac{\partial h_2}{\partial x_2} & \cdots & \dfrac{\partial h_m}{\partial x_2} \\[2mm]
\vdots & \vdots & & \vdots \\[2mm]
\dfrac{\partial h_1}{\partial x_n} & \dfrac{\partial h_2}{\partial x_n} & \cdots & \dfrac{\partial h_m}{\partial x_n}
\end{array}\right]_{\widehat{x}}
\left[\begin{array}{cccc}
\dfrac{1}{\sigma_1^2} & & & \\[2mm]
& \dfrac{1}{\sigma_2^2} & & \\[2mm]
& & \ddots & \\[2mm]
& & & \dfrac{1}{\sigma_m^2}
\end{array}\right] \times
$$

$$
\left\{\left[\begin{array}{c} z_1 \\ z_2 \\ \vdots \\ z_m \end{array}\right] - \left[\begin{array}{cccc} h_{11} & h_{12} & \cdots & h_{1n} \\ h_{21} & h_{22} & \cdots & h_{2n} \\ \vdots & \vdots & & \vdots \\ h_{m1} & h_{m2} & \cdots & h_{mn} \end{array}\right] \left[\begin{array}{c} x_1 \\ x_2 \\ \vdots \\ x_n \end{array}\right]\right\} = \left[\begin{array}{c} 0 \\ 0 \\ \vdots \\ 0 \end{array}\right]
$$

$$(14.50)$$

The error vector ζ in Eq. (14.50) is substituted by the relation given by Eq. (14.48). Equation (14.50) in compressed form can be written as

$$
\boldsymbol{H}_X^t \boldsymbol{R}^{-1} \left[\begin{array}{c} z_1 - \boldsymbol{h}_1\left(\widehat{x}_1, \widehat{x}_2, \cdots, \widehat{x}_n\right) \\ z_2 - \boldsymbol{h}_2\left(\widehat{x}_1, \widehat{x}_2, \cdots, \widehat{x}_n\right) \\ \vdots \\ z_m - \boldsymbol{h}_m\left(\widehat{x}_1, \widehat{x}_2, \cdots, \widehat{x}_n\right) \end{array}\right] = \left[\begin{array}{c} 0 \\ 0 \\ \vdots \\ 0 \end{array}\right]
$$

$$(14.51)$$

where $\boldsymbol{H}_x = \left[\begin{array}{cccc}
\dfrac{\partial h_1}{\partial x_1} & \dfrac{\partial h_2}{\partial x_1} & \cdots & \dfrac{\partial h_m}{\partial x_1} \\[2mm]
\dfrac{\partial h_1}{\partial x_2} & \dfrac{\partial h_2}{\partial x_2} & \cdots & \dfrac{\partial h_m}{\partial x_2} \\[2mm]
\vdots & \vdots & & \vdots \\[2mm]
\dfrac{\partial h_1}{\partial x_n} & \dfrac{\partial h_2}{\partial x_n} & \cdots & \dfrac{\partial h_m}{\partial x_n}
\end{array}\right]$

and $\quad z_i - h_i\left(\widehat{x}_1, \widehat{x}_2, \cdots, \widehat{x}_i, \cdots, \widehat{x}_n\right) = z_i - \left(h_{i1}\widehat{x}_1 + h_{i2}\widehat{x}_2 + \cdots + h_{i\,i}\widehat{x}_i + \cdots + h_{i\,n}\widehat{x}_n\right)$

Assume that the state variables vector is represented by its initial values given by $(x_1^0, x_2^0, x_3^0, \ldots, x_n^0)$. Equation (14.51) is linearized around its initial operating point $(x_1^0, x_2^0, x_3^0, \ldots, x_n^0)$ in the same manner as for the Newton–Raphson power flow as follows:

$$
h_1\left(x_1, x_2, \cdots, x_n\right) = h_1\left(x_1^0, x_2^0, \cdots, x_n^0\right) + \left(\Delta x_1^0\right)\dfrac{\partial h_1}{\partial x_1}\Big|^0 + \left(\Delta x_2^0\right)\dfrac{\partial h_1}{\partial x_2}\Big|^0 + \cdots + \left(\Delta x_n^0\right)\dfrac{\partial h_1}{\partial x_n}\Big|^0
$$

$$
h_2\left(x_1, x_2, \cdots, x_n\right) = h_2\left(x_1^0, x_2^0, \cdots, x_n^0\right) + \left(\Delta x_1^0\right)\dfrac{\partial h_2}{\partial x_1}\Big|^0 + \left(\Delta x_2^0\right)\dfrac{\partial h_2}{\partial x_2}\Big|^0 + \cdots + \left(\Delta x_n^0\right)\dfrac{\partial h_2}{\partial x_n}\Big|^0
$$

$$\vdots \qquad\qquad\qquad\qquad\qquad\qquad\qquad\qquad\qquad\qquad\qquad\qquad \vdots$$

$$
h_n\left(x_1, x_2, \cdots, x_n\right) = h_n\left(x_1^0, x_2^0, \cdots, x_n^0\right) + \left(\Delta x_1^0\right)\dfrac{\partial h_n}{\partial x_1}\Big|^0 + \left(\Delta x_2^0\right)\dfrac{\partial h_n}{\partial x_2}\Big|^0 + \cdots + \left(\Delta x_n^0\right)\dfrac{\partial h_n}{\partial x_n}\Big|^0
$$

$$(14.52)$$

Writing Eq. (14.52) in concise matrix form, yields

$$
\begin{bmatrix} h_1(x_1,x_2,\cdots,x_n) \\ h_2(x_1,x_2,\cdots,x_n) \\ \vdots \\ h_n(x_1,x_2,\cdots,x_n) \end{bmatrix} = \begin{bmatrix} h_1\left(x_1^0,x_2^0,\cdots,x_n^0\right) \\ h_2\left(x_1^0,x_2^0,\cdots,x_n^0\right) \\ \vdots \\ h_n\left(x_1^0,x_2^0,\cdots,x_n^0\right) \end{bmatrix} + \begin{bmatrix} \left.\dfrac{\partial h_1}{\partial x_1}\right|^0 & \left.\dfrac{\partial h_1}{\partial x_2}\right|^0 \cdots \left.\dfrac{\partial h_1}{\partial x_n}\right|^0 \\ \left.\dfrac{\partial h_2}{\partial x_1}\right|^0 & \left.\dfrac{\partial h_2}{\partial x_2}\right|^0 \cdots \left.\dfrac{\partial h_2}{\partial x_n}\right|^0 \\ \vdots \\ \left.\dfrac{\partial h_n}{\partial x_1}\right|^0 & \left.\dfrac{\partial h_n}{\partial x_2}\right|^0 \cdots \left.\dfrac{\partial h_n}{\partial x_n}\right|^0 \end{bmatrix} \begin{bmatrix} \Delta x_1^0 \\ \Delta x_2^0 \\ \vdots \\ \Delta x_n^0 \end{bmatrix}
$$

(14.53)

$$
= \begin{bmatrix} h_1\left(x_1^0,x_2^0,\cdots,x_n^0\right) \\ h_2\left(x_1^0,x_2^0,\cdots,x_n^0\right) \\ \vdots \\ h_n\left(x_1^0,x_2^0,\cdots,x_n^0\right) \end{bmatrix} + H_x^0 \begin{bmatrix} \Delta x_1^0 \\ \Delta x_2^0 \\ \vdots \\ \Delta x_n^0 \end{bmatrix}
$$

(14.54)

where $\quad H_x^0 = \begin{bmatrix} \left.\dfrac{\partial h_1}{\partial x_1}\right|^0 & \left.\dfrac{\partial h_1}{\partial x_2}\right|^0 \cdots \left.\dfrac{\partial h_1}{\partial x_n}\right|^0 \\ \left.\dfrac{\partial h_2}{\partial x_1}\right|^0 & \left.\dfrac{\partial h_2}{\partial x_2}\right|^0 \cdots \left.\dfrac{\partial h_2}{\partial x_n}\right|^0 \\ \vdots \\ \left.\dfrac{\partial h_n}{\partial x_1}\right|^0 & \left.\dfrac{\partial h_n}{\partial x_2}\right|^0 \cdots \left.\dfrac{\partial h_n}{\partial x_n}\right|^0 \end{bmatrix}$

In Eq. (14.53), Δx_i^0 represents the addition to be made to the initial estimate of the state variable x_i^0 to obtain the new estimate of the state variable i.e., $x_i^1 = x_i^0 + \Delta x_i^0$. In iterative (recursive) matrix form, this may be written as

$$x_i^{IT+1} = x_i^{IT} + \Delta x_i^{IT}.$$
$$\Delta x_i^{IT} = x_i^{IT+1} - x_i^{IT} \tag{14.55}$$

where IT is the iteration count.

Substituting the expanded form of Eq. (14.54) in Eq. (14.51), the resultant expression after re-arranging takes the following form

$$
H_X^{0t} R^{-1} \begin{bmatrix} z_1 - h_1\left(x_1^0,x_2^0,x_3^0,\cdots,x_n^0\right) \\ z_2 - h_2\left(x_1^0,x_2^0,x_3^0,\cdots,x_n^0\right) \\ \vdots \\ z_m - h_m\left(x_1^0,x_2^0,x_3^0,\cdots,x_n^0\right) \end{bmatrix} = H_X^{0t} R^{-1} H_x^0 \begin{bmatrix} \Delta x_1^0 \\ \Delta x_2^0 \\ \vdots \\ \Delta x_n^0 \end{bmatrix}
$$

or $\quad \begin{bmatrix} \Delta x_1^0 \\ \Delta x_2^0 \\ \vdots \\ \Delta x_n^0 \end{bmatrix} = \left(H_X^{0t} R^{-1} H_x^0\right)^{-1} H_X^{0t} R^{-1} \begin{bmatrix} z_1 - h_1\left(x_1^0,x_2^0,\cdots,x_n^0\right) \\ z_2 - h_2\left(x_1^0,x_2^0,\cdots,x_n^0\right) \\ \vdots \\ z_m - h_m\left(x_1^0,x_2^0,\cdots,x_n^0\right) \end{bmatrix}$ (14.56)

Substituting Eq. (14.56) in Eq. (14.55) and writing the resultant equation in a generalized iterative form leads to

$$x_i^{IT+1} - x_i^{IT} = \left(\boldsymbol{H}_X^{IT\prime} \boldsymbol{R}^{-1} \boldsymbol{H}_x^{IT} \right)^{-1} \boldsymbol{H}_X^{IT\prime} \boldsymbol{R}^{-1} \begin{bmatrix} z_1 - h_1 \left(x_1^{IT}, x_2^{IT}, \cdots, x_n^{IT} \right) \\ z_2 - h_2 \left(x_1^{IT}, x_2^{IT}, \cdots, x_n^{IT} \right) \\ \vdots \\ z_m - h_m \left(x_1^{IT}, x_2^{IT}, \cdots, x_n^{IT} \right) \end{bmatrix}$$

(14.57)

In Eq. (14.57), the Jacobian $\boldsymbol{H}_x^{IT}$ at iteration count IT is written as follows

$$\boldsymbol{H}_x^{IT} = \begin{bmatrix} \dfrac{\partial h_1}{\partial x_1}\Big|^{IT} & \dfrac{\partial h_1}{\partial x_2}\Big|^{IT} & \cdots & \dfrac{\partial h_1}{\partial x_n}\Big|^{IT} \\ \dfrac{\partial h_2}{\partial x_1}\Big|^{IT} & \dfrac{\partial h_2}{\partial x_2}\Big|^{IT} & \cdots & \dfrac{\partial h_2}{\partial x_n}\Big|^{IT} \\ & \vdots & & \\ \dfrac{\partial h_n}{\partial x_1}\Big|^{IT} & \dfrac{\partial h_n}{\partial x_2}\Big|^{IT} & \cdots & \dfrac{\partial h_n}{\partial x_n}\Big|^{IT} \end{bmatrix}$$

(14.58)

It is noted from the foregoing that Eq. (14.57) provides the iterative formula for estimating the values of the n state variables. However, it may also be observed that for estimating the new value of the state variables, the values of the Jacobian $\boldsymbol{H}_x^{IT}$ and the vector elements $z_i - h_i \left(x_1^{IT}, x_2^{IT}, \cdots, x_n^{IT} \right)$ are required to be computed for each iteration and the process continued till

$$x_i^{IT+1} - x_i^{IT} < \varepsilon$$

where ε is the specified tolerance for convergence.

The following step-wise procedure outlines the method of obtaining a solution of Eq. (14.57) iteratively and verifying the accuracy of the estimated state variables by applying the chi-square test.

Step 1: Initialize the iterative process by setting iteration count IT = 0 and guesstimating $h_i \left(x_1^{IT}, x_2^{IT}, \cdots, x_n^{IT} \right)$.

Step 2: From the weighting factors compute the diagonal variance matrix $\boldsymbol{R}$.

Step 3: Compute the elements of the Jacobian $\boldsymbol{H}_x^{IT}$ and the vector elements $z_i - h_i \left(x_1^{IT}, x_2^{IT}, \cdots, x_n^{IT} \right)$.

Step 4: Solve Eq. (14.57) and compute the corrections to be made to the state variables and therefrom calculate the new estimates of the state variables as follows

$$x_i^{IT+1} = x_i^{IT} + \Delta x_i^{IT}$$

Step 5: Check for convergence by using the expression $x_i^{IT+1} - x_i^{IT} < \varepsilon$.

Step 6: If the tolerance ε is within the specified limit, go to step 7; if not, set IT = IT + 1, replace $h_i \left(x_1^{IT}, x_2^{IT}, \cdots, x_n^{IT} \right)$ by the latest estimates of the state variables, and go to step 3.

Step 7: Compute the measurements error vector $\zeta = z - h(x_1, x_2, \cdots, x_n)$ and use the relation $\hat{f} = \sum \left(\hat{\zeta}_i^2 / \sigma_i^2\right)$ to compute the sum of the squares of the performance function $\hat{f}$. Apply the chi-square test to determine bad measurements if any.

After step 6, if convergence is obtained, the estimated values of the state variables are obtained by WLSE and may be represented by

$$x^{IT+1} = \hat{x} = [\hat{x}_1, \hat{x}_2, \cdots, \hat{x}_n]^t$$

14.7.2 Application of the State Estimator Model to Determine the State Variables of a Power System

Referring to Fig. 14.3, it is observed that in a power system, the quantities which can be measured with equipment are as follows:

- Voltage magnitude V_i at a bus i.
- Real power P_i and reactive power Q_i injected into the bus i
- Real power P_{im} and reactive power Q_{im} flowing in the line from bus i to bus m
- Real power P_{mi} and reactive power Q_{mi} flowing in the line from bus m to bus i

The line charging in Fig. 14.4 is shown lumped at the two ends of the line. In reality, the shunt capacitance is distributed along the length of the line, and therefore, the line charging MVAR cannot be measured separately. However, it forms a part of the reactive power flows Q_{pk} and Q_{kp}. As already stated, it is not economically feasible to monitor the voltage angles at the buses; hence, only the magnitude of voltage at the buses is monitored.

In a power system of N buses and L lines, the maximum numbers of possible measurements which can be made are $m = (3N + 4L)$. The digit three accounts for the voltage magnitude and the two injected powers at a bus and digit four is made up of the real and reactive power flows in both directions of a line. Similarly, if the voltage at one of the buses in the system is chosen as the reference, the number of state variables required to be estimated is given by $n = (2N - 1)$. Thus, maximum dimension of the Jacobian H matrix is $m \times n = \{(3N + 4L) \times (2N - 1)\}$.

Since m has to be greater than n, the Jacobian matrix H, unlike the Jacobian J matrix in power flow studies, is rectangular. The redundancy factor is given by

$$\frac{m}{n} = \frac{(3N + 4L)}{(2N - 1)}$$

From the preceding equation it is observed that the redundancy factor is very large in a practical electrical power system. For a line to bus ratio of 1.75, the redundancy factor when all the measurement variables are metered is 5.0.

Example 14.5 In the power system network shown in Fig. 14.9, a generator is feeding a load over a transmission line whose reactance is jX_{12}. The system is being monitored for (i) voltage magnitudes at the two buses, (ii) reactive power at bus 2, (iii) real power flow from bus 1 to 2, and (iv) reactive power flow from bus 1 to 2. Formulate the Jacobian matrix H and develop the linearized mathematical model for computing the state variables. Assume the voltage at bus 1 as the reference and the shunt reactance at bus 2 equal to jX_{20}.

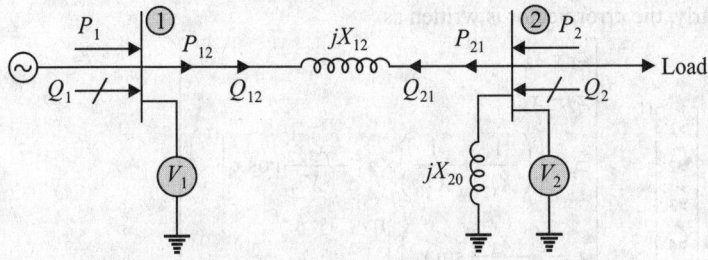

Fig. 14.9 Power system network for Example 14.5

Solution The state variables required to be estimated are: $\begin{bmatrix} x_1 = \delta_2 \\ x_2 = V_2 \\ x_3 = V_1 \end{bmatrix}$.

The measurement variables are: $\begin{bmatrix} z_1 = V_2 \\ z_2 = V_1 \\ z_3 = Q_2 \\ z_4 = P_{12} \\ z_5 = Q_{12} \end{bmatrix}$

$$Q_2 = \left(\frac{1}{X_{12}} + \frac{1}{X_{20}} \right) V_2^2 - \frac{V_1 V_2}{X_{12}} \cos(\delta_2 - \delta_1) = \left(\frac{1}{X_{12}} + \frac{1}{X_{20}} \right) x_3^2 - \frac{x_3 x_2}{X_{12}} \cos x_1$$

$$P_{12} = \frac{V_1 V_2}{X_{12}} \sin(\delta_1 - \delta_2) = -\frac{V_1 V_2}{X_{12}} \sin \delta_2 = -\frac{x_3 x_2}{X_{12}} \sin x_1$$

$$Q_{12} = \frac{V_1^2}{X_{12}} - \frac{V_1 V_2}{X_{12}} \cos(\delta_1 - \delta_2) = \frac{1}{X_{12}} \left\{ x_3^2 - x_3 x_2 \cos x_1 \right\}$$

The non-linear functions h_1, h_2, h_3, h_4, and h_5 may be written as

$$h_1(x_1, x_2, x_3, x_4, x_5) = x_2$$

$$h_2(x_1, x_2, x_3, x_4, x_5) = x_3$$

$$h_3(x_1, x_2, x_3, x_4, x_5) = \left(\frac{1}{X_{12}} + \frac{1}{X_{20}} \right) x_3^2 - \frac{x_3 x_2}{X_{12}} \cos x_1$$

$$h_4(x_1, x_2, x_3, x_4, x_5) = -\frac{x_3 x_2}{X_{12}} \sin x_1$$

$$h_5(x_1, x_2, x_3, x_4, x_5) = \frac{1}{X_{12}} \left\{ x_3^2 - x_3 x_2 \cos x_1 \right\}$$

Thus the elements of the Jacobian $\boldsymbol{H}$, for an iteration count IT, are derived from Eq. (14.58) as follows:

$$\boldsymbol{H}^{\mathrm{IT}} = \begin{bmatrix} 0 & 1 & 0 \\[6pt] 0 & 0 & 1 \\[6pt] \left. \dfrac{x_3 x_2}{X_{12}} \sin x_1 \right|^{\mathrm{IT}} & \left. -\dfrac{x_3}{X_{12}} \cos x_1 \right|^{\mathrm{IT}} & \left. \left(\dfrac{1}{X_{12}} + \dfrac{1}{X_{20}} \right) \times 2x_3 - \dfrac{x_2}{X_{12}} \cos x_1 \right|^{\mathrm{IT}} \\[10pt] \left. -\dfrac{x_3 x_2}{X_{12}} \cos x_1 \right|^{\mathrm{IT}} & \left. -\dfrac{x_3}{X_{12}} \sin x_1 \right|^{\mathrm{IT}} & \left. -\dfrac{x_2}{X_{12}} \sin x_1 \right|^{\mathrm{IT}} \\[10pt] \left. \dfrac{x_3 x_2}{X_{12}} \sin x_1 \right|^{\mathrm{IT}} & \left. -\dfrac{x_3}{X_{12}} \cos x_1 \right|^{\mathrm{IT}} & \left. \dfrac{1}{X_{12}} \left\{ 2x_3 - x_2 \cos x_1 \right\} \right|^{\mathrm{IT}} \end{bmatrix}$$

Similarly, the error vector is written as

$$
\begin{bmatrix} \zeta_1 \\ \zeta_2 \\ \zeta_3 \\ \zeta_4 \\ \zeta_5 \end{bmatrix}^{IT} = \begin{bmatrix} z_1 - x_2^{IT} \\ z_2 - x_3^{IT} \\ z_3 - \left(\left(\dfrac{1}{X_{12}} + \dfrac{1}{X_{20}} \right) \times x_3^2 - \dfrac{x_3 x_2}{X_{12}} \cos x_1 \Big|^{IT} \right) \\ z_4 - \left(-\dfrac{x_3 x_2}{X_{12}} \sin x_1 \right) \Big|^{IT} \\ z_5 - \left(\dfrac{1}{X_{12}} \left\{ x_3^2 - x_3 x_2 \cos x_1 \right\} \right) \Big|^{IT} \end{bmatrix}
$$

The linearized equation, similar to Eq. (14.57) is now written as

$$
x_i^{IT+1} - x_i^{IT} = \left(H_X^{IT \, t} R^{-1} H_x^{IT} \right)^{-1} H_X^{IT \, t} R^{-1} \begin{bmatrix} \zeta_1 \\ \zeta_2 \\ \zeta_3 \\ \zeta_4 \\ \zeta_5 \end{bmatrix}^{IT}
$$

From the preceding equation, WLSE of the state variables can be obtained.

Example 14.6   For the power system in Example 14.5, the measured variables in per unit are as follows:

$$
z_1 = V_2 = 0.95; \qquad z_2 = V_1 = 1.05; \qquad z_3 = Q_2 = 0.55;
$$
$$
z_4 = P_{12} = 0.93; \qquad z_5 = Q_{12} = 0.34
$$

With the WLSE technique, determine the values of the state variables and the confidence level of the accuracy of the results. Assume the values of the variance of the measurement errors as follows: $\sigma_1^2 = \sigma_2^2 = \sigma_3^2 = 0.0001$, $\sigma_4^2 = 0.0002$, and $\sigma_5^2 = 0.0004$. Also, assume a tolerance value of 0.0001 for convergence, and take X12 = j0.2 pu and X20 = j5 pu.

Solution

Step 1:

```
≫ x = [0; 1.0; 1.0];
```
                                    %Initialization of the iterative process by assum-
                                    ing a flat start for the three state variables,
                                    that is,
                                    δ_2 = 0 rads, V_2 = 1.0, and V_1 = 1.0

Step 2:

```
≫ R
```
                                    % Measurement matrix **R**

$$
R = \begin{bmatrix} 0.1000 & 0 & 0 & 0 & 0 \\ 0 & 0.1000 & 0 & 0 & 0 \\ 0 & 0 & 0.1000 & 0 & 0 \\ 0 & 0 & 0 & 0.2000 & 0 \\ 0 & 0 & 0 & 0 & 0.4000 \end{bmatrix} \times 10^{-3}
$$

Step 3:

```
≫ H=zeros(5,3);                 % Computation of the elements of the H matrix.
≫ H(1,2)=1;H(2,3)=1;
≫ H(3,1)=x(3)*x(2)*sin(x(1))/X12;
≫ H(3,2)=-x(3)*cos(x(1))/X12;
≫ H(3,3)=(1/X12+1/X20)*2*x(3)-x(2)*cos(x(1))/X12;
```

```
≫ H(4,1)=-x(3)*x(2)*cos(x(1))/X12;
≫ H(4,2)=-x(3)*sin(x(1))/X12;
≫ H(4,3)=-x(2)*sin(x(1))/X12;
≫ H(5,3)=x(3)*x(2)*sin(x(1))/X12;
≫ H(5,2)=-x(3)*cos(x(1))/X12;
≫ H(5,3)=(2*x(3)-x(2)*cos(x(1)))/X12;
```

$$H = \begin{bmatrix} 0 & 1.0000 & 0 \\ 0 & 0 & 1.0000 \\ 0 & -5.0000 & 5.4000 \\ -5.0000 & 0 & 0 \\ 0 & -5.0000 & 5.0000 \end{bmatrix}$$

```
≫ z = [0.95;1.05;0.55;0.93;0.34];        % computation of the elements of the vector
```

$$\zeta_i = z_i - h_i\left(x_1, x_2, x_3, x_4, x_5\right)$$

```
≫ err=[z(1)-x(2);z(2)-x(3);...
z(3)-((1/X12+1/X20)*x(3)^2-x(3)*x(2)*cos(x(1))/X12);...
z(4)-(-x(3)*x(2)*sin(x(1))/X12);...
z(5)-((x(3)^2-x(3)*x(2)*cos(x(1)))/X12)]
```

$$\text{err} = \zeta = \begin{bmatrix} -0.0500 \\ 0.0500 \\ 0.3500 \\ 0.9300 \\ 0.3400 \end{bmatrix}$$

Step 4:

```
≫ delx=inv(H'*inv(R)*H)*H'*inv(R)*err
```
% Computation of the corrections Δx_i to be made to the state variables by using Eq. (14.57) and therefrom calculate the new estimates of the state variables

$$\text{delx} = \Delta x = \begin{bmatrix} -0.1860 \\ -0.0346 \\ 0.0334 \end{bmatrix}$$

```
≫ x=x+delx
```
% Computation of new estimates of the state variables from $x_i^{IT+1} = x_i^{IT} + \Delta x_i^{IT}$

$$x = \begin{bmatrix} -0.1860 \\ 0.9654 \\ 1.0334 \end{bmatrix}$$

Step 5: Check for convergence by using the expression $x_i^{IT+1} - x_i^{IT} < \varepsilon$. Comparing the state variables, it is observed that $\begin{bmatrix} 0 \\ 1.0 \\ 1.0 \end{bmatrix} - \begin{bmatrix} 0.1860 \\ 0.9654 \\ 1.0334 \end{bmatrix} > 0.001$.

Step 6: Since the difference is greater than the specified tolerance of 0.001, go to step 3 and repeat steps 3 to 5 till convergence is achieved. The final values of the state variables after convergence are tabulated here.

Iteration count	Voltage angle δ_2	Voltage magnitude V_2	Voltage magnitude V_1
0	0	1.0	1.0
1	−0.1860	0.9654	1.0334
2	−0.1873	0.9737	1.0240
3	−0.1872	0.9740	1.0242
4	−0.1872	0.9740	1.0242

It is observed that the values of the state variables have converged in three iterations.

```
>> err=[z(1)-x(2);z(2)-x(3);...        % Computation of the measurements error
                                        vector ζ = z − h (x₁, x₂, x₃ ⋯ , xₙ)
```

vector $\zeta = z - h(x_1, x_2, x_3 \cdots, x_n)$

```
z(3)-((1/X12+1/X20)*x(3)^2-x(3)*x(2)*cos(x(1))/X12);...
z(4)-(-x(3)*x(2)*sin(x(1))/X12);...
z(5)-((x(3)^2-x(3)*x(2)*cos(x(1)))/X12)]
```

$$\text{err} = \zeta = \begin{bmatrix} -0.0240 \\ 0.0258 \\ -0.0036 \\ 0.0017 \\ -0.0039 \end{bmatrix}$$

```
>> fh=err(1)^2/0.0001+err(2)^2/0.0001+…    % Computation of the relation

err(3)^2/0.0001+err(4)^2/0.0002+…
```

$\hat{f} = \sum_{i=1}^{m} \dfrac{\hat{\zeta}_i^2}{\sigma_i^2}$ to compute the sum of the squares of the performance function $\hat{f}$ and application of the chi-square test to determine bad measurements if any.

$\text{fh} = \hat{f} = 12.6349$

With the number of degrees of freedom k equal to $5 - 3 = 2$, it is seen by referring to Table 14.2 that the computed value of $\hat{f} = 12.6349$, which is in excess of the lowest confidence level of 95% accuracy ($\alpha = 0.005$). Clearly, there is bad data in the measurements and therefore, the estimated values of the state variables are unacceptable.

The procedure for identifying the bad data is given in the next step

```
>> G=H'*inv(R)*H          % Computation of the G matrix
```

$$G = \begin{bmatrix} 1.3085 & 0.3503 & -0.8932 \\ 0.3503 & 0.3503 & -3.5946 \\ -0.8932 & -3.5946 & 4.3266 \end{bmatrix} \times 10^5$$

```
» H*inv(G)*H'*inv(R)      % Computation HG⁻¹ Hᵗ R⁻¹ for calculating the
                           diagonal elements of the product matrix.
```

Computation $HG^{-1} H^t R^{-1}$ for calculating the diagonal elements of the product matrix.

$$= \begin{bmatrix} 0.5358 & 0.4837 & -0.0316 & 0.0169 & -0.0574 \\ 0.4837 & 0.4636 & 0.1122 & -0.0176 & -0.0194 \\ -0.0316 & 0.1122 & 0.8043 & 0.0011 & 0.1896 \\ 0.0337 & -0.0352 & 0.0023 & 0.9988 & 0.0042 \\ -0.2298 & -0.0777 & 0.7584 & 0.0084 & 0.1975 \end{bmatrix}$$

```
>> diag=zeros(5,5);       % Extraction of the diagonal elements from the
                           above product matrix.
```

```
>> diag(1,1)=0.5358;diag(2,2)=0.4636;diag(3,3)=0.8043;diag(4,4)=0.9988;
>> diag(5,5)=0.1975;
>> diag
```

$$
diag = \begin{bmatrix} 0.5358 & 0 & 0 & 0 & 0 \\ 0 & 0.4636 & 0 & 0 & 0 \\ 0 & 0 & 0.8043 & 0 & 0 \\ 0 & 0 & 0 & 0.9988 & 0 \\ 0 & 0 & 0 & 0 & 0.1975 \end{bmatrix}
$$

```
>> I=eye(5,5);              %Creation of 5 × 5 identity matrix I
>> Rd=(I-diag)*R            % Computation of R' =
```
$$[\boldsymbol{I}\text{-diagonal elements of } (\boldsymbol{HG^{-1}H^t R^{-1}})]\,\boldsymbol{R}$$

$$
Rd = \boldsymbol{R'} = \begin{bmatrix} 0.0464 & 0 & 0 & 0 & 0 \\ 0 & 0.0536 & 0 & 0 & 0 \\ 0 & 0 & 0.0196 & 0 & 0 \\ 0 & 0 & 0 & 0.0002 & 0 \\ 0 & 0 & 0 & 0 & 0.3210 \end{bmatrix} \times 10^{-3}
$$

```
>> -0.0240/sqrt(Rd(1,1))    % Computation of the standardized estimates by
```
$$\text{using the relation } \frac{\zeta_i}{\sqrt{R'_{ii}}}$$

```
= -3.5226
>> 0.0258/sqrt(Rd(2,2))
= 3.5227
>> -0.0036/sqrt(Rd(3,3))
= -0.8138
>> 0.0017/sqrt(Rd(4,4))
= 3.4701
>> 0.0039/sqrt(Rd(5,5))
= 0.2177
```

From the foregoing, the standardized estimates related to measurements z_1 and z_2 are equal and the highest, therefore, any one of the two measurements can be eliminated as bad data. However, if both the measurements are grossly bad data, elimination of both of them will lead to unproductive estimates of the state variable since there is no redundancy in measurements.

14.8 COMPOSITION OF THE JACOBIAN MATRIX H AND THE MEASUREMENT VECTOR Z

The Jacobian matrix $\boldsymbol{H}$ is called the state estimation matrix and the number of columns in it always corresponds to $(2N-1)$, which is the number of state variables required to be estimated. The number of rows in $\boldsymbol{H}$ depends on the number of measurement variables being monitored. The maximum number of rows when all the measurement variables are monitored is determined by the relation $m = (3N + 4L)$. Each row in the matrix $\boldsymbol{H}$ distinctively corresponds to one of the quantities being measured.

In order to obtain reasonably good estimates of the state variables, a moderately sufficient numbers of measurements of quantities are essential so that there is redundancy in measurements and the Jacobian $\boldsymbol{H}$ matrix rectangular. In the limiting condition, the number of measurement quantities must be equal to the number of state variables and in such a situation $\boldsymbol{H}$ is a square matrix. With the number of measurements less than the number of state variables, it is not possible to solve the system. When H is a square matrix, all measurements, including bad data, have to be accepted, which puts a question mark on the confidence level of the accuracy of the estimated values of the state variables.

If it is assumed that all possible quantities are monitored and that all like measurements are organized together, that is, voltage magnitudes V_i, real and reactive power injections P_p and Q_p, real power line flows P_{pk} and P_{kp}, and reactive power line flows Q_{pk} and Q_{kp}, the configuration of the H matrix takes the form shown in Fig. 14.10.

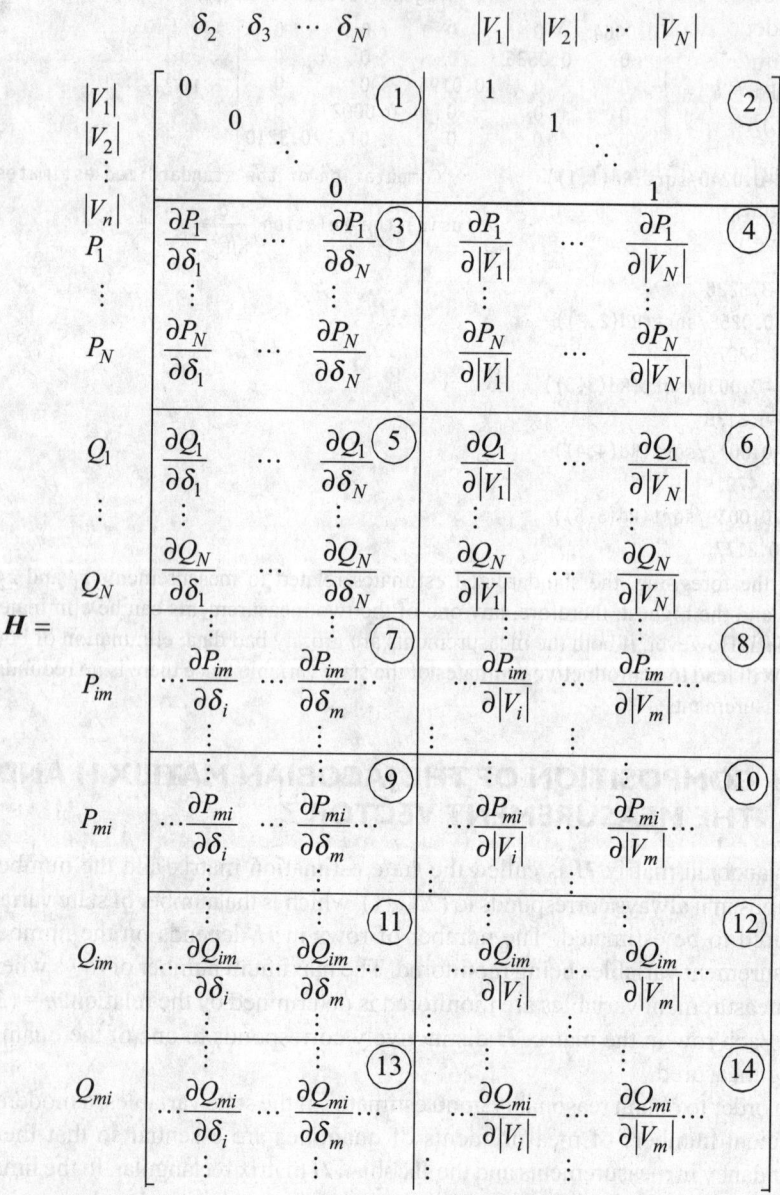

Fig. 14.10 Configuration of H matrix when all possible quantities are monitored

In the configuration in Fig. 14.10, it may be noted that there are 1, 2, ..., N columns corresponding to the N buses in the system, while the number of columns for the bus voltage angles are $(N-1)$. Since the bus voltage angles are estimated with respect to a reference angle, therefore, voltage angle at one of the buses is required to be specified. On the other hand, bus voltage magnitudes are already specified with respect to the ground, and hence, there is no need to specify another bus voltage as a reference. In Fig. 14.10, the bus voltage angle at bus 1 is chosen as the reference angle. Thus, it is observed that bus 1 is not a slack bus since the voltage need not be specified. It may also be noted that the elements in the voltage angle block 1 are zero, since measured values of the magnitudes of bus voltages are independent of bus voltage angles and also the partial derivatives of voltage magnitude with respect to voltage angles $(\partial V_i / \partial \delta_m) = 0$ for all values of i and m. In block 2 which is made up of partial derivatives of magnitudes of bus voltages $(\partial V_i / \partial \delta_m)$ the diagonal elements are unity, because $(\partial V_i / \partial \delta_m) = 1.0$ for $i = m$. Similarly, the off-diagonal elements are zero since $(\partial V_i / \partial \delta_m) = 0$ for $i \neq m$.

Equations (7.19) and (7.20), mentioned in Chapter 7, express the real and reactive powers injected into a bus and are reproduced here

$$P_i = |V_i| \sum_{k=1}^{n} |Y_{im}||V_m|\cos(\theta_{im} - \delta_i + \delta_m) \qquad \text{for } i = 1, 2, ..., n$$

$$Q_i = -|V_i| \sum_{m=1}^{n} |Y_{im}| \, |V_m|\sin(\theta_{im} - \delta_i + \delta_m) \quad \text{for } i = 1, 2, ..., n$$

The partial derivatives for computing the elements of blocks 3, 4, 5, and 6 are detailed in Eqs (7.90) to (7.97) of Chapter 7. Similarly Eq. (7.34) provides an expression for power flow in the line from bus i to bus m as follows:

$$P_{im} + jQ_{im} = V_i I_{im}^* = V_i \left(Y_{im}(V_i - V_m) + \frac{y_i}{2}V_i \right)^*$$

The complex power flow is easily separated into real and reactive line flows as follows:

$$P_{im} = G_{im}|V_i|^2 + |V_i||V_m||Y_{im}|\cos(\delta_i - \delta_m - \theta_{im})$$

$$Q_{im} = -|V_i|^2 \left(\frac{B_{im}'}{2} + B_{im} \right) - |V_i||V_m||Y_{im}|\sin(\delta_i - \delta_m - \theta_{im})$$

where $G_{im} = |Y_{im}| \cos\theta_{im}$ is the series line conductance; $B_{im}'/2$ and B_{im} are the shunt line charging susceptance and series line susceptance respectively. The derivatives for the elements of Jacobian for blocks 7 and 8 are given by

$$\frac{\partial P_{im}}{\partial \delta_i} = |V_i||V_m||Y_{im}|\sin(\delta_i - \delta_m - \theta_{im});$$

$$\frac{\partial P_{im}}{\partial V_i} = 2G_{im}|V_i| + |V_m||Y_{im}|\cos(\delta_i - \delta_m - \theta_{im})$$

$$\frac{\partial P_{im}}{\partial \delta_m} = -|V_i||V_m||Y_{im}|\sin(\delta_i - \delta_m - \theta_{im});$$

$$\frac{\partial P_{im}}{\partial V_m} = |V_i||Y_{im}|\cos(\delta_i - \delta_m - \theta_{im}) \tag{14.59a}$$

Similarly, the derivatives for the elements of the **H** matrix for blocks 11 and 12 may be written as

$$\frac{\partial Q_{im}}{\partial \delta_i} = -|V_i||V_m||Y_{im}|\cos(\delta_i - \delta_m - \theta_{im});$$

$$\frac{\partial Q_{im}}{\partial V_i} = -2|V_i|\left(\frac{B'_{im}}{2} + B_{im}\right) - |V_m||Y_{im}|\sin(\delta_i - \delta_m - \theta_{im})$$

$$\frac{\partial Q_{im}}{\partial \delta_m} = |V_i||V_m||Y_{im}|\cos(\delta_i - \delta_m - \theta_{im});$$

$$\frac{\partial Q_{im}}{\partial V_m} = -|V_i||Y_{i\,m}|\sin(\delta_i - \delta_m - \theta_{im}) \tag{14.59b}$$

Using Eq. (7.35) of Chapter 7; power flow in the line from bus *m* to bus *i* can be written as

$$P_{mi} + jQ_{mi} = V_j I_{mi}^* = V_m \left(Y_{mi} (V_m - V_i) + \frac{y_i}{2} V_m \right)^*$$

The equations for real and reactive power flows in the line from *m* to *i* can be separated and the derivatives of the **H** matrix for blocks 9, 10, 13, and 14 can be similarly derived and are as follows:

$$\frac{\partial P_{mi}}{\partial \delta_i} = -|V_i||V_m||Y_{im}|\sin(\delta_i - \delta_m - \theta_{im});$$

$$\frac{\partial P_{mi}}{\partial V_i} = |V_m||Y_{im}|\cos(\delta_i - \delta_m - \theta_{im})$$

$$\frac{\partial P_{mi}}{\partial \delta_m} = |V_i||V_m||Y_{im}|\sin(\delta_i - \delta_m - \theta_{im});$$

$$\frac{\partial P_{mi}}{\partial V_m} = 2G_{im}|V_m| + |V_i||Y_{im}|\cos(\delta_i - \delta_m - \theta_{im}) \tag{14.60a}$$

$$\frac{\partial Q_{mi}}{\partial \delta_i} = |V_i||V_m||Y_{im}|\cos(\delta_i - \delta_m - \theta_{im});$$

$$\frac{\partial Q_{mi}}{\partial V_i} = -|V_m||Y_{im}|\sin(\delta_i - \delta_m - \theta_{im})$$

$$\frac{\partial Q_{mi}}{\partial \delta_m} = -|V_i||V_m||Y_{im}|\cos(\delta_i - \delta_m - \theta_{im});$$

$$\frac{\partial Q_{mi}}{\partial V_m} = -2V_m\left(\frac{B'_{mi}}{2} + B_{mi}\right) - |V_i||Y_{im}|\sin(\delta_i - \delta_m - \theta_{im}) \tag{14.60b}$$

Readers will do well to separate the real and reactive power flows and verify the partial derivatives for the elements of the Jacobian matrix **H**.

14.8.1 Pattern of Zero and Non-zero Elements in *H* Matrix

Having formulated the full Jacobian matrix H, its 14 blocks can now be examined for the pattern of zero and non-zero elements.

Block 1: All the elements are zero.

Block 2: The matrix has a dimension equal to $N \times N$ when all the bus voltages are measured and is an identity matrix.

Blocks 3 and 5: The matrices in these two blocks relate to real and reactive power P_i and Q_i injections at the buses and have the dimensions of $N \times (N-1)$ when real and reactive power is measured at all the buses. The columns in the matrices represent the $(N-1)$ bus voltage angles. The elements of these matrices are the derivatives of real and reactive power injection with respect to δ_i and possess the same distribution pattern of zero and non-zero elements as that of a Y_{bus} matrix with the first row and column deleted.

Blocks 4 and 6: These two blocks relate to real and reactive power P_i and Q_i injections at the buses and have the dimensions of $N \times N$ when real and reactive power is measured at all the buses. The columns in the matrices represent the N bus voltage magnitudes V_i. The elements of these matrices are the derivatives of real and reactive power injection with respect to V_i and possess the same distribution pattern of zero and non-zero elements as that of Y_{bus} matrix.

Blocks 7 and 9: The elements are the derivatives of real power flows P_{im} and P_{mi}, with respect to bus voltage angles, in lines from buses i to m and m to i respectively.

Blocks 11 and 13: Similarly, the elements here are the derivatives of reactive power flows, with respect to voltage angles, in lines from buses i to m and m to i respectively.

In all the four blocks 7, 9, 11, and 13, each row in a matrix represents the real and reactive power measured. Thus, when real and reactive power flow is measured in all the L lines of a system, each matrix has a dimension of $L \times (N-1)$. The distribution pattern of zero and non-zero elements in these matrices is the same as that of the network branch-to-bus incidence matrix A, with the first column deleted. This is not unusual since, like a branch in a network, each power flow in a line is incident on two buses.

Blocks 8 and 10: The elements are the derivatives of real power flows P_{im} and P_{mi}, with respect to bus voltage magnitudes, in lines from buses i to m and m to i respectively.

Blocks 12 and 14: Similarly, the elements here are also the derivatives of reactive power flows, with respect to voltage magnitudes, in lines from buses i to m and m to i respectively.

In all the four blocks 8, 10, 12, and 14, each row in a matrix represents the real or reactive power measured. Thus, when real or reactive power flow is measured in all the L lines of a system, each matrix has a dimension of $L \times N$. The distribution pattern of zero and non-zero elements in these matrices is the same as that of the network branch-to-bus matrix A when the line measurements are numbered in the same sequence as the network branches.

When some parameters in a power system are not monitored, the rows associated with those measurements are omitted from the Jacobian matrix H. Therefore, the order of the **H** matrix will reduce by the corresponding number of rows for which the measurements are not recorded.

14.8.2 Structure of the Measurement Vector z

Making the same assumption as in the case of the **H** matrix, that all possible quantities are monitored and that all like measurements are organized together (i.e., voltage magnitudes V_p, real and reactive power injections P_p and Q_p, real power line flows P_{pk} and P_{kp}, and reactive power line flows Q_{pk} and Q_{kp}, the structure of the z matrix takes the form shown in Fig. 14.11.

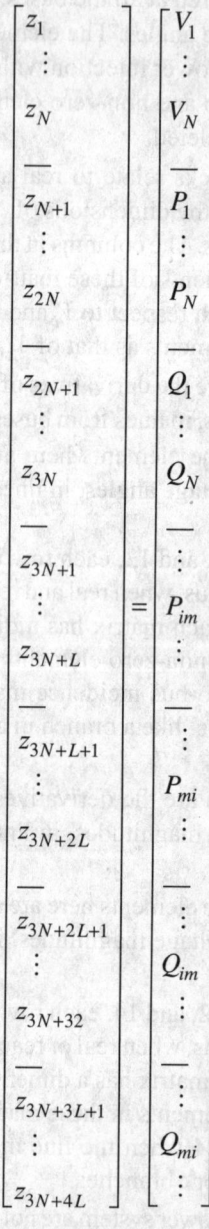

$$
\begin{bmatrix}
z_1 \\
\vdots \\
z_N \\
\hline
z_{N+1} \\
\vdots \\
z_{2N} \\
\hline
z_{2N+1} \\
\vdots \\
z_{3N} \\
\hline
z_{3N+1} \\
\vdots \\
z_{3N+L} \\
\hline
z_{3N+L+1} \\
\vdots \\
z_{3N+2L} \\
\hline
z_{3N+2L+1} \\
\vdots \\
z_{3N+32} \\
\hline
z_{3N+3L+1} \\
\vdots \\
z_{3N+4L}
\end{bmatrix}
=
\begin{bmatrix}
V_1 \\
\vdots \\
V_N \\
\hline
P_1 \\
\vdots \\
P_N \\
\hline
Q_1 \\
\vdots \\
Q_N \\
\hline
\vdots \\
P_{im} \\
\vdots \\
P_{mi} \\
\hline
\vdots \\
Q_{im} \\
\vdots \\
Q_{mi} \\
\vdots
\end{bmatrix}
$$

Fig. 14.11 Structure of the measurement vector z

As already stated, the **H** matrix is similar to the Jacobian matrix **J** employed in power flow studies, except that the former is rectangular. Therefore, the fast decoupled approximation, as described in Section 7.8.5, can be used for obtaining quick estimates of the state variables. Also, since **H** is sparse, the gain matrix $G = H^t R^{-1} H$ is also sparse. Hence, sparsity techniques (see Section 6.4.4) and optimal ordering (see Section 6.7) may be used for efficient utilization of storage space and enhanced computations. Further, explicit inversion of the **G** matrix can be avoided by resorting to triangular factorization discussed under Section 6.5.2.

Example 14.7 In the four-bus network shown in Fig. 14.12, following quantities are measured: magnitudes of voltages at buses V_1 and V_2, real and reactive powers P_1, P_3, Q_1, and Q_3 at buses 1 and 3 respectively. The real and reactive power flows are also monitored as follows:

Line 1-2	P_{12}	P_{21}	Q_{12}	Q_{21}
Line 1-4	P_{14}	P_{41}	Q_{14}	Q_{41}
Line 2-3	P_{23}	P_{32}	Q_{23}	Q_{32}

From these measurements, it is desired to compute the state variables of the network by WLSE method. What is the redundancy factor? Formulate (i) the state variables vector, (ii) the measurements vector, and (iii) the Jacobian matrix **H**. Assume bus 1 as the reference bus.

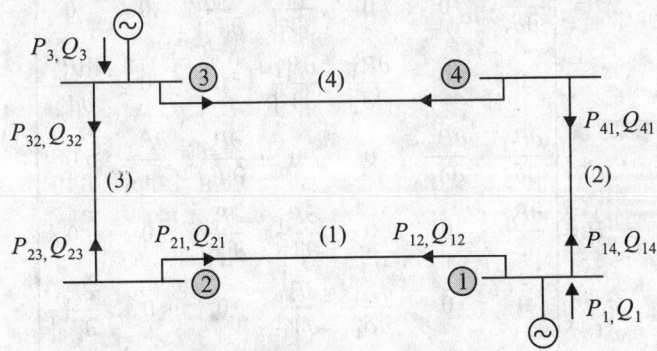

Fig. 14.12 Four-bus network of Example 14.7

Solution Number of measurements = 18
Number of state variables to be estimated = 7

Redundancy factor = $\dfrac{18}{7}$ = 2.57

(i) Since bus 1 is the reference bus, $\delta_1 = 0$

Number of state variables to be estimated = $(2N - 1) = (2 \times 4 - 1) = 7$

State vector is written as $x = [\delta_2 \quad \delta_3 \quad \delta_4 \quad |V_1| \quad |V_2| \quad |V_3| \quad |V_4|]^t$

(ii) The measurement vector is given by

$$z = [V_1\, V_2\, P_1\, P_3\, Q_1\, Q_3\, P_{12}\, P_{14}\, P_{23}\, P_{21}\, P_{41}\, P_{32}\, Q_{12}\, Q_{14}\, Q_{23}\, Q_{21}\, Q_{41}\, Q_{32}]^t$$

(iii) The branch-to-bus incidence matrix can be written by an inspection of the given four-bus network as follows:

$$[A] = \begin{matrix} & 1 & 2 & 3 & 4 \\ & \begin{bmatrix} 1 & -1 & 0 & 0 \\ 1 & 0 & 0 & -1 \\ 0 & 1 & -1 & 0 \\ 0 & 0 & 1 & -1 \end{bmatrix} \end{matrix}$$

Going by the code, that is, a column for each state variable and a row for each measurement, H matrix is written as follows:

$$
H =
\begin{matrix}
\delta_2 & \delta_3 & \delta_4 & |V_1| & |V_2| & |V_3| & |V_4| \\
\begin{bmatrix}
0 & 0 & 0 & 1 & 0 & 0 & 0 \\[4pt]
0 & 0 & 0 & 0 & 1 & 0 & 0 \\[4pt]
\dfrac{\partial P_1}{\partial \delta_2} & 0 & \dfrac{\partial P_1}{\partial \delta_2} & \dfrac{\partial P_1}{\partial |V_1|} & \dfrac{\partial P_1}{\partial |V_2|} & 0 & \dfrac{\partial P_1}{\partial |V_4|} \\[10pt]
\dfrac{\partial P_3}{\partial \delta_2} & \dfrac{\partial P_3}{\partial \delta_3} & \dfrac{\partial P_3}{\partial \delta_4} & 0 & \dfrac{\partial P_3}{\partial |V_2|} & \dfrac{\partial P_3}{\partial |V_3|} & \dfrac{\partial P_3}{\partial |V_4|} \\[10pt]
\dfrac{\partial Q_1}{\partial \delta_2} & 0 & \dfrac{\partial Q_1}{\partial \delta_2} & \dfrac{\partial Q_1}{\partial |V_1|} & \dfrac{\partial Q_1}{\partial |V_2|} & 0 & \dfrac{\partial Q_1}{\partial |V_4|} \\[10pt]
\dfrac{\partial Q_3}{\partial \delta_2} & \dfrac{\partial Q_3}{\partial \delta_3} & \dfrac{\partial Q_3}{\partial \delta_4} & 0 & \dfrac{\partial Q_3}{\partial |V_2|} & \dfrac{\partial Q_3}{\partial |V_3|} & \dfrac{\partial Q_3}{\partial |V_4|} \\[10pt]
\dfrac{\partial P_{12}}{\partial \delta_2} & 0 & 0 & \dfrac{\partial P_{12}}{\partial |V_1|} & \dfrac{\partial P_{12}}{\partial |V_2|} & 0 & 0 \\[10pt]
0 & 0 & \dfrac{\partial P_{14}}{\partial \delta_4} & \dfrac{\partial P_{14}}{\partial |V_1|} & 0 & 0 & \dfrac{\partial P_{14}}{\partial |V_4|} \\[10pt]
\dfrac{\partial P_{23}}{\partial \delta_2} & \dfrac{\partial P_{23}}{\partial \delta_3} & 0 & 0 & \dfrac{\partial P_{23}}{\partial |V_2|} & \dfrac{\partial P_{23}}{\partial |V_3|} & 0 \\[10pt]
\dfrac{\partial P_{21}}{\partial \delta_2} & 0 & 0 & \dfrac{\partial P_{21}}{\partial |V_1|} & \dfrac{\partial P_{21}}{\partial |V_2|} & 0 & 0 \\[10pt]
0 & 0 & \dfrac{\partial P_{41}}{\partial \delta_4} & \dfrac{\partial P_{41}}{\partial |V_1|} & 0 & 0 & \dfrac{\partial P_{41}}{\partial |V_4|} \\[10pt]
\dfrac{\partial P_{32}}{\partial \delta_2} & \dfrac{\partial P_{32}}{\partial \delta_3} & 0 & 0 & \dfrac{\partial P_{32}}{\partial |V_2|} & \dfrac{\partial P_{32}}{\partial |V_3|} & 0 \\[10pt]
\dfrac{\partial Q_{12}}{\partial \delta_2} & 0 & 0 & \dfrac{\partial Q_{12}}{\partial |V_1|} & \dfrac{\partial Q_{12}}{\partial |V_2|} & 0 & 0 \\[10pt]
0 & 0 & \dfrac{\partial Q_{14}}{\partial \delta_4} & \dfrac{\partial Q_{14}}{\partial |V_1|} & 0 & 0 & \dfrac{\partial Q_{14}}{\partial |V_4|} \\[10pt]
\dfrac{\partial Q_{23}}{\partial \delta_2} & \dfrac{\partial Q_{23}}{\partial \delta_3} & 0 & 0 & \dfrac{\partial Q_{23}}{\partial |V_2|} & \dfrac{\partial Q_{23}}{\partial |V_3|} & 0 \\[10pt]
\dfrac{\partial Q_{21}}{\partial \delta_2} & 0 & 0 & \dfrac{\partial Q_{21}}{\partial |V_1|} & \dfrac{\partial Q_{21}}{\partial |V_2|} & 0 & 0 \\[10pt]
0 & 0 & \dfrac{\partial Q_{41}}{\partial \delta_4} & \dfrac{\partial Q_{41}}{\partial |V_1|} & 0 & 0 & \dfrac{\partial Q_{41}}{\partial |V_4|} \\[10pt]
\dfrac{\partial Q_{32}}{\partial \delta_2} & \dfrac{\partial Q_{32}}{\partial \delta_3} & 0 & 0 & \dfrac{\partial Q_{32}}{\partial |V_2|} & \dfrac{\partial Q_{32}}{\partial |V_3|} & 0
\end{bmatrix}
\end{matrix}
$$

From the formulation of the **H** matrix, it is observed that there are two rows corresponding to the two voltages V_1 and V_2 being measured, since $\partial|V_1|/\partial|V_1| = 1$; $\partial|V_2|/\partial|V_2| = 1$. The columns corresponding to the voltage angles associated with these two rows have zero elements. The next four rows correspond to the real and reactive power injections at bus 1 and 2. The next 12 rows represent the real and reactive power flows in lines 1, 2, and 3. Readers are advised to verify that line flow matrices have the same non-zero pattern as that of the matrix **A**. It is seen that the matrix **H** is sparse.

Example 14.8 Compute the partial derivatives, in terms of the system parameters and bus voltages for the following: real and reactive power injections at bus 1 and real and reactive power flows from bus 3-2.

Solution Partial derivatives for real and reactive power injections at bus 1 are obtained from Eqs (7.19) and (7.20) of Chapter 7 as follows:

$$P_i = |V_i| \sum_{k=1}^{n} |Y_{i\,m}| \; |V_m| \cos(\theta_{i\,m} - \delta_i + \delta_m) \quad \text{for } i = 1, 2, \ldots, n \tag{7.19}$$

$$Q_i = -|V_i| \sum_{m=1}^{n} |Y_{i\,m}| \; |V_m| \sin(\theta_{i\,m} - \delta_i + \delta_m) \quad \text{for } i = 1, 2, \ldots, n \tag{7.20}$$

$$P_i = V_i \sum_{m=1}^{n} Y_{im} V_m \cos(\delta_i - \delta_m - \theta_{im})$$

$$Q_i = V_i \sum_{m=1}^{n} Y_{im} V_m \sin(\delta_i - \delta_m - \theta_{im})$$

$$P_1 = |V_1|^2 |Y_{11}| \cos\theta_{11} + |V_1||V_2||Y_{12}| \cos(\delta_1 - \delta_2 - \theta_{12})$$
$$+ |V_1||V_4||Y_{14}| \cos(\delta_1 - \delta_4 - \theta_{14})$$

$$Q_1 = |V_1|^2 |Y_{11}| \sin\theta_{11} + |V_1||V_2||Y_{12}| \sin(\delta_1 - \delta_2 - \theta_{12})$$
$$+ |V_1||V_4||Y_{14}| \sin(\delta_1 - \delta_4 - \theta_{14})$$

The derivatives corresponding to the real power injection at bus 1 are then computed as follows:

$$\frac{\partial P_1}{\partial \delta_2} = |V_1||V_2||Y_{12}| \sin(\delta_1 - \delta_2 - \theta_{12});$$

$$\frac{\partial P_1}{\partial \delta_4} = |V_1||V_4||Y_{14}| \sin(\delta_1 - \delta_4 - \theta_{14})$$

$$\frac{\partial P_1}{\partial V_1} = 2|V_1||Y_{11}| \cos\theta_{11} + |V_2||Y_{12}| \cos(\delta_1 - \delta_2 - \theta_{12}) + |V_4||Y_{14}| \cos(\delta_1 - \delta_4 - \theta_{14});$$

$$\frac{\partial P_1}{\partial V_2} = |V_1||Y_{12}| \cos(\delta_1 - \delta_2 - \theta_{12});$$

$$\frac{\partial P_1}{\partial V_4} = |V_1||Y_{14}| \cos(\delta_1 - \delta_4 - \theta_{14})$$

Similarly, the derivatives corresponding to the reactive power injection at bus 1 are computed as follows:

$$\frac{\partial Q_1}{\partial \delta_2} = -|V_1||V_2||Y_{12}| \cos(\delta_1 - \delta_2 - \theta_{12});$$

$$\frac{\partial Q_1}{\partial \delta_4} = -|V_1||V_4||Y_{14}|\cos(\delta_1 - \delta_4 - \theta_{14})$$

$$\frac{\partial Q_1}{\partial V_1} = 2|V_1||Y_{11}|\sin\theta_{11} + |V_2||Y_{12}|\sin(\delta_1 - \delta_2 - \theta_{12}) + |V_4||Y_{14}|\sin(\delta_1 - \delta_4 - \theta_{14})$$

$$\frac{\partial Q_1}{\partial V_2} = |V_1||Y_{12}|\sin(\delta_1 - \delta_2 - \theta_{12});$$

$$\frac{\partial Q_1}{\partial V_4} = |V_1||Y_{14}|\sin(\delta_1 - \delta_4 - \theta_{14})$$

The derivatives for the real and reactive power flows in line 3-2 can be written directly from Eqs (14.59a) and (14.59b) respectively, and are as follows:

$$\frac{\partial P_{32}}{\partial \delta_3} = |V_2||V_3||Y_{32}|\sin(\delta_3 - \delta_2 - \theta_{32});$$

$$\frac{\partial P_{32}}{\partial E_3} = 2|G_{32}||V_3| + |V_2||Y_{32}|\cos(\delta_3 - \delta_2 - \theta_{32})$$

$$\frac{\partial P_{32}}{\partial \delta_2} = -|V_2||V_3||Y_{32}|\sin(\delta_3 - \delta_2 - \theta_{32});$$

$$\frac{\partial P_{32}}{\partial E_2} = |V_3||Y_{32}|\cos(\delta_3 - \delta_2 - \theta_{32})$$

$$\frac{\partial Q_{32}}{\partial \delta_3} = -|V_2||V_3||Y_{32}|\cos(\delta_3 - \delta_2 - \theta_{32});$$

$$\frac{\partial Q_{32}}{\partial E_3} = -2|V_3|\left(\frac{B'_{32}}{2} + B_{32}\right) - |V_2||Y_{32}|\sin(\delta_3 - \delta_2 - \theta_{32})$$

$$\frac{\partial Q_{32}}{\partial \delta_2} = |V_2||V_3||Y_{32}|\cos(\delta_3 - \delta_2 - \theta_{32});$$

$$\frac{\partial Q_{32}}{\partial E_2} = -|V_3||Y_{32}|\sin(\delta_3 - \delta_2 - \theta_{32})$$

SUMMARY

- The estimation of the state of a power system is a continuous process which is based on online data acquisition through a centralized data acquisition system.
- The estimation of the state of a power system is based on gathering large amounts of data; in this case, bus voltage magnitudes and angles, line power flows, etc. and analysing the same.
- Weighted least square method lays the foundation for state estimation. The state estimator for power flows in transmission lines is based on balanced three-phase steady-state operation, measurement errors, and FSD of all instruments along with discretization errors in converting analogue quantities to digital signals being known.
- Since the measurement equations are non-linear, their solution employs iterative methods.
- The line power estimator has limitations such as it accounts for measurements of power flows in lines only and not accounts for measurements corrupted by noise.
- State estimation simulated on Gaussian probability density function has the ability to identify and treat bad data. The method requires the development of the Jacobian matrix *H* and solving the linearized state vector iteratively.

SIGNIFICANT FORMULAE

- Correlation between measured and state variables: $z = Hx + \zeta$

- Estimate for the noise vector: $\hat{\zeta} = \zeta - H\left(\hat{x} - x\right)$

- Weighting factor: $w_i = \dfrac{50 \times 10^{-6}}{\left[c_1 |S_i| + c_2 (FSD)\right]^2}$

- Function to minimize the sum of the squares of the errors: $H^t W \hat{\zeta} = 0$

- Gain matrix: $G = H^t W H = H^t R^{-1} H$

- Determination of state vector: $\hat{x} = G^{-1} H^t z$

- Weighting factor for line measurements: $w_i = \dfrac{50 \times 10^{-6}}{\left[c_1 |S_{li}| + c_2 (FSD)\right]^2}$

- Performance equation: $f = \left[S_m - S_c\right]^{*t} W \left[S_m - S_c\right]$

- Iterative equation for bus voltages: $V_b^{IT+1} = \left[B^t D B\right]^{-1} B^t D \left[V_m^{IT} - bV_s\right]$

- Variance: $\dfrac{1}{\sigma_i^2} = \dfrac{3}{\left[c_1 |Z_i| + c_2 (FSD)\right]^2}$

- Five-step procedure to identify and treat bad data.

   Step 1: $\hat{x} = G^{-1} H^t W z$

   Step 2: $\hat{z} = H\hat{x}$ and $\hat{\zeta}_i = z_i - \hat{z}_i$

   Step 3: $\hat{f} = \displaystyle\sum_{i=1}^{m} \dfrac{\hat{\zeta}_i^2}{\sigma_i^2}$

   Step 4: Use Table 14.2 to identify bad data.

- Step 5: Compute largest standardized error $\dfrac{\hat{\zeta}_i - 0}{\sqrt{R_i'}} = \dfrac{z_i - \hat{z}_i}{\sqrt{R_i'}}$

- Iterative equation for computing the state variables:

$$\left[x_i^{IT+1}\right] - \left[x_i^{IT}\right] = \left(H_x^{IT\,t} R^{-1} H_x^{IT}\right)^{-1} H_x^{IT\,t} R^{-1} \begin{bmatrix} z_1 - h_1\left(x_1^{IT}, x_2^{IT}, x_3^{IT} \cdots x_n^{IT}\right) \\ z_2 - h_2\left(x_1^{IT}, x_2^{IT}, x_3^{IT} \cdots x_n^{IT}\right) \\ z_3 - h_2\left(x_1^{IT}, x_2^{IT}, x_3^{IT} \cdots x_n^{IT}\right) \\ \text{-------------} \\ \text{-------------} \\ z_m - h_m\left(x_1^{IT}, x_2^{IT}, x_3^{IT} \cdots x_n^{IT}\right) \end{bmatrix}$$

EXERCISES

Review Questions

14.1 Draw a schematic block diagram of a centralized data acquisition and describe its functioning.

14.2 Describe the role of a state estimator and discuss the possible applications of state estimation.

14.3 Define and explain (i) state and measurement variables, (ii) redundancy and redundancy factor, and (iii) noise.

14.4 (a) What are the assumptions made in simulating the state estimator for power flow in transmission lines. Develop a mathematical model to iteratively compute the state variables.

(b) Draw a flow chart to show the iterative algorithm.

14.5 Plot the Gaussian probability function $p\,(z)$ versus the random variable z. Discuss the characteristics of the Gaussian probability function.

14.6 Explain chi-square distribution and chi-square test.

Numerical Problems

14.1 If $\hat{x}$ represents the estimates of variables $\hat{x}_n$ computed by the method of weighted least squares, show that

$$\hat{x} - x = \begin{bmatrix} \hat{x}_1 - x_1 \\ \hat{x}_2 - x_2 \\ \vdots \\ \hat{x}_n - x_n \end{bmatrix} = G^{-1}H^t W \zeta \text{ and } \hat{\zeta} = \left[I - HG^{-1}H^t W \right] \zeta$$

14.2 The elements of the measurement vector for the dc circuit shown in Fig. 14.13 are: $z_1 = 9.1$ A, $z_2 = 7.0$ A, $z_3 = 6.9$ A, $z_4 = 9.2$ V, and $z_5 = 7.0$ V. Using the weighted least squares estimation technique, estimate the components of the state vector. Assume that the ammeters and voltmeters have an accuracy of 100 each.

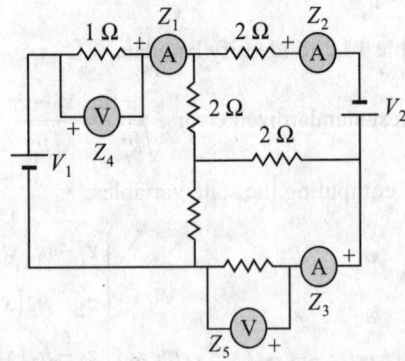

Fig. 14.13 The dc circuit for Problem 14.2

14.3 Using the state vector computed in Problem 14.2, determine the estimated measurements on the instruments and therefrom compute the errors in the measurements.

14.4 Estimate the state vector in Problem 14.2, if the voltmeter reading on the second voltmeter is 7.1 V instead of 7.0 V. What are the new estimated measurement errors? Discuss the results.

14.5 For the system shown in Fig. 14.14, estimate the state variables from the following data. Assume full-scale deflection equal to 100 both for the MW and MVAR meters, and a measurement accuracy of 2%. The analogue-to-digital conversion and transducer errors may be taken at 0.4%. All line and power flow data is in per unit. Assume tolerance of 0.0001 in absolute magnitudes of two consecutive voltages.

From bus no.	To bus no.	Complex power flow S	Series line impedance Z	Shunt admittance Y/2
1	3	$0.35 - j0.13$	$0.06 + j0.25$	0.05
1	2	$-0.80 + j0.40$	$0.05 + j0.20$	0.04
2	1	$0.76 - j0.44$	$0.05 + j0.20$	0.04
2	3	$0.15 - j0.15$	$0.03 + j0.08$	0.025
3	2	$-0.12 + j0.12$	$0.03 + j0.08$	0.025
3	1	$-0.38 + j0.10$	$0.06 + j0.25$	0.005

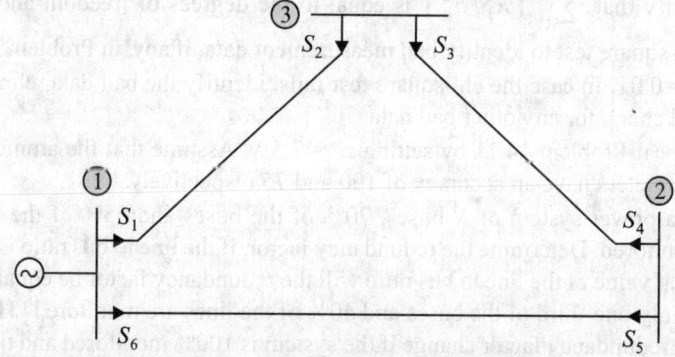

Fig. 14.14 Power system for Problem 14.7

14.6 From the best estimates of state variables computed in Problem 14.5, calculate the power line flows and the errors in between the calculated and measured values. Comment on the errors.

14.7 The weighting factors, for the power system shown in Fig. 14.15, corresponding to line measurements S_1, S_2, S_3, S_4, S_5, S_6, and S_7 are d_1, d_2, d_3, d_4, d_5, d_6, and d_7 respectively. Determine (i) $B^t D$ and (ii) $B^t DB$. Show that for $B^t D$ matrix, the weighting factor d_m is positive for a bus p when a measurement is made at the near end of the line connected to the bus and negative when the measurement is made at the far end of the line connected to the same bus. Similarly, for the matrix $B^t DB$, prove that its diagonal elements are positive weighting factors for all line measurements connected to bus p and that the off-diagonal element between bus p and bus k is the negative of the weighting factor associated with the line between the two buses.

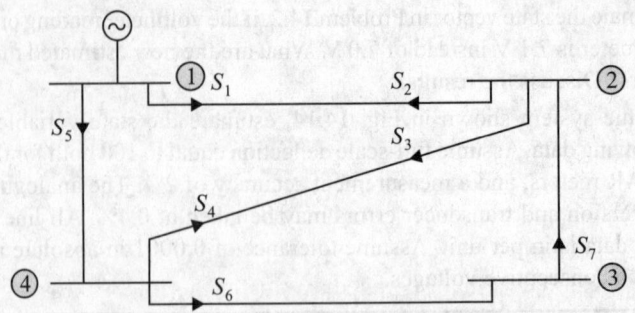

Fig. 14.15 Power system for Problem 14.7

14.8 If z varies between c_1 and c_2, derive the probability function $P_r(c_1 < z < c_2)$ in terms of $P_r(y)$ where y represents the transformation formula $y = (z - \mu)/\sigma$. If the Gaussian random variable z varies between 3σ and μ, determine its probability.

14.9 Prove that $\left[I - HG^{-1}H^t R^{-1} \right] \left[I - HG^{-1}H^t R^{-1} \right] = \left[I - HG^{-1}H^t R^{-1} \right]$. What is the product matrix called?

14.10 Show that $E\left[(x - \hat{x})(x - \hat{x})^t \right] = G^{-1} = \left(H^t R^{-1} H \right)^{-1}$.

14.11 Verify that $\sum_{i=1}^{m} \left(R_i' / \sigma_i^2 \right)$ is equal to the degrees of freedom and use the chi-square test to identify bad measurement data, if any, in Problem 14.2, for $\alpha = 0.01$. In case the chi-square test fails, identify the bad data, eliminate it, and check for any other bad data.

14.12 Repeat Problem 14.11 by setting $z_5 = 7.5$ V. Assume that the ammeters and voltmeters have an accuracy of 100 and 75 respectively.

14.13 In a power system of N buses, 70% of the buses and 55% of the lines are monitored. Determine the redundancy factor, if the line to bus ratio is 1.5. For what value of the line to bus ratio will the redundancy factor be equal to unity if only one-third of the buses and 40% of the lines are monitored? How does the redundancy factor change if the system is 100% monitored and the line to bus ratio is reduced to 1.20? Why should the redundancy factor be much in excess of unity? If N is equal to 30, determine the degrees of freedom in each case.

14.14 For a square H matrix, which is invertible, show that

$$\left(H_x^t R^{-1} H_x \right)^{-1} H_x^t R^{-1} = H_x^t$$

14.15 For the network of Example 14.6, formulate the H matrix and the z vector when the measurement quantities are as follows:

Magnitudes of voltages at buses V_3 and V_4
Real and reactive power P_3 and Q_3 at bus 3
The power flows:

Line 2-1	-	P_{21}	-	Q_{21}
Line 3-2	P_{32}	-	-	Q_{23}
Line 3-4	P_{34}	P_{43}	Q_{34}	Q_{43}

14.16 Compute the elements of H matrix for the network in Problem 14.5, in terms of the state variables and the parameters of the network for (i) power injections P_3 and Q_3 and (ii) line flow P_{21} and Q_{21}.

Multiple Choice Objective Questions

14.1 Which of the following represents the state of a system?
  (a) Frequency   (b) Voltage
  (c) Real power   (d) Reactive power

14.2 To estimate the state of a system, which of the following is required to be measured?
  (a) Bus voltages
  (b) Real and reactive bus powers
  (c) Real and reactive transmission line power flows
  (d) All of these

14.3 Which of the following is the full form of RTU?
  (a) Remote terminal unit   (b) Regional terminal unit
  (c) Remote technical unit   (d) Regional technical unit

14.4 Which of the following describes a state estimator?
  (a) Human being   (b) Analogue process
  (c) Computer program   (d) None of these

14.5 Which of the following is not an objective of state estimation?
  (a) Security analyses   (b) Fault diagnoses
  (c) Improving the quality of data   (d) None of these

14.6 Which of the following defines the redundancy factor?
  (a) $m - n$ (b) $m + n$
  (c) $m \times n$ (d) m/n

14.7 The statistical mean of noise over a large number of measurements is
  (a) -1 (b) 0
  (c) 1   (d) None of these

14.8 For good accuracy in measurements, the range of the redundancy factor should be
  (a) 0 to 0.5   (b) 0.5 to 1.0
  (c) 1.0 to 1.5   (d) 1.5 to 2.5

14.9 Which of the following represents the weighting factor w_i?
  (a) $\dfrac{50 \times 10^6}{[c_1|S_i| + c_2\,(FSD)]^2}$ (b) $\dfrac{50 \times 10^6}{[c_1|S_i| + c_2\,(FSD)]}$
  (c) $\dfrac{50 \times 10^{-6}}{[c_1|S_i| + c_2\,(FSD)]^2}$ (d) $\dfrac{50 \times 10^{-6}}{[c_1|S_i| + c_2\,(FSD)]}$

14.10 Which of the following represents the weighted least squares vector of state variables?
  (a) $G^{-1}H'Wz$ (b) $GH'Wz$
  (c) $G^{-1}HWz$ (d) $G^{-1}H^{-1}Wz$

14.11 Which of the following does not apply to the state vector $\hat{x}$ of an N bus system? It has the dimensions of
  (a) $(N-1)$ (b) $2(N-1)$
  (c) real   (d) complex

14.12 In an *N* bus system, the number of measurements is *m*. Which of the following is not true of *H* ?
  (a) It is of order $m \times m$. (b) It is of order $N \times m$.
  (c) It is diagonal. (d) All of these.

14.13 Which of the following is not a diagonal matrix?
  (a) *W* (b) *D*
  (c) *B* (d) None of these

14.14 If the number of measurements is *m* and the number of independent state variables is *n*, which of the following represents the order of the noise vector?
  (a) *m* (b) *n*
  (c) $m - n$ (d) None of these

14.15 A power system consists of *N* buses and *L* lines. Which of the following represents the number of maximum possible measurements?
  (a) $N + 2L$ (b) $2N + 3L$
  (c) $3N + 4L$ (d) $4N + 5L$

14.16 For the power system in Q. 15, which of the following is true of the Jacobian matrix *H*?
  (a) It is square. (b) It is rectangular.
  (c) It is diagonal. (d) None of these.

14.17 For the power system in Q. 15, which of the following represents the maximum number of state variables required to be estimated?
  (a) $2N$ (b) $2N - 1$
  (c) $2N + 1$ (d) None of these

14.18 A power system consists of 8 buses and 24 lines. If all the lines are monitored, which of the following is the redundancy factor?
  (a) 6 (b) 7
  (c) 8 (d) 9

Answers

14.1 (b)	14.2 (d)	14.3 (a)	14.4 (c)	14.5 (d)	14.6 (d)
14.7 (b)	14.8 (d)	14.9 (c)	14.10 (a)	14.11 (c)	14.12 (b)
14.13 (d)	14.14 (a)	14.15 (c)	14.16 (b)	14.17 (b)	14.18 (c)

CHAPTER 15

An Introduction to HVDC Power Transmission

Learning Outcomes

An engrossed study of the chapter will allow the reader to

- Know the history of development of HVDC transmission
- Make a studied comparison of ac and dc transmission systems from the perspectives of technical, economic, and environmental parameters
- Enumerate the various types of converters and explain with the help of circuit diagrams the arrangements and describe qualitatively the features of converter stations
- Explain the operation of a fully controlled three-phase bridge converter with the help of a circuit diagram and draw the voltage and current output waveforms
- Determine the average and RMS values of the voltage and current outputs
- Detail out and explain the different types of controls to regulate the flow of power in an HVDC system
- Understand the significance of HVDC system in the current power sector scenario

OVERVIEW

Interestingly, the first commercial electricity generation by Thomas Alva Edison was direct current (dc) electrical power. The first electrical transmission systems were also direct current systems. However, dc power at low voltage could not be transmitted over long distances thus giving rise to alternating current (ac) electrical systems. With the advent of transformers for lowering and raising the voltage levels, development of poly-phase circuits and the poly-phase induction motor (the workhorse of industry) led to the shift from dc electrical power to ac electrical power.

The present-day growing demand for electrical energy (in India, the demand doubles every seven years) coupled with the reduced cost of generation has led to the setting up of remote generating stations, fossil fuel fired thermal stations at pit heads, and large hydro stations at distances of hundreds of kilometres away from load centres. This in turn has led to the transmission of bulk ac power at high-voltage over long distances to the load centres. The problems associated with transmission of bulk ac power at high-voltages, over long distances has again brought focus on the development of high-voltage direct current (HVDC) systems for transmission of power. However, as generation and utilization of electric power remain at ac, the dc transmission requires conversion from ac to dc using a rectifier at the sending end while using an inverter at the receiving end for conversion from dc back to ac. Nevertheless, with the development of high-voltage valves, it was possible to once again transmit dc power at high-voltages and over long distances, giving rise to high-voltage direct current (HVDC) transmission systems. In view of liberalization and increased efforts to conserve the environment, the HVDC transmission system has become more desirable to the electricity industry owing to its environmental advantages among others. HVDC systems are economically attractive over conventional ac lines for transmission of power over long or very long distances and the power flow can be controlled rapidly and accurately with respect to both the power level and the direction. They also improve stability and power quality and this possibility is often used to improve the performance and efficiency of the connected ac networks. HVDC systems can also be used for asynchronous interconnections.

15.1 HISTORICAL PERSPECTIVE ON HVDC TRANSMISSION

The history of electrical industry documented the first commercial generation of electricity by Thomas Alva Edison in 1882 and it was dc electric power. In the year 1901, Hewitt's mercury vapour rectifier appeared. The first commercial HVDC transmission was commissioned in 1954 connecting Swedish mainland with the island of Gotland in the Baltic Sea by a submarine cable with ground return and was of rating 20 MW, 100 kV. This was the outcome of persistent research and development work with thyratrons in America and mercury arc valves in Europe before 1940. Till 1970, a number of HVDC transmission networks, using mercury arc valves with varying length and capacities to transmit power, had been commissioned all over the world. The introduction of thyristor valves in 1970 overcame some of the problems associated with mercury arc valves. The mercury arc rectifiers and thyristor valves made the design and development of line commutated current source converters possible, resulting in the growth of HVDC transmission systems throughout the world. HVDC transmission using current source converters has been in use for fifty years. Itaipu HVDC transmission project of Brazil is by far the most impressive HVDC transmission system in the world with a total rated power of 6300 MW and the highest transmission voltage of ±600 kV and the Commissioning year was 1985–1987. The longest HVDC system in operation is the 1800 km INGA-SHABA (Zaire) project at ±500 kV.

In India, the first HVDC project commissioned was Vindhyachal back-to-back link interconnecting the western and northern regional grids. The link was commissioned in 1989 and has a maximum power transfer capacity of 500 MW and operating voltage of 176 kV. The first commercial long-distance HVDC link, 814 km long, with dc voltage ±500 kV, and total rated power 1500 MW, was commissioned in 1990 between Rihand and Delhi. Since then many 500 MW lines have come up in India. The 2000 MW Talcher–Kolar link is the biggest so far and spans four states: Odisha, Andhra Pradesh, Tamil Nadu, and Karnataka.

The first microcomputer-based control equipment for HVDC was introduced in 1979. The first capacitor commutated converter (CCC) with improved commutation characteristics was used in Garabi—the Argentina–Brazil 1000-MW interconnection —in 1998. The insulated gate bipolar transistors (IGBT) with high-voltage ratings have accelerated the development of voltage source converters (VSC) for HVDC applications in the lower power range. The first voltage source converter in combination with underground dc cables was used for HVDC transmission in Gotland, Sweden, in 1999.

15.2 TRANSMISSION SYSTEMS: DC VERSUS AC

The vast majority of electric power transmissions use three-phase alternating current. It would be interesting to study the reasons behind a choice of HVDC instead of ac to transmit power. The reasons for use of HVDC instead of ac in a specific case are often numerous and complex. Each individual transmission project has its own set of reasons justifying the choice. At the planning stage, a comparison between the ac and dc transmission technology is made from the following considerations:

(a) Technical features
(b) Economic factors
(c) Environmental considerations

These will be discussed in detail here.

15.2.1 Technical Features

The comparison of ac and dc transmission systems is based on a comparison of the following technical performance parameters:

(a) Power per circuit
(b) Stability limits
(c) Voltage and current limits
(d) Reactive power control and voltage regulation
(e) Circuit breakers and short-circuit current
(f) Reliability
(g) Terminal equipment
(h) Harmonics
(i) Control of tie-line power
(j) Generating stations

Power per Circuit

In order to compare the power per circuit of ac and dc lines, it is assumed that both ac and dc insulators withstand the same peak voltage. Then,

$$E_{dc} = \sqrt{2}\, E_{ac}$$

where E_{dc} is the conductor to ground dc voltage and E_{ac} is the rms value of ac phase voltage. The current carrying capacity of both ac and dc conductors is limited by the temperature rise, that is, $I_{dc} = I_{ac}$ (rms value). Then,

$$\text{dc power per conductor} = E_{dc}I_{dc} = \sqrt{2}\, E_{ac}I_{ac}$$

Assuming a bipolar (two-conductor) dc line, the dc power per circuit P_{dc} is given by

$$P_{dc} = 2\sqrt{2}\, E_{ac}I_{ac} \tag{15.1}$$

If $\cos\varphi$ denotes the power factor, then the power per conductor $= E_{ac}I_{ac}\cos\varphi$ and the ac power per circuit P_{ac} is given by

$$P_{ac} = 3E_{ac}I_{ac}\cos\varphi \tag{15.2}$$

Dividing Eq. (15.1) by Eq. (15.2) yields the ratio of dc power to ac power per circuit. Hence,

$$\frac{P_{dc}}{P_{ac}} = \frac{2\sqrt{2}E_{ac}I_{ac}}{3E_{ac}I_{ac}\cos\varphi} = \frac{0.9428}{\cos\varphi} \tag{15.3}$$

From Eq. (15.3), it can be observed that the dc power per circuit is inversely proportional to the power factor.

Example 15.1 If an ac line has a power capacity/circuit of 1.0 per unit, determine the power factor angle at which the dc line will have the same power capacity/circuit. Plot the variation of dc power capacity/circuit when the power factor varies from 0.7 to 1.0.

Solution From Eq. (15.3), when $P_{dc} = P_{ac}$, $\cos\varphi = 0.9428$.

Therefore, power factor angle $\varphi = \cos^{-1}(09428) = 19.4712°$

In order to plot the variation of dc power capacity per circuit, the following MATLAB script file can be used.

```
>> % Script file for plotting the DC power/circuit versus power factor
>> cosphi=linspace(0.7,1.0,1000);
>> Pdc=(0.9428)./cosphi;
>> plot(cosphi,Pdc), grid
>> hold on
>> axis([0.7 1.0 0.9 1.5])
>> xlabel('Variation of power factor')
>> ylabel('DC Power per circuit')
>> title('PLOT OF POWER/CIRCUIT VS. POWER FACTOR')
>>
```

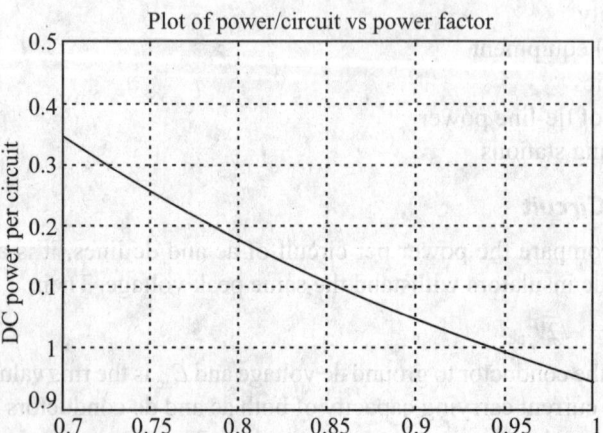

Fig. 15.1 Variation of dc power per circuit versus power factor

For the same power per circuit, a dc link is cheaper because it requires two conductors against three conductors for an ac link. A dc line would require a narrower channel of land compared to an ac link, therefore, it would be more economical since the land rights will be cheaper to acquire.

Stability Limits

The transfer of power on a loss free ac transmission line is a function of (i) the magnitudes of the terminal voltages, (ii) the difference in voltage angles at the terminals of the lines, and (iii) the line reactance, that is,

$$P = \frac{|E_S||E_R|}{X}\sin\delta$$

where $|E_S|$ and $|E_R|$ are respectively the magnitudes of sending-end and receiving-end voltages, δ is the difference in voltage angles, and X is the reactance of the line.

Further, the maximum ac power which can be transferred over a line is restricted by its steady-state and transient-state stability limits. The maximum power transfer or the steady-state limit P_{max} occurs at $\delta = 90°$ and is given by

$$P_{max} = \frac{|E_S||E_R|}{X}$$

Since both the voltage magnitudes are controlled within narrow limits, the maximum power transmitted over an ac line is inversely proportional to its reactance X, which in turn increases with distance. Therefore, the capacity of an ac line to transmit power decreases with distance. The transient stability limit is lower than the steady-state limit and as a thumb rule it may be taken as 50% (corresponding to $\delta = 30°$) of the steady-state limit. However, distance does not impose any limit on the capability of a dc line to transmit power. Figure 15.2 shows a plot of transmission capabilities of ac and dc power lines against distance.

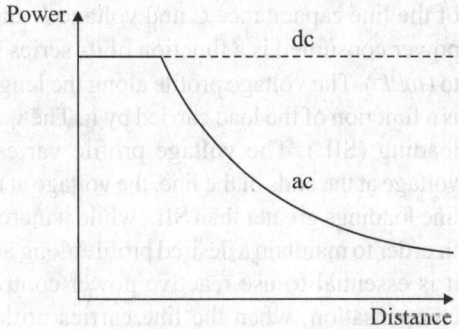

Fig. 15.2 Plot of power transmission capability as a function of distance

Additionally, a dc transmission line has no stability problem. In fact, a dc transmission line is used for asynchronous connection. It is sometimes difficult or impossible to connect two ac networks due to stability problems. In such cases, HVDC is the only way to make an exchange of power between the two networks possible. There are HVDC links between networks with different nominal frequencies (50 and 60 Hz) in Japan and South America.

Voltage and Current Limits

Lines are designed to prevent flashover or puncture of insulation at normal operational voltages and over voltages due to switching surges and lightning. In extra high-voltage (EHV) ac overhead lines, switching surges are of greater concern than lightning. Therefore, it is possible to design overhead lines which would limit the transient over voltages to 2 to 2.5 times the normal peak working voltages. The other parameters which restrict the selection of peak working voltage and minimum conductor size are radio interference (RI) and loss due to corona which increases in bad weather conditions. During bad weather, of the two parameters, RI is the more critical factor.

On the other hand, switching surges in dc links are less severe and are of the order of 1.7 times the normal voltage. Further, dc overhead lines are not as susceptible to RI in bad weather as ac lines; in fact RI in dc links marginally reduces in bad weather. In dc cables, the normal working voltage is decided by the strength of the insulation to withstand voltage stress. Generally, cable insulation is able to withstand a dc voltage which is higher than the normal peak ac voltage which is an added advantage, since the peak voltage is already $\sqrt{2}$ × (rms value of ac voltage).

Theoretically, current in overhead lines is limited by the permissible temperature rise above ambient temperature to prevent mechanical damage in the form of permanent sag in conductors. Due to other requirements such as limiting line losses and voltage drops, the current limit on overhead line is lower than the current limit based on permissible temperature for preventing permanent damage.

Reactive Power Control and Voltage Regulation

An ac overhead line, due to its inherent shunt capacitance and series inductance, produces and consumes reactive power. The reactive power generated is a function of the line capacitance C and voltage E and is equal to $(\omega C E^2)$ and the inductive power consumed is a function of its series inductance L and current I and is equal to $(\omega L I^2)$. The voltage profile along the length of a long overhead ac line, therefore, is a function of the load carried by it. The voltage profile is flat for surge impedance loading (SIL). The voltage profile varies with the line loading. For constant voltage at the ends of the line, the voltage at the mid-point of the line is decreased for line loadings greater than SIL, while it increases when line loading is less than SIL. In order to maintain a desired profile along an ac line, under varying load conditions, it is essential to use reactive power control techniques such as shunt inductive compensation, when the line carries no-load or is lightly loaded to absorb the reactive power generated due to line capacitance; and shunt capacitive compensation to generate reactive power under heavy-load conditions. The shunt capacitive current in ac cables under steady-state condition limits its length to about 40 km.

In a dc link, no reactive power is either produced or absorbed and the only voltage drop is due to the IR drop in line resistance. As such there is no restriction on distance of transmission by underground or under water cables. However, the terminal equipment (converters) at the two ends of a dc link requires reactive power. This reactive power is drawn from the ac system. The reactive power required is a function of the transmitted power but is independent of the length of the line. Synchronous condensers are normally employed to meet the requirement of reactive power.

Thus, both ac and dc transmission lines require controlled reactive power supplies. It is observed that for dc links of length greater than 400 km, the requirement of reactive power is less than those of ac lines of equivalent lengths.

Circuit Breakers and Short-circuit Current

In ac systems, under faulted conditions, circuit breaking is achieved by rapidly opening the circuit breaker contacts at zero current. Circuit breaker increases the length of the arc path and prevents re-striking of the arc.

Due to an absence of the natural zero in dc circuits, the current is forced to zero in dc circuit breaking. As such, efforts to develop circuit breakers for HVDC have not met with success. However, the absence of dc circuit breakers in modern two-terminal dc links has not been felt since faults on lines and in terminal equipment are cleared by temporarily jamming the direct current with the help of control grids in the converter valves. While research and development efforts to develop HVDC circuit breakers are going on, their absence currently is not seen as a disadvantage in the HVDC transmission development.

When power systems are interconnected through ac lines, the short-circuit fault level is raised, which often gives rise to the need to replace the existing circuit breakers with higher interrupting capacities. On the other hand, when ac systems are interconnected by a dc link, the increase in short-circuit fault level is not as much, since the dc link contributes no more than its own rated current to the ac short-circuit. Similarly, the faults on a dc link do not draw large currents from the ac system. Further, the transient current in a short-circuit on a dc link, due to a transitory discharge of the shunt capacitance of the link, is limited to twice the rated current by the automatic grid control.

Reliability

Two considerations, namely energy availability and transient reliability, determine on the whole the reliability of a system. These two parameters are defined as follows:

$$\text{Energy availability in per cent} = \left(1 - \frac{\text{Equivalent outage time}}{\text{Total time}}\right) \times 100$$

where,

Equivalent outage time

$$= (\text{Actual outage time}) \times \left(\frac{\text{System capacity lost due to outage}}{\text{Total system capacity}}\right)$$

Transient reliability is a parameter which defines the performance of HVDC systems during recordable faults on connected ac systems. A recordable fault in an ac system is defined as that fault which causes the voltage to drop below 90% of its pre-fault voltage, in one or more phases. It is defined as follows:

$$\text{Transient reliability} = \frac{\text{No. of times HVDC systems performed as designed}}{\text{No. of recordable ac faults}} \times 100$$

A two-conductor, bipolar dc link is as reliable as a double circuit ac system since in case of a fault on one conductor of the link, the supply of power is maintained via the combination of the healthy conductor and the ground as return for the period the fault is being repaired. On the other hand, operation of an ac system is not possible with ground as return due to the high impedance of the ground return and interference with telecommunication circuits. Even a monopolar dc link with ground return is equally reliable and simpler than a three-phase ac system. A monopolar link is particularly appropriate for submarine cables.

The converters at the two ends of a dc link have demonstrated their reliability but are expensive.

Terminal Equipment

Since the valves have little or no overload capacity, the converters in the terminal equipment are seen as a bottleneck to the power transmission capability. Other factors such as cut-off voltage of lightning arrestors and continuous current rating of circuit breakers on ac or dc side also impose limitations on current, voltage, and power.

Harmonics

Converters generate current and voltage harmonics in both ac and dc sides due to discrete conduction of the valves. Filters (ac, dc, and high frequency) on both sides are employed to suppress these harmonics since they interfere with the telecommunication and radio frequencies. The use of filters, however, increases the cost of terminal equipment. An added advantage of the use of filters is that they act as a source of reactive power and are able to partially meet the reactive power requirement of the converters.

Control of Tie-line Power

An operational requirement of power systems connected by tie-lines is that the latter must be managed so that the contractual commitments of all the utilities are satisfied. The need of controlling the tie-line power has acquired particular significance in view of the de-regulation of the power sector. One of the fundamental advantages with HVDC is that it is very easy to control the active power in the link.

Generating Stations

In order to improve the stability limit of a hydro generating station supplying a load centre via a long ac line, the generators are so selected as to have very low transient reactance and a high moment of inertia. These two requirements increase the unit cost of a generator. The employment of a dc connection to connect to an ac system would obviate the need for these requirements since there would be no stability issue. The other benefit of using a dc connection would be that it would be possible to vary the speed of the prime mover according to the head of water or the load. This in turn would lead to the use of a more efficient or a cheaper prime mover and it would be no longer necessary to operate the generator at the system frequency of 50 or 60 Hz; rather a generator frequency which yields the best economy could be selected. The harmonic filtering requirement in such a station might also be less stringent. It would be prudent to clarify that till date no such system exists.

15.2.2 Economic Factors

In analysing the available options to select an HVDC or ac transmission system, the focus is on the selection of a most economical method for the same transmission capacity. The cost of a transmission system depends on many factors such as power capacity to be transmitted, type of transmission medium, environmental conditions, and other safety and regulatory requirements. Investment cost of an HVDC link, can be split into two major components:

(a) Cost of terminal station equipment
(b) Cost of transmission line

The main constituents of the cost are equipment that includes valves, filters, transformers, and controls (about 53%). The balance comprises civil works, buildings, engineering, erection, commissioning, freight, insurance, etc. Typical cost structure of a converter station can be as shown in Fig. 15.3. The cost of terminal station equipment is more expensive in HVDC case due to the fact that it must perform the conversion from ac to dc and vice versa.

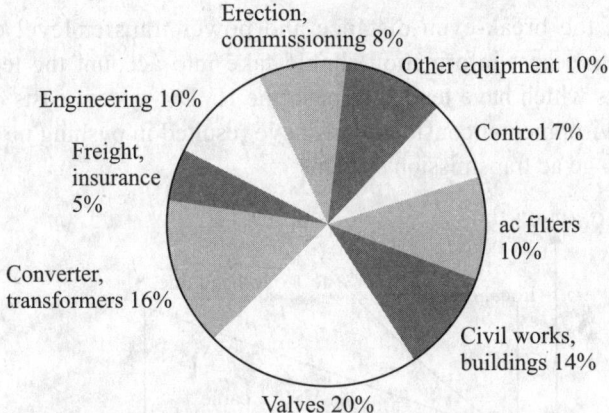

Fig. 15.3 Cost components of terminal equipment

The transmission line costs that include the costs of transmission medium (overhead lines and cables) and land acquisition/right-of-way are lower in the case of HVDC. Moreover, the operation and maintenance costs are lower in HVDC cases. Initial loss levels are higher in the HVDC system, but they do not vary with distance. In contrast, loss levels increase with distance in high-voltage ac systems. Figure 15.4 is a plot of total cost breakdown (shown with and without losses) versus distance of a high-voltage ac and HVDC transmission line for a specified equal power. From Fig. 15.4, observe that the total cost of the overhead ac power transmission is less than the HVDC line up to a certain distance, the so-called break-even distance (typically about 500–800 km). Beyond the break-even distance, the HVDC alternative will always give the lowest cost. The break-even distance is much smaller for submarine cables (typically about 50 km). The break-even distance depends on several local aspects such as permits, licences, and cost of labour.

The break-even distance also depends on several parameters, such as the type of converter technology (natural commuted, CCC based, or VSC based), type of line (overhead or cable), local costs, range of price variation in ac and dc equipment, and cost of environment conservation.

VSC based HVDC systems cater to small power applications up to 200 MW, and

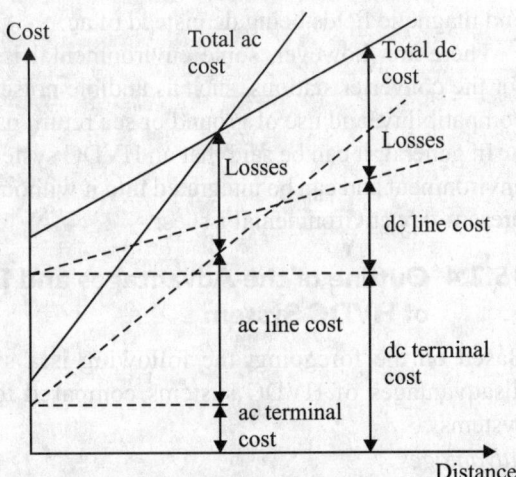

Fig. 15.4 Comparison of total costs of high-voltage ac and HVDC lines versus distance

relatively shorter distances (hundreds of km). Figure 15.5 shows that the VSC based HVDC system is economically a better alternative compared to either a high-voltage ac system or a generation source, e.g., a diesel generator local to the load centre.

However, the break-even distance and power transfer level criteria and the comparative cost information should take into account the technological developments which have tended to push the HVDC system costs downwards, while the environmental considerations have resulted in pushing up the costs of the high-voltage ac transmission systems.

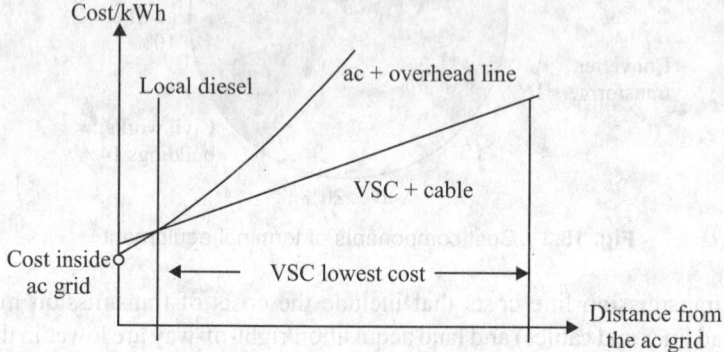

Fig. 15.5 Comparison of total costs of high-voltage ac, VSC based HVDC lines versus distance, and local diesel generator

15.2.3 Environmental Considerations

Improved energy transmission possibilities contribute to a more efficient utilization of existing power plants. The land coverage and the associated right-of-way (ROW) cost for an HVDC overhead transmission line is not as high as high-voltage ac line. Thus, an HVDC transmission line can have a lower visual profile than an equivalent ac line, and hence contributes to a lower environmental impact. There are other environmental advantages to a dc transmission line owing to the electric and magnetic fields being dc instead of ac.

There are, however, some environmental issues which must be considered for the converter stations, such as audible noise, visual impact, electromagnetic compatibility, and use of ground or sea return path in monopolar operation.

In general, it can be said that an HVDC system is highly compatible with any environment and can be integrated into it without the need to compromise on any present-day environmental issues.

15.2.4 Outline of the Advantages and Disadvantages of HVDC System

Based on the foregoing, the following is a synopsis of the advantages and disadvantages of HVDC systems compared to high-voltage ac transmission systems.

Advantages

(a) Power per conductor is more.
(b) Simple construction (fewer number of conductors, smaller transmission towers)

(c) Each conductor can be operated as an independent circuit by using ground as the return path.

(d) Since the power factor is always unity, no reactive power is needed for power factor correction.

(e) For a given rms voltage and conductor size, corona loss and RI is lower, particularly in bad weather.

(f) Stability is of no concern since synchronous operation is not required.

(g) No restriction on distance due to stability problem.

(h) Technically, operation of interconnected ac networks at different frequencies is possible.

(i) Short-circuit fault level is low.

(j) No contribution to the short-circuit fault level of an ac system.

(k) No charging current.

(l) No skin effect.

(m) Easy to control tie-line power.

(n) Bidirectional flow of power.

(o) Need for right-of-way (ROW) is lesser.

(p) High availability and reliability rate.

(q) Compatible with any environment.

Disadvantages

(a) Converter equipment is expensive.

(b) The requirement of reactive power in converters is higher.

(c) Converters act as a source of harmonics; hence, filters are required.

(d) Overload capability of converters is minimal.

(e) Difficulty of breaking current in dc network resulting in high cost of dc breakers.

(f) Complexity of control.

The advantages of HVDC transmission outweigh by far the disadvantages. Therefore, HVDC power transmission is used in the following areas:

(a) When 500 MW or more power is to be transmitted over distances above 500 km

(b) When power is to be transmitted under water bodies wider than 30 km

(c) When asynchronous operation is necessitated or when ac systems operating on different frequencies are to be interconnected

(d) When ROW for ac overhead lines is difficult to obtain, or when use of ac cables, due to lengths required become uneconomical, or when power is to be transmitted in thickly populated areas

15.3 HVDC TECHNOLOGY

The basic process which takes place in an HVDC system is the rectifier action (conversion of ac to dc) at the sending end and inverter action (conversion of dc to ac) at the receiving end. The three ways of achieving this conversion are discussed in the following sub-sections.

15.3.1 Natural Commutated Converters

The device most fundamental to the process of conversion is a *thyristor*, which is a controllable semiconductor capable of blocking high-voltages up to 10 kV and high current rating capacity of 4 kA. The thyristor valve operates at net frequency, that is, 50 or 60 Hz. It is, thus, possible to build up high-voltage blocking capability, of several hundred kV, by connecting thyristors in series. The inherent characteristic that the dc output voltage level of a thyristor bridge can be varied by controlling the firing angle provides a convenient method to efficiently and quickly vary the transmitted power.

15.3.2 Capacitor Commutated Converters (CCC)

In this arrangement, a number of capacitors, called *commutation capacitors*, are connected in series between the transformers and thyristors. This arrangement is used when converters are employed to connect weak networks, since the commutation capacitors improve the failure performance of the bridge.

15.3.3 Forced Commutated Converters

The thyristor valves in these converters are assembled with semiconductors which have the ability to turn-on and turn-off and are known as *voltage source converters* (VSC). A VSC uses a gate turn-off (GTO) thyristor or the *insulated gate bipolar transistor* (IGBT) and commutates with high frequency as against the net frequency. *Pulse width modulation* (PWM) is employed to operate a VSC which provides the facility of theoretically generating any phase and/or amplitude, up to a certain limit, by changing the PWM pattern which can be done instantaneously. A PWM VSC is analogous to a motor or generator without inertia and provides immediate control of active and reactive power independently in a transmission network. In that sense, a PWM VSC can be viewed as almost an ideal component in a transmission network.

15.4 HVDC CONVERTER OPERATION

An HVDC converter converts electric power from high-voltage alternating current (ac) to high-voltage direct current (HVDC), or vice versa. Almost all HVDC converters are inherently bidirectional; they can convert either from ac to dc (*rectification*) or from dc to ac (*inversion*). A complete HVDC transmission system always includes at least one converter operating as a rectifier (converting ac to dc) and at least one operating as an *inverter* (converting dc to ac). *Line-commutated converters* are made with electronic switches that can only be turned on. *Voltage-sourced converters* are made with switching devices that can be turned both on and off. Line-commutated converters (LCC) used mercury-arc valves until the 1970s, or thyristors from the 1970s to the present day. Voltage-source converters (VSC), which first appeared in HVDC in 1997, use transistors, usually the insulated gate bipolar transistor (IGBT).

15.4.1 Three-phase Bridge Rectifier

The basic LCC configuration of HVDC transmission invariably uses a six-thyristor bridge converter circuit. The three-phase thyristor bridge (also called Graetz circuit) is shown in Fig. 15.6. The bridge circuit is connected to a balanced three-phase voltage through a Y–Y connected transformer. The transformer connections may be Y–Δ, Δ–Δ, or Δ–Y without any change in dc output, except for a phase shift of 30° in the output. In Fig. 15.6, each thyristor connects one of the three phases to one of the two dc terminals.

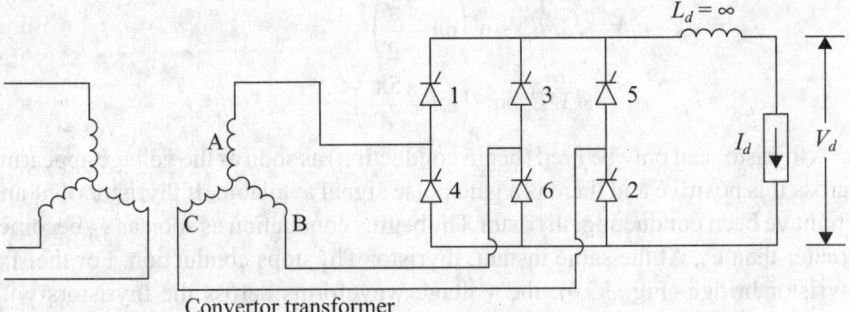

Fig. 15.6   Three-phase thyristor bridge converter (Graetz circuit)

Without Overlap

The leakage inductance of the transformer is the cause of overlap. The assumption that there is no overlap between the two thyristors in a commutation group is not correct. However, the analysis without overlap is simpler and gives an insight into the working of the converter. To simplify the analysis, it is assumed that the leakage inductance of the transformer is negligible. Further, it is assumed that the thyristors have no resistance when on (conducting) and infinite resistance when off (not conducting); and the circuit on the dc side has a high value of inductance and the dc current is constant and ripple free.

At any instant, two thyristors in the bridge, one thyristor on the upper group (Th$_1$, Th$_3$, Th$_5$) and one thyristor (from a different phase) on the lower group (Th$_2$, Th$_4$, Th$_6$), are conducting. The firing of the next thyristor in a particular group results in conduction of the thyristor that is fired and turning off of the thyristor that was already conducting in that group. The transfer of current occurs instantly as there is no overlap. The transfer of current from one thyristor to another in the same group is called commutation. The thyristors are numbered in the sequence in which they are fired. Thus, thyristor Th$_2$ is fired 60° after the firing of thyristor Th$_1$, and thyristor Th$_3$ is fired 60° after the firing of thyristor Th$_2$, and so on. Each thyristor conducts for 120°, and the interval between consecutive firing pulses is 60°. Each cycle of ac supply voltage is divided into six intervals, when a pair of valves conducts for 60°. For example, thyristors Th$_6$ and Th$_1$ conduct for 60°, thereafter thyristors Th$_1$ and Th$_2$ conduct for the next 60°, and so on.

Let the line to neutral instantaneous voltages of the supply be

$$e_A = E_m \sin \omega t$$

$$e_B = E_m \sin \left(\omega t - \frac{2\pi}{3} \right)$$

$$e_C = E_m \sin \left(\omega t + \frac{2\pi}{3} \right)$$

where E_m is the peak value of line to neutral voltage. Then, the corresponding line to line voltages are

$$e_{AB} = e_A - e_B = \sqrt{3}\ E_m \sin\left(\omega t + \frac{\pi}{6}\right)$$

$$e_{BC} = e_B - e_C = \sqrt{3}\ E_m \sin\left(\omega t - \frac{\pi}{2}\right)$$

$$e_{CA} = e_C - e_A = \sqrt{3}\ E_m \sin\left(\omega t + \frac{5\pi}{6}\right)$$

A thyristor can only be fired (begin conduction) as soon as the voltage appearing across it is positive and there is a gate pulse signal available. If thyristors Th$_1$ and Th$_2$ have been conducting, thyristor Th$_3$ begins conduction as soon as e_B becomes greater than e_A. At the same instant, thyristor Th$_1$ stops conduction. For the six-thyristor bridge (Fig. 15.6), the voltage waveforms across the thyristors with zero firing delay are shown in Fig. 15.7(a) with instant of firing of the thyristors indicated. At any given instant, one thyristor each from upper and lower groups is conducting. The sequence of conduction of the thyristors is indicated on the waveform.

The two conducting thyristors connect two of the three ac phase voltages in series to the dc terminals. Thus, the dc output voltage at any instant is given by the series combination of two ac phase voltages at that instant. For example, if thyristors Th$_1$ and Th$_2$ are conducting, the dc output voltage is given by the voltage of phase A minus the voltage of phase C, that is, line voltage e_{AC}. The instantaneous direct voltage v_d across the bridge on the thyristor side of the reactor is composed of 60° arcs of the alternating line to line voltages, as shown in Fig. 15.7(b).

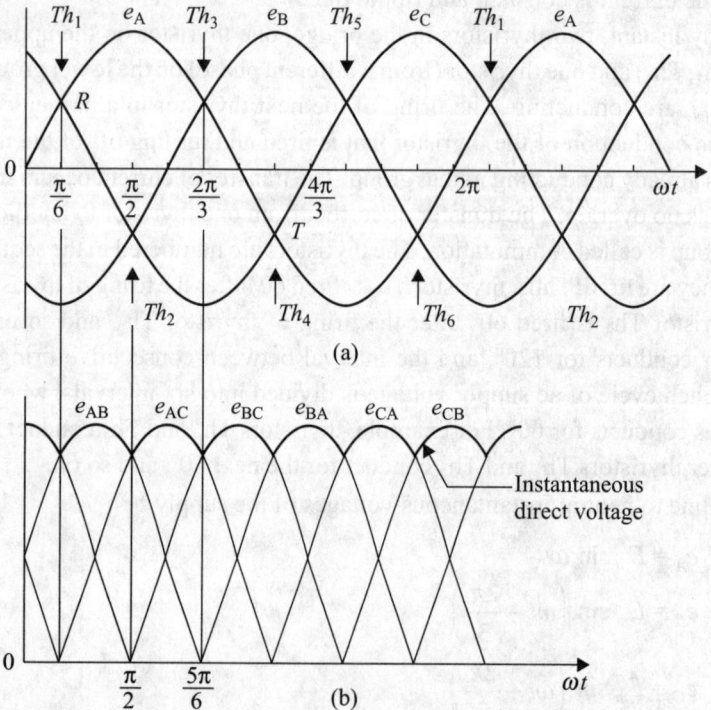

Fig. 15.7 (a) Voltage waveforms across thyristors with firing delay angle $\alpha = 0°$
(b) Instantaneous direct voltage waveform of the bridge converter with firing delay angle α and no overlap

For firing delay $\alpha = 0°$, the average dc output voltage V_{d0} is found by integrating

$$V_{d0} = \frac{\sqrt{3}\,E_m}{\pi/3} \int_{\pi/2}^{5\pi/6} \sin\left(\omega t - \frac{\pi}{6}\right) d\omega t = \frac{3\sqrt{3}\,E_m}{\pi}\left[-\cos\left(\omega t - \frac{\pi}{6}\right)\right]_{\pi/2}^{5\pi/6}$$

$$= \frac{3\sqrt{3}\,E_m}{\pi}\left[-\cos\left(\frac{5\pi}{6} - \frac{\pi}{6}\right) + \cos\left(\frac{\pi}{2} - \frac{\pi}{6}\right)\right]$$

$$= \frac{3\sqrt{3}\,E_m}{\pi}\left[-\cos\frac{2\pi}{3} + \cos\frac{\pi}{3}\right]$$

or,
$$V_{d0} = \frac{3\sqrt{3}\,E_m}{\pi} = 1.65\,E_m \qquad (15.4)$$

In terms of the rms line to neutral voltage, E_{LN}, Eq. (15.4) gets modified as under

$$V_{d0} = \frac{3\sqrt{6}}{\pi} E_{LN} = 2.34\,E_{LN} = \frac{3\sqrt{2}}{\pi}\sqrt{3}\,E_{LL} = 1.35\,E_{LL} \qquad (15.5)$$

where E_{LL} is the line to line voltage.

By delaying the firing pulse suitably by an angle α, dc voltage output can be controlled. The firing delay angle α is measured amongst two adjacent crossing points between the phase supply voltages indicated by point R in Fig. 15.7(a). The maximum delay cannot exceed 180°. For example, point T after point R in Fig. 15.7(a) marks the end of the range over which thyristor Th$_1$ can ignite. However, in practical applications due to commutation (two thyristors conducting at the same time during transfer of conduction), the extent of delay in firing angle is limited to 160°. Fig. 15.8 shows the waveforms of direct voltage on the thyristor side of the reactor when firing angle delay is α and with no commutation.

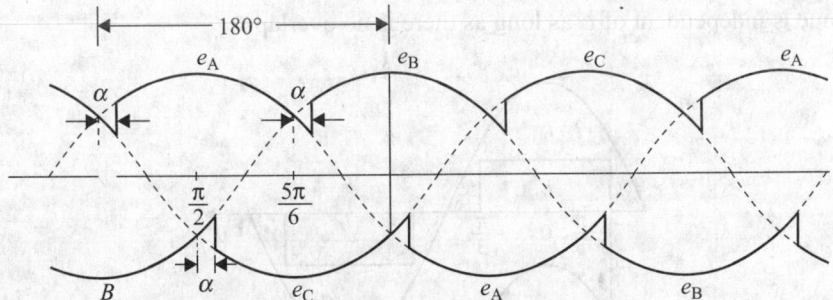

Fig. 15.8 Instantaneous direct voltage waveform of the thyristor bridge with firing delay angle α and no overlap

It can be seen from Fig. 15.8 that with a firing angle delay of α, the duration of conduction of thyristor Th$_1$ is from ($\pi/2 + \alpha$) to ($5\pi/6 + \alpha$). Therefore, the average dc voltage output V_d is given by

$$V_d = V_{d0} \int_{\pi/2+\alpha}^{5\pi/6+\alpha} \sin\left(\omega t - \frac{\pi}{6}\right) d\omega t = V_{d0}\left[-\cos\left(\omega t - \frac{\pi}{6}\right)\right]_{\pi/2+\alpha}^{5\pi/6+\alpha}$$

$$= V_{d0}\left[-\cos\left(\frac{5\pi}{6} + \alpha - \frac{\pi}{6}\right) + \cos\left(\frac{\pi}{2} + \alpha - \frac{\pi}{6}\right)\right]$$

$$= V_{d0} \left[-\cos\left(\frac{2\pi}{3} + \alpha\right) + \cos\left(\frac{\pi}{3} + \alpha\right) \right]$$

$$= \frac{3\sqrt{3}\,E_m}{\pi} \left[2\sin\left(\frac{\pi}{2} + \alpha\right) \sin\left(\frac{\pi}{6}\right) \right]$$

$$= \frac{3\sqrt{3}\,E_m}{\pi} \cos\alpha = \frac{3\sqrt{6}\,E_{LN}}{\pi} \cos\alpha = \frac{3\sqrt{2}\,E_{LL}}{\pi} \cos\alpha \qquad (15.6)$$

or $\qquad V_d = V_{d0} \cos\alpha \qquad\qquad\qquad\qquad\qquad\qquad\qquad (15.7)$

where

$$V_{d0} = \frac{3\sqrt{3}}{\pi} E_m = \frac{3\sqrt{2}\,E_{LL}}{\pi} \qquad\qquad\qquad (15.7a)$$

From Eq. (15.7) it is seen that one effect of delayed firing is to reduce the average dc output voltage by the factor $\cos\alpha$. As α varies in the range $0°$ to $180°$, $\cos\alpha$ can change from 1 to -1, and V_d can change from V_{d0} equal to 0 to $-V_{d0}$. Since the current I_d cannot reverse because of the unidirectional property of the thyristors, negative voltage V_0 in conjunction with positive current I_d represents reversed power flow, that is, conversion of dc power to ac power representing inversion. The thyristor bridge functions as a rectifier for $\alpha \le 90°$ and as an inverter for $\alpha > 90°$.

With losses in the converter neglected, the ac power must equal dc power. Thus,

$$3V I_{L1} \cos\varphi = V_d I_d = I_d V_{d0} \cos\alpha \qquad\qquad\qquad (15.8)$$

where I_{L1} is the rms value of the fundamental frequency component of alternating line current. The line current has the wave shape shown in Fig. 15.9 consisting of positive and negative rectangular pulses of height I_d and width $2\pi/3$ rad. This shape is independent of α as long as there is no overlap.

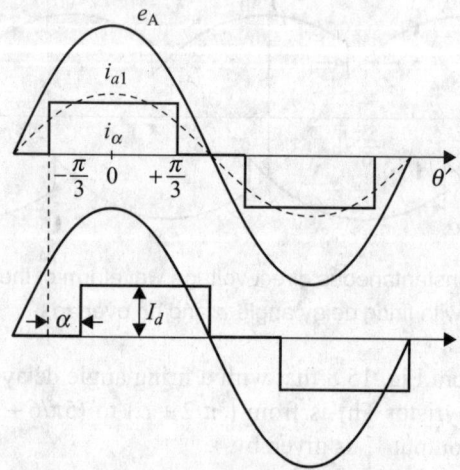

Fig. 15.9 Alternating current waveform in transformer secondary

The Fourier analysis of the peak value of the fundamental component of this wave is

$$\sqrt{2}I_{L1} = \frac{2}{\pi}\int_{-\frac{\pi}{3}}^{+\frac{\pi}{3}} I_d\cos\theta'd\theta' = \frac{2}{\pi}I_d(\sin\theta')\Big|_{-\frac{\pi}{3}}^{+\frac{\pi}{3}}$$

$$= \frac{2}{\pi}I_d\left[\sin\left(\frac{\pi}{3}\right) - \sin\left(-\frac{\pi}{3}\right)\right]$$

$$= \frac{2}{\pi}I_d\left(2\sin\frac{\pi}{3}\right) = \frac{2\sqrt{3}}{\pi}I_d = 1.11\,I_d \qquad (15.9)$$

The rms value of the fundamental component of current is

$$I_{L1} = \frac{\sqrt{6}}{\pi}I_d = 0.78\,I_d \qquad (15.10)$$

Substitution of V_{d0} from Eq. (15.5) and I_{L1} from Eq. (15.9) in Eq. (15.7) yields

$$\cos\varphi = \cos\alpha \qquad (15.11)$$

With Overlap

Because of the transformer leakage inductance L_C of the supply system, the currents in it can vary at a finite rate, and therefore, the transfer of current from one phase to another requires a finite time called the commutation time or overlap time, u/ω, where u is the overlap angle. The value of u can be different, and dependent on its value there are three modes of operation as follows:

1. Mode 1: Two-and three-thyristor conduction for $u < 60°$.
2. Mode 2: Three-thyristor conduction for $u = 60°$.
3. Mode 3: Three-and four-thyristor conduction for $u > 60°$.

The overlap angle, u, in an HVDC converter increases with the load current, but is typically around 20° at full load. Hence, the analysis of mode 1 will only be discussed here.

Figure 15.10 shows the bridge converter with thyristors Th_1, Th_2, and Th_3 conducting. Prior to this, thyristors Th_1 and Th_2 were conducting and then thyristor Th_3 is triggered, and conduction passes from thyristor Th_1 to thyristor Th_3. During commutation, all the three thyristors of the bridge conduct simultaneously. Since each commutation begins every 60° and lasts for angle u, the angular interval when two thyristors conduct is $(60° - u)$. Figure 15.11 shows the commutation from thyristor Th_1 to thyristor Th_3. During the commutation period, the dc output voltage is given by the average of the voltages of phases A and B, minus the voltage of phase C.

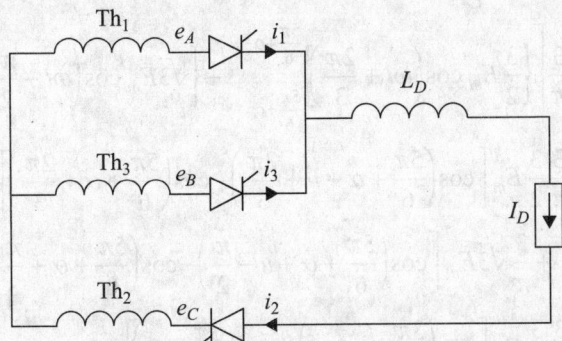

Fig. 15.10 Bridge converter circuit with thyristors Th_1, Th_2, and Th_3 conducting

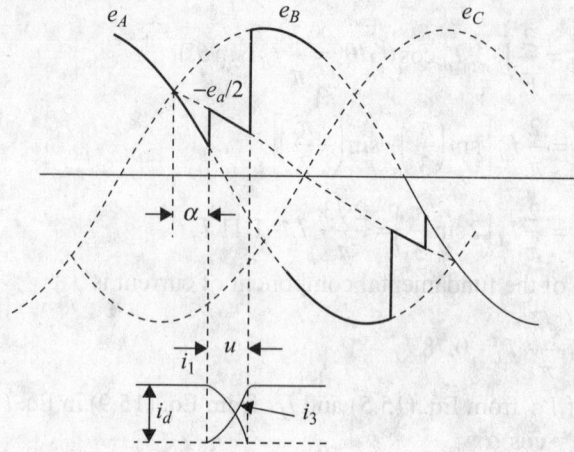

Fig. 15.11 Process of commutation between two thyristors

The instantaneous output voltage at the dc terminal of the bridge is

$$V_d = \frac{3}{\pi}\left[\int_{\alpha}^{\alpha+u}\left(\frac{e_A+e_B}{2}-e_C\right)d\omega t + \int_{\alpha+u}^{\alpha+\pi/3}(e_B-e_C)d\omega t\right] \tag{15.12}$$

Since $\dfrac{e_A+e_B}{2}=-\dfrac{e_C}{2}$

Equation (15.11) can be written as

$$V_d = \frac{3}{\pi}\left[\int_{\alpha}^{\alpha+u}\left(-\frac{3e_C}{2}\right)d\omega t + \int_{\alpha+u}^{\alpha+\pi/3}(e_B-e_C)d\omega t\right]$$

$$= \frac{3}{\pi}\int_{\frac{5\pi}{6}+\alpha}^{\frac{5\pi}{6}+\alpha+u}\left(-\frac{3e_C}{2}\right)d\omega t + \int_{\frac{5\pi}{6}+\alpha+u}^{\frac{5\pi}{6}+\alpha+\frac{\pi}{3}}e_{5C}\,d\omega t$$

$$= \frac{3}{\pi}\left[\int_{\frac{5\pi}{6}+\alpha}^{\frac{5\pi}{6}+\alpha+u}\left(-\frac{3}{2}E_m\sin\left(\omega t+\frac{2\pi}{3}\right)\right)d\omega t \right.$$
$$\left. + \int_{\frac{5\pi}{6}+\alpha+u}^{\frac{5\pi}{6}+\alpha+\frac{\pi}{3}}\sqrt{3}E_m\sin\left(\omega t-\frac{\pi}{2}\right)d\omega t\right]$$

$$= \frac{3}{\pi}\left[\left\{\frac{3}{2}E_m\cos\left(\omega t+\frac{2\pi}{3}\right)\right\}_{\frac{5\pi}{6}+\alpha}^{\frac{5\pi}{6}+\alpha+u} + \left\{\sqrt{3}E_m\cos\left(\omega t-\frac{\pi}{2}\right)\right\}_{\frac{5\pi}{6}+\alpha+\frac{\pi}{3}}^{\frac{5\pi}{6}+\alpha+u}\right]$$

$$= \frac{3}{\pi}\frac{3}{2}E_m\left[\cos\left(\frac{5\pi}{6}+\alpha+u+\frac{2\pi}{3}\right)-\cos\left(\frac{5\pi}{6}+\alpha+\frac{2\pi}{3}\right)\right]$$

$$+\frac{3}{\pi}\sqrt{3}E_m\left[\cos\left(\frac{5\pi}{6}+\alpha+u-\frac{\pi}{2}\right)-\cos\left(\frac{5\pi}{6}+\alpha+\frac{\pi}{3}-\frac{\pi}{2}\right)\right]$$

$$= \frac{3}{\pi}\frac{3}{2}E_m\left[\cos\left(\frac{3\pi}{2}+\alpha+u\right)-\cos\left(\frac{3\pi}{3}+\alpha\right)\right]$$

$$+ \frac{3}{\pi} \sqrt{3} E_m \left[\cos\left(\frac{\pi}{3} + \alpha + u\right) - \cos\left(\frac{2\pi}{3} + \alpha\right) \right]$$

$$= \frac{3}{\pi} \frac{3}{2} E_m \left[\cos\left(2\pi - \frac{\pi}{2} + \alpha + u\right) - \cos\left(2\pi - \frac{\pi}{2} + \alpha\right) \right]$$

$$+ \frac{3\sqrt{3}}{\pi} E_m \left[\begin{array}{c} \cos\dfrac{\pi}{3}\cos(\alpha+u) - \dfrac{\sin\pi}{3}\sin(\alpha+u) \\[2mm] - \cos\dfrac{2\pi}{3}\cos\alpha - \sin\dfrac{2\pi}{3}\sin\alpha \end{array} \right]$$

$$= \frac{3}{\pi} \frac{3}{2} E_m \left[\cos\left(\frac{\pi}{2} - \alpha - u\right) - \cos\left(\frac{\pi}{2} - \alpha\right) \right]$$

$$+ \frac{3\sqrt{3}}{\pi} E_m \left[\frac{1}{2}\cos(\alpha+u) - \frac{\sqrt{3}}{2}\sin(\alpha+u) + \frac{1}{2}\cos\alpha + \frac{\sqrt{3}}{2}\sin\alpha \right]$$

$$= \frac{9}{2\pi} E_m \left[\sin(\alpha - u) - \sin(\alpha) \right]$$

$$+ \frac{3\sqrt{3}}{\pi} E_m \left[\frac{1}{2}\cos(\alpha+u) - \frac{\sqrt{3}}{2}\sin(\alpha+u) + \frac{1}{2}\cos\alpha - \frac{\sqrt{3}}{2}\sin\alpha \right]$$

or
$$V_d = \frac{3\sqrt{3}}{2\pi} E_m \left[\cos(\alpha+u) + \cos\alpha \right] \tag{15.13}$$

and in terms of line to line voltage, Eq. (15.13) may be written as

$$V_d = \frac{3\sqrt{2}}{\pi} E_{LL} \left[\cos(\alpha+u) + \cos\alpha \right] \tag{15.13a}$$

Using Eqs (15.4) and (15.5) the dc voltage, V_d, may be expressed as

$$V_d = \frac{V_{d0}}{2} [\cos\alpha + \cos\delta) \tag{15.14}$$

where $\quad \delta = \alpha + u$

From Fig. 15.10, applying KVL to the closed path

$$e_A - L_C \frac{di_1}{dt} = e_B - L_C \frac{di_3}{dt}$$

or
$$L_C \frac{di_3}{dt} - L_C \frac{di_1}{dt} = e_B - e_A \tag{15.15}$$

During the commutation period, the direct current is transferred from thyristor Th_1 to thyristor Th_3. Hence, at the beginning of the period $i_1 = I_d$ and $i_3 = 0$, while at the end of the period $i_1 = 0$ and $i_3 = I_d$. Thus, at any instant during this period, $i_1 = I_d - i_3$. Integrating this yields

$$\frac{di_1}{dt} = -\frac{di_3}{dt}$$

Then, Eq. (15.15) can be written as

$$2L_C \frac{di_3}{dt} = e_B - e_A = e_{BA} = \sqrt{3}E_m \sin\left(\omega t + \pi + \frac{\pi}{6}\right) = -\sqrt{3}E_m \sin\left(\omega t + \frac{\pi}{6}\right)$$

(15.16)

or $\qquad \dfrac{di_3}{dt} = -\dfrac{\sqrt{3}}{2L_C}E_m \sin\left(\omega t + \dfrac{\pi}{6}\right)$ (15.17)

Integrating Eq. (15.17) yields

$$i_3 = \frac{\sqrt{3}}{2\omega L_C}E_m \cos\left(\omega t + \frac{\pi}{6}\right) + K$$

(15.18)

where K is an integration constant. The value of integration constant can be obtained by using initial condition at $\omega t = \dfrac{5\pi}{6} + \alpha$, $i_3 = 0$, then

$$K = \frac{\sqrt{3}}{2\omega L_C}E_m \cos \alpha$$

Thus,

$$i_3 = \frac{\sqrt{3}}{2\omega L_C}E_m \cos\left(\omega t + \frac{\pi}{6}\right) + \frac{\sqrt{3}}{2\omega L_C}E_m \cos \alpha$$

or $\qquad i_3 = I_{SC}\left[\cos\alpha + \cos\left(\omega t + \dfrac{\pi}{6}\right)\right]$ (15.19)

where $\qquad I_{SC} = \dfrac{\sqrt{3}}{2\omega L_C}E_m$ (15.20)

At $\qquad \omega t = \dfrac{5\pi}{6} + \alpha + u$, $i_3 = I_d$, then

$$I_d = I_{SC}\left[\cos\alpha + \cos\left(\frac{5\pi}{6} + \alpha + u + \frac{\pi}{6}\right)\right]$$

or, $\qquad I_d = I_{SC}\left[\cos\alpha - \cos\left(\alpha + u\right)\right] = I_{SC}\left[\cos\alpha - \cos\delta\right]$ (15.21)

Elimination of $\cos\left(\alpha + u\right)$ from Eqs (15.20) and (15.21) gives

$$V_d = V_{d0}\left[\cos\alpha - \frac{I_d}{2i_{SC}}\right] = V_{d0}\cos\alpha - R_C I_d$$

(15.22)

where R_C is called equivalent commutating reactance and is given by

$$R_C = \frac{3}{\pi}\omega L_C = \frac{3}{\pi}X_C = 6fLC$$

(15.23)

Equation (15.22) can be represented by an equivalent circuit constituting of a variable dc source with resistance R_C as shown in Fig. 15.12.

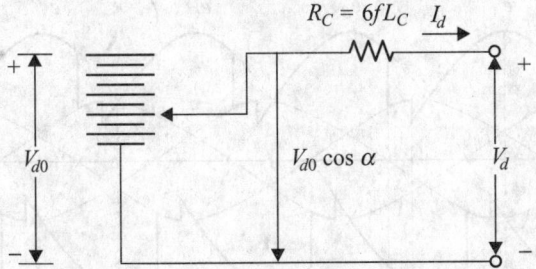

Fig. 15.12 Equivalent circuit of a rectifier

Relationship between ac and dc quantities These relationships hold between the ac quantities at the point where the voltage waves are sinusoidal and the dc quantities. The relationship between the alternating and direct voltage valid for $u < 60°$ can be obtained using Eqs (15.4) and (15.13) and is given below.

$$V_d = \frac{3\sqrt{6}}{\pi}\left[\frac{\cos\alpha + \cos(\alpha+u)}{2}\right]E_{LN} \tag{15.24}$$

Neglecting losses, active ac power P_{ac} equals dc power P_{dc}. Then

$$3\,E_{LN}I_{L1}\cos\varphi = V_d I_d \tag{15.25}$$

Substituting for V_d from Eq. (15.24) and solving for active alternating current gives

$$I_{L1}\cos\varphi = \frac{\sqrt{6}}{\pi}I_d\left[\frac{\cos\alpha+\cos(\alpha+u)}{2}\right] \tag{15.26}$$

Let it be assumed that

$$I_{L1} \cong \frac{\sqrt{6}}{\pi}I_d = I_{L10} \tag{15.27}$$

Equation (15.27) is true with a maximum error of 4.3% at $u = 60°$ and only 1.1% error for normal operating range when $u \leq 30°$. Thus,

$$\cos\varphi = \left[\frac{\cos\alpha + \cos(\alpha+u)}{2}\right] = \left[\frac{\cos\alpha + \cos\delta}{2}\right] \tag{15.28}$$

By the use of Eqs (15.21) and (15.24) Eq. (15.28) may be written as

$$\cos\varphi \cong \frac{V_d}{V_{d0}} = \cos\alpha - \frac{R_C I_d}{V_{d0}} \tag{15.29}$$

and substitution of Eq. (15.28) into Eq. (15.25) yields

$$V_d \cong \frac{3\sqrt{6}}{\pi}E_{LN}\cos\varphi \tag{15.30}$$

Equations (15.25), (15.27), and (15.30) are approximate relations, not involving α and u, for power, current, and voltage respectively.

Current and voltage waveforms of a six-pulse bridge rectifier are shown in Fig. 15.13. Direct positive and negative terminal voltages across the rectifier output are shown in Fig. 15.13(a); Fig. 15.13(b) shows direct voltage between output terminals; Fig. 15.13(c) shows the currents in the thyristors of upper group and lower group; and Fig. 15.13(d) shows the phase A line currents.

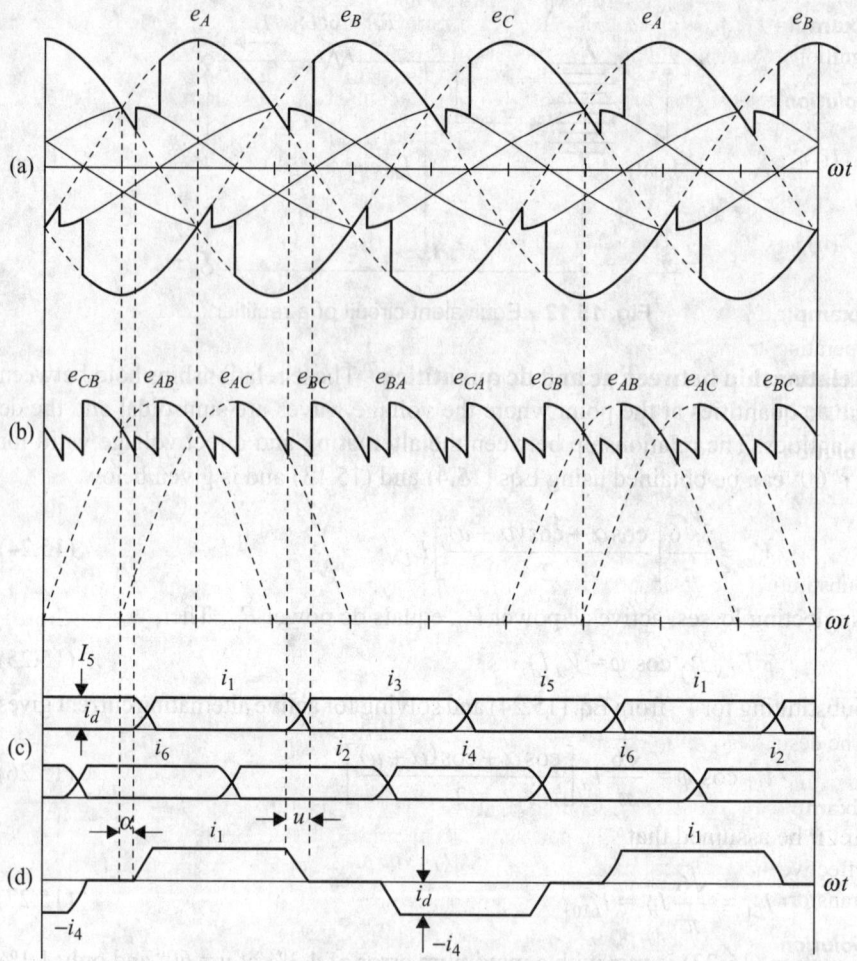

Fig. 15.13   Current and voltage waveforms of a six-pulse bridge rectifier

Example 15.2   A Graetz rectifier is connected to the 110-kV secondary winding of a 400/110 kV, three-phase transformer. If the commutation angle is 12°, determine the dc output voltage when the ignition angle is (i) 0°, (ii) 25°, (iii) 90°, and (iv) 120°.

Solution   From the given data, E_{LL} = 110 kV and u = 12°.

From Eq. (15.7a), $V_0 = (3\sqrt{2} \times 110)/\pi = 148.552$ kV

Equation (15.13) is used to compute the dc output voltage for different ignition angles.

(i) $\alpha = 0°, V_d = \dfrac{148.552}{2}[\cos(0°) + \cos(12°)] = 146.929$ kV

(ii) $\alpha = 25°, V_d = \dfrac{148.552}{2}[\cos(25°) + \cos(37°)] = 126.637$ kV

(iii) $\alpha = 90°, V_d = \dfrac{148.552}{2}[\cos(90°) + \cos(102°)] = -15.443$ kV

(iv) $\alpha = 120°, V_d = \dfrac{148.552}{2}[\cos(120°) + \cos(132°)] = -86.839$ kV

Example 15.3 Repeat Example 15.2 for commutation angle (i) 15° and (ii) 20°. The ignition angle is maintained constant at 10°.

Solution Again Eq. (15.13) is used to compute the dc output voltage with $\alpha = 10°$.

(i) $u = 15°$, $V_d = \dfrac{148.552}{2}[\cos(10°) + \cos(25°)] = 140.465\,\text{kV}$

(ii) $u = 20°$, $V_d = \dfrac{148.552}{2}[\cos(10°) + \cos(30°)] = 137.473\,\text{kV}$

Example 15.4 A six-thyristors bridge rectifier gives an output dc voltage of 90 kV when operating at $\alpha = 25°$ and $u = 20°$. Determine the effective value of the transformer line to line secondary voltage. If the transformer is rated at 220/110 kV, what is the required tap setting on the secondary side to obtain the desired dc output?

Solution Equation (15.14) is rewritten with the help of Eq. (15.7) as follows

$$E_{LN} = \frac{2\pi\,V_d}{3\sqrt{2}\,\,[\cos(\alpha) + \cos(\alpha + u)]}$$

Substituting the given data, we get

$$E_{LN} = \frac{2\pi \times 90}{3\sqrt{2}\,\left[\cos\left(25^0\right) + \cos\left(45^0\right)\right]} = 82.611\,\text{kV}$$

The desired tap setting of the transformer $= \dfrac{82.611}{110} = 0.75$

Example 15.5 A six-pulse bridge rectifier gives a dc output of 100 kV with a firing angle of 25°. What is the commutation angle? Determine (i) open circuit voltage and (ii) the effective inductance of the rectifier if it delivers 1200 A. Assume the line to line secondary transformer voltage of 110 kV and a system frequency of 50 Hz.

Solution Once again Eq. (15.14) is employed to compute u as follows

$$\cos(25° + u) = \frac{2\pi \times 100}{3\sqrt{2} \times 110} - \cos 25° = 0.44$$

or $\quad u = \cos^{-1}(0.44) - 25° = 38.89°$

(i) Open circuit voltage $= V_{d0}\cos\alpha = \dfrac{3\sqrt{2} \times 110}{\pi}\cos(25°) = 134.634\,\text{kV}$

(ii) From Eq. (15.22) it is seen that

$$L_C = \frac{\pi[V_{d0}\cos(\alpha) - V_d]}{3\omega \times I_d} = \frac{\pi\left[134.634\cos\left(25^0\right) - 100\right]}{3 \times 314 \times 1.2} = 61.2\,\text{mH}$$

15.4.2 Three-phase Bridge Inverter

A converter operates as a rectifier when $\alpha < 90°$ [see Eq. (15.7)], the dc output voltage is positive, and power flow is from the three-phase ac side to the dc side. When α exceeds 90°, polarity of the dc output voltage reverses, power flows from dc to ac side, and the converter operates as an inverter.

In the rectification process, the period from the crossover point is defined as a delay angle u and the extinction angle $\delta = \alpha + u$. Although these angles defined in the same way and having values between 90° and 180° could be used in the inverter theory, yet it is a common practice to define ignition angle β and extinction angle γ as the periods of advance as follows:

$$\beta = \pi - \alpha \tag{15.31}$$

$$\gamma = \pi - \delta \tag{15.32}$$

$$u = \delta - \alpha = \beta - \gamma \tag{15.33}$$

Inverter voltage considered negative in the general converter equations is usually taken as positive when written specifically for an inverter. General Eqs (15.14) and (15.21) for voltage and current respectively are changed to inverter equations by changing the sign of V_d and putting $\cos \alpha = -\cos \beta$ and $\cos \delta = -\cos \gamma$. These substitutions yield

$$V_d = -\frac{1}{2} V_{d0} [\cos \alpha + \cos \delta] = -\frac{1}{2} V_{d0} [-\cos \beta - \cos \gamma]$$

or $\qquad V_d = \frac{1}{2} V_{d0} [\cos \gamma + \cos \beta] \tag{15.34}$

and $\qquad I_d = I_{SC} [\cos \alpha - \cos \delta] = I_{SC} [-\cos \beta - (-\cos (\gamma)]$

or $\qquad I_d = I_{SC} [\cos \gamma - \cos (\beta)] \tag{15.35}$

Similarly Eq. (15.22) can be written as

$$V_d = -V_{d0} \cos \alpha + R_C I_d = -V_{d0} \cos (\pi - \beta) + R_C I_d$$

$$V_d = V_{d0} \cos \beta + R_C I_d \tag{15.36}$$

Inverters are commonly controlled so as to operate at constant extinction angle γ and it is useful to have relation between V_d and I_d for this condition. Elimination of $\cos \beta$ from Eqs (15.36) and (15.35) gives

$$V_d = V_{d0} \cos \gamma - R_C I_d \tag{15.37}$$

Derivation of Eq. (15.37) is left as a tutorial exercise for the reader. Equivalent circuits represented by Eqs (15.36) and (15.37) of an inverter are shown in Fig. 15.14.

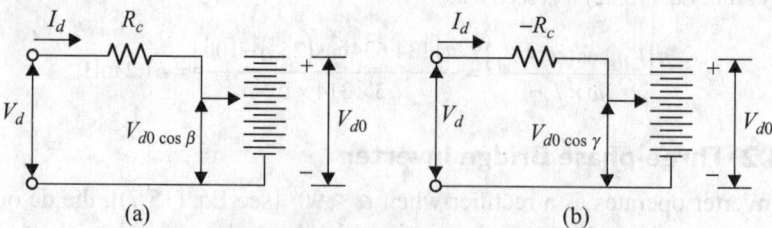

Fig. 15.14 Equivalent circuits of an inverter

Current and voltage waveforms of a six-pulse bridge inverter are shown in Fig. 15.15.

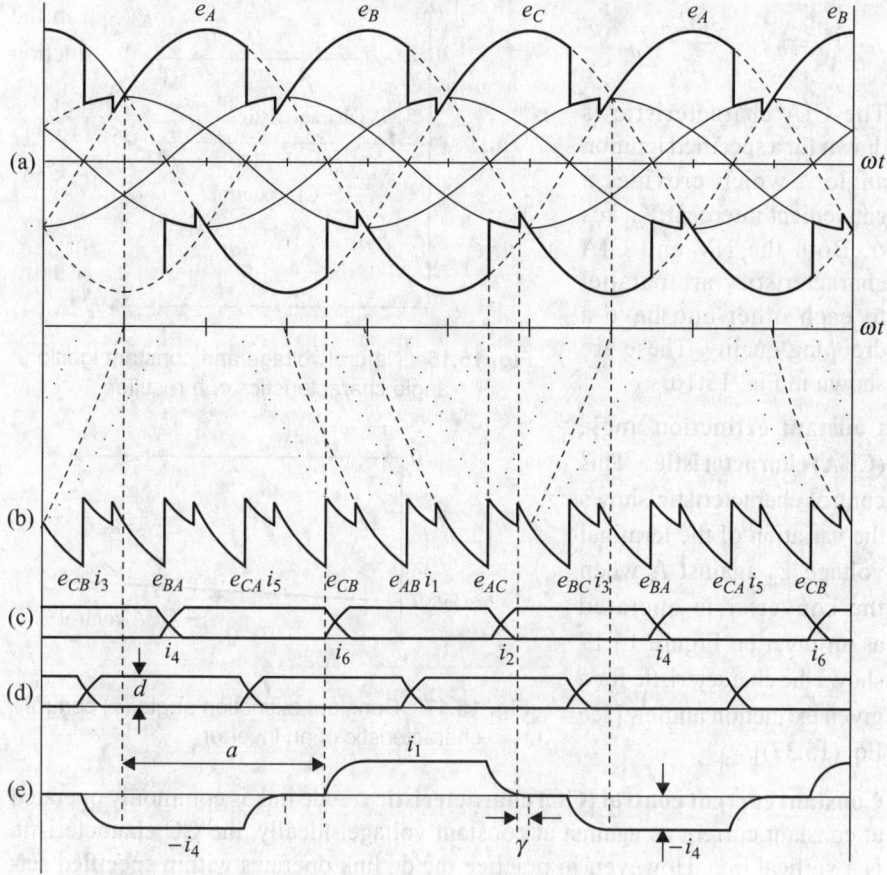

Fig. 15.15 Voltage and current waveforms of a six-pulse bridge inverter

Example 15.6 A six-thyristors bridge converter is functioning as an inverter from a 600 kV dc supply. Calculate the RMS value of the ac line voltage when the ignition angle is 22° and extinction angle is 10°.

Solution As per the data provided $\beta = 22°$ and $\gamma = 10°$. Use of Eq. (15.34) results in

$$V_{d0} = \frac{2 \times 600}{(\cos 22° + \cos 10°)} = 627.6178 \text{ kV}$$

With the help of Eq. (15.5) the RMS value of the line voltage is calculated as follows

$$E_{LN} = \frac{627.6178 \times \pi}{3\sqrt{2}} = 464.7387 \text{ kV}$$

15.4.3 Control Characteristics of Converters

Control characteristics of converters depict the variation of direct voltage V_d against direct current I_d for different operating conditions.

Natural voltage (NV) and constant ignition angle (CIA) characteristics NV characteristics are a plot of V_d versus I_d corresponding to delay angle $\alpha = 0$ and are obtained from Eq. (15.22) as

$$V_d = V_{d0} - \frac{3\omega L_C}{\pi} I_d$$

The CIA characteristic is drawn for a specified ignition angle α which provides a convenient intercept $V_{d0} \cos \alpha$. Both the NV and CIA characteristics are parallel to each other and have a drooping quality. These are shown in Fig. 15.16.

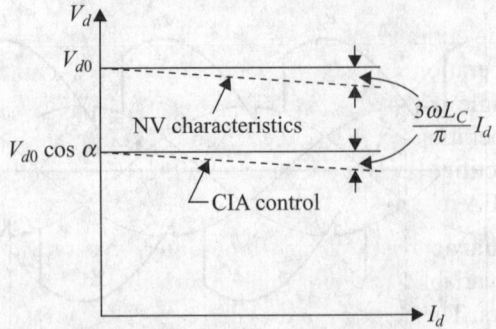

Fig. 15.16 Natural voltage and constant ignition angle characteristics of a rectifier

Constant extinction angle (CEA) characteristic This control characteristic shows the variation of the terminal voltage V_d against I_d when the converter is operated as an inverter. Figure 15.17 shows the characteristic for a given extinction angle γ [see Eq. (15.37)].

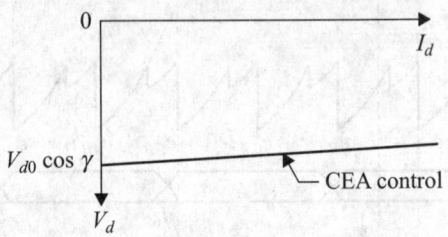

Fig. 15.17 Constant extinction angle (γ) control characteristic of an inverter

Constant current control (CC) characteristic A dc link is commonly operated at constant current as against at constant voltage. Ideally, the CC characteristic is a vertical line. However, in practice the dc link operates within specified settings I_{ds}, which is allowed to vary within a margin setting I_{dm}. As such, the CC characteristic has a small negative slope.

All-inclusive characteristics of a converter A converter is necessarily equipped with constant current (CC) and constant extinction angle (CEA) controls. The characteristics of a single converter are shown in Fig.15.18.

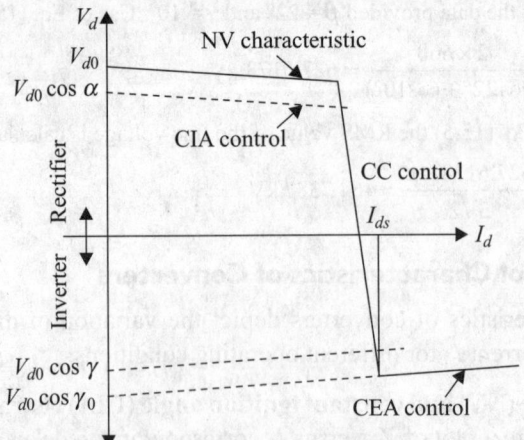

Fig. 15.18 Full characteristics of a single converter

The current I_d in the dc link is kept at a specified magnitude, both during normal operation and short-circuits, by the CC controller which adjusts the ignition angle α. The inverter is also provided with a CC control but is not called into operation since an inverter does not operate in the CC control region. In fact, a rectifier is usually operated in the CC region while an inverter functions in the CEA region.

Characteristics of compounded converters   The two sides of a HVDC link are invariably provided with two converters, each equipped with CEA and CC controllers. The controls in both the converters are kept similar to provide flexibility in either working as a rectifier or an inverter. Figure 15.19 shows the arrangement of compounded converters and the characteristics are plotted in Fig. 15.20.

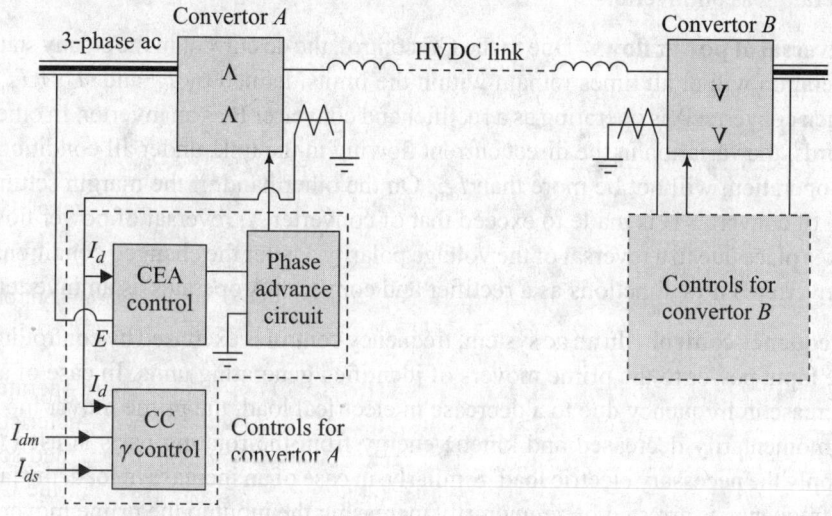

Fig. 15.19   Schematic arrangement of converters for an HVDC link

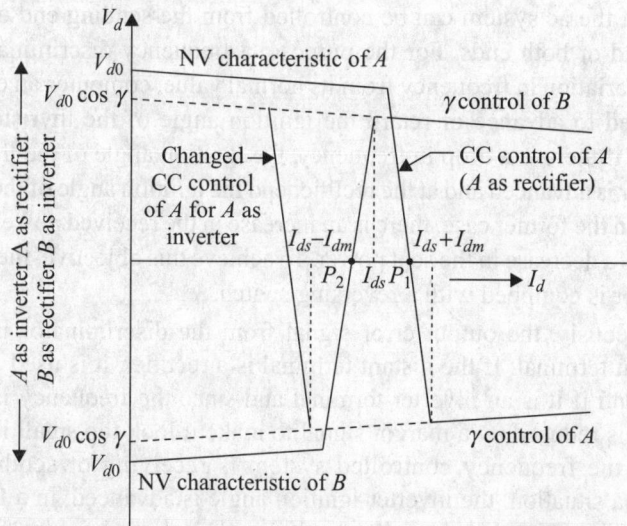

Fig. 15.20   Characteristics of compounded converters

Figure 15.20 shows the characteristics of converters A and B, with the latter characteristics plotted inverted for comparison. Power transfer takes place at the intersection of the CC control of the rectifier and the CEA control of the inverter. In Fig. 15.20, P_1 is the operating point when converter A is functioning as a rectifier and converter B is working as an inverter and P_2 represents the operating point when the operations of the two converters are reversed. The margin setting I_{dm} is around 10 to 20% of the CC control output setting I_{ds}, both for the rectifier and inverter operations. However, when the converter is operating as a rectifier the setting of I_{ds} is maintained higher than that of a rectifier by I_{dm}. Further, to ensure flow of power, the NV characteristic voltage V_0 of the converter functioning as a rectifier is maintained higher than the CEA characteristic of the converter operating as an inverter.

Reversal of power flow   Due to the CC control, the dc current in the steady state operation will at all times remain within the limits defined by I_{ds} and $(I_{ds} + I_{dm})$ when converter A is operating as a rectifier and converter B as an inverter. In other words, the variation in the direct current flowing in the link, under all conditions of operation, will not be more than I_{dm}. On the other hand, if the margin setting I_{dm} of converter B is made to exceed that of converter A, reversal of power flow takes place due to a reversal of the voltage polarity. Under the changed conditions, converter B now functions as a rectifier and converter A operates as an inverter.

Frequency control   In an ac system, frequency control is exercised by controlling the input power to the prime movers of identified generating units. In case of an increase in frequency due to a decrease in electrical load, the prime mover input is momentarily decreased and kinetic energy from the rotating parts is used to supply the necessary electric load. Similarly, in case of an increase in load, the fall in frequency is arrested by temporarily increasing the input to the prime movers.

A dc system which has a power rating comparable to or greater than the ac system to which it is connected can also be used to control the frequency. The frequency of the ac system can be controlled from the sending end or from the receiving end or both ends. For the purpose, a frequency discriminator, which detects the variation in frequency from its normal value, computes an error signal which is used to advance or retard the ignition angle of the thyristors. At the inverter end if there is a drop in frequency, the ignition angle of the thyristors of the converter is advanced and at the rectifier end the ignition angle of the thyristors is retarded. In the former case, there is an increase in the received power and in the latter there is a decrease in the sent power. To achieve this objective, the frequency discriminator is equipped with a reversing switch.

Simultaneously, the output error signal from the discriminator is also sent to the distant terminal. If the distant terminal is a rectifier, it is used as it is. On the other hand if it is an inverter terminal and since the frequency is right, the error signal is reduced by a margin signal to make it look too small irrespective of whether the frequency controlled system is receiving or sending power. Under such a situation, the inverter ignition angle is advanced, in a futile effort to raise the frequency, and consequently reaches the minimum permissible extinction angle thereby determining the direct voltage. The frequency control with a small error due to the margin signal is overtaken by the inverter when low voltage at the rectifier limits the line voltage. The process of frequency control is similar to current control; the converter with the higher voltage controls the frequency and the one with the lower voltage controls the dc line voltage.

15.5 CONFIGURATION OF HVDC CONVERTER STATIONS

The function of a converter station is to convert ac to dc (rectifier action) at the transmission end and dc to ac (inverter action) at the receiving end. Since a rectifier can be made to function as an inverter and vice versa, the converter stations at the two ends of an HVDC line are identical to each other. Figure 15.21 shows a schematic layout of a classic converter station in an HVDC link. A converter station requires considerable land space because the transformers, filters, and phase correction capacitors are placed outdoors. However, the valves and control equipment are placed in a closed air-conditioned/heated building. The various components of a converter station are discussed here.

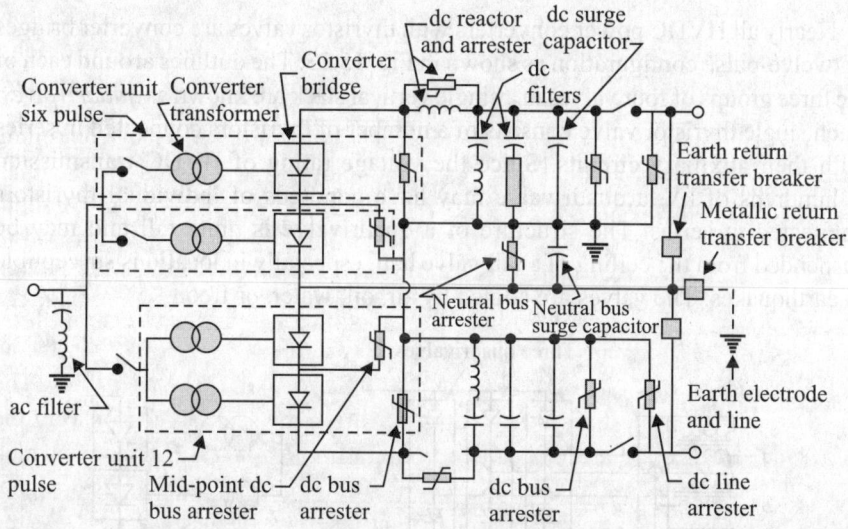

Fig. 15.21 Schematic lay out of a classic HVDC link

Converter unit A converter unit is made up of thyristor valves which can have different configurations depending upon the type of applications and the manufacturer. The valves can be organized as a single valve, double-valve cluster, or quadruple valve cluster. It is constructed from one or more thyristors in series. The graphical symbols for valves and bridge are shown in Fig. 15.22(a). The standard bridge or converter connection is defined as a full bridge connection comprising six valves or valve arms connected as shown in Fig. 15.22(b). The six-pulse valve group is connected to a three-phase ac system through a transformer in star–star connection. The six-pulse valve group was common when the valves were mercury arc.

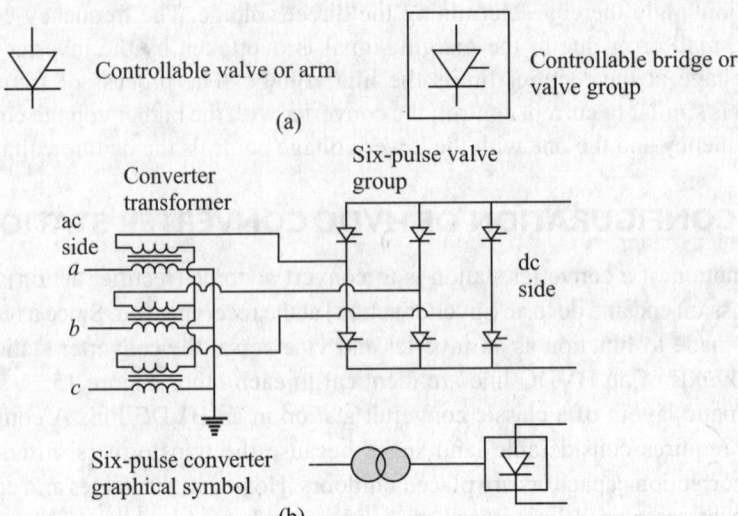

Fig. 15.22   (a) Symbols for valves and bridges (b) Six-pulse bridge converter

Nearly all HVDC power converters with thyristor valves are converter bridges of twelve-pulse configuration as shown in Fig. 15.23. The outlines around each of the three groups of four valves in a single vertical stack are known as *quadrivalves*. Each single thyristor valve consists of a number of thyristors connected in series with their auxiliary circuits. Since the voltage rating of HVDC transmission is hundreds of kV, a quadrivalve may have hundreds of individual thyristors connected in series. The structure of a quadrivalve is quite tall and may be suspended from the ceiling of a tall valve hall, especially in locations susceptible to earthquakes. The valves are cooled by air, oil, water, or freon.

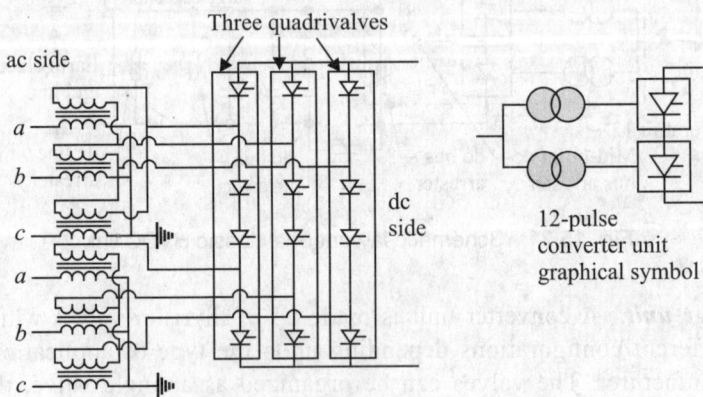

Fig. 15.23   12-pulse valve group configuration

Valve firing signals are generated in the converter at ground potential and communicated to each thyristor in the valve via an optical fibre light guide system. Gate drive amplifiers with pulse transformers convert the light signal received by the thyristor into an electrical signal. The valves are protected using snubber circuits, protective firing, and gapless surge arresters.

In case of a VSC converter bridge, a single valve is built up with a certain number of IGBTs connected in series together with auxiliary electronic circuits. VSC valves, control equipment, and cooling equipment would be in enclosures which make transport installation very easy.

The 12-pulse valve group is connected to a three-phase ac through two three-phase transformers, one in star–star connection and the other in star–delta connection.

Converter transformers The three-phase converter transformers are connected between the ac bus bars and the valves. On the valve side, the transformers are connected in star and delta with the neutral point insulated from the ground. On the ac side, the transformers are connected in parallel, in star configuration, with the neutral points earthed. Usually the converter transformers are single-phase, three-winding type. However, single-phase, two-winding or three-phase, two-winding type can also be used. The transformers have on-load tap-changing arrangement so as to cover the voltage variation of $\pm10\%$, and also for adjusting load from 10% to 100%. The short-circuit current through any valve is restricted by the magnitude of the leakage reactance of the transformer.

Converter transformers are located in the switchyard, and if the converter bridges are located in the valve hall, the connection has to be made through its wall. This is accomplished either with phase isolated busbars, where the bus conductors are housed within insulated bus ducts with oil or SF_6 as the insulating medium, or with wall bushings.

Filters Three types of filters are used in HVDC transmission systems for elimination of harmonics from ac and dc sides of the converter. These are ac filters, dc filters, and high frequency filters.

Current harmonics of 11th, 13th, 23rd, 25th, and higher order are generated on the ac side of the 12-pulse HVDC converter. Tuned ac filters are provided to suppress these harmonics to the designed level of the network. Tuning to the 5th and 7th harmonics is required if the converter is configured into six-pulse operation.

The characteristic dc voltage harmonics generated by a 12-pulse converter are of the order of $12n$, and when generated by a six-pulse converter is $6n$, where n is an integer. Harmonics generated by the HVDC converters in all their operational modes cause interference in telecommunication networks. Specially-designed dc filters are employed to minimize interference. In back-to-back HVDC stations and for completely cable transmissions, normally dc filters are not required. However, dc filters would be required if transmission is partially through an overhead line. The filters needed on the dc side are usually considerably smaller and less expensive than the filters on the ac side.

High-frequency filters are also called *radio frequency* (RF) or *power line carrier* (PLC) filters and are used to block any high-frequency currents. The high-frequency currents are located between the station ac bus and the converter transformers. Such filters might sometime become necessary to be connected to the HVDC bus between the dc filter and dc line and on the neutral side too.

With VSC converters, the current harmonics on the ac side are related directly to the PWM frequency. Therefore, the filters in this type of converters are reduced dramatically compared with natural commutated converters.

Reactive power source Converter stations require reactive power of the order of about 50% to 60% of real power supply. Luckily, a part of the reactive power requirement is met by the ac filter. For the remaining part of the requirement, switched shunt capacitors, static var compensator systems, or synchronous condensers are employed.

Smoothing reactors Linear series reactors, of sufficiently large inductance, are connected to each pole of the converter station on the dc side of the HVDC line or at an intermediate location in the dc line. The functions of a linear series reactor are prevention of intermittent current, smoothening the dc current wave form, limiting the dc fault currents, and preventing resonance in the dc circuits. The reactor is usually of oil immersed, outdoor magnetic shielded type construction.

Surge arresters The main task of surge arresters is to protect the equipment against over-voltages. Surge arresters across each valve in the converter bridge, across each converter bridge, and in the dc and ac switchyard are coordinated to protect the equipment from all over-voltages irrespective of their source. Modern HVDC substations use metal-oxide arresters and their selection is made with careful insulation coordination design.

Control systems The system control in an HVDC link is quite complex with a hierarchy of controllers. High-speed microprocessors are used for many control functions including monitoring and supervisory controls.

There are various control levels performing various functions. At *station control level*, interface with ac system, that is, either with the generating station or with the load dispatch centre is provided. The station control equipment supervises the connection of harmonic filters and may block converters during abnormal conditions.

Converter control level provides firing control and delivers the control pulses to the valve control. At this level tap-changer control is also provided.

Pole control level has a control desk where power command per rate, current command, and start and stop commands are given. At this level, measured current is compared with the current command, and the error is then passed on to the valve level firing control. This level is equipped with protection and measurement system.

Valve control level has a valve control system which supervises and distributes pulses to the valves as optical signals.

15.6 DC LINKS

The HVDC transmission system consisting of the transmission line along with its terminal and auxiliary equipment is known as the *dc link*. The three types of links used for the transmission of bulk dc power transfer are (a) monopolar link, (b) bipolar link, and (c) homopolar link.

15.6.1 Monopolar Link

Figure 15.24 shows a single-line circuit of a monopolar link. A monopolar link utilizes a single conductor which has a negative polarity. Ground is used as the positive electrode and provides the return path for the flow of current. A metallic electrode may also be used to provide the return path for the current.

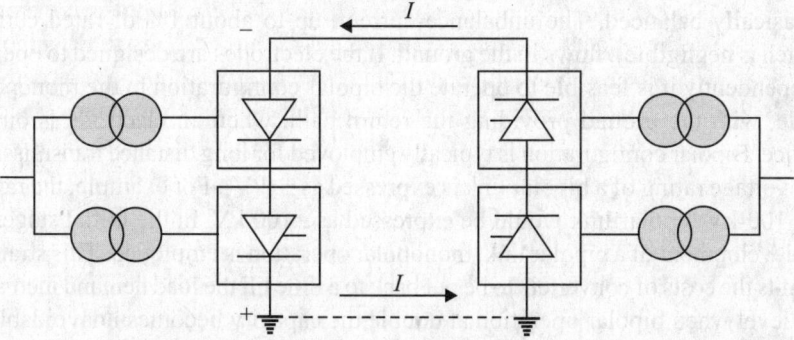

Fig. 15.24   Single-line circuit of a monopolar link

Monopolar link operation is most useful when power is to be transmitted below the sea level since it saves the cost of a second submarine cable. In such a case, the sea is used as the second electrode. Operation of a monopolar link, with ground as an electrode, over long periods of time is not advisable due to corrosion and other interference effects with the other telecommunication circuits in the region. Metallic return has the advantage that transmission of power is still possible with the transmission line pole conductor acting as a metallic return in case of a pole outage. Pole outages occur due to failure of station pole equipment.

15.6.2 Bipolar Link

A bipolar link configuration consists of two conductors, with both terminal stations grounded, one acting as the positive electrode and the other as the negative electrode. Figure 15.25 gives the schematic lay out of a bipolar link.

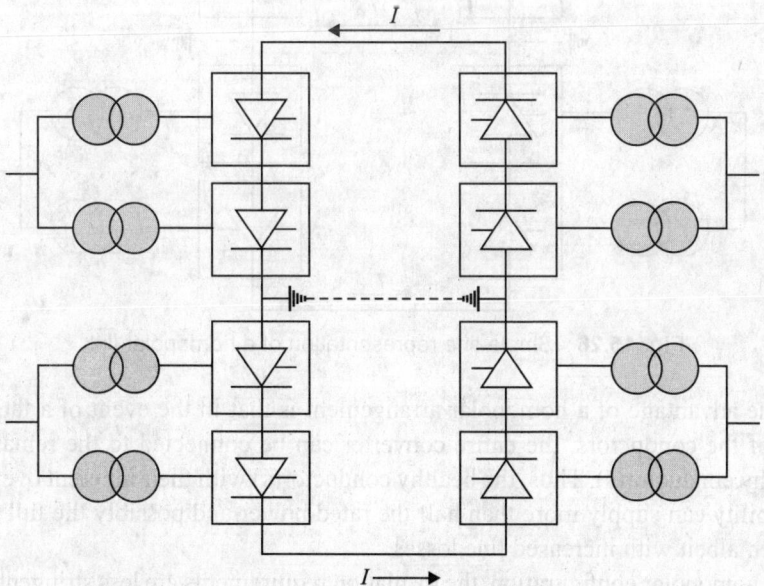

Fig. 15.25   Schematic layout of a bipolar link

The direction of flow of power in a bipolar configuration can be changed by reversing the polarities of the electrodes. The current flow in both the conductors is basically balanced. The unbalance current, up to about 1% of rated current (which is negligible), flows in the ground. If the electrodes are designed to operate independently it is feasible to operate the bipolar configuration in the monopolar mode, with the ground providing the return path, when an electrode is out of service. Bipolar configuration is typically employed for long distance transmission. The voltage rating of a bipolar link is expressed as ± (kV). For example, the rating of a 100-kV bipolar link would be expressed as ±100 kV. In the initial stages of the development of a bipolar link, monopolar operation is employed. This strategy permits the costs of converters to be put back to a time till the load demand increases to a level when bipolar operation at double the capacity becomes unavoidable.

15.6.3 Homopolar link

A homopolar link configuration is constituted of two or more conductors. It operates with all the conductors having the same polarity, normally negative, and the return path is provided by the ground or a metallic return. Thus, the ground current is twice the current in each conductor. Figure 15.26 shows a single line representation of a homopolar link.

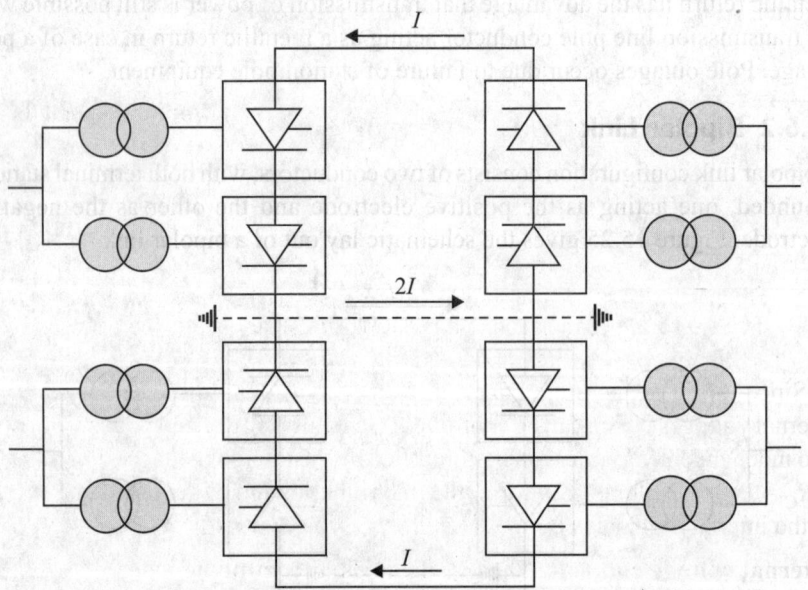

Fig. 15.26 Single-line representation of a homopolar link

The advantage of a homopolar arrangement is that in the event of a fault on one of the conductors, the entire converter can be connected to the remaining healthy conductor(s). Thus, the healthy conductor(s) with their inherent overload capability can supply more than half the rated power and possibly the full rated power, albeit with increased line losses.

In homopolar configuration, the insulation requirements are less stringent than in the bipolar arrangement which uses graded insulation. Therefore, there is a cost advantage in the former arrangement which is offset by the disadvantage of a continuous use of the ground as return. Further, due to the negative polarity of the conductors in the homopolar arrangement, the corona effects are less pronounced than compared to conductors with positive polarity. Therefore, power loss due to corona is less in a homopolar link.

15.7 SIMULATION OF POWER FLOW IN AN HVDC LINK

The equivalent circuits developed earlier (see Figs 15.14 and 15.20), are included with the line to model a HVDC link in the steady state operation. Figure 15.27 shows the equivalent circuit for simulating power flow in a HVDC link. In the figure, $R_{Cr} = (3\omega L_C)/\pi$ represents the fictitious resistance of the converter when it is working as a rectifier. Similarly, $R_{Ci} = \pm (3\omega L_C)/\pi$ symbolizes the fictitious resistance of the converter when it is operating as an inverter; the plus sign being used when constant ignition angle β representation is employed and the negative sign being used for constant extinction angle γ representation.

If I_d denotes the steady-state current in the link, then

$$I_d = \frac{V_{0r}\cos\alpha - V_{0i}\cos(\beta \text{ or } \delta)}{R_c + R_l \pm R_l} \tag{15.38}$$

where $V_{or}\cos\alpha$ and $V_{0i}\cos(\beta \text{ or } \delta)$ are internal voltages of the rectifier and inverter, respectively, and R_l is the line resistance of the link

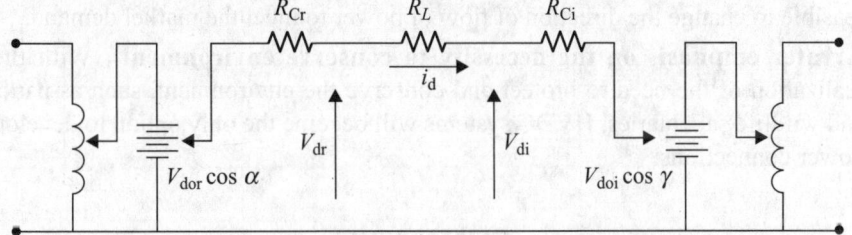

Fig. 15.27 Simulation of a HVDC link in steady-state operation

Since the resistances are constant, I_d is proportional to the difference in the internal voltages which can be controlled. Generally, power and voltage which are two independent variables can be controlled by the internal voltages. For example, if $R_C = R_i$ and the line is uniform, the midpoint voltage of the line is the average of the internal voltages.

Internal voltage control   The two methods of controlling internal voltage are (a) grid control, and (b) alternating voltage control.

The internal voltage from Eq. (15.7) can be controlled by controlling the ignition angle α; a delay in ignition angle (increase in α) will lead to a decrease in internal voltage from the ideal no-load voltage V_0 (which is directly proportional to ac voltage) by the factor $\cos\alpha$. More often than not, a tap-changing transformer is employed for ac voltage control; however, in some special cases generator excitation may be used for ac voltage control.

Both grid control and tap-changing controls are used; the former with a response time of 1 to 10 ms is initially activated for rapid action, which is followed by the slow (5 to 6 sec per step) tap changing for restoring rectifier ignition angle or inverter voltage to their typical values.

15.8   FUTURE OF HVDC TRANSMISSION IN THE MODERN POWER SECTOR

For the applications stated above, the HVDC system remains by far the most attractive option from the perspectives of both operation and economics. The following dynamics indicate that HVDC technology could become the favourite for transmission of power as against the high-voltage ac transmission technology:

Development in technology   Advances in VSC based technology and new polythene dc cables have made it feasible to transmit power up to 200 MW over transmission distances of as low as 60 km.

Restructuring of the power industry   Restructuring has placed power transmission under the management of private corporations. ROW will become a major parameter in the new environment as against its absence in the regulated structure when land acquisition and ROW was obtained in 'public interest'. Since in HVDC transmission the land acquisition and ROW requirements are much less, the HVDC system will become more cost-effective when costs required to obtain land and ROW are fully included in the economic analysis.

The cornerstone of deregulation is competitive trading   Bidirectional flow of power on an HVDC line will provide a competitive edge, since it would be feasible to change the direction of flow of power to meet the market demand.

Greater emphasis on the necessity to conserve environment   With the realization of the need to protect and conserve the environment, such as parks and wildlife sanctuaries, HVDC systems will become the only option to develop power connections.

SUMMARY

- Due to various technical and economic factors and environmental considerations HVDC is used for transmission of power.
- The basic process of rectification and inversion in an HVDC system can be accomplished through (i) natural commutated converters, or (ii) capacitor commutated converters, or (iii) forced commutated converters.
- A six-thyristors fully controlled bridge (Graetz) circuit is employed to convert ac to dc power at one end and dc to ac power at the other end of an HVDC link.
- NV, CIA, and CEA characteristics define the operation of a converter both as a rectifier and as an inverter.
- Compounded converters, one of which operates as a rectifier and the other as an inverter, are provided at the two ends of an HVDC link.
- CEA and CC controllers in the converters manage the operation of the link and power transfer takes place at the intersection of the CC control of the rectifier and CEA control of the inverter.
- Frequency control of the ac system is accomplished from both the ends with the help of a frequency discriminator.
- Power flow in an HVDC link is simulated by employing the equivalent circuits of a rectifier and an inverter. The two independent variables, power and voltage, are controlled by internal voltages which in turn are controlled by (i) grid control and (ii) ac voltage control.

- An HVDC converter station, in addition to a fully controlled six-thyristors bridge converter, consists of converter transformers, filters, a source of reactive power, smoothing reactors, surge arrestors, and control systems.
- Transmission of bulk power is undertaken through (i) monopolar, (ii) bipolar, and (iii) homopolar DC links which in addition to a transmission line is made up of terminal and auxiliary equipment.
- Flow of power is regulated by either a (i) constant current or (ii) constant voltage system.
- The various types of controls used to manage the flow of power are (i) ignition angle, (ii) extinction angle, and (iii) constant current. Other aspects associated with power flow in an HVDC link are frequency control and circuit breaking.

SIGNIFICANT FORMULAE

- Conductor to ground dc voltage: $V_{dc} = \sqrt{2}E_{ac}$
- DC power per circuit in a two conductor line: $P_{dc} = 2\sqrt{2}E_{ac}I_{ac}$
- Ratio of dc to ac power: $P_{dc}/P_{ac} = 0.9428/\cos\phi$
- Direct current voltage output, no grid control ($\alpha = 0°$):
  $V_{d0} = 1.654\, V_m = 1.351\, E_{LN}$
- Direct current voltage output:

  (i) With grid control: $V_\alpha = V_{d0}\cos\alpha$, $V_0 = \left(3\sqrt{2}E_{LN}\right)/\pi$

  (ii) With grid control and commutation: $V_d = \dfrac{1}{2}V_{d0}\left[\cos(\alpha) + \cos(\alpha+\gamma)\right]$

- Simulation of a rectifier: $V_d = V_{d0}\cos(\alpha) - \dfrac{3\omega L_C}{\pi}I_d$

- RMS value of secondary current I_s : (1) with $\alpha = 0$ and $\gamma = 0$: $0.8165\, I_d$
  (ii) with finite α and γ values: $0.78\, I_d$.

- Power factor angle: $\cos\varphi = \dfrac{1}{2}[\cos\alpha + \cos\delta]$

- Equivalent circuit of an inverter:

  $V_d = \left[V_0\cos\beta + \dfrac{3\omega L_C}{\pi}I_d\right]$ with constant γ;

  $V_d = \left[V_{d0}\cos\gamma - \dfrac{3\omega L_C}{\pi}I_d\right]$ with constant β

- Steady-state current in an HVDC link:

  $I_d = \dfrac{V_{d0r}\cos\alpha - V_{d0i}\cos(\beta \text{ or } \gamma)}{R_c + R_l \pm R_i}$

EXERCISES

Review Questions

15.1   (a) If it is assumed that the current in a dc line is equal to the effective value of the current in an ac line, what is loss of power per conductor in the two lines?

(b) What is the power loss in a dc link as a percentage of the power loss in an ac circuit? (c) Compare the losses in a dc link with those in a three-phase ac circuit when both the systems have the same percentage losses.

15.2   (a) Draw a fully labelled circuit diagram of Graetz thyristor bridge and explain its functioning.

(b) Draw and derive expressions for the dc output voltage waveforms for (i) uncontrolled, (ii) delayed ignition, and (iii) delayed ignition and overlap operations.

15.3   Develop expressions for the equivalent circuit when the converter is operating as (i) a rectifier, and (ii) an inverter. Draw the respective equivalent circuit diagrams.

15.4   Write short notes on (i) CIA control, (ii) CEA control, and (iii) CC control.

15.5   Draw a schematic diagram to show the arrangement of converters in a HVDC link. Draw the converter characteristics and explain their working.

15.6   Write concise notes on (i) frequency control, and (ii) reversal of power flow.

15.7   (a) Describe how power flow in an HVDC link can be simulated.

(b) Explain the methods of internal voltage control.

Numerical Problems

15.1   Write a MATLAB script file for plotting dc power per circuit versus ac power per circuit at a given power factor. Use the script file to plot the curves for φ ranging from 0° to 75°.

15.2   Write a MATLAB function to plot the voltage profile of an ac overhead line against distance for varying load conditions. Use the MATLAB function to plot the voltage profile, for one-twentieth wavelength of a 50-Hz, 220-kV, loss free line when it carries a three phase 300 MW load at (i) 0.8 lagging pf and (ii) 0.8 leading pf. Assume the line inductance and capacitance equal to 1.0 mH and 0.025 μF respectively and the voltage at the receiving end equal to 220 kV.

15.3   In actual practice, (i) is the length of a line greater or less than the wavelength and (ii) is the load supplied by a line greater or less than SIL? Discuss the voltage profiles obtained in Numerical Problem 15.2.

Multiple Choice Objective Questions

15.1   Which of the following is not true of the first HVDC commercial transmission link?
(a) First commissioned in 1882.
(b) It used a submarine cable.
(c) It used ground as return.
(d) All of these.

15.2   The first HVDC project in India was commissioned in
(a) 1882   (b) 1901
(c) 1979   (d) 1989

15.3   Which of the following development hastened the development of voltage source converters?
(a) Power diode
(b) Junction field effect transistor
(c) Insulated gate bipolar transistor
(d) Bipolar junction transistor

15.4 Which of the following holds for unity power factor?

    (a) $P_{dc} > P_{ac}$ (b) $P_{dc} = P_{ac}$

    (c) $P_{dc} < P_{ac}$ (d) $P_{dc} = 0$

15.5 Which of the following applies to a dc link?

    (a) It does not produce reactive power.

    (b) Terminal equipment requires reactive power.

    (c) IR is the only voltage drop.

    (d) All of these.

15.6 The break-even distance, in km, for an ac transmission line lies between

    (a) $200 - 500$ (b) $500 - 800$

    (c) $800 - 1100$ (d) above 1100

15.7 Which of the following is not possible in an HVDC transmission link?

    (a) Interconnecting ac networks operating at frequencies of 50 Hz and 60 Hz

    (b) No synchronous operation

    (c) Bidirectional flow of power

    (d) None of these

15.8 Which of the following is the most fundamental device to the process of conversion?

    (a) Thyristor   (b) Transistor

    (c) Diode   (d) None of these

15.9 Which of the following is the dc output of a controlled Graetz circuit without overlap?

    (a) $1.17 \, E_{rms}$ (b) $1.351 \, E_{rms}$

    (c) $V_0 \cos \alpha$ (d) None of these

15.10 Which of the following will switch the state of a thyristor from ON to OFF mode?

    (a) $V_{AK} = 0$ (b) $i_{AK} = 0$

    (c) $i_G = 0$ (d) All of these

15.11 Which of the following represents the requirement of reactive power, as a percentage of real power, at a converter station?

    (a) $40 - 50$ (b) $50 - 60$

    (c) $60 - 70$ (d) None of these

15.12 Which of the following is not a function of a linear series reactor?

    (a) Limiting dc fault currents

    (b) Protection against over-voltages

    (c) Prevention of intermittent current

    (d) Preventing resonance

15.13 Which of the following is a true affect on the dc output voltage of a rectifier when the ignition angle is advanced?

    (a) Decreases   (b) Increases

    (c) Unchanged   (d) None of these

15.14 Which of the following defines commutation?

    (a) Transfer of dc voltage from one valve to another in the same row

    (b) Transfer of dc voltage from one valve to another in the top and bottom row

    (c) Transfer of dc current from one valve to another in the same row

    (d) Transfer of dc current from one valve to another in the top and bottom row

15.15 The secondary transformer current during rectification is $0.78\,I_d$. Which of the following is true?

(a) $\alpha = 0, \gamma = 0$ (b) $\alpha = 0, \gamma > 0$

(c) $\alpha > 0, \gamma > 0$ (d) None of these

15.16 When a converter is operating as an inverter, which of the following is the symbol for angle of advance?

(a) α (b) β

(c) γ (d) δ

15.17 In which of the following region does an inverter operates?

(a) CEA   (b) CC

(c) Both CEA and CC   (d) None of these

15.18 From which of the following ends is the frequency of an ac system controlled in an HVDC link?

(a) Sending end   (b) Receiving end

(c) Both ends   (d) HVDC link does not control the frequency

15.19 Which of the following is used for internal voltage control?

(a) Grid control by controlling ignition angle

(b) Ac voltage control by tap changing

(c) Both (a) and (b)

(d) None of these

15.20 Which of the following is not true of a mono-polar link?

(a) The conductor has a negative polarity.

(b) Ground constitutes the return conductor.

(c) Ground is used as a positive electrode.

(d) None of these.

15.21 Which of the following is true of a homo-polar link?

(a) It has two or more conductors.

(b) All conductors possess the same polarity.

(c) It uses ground or a metallic return.

(d) All of these.

15.22 In which of the following, power can be transferred to healthy conductor(s) in case of a fault?

(a) Mono-polar   (b) Bi-polar

(c) Homo-polar   (d) All of these

15.23 Refer to Q 15.15; in which of the following direction of power can be changed?

15.24 Refer to Q 15.15; in which of the following corona loss is less?

15.25 Refer to Q 15.15; in which of the following interference with other telecommunication circuits is more?

Answers

15.1 (a)	15.2 (d)	15.3 (c)	15.4 (c)	15.5 (d)	15.6 (b)
15.7 (d)	15.8 (a)	15.9 (c)	15.10 (b)	15.11 (b)	15.12 (b)
15.13 (a)	15.14 (c)	15.15 (c)	15.16 (b)	15.17 (a)	15.18 (c)
15.19 (c)	15.20 (d)	15.21 (d)	15.22 (c)	15.23 (b)	15.24 (c)
15.25 (a)					

CHAPTER 16

Introduction to FACTS

Learning Outcomes

A focussed study of the chapter will enable the reader to
- Understand the problems arising out of operating large, interconnected, and widespread power networks in an environment characterized by continually growing demand and deregulation
- Comprehend the methodology of FACTS in enhancing transmission capability and load the lines to their maximum transmission potential
- Become familiar with the basics of power flow and stability aspects of a transmission link
- Based on an understanding that power flow in a transmission network is controlled by its parameters and it cannot self-manipulate the flow, enumerate and recognize the various methods of parametric control

OVERVIEW

The present-day electrical power systems have come a long way from the former power systems which supplied power to small local areas. Modern power systems are characterized by a maze of interconnections which are used for connecting systems at the intra-regional, inter-regional, and the national level.

Interconnections of transmission networks, in addition to supplying power, lead to the generation of energy at an optimal cost per unit and enhance the reliability of the system as well. However, the operation and control of a large and geographically widespread interconnected power system, which modern-day networks invariably are expected to be, is very complex. Failure to effectively control such a system may result in excessive reactive power in various parts of the network and large dynamic oscillations in different parts of the system. Thus, the full potential of an interconnected transmission network would remain unexploited.

In recent times, greater demands have been imposed on the transmission networks and these demands will continue to increase. Due to the deregulation of the power sector, the basic transmission challenge is to provide a network capable of delivering contracted power from power suppliers to consumers over a large geographical area. The constraints to any potential solution are cost, right-of-way, and environmental problems. Owing to these, the network must mainly be based on the existing physical transmission line structures.

The power transfer limits of a transmission line are in general determined by three levels—thermal limit, which is the inherent physical limit determined by transmission lines ampacities; uncontrolled power flow limit imposed by the natural power flow; and stability limit imposed by the requirements of a secured, stable, power transfer. Power transfer in most integrated transmission systems is constrained by transient stability and voltage stability. These constraints limit a full utilization of available transmission corridors.

Flexible alternating current transmission systems (FACTS) is a technology which provides a methodology for the utilities to effectively utilize their assets, enhance transmission capability by loading lines to their full transmission capability, and therefore, minimize the gap between the stability and the thermal limits, and improve grid reliability. The FACTS technology is based on the use of reliable high-speed power electronics, advanced control technology, advanced microcomputers, and powerful analytical tools. The key feature is the availability of power electronic switching devices at high kV and kA levels. This chapter introduces the fundamentals of the FACTS technology.

16.1 PARAMETRIC LIMITATIONS OF AC POWER FLOW IN A TRANSMISSION SYSTEM

Unlike other forms of energy, electrical energy in bulk cannot be stored. Therefore, there must always be a balance between the power generated and the load supplied. If the generation drops and is less than the load, both the frequency and voltage drop. As a consequence of a drop in frequency and voltage, the load also decreases. Since self-regulation is only feasible within a narrow margin (few percentage points), the voltage drop is arrested by injecting reactive power into the system, and thereby, raising the load. However, drop in frequency continues and the system disintegrates. If no reactive power is injected, the system fails due to drop in voltage.

When the system is generating sufficient power, real power flows through all the extra high and medium voltage lines via all the possible parallel paths to supply the load which is distributed throughout the network. However, the flow of power in a transmission network is restricted by its parameters and its incapability to manipulate the flow as desired.

16.1.1 Flow of Power in Parallel Circuits

Two parallel ac circuits interconnecting area 1 with area 2 are shown in Fig. 16.1. Let it be assumed that real power P MW flows from area 1 (the surplus generation area) to area 2 (the deficit generation area). Without any control, power flow over each circuit is based on the inverse of the transmission line impedances. If the resistances of the circuits are neglected and if the reactance of circuit 1 is X ohms and that of circuit 2 is $3X$ ohms, then power flow in the circuits 1 and 2 will be

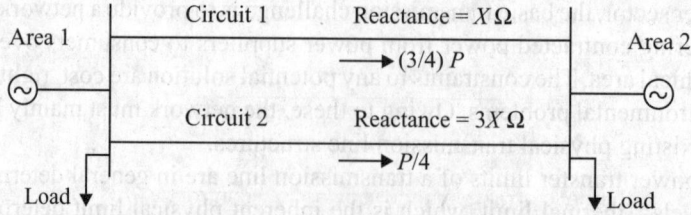

Fig. 16.1 Power flow over parallel circuit interconnecting two areas

0.75P MW and 0.25P MW, respectively. This means that circuit 1 that has the lower reactance is likely to be overloaded and circuit 2 may remain underloaded. Further, the idea of increasing the current carrying capacity of the overloaded circuit would be self-defeating, since circuit 2 with the higher reactance still remains underutilized.

Figure 16.2 shows the power system of Fig. 16.1 with the ac circuit 2 replaced by an HVDC line. The thyristor converters in the HVDC line control both the magnitude and direction of flow of power. Therefore, it is within the control of the operator to decide the loading on a line. Theoretically, it is possible to load an HVDC line to its full thermal limit. Since the converters are bidirectional, the direction of flow of power can be reversed too. Further, the fast electronic control in an HVDC line assists in the maintenance of stability in the parallel ac line.

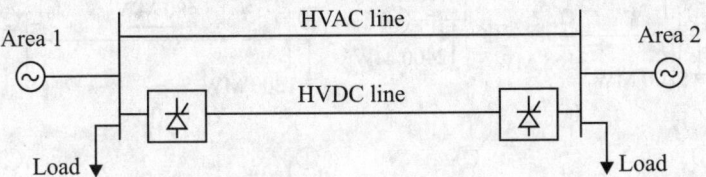

Fig. 16.2   Schematic diagram of power flow using a high-voltage ac line and an HVDC power line

An HVDC line, however, is costly and is usually considered when power is to be transmitted over long distances or when power transfer is to be done under water by submarine cables. An alternative to HVDC transmission is the use of FACTS controllers. The series FACTS controller installed in circuit 2 of Fig. 16.3 is an alternative method that can be used for controlling power flow through circuit 2. Figure 16.3(a) shows the application of series FACTS controllers, which vary the power flow in the line by changing the line reactance (impedance). Figure 16.3(b) shows the use of series FACTS controllers that control the power flow by changing the phase angle. A FACTS controller which injects an appropriate series voltage can also be employed to control the power flow. Thus, by using series FACTS controllers it is feasible to restrict overloads on ac lines by limiting loads to their rated capacity under contingency conditions, such as the loss of a parallel line.

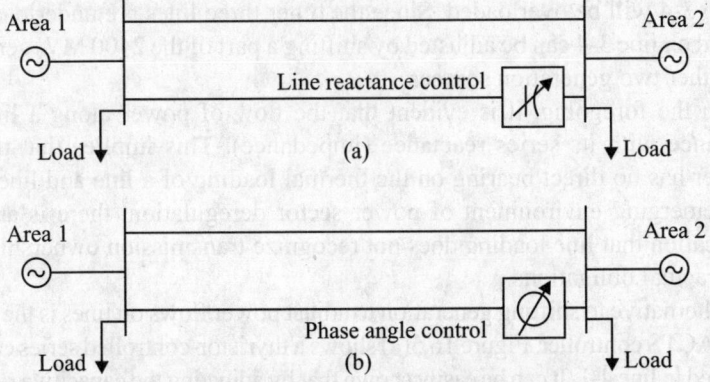

Fig. 16.3   Schematic arrangement of power flow control by FACTS technology by (a) varying line reactance and (b) varying phase angle

16.1.2 Flow of Power in Closed Loops

Figure 16.4(a) shows a hypothetical closed loop network in which a load centre is supplied from three sources of generation. For the given line reactance (impedance) indicated in Fig. 16.4(a), the flow of power to supply the load centre, is also indicated. Figures 16.4(b), (c), and (d) show the power flows in the lines due to generation sources 1, 2, and 3 respectively.

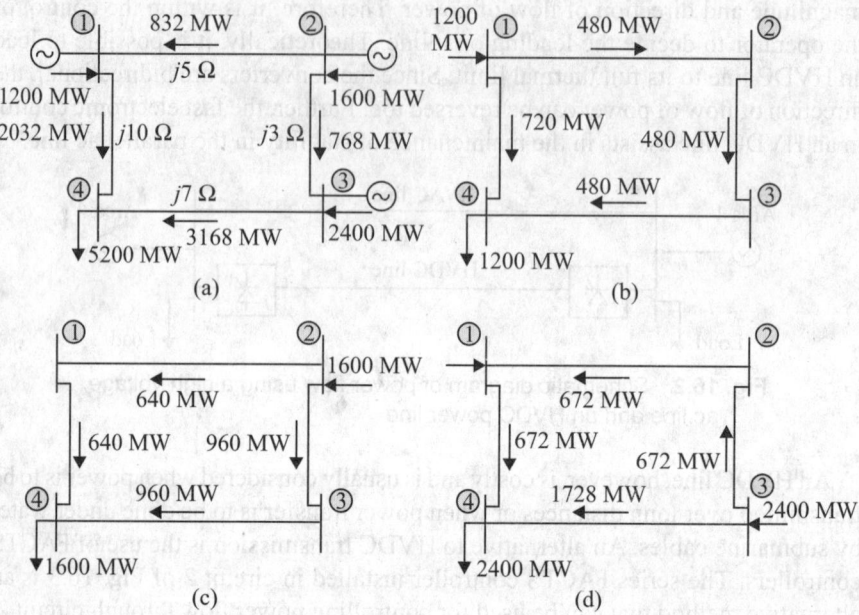

Fig. 16.4 Flow of power in closed loop network: (a) network diagram; power flow due to generation source (b) 1200 MW, (c) 1600 MW, and (d) 2400 MW

From Figs 16.4(b), (c), and (d), it can be observed that in each case there are two parallel paths and the flow of power is inversely proportional to the path reactance (impedance). Further, for the indicated line reactance (impedance), if it is assumed that the lines 1-2, 2-3, 3-4, and 4-1 are continuously rated to carry loads of 1200 MW, 1500 MW, 3000 MW, and 3000 MW respectively, it is clear that line 3-4 will be overloaded. Since the other three lines are underloaded, the overload on line 3-4 can be adjusted by shifting a part of the 2400 MW generation to the other two generation sources.

From the foregoing, it is evident that the flow of power along a line is in accordance with its series reactance (impedance). This implies that the flow of power has no direct bearing on the thermal loading of a line and line losses. In the emerging environment of power sector deregulation, there is an added complication that line loading does not recognize transmission ownership rights or contractual obligations.

An alternative to shifting generation to adjust power flows on lines is the use of a series FACTS controller. Figure 16.5(a) shows a thyristor-controlled series capacitor connected in line 4-1. It can be easily shown that by adjusting the capacitive reactance of the controller at the synchronous frequency to 5 Ω, the power flow in the lines changes and is shown in Fig. 16.5(a). The FACTS controller, since it is fully thyristor controlled, can be operated as many times as required. The other additional benefits are that it can be quickly modulated to damp out any sub-synchronous condition along with blocking low-frequency oscillations in the power flow.

The other alternative is to use a thyristor-controlled series inductor as shown in Fig. 16.5(b). Here also similar results are obtained by introducing a series inductive reactance, equal to 17.1 Ω.

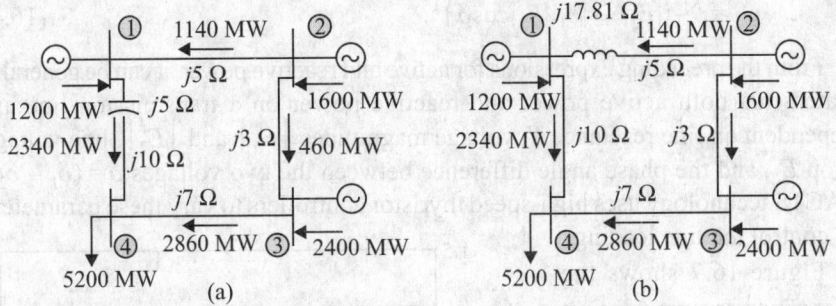

Fig. 16.5 Change in line power flows by connecting a series FACTS controller to add (a) capacitive and (b) inductive reactance

As in the case of ac parallel circuits, a thyristor-controlled phase angle regulator may be connected in series with one of the lines in the closed loop, or variable voltage may be injected in one of the lines to adjust the power flows in the steady state. It is interesting to note that in all the above cases only one type (out of several types) of a series controller is required to be connected in one line out of four lines to adjust the steady power flows in the closed loop.

16.2 POWER FLOW AND STABILITY ASPECTS OF A TRANSMISSION LINK

Consider a transmission line of reactance X pu (negligible resistance) connecting two points (stiff buses) in a power system as shown in Fig. 16.6(a). The system voltage magnitudes in per unit and phase angles are indicated in the diagram. The corresponding phasor diagram is shown in Fig. 16.6(b).

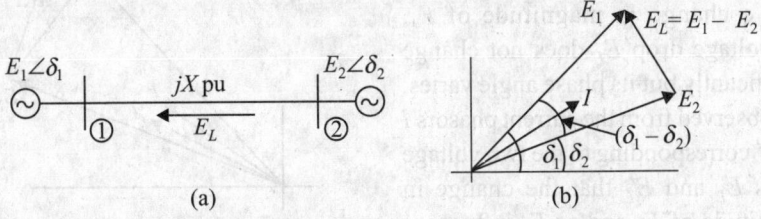

Fig. 16.6 (a) A single line diagram of an ac transmission line connecting two stiff buses in a power network (b) phasor diagram of the line

The line current lags the line voltage drop $E_L = (E_1 - E_2)$ by 90° and is given by

$$I = \frac{E_L}{X} = \frac{E_1 - E_2}{X} \tag{16.1}$$

As has already been stated, the active power flow P along the line is given by

$$P = \frac{|E_1||E_2|}{X}\sin\delta \tag{16.2}$$

where $\delta = \delta_1 - \delta_2$. Similarly reactive power flows Q_1 and Q_2 are expressed as

$$Q_1 = \frac{1}{X}\left(|E_1|^2 - |E_1||E_2|\cos\delta\right) \tag{16.3}$$

and

$$Q_2 = \frac{1}{X}\left(|E_2|^2 - |E_1||E_2|\cos\delta\right) \tag{16.4}$$

From the preceding expressions for active and reactive power, it can be generally stated that both active power and reactive power on a transmission line are dependent on line reactance X, voltage magnitudes $|E_1|$ and $|E_2|$, line voltage drop E_L, and the phase angle difference between the two voltages $\delta = (\delta_1 - \delta_2)$. FACTS technology uses high-speed thyristor controllers to vary these parameters to control the line loading.

Figure 16.7 shows the plot of power output against phase difference angle δ. Change in line reactance X changes the amplitude of the power output. At a given phase difference angle δ, the power flow on the line changes with a variation in line reactance X. For a given power flow over the line, the variation of X will correspondingly change the phase angle δ.

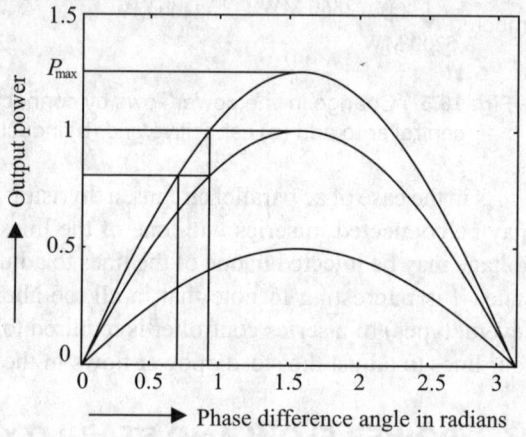

Fig. 16.7   Plot of output power versus phase difference angle δ

Variation of magnitudes of voltages E_1 and/or E_2 change the power flow. Figure 16.8 shows the phasor diagram when the magnitude of E_1 is increased to E_1'. The phasor diagram shows that with a change in magnitude of E_1, the voltage drop E_L does not change significantly but its phase angle varies. It is observed from the current phasors I and I' corresponding to the two voltage drops E_L and E_L' that the change in magnitude of E_1 and/or E_2 influences the reactive power flow more than the real power flow.

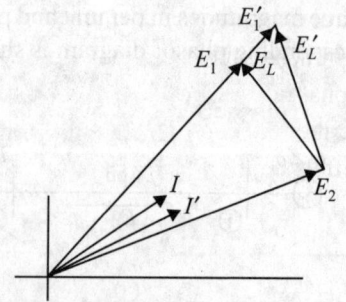

Fig. 16.8   Phasor diagram when the voltage magnitude of E_1 is increased

Power flow and current flow over the transmission line can also be varied by injecting voltage in series with the line, that is, by varying the line voltage drop E_L. Figure 16.9 shows a phasor diagram where voltage is injected in series with the line. Since the injected voltage is in series with the line voltage drop (and approximately in phase quadrature with the line current), it directly affects the line current and a small change in angle influences the reactive power flow considerably.

In case the magnitude of the voltage being injected in series and its phase angle are varied, there is variation of both the real

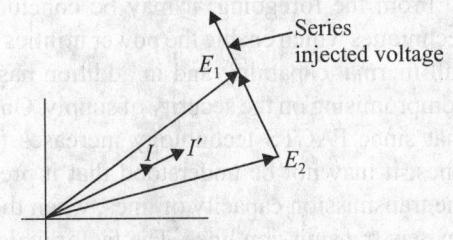

Fig. 16.9 Phasor diagram when voltage is injected in series

and reactive components of line current. Figure 16.10 shows the phasor diagram for the case of series voltage injection with variable magnitude and phase angle. It may be observed that by varying the magnitude and phase angle of the voltage injected in series, both the active and reactive power flow can be changed.

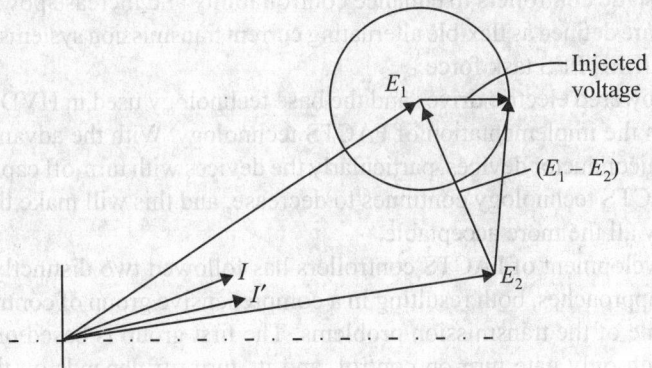

Fig. 16.10 Phasor diagram for variable magnitude and phase angle series voltage injection

FACTS technology utilizes thyristor-controlled fast acting controllers to control the following: line reactance X, magnitudes of terminal voltages E_1 and E_2, magnitude of line voltage drop or driving voltage phasor $|E_L| = (|E_1 - E_2|)$, and the phase difference angle δ to control the transmission line loading. Table 16.1 describes the control action of the thyristor-controlled fast acting controllers and its effect on the various parameters.

Table 16.1 Control action and parameters affected

Control action	Parameter affected
Control of (i) line reactance X (negligible line resistance) or (ii) angle δ	(i) Line current control (ii) Active power control (when angle is small)
Voltage injection in series with the line and perpendicular to the current	Real power control through the control of line current
Voltage injection in series with the line and any phase angle with respect to the driving voltage	Accurate control of real and reactive power flow in the line

From the foregoing, it may be concluded that FACTS technology presents techniques which enable the power utilities to load their transmission lines to their full thermal capability and in addition has the ability to control power without compromising on the security of supply. On the other hand, it may also be clarified that since FACTS technology increases the power transmission capability of lines, it may not be understood that it precludes either the need for enhancing the transmission capacity of lines, when thermal limits permit, or the need to set up new transmission lines. The factor which decides between the use of FACTS technology and the need to upgrade existing lines or add new lines is determined from an economic evaluation of the cost of losses on an existing transmission line plus the cost of FACTS technology against the cost of a new transmission line.

16.3 FACTS TECHNOLOGY

Alternating current transmission systems that employ power electronic based and other static controllers to enhance controllability and increase power transfer capability are defined as flexible alternating current transmission systems (FACTS), as per the IEEE PES task force.

High-powered electric drives and the base technology used in HVDC systems are used in the implementation of FACTS technology. With the advancement of power semiconductor devices, particularly the devices with turn-off capability, the cost of FACTS technology continues to decrease, and this will make the FACTS technology all the more acceptable.

The development of FACTS controllers has followed two distinctly different technical approaches, both resulting in a comprehensive group of controllers that solved some of the transmission problems. The first group is based on thyristor devices with only gate turn-on control, and its turn-off depends on the current reaching zero value as per circuit and system conditions. FACTS controllers in this group employ reactive elements or a tap-changing transformer with thyristor switches as controlled elements. The second group uses devices such as gate turn-off thyristors (GTO), MOS turn-off thyristors (MTO), and integrated gate-commutated thyristors (IGCT), and similar devices that have turn-on and turn-off capability. The second group uses self-commutating static converters operating as controlled voltage sources for FACTS applications. Since direct current in a voltage-sourced converter flows in either direction, the converter valves have to be bidirectional, and also, since the dc voltage does not reverse,

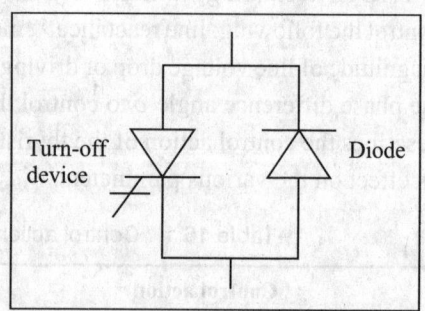

Fig. 16.11 Valve for a voltage-sourced converter

the turn-off devices need not have reverse voltage capability; such turn-off devices are known as *asymmetric turn-off devices*. Thus a voltage-sourced converter is made up of an asymmetric turn-off device such as GTO with a parallel diode connected in reverse as shown in Fig. 16.11.

16.4 BASIC TYPES OF FACTS CONTROLLERS

FACTS controllers are power electronics based systems and other static equipment that provide control of one or more transmission system parameters. A thyristor arrow inside a box, as shown in Fig. 16.12(a), is the general symbol of a FACTS controller. The four basic categories of FACTS controllers are as follows:

(a) Series controllers
(b) Shunt controllers
(c) Combined series–series controllers
(d) Combined series–shunt controllers

Figure 16.12 shows the schematic arrangements of the four types of FACTS controllers. Let us now discuss these in detail.

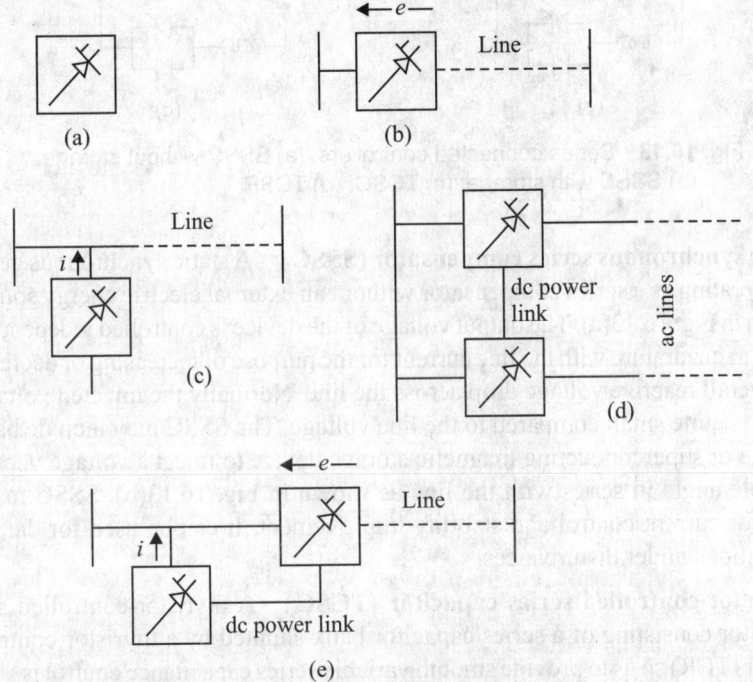

Fig. 16.12 Schematic arrangements of the basic types of FACTS controllers:
(a) general symbol, (b) series connection, (c) shunt connection,
(d) unified series–series connection, (e) unified series–shunt connection

16.4.1 Series Controller

The series controller may be a variable capacitor, inductor, or a power electronics based variable frequency source. Schematic arrangements of different types of series connected controllers are shown in Fig. 16.13. The principle of operation of a series controller is that it injects variable series voltage (current × variable reactance) in the line. When the injected series voltage is in phase quadrature with the line current, the controller absorbs or consumes reactive power only; for any other phase relationship it handles both real and reactive power. Some basic arrangements of series controllers will be discussed in the succeeding paragraphs.

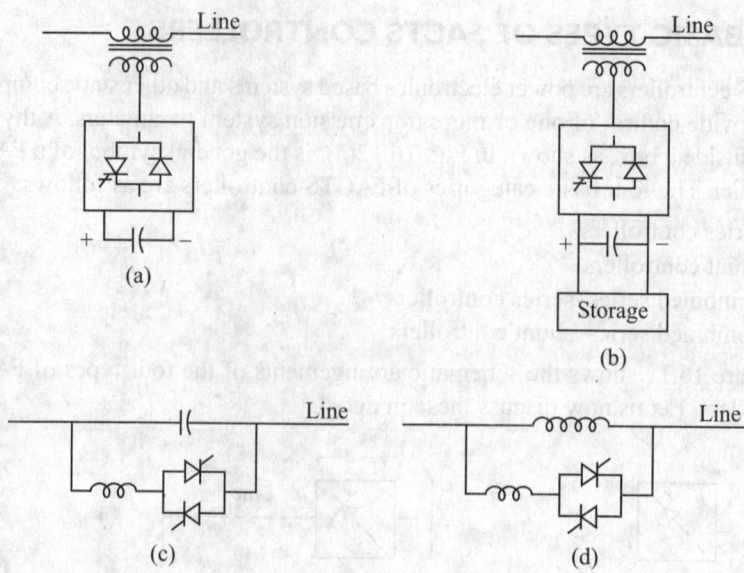

Fig. 16.13 Series connected controllers: (a) SSSC without storage;
(b) SSSC with storage; (c) TCSC; (d) TCSR

Static synchronous series compensator (SSSC) A static synchronous generator operating as a series compensator without an external electric energy source is
shown in Fig. 16.13(a). The output voltage of the device is controlled independently
and is in quadrature with the line current for the purpose of increasing or decreasing
the overall reactive voltage drop across the line. Normally the injected voltage in
series is quite small compared to the line voltage. The SSSC may include battery
storage or superconducting magnetic storage device to inject a voltage vector of
variable angle in series with the line as shown in Fig. 16.13(b). SSSC may be
used for current control and stability improvement. It can be used for damping
oscillations under disturbances.

Thyristor-controlled series capacitor (TCSC) A thyristor-controlled series
capacitor consisting of a series capacitor bank shunted by a thyristor-controlled
reactor (TCR) so as to provide smooth variable series capacitance control is shown
in Fig. 16.13(c). When the triggering delay angle of TCR is 180°, the reactor becomes non-conducting and the series capacitor has its normal impedance value.
As the triggering delay angle of TCR is reduced to less than 180°, the capacitive
reactance increases. When the triggering delay angle of TCR is 90°, the reactor
becomes fully conducting, and the total impedance becomes inductive, as the
reactor impedance is much lower than series capacitor impedance. TCSC may be
used for current control, stability improvement, damping oscillations, and fault
current limiting.

Thyristor-controlled series reactor (TCSR) A thyristor-controlled series
reactor consisting of a series reactor shunted by a thyristor-controlled reactor
(TCR) so as to provide smooth variable series reactance control is shown in
Fig. 16.13(d). When the triggering delay angle of TCR is 180°, the reactor becomes
non-conducting and the uncontrolled reactor acts as a fault current limiter. As the
triggering delay angle falls below 180°, the net inductance decreases until the

triggering angle reaches 90°, when the inductance value is the parallel combination of the two inductances. TCSR may be used for current control, and stability improvement, damping oscillations and fault current limiting.

16.4.2 Shunt Controller

Generally a shunt controller, as shown in Fig. 16.14, is similar to a series controller. It is, however, connected in parallel and injects current into the line. Thus, a variable shunt reactance connected to a transmission line injects current into the system at the point of connection. If the injected current is in phase quadrature with the line voltage, the controller handles reactive power only; for any other phase relationship of the current with the line voltage, it handles both real and reactive power. The shunt controllers may be variable impedance, variable source, or a combination of the two. Some of the basic arrangements of shunt controllers will be discussed now.

Static synchronous compensator (STATCOM)   STATCOM is one of the key FACTS controllers. It is a shunt-connected static VAR compensator and its capacitive or inductive output control is independent of the ac system voltage. A simple one line diagram of STATCOM based on voltage source converter and a current source converter is shown in Fig. 16.14(a). STATCOM may be used for voltage control, VAR compensation, and damping oscillations.

Static synchronous generator (SSG)   SSG is a combination of STATCOM and an energy source to supply or absorb power. The energy source may be a battery, large dc storage capacitor, superconducting magnet, etc. Figure 16.14(b) shows a single line diagram of a STATCOM connected to a storage device. SSG may be used for voltage control, VAR compensation, and damping oscillations under steady-state and transient conditions.

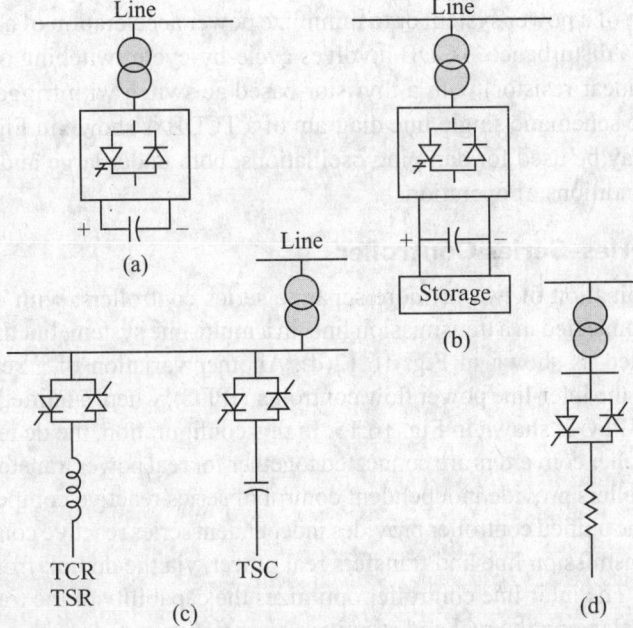

Fig. 16.14   Shunt controllers: (a) STATCOM, (b) SSG, (c) SVC/TCR, and (d) TCDB

Superconducting magnetic energy storage (SMES) Superconducting magnetic energy storage device consists of thyristor converters which can very rapidly inject/absorb both real and/or reactive power into the system to which the SMES is connected. This device can dynamically control power in an ac system, and thus can control voltage at the point of connection, and damp out transient oscillations that are caused by large power swings in the systems. As shown in the schematic diagram of Fig. 16.14(b), SMES may be connected as a storage device. This device may be used for voltage control, damping oscillations under large and small disturbances, and automatic generation control.

Static VAR compensator (SVC) A static VAR compensator is a shunt-connected, thyristor-controlled, or thyristor-switched reactor, and/or thyristor-switched capacitor, or a combination, which injects capacitive or inductive current so as to maintain the bus voltage of the electrical power system. Figure 16.14(c) shows the schematic diagram of an SVC. An SVC includes a separate thyristor-controlled or thyristor-switched reactor for absorbing reactive power and thyristor-switched capacitor for supplying the reactive power. The cost of an SVC is lower compared to that of a STATCOM. The application of an SVC can improve transmission capacity and the steady-state stability limit and can damp the sub-synchronous oscillations. SVC may be used as a stability improvement measure under small and large disturbance conditions.

Thyristor-controlled reactor (TCR) A thyristor-controlled inductor is a shunt-connected device as shown in Fig. 16.14(c). The effective reactance of a TCR is varied in a continuous manner by controlling the conduction of the thyristor valve by adjusting its triggering delay angle. TCRs can damp power oscillations.

Thyristor-controlled dynamic brake (TCDB) A thyristor-controlled dynamic brake is a shunt-connected thyristor-switched resistor, which is controlled to aid stabilization of a power system or to minimize power acceleration of a generating unit during a disturbance. TCDB involves cycle-by-cycle switching of a resistor (usually a linear resistor) with a thyristor-based ac switch with triggering angle control. The schematic single line diagram of a TCDB is shown in Fig. l6.14(d). A TCDB may be used for damping oscillations, both under large and small disturbance conditions of operation.

16.4.3 Series–Series Controller

It is a combination of two or more separate series controllers, with each series controller connected in a transmission line, in a multi-line system, but their control is coordinated as shown in Fig. 16.12(d). Another variation of a series–series controller is the inter-line power flow controller (IPFC), where a unified controller is used. An IPFC is shown in Fig. 16.15. In this configuration, the dc terminals of all the controller converters are connected together for real power transfer while the series controllers provide independent control of series reactive compensation of each line. The unified controller provides independent series reactive compensation for each transmission line and transfers real power, via the dc link from one line to the other. The inter-line controller optimizes the capability of the transmission system by balancing the real and reactive power flows.

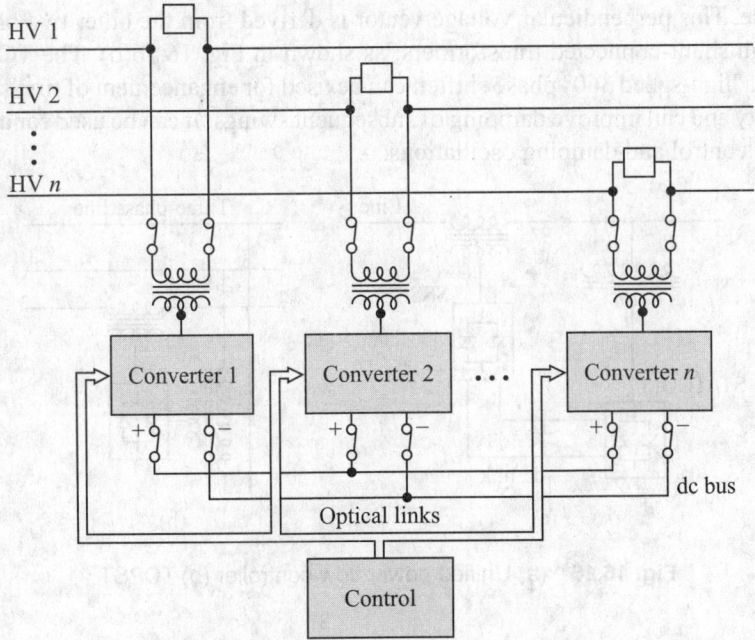

HV 1

HV 2

HV *n*

Converter 1   Converter 2   ...   Converter *n*

+   −   +   −   +   −

dc bus

Optical links

Control

Fig. 16.15   Inter-line power flow controllers

16.4.4 Combined Series–Shunt Controller

A combined series–shunt controller has separate series and shunt controllers in a transmission line whose operation is coordinated as shown in Fig. 16.12(e). Operationally, the series controller injects voltage in series with the line voltage and the shunt controller injects current into the system at the point of connection. The second type of this controller, where the shunt and series controllers are unified, is called a unified power flow controller (UPFC). A unified power flow controller is able to exchange real power via the dc power link in addition to performing the functions of a series–shunt controller.

Unified power flow controller (UPFC)   A unified power flow controller is a combination of a static synchronous compensator (STATCOM) and a static series compensator (SSSC). STATCOM and SSSC are coupled via a common dc link, to allow bidirectional flow of real power between the series output terminals of the SSC and the shunt output terminals of the STATCOM. UPFC is controlled to provide concurrently real and reactive series line compensation without an external electric energy source. The UPFC is able to control the transmission line voltage, impedance, and angle, or alternatively, the real and reactive power flow in the line. The UPFC may also provide independently controllable shunt reactive compensation. A single line diagram of a UPFC is shown in Fig. 16.16(a). A UPFC may be used for controlling active and reactive power, for voltage control, VAR compesation, damping oscillations, and fault current limiting.

Thyristor-controlled phase shifting transformer (TCPST)   TCPST is a transformer adjusted by thyristor switches to provide a rapidly variable phase angle. Phase shifting is obtained by adding a perpendicular voltage vector in series with a phase. This perpendicular voltage vector is derived from the other two phases through shunt-connected transformers, as shown in Fig. 16.16(b). The TCPST, being a high-speed 360° phase shifter, can be used for enhancement of first swing stability and can improve damping of subsequent swings. It can be used for active power control and damping oscillations.

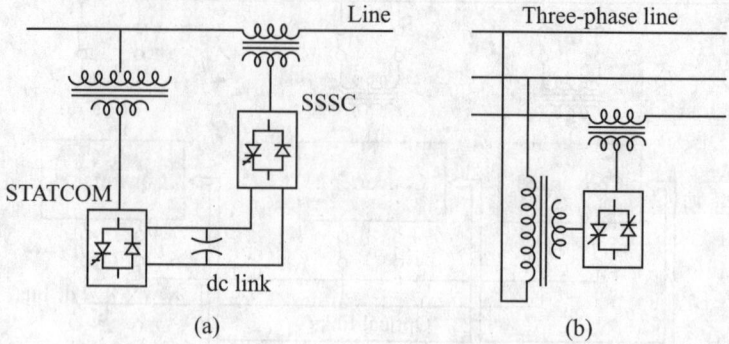

Fig. 16.16   (a) Unified power flow controller (b) TCPST

SUMMARY

- The flow of power in a transmission network is restricted by its parameters which prevent the transmission lines from exploiting them to their full transmission capabilities.
- FACTS technology which employs high-speed power electronics, advanced control technology, microcomputers, and powerful analytical tools is based on the technique of varying the reactance of a transmission line and thereby enhancing its transmission capability.
- The need to enhance transmission capability or the need to add new lines, however, is an economic issue.
- FACTS controllers are power electronics based systems, which along with their static equipment, control one or more transmission system parameters.
- The basic types of controllers are (i) series, (ii) shunt, (iii) series–series, and (iv) series–shunt.

SIGNIFICANT FORMULAE

- 90° lagging line current: $\dfrac{E_1 - E_2}{X}$

- Real power flow along a line: $\dfrac{|E_1||E_2|}{X}\sin\delta$

- Reactive power flow along a line connected between buses 1 and 2:

$$Q_1 = \frac{1}{X}\left(|E_1|^2 - |E_1||E_2|\cos\delta\right)$$

- Reactive power flow along a line connected between buses 2 and 1:

$$Q_2 = \frac{1}{X}\left(|E_2|^2 - |E_1||E_2|\cos\delta\right)$$

EXERCISES

Review Questions

16.1   Explain the importance of FACTS technology.

16.2   Describe power flow in (a) an ac parallel transmission circuit, (b) a HVAC and a HVDC line connected in parallel, and (c) an ac parallel transmission with a FACTS controller connected in series in one of the lines.

16.3   Explain how power flow takes place in a closed loop circuit. Assume a fictitious closed loop circuit of any number of buses.

16.4   Discuss, with the help of vector diagrams, the phenomena of power flow and stability in a transmission link when (a) voltage magnitude at one of the buses is increased, and (b) when the line voltage drop is varied.

16.5   Explain the basic working of FACTS technology.

16.6   Enumerate the various types of series controllers and describe their working.

16.7   List the various types of shunt controllers and describe their working.

Multiple Choice Objective Questions

16.1   Which of the following is the full form of FACTS?
   (a) Full alternating current transmission systems
   (b) Fast alternating current transmission systems
   (c) Flexible alternating current tracking systems
   (d) Flexible alternating current transmission systems

16.2   Which of the following is a key feature in the application of FACTS?
   (a) Microcomputers
   (b) Powerful analytical tools
   (c) Availability of high kV and kA rated switching devices
   (d) None of these

16.3   Which of the following is not applicable to electrical energy?
   (a) It is easily available.   (b) It can be transmitted.
   (c) It can be stored in bulk.   (d) All of these.

16.4   Two areas are interconnected by a parallel link and P pu power is transferred from area 1 to area 2. If the reactance of circuit 1 is 4 times that of circuit 2, which of the following gives the percentage of power flow in circuit 2?
   (a) 80%   (b) 60%
   (c) 40%   (d) 20%

16.5   Which of the following influences the flow of power in a transmission line?
   (a) Line losses   (b) Thermal loading
   (c) Series reactance   (d) All of these

16.6   In which of the following the load on a line cannot be varied?
   (a) Changing the reactance of a line
   (b) Changing the phase angle of a line
   (c) An HVAC line connected in parallel with an HVDC line
   (d) None of these

16.7   Which of the following occurs when the magnitudes of E_1 and/or E_2 are changed?
   (a) Maximum real power   (b) Real power line flow
   (c) Maximum reactive power   (d) All of these

16.8   Which of the following will vary both the real and reactive power flows in a line?

(a) Varying the line reactance

(b) Varying the magnitude and phase angle of the series injected voltage

(c) Varying the magnitude of the series injected voltage

(d) Varying the phase angle of the series injected voltage

16.9   Which of the following represents the reactive power flow in a line?

(a) $\dfrac{1}{X}|E_1|^2 \cos \delta$ (b) $\dfrac{1}{X}|E_2|^2 \cos \delta$

(c) $\dfrac{1}{X}\left(|E_1|^2 \cos \delta - |E_1||E_2|\cos \delta\right)$ (d) None of these

16.10   Which of the following is used in FACTS technology to control the load on a transmission line?

(a) X (b) $|E_1|$ and $|E_2|$

(c) $|E_1 - E_2|$ (d) All of these

Answers

16.1 (d)	16.2 (c)	16.3 (c)	16.4 (a)	16.5 (c)	16.6 (d)
16.7 (b)	16.8 (b)	16.9 (c)	16.10 (d)		

Matrix Operations Using MATLAB

OVERVIEW

Solving of engineering problems, using digital computers, requires the development of mathematical models to represent actual physical systems. Such simulations require translation of the characteristics of each individual component into mathematical expressions and developing a model which reflects the interconnections between the various components of the system.

Mathematical simulations of engineering problems usually lead to a series of mathematical expressions, which are a combination of non-linear equations, differential equations, and trigonometric functions. Solutions of complex mathematical equations can be achieved, using digital computers and employing numerical techniques. Matrix algebra is a very handy tool in developing mathematical models of complicated engineering problems and solving them by using digital computers.

Matrix Laboratory (commonly known as MATLAB) is a software interactive package which has been written to perform numerical computations. MATLAB is apt for performing matrix operations for engineering computations. It has a large stock of built-in functions for data analysis, linear algebra, and solution of one-dimensional equations as well as graphics and animation.

In this appendix, matrix notations and types of matrices, array, element, order of a matrix, square matrix, diagonal matrix, identity matrix, null matrix, etc. are defined. Matrix operations such as addition, subtraction and multiplication; determinants, minors, cofactors and adjoint; inverse of a matrix; matrix partitioning; sparse matrices, etc. have been discussed. Creation of different types of matrices, using MATLAB is also described. The use of MATLAB commands and functions has also been illustrated through examples.

A.1 MATRIX NOTATION AND TYPES OF MATRICES

Equation (A.1) shows the relations between the loop impedances, loop currents, and source voltages of a hypothetical network having three loop currents and voltages i_1, i_2, and i_3 and v_1, v_2, and v_3 respectively.

$$z_{11}i_1 + z_{12}i_2 + z_{13}i_3 = v_1$$

$$z_{21}i_1 + z_{22}i_2 + z_{23}i_3 = v_2 \tag{A.1}$$

$$z_{31}i_1 + z_{32}i_2 + z_{33}i_3 = v_3$$

In matrix form, Eq. (A.1) can be expressed as

$$V = ZI \tag{A.2}$$

where

$$Z = \begin{bmatrix} z_{11} & z_{12} & z_{13} \\ z_{21} & z_{22} & z_{23} \\ z_{31} & z_{32} & z_{33} \end{bmatrix} = \text{bus impedance matrix}$$

$$V = \begin{bmatrix} v_1 \\ v_2 \\ v_3 \end{bmatrix} = \text{column vector of known voltages}$$

$$I = \begin{bmatrix} i_1 \\ i_2 \\ i_3 \end{bmatrix} = \text{column vector of unknown currents}$$

Thus, a matrix is a mathematical notation to represent simultaneous equations in a concise form. It is represented by an italic boldfaced capital letter, for example, Z and A represent matrix quantities.

A matrix is constituted of an array of real or complex numbers, called *elements*, organized systematically in m rows and n columns. Such a matrix is called a *rectangular matrix*. An element in a matrix is identified by referring to its row and column numbers and is designated by lowercase letters. Thus, $z(i, j)$ in a Z matrix would refer to the impedance z in row i and column j. The dimension or order or size of a matrix is determined by the number of rows and columns in a matrix and is given by $m \times n$. When a matrix has one row ($m = 1$) and n columns, it is called a *row matrix* or a *row vector*. Similarly, when a matrix has m rows and one column ($n = 1$), it is called a *column matrix* or *column vector*. In Eq. (A.2), I and V are column vectors.

A.1.2 Types of Matrices

In a rectangular matrix when the number of rows is equal to the number of columns ($m = n$), a *square matrix* is realized. The dimension of such a matrix is equal to m or n. The elements lying on the diagonal of a square matrix are called the *diagonal elements* and all other elements are called the *off-diagonal* elements. Thus, the dimension of matrix Z of Eq. (A.2) is three and z_{11}, z_{22}, z_{33} are the diagonal elements of the matrix. All other elements in the matrix are the off-diagonal elements.

A square matrix is symmetric when the corresponding elements above and below the diagonal are equal. The Z matrix of Eq. (A.2) would be a symmetric matrix if $z_{ij} = z_{ji}$ for all values of i and j.

A *diagonal matrix* is a square matrix which has all the off-diagonal elements equal to zero. The following matrix is a diagonal matrix.

$$Z = \begin{bmatrix} z_{11} & 0 & 0 \\ 0 & z_{22} & 0 \\ 0 & 0 & z_{33} \end{bmatrix}$$

A *unity* or *identity matrix* is a square matrix whose diagonal elements are unity and the off-diagonal elements are zero. A unity matrix is represented by U. For example,

$$U = \begin{bmatrix} 1 & 0 & 0 \\ 0 & 1 & 0 \\ 0 & 0 & 1 \end{bmatrix}$$

A *null* or *zero matrix* is defined as a matrix having all its elements equal to zero. A *lower triangular matrix* is a square matrix with all the elements above the diagonal ($i < j$) equal to zero. For example,

$$Z = \begin{bmatrix} z_{11} & 0 & 0 \\ z_{21} & z_{22} & 0 \\ z_{31} & z_{32} & z_{33} \end{bmatrix}$$

Similarly an *upper triangular matrix* is a square matrix with all the elements below the diagonal $(i > j)$ equal to zero. For example,

$$Z = \begin{bmatrix} z_{11} & z_{12} & z_{13} \\ 0 & z_{22} & z_{23} \\ 0 & 0 & z_{33} \end{bmatrix}$$

Transpose of a rectangular matrix is obtained when its rows are interchanged with its columns. In other words, if a matrix is of order $m \times n$, then its transpose will be a matrix of order $n \times m$. The following matrix is of the order 2×3.

$$Y = \begin{bmatrix} y_{11} & y_{12} & y_{13} \\ y_{21} & y_{22} & y_{23} \end{bmatrix}$$

The transpose of the preceding matrix will be of order 3×2 and is designated by Y^t. Thus,

$$Y^t = \begin{bmatrix} y_{11} & y_{21} \\ y_{12} & y_{22} \\ y_{13} & y_{23} \end{bmatrix}$$

For a square matrix, constituted of elements of real numbers, if $YY^t = U = Y^tY$, then Y is called an *orthogonal matrix*.

A square matrix consisting of elements of real numbers is *skew-symmetric* when its diagonal elements are zero and the corresponding off-diagonal elements are equal but have the opposite signs. The following matrix is an example of a skew-symmetric matrix.

$$Z = \begin{bmatrix} 0 & 1 & -2 \\ -1 & 0 & -4 \\ 2 & 4 & 0 \end{bmatrix}$$

Another property of a skew-symmetric matrix is that $Z = -Z^t$. A *conjugate matrix* is obtained when all the elements of a rectangular matrix are replaced by their conjugates, that is, $r + jx$ is replaced by $r - jx$. The following matrix represents a 3×2 complex matrix.

$$Z = \begin{bmatrix} 1 + j2 & 4 \\ 2 - j2 & -j4 \\ 3 & 1 - j2 \end{bmatrix}$$

The conjugate of Z is designated by Z^* and is expressed as follows:

$$Z^* = \begin{bmatrix} 1 - j2 & 4 \\ 2 + j2 & +j4 \\ 3 & 1 + j2 \end{bmatrix}$$

A *square complex matrix* Z is defined as a *Hermitian matrix*, if all the diagonal elements are real and $Z = (Z^*)^t$. An example of a Hermitian matrix is as follows:

$$Z = \begin{bmatrix} 8 & 3 + j5 \\ 3 - j5 & 6 \end{bmatrix}$$

A *skew-Hermitian* matrix is defined as a square complex matrix whose all diagonal elements are either zero or complex and $Z = (Z^*)^t$. An example of a skew-Hermitian matrix is as follows:

$$Z = \begin{bmatrix} j4 & (4-j3) \\ (-4-j3) & j6 \end{bmatrix}$$

If a square complex matrix Z satisfies the relation $Z(Z^*)^t = U = (Z^*)^t Z$, then Z is defined as a *unitary matrix*. If the elements of a unitary matrix are real, then the matrix is an orthogonal matrix.

Using MATLAB

MATLAB is an interactive software package. The name MATLAB stands for Matrix Laboratory. It has hundreds of built-in functions for computation, graphics, and animation. Matrix dimensions are determined automatically by MATLAB, or in other words, no implicit declarations for dimensions are required.

A matrix can be created by entering the elements row wise, each element separated by a comma or a space, and the rows are separated from each other by a semi-colon (;), or a carriage return. The elements are enclosed within a square bracket. Row and column vectors are entered in the same manner as a matrix. However, a column vector will be created when the elements are separated by semi-colon. The following example shows how matrices and vectors are created with MATLAB.

Example A.1 For a network impedance matrix given by

$$Z = \begin{bmatrix} 6 & -2 & -3 \\ -2 & 5 & -1 \\ -3 & -1 & 5 \end{bmatrix}$$

write the MATLAB commands to create the impedance matrix and obtain a print out.

Solution The following MATLAB commands will give the impedance matrix Z.

```
≫ Z = [6 -2 -3; -2 5 -1; -3 -1 5]
Z =
     6      -2      -3
    -2       5      -1
    -3      -1       5    .
```

Example A.2 Write the MATLAB commands for creating column vectors for the known voltage sources $V^t = [20\ -10\ 0]$ and the unknown loop current vector $I^t = [I_1\ I_2\ I_3]$.

Solution The following MATLAB commands will give the voltage vector V and current vector I.

```
≫ V = [20; -10; 0]
≫ V =
    20
   -10
     0
≫ I = [I_1; I_2; I_3]
    I = I_1
        I_2
        I_3
```

Additionally, in order to ease generation and manipulation of matrices, several utility matrices have been provided in MATLAB. A few examples, related to the types of matrices described in this section are as follows:

MATLAB Command	Output
B = A'	Produces the transpose of matrix A if A is real, and produces the conjugate transpose of matrix A if A is complex.
eye (m, n)	Produces a matrix of dimension $m \times n$ with unity on the main diagonal
eye(m)	Produces a $m \times m$ identity matrix
zeros (m, n)	Produces a $m \times n$ matrix of zeros
ones (m, n)	Produces a $m \times n$ matrix of ones
ones(m)	Produces a square matrix of order m
tril (Z)	Extracts the lower triangular part of the Z matrix
triu (Z)	Extracts the upper triangular part of the Z matrix

A.2 PERFORMING MATHEMATICAL OPERATIONS ON MATRICES AND ARRAYS

Arithmetic operations (addition, subtraction, multiplication, and division) can be carried out on matrices and arrays and the laws governing them have been discussed in this section.

A.2.1 Addition and Subtraction of Matrices

Conformable matrices, that is, matrices having the same dimensions, can be added or subtracted. For example, two matrices X and Y each of order $m \times n$ can be added or subtracted to obtain a third matrix Z of order $m \times n$. Each element of Z is given by

$$z_{ij} = x_{ij} \pm y_{ij}$$

Addition and subtraction of matrices obey the following associative and commutative laws of arithmetic:

$$X \pm Y \pm Z = X \pm (Y \pm Z) = (X \pm Y) \pm Z$$

$$X \pm Y = Y \pm X$$

A.2.2 Multiplication and Division of Matrices by Scalar Quantities

Multiplication of a matrix X, by a scalar quantity k, is performed by multiplying each element of the matrix by k as follows:

$Y = kX$ where $y_{ij} = kx_{ij}$ for all values of i and j.

Similarly, $Y = X/k$ where $y_{ij} = x_{ij}/k$ for all values of i and j. Multiplication of matrices by scalar quantities obey the commutative and distributive laws, that is,

$$kX = Xk$$

$$k(X \pm Y) = (X \pm Y)k = kX \pm kY$$

A.2.3 Multiplication of Matrices

Multiplication of two matrices can be performed only if the dimensions of the two matrices are in conformity with each other. In other words, the product of the two matrices is defined if the number of columns of the first matrix is equal to the number of rows of the second matrix. Thus, if a matrix X of dimension $m \times q$ is to be multiplied with a matrix Y, then the dimension of Y must be $q \times n$. The product of the two matrices is a matrix of order $m \times n$.

$$XY = \begin{bmatrix} x_{11} & x_{12} \\ x_{21} & x_{22} \\ x_{31} & x_{32} \end{bmatrix} \begin{bmatrix} y_{11} & y_{12} \\ y_{21} & y_{22} \end{bmatrix} = \begin{bmatrix} x_{11}y_{11} + x_{12}y_{21} & x_{11}y_{12} + x_{12}y_{22} \\ x_{21}y_{11} + x_{22}y_{21} & x_{21}y_{12} + x_{12}y_{22} \\ x_{31}y_{11} + x_{32}y_{21} & x_{31}y_{12} + x_{32}y_{22} \end{bmatrix}$$

If it is assumed that $Z = XY$, it can be observed from the preceding matrix multiplication that the dimension of matrix Z is 3×2. Also, a general mathematical relation can be written for each element of Z as follows:

$$z_{ij} = \sum_{k=1}^{q} x_{ik} y_{kj} \text{ for } i = 1, 2, 3, \ldots, m \text{ and } j = 1, 2, 3, \ldots, n$$

Matrix multiplication can be accomplished by the following laws.

- Observe that the product YX is not defined since the number of columns of Y is not equal to the number of rows of X. Thus, generally, $Z = XY \neq YX$. Hence matrix multiplication is not commutative.
- However, matrix multiplication obeys the associative, distributive, and reversal laws, when the dimensions of the matrices conform to the laws of multiplication and addition.

Example A.3   Demonstrate the *associative law* $X(YZ) = (XY)Z$.

Solution

```
>> X=[1 2 3;4 5 6;7 8 9], Y=[1 3 5;2 4 6; 3 5 7], Z=[4 6 8;5 7 9;7 9 10]
X =

      1       2       3
      4       5       6
      7       8       9
Y =

      1       3       5
      2       4       6
      3       5       7
Z =

      4       6       8
      5       7       9
      7       9       10
>> X*(Y*Z)
ans =
    452     608     726
   1082    1454    1734
   1712    2300    2742
>> (X*Y)*Z
ans =
    452     608     726
   1082    1454    1734
   1712    2300    2742
```

Example A.4   Demonstrate the *distributive law* $X(Y + Z) = XY + XZ$.

Solution

```
>> X=[1 2;3 4], Y=[10 9;11 13], Z=[6 -2; -2 8]
X =

      1       2
      3       4
Y =

     10       9
     11      13
```

```
Z =
      6       -2
     -2        8
≫ X*(Y+Z)
ans =
     34       49
     84      105
≫ X*Y+X*Z
ans =
     34       49
     84      105
```

Reversal law If $Z = XY$, then $Z^t = Y^t X^t$. The transpose of a real matrix A is obtained by typing A', that is, the name of the matrix followed by a single quote (prime). For a complex matrix A, $B = A'$ produces matrix B that is a conjugate transpose of matrix A.

Example A.5 Demonstrate the reversal law $Z = XY$ and then $Z^t = Y^t X^t$.

Solution

```
≫ X = [12 10 8;10 8 9;5 9 11;];Y=[2 5 9;6 10 2;8 15 4];
≫ Z=X*Y
Z =   148       280       160
      140       265       142
      152       280       107
≫ X' = 12 10 5
         10    8     9
          8    9    11
≫ Y' = 2 6 8
         5   10   15
         9    2    4
≫ Y'* X' = 148 140 152
                280      265      280
                160      142      107
```

It can be observed that $Y'X'$ is the transpose of Z, thereby verifying the reversal law. Two cases are discussed here.

Case (a) If $XY = 0$, it is not necessary that either $X = 0$ or $Y = 0$.

Example A.6 Demonstrate case (a)

```
≫ X=[1 2 3;4 5 9;7 8 15];
≫ Y=[1;1;-1];
≫ Z =X*Y
 = 0
   0
   0
```

Case (b) If $XY = XZ$, it is not necessary that $Y = Z$. The transpose of a column vector is a row vector and the multiplication of the transpose of the vector with the column vector is the sum of the square of each element.

A.2.4 Determinant of a Matrix

Only the determinant of a square matrix is defined and it has a single value. Determinant $|Z|$ for the square matrix Z is

$$|Z| = \begin{vmatrix} z_{11} & z_{12} & z_{13} \\ z_{21} & z_{22} & z_{23} \\ z_{31} & z_{32} & z_{33} \end{vmatrix}$$

The order of the determinant is the same as that of the matrix.

Minor M_{ij} of the element z_{ij} is defined as that determinant which is obtained by striking out the ith row and jth column of a determinant $|Z|$ of order n. The order of the minor is $n-1$. The minor of z_{11} is given by

$$M_{11} = \begin{vmatrix} z_{22} & z_{23} \\ z_{32} & z_{33} \end{vmatrix}$$

Cofactor α_{ij} of an element z_{ij} is given by

$$\alpha_{ij} = (-1)^{i+j} M_{ij}$$

where M_{ij} is the minor of z_{ij}. The cofactor of element z_{31} of $|Z|$ is written as

$$\alpha_{31} = (-1)^{3+1} M_{31} = \begin{vmatrix} z_{12} & z_{13} \\ z_{22} & z_{23} \end{vmatrix} = (z_{12}z_{23} - z_{22}z_{13})$$

Generally the value of an nth order determinant can be obtained from the following formulae

$$|Z| = \sum_{i=1}^{n} z_{ij}\alpha_{ij} \text{ with } j \text{ chosen for one column} \tag{A.3a}$$

$$|Z| = \sum_{j=1}^{n} z_{ij}\alpha_{ij} \text{ with } i \text{ chosen for one row} \tag{A.3b}$$

Example A.7   Demonstrate the application of the above formulae for computing the determinant $|A|$ of the given matrix $[A]$.

$$[A] = \begin{bmatrix} 1 & 2 & 3 \\ 2 & 4 & 0 \\ 3 & 1 & 1 \end{bmatrix} = 1\begin{vmatrix} 4 & 0 \\ 1 & 1 \end{vmatrix} - 2\begin{vmatrix} 2 & 0 \\ 3 & 1 \end{vmatrix} + 3\begin{vmatrix} 2 & 4 \\ 3 & 1 \end{vmatrix}$$

$$= 1(4) - 2(2) + 3(2 - 12) = -30$$

Solution   The determinant of any square matrix can be obtained by using the MATLAB library function det as follows:

```
≫ A= [1 2 3;2 4 0;3 1 1];
≫ det (A)
ans =
-30
```

It may be noted that a matrix is said to be singular if the value of its determinant is zero. If the elements of a square matrix are replaced by their cofactors and transposed, then the square matrix so obtained is defined as the *adjoint matrix*. Thus,

$$\text{adjoint } Z = \begin{bmatrix} \alpha_{11} & \alpha_{12} & \alpha_{13} \\ \alpha_{21} & \alpha_{22} & \alpha_{23} \\ \alpha_{31} & \alpha_{32} & \alpha_{33} \end{bmatrix}^t = \begin{bmatrix} \alpha_{11} & \alpha_{21} & \alpha_{31} \\ \alpha_{12} & \alpha_{22} & \alpha_{32} \\ \alpha_{13} & \alpha_{23} & \alpha_{33} \end{bmatrix}$$

In MATLAB, the statement $A(m:n, k:l)$ specifies rows m to n and columns k to l of the matrix A. This statement is useful for determining the cofactors of a matrix.

A.2.5 Inversion of a Matrix

In order to obtain the unknown loop currents in Eq. (A.2) it is necessary to divide the known voltage vector E by the coefficient matrix Z. However, in matrix operations, division does not exist except by a scalar quantity as explained in Section A.2.2. Division in Eq. (A.2) is performed by computing the inverse of the coefficient matrix Z.

The inverse of a matrix is designated by Z^{-1} and exhibits the following property:

$$ZZ^{-1} = Z^{-1}Z = U$$

The inverse of a matrix is given by the following relation

$$Z^{-1} = \frac{\text{adjoint } Z}{|Z|} \tag{A.4}$$

If the determinant $|Z| = 0$, the matrix is singular and inverse does not exist.

Example A.8 Demonstrate the various steps to compute the inverse of a matrix using Eq. (A.4).

Solution MATLAB commands have been used in finding the inverse of a matrix.

Step 1: Input matrix and compute its determinant

```
≫ A= [1 3 5;0 2 4;-2 0 6]
A =
    1      3      5
    0      2      4
   -2      0      6
≫ det(A)
ans =
    8
```

Step 2: Initialize a null matrix for cofactors

```
≫ cofac = zeros(3,3)
cofac =
    0      0      0
    0      0      0
    0      0      0
```

Step 3: Compute the cofactors

```
≫ tem = A(2:3,2:3)
tem =
    2      4
    0      6
    » cofac(1,1)=det(tem)
cofac =
   12      0      0
    0      0      0
    0      0      0
≫ tem=A(2:3,1:2:3)
tem =
    0      4
   -2      6
≫ cofac(1,2)=-3*det(tem)
cofac =
   12    -24      0
    0      0      0
    0      0      0
≫ tem = A(2:3,1:2)
```

```
tem =
    0        2
   -2        0
≫ cofac(1,3) = 5*det(tem)
cofac =
   12      -24       20
    0        0        0
    0        0        0
≫ tem=A(1:2:3,1:2:3)
tem =
    1        5
   -2        6
≫ cofac(2,2) = 2*det(tem)
cofac =
   12      -24       20
    0       32        0
    0        0        0
≫ tem=A(2:3,1:2)
tem =
    0        2
   -2        0
≫ cofac(2,3) = -4*det(tem)
cofac =
   12      -24       20
    0       32      -24
    0        0        0
≫ tem=A(1:2,2:3)
tem =
    3        5
    2        4
≫ cofac(3,1) = -2*det(tem)
cofac =
   12      -24       20
    0       32      -24
   -4        0        0
≫ tem=A(1:2,1:2)
tem =
    1        3
    0        2
≫ cofac(3,3) = 6*det(tem)
cofac =
   12      -24       20
    0       32      -24
   -4        0       12
```

Step 3: Compute the adjoint matrix

```
≫ adjoint=cofac'
adjoint =
   12        0       -4
  -24       32        0
   20      -24       12
```

Step 4: Compute the matrix inverse

```
≫ adjoint/det(A)
ans =
  1.5000        0      -0.5000
 -3.0000    4.0000         0
  2.5000   -3.0000     1.5000
```

A.2.6 Solving Linear Simultaneous Equations

As already stated, division of matrices, except by a scalar quantity, does not exist. Matrix inversion is employed to obtain solutions of linear simultaneous equations as explained here.

Multiply both sides of Eq. (A.1) by Z^{-1}. Thus,

$$Z^{-1}V = Z^{-1}ZI$$

$$Z^{-1}V = UI = I \qquad (A.5)$$

Equation (A.5) shows that Eq. (A.2) can be solved by pre-multiplying the known column vector of voltages by the inverse of Z matrix.

Example A.9 A network has two independent meshes and the mesh currents in each are I_1 and I_2. The relation between the mesh currents, source voltages, and network parameters are given by the following matrix equations:

$$\begin{bmatrix} 14 + j20 & -8 - j12 \\ -8 - j12 & 18 + j6 \end{bmatrix} \begin{bmatrix} I_1 \\ I_2 \end{bmatrix} = \begin{bmatrix} 20 \\ j30 \end{bmatrix}$$

Determine the mesh currents.

Solution From Eq. (A.5), it is clear that

$$\begin{bmatrix} I_1 \\ I_2 \end{bmatrix} = \begin{bmatrix} 14 + j20 & -8 - j20 \\ -8 - j12 & 18 + j6 \end{bmatrix}^{-1} \begin{bmatrix} 20 \\ j30 \end{bmatrix}$$

By the method of determinants and employing MATLAB, the mesh currents are computed as follows:

```
>> Z= [(14+20i) (-8-12i);(-8-12i) (18+6i)]
Z =
   14.0000 +20.0000i -8.0000 -12.0000i
   -8.0000 -12.0000i 18.0000 + 6.0000i
>> Z1=[20 (-8-12i);30i (18+6i)]
Z1 =
   20.0000      -8.0000 - 12.0000i
   0 + 30.0000i 18.0000 + 6.0000i
>> det (Z)
ans =
   2.1200e+002 + 2.5200e+002i
>> det(Z1)
ans =
   0 + 3.6000e + 002i
>> Z2= [(14+20i) 20;( -8-12i) 30i]
Z2 =
   14.0000 + 20.0000i     20.0000
   -8.0000 - 12.0000i     0 +30.0000i
>> det(Z2)
ans =
   -4.4000e + 002 + 6.6000e+002i
>> I2 = det(Z2)/det(Z)
I2 =
   0.6735 + 2.3126i
>> I1 = det(Z1)/det(Z)
I1 =
   0.8365 + 0.7037i
```

Alternatively, the mesh currents can be calculated by using the MATLAB library function of inv.

```
>> Z= [(14+20i) (-8-12i);( -8- -2i) (18+6i)]
Z =
      14.0000 +20.0000i -8.0000 -12.0000i
      -8.0000 -12.0000i 18.0000 + 6.0000i
>> V= [20;30i]
V =
  20.0000
   0 + 30.0000i
>> inv(Z)
ans =
    0.0491 - 0.0301i 0.0435 + 0.0049i
    0.0435 + 0.0049i 0.0738 + 0.0066i
>> I=inv (Z)*V
I =
    0.8365 + 0.7037i
    0.6735 + 2.3126i
```

It may be noted that MATLAB accepts complex operator i instead of the operator j, normally employed by engineers, for performing complex mathematical computations. MATLAB also permits right division. Right division is employed to solve a matrix equation and can be performed by using the backward stroke (\). The use of the backward stroke(\) is demonstrated here for solving the matrix equation of Example A.9.

```
>> I=Z\V
I =
  0.8365 + 0.7037i
  0.6735 + 2.3126i
```

A.3 MATRIX PARTITIONING

In order to perform matrix operations on matrices, representing large electrical networks, it is convenient to break up the matrix into smaller matrices. The process of dividing a large square matrix into smaller matrices is called *partitioning*. Each part of the matrix is called a sub-matrix. The rule for partitioning is that all the resultant sub-matrices should be conformable to mathematical operations. The rule for partitioning a matrix is that if the horizontal partition line is drawn between the r and $r + 1$ rows, then the vertical partition line should be drawn between the r and $r + 1$ columns. The mathematical operations such as addition, subtraction, and multiplication can then be performed by treating each sub-matrix as a single element. Equation (A.6) shows a 4×4 matrix A partitioned between the third and fourth rows and columns.

$$A = \begin{array}{|ccc|c|} a_{11} & a_{12} & a_{13} & a_{14} \\ a_{21} & a_{22} & a_{23} & a_{24} \\ a_{31} & a_{32} & a_{33} & a_{34} \\ \hline a_{41} & a_{42} & a_{43} & a_{44} \end{array} \qquad (A.6)$$

If

$$A_1 = \begin{array}{ccc} a_{11} & a_{12} & a_{13} \\ a_{21} & a_{22} & a_{23} \\ a_{31} & a_{32} & a_{33} \end{array} \qquad A_2 = \begin{array}{c} a_{14} \\ a_{24} \\ a_{34} \end{array} \qquad A_3 = \begin{array}{|ccc|} a_{14} & a_{24} & a_{32} \end{array}$$

and $\quad A_4 = \boxed{a_{44}}$

then

$$A = \begin{array}{|c|c|} \hline A_1 & A_2 \\ \hline A_3 & A_4 \\ \hline \end{array}$$

The mathematical operations can now be performed by treating each sub-matrix as an element as follows:

$$\begin{array}{|c|c|} \hline A_1 & A_2 \\ \hline A_3 & A_4 \\ \hline \end{array} \pm \begin{array}{|c|c|} \hline B_1 & B_2 \\ \hline B_3 & B_4 \\ \hline \end{array} = \begin{array}{|c|c|} \hline C_1 & C_2 \\ \hline C_3 & C_4 \\ \hline \end{array}$$

where

$$C_1 = A_1 \pm B_1$$
$$C_2 = A_2 \pm B_2$$
$$C_3 = A_3 \pm B_3$$
$$C_4 = A_4 \pm B_4$$

In case the matrices A and B are to be multiplied, that is, $C = A*B$, then

$$C_1 = A_1*B_1 + A_2*B_3$$
$$C_2 = A_1*B_2 + A_2*B_4$$
$$C_3 = A_3*B_1 + A_4*B_4$$
$$C_4 = A_3*B_2 + A_4*B_4$$

Similarly, the transpose of matrix $[A]$ is given by

$$A^t = \begin{array}{|c|c|} \hline A_1^t & A_3^t \\ \hline A_2^t & A_4^t \\ \hline \end{array}$$

Assume that

$$A^{-1} = \begin{array}{|c|c|} \hline B_1 & B_2 \\ \hline B_3 & B_4 \\ \hline \end{array}$$

then

$$B_1 = [A_1 - A_2 A_4^{-1} A_3]^{-1} \qquad (A.7a)$$
$$B_2 = -B_1 A_2 A_4^{-1} \qquad (A.7b)$$
$$B_3 = -A_4^{-1} A_3 B_1 \qquad (A.7c)$$
$$B_4 = A_4^{-1} - A_4^{-1} A_3 B_2 \qquad (A.7d)$$

The essential condition to compute A^{-1} is that sub-matrices A_1 and A_4 must be square. The advantage of matrix partitioning and the orderliness with which a matrix inverse can be obtained is shown in the following example.

Example A.10 The matrix Y is the admittance matrix of a six-node network shown in Fig. A.1. It is desired to reduce the network by eliminating the nodes 4, 5, and 6. Determine the admittance matrix of the resultant reduced network.

$$Y = \begin{array}{cccccc} & 1 & \quad 2 & \quad 3 & \quad 4 & \quad 5 & \quad 6 \end{array}$$

$$Y = \begin{bmatrix} -j0.9 & 0 & 0 & j0.9 & 0 & 0 \\ 0 & -j1.1 & 0 & 0 & j1.1 & 0 \\ 0 & 0 & -j1.0 & 0 & 0 & j1.0 \\ j0.9 & 0 & 0 & -j6.9 & j1.0 & j5.0 \\ 0 & -j1.1 & 0 & j1.0 & j7.1 & j5.0 \\ 0 & 0 & -j1.0 & j5.0 & j5.0 & -j11.0 \end{bmatrix}$$

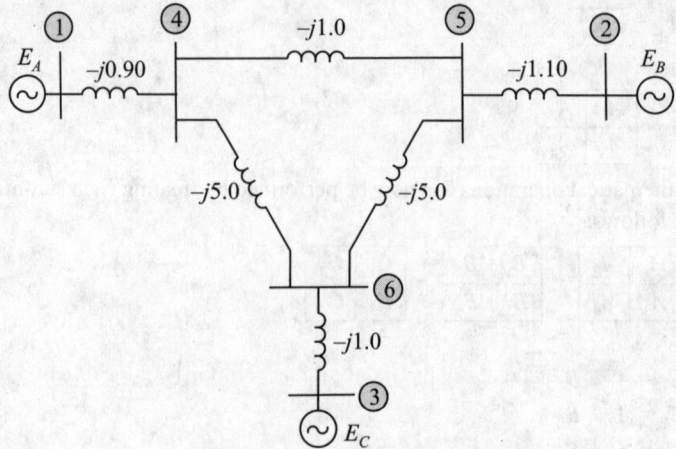

Fig. A.1   Six-node network

Solution   To eliminate nodes 4, 5, and 6, the Y matrix is partitioned between rows and columns 3 and 4 as shown.

$$Y = \begin{array}{|ccc|ccc|}
\hline
-j0.9 & 0 & 0 & -j0.9 & 0 & 0 \\
0 & -j1.1 & 0 & 0 & j1.1 & 0 \\
0 & 0 & -j1.0 & 0 & 0 & j1.0 \\
\hline
j0.9 & 0 & 0 & -j6.9 & j1.0 & j5.0 \\
0 & j1.1 & 0 & j1.0 & -j7.1 & j5.0 \\
0 & 0 & j1.0 & j5.0 & 5.0 & -j11.0 \\
\hline
\end{array}$$

Using MATLAB the nodes are eliminated as follows:

Step 1: Input matrix Y

≫ Y=[–0.9i 0 0 0.9i 0 0;0 –1.1i 0 0 1.1i 0;0 0 –1.0i 0 0 1.0i;0.9i 0 0 –
6.9i 1.0i 5.0i;0 1.1i 0 1.0i
–7.1i 5.0i;0 0 1.0i 5.0i 5.0i –11.0i];

Step 2: Partition the network admittance matrix Y between rows and columns 3 and 4 to eliminate nodes 4, 5, and 6. Extract sub-matrices A1, A2, A3, and A4.

≫ A1=[Y(1:3,1:3)]

A1 =

    0 – 0.9000i   0   0
         0   0 – 1.1000i   0
         0   0   0 – 1.0000i

≫ A2=[Y(1:3,4:6)]

A2 =

    0 + 0.9000i   0   0
         0   0 + 1.1000i   0
         0   0   0 + 1.0000i

≫ A3=[Y(4:6,1:3)]

A3 =

    0 + 0.9000i   0   0
         0   0 + 1.1000i   0
         0   0   0 + 1.0000i

```
>> A4 = [Y(4:6,4:6)]
A4 =
   0 - 6.9000i   0 + 1.0000i   0 + 5.0000i
   0 + 1.0000i   0 - 7.1000i   0 + 5.0000i
   0 + 5.0000i   0 + 5.0000i   0 -11.0000i
```

Step 3: Compute the admittance matrix of the reduced network.

```
>> A1-A2*inv(A4)*A3
ans =
   0 - 0.5637i   0 + 0.2787i   0 + 0.2850i
   0 + 0.2787i   0 - 0.6184i   0 + 0.3397i
   0 + 0.2850i   0 + 0.3397i   0 - 0.6248i
```

It can be observed that the reduced matrix gives the admittance of the three-node network.

A.4 SPARSITY IN MATRICES

A matrix is said to be sparsely filled (or simply sparse), if the number of zero elements in a matrix is in excess of the number of non-zero elements. If the number of non-zero elements is less than 15%, the matrix is called a *sparse matrix*.

In electrical power networks, the bus admittance matrix has a large number of zero elements compared to the non-zero elements. This is due to the property of bus admittance matrix which reflects only the direct connection between the buses. Also, the bus admittance matrix is symmetrical about the major diagonal. An example of a six-bus admittance matrix of a hypothetical network is given below. Crosses (×) are the only non-zero elements in the network.

	1	2	3	4	5	6
	×	×		×	×	
	×	×				×
			×		×	
	×				×	
	×	×		×		
		×				×

It can be observed that the number of elements required to be stored (along with the diagonal elements) for a symmetric square matrix of order n is equal to $n(n + 1)/2$. The sparsity in bus admittance matrices in electrical power networks can further be exploited to reduce storage requirements and increase speed of computation with the help of sparsity techniques discussed in Chapter 6 of this book.

EXERCISES

Numerical Problems

A.1   Determine (i) $A + B$, (ii) $A*B$ (iii) A^{-1}, (iv) A^2, (v) A^t, (vi) $A^t B^t$, (vii) $A^{0.5}$ for the matrices given below. Also state in each case the type of the resultant matrix.

$$A = \begin{bmatrix} 4 & -6 & 3+j6 \\ 1 & 2-j4 & 8-j9 \\ 2+j1 & 0 & 14 \end{bmatrix}; B = \begin{bmatrix} 1+j2 & 2.5-j3.4 & 0 \\ 6 & 1.2+j4.5 & j4 \\ 2-j0.5 & 3.2 & j0.8 \end{bmatrix}$$

A.2 Verify the reversal rule for Problem A.1

A.3 For the matrices of Problem A.1, calculate $(A^*)^t A$ and give the type of the resultant matrix.

A.4 Write the following simultaneous equations in matrix form.

$$2x + 4y - 3z = 10$$
$$-5y + z = 6$$
$$4x + 8y = 0$$

A.5 If the equations in Problem A.4 are represented by $AX = Y$, state the type of each matrix. Compute A^{-1} and therefrom calculate the values of x, y, and z.

A.6 What is the rule for multiplying two matrices? Two matrices of order $m \times n$ and $n \times k$ are multiplied. What is the size of the product matrix? Find AB for the matrices given below. What is the type of the resultant matrix?

$$A = \begin{bmatrix} 1 & 3 & 5 & 7 \\ 2 & 4 & 6 & 8 \\ 9 & 5 & 0 & 7 \\ 1 & 2 & 3 & 4 \end{bmatrix}; B = \begin{bmatrix} 20 + j10 \\ 0 \\ -j6 \\ 10 \end{bmatrix}$$

A.7 Explain the following with the help of examples: (i) determinant, (ii) minor and cofactor, and (iii) adjoint matrix. What is the significance of a singular and a non-singular matrix? Compute the determinant of matrix A given in Problem A.6. Does the inverse of the matrix exist?

A.8 For the matrix A given below, compute (i) the cofactors and (ii) the adjoint matrix. Then compute the inverse of A.

$$[A] = \begin{bmatrix} 2 & 5 & 6 & 0 \\ 10 & -6 & 2 & 1 \\ 3 & 7 & 9 & 2 \\ 4 & 0 & 3 & 1 \end{bmatrix}$$

A.9 Determine $D = A - C^t BC$ and the type of matrix D when

$$A = \begin{bmatrix} -2 & -1 & 1 \\ -1 & 2 & 3 \\ 1 & 3 & 4 \end{bmatrix}; B = \begin{bmatrix} 3 & -3 \\ -3 & 5 \end{bmatrix}; \text{ and } C = \begin{bmatrix} 3 & -1 & 0 \\ -1 & 0 & 3 \end{bmatrix}$$

A.10 Explain the significance of matrix partitioning. State the rule for partitioning matrices. Determine the inverse of the partitioned matrix given below.

4	2	-4	6
3	7	0	3
2	0	-3	4
1	3	4	8

A.11 Given the following partitioned matrices. Determine AA_2.

AA_1	3	4	1	7	0	2
	4	5	0	2	6	3
=	1	1	5	10	3	2
AA_2	0	3	10	8	0	3
	2	5	3	0	9	0

A.12 What is a skew-symmetric matrix? Choose a 4×4 skew-symmetric matrix A and compute $B = 0.2A + 0.5U$. What is the type of matrix B?

A.13 For the five-node network shown in Fig. A.2, write the admittance matrix. What is the size of the matrix? Eliminate nodes 3, 4, and 5 by the method of partitioning. What is the order of the matrix? Draw the circuit diagram represented by the reduced matrix.

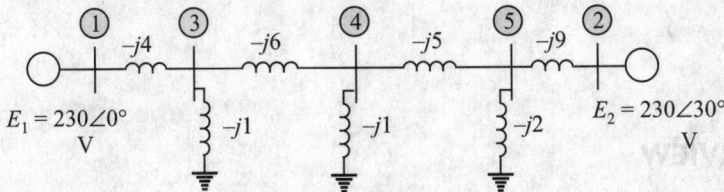

Fig. A.2 Five-node network

A.14 For the network shown in Fig. A.2, compute the impedance matrix of the (i) complete network using the method of partitioning and (ii) the reduced two-node network.

A.15 For the source voltages shown in Fig. A.2, determine the currents at nodes 1 and 2.

Test System Data

OVERVIEW

In this appendix, system data and load flow results for IEEE 14 bus, IEEE 30 bus, and IEEE 57 bus test systems are given. The load flow results have also been verified by actual programming. The data is presented in the following form:

- Bus data and load flow results
- Line data
- Transformer data
- Shunt capacitor data
- Bus data for voltage-controlled buses

B.1 IEEE 14 BUS TEST SYSTEM

Bus data and load flow results						
Bus no.	Bus voltage		Generation		Load	
	Magnitude per unit	Phase angle degrees	Real (MW)	Reactive (MVAR)	Real (MW)	Reactive (MVAR)
1	1.060	0.0	232.4	−16.9	0.0	0.0
2	1.045	−4.98	40.0	42.4	21.7	12.7
3	1.010	−12.72	0.0	23.4	94.2	19.0
4	1.019	−10.33	0.0	0.0	47.8	3.9
5	1.020	−8.78	0.0	0.0	7.6	1.6
6	1.070	−14.22	0.0	12.2	11.2	7.5
7	1.062	−13.37	0.0	0.0	0.0	0.0
8	1.090	−13.36	0.0	17.4	0.0	0.0
9	1.056	−14.94	0.0	0.0	29.5	16.6
10	1.051	−15.10	0.0	0.0	9.0	5.8
11	1.057	−14.79	0.0	0.0	3.5	1.8
12	1.055	−18.07	0.0	0.0	6.1	1.6
13	1.050	−15.16	0.0	0.0	13.5	5.8
14	1.036	−16.04	0.0	0.0	14.9	5.0

Line data				
Line no.	Between buses	Line impedance		Half line charging susceptance (pu)
		R (pu)	X (pu)	
1	1-2	0.01938	0.05917	0.02640
2	2-3	0.04699	0.19797	0.02190
3	2-4	0.05811	0.17632	0.01870
4	1-5	0.05403	0.22304	0.02460
5	2-5	0.05695	0.17388	0.01700
6	3-4	0.06701	0.17103	0.01730
7	4-5	0.01335	0.04211	0.00640
8	5-6	0.0	0.25202	0.0
9	4-7	0.0	0.20912	0.0
10	7-8	0.0	0.17615	0.0
11	4-9	0.0	0.55618	0.0
12	7-9	0.0	0.11001	0.0
13	9-10	0.03181	0.08450	0.0
14	6-11	0.09498	0.19890	0.0
15	6-12	0.12291	0.25581	0.0
16	6-13	0.06615	0.13027	0.0
17	9-14	0.12711	0.27038	0.0
18	10-11	0.08205	0.19207	0.0
19	12-13	0.22092	0.19988	0.0
20	13-14	0.17093	0.34802	0.0

Transformer data		
Transformer	Between buses	Tap setting
1	4-7	0.978
2	4-9	0.969
3	5-6	0.932

Shunt capacitor data	
Bus no.	Susceptance (pu)
9	0.190

Voltage controlled bus data			
Bus no.	Voltage magnitude (pu)	Reactive power limits	
		Minimum (MVAR)	Maximum (MVAR)
2	1.045	− 40.0	50.0
3	1.010	0.0	40.0
6	1.070	− 6.0	24.0
8	1.090	− 6.0	24.0

B.2 IEEE 30 BUS TEST SYSTEM

Bus data and load flow results						
Bus no.	Bus voltage		Generation		Load	
	Magnitude per unit	Phase angle degrees	Real (MW)	Reactive (MVAR)	Real (MW)	Reactive (MVAR)
1	1.05	0.0	138.48	−2.79	0.0	0.0
2	1.0338	−2.73	57.56	2.47	21.7	12.7
3	1.0313	−4.68	0.0	9.0	2.4	1.2
4	1.0263	−5.61	0.0	0.0	7.6	1.6
5	1.0058	−8.99	24.56	22.57	94.2	19.0
6	1.0208	−6.45	0.0	0.0	0.0	0.0
7	1.0069	−8.02	0.0	0.0	22.8	10.9
8	1.0230	−6.47	35.0	34.84	30.0	30.0
9	1.0332	−8.03	0.0	0.0	0.0	0.0
10	1.0183	−9.93	0.0	0.0	5.8	2.0
11	1.0913	−6.13	17.93	30.78	0.0	0.0
12	1.0399	−9.40	0.0	0.0	11.2	7.5
13	1.0883	−8.20	16.91	37.83	0.0	0.0
14	1.0236	−10.31	0.0	0.0	6.2	1.6
15	1.0179	−10.36	0.0	0.0	8.2	2.5
16	1.0235	−9.90	0.0	0.0	3.5	1.8
17	1.0144	−10.14	0.0	0.0	9.0	5.8
18	1.0057	−10.93	0.0	0.0	3.2	0.9
19	1.0017	−11.06	0.0	0.0	9.5	3.4
20	1.0051	−10.83	0.0	0.0	2.2	0.7
21	1.0061	−10.40	0.0	0.0	17.5	11.2
22	1.0069	−10.39	0.0	0.0	0.0	0.0
23	1.0053	−10.72	0.0	0.0	3.2	1.6
24	0.9971	−10.85	0.0	0.0	8.7	6.7
25	1.0086	−10.91	0.0	0.0	0.0	0.0
26	0.9908	−11.33	0.0	0.0	3.5	2.3
27	1.0245	−10.66	0.0	0.0	0.0	0.0
28	1.0156	−6.87	0.0	0.0	0.0	0.0
29	1.0047	−11.89	0.0	0.0	2.4	0.9
30	0.9932	−12.77	0.0	0.0	10.6	1.9

Line data				
Line no.	Between buses	Line impedance		Half line charging susceptance (pu)
		R (pu)	X (pu)	
1	1-2	0.0192	0.0575	0.0264
2	1-3	0.0452	0.1852	0.0204
3	2-4	0.0570	0.1737	0.0184
4	3-4	0.0132	0.0379	0.0042

(Contd)

(Contd)

Line data				
Line no.	Between buses	Line impedance		Half line charging susceptance (pu)
		R (pu)	X (pu)	
5	2-5	0.0472	0.1983	0.0209
6	2-6	0.0581	0.1763	0.0187
7	4-6	0.0119	0.0414	0.0045
8	5-7	0.0460	0.1160	0.0102
9	6-7	0.0267	0.0820	0.0085
10	6-8	0.0120	0.0420	0.0045
11	6-9	0.0	0.2080	0.0
12	6-10	0.0	0.5560	0.0
13	9-11	0.0	0.2080	0.0
14	9-10	0.0	0.1100	0.0
15	4-12	0.0	0.2560	0.0
16	12-13	0.0	0.1400	0.0
17	12-14	0.1231	0.2559	0.0
18	12-15	0.0662	0.1304	0.0
19	12-16	0.0945	0.1987	0.0
20	14-15	0.2210	0.1997	0.0
21	16-17	0.0824	0.1932	0.0
22	15-18	0.1070	0.2185	0.0
23	18-19	0.0639	0.1292	0.0
24	19-20	0.0340	0.0680	0.0
25	10-20	0.0936	0.2090	0.0
26	10-17	0.0324	0.0845	0.0
27	10-21	0.0348	0.0749	0.0
28	10-22	0.0727	0.1499	0.0
29	21-22	0.0116	0.0236	0.0
30	15-23	0.1000	0.2020	0.0
31	22-24	0.1150	0.1790	0.0
32	23-24	0.1320	0.2700	0.0
33	24-25	0.1885	0.3292	0.0
34	25-26	0.2544	0.3800	0.0
35	25-27	0.1093	0.2087	0.0
36	26-27	0.0	0.3960	0.0
37	27-29	0.2198	0.4153	0.0
38	27-30	0.3202	0.6027	0.0
39	29-30	0.2399	0.4533	0.0
40	8-28	0.0636	0.2000	0.0214
41	6-28	0.0169	0.0599	0.0065

Transformer data		
Transformer no.	Between buses	Tap setting
1	6-9	1.0155
2	6-10	0.9629
3	4-12	1.0129
4	28-27	0.9581

Shunt capacitor data	
Bus no.	Susceptance (pu)
10	0.19
24	0.04

Voltage controlled bus data			
Bus no.	Voltage magnitude (pu)	Reactive power limits	
		Minimum (MVAR)	Maximum (MVAR)
2	1.0338	−20.0	60.0
5	1.0058	−15.0	62.5
8	1.0230	−15.0	50.0
11	1.0913	−10.0	40.0
13	1.0883	−15.0	45.0

B.3 IEEE 57 BUS TEST SYSTEM

Bus data and load flow results						
Bus no.	Bus voltage		Generation		Load	
	Magnitude per unit	Phase angle degrees	Real (MW)	Reactive (MVAR)	Real (MW)	Reactive (MVAR)
1	1.040	0.0	478.0	128.9	55.0	17.0
2	1.010	−1.18	0.0	−0.8	3.0	88.0
3	0.985	−5.97	40.0	−1.0	41.0	21.0
4	0.981	−7.32	0.0	0.0	0.0	0.0
5	0.976	−8.52	0.0	0.0	13.0	4.0
6	0.980	−8.65	0.0	0.8	75.0	2.0
7	0.984	−7.58	0.0	0.0	0.0	0.0
8	1.005	−4.45	450.0	62.1	150.0	22.0
9	0.980	−9.56	0.0	2.2	121.0	26.0
10	0.986	−11.43	0.0	0.0	5.0	2.0
11	0.974	−10.17	0.0	0.0	0.0	0.0
12	1.015	−10.46	310.0	128.5	377.0	24.0
13	0.979	−9.79	0.0	0.0	18.0	2.3
14	0.970′	−9.33	0.0	0.0	10.5	5.3

(Contd)

(Contd)

Bus data and load flow results						
Bus no.	Bus voltage		Generation		Load	
	Magnitude per unit	Phase angle degrees	Real (MW)	Reactive (MVAR)	Real (MW)	Reactive (MVAR)
15	0.988	− 7.18	0.0	0.0	22.0	5.0
16	1.013	− 8.85	0.0	0.0	43.0	3.0
17	1.017	− 5.39	0.0	0.0	42.0	8.0
18	1.001	−11.71	0.0	0.0	27.2	9.8
19	0.970	−13.20	0.0	0.0	3.3	0.6
20	0.964	−13.41	0.0	0.0	2.3	1.0
21	1.008	−12.89	0.0	0.0	0.0	0.0
22	1.010	−12.84	0.0	0.0	0.0	0.0
23	1.008	−12.91	0.0	0.0	6.3	2.1
24	0.999	−13.25	0.0	0.0	0.0	0.0
25	0.982	−18.13	0.0	0.0	6.3	3.2
26	0.959	−12.95	0.0	0.0	0.0	0.0
27	0.982	−11.48	0.0	0.0	9.3	0.5
28	0.997	−10.45	0.0	0.0	4.6	2.3
29	1.010	− 9.75	0.0	0.0	17.0	2.6
30	0.962	−18.68	0.0	0.0	3.6	1.8
31	0.936	−19.34	0.0	0.0	5.8	2.9
32	0.949	−18.46	0.0	0.0	1.6	0.8
33	0.947	−18.50	0.0	0.0	3.8	1.9
34	0.959	−14.10	0.0	0.0	0.0	0.0
35	0.966	−13.86	0.0	0.0	6.0	3.0
36	0.976	−13.59	0.0	0.0	0.0	0.0
37	0.985	−13.41	0.0	0.0	0.0	0.0
38	1.013	−12.71	0.0	0.0	14.0	7.0
39	0.983	−13.46	0.0	0.0	0.0	0.0
40	0.973	−13.62	0.0	0.0	0.0	0.0
41	0.996	−14.05	0.0	0.0	6.3	3.0
42	0.966	−15.50	0.0	0.0	7.1	4.0
43	1.010	−11.33	0.0	0.0	2.0	1.0
44	1.017	−11.83	0.0	0.0	12.0	1.8
45	1.036	− 9.25	0.0	0.0	0.0	0.0
46	1.060	−11.09	0.0	0.0	0.0	0.0
47	1.033	−12.49	0.0	0.0	29.7	11.6
48	1.027	−12.57	0.0	0.0	0.0	0.0
49	1.036	−12.92	0.0	0.0	18.0	8.5
50	1.023	−13.39	0.0	0.0	21.0	10.5

(Contd)

(Contd)

<table>
<tr><th colspan="7">Bus data and load flow results</th></tr>
<tr><th rowspan="2">Bus no.</th><th colspan="2">Bus voltage</th><th colspan="2">Generation</th><th colspan="2">Load</th></tr>
<tr><th>Magnitude per unit</th><th>Phase angle degrees</th><th>Real (MW)</th><th>Reactive (MVAR)</th><th>Real (MW)</th><th>Reactive (MVAR)</th></tr>
<tr><td>51</td><td>1.052</td><td>−12.52</td><td>0.0</td><td>0.0</td><td>18.0</td><td>5.3</td></tr>
<tr><td>52</td><td>0.980</td><td>−11.47</td><td>0.0</td><td>0.0</td><td>4.9</td><td>2.2</td></tr>
<tr><td>53</td><td>0.971</td><td>−12.23</td><td>0.0</td><td>0.0</td><td>20.0</td><td>10.0</td></tr>
<tr><td>54</td><td>0.996</td><td>−11.69</td><td>0.0</td><td>0.0</td><td>4.1</td><td>1.4</td></tr>
<tr><td>55</td><td>1.031</td><td>−10.78</td><td>0.0</td><td>0.0</td><td>6.8</td><td>3.4</td></tr>
<tr><td>56</td><td>0.968</td><td>−16.04</td><td>0.0</td><td>0.0</td><td>7.6</td><td>2.2</td></tr>
<tr><td>57</td><td>0.965</td><td>−16.56</td><td>0.0</td><td>0.0</td><td>6.7</td><td>2.0</td></tr>
</table>

<table>
<tr><th colspan="5">Line data</th></tr>
<tr><th rowspan="2">Line no.</th><th rowspan="2">Between buses</th><th colspan="2">Line impedance</th><th rowspan="2">Half line charging susceptance (pu)</th></tr>
<tr><th>R (pu)</th><th>X (pu)</th></tr>
<tr><td>1</td><td>1-2</td><td>0.0083</td><td>0.0280</td><td>0.0645</td></tr>
<tr><td>2</td><td>2-3</td><td>0.0298</td><td>0.0850</td><td>0.0409</td></tr>
<tr><td>3</td><td>3-4</td><td>0.0112</td><td>0.0366</td><td>0.0190</td></tr>
<tr><td>4</td><td>4-5</td><td>0.0625</td><td>0.1320</td><td>0.0129</td></tr>
<tr><td>5</td><td>4-6</td><td>0.0430</td><td>0.1480</td><td>0.0174</td></tr>
<tr><td>6</td><td>6-7</td><td>0.0200</td><td>0.1020</td><td>0.0138</td></tr>
<tr><td>7</td><td>6-8</td><td>0.0339</td><td>0.1730</td><td>0.0235</td></tr>
<tr><td>8</td><td>8-9</td><td>0.0099</td><td>0.0505</td><td>0.0274</td></tr>
<tr><td>9</td><td>9-10</td><td>0.0369</td><td>0.1679</td><td>0.0220</td></tr>
<tr><td>10</td><td>9-11</td><td>0.0258</td><td>0.0848</td><td>0.0109</td></tr>
<tr><td>11</td><td>9-12</td><td>0.0648</td><td>0.2950</td><td>0.0386</td></tr>
<tr><td>12</td><td>9-13</td><td>0.0481</td><td>0.1580</td><td>0.0203</td></tr>
<tr><td>13</td><td>13-14</td><td>0.0132</td><td>0.0434</td><td>0.0055</td></tr>
<tr><td>14</td><td>13-15</td><td>0.0269</td><td>0.0869</td><td>0.0115</td></tr>
<tr><td>15</td><td>1-15</td><td>0.0178</td><td>0.0910</td><td>0.0494</td></tr>
<tr><td>16</td><td>1-16</td><td>0.0454</td><td>0.2060</td><td>0.0273</td></tr>
<tr><td>17</td><td>1-17</td><td>0.02380</td><td>0.1080</td><td>0.0143</td></tr>
<tr><td>18</td><td>3-15</td><td>0.0162</td><td>0.053</td><td>0.0272</td></tr>
<tr><td>19</td><td>4-18</td><td>0.0</td><td>0.555</td><td>0.0</td></tr>
<tr><td>20</td><td>4-18</td><td>0.0</td><td>0.430</td><td>0.0</td></tr>
<tr><td>21</td><td>5-6</td><td>0.0302</td><td>0.0641</td><td>0.0062</td></tr>
<tr><td>22</td><td>7-8</td><td>0.0139</td><td>0.0712</td><td>0.0097</td></tr>
<tr><td>23</td><td>10-12</td><td>0.0277</td><td>0.1262</td><td>0.0164</td></tr>
<tr><td>24</td><td>11-13</td><td>0.0223</td><td>0.0732</td><td>0.0094</td></tr>
</table>

(Contd)

		Line data		
Line no.	Between buses	Line impedance		Half line charging susceptance (pu)
		R (pu)	X (pu)	
25	12-13	0.0178	0.0580	0.0302
26	12-16	0.0180	0.0813	0.0108
27	12-17	0.0397	0.1790	0.0238
28	14-15	0.0171	0.0547	0.0074
29	18-19	0.4610	0.6850	0.0
30	19-20	0.2830	0.4340	0.0
31	20-21	0.0	0.7767	0.0
32	21-22	0.0736	0.1170	0.0
33	22-23	0.0099	0.0152	0.0
34	23-24	0.1660	0.2560	0.0042
35	24-25	0.0	1.182	0.0
36	24-25	0.0	1.230	0.0
37	24-26	0.0	0.0473	0.0
38	26-27	0.165	0.2540	0.0
39	27-28	0.0618	0.0954	0.0
40	28-29	0.0418	0.0587	0.0
41	7-29	0.0	0.0648	0.0
42	25-30	0.135	0.202	0.0
43	30-31	0.326	0.497	0.0
44	31-12	0.507	0.755	0.0
45	32-33	0.0392	0.0360	0.0
46	32-34	0.0	0.953	0.0
47	34-35	0.052	0.0780	0.0016
48	35-36	0.043	0.0537	0.0008
49	36-37	0.0290	0.0366	0.0
50	37-38	0.0651	0.1009	0.0010
51	37-39	0.0239	0.0379	0.0
52	36-40	0.0300	0.0466	0.0
53	22-38	0.0192	0.0295	0.0
54	11-41	0.0	0.7490	0.0
55	41-42	0.207	0.3520	0.0
56	41-43	0.0	0.412	0.0
57	38-44	0.0289	0.0585	0.0010
58	15-45	0.0	0.1042	0.0
59	14-46	0.0	0.0735	0.0
60	46-47	0.023	0.068	0.0016

(Contd)

(Contd)

Line data				
Line no.	Between buses	Line impedance		Half line charging susceptance (pu)
		R (pu)	X (pu)	
61	47-48	0.0182	0.0233	0.0
62	48-49	0.0834	0.1290	0.0024
63	49-50	0.0801	0.1280	0.0
64	50-51	0.1386	0.2200	0.0
65	10-51	0.0	0.0712	0.0
66	13-49	0.0	0.1910	0.0
67	29-52	0.1442	0.1870	0.0
68	52-53	0.0762	0.0984	0.0
69	53-54	0.1878	0.2320	0.0
70	54-55	0.1732	0.2265	0.0
71	11-43	0.0	0.1530	0.0
72	44-45	0.0624	0.1242	0.0020
73	40-56	0.0	1.1950	0.0
74	56-41	0.5530	0.5490	0.0
75	56-42	0.2125	0.3540	0.0
76	39-57	0.0	1.3550	0.0
77	57-56	0.1740	0.2600	0.0
78	38-49	0.1150	0.1770	0.0030
79	38-48	0.0312	0.0482	0.0
80	9-55	0.0	0.1205	0.0

Transformer data		
Transformer no.	Between buses	Tap setting
1	4-18	0.970
2	4-18	0.978
3	7- 9	0.967
4	9-55	0.940
5	10-51	0.930
6	11-41	0.955
7	11-43	0.958
8	3-49	0.895
9	14-46	0.900
10	15-45	0.955
11	21-20	1.043
12	24-25	1.000
13	24-25	1.000

(Contd)

(Contd)

Transformer data		
Transformer no.	Between buses	Tap setting
14	24-26	1.043
15	34-32	0.975
16	39-57	0.980
17	40-56	0.958

Shunt capacitor data	
Bus no.	Susceptance (pu)
18	0.100
25	0.059
53	0.063

Voltage controlled bus data			
Bus no.	Voltage magnitude (pu)	Reactive power limits	
		Minimum (MVAR)	Maximum (MVAR)
2	1.010	− 17.0	50.0
3	0.985	− 10.0	60.0
6	0.980	− 8.0	25.0
8	1.005	− 140.0	200.0
9	0.980	− 3.0	9.0
12	1.015	− 50.0	155.0

Answers

Chapter 1: Power Sector Outlook

1.1 1.2

1.2 90 MW

1.3 0.35

1.4 4.05%

1.5 375 GW from the plot

1.6 7.32 MW, 52.25%

Chapter 2: Basic Concepts

2.1 Transmission line: $Z_{pu} = (2.2957 + j4.5914)$, Generator: $X_{pu} = 0.0540$,
Generator end transformer: $X_{pu} = 0.08$, Motor end transformer: $X_{pu} = 0.0833$,
Motor: $X_{pu} = 0.0625$

2.2 $P = 96.8$ kW, $Q = 48.4$ kVAR, $S = 108.23$ kVA, (i) $R = 5.0$ pu;
$X = -j2.5$ pu (ii) R = 5.0 pu; X = $-j$ 2.5 pu

2.3 Generator 1: $X = 0.3750$ pu, Generator 2: $X = 0.40$ pu,
Equivalent generator: $X = 0.1935$

2.4 Voltage at generator terminals = $0.9954 \angle 20.7283°$ pu

2.5 $X_p = 0.0397$ pu, $X_s = 0.0017$ pu, $X_t = 0.0083$ pu

2.6 (a) $1.0 \angle -60.0°$, (b) $1.7321 \angle 30.0°$, (c) $1.7321 \angle 150.0°$,
(d) $1.7321 \angle 90.0°$, (e) $1.7321 \angle -90.0°$

Chapter 3: Three-phase Transmission Line Parameters

3.1 $R_2 = R_1 \dfrac{(1 + \alpha_0 t_2)}{(1 + \alpha_0 t_1)}$

3.2 $l = \sqrt{\dfrac{Rv}{\rho}}$ $d = \left(\dfrac{16\rho v}{\pi^2 R}\right)^{\frac{1}{4}}$

3.3 $P_{20} = 0.0173\mu$ Ω-m, $R_{60} = 0.5156$ Ω

3.4 28.49° C

3.5 11.8836 Ω

3.6 1.9271 cm

3.7 0.3773 mH

3.8 42.9264 mH

3.9 (i) $3.1152r$ (ii) $12.4608r$ (iii) $24.9216r$ (iv) $8.8111r$ (v) $70.489r$

3.10 (i) 0.4414 cm (ii) 11.9784 cm

3.11 $(2 \times 10^{-7}) \times 1.1331[k^2(1 + k^2)]^{\frac{1}{4}}\sqrt{\dfrac{H}{r}}$ H/m, 2.7486 Ω

3.12 1.0433e-011 F/m, 0.9585

3.13 0.2673 Ω/km, 0.3057 Ω/km, 3.6546e-006 S/km, 17.8200 Ω/conductor, 61.1424 Ω,
7.3092e-004 S

3.14 $j\omega \left[2 \times 10^{-7} \times I_R \angle 120° \left(\ln \dfrac{D_{R-2}}{D_{R-1}} + \ln \dfrac{D_{B-2}}{D_{B-1}} \right) \right]$, 3.3920∠60.0° V/km

Chapter 4: Transmission Line Model and Performance

4.1 (i) Line to line: 502.40∠16.4709° kV, (ii) 942.8550∠−21.1336° A,

(iii) Three-phase power = 820.46∠37.6045° MVA,

(iv) 25.6%, Z_c = 251.6611 Ω, β = 0.0012 rads/km, λ = 5298.1 km,

  v = 2.9986e + 005 km/sec

4.2 18125 kVAR, 0.9219

4.3 42.3910 MVAR, 35.8954 MW

4.5 Voltage at the junction of long line and cable at the other end at 10 μs is
49.95 kV.

Chapter 5: Simulation of Power System Elements

5.1 (i) 1.0633∠23.1522°pu, 1.0533∠10.7774° pu

(ii) 425.3077 kV, 421.3235 kV

(iii) 96 MW, 14.64 MVAR

5.2 (i) 1.6125∠23.3852° pu   (ii) 1.6012∠16.6012° pu

5.4 (i) 15.8262°, 1.3200∠15.8262° pu   (ii) 2.20 pu

(iii) 1.5022 pu, 0.9223 (lagging),   (iv) 1.2911 pu, 0.9958 (lagging)

5.5 (i) 0.6508∠11.4793° pu,   (ii) S = (1.7000 − j4.7708) pu   (iii) 11.4793°

(iv) 1.1204 pu

5.6 (i) 0.9844 pu, ∠21.7668°   (ii) 1.6180 pu,   (iii) 1.5371 pu

(iv) Load reactive power = 0.45 pu

5.7 (i) 300 kVA,   (ii) 99.1304%

5.8 Tap setting at (i) sending end = 1.1112 pu and

(ii) receiving end = 0.90 pu

Chapter 6: Formulation of Network Matrices

6.2

$$
\text{node} \rightarrow \quad 0 \;\; 1 \;\; 2 \;\; 3 \;\; 4
$$
$$
\text{element} \downarrow
$$

$$
\bar{A} = \begin{matrix} 1 \\ 2 \\ 3 \\ 4 \\ 5 \\ 6 \\ 7 \end{matrix}
\begin{bmatrix}
1 & -1 & 0 & 0 & 0 \\
1 & 0 & -1 & 0 & 0 \\
0 & 1 & -1 & 0 & 0 \\
0 & 1 & 0 & -1 & 0 \\
0 & 1 & 0 & 0 & -1 \\
0 & 0 & 0 & -1 & 1 \\
0 & 0 & -1 & 0 & 1
\end{bmatrix}
$$

$$
\text{bus} \rightarrow \quad 1 \;\; 2 \;\; 3 \;\; 4
$$
$$
\text{element} \downarrow
$$

$$
A = \begin{matrix} 1 \\ 2 \\ 3 \\ 4 \\ 5 \\ 6 \\ 7 \end{matrix}
\begin{bmatrix}
-1 & 0 & 0 & 0 \\
0 & -1 & 0 & 0 \\
1 & -1 & 0 & 0 \\
1 & 0 & -1 & 0 \\
1 & 0 & 0 & -1 \\
0 & 0 & -1 & 1 \\
0 & -1 & 0 & 1
\end{bmatrix}
$$

6.3 $node \rightarrow$ 0   2   3   4
    element $\downarrow$

$$\bar{A} = \begin{array}{c} 1 \\ 2 \\ 3 \\ 4 \\ 5 \\ 6 \\ 7 \end{array} \begin{bmatrix} 1 & 0 & 0 & 0 \\ 1 & -1 & 0 & 0 \\ 0 & -1 & 0 & 0 \\ 0 & 0 & -1 & 0 \\ 0 & 0 & 0 & -1 \\ 0 & 0 & -1 & 1 \\ 0 & -1 & 0 & 1 \end{bmatrix}$$

6.4

$$Y_{\text{bus}} = \begin{bmatrix} 21.0000 & -2.0000 & -2.5000 & -4.0000 \\ -2.0000 & 16.0000 & 0 & -4.0000 \\ -2.5000 & 0 & 7.5000 & -5.0000 \\ -4.0000 & -4.0000 & -5.0000 & 13.0000 \end{bmatrix}$$

$$Z_{\text{bus}} = \begin{bmatrix} 0.0644 & 0.0195 & 0.0520 & 0.0458 \\ 0.0195 & 0.0756 & 0.0350 & 0.0427 \\ 0.0520 & 0.0350 & 0.2266 & 0.1139 \\ 0.0458 & 0.0427 & 0.1139 & 0.1480 \end{bmatrix}$$

6.5   *bus* $\rightarrow$ 1   2   3   4   5   *bus* $\rightarrow$ 1   2   3   5
    element $\downarrow$ *element* $\downarrow$

$$\bar{A} = \begin{array}{c} 1 \\ 2 \\ 3 \\ 4 \\ 5 \\ 6 \end{array} \begin{bmatrix} 1 & -1 & 0 & 0 & 0 \\ 0 & 1 & -1 & 0 & 0 \\ 0 & 1 & -1 & 0 & 0 \\ 0 & 0 & -1 & 1 & 0 \\ 0 & 0 & 0 & 1 & -1 \\ -1 & 0 & 0 & 0 & 1 \end{bmatrix} \qquad A = \begin{array}{c} 1 \\ 2 \\ 3 \\ 4 \\ 5 \\ 6 \end{array} \begin{bmatrix} 1 & -1 & 0 & 0 \\ 0 & 1 & -1 & 0 \\ 0 & 1 & -1 & 0 \\ 0 & 0 & -1 & 0 \\ 0 & 0 & 0 & -1 \\ -1 & 0 & 0 & 1 \end{bmatrix}$$

$$Z_{\text{bus}} = \begin{bmatrix} 0.1823i & 0.1497i & 0.1088i & 0.1139i \\ 0.1497i & 0.1721i & 0.1252i & 0.0935i \\ 0.1088i & 0.1252i & 0.1456i & 0.0680i \\ 0.1139i & 0.0935i & 0.0680i & 0.1650i \end{bmatrix}$$

$$Y_{\text{bus}} = \begin{bmatrix} -23.3333i & +16.6667i & 0 & +6.6667i \\ +16.6667i & -30.0000i & +13.3333i & 0 \\ 0 & +13.3333i & -18.3333i & 0 \\ +6.6667i & 0 & 0 & -10.6667i \end{bmatrix}$$

6.6

$$Y_{\text{bus}} = \begin{bmatrix} -37.5000i & +12.5000i & -20.0000i \\ +12.5000i & -47.8333i & +33.3333i \\ +20.0000i & +33.3333i & -53.3333i \end{bmatrix}$$

6.8 $V_1 = 1.0693$, $V_2 = 1.0343$, and $V_3 = 1.0562$

6.9 Same as in **6.8**

6.11

$$V_{\text{bus}} = \begin{bmatrix} 1.0694 \\ 1.0343 \\ 1.0563 \end{bmatrix}$$

6.12

$$Y_{\text{bus}} = \begin{bmatrix} -80.0i & 50.0i & 0 & 0 & 0 & 25.0i \\ 50.0i & -133.33i & 50.0i & 0 & 33.33i & 0 \\ 0 & 50.0i & -70.0i & 20.0i & 0 & 0 \\ 0 & 0 & 20.0i & -80.0i & 40.0i & 20.0i \\ 0 & 33.33i & 0 & 40.0i & -123.33i & 50.0i \\ 25.0i & 0i & 0i & 20.0i & 50.0i & -95.0i \end{bmatrix}$$

6.13

$$Y_{\text{bus}} = \begin{bmatrix} -80i & 50i & 0 & 0 & 0 & 25i \\ 50i & -100i & 50i & 0 & 0 & 0 \\ 0 & 50i & -120i & 20i & 0 & 50i \\ 0 & 0 & 20i & -60i & 40i & 0 \\ 0 & 0 & 0 & 40i & -90i & 50i \\ 25i & 0 & 50i & 0 & 50i & -125i \end{bmatrix}$$

6.14

$$Y_{\text{bus}} = \begin{bmatrix} 65.0i & 40.0i & 0 & 25.0i \\ 40.0i & -110.0i & 20.0i & 0 \\ 0 & 20.0i & -160.0i & 40.0i \\ 25.0i & 0 & 40.0i & 65.0i \end{bmatrix}$$

$$TOF = \begin{bmatrix} 65.0i & -0.62 & 0 & -0.38 \\ 40.0i & -85.38i & -0.23 & -0.18 \\ 0 & 20.0i & -155.32i & -0.28 \\ 25.0i & 15.38i & 43.60i & 40.37i \end{bmatrix}$$

6.15

$$\begin{bmatrix} V_1 \\ V_2 \\ V_3 \\ V_4 \end{bmatrix} = \begin{bmatrix} 0.0557i \\ 0.0336i \\ 0.0232i \\ 0.0511i \end{bmatrix}$$

6.17

$$Y_{\text{bus}} = \begin{bmatrix} 71.64i & 54.52i & 0 & 12.12i \\ 54.52i & -12.188i & 50.0i & 17.36i \\ 0 & 50.0i & -70.0i & 20.0i \\ 12.12i & 17.36i & 20.0i & -49.48i \end{bmatrix}$$

6.19

$$Z_{\text{bus}} = j \begin{bmatrix} 0.0084 & 0.0032 & 0.0047 & 0.0070 \\ 0.0032 & 0.0136 & 0.0107 & 0.0061 \\ 0.0047 & 0.0107 & 0.0271 & 0.0133 \\ 0.0070 & 0.0061 & 0.0133 & 0.0248 \end{bmatrix}$$

6.20

$$Z_{bus} = \begin{bmatrix} 0.2009i & 0.0981i & 0.1542i \\ 0.0981i & 0.3037i & 0.1916i \\ 0.1542i & 0.1916i & 0.4439i \end{bmatrix}$$

6.21

$$Z_{bus} = \begin{bmatrix} 0.2162i & 0.0675i & 0.1486i \\ 0.0675i & 0.3648i & 0.2027i \\ 0.1486i & 0.2027i & 0.4459i \end{bmatrix}$$

Chapter 7: Power Flow Studies

7.1

$$Y = \begin{bmatrix} 0.0017 - 0.0077i & -0.0017 + 0.0081i \\ -0.0017 + 0.0081i & 0.0017 - 0.0077i \end{bmatrix}$$

$$Y_{pu} = \begin{bmatrix} (0.1854 - 0.8314i) & (-0.1854 + 0.8736i) \\ (-0.1854 + 0.8736i) & (0.1854 - 0.8314i) \end{bmatrix}$$

7.2 P (lagging pf) = 0.0466 pu, P (leading pf) = 0.6786 pu

7.3 For lagging pf: $S_1 = (0.0474 + 0.0387i)$ pu, $S_2 = (0.0466 + 0.0350i)$ pu

For leading pf: $S_1 = (0.8457 + 0.2790i)$ pu, $S_2 = (0.6785 - 0.5088i)$ pu,

$|V_2| = 1.158$ pu, $P_2 = 0.1659$ pu

7.4 (i) 35.2189°, (ii) 9.5220 pu, (iii) At bus 1: (23.0000 + 17.5220i) pu, 0.7955 lagging; at bus 2: (7.0000 + 24.5220i) pu, 0.2745 lagging

7.5 0.9472 pu, Slack bus power = 1.0560 pu

7.7 (i) 0.8286 pu (ii) Line 1-2: 0.3870 pu, Line 1-3: 0.4025 pu,

Line 2-3: – 0.1083 pu, (iii) 0.0286 pu

7.8 1.0708∠–9.5°

7.10 (i) (1.9425 + 0.1876i) pu,

(ii) From bus 1 to bus 2: (–0.8425 – 0.2573i) pu, (iii) 0.0425 pu

7.11

$$Y_{bus} = \begin{bmatrix} (13.1655 - 30.4117i) & (-7.6923 + 11.5385i) \\ (-7.6923 + 11.5385i) & (9.6923 - 17.4885i) \\ (-0.4732 + 3.9432i) & (-2.0000 + 6.0000i) \\ (-5.0000 + 15.0000i) & 0 \end{bmatrix}$$

$$\begin{bmatrix} (-0.4732 + 3.9432i) & (-5.0000 + 15.0000i) \\ (-2.0000 + 6.0000i) & 0 \\ (6.8850 - 17.2512i) & (-4.4118 + 7.3529i) \\ (-4.4118 + 7.3529i) & (9.4118 - 22.3179i) \end{bmatrix}$$

7.12 $V_1 = 1.04 + j0$, $V_2 = 1.0065 - j0.0044$, $V_3 = 0.9954 + j0.0138$,

$V_4 = 1.0076 - j0.0171$

7.13 $V_1 = 1.040∠0$, $V_2 = 1.025∠-0.0051$, $V_3 = 1.052∠0.020$,

$V_4 = 1.040 ∠0.0296$

[Note voltage magnitude is in pu and voltage angles are in radians]

7.14 (i) $(0.7214 + 0.9216i)$ pu, (ii) Line 1-2: S12 = $(0.3208 + 0.3344i)$ pu, Line 1-3: S13 = $(-0.0346 + 0.1735i)$ pu,

Line 1-4: S14 = $(0.4352 + 0.4138i)$ pu,

Line 2-3: S23 = $(0.0871 + 0.0838i)$ pu,

Line 3-4: S34 = $(0.1757 - 0.2424i)$ pu. System losses $P = 0.0214$ pu

7.16

$$Z_{bus} = \begin{bmatrix} (0.0080 + 0.0164i) & (0.0040 + 0.0091i) \\ (0.0040 + 0.0091i) & (0.0121 + 0.0273i) \end{bmatrix}$$

7.17 (i) System loss $P = 0.3438$ pu,

(ii) Line 1-2: S12 = $(0.3789 - 0.2962i)$ pu,

Line 2-1: S21 = $-(0.3747 - 0.3046i)$ pu,

Line 1-3: S13 = $(1.2649 - 0.9949i)$ pu,

Line 3-1: S31 = $-(1.2414 - 1.0419i)$ pu,

Line 2-3: S23 = $(0.2218 - 0.1539i)$ pu,

Line 3-2: S32 = $-(0.2205 - 0.1572i)$ pu

(iii) $P = 0.3438$ pu

7.18 (i) $V_1 = 1.02\angle 0°$, $V_2 = 0.9847\angle -2.7392°$ pu, $V_3 = 0.9481\angle -5.6861°$ pu

(ii) Line 1-2: $Q = 0.6$ Line 1-2: $Q = 0$, Line 2-3: $Q = 0.6$. $V_1 = 1.02\angle 0°$,

$V_2 = 1.02\angle -2.6443°$ pu, $V_3 = 0.9847\angle -5.3835°$ pu.

Reactive compensation $Q_2 = 0.6$ pu

7.19 (i) $(0.6972 + 0.4560i)$ pu, (ii) $(0.3192 + 0.0699i)$ pu

(iii) $0.9180\angle -1.6759°$ pu, (iv) $(0.1266 + 0.1445i)$ pu; power flow from 1 to 2: = $(0.1882 + 0.0486i)$ pu, power flow into 2 from line 1-2: = $(0.1868 + 0.0465i)$ pu, power flow from 3 to 2 = $(0.1333 + 0.0260i)$ pu, power flow into 2 from 3: = $(0.1324 + 0.0234i)$ pu, power flow from 1 to 3: = $(-0.0372 + 0.0046i)$ pu, Power flow into 3 from 1-3: = $(-0.0372 + 0.0043i)$ pu, power flow out of 3 into 4: = $(0.8295 + 0.7783i)$ pu, power into 4 from 3-4: $(0.7577 + 0.6587i)$ pu, power into 4 from 1-4: $(0.4923 + 0.3413i)$ pu, power flow from 1 to 4: = $(0.5462 + 0.4028i)$ pu. $\eta = 81.45\%$

Chapter 8: Power System Control

8.1 (i) $\dfrac{K_A}{\left(1 + 2.45s + 1.1s^2 + 0.4s^3\right)}$

(ii) $\dfrac{K_A(1 + 0.1s)}{K_A + 1 + 2.55s + 1.345s^2 + 0.52s^3 + 0.04s^4}$

(iii) $0.04s^4 + 0.52s^3 + 1.345s^2 + 2.55s + 1 + K_A = 0$

8.2 Maximum gain $K_A = 4.0$

8.3 $G(s) = \dfrac{Ms}{R + Ls}$, $G(s) = \dfrac{M}{R}s$

8.4 $\dfrac{K\left(0.016s^2 + 0.44s + 1\right)}{0.0064s^4 + 0.2416s^3 + 10.22s^2 + 6.54s + 1}$,

$$0.0064s^4 + 0.02416s^3 + (10.22 + 0.016K)s^2 + (6.54 + 0.44K)s + (1+K) = 0$$

8.5 (i) 24 MW, 96 MW   (ii) 0.0250 Hz/MW,0.0062 Hz/MW

8.6 6.67MW

8.7 (a) $\dfrac{1}{(1+0.4s)(1+0.1s)}$ (b) $\Delta P(t) = \left[1 + \dfrac{1}{3}e^{-10t} - \dfrac{4}{3}e^{-2.5t}\right]$

8.8 (i) – 0.136 Hz, 49.864 Hz, 45.5 MW, 54.5 MW

(ii) – 0.132 Hz, 49.868 Hz, 44.0 MW, 52.8 MW

8.9 (i) Change in tie-line flow = 63.8 MW, frequency deviation = – 0.0033 Hz, change in turbine power generation in area 1 = 82.5 MW, load reduction in area 1 due a drop in frequency = – 2.5 MW,   (ii) change in line flow = 86.2 MW, frequency deviation = – 0.0033 Hz, change in turbine power generation in area 2 = 60 MW, load reduction in area 2 due a drop in frequency = 3.0 MW

8.10 $\alpha = 0.7937$, $\omega = 3.7367$ rads/sec

Chapter 9: Symmetrical Fault Analyses

9.1 $I_f = (0 - 9.8327i)$ pu; $V_1 = 0.5167$ pu, $V_2 = 0$ pu, $V_3 = 1.6914$ pu, $V_4 = 1.3569$ pu

9.2 $I_f = (0 - 13.2626i)$ pu; $V_1 = 0$ pu, $V_2 = 0.3488$ pu, $V_3 = -0.7931$ pu, $V_4 = -0.5663$ pu, Line currents: $I_{1-2} = (0 + 17.4403i)$ pu, $I_{1-4} = (0 - 14.1578i)$ pu, $I_{2-3} = (0 - 11.4191i)$ pu, $I_{2-4} = (0 - 36.6048i)$ pu

9.3 $I_f = (0 - 5.7013i)$ pu; $V_1 = 0.5701$ pu, $V_2 = 0.7201$ pu, $V_3 = 0.2292$ pu, $V_4 = 0.3267$ pu, Line currents: $I_{1-2} = (0 + 7.4971i)$ pu, $I_{1-4} = (0 - 6.0861i)$ pu, $I_{2-3} = (0 - 4.9088i)$ pu, $I_{2-4} = (0 - 15.7355i)$ pu

9.4 Per unit values: $G1 = 0.00011$, $T1 = 0.08$, $G2 = 0.00011$, $T2 = 0.1667$, $G3 = 0.00013$, $T3 = 0.00050$, $X14 = 0.1250$, $X24 = 0.1563$, $X34 = 0.0938$. (i) $-22.1355i$ kA,   (ii) $V_2 = 327.04$ kV, $V_3 = 403$ kV, $V_4 = 254.96$ kV,   (iii) $I_{1-4} = 6.3739$ kA, $I_{2-4} = -1.4415i$ kA, $I_{3-4} = 4.9324i$ kA

9.5 (i) $-23.4326i$ kA,   (ii) $V_1 = 246.24$ kV, $V_2 = 195.44$ kV, $V_3 = 401.32$ kV, (iii) $I_{1-4} = -6.1555$ kA, $I_{2-4} = -3.9075i$ kA, $I_{3-4} = -13.3696i$ kA

Chapter 10: Symmetrical Components and Unsymmetrical Fault Analyses

10.1 $I_{a0} = 4.4096\angle-139.11°$ A, $I_{a1} = 4.4096\angle139.11°$ A, $I_{a2} = 21.6667\angle0°$ A

10.2 $V_{an} = 100\angle90°$ V, $V_{bn} = 13.3975\angle-30°$ V, $V_{cn} = 186.6025\angle-150°$ V

10.3 Symmetrical components of line voltages: $V_{ab1} = 199.3220\angle63.8767°$ V, $V_{ab2} = 23.1823\angle155.5361°$ V, Phase to neutral voltages: $V_{an} = 201.3313\angle147.2675°$V, $V_{bn} = 179.2564\angle37.3964°$ V, $V_{cn} = 205.9470\angle-9.1282°$ V, Phase currents: $I_{an} = 20.1331\angle147.2675°$A, $I_{bn} = 17.9256\angle37.3964°$ A, $I_{cn} = 20.5947\angle-99.1282°$ A

10.4 12.08 kW

10.5 Generator reactances: $X_{g1} = 0.2229$ pu, $X_{g2} = 0.3333$ pu; transformer reactances: $X_{t1} = 0.20$ pu, $X_{t2} = 0.30$pu, $X_{t3} = 0.50$ pu; motor reactance $X_m = 0.2645$ pu; line impedances: line 1 $(0.1033 + 0.2066i)$ pu, line 2 = $(0.0517 + 0.1033i)$ pu, line 3 = $(0.0930 + 0.1240i)$ pu; load: as a parallel R-X combination $R = 4.1667$ pu, $X = 5.5556$ pu

10.6
$$\begin{bmatrix} Z_0 \\ Z_1 \\ Z_2 \end{bmatrix} = \frac{1}{3} \begin{bmatrix} (Z_a + 2Z_b) & (Z_a - Z_b) & (Z_a - Z_b) \\ (Z_a - Z_b) & (Z_a + 2Z_b) & (Z_a - Z_b) \\ (Z_a - Z_b) & (Z_a - Z_b) & (Z_a + 2Z_b) \end{bmatrix}$$

10.7 Fault current = $11.6708\angle - 66.5856°$ kA, $V_{a0} = 0.3542\angle - 176.0837°$ pu $V_{a1} = 0.6999\angle - 3.1862°$ pu, $V_{a2} = 0.3037\angle - 17.6410°$ pu Line voltages:

$$\begin{bmatrix} V_a \\ V_b \\ V_c \end{bmatrix} = \begin{bmatrix} 0.1112\angle - 66.5603 \\ 1.0148\angle - 122.97350° \\ 1.0395\angle 122.0932 \end{bmatrix} \text{pu}$$

10.8 (i) Sequence currents: $I_{a1} = 1.1982\angle - 70.6445°$ pu,

$I_{a2} = 1.1982\angle 109.3555°$ pu,

Phase voltages: $\begin{bmatrix} V_a \\ V_b \\ V_c \end{bmatrix} = \begin{bmatrix} 1.0000\angle 0° \\ 0.5492\angle - 178.2058° \\ 0.4514\angle 177.8167° \end{bmatrix}$ pu,

Line voltages: $V_{ab} = 1.5490\angle 0.6361°$ pu, $V_{bc} = 0.1038\angle - 160.6445°$ pu,

$V_{ca} = 1.4511\angle 179.3210°$ pu, Line currents: $I_b = 2.0754\angle - 160.6445°$ pu,

$I_c = 2.0754\angle 19.3555°$pu (ii) $I_{a1} = (0.5939 - 1.4048i)$ pu,

$I_{a2} = (-0.3050 + 0.7624i)$ pu, $I_{a0} = (-0.2889 + 0.6424i)$ pu,

Currents: $\begin{bmatrix} I_a \\ I_b \\ I_c \end{bmatrix} = \begin{bmatrix} 0.0000 - 0.0000i \\ -2.3101 + 0.1851i \\ 1.4435 + 1.7420i \end{bmatrix}$,

Sequence voltages: $V_{a0} = (0.3355 + 0.0277i)$ pu,

$V_{a1} = (0.3797 - 0.0747i)$ pu, $V_{a2} = (0.3347 + 0.0337i)$ pu,

Phase voltages: $\begin{bmatrix} V_a \\ V_b \\ V_c \end{bmatrix} = \begin{bmatrix} 1.0500\angle - 0.7226° \\ 0.1159\angle 175.4195° \\ 0.1131\angle 50.3532° \end{bmatrix}$,

Line voltages: $V_{ab} = 1.1656\angle - 1.1058°$ pu,

$V_{bc} = 0.2032 \angle -157.4722°$ pu, $V_{ca} = 0.9829\angle 174.1404°$ pu, actual line currents:

$\begin{bmatrix} I_a \\ I_b \\ I_c \end{bmatrix} = \begin{bmatrix} 0.0000 \\ 12.1637\angle 175.4195° \\ 11.8743\angle 50.3532° \end{bmatrix}$ kA

(iii) Currents: $I_a = I_{a1} = 2.3462\angle - 67.4722°$ pu $= 12.3142\angle - 67.4722°$ kA,

$I_b = \angle -187.4722°$, $I_c = \angle 52.5278°$; Voltages: $V_a = 0.1173\angle - 67.4722°$ pu,

$V_b = 0.1173\angle - 187.4722°$, $V_c = 0.1173\angle 52.5278°$

10.9 $V_{ab} = 2.2332\angle 23.4324°$ pu, $V_{bc} = 1.7321\angle -90.0000°$ pu,

$V_{ca} = 2.2160\angle 157.6129°$ pu

10.10 $I_a = I_{a1} = 8.6197\angle -65.5699°$ pu $= 75.403$ kA, $I_b = 8.6197\angle -185.5699°$ pu,

$I_c = 8.6197\angle 54.4301°$, $V_a = 0.4310\angle -65.5699°$ pu, $V_b = 0.4310\angle -185.5699°$,

$V_c = 0.4310\angle 54.4301°$

10.11 Currents: Generator side $= (3.1392 - 7.6860i)$ pu $= 72.6266\angle -67.7833°$ kA,

Line-1 $= (0.2606 - 0.1003i)$ pu $= 2.4430\angle -21.0420°$ kA,

Line-2 $= (0.1651 - 0.0618i)$ pu $= 1.5423\angle -20.5063°$ kA;

Voltages: node 2 $= 0.4068\angle -62.0658°$ pu, node 3 $= 0.3795\angle -48.6134°$ pu, node 4 $= 0.3642\angle -43.7142°$ pu, node 5 $= 0.3642\angle -43.7142°$ pu, node 6 $= 0.4040\angle -54.7471°$ pu

10.12

$$\begin{bmatrix} I_a \\ I_b \\ I_c \end{bmatrix} = \begin{bmatrix} 0.0000 \\ 4.1093\angle 163.6605° \\ 3.3023\angle 51.9268° \end{bmatrix}, \begin{bmatrix} V_a \\ V_b \\ V_c \end{bmatrix} = \begin{bmatrix} 0.8856\angle -3.8380° \\ 0.1202\angle 131.8273° \\ 0.1061\angle 99.9696° \end{bmatrix},$$

Line voltages: $V_{ab} = 0.9752\angle -8.7779°$ pu, $V_{bc} = 0.0635\angle -166.4276°$ pu, $V_{ca} = 0.9167\angle 169.7115°$ pu

10.13

Line currents: $\begin{bmatrix} I_a \\ I_b \\ I_c \end{bmatrix} = \begin{bmatrix} 0 \\ 2.3606\angle -169.2804° \\ 2.3606\angle 10.7196° \end{bmatrix}$ pu,

Line voltages: $\begin{bmatrix} V_a \\ V_b \\ V_c \end{bmatrix} = \begin{bmatrix} 1.0000\angle 0° \\ 0.5333\angle -179.3232° \\ 0.4668\angle 179.2267° \end{bmatrix}$ pu

Chapter 11: Power System Stability

11.1 (i) 6.05 MVJ/MVA,　(ii) 0.6722 MJ s/elec. degrees

11.2 (i) $3.7011e + 003$ MJ,　(ii) 14.8044 MJ/MVA,　(iii) 0.4112 MJ s/elec. degrees

11.3 (i) 24 MW,　(ii) $1.2343e+003$ elec. degrees/s^2

11.4 16.1854°,　(i) 1566.6290 rpm,　(ii) 28.5283°

11.5 $P_{eqi} = \dfrac{H_2 P_{1i} - H_1 P_{2i}}{H_1 + H_2} = 0.2429$ pu, $P_{equ} = \dfrac{H_2 P_{1u} - H_1 P_{2u}}{H_1 + H_2}$,

$H_{eq} = \dfrac{H_1 H_2}{H_1 + H_2} = 2.5714$ MJ/MVA, max. power $= 1.9714$ pu

11.6 $E' = 1.2403\angle 14.9514°$, $\dfrac{d^2\delta}{dt^2} = 31.3725 - 42.2941\sin\delta$, accelerating power $=$ 0.5217 pu, acceleration $= 20.4886$ elec. rads/s^2

11.7 $\delta_m = 66.5220°$, $t_c = 16$ cycles

11.8 $\delta_{\text{crit}} = 96.29°$

11.9 (i) $\delta = 30.0°$, (ii) $P_u = 0.5844$ pu, $P_{\text{acc.}} = 0.4156$ pu, $\alpha = 18.6522$ rads/s^2,

(iii) $\delta = 78.0912°$, $P_{\text{acc.}} = -0.1436$ pu, $\alpha = -6.4395$ rads/s^2

11.11

$$Z = \begin{bmatrix} Z_1 \\ Z_2 \end{bmatrix} = \begin{bmatrix} \delta \\ \dot{\delta} \end{bmatrix}, \ \dot{Z} = \begin{bmatrix} \dot{Z}_1 \\ \dot{Z}_2 \end{bmatrix} = \begin{bmatrix} Z_2 \\ \dot{Z}_2 \end{bmatrix} = \begin{bmatrix} \dot{Z}_2 \\ P_a \end{bmatrix} \text{ when } Z_1 = \delta \text{ and } Z_2 = \dot{\delta}$$

11.12 (i) 2.9452, (ii) 6.4693 rads/s

Chapter 13: Contingency Analysis

13.1 (i) Currents: $(-0.3953 - 0.1454i)$ pu, $(0.0212 + 0.0009i)$ pu.

$V_1 = 1.0\angle 0°$ pu, $V_2 = 0.9590\angle 8.87°$ pu, $V_3 = 0.9804\angle 3.00°$ pu,

$V_4 = 0.9615\angle 4.59°$ pu, $V_5 = 0.9869\angle 2.67°$ pu

(ii) Currents: $(0.1396 + 0.0523i)$ pu, $(-0.0891 - 0.0350i)$ pu,

$(0.0325 + 0.0273i)$ pu; $V_1 = 1.0\angle 0°$ pu, $V_2 = 0.9673\angle 6.65°$ pu,

$V_3 = 1.0303\angle - 4.35°$ pu, $V_4 = 0.9582\angle 5.32°$ pu, $V_5 = 0.9930\angle 2.48°$ pu

13.2 (i) $(0 + 0.1504i)$ pu, (ii) $L_{12\text{-}14} = 0.6364$, $L_{13\text{-}14} = 0.3636$, $L_{23\text{-}14} = 0.2273$,

$L_{24\text{-}14} = 0.4091$, $L_{34\text{-}14} = 0.5909$, (iii) $I'_{12} = 0.2434\angle -16.6424°$ pu,

$I'_{13} = 0.0868\angle 147.2053°$ pu, $I'_{23} = 1.0324\angle 156.6423°$ pu,

$I'_{24} = 0.2239\angle 132.1526°$ pu, $I'_{34} = 0.6518\angle - 6.8094°$ pu

(iv) $V_1 = 1.0 \angle 0°$ pu, $V_2 = 0.9757\angle -2.6009°$ pu, $V_3 = 1.0159\angle 2.9625°$ pu,

$V_4 = 1.0039\angle - 0.4001°$ pu

13.3 $I'_{12} = 0.2029\angle - 20.5660°$ pu, $I'_{13} = 0.0948\angle 157.2656°$ pu,

$I'_{14} = 0.0590\angle 11.6516°$ pu, $I'_{23} = 0.9811\angle 158.3863°$ pu,

$I'_{24} = 0.0831\angle 137.1839°$ pu, $I'_{34} = 0.6820\angle -11.4663°$ pu,

$V_1 = 1.0000\angle 0°$ pu, $V_2 = 0.9833\angle -2.7677°$ pu, $V_3 = 1.0193\angle 2.4590°$ pu,

$V_4 = 1.0050\angle -1.3176°$ pu

13.4

$$\mathbf{Z}_{th-45} = \begin{array}{c} 4 \\ 5 \end{array}\begin{bmatrix} \overset{4}{0.2543i} & \overset{5}{0.4181i} \\ 0.4181i & 0.1638i \end{bmatrix} \text{pu} \quad \mathbf{Z}_{th-45} = \begin{array}{c} 6 \\ 7 \end{array}\begin{bmatrix} \overset{6}{0.0319i} & \overset{7}{0.0688i} \\ 0.0688i & 0.0369i \end{bmatrix} \text{pu}$$

13.5

(i) $\begin{bmatrix} 4.7630i & 4.5087i \\ 4.5087i & 4.6725i \end{bmatrix}$ pu

(ii) $\begin{bmatrix} 4.1454i & 4.1135i \\ 4.1135i & 4.1504i \end{bmatrix}$ pu $\quad \mathbf{Z}_{\text{loop}} = \begin{bmatrix} 9.1084i & 8.6222i \\ 8.6222i & 9.2228i \end{bmatrix}$ pu

13.6

$$V'_A = \begin{bmatrix} V'_1 \\ V'_2 \\ V'_3 \\ V'_4 \\ V'_5 \end{bmatrix} = \begin{bmatrix} 1.0078\angle - 0.2599° \\ 0.9840\angle 4.7550° \\ 1.0305\angle - 3.1159° \\ 1.0009\angle 2.2666° \\ 1.0214\angle 0.3814° \end{bmatrix} \text{pu}$$

$$V_B' = \begin{bmatrix} V_1' \\ V_2' \\ V_3' \\ V_4' \end{bmatrix} = \begin{bmatrix} 1.0276\angle 2.2262° \\ 1.0120\angle -1.5818° \\ 1.0031\angle -0.5199° \\ 0.9839\angle -3.5014° \end{bmatrix} \text{pu}$$

13.7

At boundary A $\begin{bmatrix} -2.4250i & 2.3400i \\ 2.3400i & -2.4720i \end{bmatrix}$ pu,

At boundary B $\begin{bmatrix} -14.6057i & 14.4759i \\ 14.4759i & -14.5882i \end{bmatrix}$ pu

13.10 (i)

Distribution factor →	1-2	1-5	2-3	3-4	3-5	4-5
Line outage ↓						
1-2	0	1.0000	−1.0000	−0.3478	−0.6522	−0.3478
1-5	1.0000	0	1.0000	0.3478	0.6522	0.3478
2-3	−1.0000	1.0000	0	−0.3478	−0.6522	−0.3478
3-4	−0.2520	0.2520	−0.2520	0	0.7480	−1.0000
3-5	−0.3871	0.3871	−0.3871	0.6129	0	0.6129
4-5	−0.2520	0.2520	−0.2520	−1.0000	0.7480	0

(ii) Line 5-3: $K531 = 0.1520$, $K535 = 0.2645$;

Line 5-4: $K541 = 0.0811$, $K545 = 0.1411$

13.11 $I_{53}' = 0.9383\angle -1.94°$ pu

13.12 0.5561 pu (reduction)

Chapter 14: State Estimation Techniques

14.2 $\begin{bmatrix} V_1 \\ V_2 \end{bmatrix} = \begin{bmatrix} 20.1821 \\ 9.9415 \end{bmatrix}$ V

14.3

Measurements: $\begin{bmatrix} 9.1642 \\ 7.0189 \\ 6.9217 \\ 9.1642 \\ 6.9717 \end{bmatrix}$ Error: $\begin{bmatrix} -0.0642 \\ -0.0189 \\ -0.0217 \\ 0.0358 \\ 0.0783 \end{bmatrix}$

14.4

$\begin{bmatrix} V_1 \\ V_2 \end{bmatrix} = \begin{bmatrix} 20.1906 \\ 10.0113 \end{bmatrix}$ V, Error: $\begin{bmatrix} -0.0811 \\ -0.0415 \\ -0.0377 \\ 0.0189 \\ 0.1623 \end{bmatrix}$

14.5

$\begin{bmatrix} V_1 \\ V_2 \\ V_3 \end{bmatrix} = \begin{bmatrix} 1.0500 + j0.0000 \\ 0.9842 + j0.0005 \\ 0.9917 - j0.0186 \end{bmatrix}$

14.6 $S_1 = (0.1293 + 0.1587i)$ pu, $S_2 = (-0.1259 - 0.2489i)$ pu,

$S_3 = (-0.1735 + 0.1381i)$ pu, $S_4 = (0.1753 - 0.1823i)$ pu,

$S_5 = (-0.0737 - 0.3442i)$ pu, $S_6 = (0.0788 + 0.2817i)$ pu

14.7

(i)
$$\begin{bmatrix} -d_1 & d_2 & d_3 & -d_4 & 0 & 0 & -d_7 \\ 0 & 0 & 0 & 0 & 0 & -d_6 & d_7 \\ 0 & 0 & -d_3 & d_4 & -d_5 & d_6 & 0 \end{bmatrix}$$

(ii)
$$\begin{bmatrix} (d_1+d_2+d_3+d_4+d_7) & -d_7 & (-d_3-d_4) \\ -d_7 & (d_6+d_7) & -d_6 \\ (-d_3-d_4) & -d_6 & (d_3+d_4+d_5+d_6) \end{bmatrix}$$

14.8 $Pr(c_1 < z < c_2) = \dfrac{1}{\sqrt{2\pi}} \displaystyle\int_{y_1}^{y_2} \varepsilon^{-\frac{1}{2}y^2} \, dy$, 99%

14.11 Computed value of $\hat{f}$ is 1.2358 against its critical value of 11.35. Hence, there is no bad data.

14.12 Computed value of $\hat{f}$ is 30.2943 and its critical value is 11.35. There is bad data. Measurement corresponding to instrument 5 is identified as bad data and eliminated. Computed value of $\hat{f}$ now works out to 0.43 which is less than the critical value of 9.21.

14.13 Redundancy factor = 2.7458, Degrees of freedom = 103, Line to bus ratio = 0.6042, Degrees of freedom = 0, Redundancy factor = 3.9661, Degrees of freedom = 175

14.15

$$[\mathbf{H}] = \begin{bmatrix} & \delta_2 & \delta_3 & \delta_4 & V_1 & V_2 & V_3 & V_4 \\ & 0 & 0 & 0 & 0 & 0 & 1 & 0 \\ & 0 & 0 & 0 & 0 & 0 & 0 & 1 \\ & \frac{\partial P_3}{\partial \delta_2} & \frac{\partial P_3}{\partial \delta_3} & \frac{\partial P_3}{\partial \delta_4} & 0 & \frac{\partial P_3}{\partial V_2} & \frac{\partial P_3}{\partial V_3} & \frac{\partial P_3}{\partial V_4} \\ & \frac{\partial Q_3}{\partial \delta_2} & \frac{\partial Q_3}{\partial \delta_3} & \frac{\partial Q_3}{\partial \delta_4} & 0 & \frac{\partial Q_3}{\partial V_2} & \frac{\partial Q_3}{\partial V_3} & \frac{\partial Q_3}{\partial V_4} \\ & \frac{\partial P_{32}}{\partial \delta_2} & \frac{\partial P_{32}}{\partial \delta_3} & 0 & 0 & \frac{\partial P_{32}}{\partial V_2} & \frac{\partial P_{32}}{\partial V_3} & 0 \\ & 0 & \frac{\partial P_{34}}{\partial \delta_3} & \frac{\partial P_{34}}{\partial \delta_4} & 0 & 0 & \frac{\partial P_{34}}{\partial V_3} & \frac{\partial P_{34}}{\partial V_4} \\ & \frac{\partial P_{21}}{\partial \delta_2} & 0 & 0 & \frac{\partial P_{21}}{\partial V_1} & \frac{\partial P_{21}}{\partial V_2} & 0 & 0 \\ & 0 & \frac{\partial P_{43}}{\partial \delta_3} & \frac{\partial P_{43}}{\partial \delta_4} & 0 & 0 & \frac{\partial P_{43}}{\partial V_3} & \frac{\partial P_{43}}{\partial V_4} \\ & 0 & \frac{\partial Q_{34}}{\partial \delta_3} & \frac{\partial Q_{34}}{\partial \delta_4} & 0 & 0 & \frac{\partial Q_{34}}{\partial V_3} & \frac{\partial Q_{34}}{\partial V_4} \\ & \frac{\partial Q_{21}}{\partial \delta_2} & 0 & 0 & \frac{\partial Q_{21}}{\partial V_1} & \frac{\partial Q_{21}}{\partial V_2} & 0 & 0 \\ & \frac{\partial Q_{23}}{\partial \delta_2} & \frac{\partial Q_{23}}{\partial \delta_3} & 0 & 0 & \frac{\partial Q_{23}}{\partial V_2} & \frac{\partial Q_{23}}{\partial V_3} & 0 \\ & 0 & \frac{\partial Q_{43}}{\partial \delta_3} & \frac{\partial Q_{43}}{\partial \delta_4} & 0 & 0 & \frac{\partial Q_{43}}{\partial V_3} & \frac{\partial Q_{43}}{\partial V_4} \end{bmatrix}$$

$$[z] = [V_3 \ V_4 \ P_3 \ Q_3 \ P_{32} \ P_{34} \ P_{21} \ P_{43} \ Q_{34} \ Q_{21} \ Q_{23} \ Q_{43}]^t$$

14.16

(i) $\dfrac{\partial P_3}{\partial \delta_2} = |Y_{32}||V_3||V_2|\sin(\delta_3 - \delta_2 - \theta_{32}),$

$\dfrac{\partial P_3}{\partial \delta_3} = -|Y_{32}||V_3||V_2|\sin(\delta_3 - \delta_2 - \theta_{32}) - |Y_{34}||V_3||V_4|\sin(\delta_3 - \delta_4 - \theta_{34})$

$\dfrac{\partial P_3}{\partial \delta_4} = |Y_{34}||V_3||V_4|\sin(\delta_3 - \delta_4 - \theta_{34}), \quad \dfrac{\partial P_3}{\partial V_2} = |Y_{32}||V_3|\cos(\delta_3 - \delta_2 - \theta_{32}),$

$\dfrac{\partial P_3}{\partial V_3} = Y_{32}|V_2|\cos(\delta_3 - \delta_2 - \theta_{32}) + 2|Y_{33}||V_3|\cos\theta_{33} + |Y_{34}||V_4|\cos(\delta_3 - \delta_4 - \theta_{34}),$

$\dfrac{\partial P_3}{\partial V_4} = |Y_{34}||V_3|\cos(\delta_3 - \delta_4 - \theta_{34}), \quad \dfrac{\partial Q_3}{\partial \delta_2} = -|Y_{32}||V_3||V_2|\cos(\delta_3 - \delta_2 - \theta_{32}),$

$\dfrac{\partial Q_3}{\partial \delta_3} = |Y_{32}||V_3||V_2|\cos(\delta_3 - \delta_2 - \theta_{32}) - |Y_{34}||V_3||V_4|\cos(\delta_3 - \delta_4 - \theta_{34})$

$\dfrac{\partial Q_3}{\partial \delta_4} = -|Y_{34}||V_3||V_4|\cos(\delta_3 - \delta_4 - \theta_{34}), \quad \dfrac{\partial Q_3}{\partial V_2} = |Y_{32}||V_3|\sin(\delta_3 - \delta_2 - \theta_{32})$

$\dfrac{\partial Q_3}{\partial V_3} = |Y_{32}||V_2|\sin(\delta_3 - \delta_2 - \theta_{32}) + 2|Y_{33}||V_3|\sin\theta_{33} + |Y_{34}||V_4|\sin(\delta_3 - \delta_4 - \theta_{34}),$

$\dfrac{\partial Q_3}{\partial V_4} = |Y_{34}||V_3|\sin(\delta_3 - \delta_4 - \theta_{34})$

(ii) Line flows

$\dfrac{\partial P_{21}}{\partial \delta_2} = |V_2||V_1||Y_{21}|\sin(\delta_2 - \delta_1 - \theta_{21}),$

$\dfrac{\partial P_{21}}{\partial V_2} = 2G_{21}|V_2| + |V_1||Y_{21}|\cos(\delta_2 - \delta_1 - \theta_{21})$

$\dfrac{\partial P_{21}}{\partial V_1} = |V_2||Y_{21}|\cos(\delta_2 - \delta_1 - \theta_{21}), \quad \dfrac{\partial Q_{21}}{\partial \delta_2} = -|V_2||V_1||Y_{21}|\cos(\delta_2 - \delta_1 - \theta_{21})$

$\dfrac{\partial Q_{21}}{\partial V_2} = -2|V_2|\left(\dfrac{B'_{21}}{2} + B_{21}\right) - |V_1||Y_{21}|\sin(\delta_2 - \delta_1 - \theta_{21})$

$\dfrac{\partial Q_{21}}{\partial V_1} = -|V_2||Y_{21}|\sin(\delta_2 - \delta_1 - \theta_{21})$

Chapter 15: An Introduction to HVDC Power Transmission

15.1

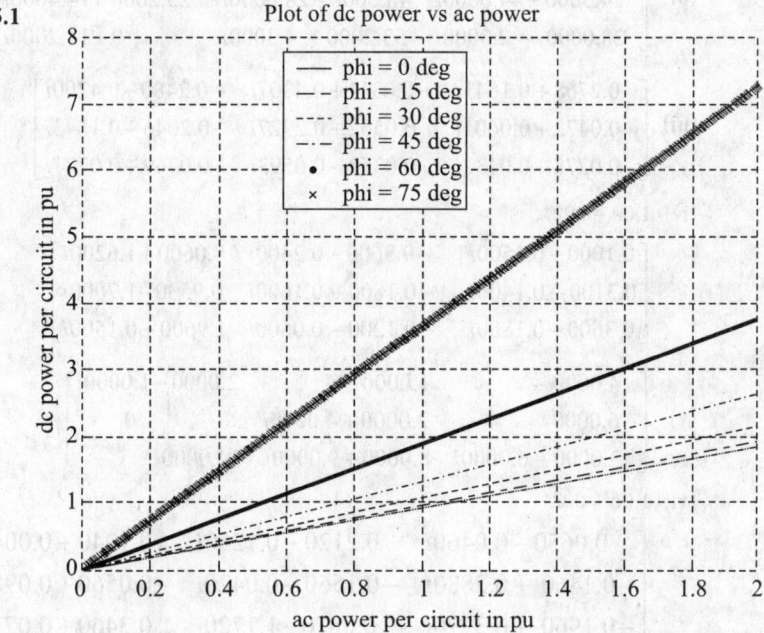

Plot of dc power vs ac power

15.2

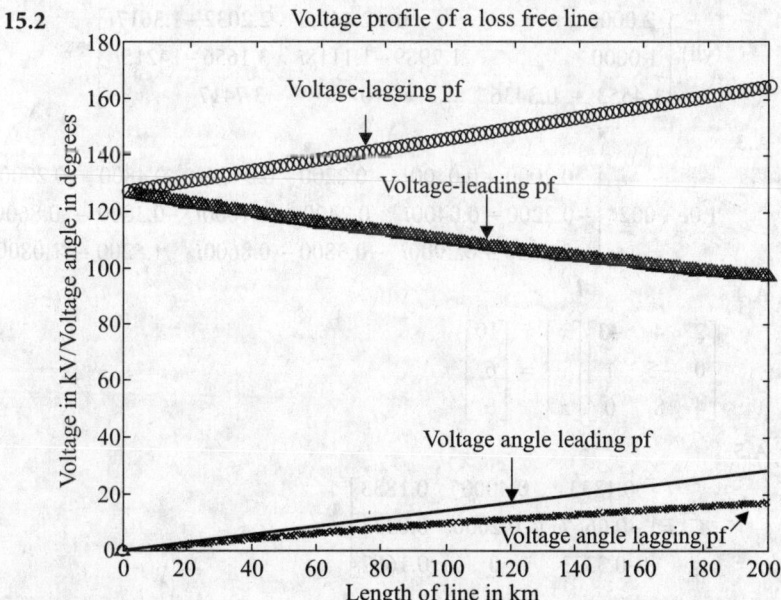

Voltage profile of a loss free line

Appendix A

A.1

(i)
$$\begin{bmatrix} 5.0000 + 2.0000i & -3.5000 - 3.4000i & 3.0000 + 6.0000i \\ & 7.0000 & 3.2000 + 0.5000i & 8.0000 - 5.0000i \\ 4.0000 + 0.5000i & 3.2000 & 14.0000 + 0.8000i \end{bmatrix}$$

(ii)
$$
\begin{bmatrix}
-23.0000+18.5000i & 12.4000-21.4000i & -4.8000-21.6000i \\
24.5000-44.0000i & 48.5000-28.0000i & 23.2000+14.4000i \\
28.0000-2.0000i & 53.2000-4.3000i & 0+11.2000i
\end{bmatrix}
$$

(iii)
$$
\begin{bmatrix}
0.2768+0.1643i & -0.0311+0.4307i & -0.2480-0.4200i \\
0.0472+0.0601i & 0.0562+0.2927i & -0.2046-0.1643i \\
-0.0278-0.0432i & 0.0352-0.0593i & 0.0769+0.0777i
\end{bmatrix}
$$

(iv) $1.0e+002*$
$$
\begin{bmatrix}
0.1000+0.1500i & -0.3600+0.2400i & 0.0600+1.6200i \\
0.3100-0.1400i & -0.1800-0.1600i & 0.9500-1.7000i \\
0.3600+0.1800i & -0.1200-0.0600i & 1.9600+0.1500i
\end{bmatrix}
$$

(v)
$$
\begin{bmatrix}
4.0000 & 1.0000 & 2.0000-1.0000i \\
-6.0000 & 2.0000+4.0000i & 0 \\
3.0000-6.0000i & 8.0000+9.0000i & 14.0000
\end{bmatrix}
$$

(vi) $1.0e+002*$
$$
\begin{bmatrix}
0.0650-0.0460i & 0.2120-0.1250i & 0.1040+0.0040i \\
-0.1460+0.2880i & -0.1560-0.0420i & -0.0560+0.0980i \\
-0.1960+0.3770i & 0.6810-1.1720i & 0.3460+0.0710i
\end{bmatrix}
$$

(vii)
$$
\begin{bmatrix}
2.0000 & 0-2.4495i & 2.2032+1.3617i \\
1.0000 & 1.7989-1.1118i & 3.1656-14215i \\
1.4553+0.3436i & 0 & 3.7417
\end{bmatrix}
$$

A.3
$$
1.0e+002*
\begin{bmatrix}
0.2000+0.0400i & -0.2200-0.0400i & 0.4800+0.2900i \\
-0.2200-0.0400i & 0.2400-0.1600i & -0.3800-0.8600i \\
0.4800+0.2900i & -0.3800-0.8600i & 1.5200-1.0800i
\end{bmatrix}
$$

A.4
$$
\begin{bmatrix}
2 & 4 & -3 \\
0 & -5 & 1 \\
4 & 8 & 0
\end{bmatrix}
\begin{bmatrix}
x \\ y \\ z
\end{bmatrix}
=
\begin{bmatrix}
10 \\ 6 \\ 0
\end{bmatrix}
$$

A.5
$$
A^{-1} =
\begin{bmatrix}
0.1333 & 0.4000 & 0.1833 \\
-0.0667 & -0.2000 & 0.0333 \\
-0.3333 & 0 & 0.1667
\end{bmatrix},
$$

$x = 3.7333, y = -1.8667, z = -3.3333$

A.6
$$
1.0e+002*
\begin{bmatrix}
0.9000-0.2000i \\
1.2000-0.1600i \\
2.5000+0.9000i \\
0.6000-0.0800i
\end{bmatrix}
$$

A.7 0, No

A.8 cofac =   adjoint =

$$
\begin{array}{rrrr}
-11 & -13 & 12 & 8 \\
27 & 36 & -39 & 9 \\
31 & 56 & -42 & 2 \\
-89 & -112 & 123 & -28
\end{array}
\qquad
\begin{array}{rrrr}
-11 & 27 & 31 & -89 \\
-13 & 36 & 56 & -112 \\
12 & -39 & -42 & 123 \\
8 & 9 & 2 & -28
\end{array}
$$

$$
A^{-1} = \begin{bmatrix}
0.7333 & -1.8000 & -2.0667 & 5.9333 \\
0.8667 & -2.4000 & -3.7333 & 7.4667 \\
-0.8000 & 2.6000 & 2.8000 & -8.2000 \\
-0.5333 & -0.6000 & -0.1333 & 1.8667
\end{bmatrix}
$$

A.9
$$
\begin{bmatrix}
-52 & 11 & 43 \\
11 & -1 & -6 \\
43 & -6 & -41
\end{bmatrix}
$$

A.10
$$
\begin{bmatrix}
1.9462 & -0.5692 & -2.5538 & 0.0308 \\
-0.6692 & 0.3538 & 0.8308 & -0.0462 \\
0.7846 & -0.2769 & -1.2154 & 0.1231 \\
-0.3846 & 0.0769 & 0.6154 & 0.0769
\end{bmatrix}
$$

A.11
$$
\begin{bmatrix}
32.0000 + 13.0000i \\
24.0000 + 29.0000i \\
4.0000 + 21.0000i
\end{bmatrix}
$$

A.12
$$
\begin{bmatrix}
0.5000 & 0.4000 - 0.6000i & 0.6000 - 0.8000i & 0.8000 - 1.0000i \\
-0.4000 - 0.6000i & 0.5000 & 1.0000 - 1.2000i & 1.2000 - 1.4000i \\
-0.6000 - 0.8000i & -1.0000 - 1.2000i & 0.5000 & 1.4000 - 1.6000i \\
-0.8000 - 1.0000i & -1.2000 - 1.4000i & -1.4000 - 1.6000i & 0.5000
\end{bmatrix}
$$

A.13
$$
\left[
\begin{array}{cc|ccc}
0 - 4.0000i & 0 & 0 + 4.0000i & 0 & 0 \\
0 & 0 - 9.0000i & 0 & 0 & 0 + 9.0000i \\
\hline
0 + 4.0000i & 0 & 0 - 11.0000i & 0 + 6.0000i & 0 \\
0 & 0 & 0 + 6.0000i & 0 - 12.0000i & 0 + 5.0000i \\
0 & 0 + 9.0000i & 0 & 0 + 5.0000i & 0 - 16.0000i
\end{array}
\right] S
$$

$$
\begin{bmatrix}
0 - 1.8810i & 0 + 0.8565i \\
0 + 0.8565i & 0 - 2.8335i
\end{bmatrix} S
$$

A.14 (i)
$$
\begin{bmatrix}
0 + 0.6165i & 0 + 0.1863i & 0 + 0.3665i & 0 + 0.2609i & 0 + 0.1863i \\
0 + 0.1863i & 0 + 0.4092i & 0 + 0.1863i & 0 + 0.2174i & 0 + 0.2981i \\
0 + 0.3665i & 0 + 0.1863i & 0 + 0.3665i & 0 + 0.2609i & 0 + 0.1863i \\
0 + 0.2609i & 0 + 0.2174i & 0 + 0.2609i & 0 + 0.3043i & 0 + 0.2174i \\
0 + 0.1863i & 0 + 0.2981i & 0 + 0.1863i & 0 + 0.2174i & 0 + 0.2981i
\end{bmatrix} \Omega
$$

(ii)
$$
\begin{bmatrix}
0 + 0.6165i & 0 + 0.1863i \\
0 + 0.1863i & 0 + 0.4092i
\end{bmatrix} \Omega
$$

A.15 $1.0e + 002 * \begin{bmatrix} -0.9849 - 2.6204i \\ 3.2585 - 3.6741i \end{bmatrix} A$

Bibliography

A. Brameller, R.N. Allan, and Y.M. Hamam, *Sparsity*, 1st edn, Pitman Publishing, NY, 1976.

A. Greenwood, *Electrical Transients in Power Systems*, Wiley Interscience, NY, 1971.

A. Woodford, 'HVDC Transmission', www.hvdc.ca/pdf_misc/dcsum.pdf .

A.E. Fitzgerald, C. Kingsley Jr, and S.D. Umans, *Electric Machinery*, 5th edn, McGraw Hill, 1992.

A.H. El-Abiad and G.W. Stagg, 'Automatic evaluation of power system performance—Effects of line and transformer outages', *AIEE Transactions*, 1963, pp. 712–16.

A.H. El-Abiad and K. Nagappan, 'Transient stability regions of multimachine power systems' *IEEE Transactions*, PAS, vol. 85, 1966, pp. 169–79.

A.H. El-Abiad, 'Digital calculation of line to ground short circuit by matrix method', *IEEE Transactions*, PAS, vol. 79, Part III, 1960, pp. 323–32.

A.H. El-Abiad, R. Guidone, and G.W. Stagg, 'Digital calculation of short circuits using a high speed digital computer', *IEEE Transactions*, PAS, vol. 80, Part III, 1961, pp. 702–7.

A.R. Abhyankar and S.A. Khaparde, 'Introduction to deregulation in power industry', *Proceedings of Workshop on Deregulation in Power Industry*, IIT, Mumbai, 2002.

A.R. Bergen and V. Vittal, *Power System Analysis*, 2nd edn, Pearson Education Asia, New Delhi, 2001.

B. Stott and O. Alasc, 'Fast decoupled load flow', *IEEE Transactions*, vol. PAS-93, 1974, pp. 859–67.

B. Stott, 'Effective starting process for Newton–Raphson load flows', *Proceedings IEE*, vol. 118, 1974.

B. Stott, 'Review of load flow calculation methods', *Proceedings IEEE*, vol. 62, no. 7, 1974, pp. 916–29.

B.M. Weedy, *Electric Power Systems*, 3rd edn, John Wiley, London, 1979.

Carson W. Taylor, *Power System Voltage Stability*, McGraw-Hill Inc. International Edition, 1994.

C.A. Gross, *Power System Analysis*, 2nd edn, John Wiley, NY, 1983.

C.S. Rao and T.K. Nagsarkar, 'Half wave thyristor controlled dynamic brake to improve transient stability', *IEEE Transactions on Power Apparatus and Systems*, PAS-103, vol. 5, 1984, pp. 1077–83.

C.S. Rao and T.K. Nagsarkar, 'Transient stability improvement with thyristor controlled braking device', *IEEE Power Engineering Society Winter Meeting*, Paper No. A90 079-4, NY, 1980.

C.S. Rao and T.K. Nagsarkar, 'Transient stability investigations of multimachine power systems employing thyristor controlled dynamic brake', *Proceedings of AMSE*, Paris, March 1-3, 1982.

Central Electricity Authority, 'Chapter 3: Transmission planning philosophy', *Draft National Electricity Plan*, 2005.

Central Electricity Authority, 'Chapter 5: Electric power supply and system losses', *General Review*, 2005.

Central Electricity Authority, 'Chapter 7: Electricity transmission and distribution system', *General Review*, 2005.

Central Electricity Authority, 'Review of growth of Indian electricity sector', *General Review*, 2005.

D.P. Kothari and J.J. Nagrath, *Modern Power System Analysis*, 3rd edn, Tata McGraw Hill, New Delhi, 2003.

E. Handschin, F.C. Schweppe, J. Kohlas, and A. Fiechter, 'Bad data analysis of power system state estimation', *IEEE Transactions*, PAS, vol. 94, no. 2, 1975, pp. 329–37.

E.E. Fetzer and P.M. Anderson, 'Observability in the state estimation of power systems', *IEEE Transactions*, PAS, vol. 94, no. 6, 1975.

E.R.D. Begamudre, *Energy Conversion Systems*, New Age International (P) Ltd, Publishers, New Delhi, 2000.

E.W. Kimbark, 'Improvement of power system stability by changes in the network', *IEEE Transactions*, PAS, vol. 88, 1969, pp. 773–81.

E.W. Kimbark, 'Improvement of power system stability by switched series capacitors', *IEEE Transactions*, PAS, vol. 85, 1966, pp. 180–8.

E.W. Kimbark, *Direct Current Transmission*, vol. 1, Wiley, NY, 1971.

E.W. Kimbark, *Power System Stability*, vol. 1, John Wiley, NY, 1948.

E.W. Kimbark, *Power System Stability*, vol. 3, John Wiley, NY, 1956.

F. Broussolle, 'State estimation in power systems: Detecting bad data through the sparse inverse matrix method', *IEEE Transactions*, PAS, vol. 97, no. 3, 1978.

F.C. Schweppe, J. Wildes, and D.B. Rom, 'Power system static-state estimation— Part I: Exact model, Part II: Approximate model, Part III: Implementation', *IEEE Transactions*, PAS, vol. 89, no. 1, 1970, pp. 120–35.

G. Singh, T.K. Nagsarkar, and C.S. Rao, 'Selection of stabilizing signal for Bhaba hydro generators', *Modelling Simulation and Control A*, AMSE Press, vol. 1, no. 4, 1985, pp. 17–25.

G.K. Brown, H.H. Happ, C.E. Person, and C.C. Young, 'Transient stability solution by an impedance matrix method', *IEEE Transactions*, PAS, vol. 84, 1965, pp. 1204–14.

G.L. Kusic, *Computer Aided Power System Analysis*, Prentice Hall, 1986.

G.W. Stagg and A.H. El-Abiad, *Computer Methods in Power System Analysis*, McGraw Hill Kogakusha Ltd, Tokyo, 1968.

H. Saadat, *Power System Analysis*, Tata McGraw Hill, New Delhi, 2002.

H.E. Brown, 'Contingencies evaluated by Z-matrix method', *IEEE Transactions*, PAS, vol. 88, 1969, pp. 1877–82.

H.E. Brown, 'Low voltages caused by contingencies predicted by a high speed Z-matrix method', *IEEE Transactions*, PAS, vol. 96, 1977, pp. 44–51.

H.E. Brown, C.E. Person, L.K. Kirchmeyer, and G.W. Stagg, 'Digital calculation of three-phase short circuit by matrix method', *AIEE Transactions*, PAS, vol. 79, 1961, pp. 1277–82.

H.E. Brown, G.K. Carter, H.H. Happ, and C.E. Person, 'Power flow solution by impedance matrix method', *Transactions AIEE*, vol. 82, part 3, 1963, pp. 1–10.

H.E. Brown, *Solution of Large Networks by Matrix Methods*, John Wiley, NY, 1975.

H.P. Horisberger, J.C. Richard, and C. Rossier, 'Fast decoupled state estimation', *IEEE Transactions*, PAS, vol. 95, no. 1, 1976, pp. 208–15.

H.W. Dommel and N. Sato, 'Fast transient stability solutions', *IEEE Transactions*, PAS, 1972, pp. 253–60.

I.O. Hebibellah, and V.H. Quintana, 'Exact decoupled rectangular coordinate state estimation with effective data structure management', *IEEE Transactions*, PAS, vol. 7, no. 1, 1992.

ICPCI, 'Does India Really have a Power Shortage?', www.indiainfoline.com.

IEEE Task Force, 'A description of discrete supplementary controls for stability', *IEEE Transactions*, PAS, vol. 97, 1978, pp. 149–65.

India Core 'Blueprint for Power Sector Development', *Overview of Power Sector in India*, Annexure XI, IndiaCore, 2005.

J. Arillaga, C.P. Arnold, and B.J. Harker, *Computer Modeling of Electrical Power Systems*, John Wiley, NY, 1986.

J. Arillaga, *High Voltage Direct Current Transmission*, IEE Power Engineering Series 6, Peter Peregrinus Ltd., London, 1983.

J.B. Ward and H.W. Hale, 'Digital computer solution of power flow problem', *AIEE Transactions*, vol. PAS-75, 1956, pp. 398–404.

J.D. Glover and M.S. Sarma, *Power System Analysis and Design*, 3rd edn, Thomson Asia Pvt Ltd, Bangalore, 2003.

J.E. Van Ness and J.H. Griffin, 'Elimination Methods for Load Flow Studies', *AIEE Transactions*, vol. 80, Part III, 1961, pp. 299–304.

J.F. Dopazo, G.W. Stagg, and L.S. Van Slyck, 'State calculations of power systems from line flow measurements—Part II', *IEEE Transactions*, PAS, vol. 91, no. 1, 1972, pp. 145–51.

J.F. Dopazo, GW. Stagg, and L.S. Van Slyck, 'State calculations of power systems from line flow measurements—Part I', *IEEE Transactions*, PAS, vol. 89, no. 1, 1970, pp. 1698–1708.

J.J. Grainger and W.D. Stevenson Jr, *Power System Analysis*, Tata McGraw Hill, 2003.

K. Prabhashankar and W. Janischewsyj, 'Digital simulation of multimachine power systems for stability studies', *IEEE Transactions*, PAS, vol. 87, 1968, pp. 3–80.

K.L. Lo, P.S. Ong, R.D. McColl, A.M. Moffatt, and J.L. Suley, 'Development of static state estimator—Part I and Part II', *IEEE Transactions*, PAS, vol. 102, no. 8, 1983.

K.R. Padiyar, *Power System Dynamics: Stability and Control*, Interline Publishing Pvt Ltd, Bangalore, 2000.

L.P. Singh, *Advanced Power System Analysis and Dynamics*, 2nd edn, Wiley Eastern Ltd, New Delhi, 1986.

L.V. Bewley, *Traveling Waves on Transmission Systems*, 2nd edn, Dover Publications Inc., NY, 1963.

L.W. Coombe and D.G. Lewis, 'Digital calculation of short circuit currents in large complex impedance network', *IEEE Transactions*, PAS, vol. 75, 1956, pp. 1394–97.

M.A. Pai, *Computer Techniques in Power System Analysis*, Tata McGraw Hill, New Delhi, 1980.

M.G. Say, *Design of Alternating Current Machinery*, 5th edn, ELBS, 1983.

M.L. Shelton, P.F. Winkleman, W.A. Mittlestadt, and W.J. Belleroy, 'Bonneville power administration 1400 MW braking resistor', *IEEE Transactions*, PAS 94, 1965, pp. 602–11.

M.S. Sukhija and B. Thapar, 'A criterion for reducing power systems for stability', *All India Paper Meeting of the Institution of Engineers (India)*, Bombay, 1978.

M.S. Sukhija and B. Thapar, 'A study of power oscillations in Dehar machines with static exciters', *Symposium on Power Plant Dynamics and Control*, Bhabha Atomic Research Centre (BARC), Bombay, 1976.

M.S. Sukhija and B. Thapar, 'Computer techniques for solving transient stability problems', *Symposium on Power Systems Operation and Control*, IIT, Delhi, June 1972.

M.S. Sukhija and B. Thapar, 'Electrical equivalents for stability studies', *Journal of the Institution of Engineers (India)*, vol. 55, 1975.

M.S. Sukhija and B. Thapar, 'Some important aspects in the development of stability equivalents', *All India Seminar on Computer Applications in Power Systems*, Calcutta, 1978.

M.S. Sukhija, 'Fundamentals of Sensitivity Analysis', *The Institution of Engineers (India)*, Punjab, Haryana, and Himachal Pradesh Centre, vol. XI, Jan. and Feb., 1972.

M.S. Sukhija, 'Theoretical aspects of increasing aluminum to steel ratio in ACSR conductors', *Seminar on Recent Developments in Cable & Conductor Technology*, Raghvendra Industrial Research Foundation, Rajpura, 1978.

M.V. Deshpande, *Electrical Power System Design*, Tata McGraw Hill, New Delhi, 1984.

N. Sato and W.F. Tinney, 'Techniques for exploiting sparsity of the Newton admittance matrix', *IEEE Transactions*, PAS, vol. 82, 1963, pp. 944–50.

N.G. Hingorani and L. Gyugyi, *Understanding FACTS*, IEEE Press, Picataway, NJ, 2000.

O.I. Elgerd, *Electric Energy Systems Theory: An Introduction*, 2nd edn, Tata McGraw Hill, 1989.

O.I. Elgerd, *Electric Energy Systems Theory: An Introduction*, 1st edn, McGraw Hill, 1971.

P. Kundur, *Power System Stability and Control*, McGraw Hill, NY, 1994.

P.M. Anderson and A.A. Fouad, *Power System Control and Stability*, Iowa State University Press, Ames, Iowa, 1977.

P.M. Anderson, *Analysis of Faulted Power Systems*, IEEE Press, Piscataway, NJ, 1995.

R. Bakshi, 'Wind energy in India', *IEEE Power Engineering Review*, vol. 22, no. 9, 2002, pp. 16-17.

R. Khanna, T.K. Nagsarkar, and C.S. Rao, 'Generalised fast decoupled load flow for high R/X ratios', *Journal of AMSE*, vol. 71, no. 1, Paris, France, 1998, pp. 1–9.

R. Pratap, *Getting Started with MATLAB*, Oxford University Press, NY, 2004.

R. Rudervall, J.P. Carpentier, and R. Sharma, 'High Voltage Direct Current (HVDC) Transmission Systems Technology Review Paper', www.worldbank.org/html/fpd/em/transmission/technology_abb.pdf.

R.B. Shipley, N. Sato, D.W. Coleman, and C.F. Watts, 'Direct calculation of power system stability using the impedance matrix', *IEEE Transactions*, PAS, vol. 85, no. 7, 1966.

R.E. Larson, W.F. Tinney, et al., 'State estimation in power systems—Part I: Theory and feasibility; Part II: Implementation and applications', *IEEE Transactions*, PAS, vol. 89, no. 3, 1970, pp. 349–59.

R.G. Andretich, G.K. Brown, H.H. Happ, and C.E. Person, 'Piece-wise solution of the impedance matrix load flow', *IEEE Transactions*, PAS, vol. 87, 1968, pp. 1877–82.

R.T. Byrely, R.W. Long, C.J. Baldwin, C.W. King, 'Digital calculations of power system network under faulted conditions', *IEEE Transactions*, PAS, vol. 77, Part III, 1958, pp. 1296–1307.

S. Rahman, 'Green Power: What is it and where can we find it', *IEEE Power and Energy*, vol. 1, no. 1, 2003, pp. 30–37.

S.M.C. Pillai and R. Krishnamurthy, 'Problems and prospects of privatisation and regulation in India's power sector', *Energy for Sustainable Development*, vol. III, no. 6, 1997.

T. Gonen, *Electric Power Distribution System Engineering*, McGraw Hill, Singapore, 1986.

T.K. Nagsarkar and A. Singh, 'Dynamic simulation of HVDC transmission systems on PC-AT—Part I: Mathematical Model', *Journal of AMSE*, vol. 32, no. 1, 1997, p. 35.

T.K. Nagsarkar and A. Singh, 'Dynamic simulation of HVDC transmission systems on PC-AT—Part II: Computational algorithm and test results', *Journal of AMSE*, vol. 32, no. 1, 1997, p. 46.

T.K. Nagsarkar and A.K. Jindal, 'Transient stability improvement using dynamic brake with shunt impedance elements', *Journal of AMSE*, vol. 74, no. 34, 2001, pp. 17–30.

T.K. Nagsarkar and C.S. Rao, 'Performance comparison of full wave and half wave thyristor controlled dynamic brake in augmenting transient stability', *Modelling Simulation and Control A*, AMSE Press, vol. 9, no. 4, 1987, pp. 1–12.

T.K. Nagsarkar and C.S. Rao, 'Some aspects of transient stability improvement with thyristor controlled dynamic brake', *IEEE Power Engineering Society Winter Meeting*, Paper no. A-80 004-2, NY, 1980.

T.K. Nagsarkar and C.S. Rao, Written Discussion on the paper 'Augmentation of transient stability of a power system by braking', by S.S. Joshi and D.G. Tamaskar, *IEEE Transactions on Power Apparatus and Systems*, PAS-104, no. 11, 1985, pp. 3011-3012.

T.K. Nagsarkar and C.S. Rao, Written Discussion on the paper 'Closed loop quasi-optimal dynamic braking and shunt reactive control strategy for transient stability', by A.H.M.A. Rahim and D.A.A. Alamgir, *IEEE Transactions on Power Applications and Systems*, vol. 3, 1988, p. 886.

T.K. Nagsarkar and M.S. Sukhija, *Basic Electrical Engineering*, Oxford University Press India, New Delhi, 2005.

T.K. Nagsarkar and N.K. Dutta, 'A mathematical model for effective inductance of transmission line tower', *Journal of AMSE*, vol. 9, no. 3, 1987, p. 24.

T.K. Nagsarkar and R. Gupta, 'Some aspects of optimal reactive power flow for real power loss minimization' *Journal of Modeling Simulation & Control*, AMSE Press, vol. 20, no. 2, 1989, p. 29.

T.K. Nagsarkar and R. Nijhawan, 'Newton–Raphson power flow algorithm using bi-factorization technique', *Journal of AMSE*, vol. 31, no. 1, 1989, p. 15.

T.K. Nagsarkar and R. Segal, 'Fast dynamic response of power system to large disturbances', *Journal of AMSE*, vol. 54, no. 1, 1999, p. 1.

T.K. Nagsarkar and R. Singh, 'Optimal transformer tap adjustment for minimum real power loss', *Journal of AMSE*, vol. 50, no. 4, 1992, p. 15.

T.K. Nagsarkar, G. Singh, and G. Sharma, 'Transient stability improvement of Bhakra left bank generators with static excitation system', *Journal of AMSE*, vol. 192, no. 2, 1988, p. 1.

T.K. Nagsarkar, S. Chatterji, and B.D. Chaudhary, 'A modified interface for serial data transmission', *Proceedings of the International AMSE Conference*, vol. 3, 1992, p. 57.

V.A. Venikov, *Transient Processes in Electrical Power System*, Mir Publishers, Moscow, 1977.

W.A. Mittlestadt and J.L. Saugen, 'A method of improving power system transient stability using controllable parameters', *IEEE Transactions*, PAS, vol. 90, 1970, pp. 23–7.

W.D. Stevenson Jr, *Elements of Power System Analysis*, 4th edn, McGraw Hill, Singapore, 1982.

W.F. Tinney and C.E. Hart, 'Power flow solution by Newton's method', *IEEE Transactions*, vol. PAS-86, 1967, pp. 1449–56.

W.F. Tinney and J.W. Walker, 'Direct solution sparse network equations by optimally ordered triangular factorization', *Proceedings IEEE*, vol. PAS-55, 1967, pp. 1801–9.

Westinghouse Electric Corporation, *Electrical Transmission and Distribution Reference Book*, 1st edn, Oxford Book Company, Calcutta, 1964.

Y.H. Song and A.T. Johns, 'Flexible ac transmission systems (FACTS)', *IEE Power Energy Series 30*, UK, 1999.

Web References

Chapter 1

http://web.mit.edu/taalebi/www/scitech/pvtutorial.pdf

http://www.nfcrc.uci.edu/EnergyTutorial/energy.html

http://cseindia.org/userfiles/tarun_kapoor.pdf

http://www.energetica-india.net/news/status-of-grid-connected-solar-power-in-india-5039-mw--mnre

http://www.nrel.gov/docs/fy11osti/48948.pdf

http://books.google.co.in/books?id=oaB4vzI-UdwC&pg=PA11&source=gbs_toc_r&cad=4#v=onepage&q&f=false

http://supernovasolarengineering.com/solar-energy-a-photovoltaic-basics

http://www.indiaworldenergy.org/pdf/T&D%20Report_PGCIL.pdf

https://sites.google.com/site/ufases/how-pv-works

http://www.cea.nic.in/reports/yearly/annual_rep/2010-11/ar_10_11.pdf

www.mpoweruk.com/**mhd_generator**.htm

www.thermopedia.com/content/934/

www.buzzle.com/articles/how-do-**mhd-generators**-work.html

www.ustudy.in › ... › Power Generation

www.msubbu.in/ln/energy/Lecture-18-**MHD**.pdf

www.**cea**.nic.in/reports/monthly/executive_rep/**dec12/1-2**.pdf

www.powermin.nic.in/bharatnirman/bharatnirman.asp

Chapter 2

http://books.google.co.in/books?id=REww2ZF2RwwC&pg=PA1&source=gbs_toc_r&cad=4#v=onepage&q&f=false

http://books.google.co.in/books?id=xgSXRWvVwaEC&pg=PA301&source=gbs_toc_r&cad=4#v=onepage&q&f=false

http://www.ahmetkucuker.com/wp-content/uploads/2011/05/Three-Phase_Circuits.pdf

Chapter 5

http://ewh.ieee.org/cmte/psace/CAMS_taskforce/software.htm

Chapter 6

http://uqu.edu.sa/files2/tiny_mce/plugins/filemanager/files/4300303/EPS/Electrical_Power_System_Analysis_4._The_Impedance_Model_And_Network_Calculations.pdf

http://books.google.co.in/books?id=Ekfgjd0qADYC&pg=PA167&lpg=PA167&dq=computer+algorithm+for+formulation+of+bus+impedance+matrix+for+power+system+networks&source=bl&ots=oxVkGzjUGI&sig=9lTQXHSVI48k96waavd8sCNOy90&hl=en&sa=X&ei=aT39UMaTMY_prQee7YCoDA&redir_esc=y

Chapter 7

http://books.google.co.in/books?id=v3RxH_GkwmsC&pg=PA267&lpg=PA267&dq=comparison+of+gauss+seidel+and+newton+raphson+method&source=bl&ots=k0IgpaLBm1&sig=V40jjRylMgsvpi3gUqmikvlf9No&hl=en&sa=X&ei=ikf5UKGyGsrprAec3IHQDw&redir_esc=y#v=onepage&q=comparison%20of%20gauss%20seidel%20and%20newton%20raphson%20method&f=false

http://electricalquestionsguide.blogspot.in/2011/05/gauss-seidel-newton-raphson-methods.html

http://books.google.co.in/books?id=z_DF3lJ1gF8C&pg=PA219&lpg=PA219&dq=comparison+of+gauss+seidel+and+newton+raphson+method&source=bl&ots=6Tx-3wpMZw&sig=6NjF43XvLsEiQtMjqzsYu0PPgHk&hl=en&sa=X&ei=ikf5UKGyGsrprAec3IHQDw&redir_esc=y#v=onepage&q=comparison%20of%20gauss%20seidel%20and%20newton%20raphson%20method&f=false

http://www.uclm.es/area/gsee/Archivos%20Pag- web/docencia/aelect/01_LoadFlow_R1.pdf

http://www.esat.kuleuven.be/electa/publications/fulltexts/pub_1536.pdf

http://home.eng.iastate.edu/~jdm/ee553/FastPowerFlow.pdf

http://www.esat.kuleuven.be/electa/publications/fulltexts/pub_1456.pdf

http://home.eng.iastate.edu/~jdm/ee553/DCPowerFlowEquations.pdf

Chapter 10

http://opencourseware.kfupm.edu.sa/colleges/ces/ee/ee360/files%5C3-Lesson_Notes_Lec_11_3_phase_traqnsformers.pdf

http://books.google.co.in/books?id=XXkjEoKeWkIC&pg=PA59&lpg=PA59&dq=phase+shift+through+a+transformer&source=bl&ots=CkwQYFc33i&sig=Z6pt4o7B807arZySPChnRQyfIiU&hl=en&sa=X&ei=ws0LUdKBLcjZrQe-_YC4AQ&redir_esc=y#v=onepage&q=phase%20shift%20through%20a%20transformer&f=false

http://www.allaboutcircuits.com/vol_2/chpt_9/4.html

http://electrical-engineering-portal.com/understanding-vector-group-transformer-1

http://www.transtutors.com/homework-help/electrical-engineering/transformer/three-phase-transformer-connections.aspx

http://www.ee.up.ac.za/main/_media/en/undergrad/subjects/eeo420/power_transformers_2.pdf

Chapter 12

http://home.iitk.ac.in/~saikatc/EE632_files/VS_SC.pdf

http://www.iea.lth.se/~ielolof/stability/Seminar5.pdf

http://books.google.co.in/books?id=m3yB1mnlq24C&pg=PA22&source=gbs_toc_r&cad=3#v=onepage&q&f=false

http://ebcano.files.wordpress.com/2008/07/microsoft-word-ebcano_vs_0908.pdf   *

http://www.slideshare.net/Shahabkhan/definition-classification-of-power-system-stability   * http://books.google.co.in/books?id= v3RxH_GkwmsC&pg=PA1022&lpg=PA1022&dq = Tutorial+on+voltage+stability+in+electrical+power+systems&source=bl&ots=k0J9qc HEn1&sig=Ash8P6xeFd0a2_TrRI6h8KTKr9Q&hl=en&sa=X&ei=n9QoUZf0MsvNrQ ezvoGIAw&redir_esc=y#v=onepage&q=Tutorial%20on%20voltage%20stability%20 in%20electrical%20power%20systems&f=false

http://www.youtube.com/watch?v=xRvFTK7Xu8E

http://www.google.com/search?q=history+of+voltage+stability+in+power+systems&hl= en&ei=841ZUeCUMsTusgb6nIH4Ag&start=10&sa=N&biw=1350&bih=611

http://books.google.co.in/books?id=FsQIc3GxpZ0C&pg=PA64&source=gbs_ toc_r&cad=4#v=onepage&q&f=false

Chapter 15

http://www.sari-energy.org/PageFiles/What_We_Do/activities/HVDC_Training/ Materials/1JNL100020-842%20-%20PDF%20-%20Rev.%2000.PDF

http://www.cdeep.iitb.ac.in/nptel/Electrical%20Engineering/Power%20System%20Opera-tion%20and%20Control/Course_home_L18a.html

http://freevideolectures.com/Course/3076/High-Voltage-DC-Transmission

http://www.elect.mrt.ac.lk/HV_Chap11.pdf *

http://web.ing.puc.cl/~power/paperspdf/dixon/21.pdf

http://nptel.iitm.ac.in/courses/Webcourse-contents/IIT%20Kharagpur/Power%20Electron-ics/PDF/L-13(DK)(PE)%20((EE)NPTEL)%20.pdf

http://faculty.ksu.edu.sa/eltamaly/Documents/Courses/EE%20435/notes3.pdf

http://books.google.co.in/books?id=eS1z95mzi28C&pg=PA205&lpg=PA205&dq=three+ phase+thyristor+controlled+rectifier&source=bl&ots=G0Msd0zGIw&sig=jyhGDWif rZauuTCy530qqjZlJ70&hl=en&sa=X&ei=VsAwUZzLCoLOmgWY4ICYCg&redir_ esc=y#v=onepage&q=three%20phase%20thyristor%20controlled%20rectifier&f=false

http://pel-course.xjtu.edu.cn/courseware/pdf_en/full/chapter2.pdf

http://books.google.co.in/books?id=gSoDaumDrjoC&pg=PA42&source=gbs_ toc_r&cad=4#v=onepage&q&f=false   Padiyar

http://www.wiziq.com/tutorial/64397-HVDC

http://home.iitk.ac.in/~saikatc/EE632_files/VS_SC.pdf

http://www.iea.lth.se/~ielolof/stability/Seminar5.pdf

http://books.google.co.in/books?id=m3yB1mnlq24C&pg=PA22&source=gbs_ toc_r&cad=3#v=onepage&q&f=false

http://ebcano.files.wordpress.com/2008/07/microsoft-word-ebcano_vs_0908.pdf   *

http://www.slideshare.net/Shahabkhan/definition-classification-of-power-system-stabili-ty   *

http://books.google.co.in/books?id=v3RxH_GkwmsC&pg=PA1022&lpg=PA1022&dq = Tutorial+on+voltage+stability+in+electrical+power+systems&source=bl&ots=k0J9qc HEn1&sig=Ash8P6xeFd0a2_TrRI6h8KTKr9Q&hl=en&sa=X&ei=n9QoUZf0MsvNrQ ezvoGIAw&redir_esc=y#v=onepage&q=Tutorial%20on%20voltage%20stability%20 in%20electrical%20power%20systems&f=false

http://www.youtube.com/watch?v=xRvFTK7Xu8E

http://www.google.com/search?q=history+of+voltage+stability+in+power+systems&hl= en&ei=841ZUeCUMsTusgb6nIH4Ag&start=10&sa=N&biw=1350&bih=611

http://books.google.co.in/books?id=FsQIc3GxpZ0C&pg=PA64&source=gbs_ toc_r&cad=4#v=onepage&q&f=false

Index

Related Titles

PROTECTION AND SWITCHGEAR | 9780198075509

Bhavesh Bhalja, *AD Patel Institute of Technology, Gujarat*;
R.P. Maheshwari, *IIT Roorkee; Nilesh G. Chothani, AD Patel Institute of Technology, Gujarat*

   The book provides in-depth coverage of apparatus protection in an easily comprehensible manner.

Key Features

- Discusses circuit-breaking fundamentals and selection & testing of circuit breakers using actual field data
- Provides the International code list for protective relaying scheme with description of each device
- Discusses the latest development of digital/numerical relaying scheme for line and equipment protection in each chapter

MICROPROCESSORS AND
MICROCONTROLLERS | 9780198066477

N. Senthil Kumar, *Mepco Schlenk Engineering College, Sivakasi;*
M. Saravanan, *Thiagarajar College of Engineering, Madurai;*
S. Jeevananthan, *Pondicherry Engineering College, Puducherry*

The book is designed as a comprehensive textbook which provides exhaustive coverage of basic principles, functioning, and applications of microprocessors and microcontrollers.

Key Features

- Discusses advanced processors such as 80186, 80286, 80386, 80486, Pentium, PowerPC, and PIC 16F877
- Includes case studies on traffic light control, washing machine control, and elevator control to enable students to appreciate the applications of processors
- Provides numerous programming examples (in assembly language) which can also be used for conducting lab sessions

COMMUNICATION SYSTEMS | 9780198078050

V. Chandra Sekar, *Formerly, Srinivasa Ramanujan Centre, Kumbakonam*

This textbook is designed specifically for students of computer science, information technology, and electrical engineering disciplines.

Key Features

- Includes practical circuits for topics such as FM detection and generation, transmitters and receivers, and phase-locked loop (PLL)
- Provides application case studies on PLL frequency synthesizer, satellite instructional television experiment (SITE), and digital microwave radio
- Includes an appendix on MATLAB programs with their representative outputs for various modulation techniques

Other Related Titles

9780198066798 Tarun Kumar Rawat: *Signals and Systems*
9780198087229 R N Mutagi: *Digital Communication*
9780198067665 Debaprasad Das: *VLSI Design*
9780195670929 V.R. Moorthi: *Power Electronics*

Visit us at www.oup.co.in and www.oupinheonline.com